Relativistic Quantum Mechanics

Relativistic Quantum Mechanics

J.P. Singh
Department of Management Studies
Indian Institute of Technology Roorkee
Roorkee, Uttarakhand

I.K. International Publishing House Pvt. Ltd.

NEW DELHI • BANGALORE

Published by

I.K. International Publishing House Pvt. Ltd.
S-25, Green Park Extension
Uphaar Cinema Market
New Delhi–110 016 (India)
E-mail: info@ikinternational.com
Website: www.ikbooks.com

ISBN: 978-93-82332-63-3

Published by Krishan Makhijani for I.K. International Publishing House Pvt. Ltd., S-25, Green Park Extension, Uphaar Cinema Market, New Delhi–110 016 and Printed by Rekha Printers Pvt. Ltd., Okhla Industrial Area, Phase II, New Delhi–110 020

Preface

The primary objective of writing this book on relativistic quantum mechanics is to address the needs of postgraduate students who have had a first course in non-relativistic quantum mechanics. Traditional literature on quantum physics can be classified almost exclusively into: (i) books that provide a comprehensive coverage of non-relativistic quantum mechanics (Schiff, Davydov, Merzbacher, Messiah, etc.) and develop the subject as dynamics of particles in a non-relativistic setting and (ii) books that unify quantum mechanics with the special theory of relativity leading into the formulation of quantum field theory (Zee, Ryder, Peskin & Schroeder, etc.). The latter segment, however, almost exclusively, gets straight into the field theoretic description of microscopic physics. The fallout of this mutually exclusive segmentation is that both segments cover relativistic particle physics only marginally, i.e. in the asymptote rather than as a mainstream course. The coverage is, nowhere near adequate for a student to get acquainted with the intricacies of relativistic quantum mechanics, and hence, appreciate the nuances that have led to the origin and development of the field theoretic description in supersession of the particle-based version as the unifying theory of quantum mechanics and special theory of relativity.

This book fights that trend by covering in detail the topics that are thrown by the wayside in the traditional coverage. Admittedly, a fully relativistic treatment of the quantum mechanics of particles needs to be evolved using the field theoretic tools, yet, the study of single particle dynamics through more conventional wave equations is certainly, not without value. It provides valuable insights into the underlying physics. This book aims at addressing this gap by providing a lucid description of single particle relativistic dynamics at the quantum scale adapting, in a sense, the Schrödinger wave equations. It gives a thorough analysis of the Klein-Gordon and Dirac equations highlighting their properties.

Dynamics of physical systems are represented by differential equations. A study of the symmetry groups of such equations provides valuable insights into the system dynamics. The Lorentz and Poincare groups constitute the underlying symmetry groups of special relativity and, thus, of the relativistic wave equations. To put these issues in perspective and realizing the cardinal role played by "symmetry" in modern physics, Chapter 1 is dedicated to a study of the Lorentz and Poincare groups highlighting their salient features. Chapter 2 provides an overview of the representation theory of these two symmetry groups, covering both the finite dimensional (non-unitary) and unitary representations of the Poincare group, adopting the method of "induced representations".

The reader gets an overview of relativistic quantum mechanics in Chapter 3, wherein the dynamics of spinless particles is introduced through the Klein-Gordon equation. In addition to a comprehensive coverage of the conventional topics, this chapter also introduces the Feshbach Villars two-component representation of the KG equation and its properties.

The Dirac equation for spin-½ particles is acknowledged by the fraternity as a work of sheer ingenuity. Its remarkable features are not confined only to the revolutionary physical insights that it provides – but also extend to some purely extraordinary mathematical ramifications. While a complete coverage of the physics and mathematics of the Dirac equation would warrant a treatise of its own, Chapters 4-8 of the book embrace most of its physical aspects. Further, Chapters 11 and 12 provide solutions of the Dirac equation for the two-body central force problem of the hydrogen atom in configuration and momentum space respectively. Dirac's "hole theory" to explain the negative energy solutions also finds coverage here.

Chapter 9 covers the salient features of the relativistic dynamics of spin ½ massless particles and show how such dynamics can be represented by two-component as well as four-component wavefunctions. The important issue of polarization of massless particles also finds a discussion in this chapter. Chapter 10 takes up the issue of Dirac "wavepackets".

On the pedagogical plane, the book carries a few innovations despite being a conventional textbook. Efforts have been made to provide extended derivations of all relevant physical results without missing any intermediate steps to make the book as student-friendly as possible. Matter that would, probably, be relegated to chapter-end problems or appendices by most authors has been incorporated / retained in the mainstream text to facilitate a superior appreciation of results.

Needless to say, some notational/typographical errors would show up in the final script despite the most scrupulous of proofreadings. The author would be privileged and would owe a sincere debt of gratitude if such errors are brought to his notice. Indeed, the author would not only welcome but also be sincerely obliged for suggestions for adding value to the text.

A work of these dimensions would never have been accomplished without the blessings of well-wishers. It is, now, the pleasant duty of the author to acknowledge their angelic contributions to the project. Eminent personalities like Prof David Deutsch of Oxford, Prof Jose Luis Lopez, Prof Ignazio Licata & Prof Ammar Sakaji of the EJTP Editorial Team warrant specific mention in this context.

J.P. Singh

Contents

CHAPTER 0

Notation and Conventions

0.1 4-VECTORS OF SPECIAL RELATIVITY

Special relativity places space and time coordinates on an equal footing. It, therefore, becomes necessary to represent all vectorial quantities by 4-vectors in four-dimensional spacetime consisting of one time coordinate and the three conventional spatial coordinates. The temporal part of the 4-vector is usually ascribed a super (sub) script index of 0 and the three spatial coordinates carry the indices 1, 2, 3 – whether the index takes the form of a superscript or a subscript depends on whether the physical quantity under reference transforms contravariantly or covariantly under a coordinate transformation. Thus, a point in spacetime is represented by the 4-vector $x \equiv (x^0, \mathbf{x}) \equiv (x^0, x^1, x^2, x^3)$. Any arbitrary 4-vector A will be expressible in four components A^μ or A_μ, $\mu = 0, 1, 2, 3$ with the index positioning being upper or lower depending on whether the vector transforms contravariantly or covariantly under coordinate transformations.

0.2 CONTRAVARIANT AND COVARIANT VECTORS AND TENSORS

Given a four-dimensional space-time, let $x \equiv (x^0, \mathbf{x}) \equiv (x^0, x^1, x^2, x^3)$ and $x' \equiv (x'^0, \mathbf{x}')$ $\equiv (x'^0, x'^1, x'^2, x'^3)$ be the 4-coordinates of the same space-time point P with reference to different frames of reference I and I' respectively. Let us, now, consider a physical entity represented by a 4-vector A. The value of the physical entity is obviously a function of the space-time points, i.e. $A \equiv A(x)$ as well as $A \equiv A(x')$. Let A^i, $i = 0, 1, 2, 3$ be the components of A with reference to the unprimed frame I and let A'^α, $\alpha = 0, 1, 2, 3$ be the corresponding components with reference to the primed frame I'. The physical quantity and the corresponding vector A is called a contravariant vector if the components of A transform under coordinate transformations between the frames I and I' according to the transformation law:

$$A'^\alpha = \sum_{i=0}^{3} \frac{\partial x'^\alpha}{\partial x^i} A^i$$

The vector A is a covariant vector if the components of A transform under coordinate transformations between I and I' as:

$$A'^{\alpha} = \sum_{i=0}^{3} \frac{\partial x^i}{\partial x'^{\alpha}} A_i$$

Similarly, a second rank tensor A is called a contravariant tensor of rank two if its 16 components A^{ij} transform according to the transformation law

$$A'^{\alpha\beta} = \sum_{i,j=0}^{3} \frac{\partial x'^{\alpha}}{\partial x^i} \frac{\partial x'^{\beta}}{\partial x^j} A^{ij}$$

to the corresponding components $A'^{\alpha\beta}$ in the primed system under the coordinate transformation $x \to x'$. A covariant tensor of rank two obeys the transformation law

$$A'_{\alpha\beta} = \sum_{i,j=0}^{3} \frac{\partial x^i}{\partial x'^{\alpha}} \frac{\partial x^j}{\partial x'^{\beta}} A_{ij}$$

Mixed tensors also exist. A tensor of contravariant rank p and covariant rank q has 4^{p+q} components designated $A^{i_1 i_2 \dots i_p}_{j_1 j_2 \dots j_q}$ if they transform under coordinate transformation as

$$A'^{\alpha_1 \alpha_2 \dots \alpha_p}_{\beta_1 \beta_2 \dots \beta_q} = \sum_{i_1,\dots,i_p, j_1,\dots,j_q=0}^{3} \frac{\partial x'^{\alpha_1}}{\partial x^{i_1}} \cdots \frac{\partial x'^{\alpha_p}}{\partial x^{i_p}} \frac{\partial x^{j_1}}{\partial x'^{\beta_1}} \cdots \frac{\partial x^{j_p}}{\partial x'^{\beta_q}} A^{i_1 i_2 \dots i_p}_{j_1 j_2 \dots j_q}$$

with α_r, β_s both being free indices, each taking any value 0, 1, 2, 3 respectively.

The coordinates themselves transform as contravariant vectors for $dx'^{a} = \sum_{i=0}^{3} \frac{\partial x'^{\alpha}}{\partial x^i} dx^i$ so that we represent the space-time vector $x \equiv x^{\mu} \equiv (x^0, x^1, x^2, x^3) \equiv (t, x, y, z)$. The covariant space-time vector will be $x_{\mu} \equiv g_{\mu\nu} x^{\nu} \equiv (x^0, -x^1, -x^2, -x^3) \equiv (t, -x, -y, -z)$.

0.3 EINSTEIN'S SUMMATION CONVENTION

(a) If a Greek index is repeated twice in the same term, then summation over it over the values 0, 1, 2, 3 is implied.

(b) If a Latin index is repeated twice in the same term, then summation over it over the values 1, 2, 3 is implied.

(c) If an index appears only once in any term, it carries a definite value and is called a "free index".

(d) An index which is repeated and over which summation is implied as per (a) above, is called a "dummy index". Obviously, one can replace a dummy index by any other index that does not appear in the same term.

The aforesaid summation convention (due to Einstein) shall be followed throughout the book unless otherwise explicitly stated.

0.4 RAISING AND LOWERING OF INDICES

Contravariant (covariant) indices can be lowered (raised) by use of the metric tensor, which we define as $g \equiv (g_{\mu\nu}) = (g^{\mu\nu}) = diag\ (1, -1, -1, -1)$ which corresponds to the four-dimensional Minkowski space with signature (1,–1,–1,–1). For example, $A_\nu = g_{\mu\nu}A^\mu$, $A^\nu = g^{\mu\nu} A_\mu$, etc. Considering a tensor of second rank with components labelled $A^{\mu\nu}$ and taking the scalar product of $A^{\mu\nu}$ with the metric tensor $g_{\mu\nu}$ successively, we obtain (i) $A^{\mu\nu} g_{\nu\lambda} = A^\mu_\lambda$, (ii) $A^{\mu\nu} g_{\mu\lambda} = A^\nu_\lambda$, (iii) $A^\mu_\nu g_{\mu\lambda} = A^\rho_\lambda g_{\rho\nu} = A^{\mu\rho} g_{\mu\lambda} g_{\rho\nu} = A_{\lambda\nu}$. Thus, a second rank contravariant tensor can be manipulated to yield two mixed tensors. A^μ_λ and A^ν_λ and a fully covariant tensor $A_{\lambda\nu}$ by lowering of the index. Similarly, the index of a covariant second rank tensor can be raised to yield two mixed tensors and one fully contravariant second rank tensor. The reciprocal relationship also holds, viz. $A^\mu{}_\lambda g^{\lambda\mu} = A^{\mu\nu}$, $A_{\lambda.}{}^\mu g^{\lambda\nu} = A^{\mu\nu}$, $A_{\mu\nu} g^{\mu\lambda} g^{\nu\rho} = A^{\lambda\rho}$. In tensor usage, the order of the indices is extremely significant and must not be ignored. We, therefore, illustrate the raising and lowering of indices by another example. For this purpose, we consider a fourth rank contravariant tensor $A^{\mu\nu\lambda\rho}$ and successively lower its indices to get (i) $A_{\mu\dots}^{\nu\lambda\rho} = A^{\sigma\nu\lambda\rho} g_{\sigma\mu}$, (ii) $A^{\mu\ \lambda\rho}_{\ \nu..} = A^{\mu\sigma\lambda\rho} g_{\sigma\nu}$, (iii) $A^{\mu\nu\ \rho}_{..\lambda.} = A^{\mu\nu\sigma\rho} g_{\sigma\lambda}$ & (iv) $A^{\mu\nu\lambda}_{\ \ \ \dots\rho} = A^{\mu\nu\lambda\sigma} g_{\sigma\rho}$ (v) $A^{\lambda\rho}_{\mu\nu} = A^{\sigma\omega\lambda\rho} g_{\sigma\nu} g_{\omega\nu}$ (vi) $A^{\nu\ \rho}_{\mu\cdot\lambda} = A^{\sigma\nu\omega\rho} g_{\sigma\mu} g_{\omega\lambda}$, (vii) $A_{\rho\,.\,\mu} = \nu\lambda^{\sigma\nu\lambda\omega} g_{\sigma\mu}\, g_{\omega\rho}$, (viii). $A^{\mu\ \lambda}_{.\nu.\rho} = A^{\mu\sigma\lambda\omega} g_{\sigma\nu}\, g_{\omega\rho}$, and so on.

Elements of the metric tensor also satisfy (i) $g^{\mu\nu} = g_{\mu\nu}$, (ii) $g^\mu_\nu = g^\nu_\mu = \delta_{\mu\nu}$ (iii) $g_{\mu\nu}\, g^{\nu\lambda} = \delta^\lambda_\mu$. It is to be noted that while expressing matrix elements with two indices the first index (irrespective of whether it is a superscript or subscript) indicates the row number while the second index indicates the column number. For instance, in both $g^\mu_{\ \nu}$ as well as $g_\mu{}^\nu$, μ is the row indicator and ν is the column number.

0.5 INNER PRODUCT OF 4-VECTORS

Given two 4-vectors $A \equiv (A^0, \mathbf{A})$ and $B \equiv (B^0, \mathbf{B})$, we define their inner product by $AB \equiv A^0 B^0 - \mathbf{A.B}$ and the norm by $|A|^2 = A^2 = (A^0)^2 - \mathbf{A}^2$. Following the summation convention, we can write

$$
\begin{aligned}
A.B &= A^0B^0 - \mathbf{A.B} = A^0B^0 - A^1B^1 - A^2B^2 - A^3B^3 \\
&= g^{00}\, A_0B^0 - g^{11}\, A_1B^1 - g^{22}\, A_2B^2 - g^{33}A_3B^3 \\
&= g^{00}\, A_0B^0 + A_1\, B^1 + A_2\, B^2 + A_3\, B^3 \\
&= A^\mu\, B_\mu = A_\mu\, B^\mu = A_\mu\, g^{\mu\nu}\, B_\nu = A^T gB
\end{aligned}
$$

CHAPTER 1

Symmetry in Relativistic Quantum Mechanics

A "symmetry" may be defined as the equivalence of different physical situations. It implies the impossibility of acquiring additional knowledge about the physical system by subjecting it to symmetric transformations. For example, the laws of nature that govern the occurrence of physical phenomena are invariant under translations, rotations and uniform relative motion of frames of reference.

1.1 ACTIVE AND PASSIVE TRANSFORMATIONS

Let us consider the operator equation

$$\mathbf{y} = \hat{A}\mathbf{x} \tag{1}$$

where **x**, **y** are vectors in a Hilbert space H and the operator $\hat{A}$ represents a linear transformation in the said space. If $\{e_i, i = 1, 2, ..., n\}$ constitutes a basis in H, then we have (in the Einstein's summation convention that is followed throughout the text, unless explicitly stated otherwise):

$$y^i e_i = \mathbf{y} = \hat{A}\mathbf{x} = \hat{A}x^k e_k = x^k \hat{A} e_k = x^k a^i_k e_i \tag{2}$$

whence the linear independence of e_i implies

$$y^i = x^k a^i_k \tag{3}$$

or, in matrix notation

$$\mathbf{y} = \mathbf{A}\mathbf{x} \tag{4}$$

where **A** is the matrix representation of the operator $\hat{A}$. Thus, the action of the operator $\hat{A}$ on the vector **x** with components $\{x^k\}$ takes it to another vector **y** with components $\{y^i\}$ in the same vector space. This is the active point of view of looking at the transformation embedded in the operator $\hat{A}$.

The operator eq. (1) may as well be written as

$$\hat{A}^{-1}\mathbf{y} = \hat{A}^{-1}y^i e_i = y^i \hat{A}^{-1} e_i = y^i e'_i = \mathbf{x} = x^k e_k \qquad (5)$$

Now, e_i s are also linearly independent, for, considering the expression , $\sum_i \alpha^i e'_i = 0$, we have

$$0 = \sum_i \alpha^i \hat{A}^{-1} e_i = \hat{A}^{-1} \sum_i \alpha^i e_i = \sum_i \alpha^i e_i \Leftrightarrow \alpha^i = 0 \qquad (6)$$

where the last step follows from the linear independence of the original basis e_i. Hence, e'_i also constitutes a basis of our Hilbert space. In the new basis e'_i, we can write, using eq. (5), $y^i e'_i = \mathbf{x} = x^k e_k$ so that $\{x^k\}$ and $\{y^i\}$ may be regarded as components of the same vector $\mathbf{x}$ with respect to the different bases e_i and e'_i. The new basis can, of course, be obtained from the old one by operating on the latter with the operator $\hat{A}^{-1}$. This is the passive point of view. Thus, in the passive point of view the vector $\mathbf{x}$ remains fixed while the transformation is achieved by transforming the frame of reference with the inverse transformation. The important point here is that, since the passive transformation entails the choice of a new basis, it may be achieved through a unitary transformation. On the other hand, an active transformation involves an actual change in the state of the quantum system.

A symmetry is deemed to exist if there exists a correspondence between the active and passive transformations. In other words, if a new basis is identified through a passive transformation then there should exist a corresponding active transformation such that the new state vector created by operating this active transformation on the original state vector possesses components, with respect to the new basis, that are identical with those of the original state vector with respect to the original basis. There is, however, a rider to this simple proposition, viz. that the vector space in quantum mechanics is a complex space, as a result of which, a one-to-one correspondence fails to exist between the physical states and the set of vector components. In actual fact, the basis vectors may be multiplied by complex numbers of unit modulus without any change in the corresponding vector components representing a given physical state. Thus, there exists a certain phase arbitrariness in defining the components of a state vector that represent a given physical state because of the Hilbert space being a complex vector space. This phase arbitrariness, however, does not impact the transition probability between two physical states as represented by $P_{\alpha\beta} = |\langle\alpha|\beta\rangle|^2$. This leads us to the much celebrated Wigner's theorem.

1.2 WIGNER'S THEOREM (WEINBERG, 1995)

Wigner's theorem provides that "Any symmetry transformation can be represented on the Hilbert space of physical states by an operator that is either linear and unitary or antilinear and antiunitary".

Proof

As mentioned earlier, we define a symmetry transformation as a transformation $\hat{T}: H \rightarrow H$ of state vectors (representing physical states in the Hilbert space H) that preserves transition probabilities in the sense that if ψ_1 and ψ_2 are state vectors belonging to quantum $\Re_1$ and $\Re_2$ rays respectively and representing the physical states of a quantum system and $\psi_1' = \hat{T}\psi_1$ and $\psi_2' = \hat{T}\psi_2$ are the corresponding states after operating with $\hat{T}$ that belong to the transformed rays $\mathfrak{R}'_1 = \hat{T}\mathfrak{R}_1$ and $\mathfrak{R}_2 = \hat{T}\mathfrak{R}_2$, then

$$|\langle\psi_1'|\psi_2'\rangle|^2 = |\langle\psi_1|\psi_2\rangle|^2 \tag{7}$$

In this context, we define a "quantum ray" as a set of normalized states, i.e., $|\langle\psi|\psi\rangle = 1$ such that any two elements ψ, ϕ of the same ray can be represented as $\psi = \lambda\phi$, where λ is an arbitrary complex number of unit modulus, i.e., ψ, ϕ differ only in phase, $\psi = \lambda\phi = e^{i\varphi}\phi$ since $|\lambda| = 1 \Rightarrow 1 = e^{i\varphi}$ with φ real. Furthermore, a symmetry transformation must also be invertible and the inverse should also preserve the transition probabilities.

Let ψ_k, ψ_l belong to some complete orthonormal set of state vectors with $\mathfrak{R}_k$ and $\mathfrak{R}_l$ being the corresponding rays respectively, so that

$$\langle\psi_k|\psi_l\rangle = \delta k_l \tag{8}$$

Let ψ_k', ψ_l' be the corresponding transformed state vectors belonging to the transformed rays $\mathfrak{R}'_k$ and $\mathfrak{R}'_l$ respectively, so that from eq. (7), we have

$$|\langle\psi_k'|\psi_l'\rangle|^2 = |\langle\psi_k|\psi_l\rangle|^2 = \delta k_l \tag{9}$$

Now $|\langle\psi_k'|\psi_k'\rangle|$, being a norm must necessarily be real and positive. Hence, $\langle\psi_k'|\psi_k'\rangle = 1$. This implies that

$$|\langle\psi_k'|\psi_l'\rangle| = \delta k_l \tag{10}$$

We claim that the set of transformed states ψ_i' also forms a complete set. For, if it is not so, then let ψ_m' belonging to the ray $\mathfrak{R}'_m$ be a non-zero state vector that is orthogonal to all of the ψ_i'. Then, the inversely transformed ray $\mathfrak{R}''_m = \hat{T}^{-1}\mathfrak{R}'_m$ shall consist of non-zero state vectors ψ''_m which are orthogonal to ψ_i for all i, that is, for which the following holds

$$|\langle\psi_i'|\psi''_m\rangle|^2 = |\langle\psi_i'|\psi_m'\rangle|^2 = 0 \tag{11}$$

However, this is a contradiction because we have assumed $\{\psi_i\}$ to form a complete set. It, therefore, follows that the transformed states also form a complete set.

Our next step is to establish a phase convention for the transformed states ψ_i'. For the purpose, we identify a particular state vector, say ψ_1, and then constitute the new vector

$$\Upsilon_k \equiv \frac{1}{\sqrt{2}}(\psi_1 + \psi_k) \tag{12}$$

with $k \neq 1$. Since ψ_1, ψ_k are the constituents of an orthonormal set of vectors, both of them are unit vectors so that

$$\Upsilon_k \equiv \frac{1}{\sqrt{2}}(\psi_1 + \psi_k)$$

is also normalized. Hence, it belongs to a ray, say $\mathfrak{I}_k$, $k \neq 1$. Now, any state vector Υ_k' belonging to the transformed ray $\mathfrak{I}'_k = \hat{T}\mathfrak{I}_k$ may be expanded in the basis of the state vectors ψ_i' as

$$\Upsilon'_k = \sum_l c_{kl}\,\psi'_l \tag{13}$$

Then, from eq. (7), we have

$$|c_{kk}| = |c_{k1}| = \frac{1}{\sqrt{2}} \tag{14}$$

with $|c_{kl}| = 0$ for $l \neq k$, $l \neq 1$ as can be seen by computing $\langle\Upsilon'_k|\psi'_i\rangle$ and $\langle\Upsilon_k|\psi_i\rangle$ successively for each member of the two bases ψ'_i and ψ_i respectively. We can, therefore, write

$$\Upsilon'_k = \frac{1}{\sqrt{2}}\ \text{sgn}\ (c_{k1})\psi'_1 + \frac{1}{\sqrt{2}}\ \text{sgn}\ (c_{kk})\ \psi'_k \tag{15}$$

Now, by choosing the phase of the two state vectors Υ'_k and ψ'_k appropriately, we can manipulate the phases of the two non-zero coefficients c_{kk}, c_{k1} such that $c_{kk} = c_{k1} = \frac{1}{\sqrt{2}}$. For example, let us divide eq. (15) throughout by sgn (c_{k1}). We, then, absorb $\frac{\text{sgn}(c_{kk})}{\text{sgn}(c_{k1})}$ in ψ'_k by suitably adjusting its phase. It may be noted that a change in phase of a state vector does not change its ray so that even after the change in phase ψ'_k will continue to belong to the same ray $\mathfrak{R}'_k$. Similarly, by a phase transformation, we absorb $\frac{1}{\text{sgn}(c_{k1})}$ in the state vector Υ'_k without altering its ray. This can be done for any value k. Let us designate the vectors corresponding to Υ'_k and ψ'_k that are so phase-adjusted as $\Upsilon'''_k \equiv \hat{U}\,\Upsilon_k$ and $\psi'''_k \equiv \hat{U}\,\psi_k$ respectively.

Further, we write $\psi'''_1 = \hat{U}\,\psi_1 = \psi'_1$ by choosing ψ_1 accordingly. Then, we have

$$\Upsilon'''_k = \frac{1}{\sqrt{2}}(\psi'''_1 + \psi'''_k) \tag{16}$$

and so

$$\hat{U}\,\frac{1}{\sqrt{2}}(\psi_k + \psi_1) = \hat{U}\,\Upsilon_k = \Upsilon'''_k = \frac{1}{\sqrt{2}}\left(\hat{U}\psi_k + \hat{U}\psi_1\right) \tag{17}$$

It is emphasized that since ψ'''_k's are obtained from ψ'_k purely by phase transformations and the orthogonality structure is not affected by such transformations, ψ'''_k will continue to constitute a complete orthogonal set if ψ'_k constitutes a complete orthogonal set.

To summarize, we have

(i) $\psi'''_k \equiv \hat{U}\,\psi_k$, being simply a phase adjusted version of ψ'_k, belongs to the same ray as ψ'_k.

(ii) Since $\{\psi'_k\}$ has been shown to be an orthonormal basis (on the assumption that $\{\psi_k\}$ is so), it follows that $\{\psi'''_k\}$ also constitutes an orthonormal basis, as phase transformations affect neither the orthogonality nor the normality of the state vectors.

Our next step is to define the transformation $\hat{U}\,\psi$ for an arbitrary state vector ψ. Let ψ belong to an arbitrary ray $\Re$. We expand ψ in the original (unprimed) basis $\{\psi_i\}$ as

$$\psi = \sum_k C_k \, \psi_k \tag{18}$$

Similarly, we can expand any statec ψ' belonging to the transformed ray $\Re' \equiv \hat{T}\Re$ in the complete orthogonal set $\psi'''_k \equiv \hat{U}\,\psi_k$ as

$$\psi' = \sum_k C'_k \, \hat{U}\,\psi_k \tag{19}$$

Then $|\langle\psi_k|\psi\rangle|^2 = |\langle\hat{U}\,\psi_k|\psi'\rangle|^2$ mandates that for every value of k, we must have

$$|C_K|^2 = |C'_k|^2 \tag{20}$$

Similarly, it follows from the equality of $|\langle\Upsilon_k|\psi\rangle|^2 = |\langle\hat{U}\,\Upsilon_k|\psi'\rangle|^2$ that for every value of $k \neq 1$, we must have

$$|C_k + C_1|^2 = |C'_k + C'_1|^2 \tag{21}$$

Let us write $C_j = a_j + ib_j$ and $C'_j = a'_j + ib'_j$. Then eqs. (20) and (21) may be written as

$$a_k^2 + b_k^2 = a_k'^2 + b_k'^2 \; \forall k \tag{22}$$

and

$$\begin{aligned} &a_k^2 + a_1^2 + 2a_k a_1 + b_k^2 + b_1^2 + 2b_k b_1 \\ &= a_k'^2 + a_1'^2 + 2a'_k a'_1 + b_k'^2 + b_1'^2 + 2b'_k b'_1 \quad \forall\, k \neq 1 \end{aligned} \tag{23}$$

whence

$$a_k a_1 + b_k b_1 = a'_k \, a'_1 + b'_k \, b'_1 \quad \forall k \tag{24}$$

which can be written as

$$\frac{a_k\, a_1 + b_k\, b_1}{a_1^2 + b_1^2} = \frac{a'_k\, a'_1 + b'_k\, b'_1}{a_1'^2 + b_1'^2} \quad \forall k \tag{25}$$

or

$$\mathrm{Re}\left(\frac{C_k}{C_1}\right) = \mathrm{Re}\left(\frac{C'_k}{C'_1}\right) \tag{26}$$

Similarly, eqs. (22) and (24) provide, in conjunction

$$\left(\frac{a_1\, b_k - a_k\, b_1}{a_1^2 + b_1^2}\right)^2 = \frac{a_1^2\, b_k^2 + a_k^2\, b_1^2 - [(a_k\, a_1 + b_k\, b_1)^2 - (a_1^2\, a_k^2 + b_k^2\, b_1^2)]}{(a_1^2 + b_1^2)^2}$$

$$= \frac{(a_1^2 + b_1^2)(a_k^2 + b_k^2) - (a_k\, a_1 + b_k\, b_1)^2}{(a_1^2 + b_1^2)^2}$$

$$= \frac{(a_1'^2 + b_1'^2)(a_k'^2 + b_k'^2) - (a_k' a_1' + b_k' b_1')^2}{(a_1'^2 + b_1'^2)^2} \tag{27}$$

whence

$$\left[\mathrm{Im}\left(\frac{C_k}{C_1}\right)\right]^2 = \left[\mathrm{Im}\left(\frac{C_k'}{C_1'}\right)\right]^2 \tag{28}$$

Equations (26) and (28) together imply that either

$$\frac{C_k}{C_1} = \frac{C_k'}{C_1'} \text{ or } \frac{C_k}{C_1} = \left(\frac{C_k'}{C_1'}\right)^* \tag{29}$$

Our next step is to show that the same choice, i.e., either

$$\frac{C_k}{C_1} = \frac{C_k'}{C_1'} \text{ or } \frac{C_k}{C_1} = \left(\frac{C_k'}{C_1'}\right)^*$$

must be made for each k.

Let, if possible, for some $k = \alpha$ we have $\frac{C_\alpha}{C_1} = \frac{C_\alpha'}{C_1'}$ while for some other $k = \beta \neq \alpha$ we have $\frac{C_\beta}{C_1} = \left(\frac{C_\beta'}{C_1'}\right)^*$. Let us further assume that the ratio $\frac{C_\alpha}{C_1}$ is complex so that the two cases $\frac{C_k}{C_1} = \frac{C_k'}{C_1'}$ and $\frac{C_k}{C_1} = \left(\frac{C_k'}{C_1'}\right)^*$ are different. Obviously, we also have to assume then that $\alpha \neq 1$, $\beta \neq 1$, $\alpha \neq \beta$.

Let us define a state vector $\xi = \frac{1}{\sqrt{3}}(\psi_1 + \psi_\alpha + \psi_\beta)$. Let the corresponding vector belonging to the transformed ray be ξ'. Since all the ratios of the form $\frac{C_k}{C_1}$ in the state vector ξ are real, the ratios $\frac{C_k'}{C_1'}$ for the corresponding transformed vector ξ' will also be the same and so we can write $\xi' = \frac{\omega}{\sqrt{3}}(\hat{U}\psi_1 + \hat{U}\psi_\alpha + \hat{U}\psi_\beta)$, where ω is a phase factor with $|\omega| = 1$. Equality of the transition probabilities $|\langle\xi|\psi\rangle|^2 = |\langle\xi'|\psi'\rangle|^2$, where ψ' is given by eq. (19) and use of eq. (20) mandates that

$$\left|1 + \frac{C_\alpha}{C_1} + \frac{C_\beta}{C_1}\right|^2 = \left|1 + \frac{C_\alpha'}{C_1'} + \frac{C_\beta'}{C_1'}\right|^2 \tag{30}$$

since $\frac{C_\alpha}{C_1} = \frac{C'_\alpha}{C'_1}$ and $\frac{C_\beta}{C_1} = \left(\frac{C'_\beta}{C'_1}\right)^*$, we have

$$\left|1 + \frac{C_\alpha}{C_1} + \frac{C_\beta}{C_1}\right|^2 = \left|1 + \frac{C_\alpha}{C_1} + \frac{C^*_\beta}{C^*_1}\right|^2 \tag{31}$$

Let $\frac{C_\alpha}{C_1} = \frac{C'_\alpha}{C'_1} = c + id$ and $\frac{C_\beta}{C_1} = \left(\frac{C'_\beta}{C'_1}\right)^* = m + in$, then eq. (31) implies that

$$(1 + c + m)^2 + (d + n)^2 = (1 + c + m)^2 + (d - n)^2 \tag{32}$$

whence either $d = 0$ or $n = 0$. Therefore, either $\frac{C_\alpha}{C_1}$ or $\frac{C_\beta}{C_1}$ must necessarily be real, thereby contradicting our assumption.

It follows that for a given symmetry transformation $\hat{T}$ applied to an arbitrary state vector $\psi = \sum_k C_k \psi_k$ we must either have $\frac{C_k}{C_1} = \frac{C'_k}{C'_1}$ for all k or $\frac{C_k}{C_1} = \left(\frac{C'_k}{C'_1}\right)^*$ for all k.

We have, thus, established that given a state vector ψ, all its expansion coefficients C_k in the expansion $\psi = \sum_k C_k \psi_k$ and their respective transformed counterparts C'_k must uniformly obey either of the two rules, viz. $\frac{C_k}{C_1} = \frac{C'_k}{C'_1}$ for all k or $\frac{C_k}{C_1} = \left(\frac{C'_k}{C'_1}\right)^*$ for all k. We are, however, yet to establish that for a given symmetry transformation, the same rule must apply to all state vectors in the given space. We establish this in the sequel.

For the purpose, depending on which of the two rules is realized, we define $\psi' = \hat{U}\psi$ belonging to the transformed ray $\hat{T}\Re$ with the phase so chosen that $C_1 = C'_1$ or $C_1 = C'^*_1$ so as to conform to the appropriate rule applicable to the coefficients, i.e. $\frac{C_k}{C_1} = \frac{C'_k}{C'_1}$ or $\frac{C_k}{C_1} = \left(\frac{C'_k}{C'_1}\right)^*$ respectively. Then, either

$$\hat{U}\left(\sum_k C_k \psi_k\right) = \sum_k C_k \hat{U} \psi_k \tag{33}$$

or

$$\hat{U}\left(\sum_k C_k \psi_k\right) = \sum_k C^*_k \hat{U} \psi_k \tag{34}$$

Now, let us suppose, if possible, that the rule envisaged in eq. (33) applies to a state vector $A = \sum_k A_k \psi_k$ and the rule envisaged in eq. (34) applies to another state vector $B = \sum_k B_k \psi_k$. Then, by the invariance of the transition probabilities $|\langle B|A\rangle|^2 = |\langle B'|A'\rangle|^2$, we have, on using the orthonormality of the basis vectors

$$\left|\sum_k B_k^* A_k\right|^2 = \left|\sum_k B_k A_k\right|^2 \tag{35}$$

Writing $A_k = c_k + id_k$, $B_k = m_k + in_k$, eq. (35) gives

$$\left|\sum_k (m_k - in_k)(c_k + id_k)\right|^2 = \left|\sum_k (m_k + in_k)(c_k + id_k)\right|^2$$

or

$$\left|\sum_k [(m_k c_k + n_k d_k) + i(m_k d_k - n_k c_k)]\right|^2 = \left|\sum_k [(m_k c_k - n_k d_k) + i(m_k d_k + n_k c_k)]\right|^2$$

which implies

$$\left[\sum_k (m_k c_k + n_k d_k)\right]^2 + \left[\sum_k (m_k d_k - n_k c_k)\right]^2$$

$$= \left[\sum_k (m_k c_k - n_k d_k)\right]^2 + \left[\sum_k (m_k d_k - n_k c_k)\right]^2$$

or

$$\left[\left(\sum_k m_k c_k\right)^2 + \left(\sum_l n_l d_l\right)^2 + 2\left(\sum_k m_k c_k\right)\left(\sum_l n_l d_l\right)\right]$$

$$+ \left[\left(\sum_k m_k d_k\right)^2 + \left(\sum_l n_l c_l\right)^2 - 2\left(\sum_k m_k d_k\right)\left(\sum_l n_l c_l\right)\right]$$

$$= \left[\left(\sum_k m_k c_k\right)^2 + \left(\sum_l n_l d_l\right)^2 - 2\left(\sum_k m_k c_k\right)\left(\sum_l n_l d_l\right)\right]$$

$$+ \left[\left(\sum_k m_k d_k\right)^2 + \left(\sum_l n_l c_l\right)^2 + 2\left(\sum_k m_k d_k\right)\left(\sum_l n_l c_l\right)\right]$$

whence $\sum_{\substack{k,l\\k\neq l}}(m_k n_l - m_l n_k)(c_k d_l - c_l d_k) = 0$, which gives

$$\sum_{\substack{k,l\\k\neq l}} \text{Im}(A_k^* A_l)\ \text{Im}(B_k^* B_l) = 0 \tag{36}$$

However, since for $k = l$, $A_k^* A_k$, $B_k^* B_k$ are both real quantities, $\text{Im}(A_k^* A_k) = \text{Im}(B_k^* B_k) = 0$ so that $k \neq l$ constraint in the summation index of eq. (36) may be dropped and we may write

$$\sum_{k,l} \text{Im}(A_k^* A_l)\,\text{Im}(B_k^* B_l) = 0 \tag{37}$$

Let us assume that eq. (37) is satisfied for a pair of state vectors $A = \sum_k A_k \psi_k$ and $B = \sum_k B_k \psi_k$ belonging to different rays. However, in such a case, when neither A_k nor B_k are all of the same phase, we can find a third vector $Q = \sum_k Q_k \psi_k$ for which

$$\sum_{kl} \text{Im}(Q_k^* Q_l)\ \text{Im}(A_k^* A_l) \neq 0 \tag{38}$$

and also

$$\sum_{kl} \text{Im}(Q_k^* Q_l)\ \text{Im}(B_k^* B_l) \neq 0 \tag{39}$$

This is proved as follows:

(a) Let for some k, l both $A_k^* A_l$, $B_k^* B_l$ be complex. Then, we choose the state vector Q such that all the Q_i's vanish except for Q_k, Q_l. Further, these two coefficients Q_k, Q_l are chosen to be in different phases. Then, writing $A_k \equiv a_k(\cos\theta_k^A + i\sin\theta_k^A)$ with similar notations for the other coefficients, we have

$$\text{Im}(Q_k^* Q_l)\ \text{Im}(A_k^* A_l) = q_k q_l a_k a_l \sin(\theta_l^Q - \theta_k^Q)\sin(\theta_l^A - \theta_k^A) \tag{40}$$

which can only vanish if either of $\theta_l^A = \theta_k^A$ or $\theta_l^Q = \theta_k^Q$ holds. The latter possibility is ruled out by construction for Q_k, Q_l are chosen to be in different phases whereas the former implies that $A_k^* A_l$ is real, a contradiction. Hence,

$$\sum_{kl} \text{Im}(Q_k^* Q_l)\ \text{Im}(A_k^* A_l) = q_k q_l q_k a_l \sin(\theta_l^Q - \theta_k^Q)\sin(\theta_l^A - \theta_k^A) \neq 0 \tag{41}$$

Similarly, we can establish eq. (39).

(b) If for some k, l, $A_k^* A_l$ is complex but $B_k^* B_l$ is real, then let, for some other pair of indices m, n (where, possibly, but not necessarily, $m = k$ or $n = l$ but not both) $B_m^* B_n$ be complex. If the corresponding $A_m^* A_n$ is complex, then, as in case (a), we choose all the Q_i's vanish except for Q_m, Q_n. Further, these two coefficients Q_m, Q_n are chosen to be in different phases. If, however, corresponding to $B_m^* B_n$ complex, $A_m^* A_n$ is real, then we choose all the Q_i's vanish except for Q_k, Q_l, Q_m, Q_n. These four coefficients Q_k, Q_l, Q_m, Q_n are chosen to be in different phases. Now, if both $B_m^* B_n$ and $A_m^* A_n$ are complex, the case can be handled as in (a). If $B_m^* B_n$ is complex but $A_m^* A_n$ is real and $A_k^* A_l$ is complex but $B_k^* B_l$ is real, then we have

$$\text{Im}(Q_k^* Q_l)\ \text{Im}(B_k^* B_l) = 0 \tag{42}$$

$$\text{Im}(Q_k^* Q_l)\ \text{Im}(A_k^* A_l) \neq 0 \tag{43}$$

$$\text{Im}(Q_m^* Q_n)\ \text{Im}(B_m^* B_n) \neq 0 \tag{44}$$

$$\text{Im}(Q_m^* Q_n)\ \text{Im}(A_m^* A_n) = 0 \tag{45}$$

whence eqs. (38) and (39) both hold.

(c) The case of $A_k^* A_l$ real, $B_k^* B_l$ complex, for some k, l is handled as in case (b).

From eq. (38), it follows that the same rule, viz. (i) $\hat{U}\left(\sum_k C_k \psi_k\right) = \sum_k C_k \hat{U} \psi_k$ or (ii) $\hat{U}\left(\sum_k C_k \psi_k\right) = \sum_k C_k^* \hat{U} \psi_k$ must be followed for $A = \sum_k A_k \psi_k$ and $Q = \sum_k Q_k \psi_k$. Similarly, from eq. (39) it follows that the same rule of (i) or (ii) must necessarily be followed for $B = \sum_k B_k \psi_k$ and $Q = \sum_k Q_k \psi_k$. These two conditions imply that the same rule (i) or (ii) must be followed for both the state vectors $A = \sum_k A_k \psi_k$ and $B = \sum_k B_k \psi_k$ which is what we set out to prove. We have, therefore, established that for a given symmetry transformation $\hat{T}$, all the state vectors must necessarily obey rule

(i) $\hat{U}\left(\sum_k C_k \psi_k\right) = \sum_k C_k \hat{U} \psi_k$ or all the state vectors must obey rule

(ii) $\hat{U}\left(\sum_k C_k \psi_k\right) = \sum_k C_k^* \hat{U} \psi_k$.

We show, as the final step of this proof that the transformation $\hat{U}$ is either (a) linear and unitary or (b) antilinear and antiunitary.

Let us assume that all the state vectors obey rule (i) $\hat{U}\left(\sum_k C_k \psi_k\right) = \sum_k C_k \hat{U} \psi_k$. Consider any two arbitrary state vectors $\psi = \sum_k A_k \psi_k$ and $\phi = \sum_k B_k \psi_k$. By rule (i), we then have

$$\hat{U}(\alpha\psi + \beta\phi) = \hat{U}\sum_k (\alpha A_k + \beta B_k)\psi_k = \sum_k (\alpha A_k + \beta B_k)\hat{U}\psi_k$$

$$= \alpha\sum_k A_k \hat{U}\psi_k + \beta\sum_k B_k \hat{U}\psi_k = \alpha\,\hat{U}\psi + \beta\,\hat{U}\phi \tag{46}$$

Equation (46) establishes the linearity of $\hat{U}$.

We now show that it is unitary. We have

$$\langle\hat{U}\psi|\hat{U}\phi\rangle = \sum_{kl} A_k^* B_l \langle\hat{U}\psi_k|\hat{U}\psi_l\rangle = \sum_k A_k^* B_l = \langle\psi|\phi\rangle \tag{47}$$

so that $\hat{U}$ is unitary.

The case of $\hat{U}$ being antilinear and antiunitary can be resolved in an identical manner. Let us assume that all state vecters $C = \sum_k C_k\,\psi_k$ obey rule (ii), viz. $\hat{U}\left(\sum_k C_k\,\psi_k\right) = \sum_k C_k^*\,\hat{U}\psi_k$. We expand any two arbitrary vectors $\psi = \sum_k A_k\,\psi_k$ and $\phi = \sum_k B_k\,\psi_k$ as in the previous case and have

$$\hat{U}(\alpha\psi + \beta\phi) = \hat{U}\sum_k (\alpha A_k + \beta B_k)\psi_k = \sum_k (\alpha^* A_k^* + \beta^* B_k^*)\,\hat{U}\psi_k$$

$$= \alpha^*\sum_k A_k^*\,\hat{U}\psi_k + \beta^*\sum_k B_k^*\,\hat{U}\psi_k = \alpha^*\,\hat{U}\psi + \beta^*\,\hat{U}\phi \tag{48}$$

thereby establishing antilinearity.

Furthermore,

$$\langle\hat{U}\psi|\hat{U}\phi\rangle = \sum_{kl} A_k B_l^* \langle\hat{U}\psi_k|\hat{U}\psi_l\rangle = \sum_k A_k B_l^* = \langle\psi|\phi\rangle^* \tag{49}$$

showing antiunitarity.

In the context of symmetry transformations being unitary or antiunitary, the physical nature of the transformation would determine this characteristic. For example, transformations such as translations and rotations that belong to a continuous group must necessarily be unitary because, in the case of unitary transformations, finite transformations can be generated by a sequence of infinitesimal transformations as is required in the case of continuous groups. When we perform such infinitesimal transformations, the transformed basis vectors e_i^t are arbitrarily close to e_i. If the transformed state vector $\psi^t = a^{it}\,e_i^t$ is also to be arbitrarily close to the original state vector $\psi = a^i e_i$, then we must mandate $a^{it} = a^i$ and not $a^{it} = a^{i*}$. It follows that antiunitary transformations do not meet this criterion of continuous groups.

However, it is emphasized that the above argument debarring antiunitary transformations is specific only to the continuous case and is not applicable to discrete transformations like space reflection (which is unitary) and time reversal (which is antiunitary).

1.3 RELATIVISTIC INVARIANCE (PERES, 1995)

It is now well accepted that the laws of physics are not invariant under a Galelian transformation defined by $\mathbf{r} \to \mathbf{r}' = \mathbf{r} + \mathbf{v}t, t \to t' = t$, even in the non-relativistic limit $v << c$. The reason is that the second of the above equations implies the existence of a universal time that is independent of the motion of the measuring apparatus. This goes against the tenets of special relativity.

Let us consider a hypothetical situation wherein we have two distant clocks in arbitrary motion in respective reference frames I and I' and we want to synchronize the times indicated by them. To synchronize these clocks, we need to convey signals from one to the other. This cannot be done instantaneously since there is no physical agent that propagates with infinite speed. The best we can do is to convey optical signals that travel with the velocity of light. Furthermore, these signals attain the same velocity in every inertial frame. Thus, the times t and t' associated with the coordinate systems I and I', that are in uniform relative motion with velocity $\mathbf{v}$, are related to the space vectors as $c^2t^2 - \mathbf{r}^2 = c^2t'^2 - \mathbf{r}'^2$.

The above equation implies that along with the infinitesimal space transformation $\mathbf{r} \to \mathbf{r}' = \mathbf{r} + \mathbf{v}t$, there must necessarily be a time transformation given by $t \to t' = t + \frac{\mathbf{v}.\mathbf{r}}{c^2}$. Thus, the Lorentz transformations that constitute the underlying symmetry transformations of relativistic quantum mechanics comprises the aforesaid pair of infinitesimal transformations. It may be noted that the time transformation $t \to t' = t + \frac{\mathbf{v}.\mathbf{r}}{c^2}$ does not reduce to the Galilean version $t \to t' = t$ even for non-relativistic velocities $v << c$. This is because $t \to t' = t + \frac{\mathbf{v}.\mathbf{r}}{c^2}$ must hold for arbitrarily large $\mathbf{r}$ and even for $v << c$, we can, at best, neglect terms of second order or higher in $\frac{|\mathbf{v}|}{c}$, but must retain terms of the first order.

The explicit involvement of the position vector $\mathbf{r}$ in the relativistic transformation for time creates complications in applying the canonical (unitary) formalism for, in the canonical (unitary) formalism we cannot have $t \to t' = t + \frac{\mathbf{v}.\mathbf{r}}{c^2}$, because both t, t' are c-numbers whereas $\mathbf{r}$, the position vector, must, in the canonical (unitary) formalism, correspond to the coordinate operator $\hat{q}$.

In essence, the special theory of relativity propounded by Einstein, envisages placing the spatial and temporal coordinates on an identical footing and, thereby, merging space and time into a single concept called "spacetime". Dynamical laws may, then, be expressed conveniently in terms of derivatives of "proper time", i.e., $d\tau^2 = dt^2 - \frac{1}{c^2}d\mathbf{r}^2$. However, while this is an elegant formulation for classical relativistic dynamics, it does not lend itself very well to canonical quantization. A multi-particle system would involve several proper times (one corresponding to each particle) to be assimilated in a single wave function that is used to describe the system.

The canonical quantization of a classical system envisages the following steps:

(a) Let (x^i, p_i) be a set of generalized coordinates of a classical system. From *each* coordinate function, we obtain an operator that acts on state vectors in a Hilbert space H consisting of square integrable functions by making the following substitutions:

$$x^i \mapsto \hat{x}^i : \hat{x}^i|x^i\rangle = x^i|x^i\rangle \text{ and } p_i \mapsto \hat{p}_i \equiv -i\hbar \frac{\partial}{\partial x^i}$$

(b) For $x^i p_i$, the above rule creates a problem since $x^i p_i = p_i x^i$ whereas $\hat{x}^i \hat{p}_i \neq \hat{p}_i \hat{x}^i$. This problem is referred to as the "operator ordering problem". One approach to resolving this is to take $x^i p_j \mapsto \frac{1}{2}\left(\hat{x}^i \hat{p}_j + \hat{p}_j \hat{x}^i\right)$.

(c) This process is called canonical quantization since it preserves the canonical Poisson brackets, e.g., $\left[\hat{x}^i, \hat{p}_j\right] = \left(\hat{x}^i \hat{p}_j - \hat{p}_j \hat{x}^i\right) = i\hbar\delta^i_j$ which is analogous to the classical Poisson bracket $\{x^i, p_j\} = \delta^i_j$.

Thus, compatibility of the special theory of relativity with the canonical formalism mandates the following conditions:

(a) The dynamical laws, in the relativistic framework, be invariant under the relevant symmetry transformation of coordinates in four-dimensional spacetime, in analogy with the invariance of the dynamical laws under three-dimensional rotations and translations in the non-relativistic framework.

(b) There should be a canonical (unitary) transformation of the dynamical variables or the corresponding operators $\hat{q}^i$, $\hat{p}_i$ for each particle such that $\hat{q}^i$ transforms as the position vector **r** under the above symmetry transformations.

Towards resolving these issues Dirac postulated that the Lorentz transformations of special relativity constitute canonical (unitary) transformations in the relativistic framework just as spatial rotations constitute the canonical (unitary) transformations in the non-relativistic framework. His argument was based on the premises that if a canonical formulation is possible in one Lorentz frame, it should be equally possible in every Lorentz frame.

We now establish the existence of a canonical (unitary) representation of the Lorentz transformation for a single free particle with canonical variables $\mathbf{q}(t)$, $\mathbf{p}(t)$.

Since a classical canonical transformation or a quantum unitary transformation does not modify time, t not being a dynamical variable, we can write $\mathbf{q}(t) \to \mathbf{q}'(t) \equiv \mathbf{q}(t) + \delta\mathbf{q}(t)$.

The effect of an infinitesimal Lorentz transformation of time given by $t \to t' = t + \frac{\mathbf{v.r}}{c^2}$ on $\mathbf{q}(t)$ is

$$\delta\mathbf{q}(t) = \mathbf{q}'(t) - \mathbf{q}(t) = \mathbf{q}'\left(t' - \frac{1}{c^2}\mathbf{v.r}\right) - \mathbf{q}(t) = \mathbf{q}'(t') - \frac{1}{c^2}\,\mathbf{v.r}\dot{\mathbf{q}}'(t') - \mathbf{q}(t)$$

to the first order in **v**.

Now, if we require that the canonical variable (operator) $\mathbf{q}(t)$ must transform in the same manner as the coordinate vector **r**, then, we must also have $\mathbf{q}'(t') = \mathbf{q}(t) + \mathbf{v}t$ so that $\dot{\mathbf{q}}'(t') = \dot{\mathbf{q}}(t) + \mathbf{v}$, whence $\frac{1}{c^2}\mathbf{v}\dot{\mathbf{q}}'(t') = \frac{1}{c^2}\mathbf{v}\dot{\mathbf{q}}(t) + \frac{1}{c^2}\mathbf{v}^2 \approx \frac{1}{c^2}\mathbf{v}\dot{\mathbf{q}}(t)$. Substituting this and also $\mathbf{r} \equiv \mathbf{q}$ in the above equation, we get, on using the Poisson bracket equation of motion,

$$\delta\mathbf{q}(t) = \mathbf{v}t - \frac{1}{c^2}\,\mathbf{v}.\mathbf{q}(t)\;\dot{\mathbf{q}}(t) = [\mathbf{q}, \mathbf{v}t.\mathbf{p}]_{PB} - \left[\mathbf{q}, \frac{1}{c^2}(\mathbf{v}.\mathbf{q})H\right]_{PB}$$

where H is the generator of time translations corresponding to the system's Hamiltonian and satisfies $d\dot{\mathbf{q}}/dt = \partial H/\partial\mathbf{p}$. It, therefore, follows that the infinitesimal transformation $\mathbf{q}(t) \to \mathbf{q}'(t)$ is a canonical transformation. This transformation is generated by $\mathbf{v}.\mathbf{K} = \mathbf{v}.\left(t.\mathbf{p} - \frac{1}{c^2}\mathbf{q}H\right)$ whose quantum analogue is $\hat{K}_i = t.\hat{\mathbf{p}}_i - \frac{1}{2c^2}\left(\hat{H}\,\hat{\mathbf{x}}_i + \hat{\mathbf{x}}_i\hat{H}\right)$.

1.4 THE LORENTZ AND POINCARE GROUPS (WIGNER, 1939,1959; OHNUKI, 1976; TUNG, 1984; RAO, 1988)

As mentioned earlier, the homogeneity and isotropy of the three-dimensional space was generalized to include the time dimension by the special theory of relativity to a four-dimensional spacetime. The symmetry groups of four-dimensional spacetime are the Lorentz group and the Poincare group.

The postulated symmetry between the three-dimensional space and the time dimension manifests itself as the constancy of the speed of light in all coordinate frames (which is one of the postulates of special relativity).

In addition to this constancy of the speed of light, special relativity also mandates that the basic laws of physics are invariant with respect to:

(a) Translations in all four coordinates.
(b) All homogeneous linear transformations on the spacetime coordinates that leave the length of the 4-vector given by $|x| = [(x^0)^2 - (\mathbf{x})^2]^{1/2} = (g_{\mu\nu}x^\mu x^\nu)^{1/2}$ invariant, where $(g_{\mu\nu})$ is the metric tensor given by $(g_{\mu\nu}) = \text{diag}\,(1\ -1\ -1\ -1)$ corresponding to the Minkowski metric with signature $(1\ \ -1\ \ -1\ \ -1)$. We shall follow natural units in which $\hbar = c = 1$ and $x^0 \equiv t$ unless explicitly stated otherwise.

1.4.1 Minkowski Spacetime and 4-vectors

The Minkowski space is a four-dimensional real vector space endowed with an indefinite scalar product $\langle x|y\rangle$ that has the following properties:

(a) The scalar product has an indefinite norm, i.e., $|x|^2 \equiv \langle x|x\rangle$ may take positive, negative or nil values.
(b) $\langle x|y\rangle = \langle y|x\rangle^* = \langle y|x\rangle$ since the Minkowski space is a real vector space.

(c) $\langle \alpha x + \beta y | \gamma z \rangle = \alpha\gamma \langle x|z\rangle + \beta\gamma\langle y|z\rangle$, $\alpha, \beta, \gamma \in \mathbb{R}$.

(d) In a basis of the following linearly independent tetrad of unit vectors $e_0 = (1\ 0\ 0\ 0)^T$, $e_1 = (0\ 1\ 0\ 0)^T$, $e_2 = (0\ 0\ 1\ 0)^T$, $e_3 = (0\ 0\ 0\ 1)^T$ the scalar product of the basis vectors is defined by $\langle e_\mu | e_\nu \rangle = g_{\mu\nu}$ so that if $x = x^\mu e_\mu$, $y = y^\nu e_\nu$, then

$$\left\langle \langle x|y\rangle = x^0\, y^0 - \sum_{i=1}^{3} x^i\, y^i \text{ and } \langle x|x\rangle = (x^0)^2 - \sum_{i=1}^{3} (x^i)^2 \right\rangle$$

Any arbitrary 4-vector in Minkowski space can be classified as:

(a) Spacelike if $\langle x|x\rangle < 0$

(b) Null vector if $\langle x|x\rangle = 0$

(c) Timelike if $\langle x|x\rangle > 0$

No two linearly independent null vectors are orthogonal for let, if possible, $x \equiv (x^0, \mathbf{x})$, $y \equiv (y^0, \mathbf{y})$, be two null vectors so that $|\mathbf{x}| = |x^0|$ and $|\mathbf{y}| = |y^0|$. Further, let x, y be orthogonal to each other so that $0 = \langle x|y\rangle = x^0 y^0 - \mathbf{x}.\mathbf{y} = x^0 y^0 - |\mathbf{x}||\mathbf{y}| \cos\theta$. Hence, we have $0 = x^0 y^0 - .|\mathbf{x}||\mathbf{y}| \cos\theta = x^0 y^0 - |x^0||y^0| \cos\theta = x^0 y^0 (1 \mp \cos\theta)$, whence $\theta = 0, \pi$ implying that $\mathbf{x}$, $\mathbf{y}$ are parallel and, therefore, $\mathbf{y} = \lambda\mathbf{x}$, whence $0 = \langle x|y\rangle = x^0 y^0 - \mathbf{xy} = x^0 y^0 - \lambda\mathbf{xx} = x^0 y^0 - \lambda|\mathbf{x}|^2 = x^0 y^0 - \lambda(x^0)^2$ so that $y^0 = \lambda x^0$. Hence, $y = \lambda x$ so that x, y are not linearly independent.

Furthermore, any vector orthogonal to a null vector cannot be timelike for if y is orthogonal to null vector x, then

$$0 = x^0 y^0 - |\mathbf{x}||\mathbf{y}| \cos\theta = x^0 y^0 - |x^0||\mathbf{y}| \cos\theta = |x^0|(\pm y^0 - |\mathbf{y}| \cos\theta)$$

whence $|\mathbf{y}| \geq |y^0|$ so that y can either be null or spacelike only.

No two timelike vectors are orthogonal for let x be a timelike vector whence $|\mathbf{x}| < |x^0|$ and let y be orthogonal to x. Then

$$0 = x^0 y^0 - |\mathbf{x}||\mathbf{y}| \cos\theta > x^0 y^0 - |x^0||\mathbf{y}| \cos\theta = |x^0|\, (\pm y^0 - |\mathbf{y}| \cos\theta)$$

whence $0 > (\pm y^0 - |\mathbf{y}|\cos\theta)$ or $|\mathbf{y}| > |y^0|$ implying that y can only be spacelike.

However, a vector that is orthogonal to a spacelike vector can be spacelike, null or timelike for, let, x be a spacelike vector, whence $|\mathbf{x}| > |x^0|$ and let y be orthogonal to x.

Then

$$0 = x^0 y^0 - |\mathbf{x}||\mathbf{y}| \cos\theta > x^0 y^0 - |x^0||\mathbf{y}|\cos\theta = |x^0|\, (\pm y^0 - |\mathbf{y}|\cos\theta)$$

whence $0 < (\pm y^0 - |\mathbf{y}|\cos\theta)$ which does not impose any restriction between $|\mathbf{y}|$ and $|y^0|$ implying that y can be spacelike, null or timelike.

In a Minkowski space, orthogonality does not imply linear independence. In fact, the case of two orthogonal null vectors being linearly dependent has already been discussed in the preceding paragraphs.

The 4-vectors share another very important property, viz. any linearly independent tetrad of mutually orthogonal vectors must necessarily consist of three spacelike and one timelike vectors. To establish this proposition, we consider a tetrad of orthogonal and linearly independent vectors $\{|e_\mu\rangle, \mu = 0, 1, 2, 3\}$. In view of the orthogonality of these vectors, we have $\langle e_\mu | e_\nu \rangle = g_{\mu\nu}$, $\mu, \nu = 0,$

1, 2, 3. Since these vectors are linearly independent, we can expand an arbitrary vector $|x\rangle$ in our Minkowski space as $|x\rangle = x^{\mu}|e_{\mu}\rangle$, whence we have $|x|^2 = \langle x|x\rangle = g_{00}(x^0)^2 + g_{11}(x^1)^2 + g_{22}(x^2)^2 + g_{33}(x^3)^2$. In view of the above results, on orthogonality of null and timelike vectors, we need only consider the following possible combinations for our tetrad of orthogonal linearly independent vectors:

(a) All $\{|e_{\mu}\rangle, \mu = 0, 1, 2, 3\}$ are spacelike vectors.

(b) Three of the $|e_{\mu}\rangle$ are spacelike and one is null.

(c) Three of the $|e_{\mu}\rangle$ are spacelike and one is timelike.

In case (a), $\langle e_{\mu}|e_{\mu}\rangle = g_{\mu\mu} < 0 \ \forall \ \mu = 0, 1, 2, 3$ so that $|x|^2 < 0$ for every arbitrary vector $|x\rangle$ in the Minkowski space which is unacceptable. Similarly, in case (b), assuming, without loss of generality, $|e_0\rangle$ as the null vector and the other three $|e_i\rangle$ as spacelike, we have $g_{00} = 0$, $g_{kk} < 0$, $\forall\ k = 1, 2, 3$ which implies that $|\,x\,|^2 \leq 0$ for every arbitrary vector $|x\rangle$ in the Minkowski space which is another unacceptable situation. Hence, orthogonal linearly independent tetrads consisting of four spacelike or three spacelike and one null vectors are forbidden in a Minkowski space. The situation (c) does not lead to any contradiction. Further, such vectors have been proved to exist in the preceding paragraphs. Hence, we can have orthogonal linearly independent tetrads of three spacelike vectors and one timelike vector only.

Thus, we conclude that:

(a) The maximum number of mutually orthogonal spacelike vectors in a Minkowski space is three.

(b) There can be a maximum of only two orthogonal spacelike vectors that are orthogonal to the same null vector.

(c) Every orthonormal basis in a Minkowski space shall consist of three spacelike vectors and one timelike vector. However, this restriction does not apply to non-orthogonal bases.

(d) A Minkowski space admits orthogonal pairs only of the types: (i) spacelike-spacelike, (ii) spacelike-null and (iii) spacelike-timelike.

1.4.2 Lorentz Group

A subgroup of the Poincare group is the Lorentz group. The Lorentz group consists of all homogeneous linear transformations on the spacetime coordinates that leave the length of the 4-vector given by $|\,x\,| = [(x^0)^2 - (\mathbf{x})^2]^{1/2} = (g_{\mu\nu}\, x^{\mu} x^{\nu})^{1/2}$ invariant, where $(g_{\mu\nu})$ is the metric tensor given by $(g_{\mu\nu}) = \text{diag}\,(1\ -1\ -1\ -1)$ corresponding to the Minkowski metric with signature $(1\ -1\ -1\ -1)$. As mentioned above, the Lorentz transformation is the spacetime transformation of special relativity connecting two inertial frames of reference I and I' that are in uniform relative motion.

In its simplest form, we consider two inertial frames with an orientation such that the respective spatial axes are parallel to each other and the primed frame I' is moving with a relative velocity $\mathbf{v} \equiv (v, 0, 0)$ along the direction of x^1 axis as observed from the unprimed frame I. We further assume that the origins of the two frames coincide at time $x^0 = 0$. The transformation equations, in this case, in natural units, will take the form

$$x'^1 = \frac{x^1 - vx^0}{\sqrt{1 - v^2}} \tag{50}$$

$$x'^2 = x^2 \tag{51}$$

$$x'^3 = x^3 \tag{52}$$

$$x'^0 = \frac{x^0 - vx^1}{\sqrt{1 - v^2}} \tag{53}$$

We can write the above transformation in matrix form as

$$x' = \Lambda x \tag{54}$$

where $x' \equiv \begin{pmatrix} x'^0 \\ x'^1 \\ x'^2 \\ x'^3 \end{pmatrix}$, $\Lambda \equiv \begin{pmatrix} \gamma & -\gamma\beta & 0 & 0 \\ -\gamma\beta & \gamma & 0 & 0 \\ 0 & 0 & 1 & 0 \\ 0 & 0 & 0 & 1 \end{pmatrix}$, $x \equiv \begin{pmatrix} x^0 \\ x^1 \\ x^2 \\ x^3 \end{pmatrix}$, $\beta \equiv v$ and $\gamma \equiv \sqrt{1 - v^2}$, respectively.

In terms of matrix components, we can write eq. (54) as $x'^\mu = \Lambda^\mu_\nu x^\nu$ with summation over repeated indices. A simple calculation shows that

$$|x'|^2 = (x'^0)^2 - \mathbf{x}'^2 = (x^0)^2 - \mathbf{x}^2 = |x|^2 \tag{55}$$

Hence, we have

$$\langle x'|x'\rangle \equiv \langle \Lambda x|\Lambda x\rangle = \langle x|x\rangle \tag{56}$$

It also follows from eq. (56) that if x, y are any two spacetime vectors representing any two spacetime points with x', y' being their images under the Lorentz transformation Λ, then

$$\langle x'|y'\rangle \equiv \langle \Lambda x|\Lambda y\rangle = \langle x|y\rangle \tag{57}$$

thereby establishing the invariance of the Minkowski product under Lorentz transformation.

This enables us to define a Lorentz transformation as "a real linear mapping of the Minkowski spacetime onto itself that preserves the scalar product $\langle x|y\rangle = x^0y^0 - \mathbf{x}\cdot\mathbf{y}$ for any two Minkowski vectors x, y".

Thus, for a Lorentz transformation, $x \in M \to x' \equiv \Lambda x \in M$, where $\Lambda \equiv (\Lambda^\mu_\nu)$ is a Lorentz matrix that satisfies eq. (57). Using the orthogonality of the basis vectors in Minkowski space $\langle e_\mu|e_\nu\rangle = g_{\mu\nu}$, we can define the scalar product as

$$\langle x|y\rangle = g_{\mu\nu}x^\mu y^\nu = x^T G y \tag{58}$$

where $G = (g_{\mu\nu}) = \text{diag}\,(1\ -1\ -1\ -1)$. Therefore, the Lorentz matrix must satisfy

$$\langle x'|y'\rangle = \langle \Lambda x|\Lambda y\rangle = (\Lambda x)^T G(\Lambda y) = x^T\Lambda^T G\Lambda y = \langle x|y\rangle = x^T G y \tag{59}$$

whence

$$\Lambda^T G\,\Lambda = G \tag{60}$$

Taking determinants of both sides, we get det Λ^T det G det Λ = det G, whence (det $\Lambda)^2 = 1$. This shows that Λ is non-singular and Λ^{-1} exists. Furthermore, since Λ is non-singular, we also get $(\Lambda^T)^{-1}G\Lambda^{-1} = (\Lambda^{-1})^T G\Lambda^{-1} = G$, whence Λ^{-1} is also a Lorentz transformation. Again, if Λ_1 and Λ_2 are two Lorentz transformations, then $(\Lambda_1\Lambda_2)^T G(\Lambda_1\Lambda_2) = \Lambda_2^T\Lambda_1^T G\Lambda_1\Lambda_2 = \Lambda_2^T G\Lambda_2 = G$ so that $\Lambda_1\Lambda_2$ is also a Lorentz transformation. Similarly, $\Lambda_2\Lambda_1$ is also a Lorentz transformation. The 4 × 4 identity matrix constitutes the identity element of the Lorentz group. Furthermore, from eq. (60), we have

$$\Lambda^{-1} = G^{-1}\Lambda^T G = G\Lambda^T G = \begin{pmatrix} \Lambda^0_0 & -\Lambda^1_0 & -\Lambda^2_0 & -\Lambda^3_0 \\ -\Lambda^0_1 & \Lambda^1_1 & \Lambda^2_1 & \Lambda^3_1 \\ -\Lambda^0_2 & \Lambda^1_2 & \Lambda^2_2 & \Lambda^3_2 \\ -\Lambda^0_3 & \Lambda^1_3 & \Lambda^2_3 & \Lambda^3_3 \end{pmatrix} \tag{61}$$

corresponding to a Lorentz transformation $\Lambda = (\Lambda^\mu_\nu)$. Thus, we have proved all the group properties of the group of Lorentz transformations. The Lorentz group corresponds to the group $O(3, 1)$ of 4 × 4 real, orthogonal matrices that preserve a quadratic form with three signs of one type and one of the other. The unconstrained 4 × 4 matrices have sixteen degrees of freedom. However, eq. (60) imposes ten constraints on the Lorentz matrices, thereby leaving only six degrees of freedom. Hence, the Lorentz group is a six real parameter group. Several interesting conclusions can be drawn by writing eq. (60) explicitly in terms of the matrix components. We have

$$\Lambda^T G\Lambda = \begin{pmatrix} (\Lambda^0_0)^2 - \sum_{i=1}^3(\Lambda^i_0)^2 & \Lambda^0_0\Lambda^0_1 - \sum_{i=1}^3\Lambda^i_0\Lambda^i_1 & \Lambda^0_0\Lambda^0_2 - \sum_{i=1}^3\Lambda^i_0\Lambda^i_2 & \Lambda^0_0\Lambda^0_3 - \sum_{i=1}^3\Lambda^i_0\Lambda^i_3 \\ \Lambda^0_0\Lambda^0_1 - \sum_{i=1}^3\Lambda^i_0\Lambda^i_1 & (\Lambda^0_1)^2 - \sum_{i=1}^3(\Lambda^i_1)^2 & \Lambda^0_1\Lambda^0_2 - \sum_{i=1}^3\Lambda^i_1\Lambda^i_2 & \Lambda^0_1\Lambda^0_3 - \sum_{i=1}^3\Lambda^i_1\Lambda^i_3 \\ \Lambda^0_0\Lambda^0_2 - \sum_{i=1}^3\Lambda^i_0\Lambda^i_2 & \Lambda^0_1\Lambda^0_2 - \sum_{i=1}^3\Lambda^i_1\Lambda^i_2 & (\Lambda^0_2)^2 - \sum_{i=1}^3(\Lambda^i_2)^2 & \Lambda^0_2\Lambda^0_3 - \sum_{i=1}^3\Lambda^i_2\Lambda^i_3 \\ \Lambda^0_0\Lambda^0_3 - \sum_{i=1}^3\Lambda^i_0\Lambda^i_3 & \Lambda^0_1\Lambda^0_3 - \sum_{i=1}^3\Lambda^i_1\Lambda^i_3 & \Lambda^0_2\Lambda^0_3 - \sum_{i=1}^3\Lambda^i_2\Lambda^i_3 & (\Lambda^0_3)^2 - \sum_{i=1}^3(\Lambda^i_3)^2 \end{pmatrix} \tag{62}$$

Since G = diag (1 –1 –1 –1) the following constraints result from eqs. (60) and (62):

$$(\Lambda^0_j)^2 - \sum_{i=1}^3(\Lambda^i_j)^2 = -1, j = 1, 2, 3 \text{ (3 constraints)} \tag{63}$$

$$(\Lambda_0^0)^2 - \sum_{i=1}^{3} (\Lambda_0^i)^2 = +1 \text{ (1 constraint)} \tag{64}$$

$$\Lambda_j^0 \Lambda_k^0 - \sum_{i=1}^{3} \Lambda_j^i \Lambda_k^i = 0, j < k; j, k = 0, 1, 2, 3 \text{ (6 constraints)} \tag{65}$$

Equations (63)-(65) also imply that the four columns of a Lorentz matrix constitute an orthonormal tetrad with the first column being the timelike basis and the next three columns being the spacelike basis vectors. Similarly, since Λ^T is also a Lorentz matrix, it follows that the four rows of a Lorentz matrix also constitute an orthonormal tetrad with the first row being timelike and the other three rows being spacelike.

1.4.3 Classification of Lorentz Transformations

From eq. (60), $(\det \Lambda)^2 = 1$, whence $(\det \Lambda) = \pm 1$. Lorentz transformations with $(\det \Lambda) = +1$ are classified as proper Lorentz transformations while those with $(\det \Lambda) = -1$ are classified as improper Lorentz transformations.

Furthermore, from eq. (64), we have $(\Lambda_0^0)^2 = \sum_{i=1}^{3} \left(\Lambda_0^i\right)^2 + 1$, whence either $\Lambda_0^0 \geq +1$ or $\Lambda_0^0 \leq -1$. Transformations with $\Lambda_0^0 \geq +1$ are called orthochronous Lorentz transformations while those with $\Lambda_0^0 \leq -1$ are called antichronous.

The orthochronous Lorentz transformations, i.e., those with $\Lambda_0^0 \geq 1$ preserve the sign of every timelike vector while antichronous ones reverse it. To establish this, we note that the zeroth component of any arbitrary vector in the primed frame can be obtained as $x'^0 = \Lambda_\mu^0 x^\mu = \Lambda_i^0 x^i + \Lambda_0^0 x^0$.Now, if x is timelike then $|\mathbf{x}| < |x^0|$. Further, from $(\Lambda_0^0)^2 = \sum_{i=1}^{3} \left(\Lambda_0^i\right)^2 + 1$, we have $1 = \frac{1}{\left(\Lambda_0^0\right)^2} \sum_{i=1}^{3} \left(\Lambda_0^i\right)^2 + \frac{1}{\left(\Lambda_0^0\right)^2}$. Since $\Lambda_0^0 \geq +1$, we have $0 < \frac{1}{\left(\Lambda_0^0\right)^2} \leq 1$ so that $0 \leq \frac{1}{\left(\Lambda_0^0\right)^2} \sum_{i=1}^{3} \left(\Lambda_0^i\right)^2 < 1$. This gives $\frac{1}{\Lambda_0^0 x^0} x'^0 = \frac{\Lambda_i^0 x^i}{\Lambda_0^0 x^0} + 1 > 0$ so that if Λ is orthochronous, then $\Lambda_0^0 \geq +1$ and the sign of the zeroth vector remains unchanged. On the other hand, if Λ is antichronous, then $\Lambda_0^0 \leq -1$ so that the zeroth component must necessarily change sign in order that the above inequality be satisfied.

On the basis of the above analysis, we can split the Lorentz group into four disjoint sets as follows:

(a) $L_+^\uparrow$ with $\Lambda_0^0 \geq +1$ and $\det \Lambda = +1$ (66)

(b) $L_+^{\downarrow}$ with $\Lambda_0^0 \le -1$ and $\det \Lambda = +1$ (67)

(c) $L_-^{\uparrow}$ with $\Lambda_0^0 \ge +1$ and $\det \Lambda = -1$ (68)

(d) $L_-^{\downarrow}$ with $\Lambda_0^0 \le -1$ and $\det \Lambda = -1$ (69)

Of these four subsets, only (a) constitutes a subgroup of the Lorentz group since only (a) contains the identity element of the Lorentz group. It is called the orthochronous proper Lorentz group (OPLG). The other three, although subsets of the Lorentz group, are not subgroups of the Lorentz group since none contains the identity element which has $\Lambda_0^0 \ge +1$ and $\det \Lambda = +1$ and is, therefore, a member of (a) only.

Subgroups of the Lorentz group may, however, be constructed using any of (b), (c) or (d) by adjoining them with (a). We have

(i) The proper Lorentz group $L_+^{\uparrow} \cup L_+^{\downarrow}$

(ii) The orthochronous Lorentz group $L_+^{\uparrow} \cup L_-^{\uparrow}$

(iii) The antichronous Lorentz group $L_+^{\uparrow} \cup L_-^{\downarrow}$

To establish that (i) is a subgroup of $L = \{\Lambda\}$, we need to establish that the composition of two proper Lorentz transformations is a proper Lorentz transformation, that the inverse of a proper Lorentz transformation is a proper Lorentz transformation and that the identity is a proper Lorentz transformation. We have, if Λ_1 and Λ_2 are two proper Lorentz transformations then, $\det(\Lambda_1\Lambda_2) = \det(\Lambda_1)\det(\Lambda_2) = 1$ showing that $\Lambda_1\Lambda_2$ is a proper Lorentz transformation. Secondly, for a proper Lorentz transformation Λ, we have $\det(\Lambda^{-1}) = [\det(\Lambda)]^{-1} = 1$ so that Λ^{-1} is proper. Finally, $\det(\mathbf{I}_{4\times 4}) = 1$ so that the identity is also a proper Lorentz transformation and hence, a member of the subgroup.

To prove that (ii) is a subgroup of $\Lambda = \{\Lambda\}$, we need to establish that the composition of two orthochronous Lorentz transformations is an orthochronous Lorentz transformation, that the inverse of an orthochronous Lorentz transformation is an orthochronous Lorentz transformation and that the identity is an orthochronous Lorentz transformation. We need to show that, if Λ_1 and Λ_2 are two orthochronous Lorentz transformations so that $(\Lambda_1)_0^0 \ge 1$ and $(\Lambda_2)_0^0 \ge 1$, then $(\Lambda_1\Lambda_2)_0^0 \ge 1$. We have, $(\Lambda_1\Lambda_2)_0^0 = (\Lambda_1)_\mu^0(\Lambda_2)_0^\mu$. Given two quadruples $(1\ (\Lambda_1)_1^0\ (\Lambda_1)_2^0\ (\Lambda_1)_3^0)$ and $(1 - (\Lambda_2)_1^0 - (\Lambda_2)_2^0 - (\Lambda_2)_3^0)$, by applying Schwarz's inequality, we have

$$1 - \sum_{i=1}^{3}(\Lambda_1)_i^0(\Lambda_2)_0^i \le \left|\left\{1 + \sum_{i=1}^{3}\left[(\Lambda_1)_i^0\right]^2\right\}^{\frac{1}{2}}\right| \cdot \left|\left\{1 + \sum_{i=1}^{3}\left[(\Lambda_2)_i^0\right]^2\right\}^{1/2}\right| \tag{70}$$

Since eq. (64) holds equally well for summation of elements across columns, we also have $(\Lambda_0^0)^2 - \sum_{i=1}^{3}\left(\Lambda_i^0\right)^2 = +1$, whence the right-hand side of eq. (70) reduces to $(\Lambda_1)_0^0(\Lambda_2)_0^0$. Thus,

$1 \le \sum_{i=1}^{3} (\Lambda_1)_i^0 (\Lambda_2)_0^i + (\Lambda_1)_0^0 (\Lambda_2)_0^0 = (\Lambda_1\Lambda_2)_0^0$, as desired. The fact that the inverse of an orthochronous Lorentz transformation is orthochronous follows from the explicit expression of the inverse given by eq. (61) and the orthochronous nature of identity is trivial.

We now establish that $L_+^\uparrow \cup L_-^\downarrow$ also constitutes a subgroup of the Lorentz group. The set $L_+^\uparrow \cup L_-^\downarrow$ consists of elements of the type Λ or $-G\Lambda$, where $\Lambda \in L_+^\uparrow$. Consider any two elements of this set, say Ξ_1 and Ξ_2. If both $\Xi_1, \Xi_2 \in L_+^\uparrow$ then their composition also belongs to $L_+^\uparrow$ and, hence, to $L_+^\uparrow \cup L_-^\downarrow$. If Ξ_1 is of the type $-G\Lambda$, where $\Lambda \in L_+^\uparrow$, and $\Xi_2 \in L_+^\uparrow$ then $\Xi_1\Xi_2 = -G\Lambda\Xi_2 = -G\Lambda'$ where $\Lambda' \in L_+^\uparrow$ whence $\Xi_1 \Xi_2 \in L_+^\uparrow \cup L_-^\downarrow$. If Ξ_2 is of the type $-G\Lambda$, where $\Lambda \in L_+^\uparrow$ and $\Xi_1 \in L_+^\uparrow$, then $\Xi_1 \Xi_2 = -\Xi_1 G\Lambda = -\Xi_1(\Lambda^T)^{-1} G = -\Xi_1(\Lambda^{-1})^T G = -\Lambda' G = -G(\Lambda'^T)^{-1} \in L_+^\uparrow \cup L_-^\downarrow$ where $\Lambda' \in L_+^\uparrow$.

If both Ξ_1, Ξ_2 are of the type $-G\Lambda$, where $\Lambda \in L_+^\uparrow$, then we have $\Xi_1\Xi_2 = (-G\Lambda_1)(-G\Lambda_2) = (\Lambda_1^T)^{-1} GG\Lambda_2 = (\Lambda_1^T)^{-1} \Lambda_2 \in L_+^\uparrow \cup L_-^\downarrow$. These proofs make use of the fact that the transpose as well as the inverse of an OPLT is also an OPLT. Cases with factors reversed can be proved similarly.

Also, if $\Xi \in L_+^\uparrow$, then $\Xi^{-1} \in L_+^\uparrow \subset L_+^\uparrow \cup L_-^\downarrow$. If Ξ is of the form $-G\Lambda$, where $\Lambda \in L_+^\uparrow$, then $\Xi^{-1} = (-G\Lambda)^{-1} = -\Lambda^{-1}G^{-1} = -\Lambda^{-1}G = -G\Lambda^T \in L_+^\uparrow \cup L_-^\downarrow$. The identity is obviously a member of $L_+^\uparrow \cup L_-^\downarrow$ confirming that it is a subgroup.

It is interesting to observe that elements of $L_+^\uparrow$, $L_-^\downarrow$ and $L_-^\downarrow$ can be obtained from the elements of $L_+^\uparrow$ by operating on the latter by $-\mathbf{I}$, $+G$ and $-G$ respectively, so that $L_+^\uparrow = -\mathbf{I}L_+^\uparrow$, $L_-^\downarrow = GL_+^\uparrow$ and $L_-^\downarrow = -GL_+^\uparrow$ respectively.

1.4.4 Generators of Lorentz Transformations

We now show that every Lorentz transformation in the neighbourhood of the identity is uniquely expressible in the form $\Lambda = e^S$, where S is a 4×4 matrix satisfying

$$S^T G + GS = 0 \tag{71}$$

We first prove that there does exist a unique S such that $\Lambda = e^S$. Let us assume that there does exist such a mapping. Then, we have

$$(\Lambda)_\rho^\mu = (e^s)_\rho^\mu = \delta_\rho^\mu + (\mathrm{S})_\rho^\mu + \frac{(S^2)_\rho^\mu}{2!} + \ldots \tag{72}$$

Now

$$\frac{\partial \delta_\rho^\mu}{\partial (S)_\sigma^\nu} = 0 \tag{73}$$

$$\frac{\partial(S^2)^{\mu}_{\rho}}{\partial(S)^{\nu}_{\sigma}} = \frac{\partial}{\partial(S)^{\nu}_{\sigma}} [(S)^{\mu}_{\lambda} (S)^{\lambda}_{\rho}]$$

$$= \frac{\partial(S)^{\mu}_{\lambda}}{\partial(S)^{\nu}_{\sigma}} (S)^{\lambda}_{\rho} + (S)^{\mu}_{\lambda} \frac{\partial(S)^{\lambda}_{\rho}}{\partial(S)^{\nu}_{\sigma}} \tag{74}$$

The right-hand side goes to zero at $S = 0$ because the coefficients of the derivatives will vanish at $S = 0$. Derivatives of all higher powers of S will also vanish on account of the same reasoning. However,

$$\frac{\partial(S)^{\mu}_{\rho}}{\partial(S)^{\nu}_{\sigma}} = \delta^{\mu}_{\nu}\, \delta^{\rho}_{\sigma} = (\mathbf{I})^{\mu}_{\nu}\, (\mathbf{I})^{\rho}_{\sigma} = (\mathbf{I} \times \mathbf{I})^{\mu\rho}_{\nu\sigma} \tag{75}$$

Therefore,

$$\frac{\partial(\Lambda)^{\mu}_{\rho}}{\partial(S)^{\nu}_{\sigma}} = (\mathbf{I} \times \mathbf{I})^{\mu\rho}_{\nu\sigma} + \text{terms tending to 0 as } S \to 0 \tag{76}$$

From eq. (76), it follows that

$$\det \left.\frac{\partial(\Lambda)^{\mu}_{\rho}}{\partial(S)^{\nu}_{\sigma}}\right|_{S=0} = \det\,(\mathrm{I} \times \mathrm{I})|_{S=0} = 1 \neq 0 \tag{77}$$

Hence, the functional determinant of the derivatives of the elements of Λ with respect to the elements of S at $S = 0$ does not vanish thereby proving that there exists a unique S corresponding to the given Lorentz transformation Λ such that $S \to \Lambda = e^S$ maps S in a neighbourhood at $S = 0$ onto Λ in the neighbourhood of the identity matrix. Now, $G = \Lambda^T G \Lambda = e^{S^T} G e^S \Rightarrow e^{-S^T} G = G e^S \Rightarrow G e^{-S^T} G = e^S$, whence $e^{-GS^TG} = e^S$ whereby the uniqueness of S implies $-GS^TG = S \Rightarrow S^TG + GS = 0$ where we have used

$$e^{-GS^TG} = \mathbf{I} + (-GS^TG) + \frac{1}{2!}(-GS^TG)^2 + \frac{1}{3!}(-GS^TG)^3 + \dots$$

$$= G\mathbf{I}G + G(-S^T)G + \frac{1}{2!}(-GS^TG)(-GS^TG) + \frac{1}{3!}(-GS^TG)(-GS^TG)(-GS^TG) + \dots$$

$$= G\mathbf{I}G + G(-S^T)G + \frac{1}{2!}G(-S^T)^2G + \frac{1}{3!}G(-S^T)^3G + \dots = Ge^{-S^T}G \tag{78}$$

1.4.5 Spectrum of a Lorentz Matrix

To study the eigenvalue spectrum of OPLTs, it is more convenient to introduce Minkowski coordinates by the substitution $\mathbf{x} \mapsto \mathbf{x}$ and $ix^0 \mapsto x^0$. This is achieved by the transformation matrix $H = \text{diag}\,(i\ 1\ 1\ 1)$ which takes real coordinates into the Minkowski coordinates. This matrix has the following properties:

(i) $H^{-1} \equiv \text{diag}\,(-i\ 1\ 1\ 1)$ (79)

(ii) $H^T = H$ (80)

(iii) $(H^T)^{-1} = H^{-1}$ (81)

(iv) $HGH = H^{-1}GH^{-1} = -\mathbf{I}$ (82)

Let Λ_m represent the Lorentz transformation in Minkowski coordinates. Then, $\Lambda_m = H\Lambda H^{-1}$, whence $\Lambda = H^{-1}\Lambda_m H$. Similarly, the S matrix gets transformed as $S_m = HSH^{-1}$. We have, from eq. (71)

$$S^T = -GSG$$

$$(H^{-1} S_m H)^T = -G(H^{-1}S_m H)G \Rightarrow$$

$$\Rightarrow \quad H^T S_m^T (H^{-1})^T = -G(H^{-1}S_m H)G$$

$$HS_m^T H^{-1} = -G(H^{-1}S_m H)G \Rightarrow$$

$$\Rightarrow \quad S_m^T = -H^{-1}G(H^{-1}S_m H)\, GH = -S_m \tag{83}$$

Thus, in the Minkowski representation, the S_m matrix is a skew-symmetric matrix.

Furthermore, using

$$\Lambda^T G\Lambda = G \Rightarrow (H^{-1}\Lambda_m H)^T\, GH^{-1}\Lambda_m H = G$$

$$\Rightarrow \quad H^T(\Lambda_m)^T(H^{-1})^T GH^{-1}\Lambda_m H = G$$

$$\Rightarrow \quad H(\Lambda_m)^T H^{-1}GH^{-1}\, \Lambda_m H = G \Rightarrow -H(\Lambda_m)^T \Lambda_m H = G$$

$$\Rightarrow \quad (\Lambda_m)^T \Lambda_m = -H^{-1}GH^{-1} = \mathbf{I} \tag{84}$$

Thus, in Minkowski coordinates, $\Lambda_m = e^{S_m}$, where S_m is a skew-symmetric matrix and also $(\Lambda_m)^T \Lambda_m = \mathbf{I}$. Furthermore, since in Minkowski coordinates the $(0k)$ and the $(k0)$ elements of $\Lambda_m = e^{S_m}$ are purely imaginary (this is because the 4-vectors in Minkowski coordinates have their first components purely imaginary and the other three components real, so that a Lorentz transformation that is required to send 4-vectors to 4-vectors, must have Λ^i_j real, Λ^i_0 and Λ^0_j purely imaginary and Λ^0_0 real), it follows that the $(0k)$ and the $(k0)$ elements of S_m must also necessarily be purely imaginary.

Furthermore, since $\Lambda_m = e^{S_m}$, the eigenvalues of Λ_m will be of the form e^λ where λ is the eigenvalue of S_m. Let

$$S_m = \begin{pmatrix} 0 & ie^1 & ie^2 & ie^3 \\ -ie^1 & 0 & -h^3 & h^2 \\ -ie^2 & h^3 & 0 & -h^1 \\ -ie^3 & -h^2 & h^1 & 0 \end{pmatrix} \tag{85}$$

Solving the eigenvalue equation

$$|S_m - \lambda\mathbf{I}| = 0 \tag{86}$$

we get

$$\sum_{k=0}^{4} (-1)^k \lambda^{4-k} p_k = 0 \tag{87}$$

where p_k represents the sum of the principal minors of order k.

Now,

(i) p_1 represents the trace of S_m which is zero since S_m is skew symmetric.

(ii) p_2 is the sum of the principal minors of order 2 which can be shown to be

$$\sum_{i=1}^{3}(h^i)^2 - \sum_{i=1}^{3}(e^i)^2 \equiv \mathbf{h}^2 - \mathbf{e}^2 \tag{88}$$

(iii) p_3 is the sum of the third order determinants. However, all determinants of odd dimensions of a skew-symmetric matrix are zero so that p_3 being their sum also vanishes.

(iv) $p_4 = \det(S_m) = -(\mathbf{h}.\mathbf{e})^2$ (89)

where $\mathbf{h} \equiv (h^1, h^2, h^3)$ and $\mathbf{e} \equiv (e^1, e^2, e^3)$.

Using these results, eq. (86) gives

$$\lambda^4 + \lambda^2(\mathbf{h}^2 - \mathbf{e}^2) - (\mathbf{h}.\mathbf{e})^2 = 0 \tag{90}$$

whence

$$\lambda^2 = \frac{-(\mathbf{h}^2 - \mathbf{e}^2) \pm [(\mathbf{h}^2 - \mathbf{e}^2)^2 + 4(\mathbf{h}.\mathbf{e})^2]^{1/2}}{2} \tag{91}$$

Now

$$|(\mathbf{h}^2 - \mathbf{e}^2)| \leq |[(\mathbf{h}^2 - \mathbf{e}^2)^2 + 4(\mathbf{h}.\mathbf{e})^2]^{1/2}| \tag{92}$$

so that, we may write

$$(\theta_b)^2 = \frac{-(\mathbf{h}^2 - \mathbf{e}^2) + [(\mathbf{h}^2 - \mathbf{e}^2)^2 + 4(\mathbf{h}.\mathbf{e})^2]^{1/2}}{2} \tag{93}$$

and

$$(i\theta_r)^2 = \frac{-(\mathbf{h}^2 - \mathbf{e}^2) - [(\mathbf{h}^2 - \mathbf{e}^2)^2 + 4(\mathbf{h}.\mathbf{e})^2]^{1/2}}{2} \tag{94}$$

whence the eigenvalues of S_m are $\pm\,\theta_b$ and $\pm\,i\theta_r$ and those of Λ_m are exponentials of these.

Expressions for the eigenvalues in terms of the matrix elements of the Lorentz matrix can be obtained by solving the eigenvalue equation analogous to eq. (86), i.e.,

$$|\Lambda_m - \mu\mathbf{I}| = 0 \tag{95}$$

which expands to

$$\mu^4 - q_1\mu^3 + q_2\mu^2 - q_3\mu + q_4 = 0 \tag{96}$$

where

(i) q_1 represents the trace of Λ_m.

(ii) q_2 is the sum of principal minors of the second order.

(iii) q_3 is the sum of all principal minors of third order. Now, since Λ_m is orthogonal, each element is equal to its own cofactor. Thus, every principal minor of the third order which is a cofactor of the corresponding diagonal element becomes equal to that diagonal element. In other words, the sum of the principal minors of the third order is equal to the trace of the matrix, i.e., $q_3 = q_1$.

(iv) q_4 is equal to the determinant of the Lorentz matrix which, for OPLT, is unity.

Hence, we may write eq. (96) as

$$\mu^4 - q_1\mu^3 + q_2\mu^2 - q_1\mu + 1 = 0 \tag{97}$$

The solution of this equation are the eigenvalues of Λ_m which, we know, are of the form $e^{\pm i\theta_r}$ and $e^{\pm\theta_b}$, i.e.,

$$\mu = e^{\pm i\theta_r}, e^{\pm\theta_b} \tag{98}$$

so that, by the theory of equations, we have

$$q_1 = e^{i\theta_r} + e^{-i\theta_r} + e^{\theta_b} + e^{-\theta_b} \tag{99}$$

and

$$q_2 = e^{i\theta_r} e^{-i\theta_r} + e^{i\theta_r} e^{\theta_b} + e^{i\theta_r} e^{-\theta_b} + e^{-i\theta_r} e^{\theta_b} + e^{-i\theta_r} e^{-\theta_b} + e^{\theta_b} e^{-\theta_b} \tag{100}$$

Equations (99) and (100) can be put in the form

$$q_1 = 2 \cosh \theta_b + 2 \cos \theta_r \tag{101}$$

and

$$q_2 = 4 \cosh \theta_b \cos \theta_r + 2 \tag{102}$$

whence

$$\cosh \theta_b = \frac{1}{4}[q_1 + (q_1^2 - 4q_2 + 8)^{1/2}] \tag{103}$$

and

$$\cos \theta_r = \frac{1}{4}[q_1 - (q_1^2 - 4q_2 + 8)^{1/2}] \tag{104}$$

The eigenvalues obtained in Minkowskian coordinates will hold equally well in real coordinates since eigenvalues are unaffected by coordinate transformations.

1.5 THE LIE ALGEBRA OF THE OPLG (TUNG, 1984; RAO, 1988; WEINBERG, 1995)

We outline here the general theory behind infinitesimal linear transformations, the concept of generators of linear groups and the related concept of Lie algebras.

For this purpose, we consider a linear group G of n-dimensional matrices $A = (a_k^i)$. Each element a_k^i is a differentiable function of r parameters identified as $\alpha_1, \alpha_2, \ldots, \alpha_r$. Then r is construed as the dimension of the group G. Let us assume that the identity element of G is identified with the values $\alpha_1^0, \alpha_2^0, \ldots, \alpha_r^0$ of the r parameters. If we write $\alpha_k' = \alpha_k - \alpha_k^0$ for $k = 1, 2, \ldots, r$, then the identity element of G will be identified with the values $\alpha_k' = 0$ for $k = 1, 2, \ldots, r$. In view of this reasoning, we can assume, without loss of generality, that the identity element of G is identified with the values $\alpha_k = 0$ for $k = 1, 2,..., r$. In such a situation, we have

$$a_k^i\,(0, 0,..., 0) = \delta_k^i \tag{105}$$

and

$$A(0, 0,..., 0) = \mathbf{I} \tag{106}$$

where $i, k = 1, 2,..., n$.

Now, the matrix A which is an element of the group G, is completely and uniquely specified by the values of the parameters $\alpha_1, \alpha_2,..., \alpha_r$. Hence, we can map each A into a Euclidean space of r dimensions. This r-dimensional space is called the parameter space of group G. The identity element of G is mapped into the origin of the parameter space. Since, we require a one-one continuous mapping between points in the parameter space and the elements of the group G in the neighbourhood of the identity (origin in the parameter space), the Jacobian matrix $\left(\dfrac{\partial a_k^i}{\partial \alpha_j}\right)_{\alpha_1=\alpha_2=...=\alpha_r=0}$, for $i, k = 1, 2, ..., n$ evaluated at the origin must be of rank r so that the r rows of the Jacobian matrix are linearly independent and we can solve for α_k from a_k^i. The linear independence of the r rows of the Jacobian matrix implies that the r matrices defined by $u_k = \left(\dfrac{\partial A}{\partial \alpha_k}\right)_{\alpha_1=...=\alpha_r=0}$ for $k = 1, 2,...,r$ are also linearly independent. Since, the right-hand side of $u_k = \left(\dfrac{\partial A}{\partial \alpha_k}\right)_{\alpha_1=...=\alpha_r=0}$ is the value of $\dfrac{\partial A}{\partial \alpha_k}$ computed at the origin of the parameter space, i.e., it is the matrix whose entries are, respectively, the values of $\dfrac{\partial a_j^i\,(\alpha_1,...\alpha_r)}{\partial \alpha_k}$ computed at $\alpha_1 = ... = \alpha_r = 0$, it follows that all the elements of u_k will be constants.

Now, consider the matrix $A(\alpha_1(\theta),..., \alpha_r(\theta)) \equiv A(\theta)$, where $\alpha_k(\theta)$ are differentiable functions of θ with $\alpha_k(0) = 0$. Then, $A(\theta)$ is a one parameter family of matrices of G and $A(0) = \mathbf{I}$. The matrix $\left.\dfrac{\partial A(\theta)}{\partial \theta}\right|_{\theta=0}$ is called the infinitesimal transformation of the transformations represented by the family $A(\theta)$.

The set of all such infinitesimal transformations constitutes a vector space, say V, for

(i) Given two one parameter families of matrices $A(\theta)$, $B(\theta)$ of G and infinitesimal generators $u = \left.\dfrac{\partial A(\theta)}{\partial \theta}\right|_{\theta=0}$ and $v = \left.\dfrac{\partial B(\theta)}{\partial \theta}\right|_{\theta=0}$, respectively, we have

$$\left.\frac{\partial A(\theta)B(\theta)}{\partial \theta}\right|_{\theta=0} = \left.\frac{\partial A(\theta)}{\partial \theta}.B(\theta)\right|_{\theta=0} + A(\theta)\,.\left.\frac{\partial B(\theta)}{\partial \theta}\right|_{\theta=0}$$

$$= \left.\frac{\partial A(\theta)}{\partial \theta}\right|_{\theta=0} B(0) + A(0) \,.\, \left.\frac{\partial B(\theta)}{\partial \theta}\right|_{\theta=0} = \left.\frac{\partial A(\theta)}{\partial \theta}\right|_{\theta=0} \mathbf{I} + \mathbf{I} \left.\frac{\partial B(\theta)}{\partial \theta}\right|_{\theta=0}$$

$$= \left.\frac{\partial A(\theta)}{\partial \theta}\right|_{\theta=0} + \left.\frac{\partial B(\theta)}{\partial \theta}\right|_{\theta=0} = u + v \tag{107}$$

so that $u, v \in V \Rightarrow u + v \in V$.

(ii) Given a one parameter family of matrices $A(\theta)$ of G and infinitesimal generators $u = \left.\frac{\partial A(\theta)}{\partial \theta}\right|_{\theta=0}$, we have

$$\left.\frac{\partial A(\lambda\theta)}{\partial \theta}\right|_{\theta=0} = \lambda \left.\frac{\partial A(\theta)}{\partial \theta}\right|_{\theta=0} = \lambda u \tag{108}$$

whence λu is the infinitesimal generator corresponding to the family $A(\lambda\theta)$ so that

$$u \in V \Rightarrow \lambda u \in V \tag{109}$$

Hence, V is a vector space. Now, consider an arbitrary $u \in V$. Then, we have, for some $A(\theta) \in G$

$$u = \left.\frac{\partial A(\theta)}{\partial \theta}\right|_{\theta=0} = \sum_{k=1}^{r} \left(\frac{\partial A(\theta)}{\partial \alpha_k}\right)_0 \left(\frac{\partial \alpha_k}{\partial \theta}\right)_0 = \sum_{k=1}^{r} u_k \beta^k \tag{110}$$

Since $u \in V$ is arbitrary, the r linearly independent matrices $u_k = \left(\frac{\partial A(\theta)}{\partial \alpha_k}\right)_0$ constitute a basis for V which is a r-dimensional vector space. Now, consider $A(\phi)$ to be an arbitrary one parameter family of matrices of G and let $B(\theta)$ be another such one parameter family characterized by a different parameter θ. Then, $v = \left.\frac{\partial B(\theta)}{\partial \theta}\right|_{\theta=0} \in V$. Further,

$$w = A(\phi)\, vA^{-1}(\phi) = A(\phi) \left.\frac{\partial B(\theta)}{\partial \theta} A^{-1}(\phi)\right|_{\theta=0} = \left.\frac{\partial [A(\phi) B(\theta) A^{-1}(\phi)]}{\partial \theta}\right|_{\theta=0} \in V$$

Since, V is a vector space, it follows that $\dfrac{A(\phi)\left.\dfrac{\partial B(\theta)}{\partial \theta} A^{-1}(\phi)\right|_{\theta=0} - \left.\dfrac{\partial B(\theta)}{\partial \theta}\right|_{\theta=0}}{\phi} \in V$ for $\phi \neq 0$

because both terms of the numerator belong to V and the denominator is a scalar. Continuity of $A(\phi)$ in the neighbourhood of the identity implies that

$$\lim_{\phi\to 0} \frac{A(\phi)\frac{\partial B(\theta)}{\partial \theta}A^{-1}(\phi)\Big|_{\theta=0} - \frac{\partial B(\theta)}{\partial \theta}\Big|_{\theta=0}}{\phi} = \frac{\partial}{\partial \phi}\left[A(\phi)\frac{\partial B(\theta)}{\partial \theta}A^{-1}(\phi)\right]\Bigg|_{\theta,\,\phi=0} \in V \tag{111}$$

Identifying $u = \frac{\partial A(\phi)}{\partial \phi}\Big|_{\phi=0}$ and $v = \frac{\partial B(\theta)}{\partial \theta}\Big|_{\theta=0}$ so that

$$\frac{\partial A^{-1}(\phi)}{\partial \phi}\Bigg|_{\phi=0} = -\frac{1}{A^2(\phi)}\frac{\partial A(\phi)}{\partial \phi}\Bigg|_{\phi=0} = -\frac{1}{A^2(0)}\frac{\partial A(\phi)}{\partial \phi}\Bigg|_{\phi=0} = -\frac{\partial A(\phi)}{\partial \phi}\Bigg|_{\phi=0} = -u \tag{112}$$

we have

$$\frac{\partial}{\partial \phi}\left[A(\phi)\frac{\partial B(\theta)}{\partial \theta}A^{-1}(\phi)\right]_{\theta,\phi=0} = \left[\frac{\partial A(\phi)}{\partial \phi}.\frac{\partial B(\theta)}{\partial \theta}A^{-1}(\phi) + A(\phi)\frac{\partial B(\theta)}{\partial \theta}\frac{\partial A^{-1}(\phi)}{\partial \phi}\right]\Bigg|_{\theta,\phi=0}$$

$$= \left[\frac{\partial A(\phi)}{\partial \phi}.\frac{\partial B(\theta)}{\partial \theta}\right]\Bigg|_{\theta,\phi=0} A^{-1}(0) + A(0)\left[\frac{\partial B(\theta)}{\partial \theta}.\frac{\partial A^{-1}(\phi)}{\partial \phi}\right]\Bigg|_{\theta,\phi=0}$$

$$= \left[\frac{\partial A(\phi)}{\partial \phi}.\frac{\partial B(\theta)}{\partial \theta}\right]\Bigg|_{\theta,\phi=0} + \left[\frac{\partial B(\theta)}{\partial \theta}.\frac{\partial A^{-1}(\phi)}{\partial \phi}\right]\Bigg|_{\theta,\phi=0} = uv - vu \in V$$

making V into a Lie algebra.

Further, since $uv - vu \in V$, it can be expressed as a linear combination of the basis vectors u_k.

Identifying u, v with the basis vectors u_k themselves, we also have

$$u_i u_j - u_j u_i = \sum_{k=1}^{r} C_{ij}^k\, u_k \tag{113}$$

C_{ij}^k are called structure constants of the group G. Equation (113) constitutes the Lie-Cartan integrability conditions.

We now consider the case of infinitesimal transformations belonging to the OPLG. In Minkowskian coordinates, any element Λ_m of the OPLG $L_+^{\uparrow}$ satisfies $(\Lambda_m)^T(\theta)\,\Lambda_m(\theta) = \mathbf{I}$, whence

$$\frac{\partial \Lambda_m^T(\theta)\Lambda_m(\theta)}{\partial \theta}\Bigg|_{\theta=0} = \frac{\partial \Lambda_m^T(\theta)}{\partial \theta}\Lambda_m(\theta)\Bigg|_{\theta=0} + \Lambda_m^T(\theta)\frac{\partial \Lambda_m(\theta)}{\partial \theta}\Bigg|_{\theta=0}$$

$$= \left.\frac{\partial \Lambda_m^T(\theta)}{\partial \theta}\right|_{\theta=0} \Lambda_m(0) + \Lambda_m^T(0) \left.\frac{\partial \Lambda_m(\theta)}{\partial \theta}\right|_{\theta=0} = \left.\frac{\partial \Lambda_m^T(\theta)}{\partial \theta}\right|_{\theta=0} + \left.\frac{\partial \Lambda_m(\theta)}{\partial \theta}\right|_{\theta=0}$$

$$= u^T + u = \left.\frac{\partial \mathbf{I}}{\partial \theta}\right|_{\theta=0} = 0, \quad \text{where } u = \left.\frac{\partial \Lambda_m(\theta)}{\partial \theta}\right|_{\theta=0} \tag{114}$$

showing that the infinitesimal generators of OPLTs in Minkowskian coordinates are skew symmetric matrices.

Now, in Minkowskian coordinates $(0k)$ and $(k0)$ elements of Λ_m are purely imaginary. It, therefore, follows that $(0k)$ and $(k0)$ elements of $u = \left.\frac{\partial \Lambda_m(\theta)}{\partial \theta}\right|_{\theta=0}$ will also be purely imaginary. Thus, the Lie algebra of OPLG is the algebra of skew symmetric matrices that additionally satisfy the condition that in Minkowskian coordinates, $(0k)$ and $(k0)$ elements are purely imaginary.

We have also shown above that in the neighbourhood of identity, in Minkowski coordinates, $\Lambda_m = e^{S_m}$, where S_m is a skew-symmetric matrix with imaginary $(0k)$ and $(k0)$ elements. It follows that the matrix S_m belongs to the vector space of u. Further, S_m has been shown to have the form

$$S_m = \begin{pmatrix} 0 & ie^1 & ie^2 & ie^3 \\ -ie^1 & 0 & -h^3 & h^2 \\ -ie^2 & h^3 & 0 & -h^1 \\ -ie^3 & -h^2 & h^1 & 0 \end{pmatrix}$$

whence we can write

$$S_m = \sum_{i=1}^{3} h^i u_i + \sum_{i=1}^{3} e^i v_i \equiv \sum_{i=1}^{3} h^i \hat{J}_i + \sum_{i=1}^{3} e^i \hat{K}_i \tag{115}$$

where

$$u_1 \equiv \hat{J}_1 = \begin{pmatrix} 0 & 0 & 0 & 0 \\ 0 & 0 & 0 & 0 \\ 0 & 0 & 0 & -1 \\ 0 & 0 & 1 & 0 \end{pmatrix}, u_2 \equiv \hat{J}_2 = \begin{pmatrix} 0 & 0 & 0 & 0 \\ 0 & 0 & 0 & 1 \\ 0 & 0 & 0 & 0 \\ 0 & -1 & 0 & 0 \end{pmatrix}, u_3 \equiv \hat{J}_3 = \begin{pmatrix} 0 & 0 & 0 & 0 \\ 0 & 0 & -1 & 0 \\ 0 & 1 & 0 & 0 \\ 0 & 0 & 0 & 0 \end{pmatrix}$$

$$v_1 \equiv \hat{K}_1 = \begin{pmatrix} 0 & i & 0 & 0 \\ -i & 0 & 0 & 0 \\ 0 & 0 & 0 & 0 \\ 0 & 0 & 0 & 0 \end{pmatrix}, v_2 \equiv \hat{K}_2 = \begin{pmatrix} 0 & 0 & i & 0 \\ 0 & 0 & 0 & 0 \\ -i & 0 & 0 & 0 \\ 0 & 0 & 0 & 0 \end{pmatrix}, v_3 \equiv \hat{K}_3 = \begin{pmatrix} 0 & 0 & 0 & i \\ 0 & 0 & 0 & 0 \\ 0 & 0 & 0 & 0 \\ -i & 0 & 0 & 0 \end{pmatrix} \tag{116}$$

and, in accordance with convention, we have designated $u_i \equiv \hat{J}_i$ and $v_i \equiv \hat{K}_i$ as the six generators of the OPLG.

These six matrices also constitute a basis for the vector space constituted by the infinitesimal generators of $L_+^{\uparrow}$.

For $S_m = \hat{J}_1$, we get the corresponding finite Lorentz transformation by exponentiation as

$$\Lambda_{r_1}(\theta) = \begin{pmatrix} 1 & 0 & 0 & 0 \\ 0 & 1 & 0 & 0 \\ 0 & 0 & \cos\theta & -\sin\theta \\ 0 & 0 & \sin\theta & \cos\theta \end{pmatrix} \tag{117}$$

which may be construed as a rotation in the $x^2 - x^3$ plane. It is obvious that $\left.\dfrac{\partial \Lambda_{r_1}(\theta)}{\partial\theta}\right|_{\theta=0} = \hat{J}_1$.

Similarly, for $S_m = \hat{K}_1$, we get the corresponding finite Lorentz transformation by exponentiation as

$$\Lambda_{b_1}(\theta) = \begin{pmatrix} \cosh\theta & i\sinh\theta & 0 & 0 \\ -i\sinh\theta & \cosh\theta & 0 & 0 \\ 0 & 0 & 1 & 0 \\ 0 & 0 & 0 & 1 \end{pmatrix}$$

which is a Lorentz transformation from one three-dimensional coordinate frame to another frame that is moving with a velocity $v = c \tanh\theta$ along the common x^1 axis.

The transformation of the S_m matrix and the generators of eq. (116) from Minkowskian coordinates to real coordinates is achieved by the inverse transformation $H^{-1} \equiv \text{diag}(-i\ 1\ 1\ 1)$ given by eq. (79). For example, the S_m matrix is transformed as

$$S_m \to H^{-1} S_m H = S = \begin{pmatrix} 0 & e^1 & e^2 & e^3 \\ e^1 & 0 & -h^3 & h^2 \\ e^2 & h^3 & 0 & -h^1 \\ e^3 & -h^2 & h^1 & 0 \end{pmatrix} \tag{118}$$

Similarly, the infinitesimal generators may also be transformed to real coordinates, e.g.,

$$\hat{J}_1 \to H^{-1} \hat{J}_1 H = \begin{pmatrix} 0 & 0 & 0 & 0 \\ 0 & 0 & 0 & 0 \\ 0 & 0 & 0 & -1 \\ 0 & 0 & 1 & 0 \end{pmatrix} \tag{119}$$

and

$$\hat{K}_1 \to H^{-1}\,\hat{K}_1\,H = \begin{pmatrix} 0 & 1 & 0 & 0 \\ 1 & 0 & 0 & 0 \\ 0 & 0 & 0 & 0 \\ 0 & 0 & 0 & 0 \end{pmatrix} \text{ etc.} \tag{120}$$

Equation (115) holds in real coordinates as well for the left-hand side gets transformed as $S_m \to H^{-1} S_m H$ whereas the right-hand side becomes

$$\sum_{i=1}^{3} h^i H^{-1}\,\hat{J}_i\,H + \sum_{i=1}^{3} e^i H^{-1}\,\hat{K}_i H = H^{-1}\left(\sum_{i=1}^{3} h^i \hat{J}_i + \sum_{i=1}^{3} e^i \hat{K}_i\right) H = H^{-1}\,S_m\,H \text{ as desired.}$$

As we have already seen, the permissible range of values of Λ_0^0 are either (i) $\Lambda_0^0 \geq 1$ or (ii) $\Lambda_0^0 \leq -1$. In the present case, since we are considering transformations in the neighbourhood of the identity element (which has $\Lambda_0^0 = 1$) that can be obtained by a continuous change of parameters from their zero values (corresponding to the identity element), we can only obtain Lorentz transformations corresponding to $\Lambda_0^0 \geq 1$. Transformations corresponding to $\Lambda_0^0 \leq -1$ cannot be obtained by this method since, in this case, a jump in the parameter value is necessitated due to a jump in Λ_0^0 from the value 1 (that corresponds to the zero value of the parameter) to $\Lambda_0^0 \leq -1$, i.e., a discontinuity in the value of the parameter is involved which goes against our *ab initio* assumption about the continuity of the parameter.

We summarize our results of this section, since they are of significant importance:

(a) Every orthochronous proper Lorentz transformation Λ_m (expressed in Minkowski coordinates) in a sufficiently small neighbourhood of the identity element **I** of the OPLG, of which the group $SO(3, 1)$ provides a representation, may be expressed uniquely as

$$\Lambda_m = \exp(S_m) = \exp(h.\hat{J} + e.\hat{K}) \tag{121}$$

where the skew-symmetric matrix S_m is called the infinitesimal Lorentz transformation (ILT) associated with the OPLT and $\hat{\mathbf{J}} \equiv \left(\hat{J}_1, \hat{J}_2, \hat{J}_3\right)$, $\hat{\mathbf{K}} \equiv \left(\hat{K}_1, \hat{K}_2, \hat{K}_3\right)$ are known the generators of the group $SO(3, 1)$. They are defined in terms of eq. (116) or eqs. (119) and (120).

(b) The infinitesimal Lorentz transformation S_m is an element of the Lie algebra $sO(3, 1)$ of the linear group $SO(3, 1)$.

(c) The six generators $\hat{J}_i$ and $\hat{K}_i$, $i = 1, 2, 3$ form a basis for the Lie algebra $sO(3, 1)$ while the six real variables h^i and e^i provide a parameterization for the Lie algebra $sO(3, 1)$ and the corresponding group $SO(3, 1)$. **h**,**e** are sometimes referred to as the parameter vectors.

(d) The generators $\hat{J}_i$ and $\hat{K}_i$, $i = 1, 2, 3$, satisfy the following relations as may be explicitly verified from their above representations:

$$\text{(i) } \hat{J}_i^T = -\hat{J}_i \tag{122}$$

(ii) $\hat{K}_i^T = -\hat{K}_i$ (123)

(iii) $\hat{J}_i^3 = -\hat{J}_i$ (124)

(iv) $\hat{K}_i^3 = \hat{K}_i$ (125)

(v) $\left[\hat{J}_i, \hat{J}_j\right] = \varepsilon_{ijk}\hat{J}_k$ (126)

(vi) $\left[\hat{K}_i, \hat{K}_j\right] = -\varepsilon_{ijk}\hat{J}_k$ (127)

(vii) $\left[\hat{J}_i, \hat{K}_j\right] = \varepsilon_{ijk}\hat{K}_k$ (128)

where ε_{ijk} is totally anti-symmetric in i, j, k and $\varepsilon_{123} = 1$.

(e) We have, from eq. (124)

$$\exp \quad \hat{J}_i\theta = \mathbf{I} + \hat{J}_i\left(\theta - \frac{\theta^3}{3!} + \frac{\theta^5}{5!} - \ldots\right) + \hat{J}_i^2\left(\frac{\theta^2}{2!} - \frac{\theta^4}{4!} + \ldots\right)$$

$$= \mathbf{I} + \hat{J}_i \sin\theta + \hat{J}_i^2(1 - \cos\theta) \tag{129}$$

Similarly,

$$\exp\left(\hat{K}_i\theta\right) = \mathbf{I} + \hat{K}_i \sinh\theta + \hat{K}_i^2(\cosh\theta - 1) \tag{130}$$

1.6 TRANSFORMATION OF AN ARBITRARY INFINITESIMAL LORENTZ TRANSFORMATION INTO CANONICAL FORM (RAO, 1988)

The objective here is to obtain from a given matrix S_m that represents an infinitesimal Lorentz transformation in Minkowski coordinates, another matrix S_{mc}, by performing an OPLT transformation, say T_m, on S_m such that S_{mc} has the simplest possible structure, i.e., we look for S_{mc} that is related to S_m through a similarity OPLT given by T_m, such that S_{mc} has a canonical structure. We have $S_{mc} = T_m S_m T_m^{-1} = T_m S_m T_m^T$ (since $T_m^{-1} = T_m^T$ for an OPLT in Minkowski coordinates).

We consider the non-null case first. It involves the following steps.

Step 1

We perform a spatial rotation represented by the matrix R on the parameter vectors $\mathbf{e}$, $\mathbf{h}$ to get $\mathbf{e}'$, $\mathbf{h}'$, i.e., $\mathbf{e} \to \mathbf{e}' = R\mathbf{e}$ and $\mathbf{h} \to \mathbf{h}' = R\mathbf{h}$. The elements of the rotation matrix R are given by

$$(R_1^1, R_2^1, R_3^1) = \frac{(\mathbf{e} \times \mathbf{h})}{|\mathbf{e} \times \mathbf{h}|} \tag{131}$$

$$(R_1^2, R_2^2, R_3^2) = \frac{\mathbf{h} \times (\mathbf{e} \times \mathbf{h})}{|\mathbf{h}||\mathbf{e} \times \mathbf{h}|} \tag{132}$$

$$(R_1^3, R_2^3, R_3^3) = \frac{\mathbf{h}}{|\mathbf{h}|} \tag{133}$$

The orthogonality of this matrix can easily be verified by making use of the fact that the three vectors $\frac{(\mathbf{e}\times\mathbf{h})}{|\mathbf{e}\times\mathbf{h}|}$, $\frac{\mathbf{h}\times(\mathbf{e}\times\mathbf{h})}{|\mathbf{h}||\mathbf{e}\times\mathbf{h}|}$ and $\frac{\mathbf{h}}{|\mathbf{h}|}$ are orthogonal whereas the fact that $|R| = +1$ can be verified by explicit computation.

The new (rotated) components of the parameter vectors **e**, **h** are obtained as $e'^i = R^i_k e^k$ and $h'^i = R^i_k h^k$ respectively. We have

$$e'^1 = R^1_1 e^1 + R^1_2 e^2 + R^1_3 e^3 = \frac{1}{|\mathbf{e}\times\mathbf{h}|}[(\mathbf{e}\times\mathbf{h})_1 e^1 + (\mathbf{e}\times\mathbf{h})_2 e^2 + (\mathbf{e}\times\mathbf{h})_3 e^3]$$

$$= \frac{1}{|\mathbf{e}\times\mathbf{h}|}(\mathbf{e}\times\mathbf{h})\,.\,\mathbf{e} = 0 \tag{134}$$

$$e'^2 = \frac{1}{|\mathbf{h}||\mathbf{e}\times\mathbf{h}|}[\mathbf{h}\times(\mathbf{e}\times\mathbf{h})]\,.\,\mathbf{e} = \frac{1}{|\mathbf{h}||\mathbf{e}\times\mathbf{h}|}[(\mathbf{h}\times\mathbf{e})\,.\,(\mathbf{h}\times\mathbf{e})] = \frac{|\mathbf{e}\times\mathbf{h}|^2}{|\mathbf{h}||\mathbf{e}\times\mathbf{h}|} = \frac{|\mathbf{e}\times\mathbf{h}|}{|\mathbf{h}|} \tag{135}$$

$$e'^3 = \frac{\mathbf{h}}{|\mathbf{h}|}\,.\,\mathbf{e} = \frac{1}{|\mathbf{h}|}\,\mathbf{he} \tag{136}$$

Similarly,

$$h'^1 = \frac{1}{|\mathbf{e}\times\mathbf{h}|}(\mathbf{e}\times\mathbf{h})\,.\,\mathbf{h} = 0 \tag{137}$$

$$h'^2 = \frac{1}{|\mathbf{h}||\mathbf{e}\times\mathbf{h}|}[\mathbf{h}\times(\mathbf{e}\times\mathbf{h})].\mathbf{h} = 0 \tag{138}$$

$$h'^3 = \frac{\mathbf{h}}{|\mathbf{h}|}\,.\,\mathbf{h} = |\mathbf{h}| \tag{139}$$

Therefore, S_m matrix after incorporating the effect of this rotation, is given by

$$S'_m = \begin{pmatrix} 0 & ie'^1 & ie'^2 & ie'^3 \\ -ie'^1 & 0 & -h'^3 & h'^2 \\ -ie'^2 & h'^3 & 0 & -h'^1 \\ -ie'^3 & -h'^2 & h'^1 & 0 \end{pmatrix} = \begin{pmatrix} 0 & 0 & ie'^2 & ie'^3 \\ 0 & 0 & -h'^3 & 0 \\ -ie'^2 & h'^3 & 0 & 0 \\ -ie'^3 & 0 & 0 & 0 \end{pmatrix} \tag{140}$$

with the e'^i, h'^i being given by eqs. (134)-(139).

Step 2

The next step is to transform S'_m by a pure boost B_m along x^1 axis with the matrix representation

$$B_m = \begin{pmatrix} \cosh\theta & i\sin\theta & 0 & 0 \\ -i\sinh\theta & \cosh\theta & 0 & 0 \\ 0 & 0 & 1 & 0 \\ 0 & 0 & 0 & 1 \end{pmatrix} \tag{141}$$

Under this transformation S_m' transforms as

$$S'_m \rightarrow S''_m = B_m S'_m B_m^{-1} = B_m S'_m B_m^T \tag{142}$$

which gives

$$S''_m = \begin{pmatrix} 0 & 0 & ie''^2 & ie''^3 \\ 0 & 0 & -h''^3 & h''^2 \\ -ie''^2 & h''^3 & 0 & 0 \\ -ie''^3 & -h''^2 & 0 & 0 \end{pmatrix} \tag{143}$$

with the identification

$$\mathbf{e}'' \equiv (e''^1, e''^2, e''^3) = (0, e'^2 \cosh\theta - h'^3 \sinh\theta, e'^3 \cosh\theta) \tag{144}$$

$$\mathbf{h}'' \equiv (h''^1, h''^2, h''^3) = (0, e'^3 \sinh\theta, h'^3 \cosh\theta - e'^2 \sinh\theta) \tag{145}$$

A complete description of the boost involves fixing the value of θ. To achieve this, we mandate that the transformed parameter vectors $\mathbf{e}''$, $\mathbf{h}''$ be parallel to each other, whence

$$\frac{e''^2}{e''^3} = \frac{h''^2}{h''^3} = \lambda \tag{146}$$

Using eqs. (144)-(146), we get

$$\frac{e'^2 \cosh\theta - h'^3 \sinh\theta}{e'^3 \cosh\theta} = \frac{e'^3 \sinh\theta}{h'^3 \cosh\theta - e'^2 \sinh\theta} = \lambda \tag{147}$$

Again, using eqs. (134)-(139), we get

$$\frac{\frac{|\mathbf{e}\times\mathbf{h}|}{|\mathbf{h}|}\cosh\theta - |\mathbf{h}|\sinh\theta}{\frac{1}{|\mathbf{h}|}\mathbf{he}\cosh\theta} = \frac{\frac{1}{|\mathbf{h}|}\mathbf{he}\sinh\theta}{|\mathbf{h}|\cosh\theta - \frac{|\mathbf{e}\times\mathbf{h}|}{|\mathbf{h}|}\sinh\theta}$$

or

$$|\mathbf{e}\times\mathbf{h}|\,(\cosh^2\theta + \sinh^2\theta) = \left[\frac{1}{|\mathbf{h}|^2}(\mathbf{e}.\mathbf{h})^2 + \frac{1}{|\mathbf{h}|^2}(|\mathbf{e}\times\mathbf{h}|)^2 + |\mathbf{h}|^2\right]\sinh\theta\cosh\theta$$

or

$$|\mathbf{e}\times\mathbf{h}|\cosh 2\theta = \frac{1}{2}\left[\frac{1}{|\mathbf{h}|^2}\mathbf{e}^2\,\mathbf{h}^2 + |\mathbf{h}|^2\right]\sinh 2\theta$$

whence

$$\tanh 2\theta = \frac{2|\mathbf{e}\times\mathbf{h}|}{\mathbf{e}^2 + \mathbf{h}^2} \tag{148}$$

Further,

$$\cosh 2\theta = \frac{\varepsilon(\mathbf{e}^2 + \mathbf{h}^2)}{\sqrt{(\mathbf{e}^2 + \mathbf{h}^2)^2 - 4(\mathbf{e} \times \mathbf{h})^2}} \tag{149}$$

$$\sinh 2\theta = \frac{2\varepsilon|\mathbf{e} \times \mathbf{h}|}{\sqrt{(\mathbf{e}^2 + \mathbf{h}^2)^2 - 4(\mathbf{e} \times \mathbf{h})^2}} \tag{150}$$

with $\varepsilon = \pm 1$. The above equations determine θ subject to

$$(\mathbf{e}^2 + \mathbf{h}^2)^2 - 4(\mathbf{e} \times \mathbf{h})^2 \neq 0 \tag{151}$$

or equivalently

$$(\mathbf{h}^2 - \mathbf{e}^2)^2 + 4(\mathbf{e}.\mathbf{h})^2 \neq 0 \tag{152}$$

which holds for the non-null cases and is violated only when $(\mathbf{h}^2 - \mathbf{e}^2) = 0$ and $(\mathbf{e}.\mathbf{h}) = 0$. This happens to be the very definition of the null transformation. In other words, eqs. (148)-(150) determine θ uniquely upto a $\pm$ sign for all non-null transformations.

Step 3

The final step to obtain the canonical form is to perform a rotation about x^1 axis characterized by angle ϕ and represented by the matrix

$$R'(\phi) = \Lambda_{r_1}(\phi) = \begin{pmatrix} 1 & 0 & 0 & 0 \\ 0 & 1 & 0 & 0 \\ 0 & 0 & \cos\phi & -\sin\phi \\ 0 & 0 & \sin\phi & \cos\phi \end{pmatrix} \tag{153}$$

The transformed S''_m is obtained as

$$S''_m \to S'''_m = R'(\phi)\, S''_m\, R'(\phi)^{-1} = R'(\phi)\, S''_m\, R'(\phi)^T \tag{154}$$

giving

$$S'''_m = \begin{pmatrix} 0 & 0 & ie'''^2 & ie'''^3 \\ 0 & 0 & -h'''^3 & h'''^2 \\ -ie'''^2 & h'''^3 & 0 & 0 \\ -ie'''^3 & -h'''^2 & 0 & 0 \end{pmatrix} \tag{155}$$

with the identification

$$\mathbf{e}''' \equiv (e'''^1, e'''^2, e'''^3) = (0, e''^2 \cos\phi - e''^3 \sin\phi, e''^2 \sin\phi + e''^3 \cos\phi \tag{156}$$

$$\mathbf{h}''' \equiv (h'''^1, h'''^2, h'''^3) = (0, -h''^3 \sin\phi + h''^2 \cos\phi, h''^3 \cos\phi + h''^2 \sin\phi \tag{157}$$

It may be noted that ϕ is still arbitrary. We can further simplify the matrix of eq. (155) read with eqs. (156)-(157) by selecting an appropriate value of ϕ that will take two of the nonzero elements of S'''_m to zero. For example, we may select ϕ such that $\tan\phi = \frac{e''^2}{e''^3} = \frac{h''^2}{h''^3} = \lambda$ whence both e'''^2, h'''^2 vanish.

Thus, with this choice of the angle of rotation, we obtain the canonical form corresponding to any arbitrary infinitesimal Lorentz transformation as

$$S'''_m = \begin{pmatrix} 0 & 0 & 0 & ie'''^3 \\ 0 & 0 & -h'''^3 & 0 \\ 0 & h'''^3 & 0 & 0 \\ -ie'''^3 & 0 & 0 & 0 \end{pmatrix} \tag{158}$$

An interesting relationship exists between the components of this matrix S'''_m and the eigenvalues $\pm i\theta_r$ and $\pm\theta_b$ of the infinitesimal Lorentz transformation S_m. The eigenvalues of S'''_m are found to be $\pm ih'''^3$ and $\pm e'''^3$. Identifying these with the eigenvalues $\pm i\theta_r$ and $\pm\theta_b$ of the infinitesimal Lorentz transformation S_m (since the transformations envisaged in this procedure of conversion to canonical form are similarity transformations that do not change the eigenvalues of the original matrix) and noting that in our formulation both $\pm h'''^3$ and $\pm e'''^3$ must necessarily be real, we make the correspondence $h'''^3 \equiv \theta_r$ and $e'''^3 \equiv \theta_b$ so that the canonical form of an infinitesimal Lorentz transformation S'''_m can also be expressed in terms of the eigenvalues of the original infinitesimal Lorentz transformation as

$$S'''_m = \begin{pmatrix} 0 & 0 & 0 & i\theta_b \\ 0 & 0 & -\theta_r & 0 \\ 0 & \theta_r & 0 & 0 \\ -i\theta_b & 0 & 0 & 0 \end{pmatrix} \tag{159}$$

Now, consider the canonical form of a null infinitesimal Lorentz transformation. The null transformation is characterized by $\mathbf{h}^2 - \mathbf{e}^2 = 0$, $\mathbf{e.h} = 0$ and has all the four eigenvalues equal to zero. The relations $\mathbf{h}^2 - \mathbf{e}^2 = 0$ and $\mathbf{e.h} = 0$ imply that the parameter vectors **e**, **h** are equal in magnitude and also mutually perpendicular.

Furthermore, there does not exist any θ that satisfies eqs. (149) and (150) because eq. (152) is not satisfied. It follows that these parameter vectors **e**, **h** can never be made parallel by any OPLT. Thus, the maximum simplification of the "null transformation" matrix that can be achieved is to bring these two parameter vectors **e**, **h** along any two space coordinate axes through a pure rotation R_n. If we choose x^1 and x^3 axis for this purpose, we get the canonical form of a null ILT as

$$S'_m = R_n S_m R_n^{-1} = R_n\, S_m\, R_n^T = \begin{pmatrix} 0 & i\theta & 0 & 0 \\ -i\theta & 0 & -\theta & 0 \\ 0 & \theta & 0 & 0 \\ 0 & 0 & 0 & 0 \end{pmatrix} \tag{160}$$

where $|\mathbf{h}| = |\mathbf{e}| = \theta$.

1.6.1 Classification of OPLTs

The aforesaid canonical form of infinitesimal generators of OPLTs enable us to classify OPLTs on a rational basis as follows:

(a) Screw like OPLTs corresponding to the eigenvalues $\pm\,\theta_r \neq 0$, $\pm\,\theta_b \neq 0$ of the infinitesimal generators.

(b) Rotation like OPLTs corresponding to eigenvalues $\pm\,\theta_r \neq 0$, $\pm\,\theta_b = 0$ of the infinitesimal generators.

(c) Boost like OPLTs corresponding to eigenvalues $\pm\,\theta_r = 0$, $\pm\,\theta_b \neq 0$ of the infinitesimal generators.

(d) Null OPLTs corresponding to eigenvalues $\pm\,\theta_r = 0$, $\pm\,\theta_b = 0$ of the infinitesimal generator.

OPLTs may also be characterized in terms of the magnitude and orientation of the parameter vectors $\mathbf{e}$, $\mathbf{h}$ as (i) $\mathbf{e}, \mathbf{h} \neq 0$ with $\mathbf{e}^2 - \mathbf{h}^2$ unrestricted for screw like OPLTs, (ii) $\mathbf{e}, \mathbf{h} = 0$ with $\mathbf{e}^2 - \mathbf{h}^2 < 0$ for rotation like OPLTs, (iii) $\mathbf{e}, \mathbf{h} = 0$ with $\mathbf{e}^2 - \mathbf{h}^2 > 0$ for boost like OPLTs and (iv) $\mathbf{e}, \mathbf{h} = 0$ with $\mathbf{e}^2 - \mathbf{h}^2 = 0$ for null OPLTs.

1.7 THE POINCARE GROUP (WIGNER, 1939,1959; OHNUKI, 1976; TUNG, 1984; RAO, 1988)

The Poincare group is, essentially, an extension of the Lorentz group to include therein translations in all the four dimensions. In other words, the Poincare group consists of:

(a) Translations in all four coordinates.

(b) All homogeneous linear transformations on the spacetime coordinates that leave the length of the 4-vector given by $|x| = [(x^0)^2 - (\mathbf{x})^2]^{1/2} = (g_{\mu\nu}\, x^\mu\, x^\nu)^{1/2}$ invariant where $(g_{\mu\nu})$ is the metric tensor given by $(g_{\mu\nu}) = \text{diag}\,(+1\ -1\ -1\ -1)$ corresponding to the Minkowski metric with signature $(+1\ -1\ -1\ -1)$.

Thus, if $a \in M$ is a given 4-vector and Λ is a Lorentz transformation, then a general element of the Poincare group is a transformation that operates on an arbitrary 4-vector $x \in M$ as

$$x \in M \to x' = \Lambda x + a \in M \tag{161}$$

An element of the Poincare group $P(\Lambda;\, a)$ may be represented by five-dimensional matrices as

$$P(\Lambda;\, a) = \left(\begin{array}{c|c} \Lambda & a \\ \hline \mathbf{0} & 1 \end{array}\right) \tag{162}$$

where $\mathbf{0} \equiv (0\ \ 0\ \ 0\ \ 0)$.

Since the Poincare group consists of the Lorentz group that is described by six parameters and translations in four directions (which are linearly independent) corresponding to another four parameters, the Poincare group is a ten-parameter group with six parameters corresponding to the Lorentz group and four parameters to the four-dimensional translation group. The group nature of the Poincare group is easily established by making use of the representation given by eq. (162). If $P(\Lambda_1;\, a_1)$ and $P(\Lambda_2;\, a_2)$ are two arbitrary elements of the group, designated P, then we have

$$P(\Lambda_1;\, a_1)\; P(\Lambda_2;\, a_2) = \left(\begin{array}{c|c} \Lambda_1 & a_1 \\ \hline \mathbf{0} & 1 \end{array}\right)\left(\begin{array}{c|c} \Lambda_2 & a_2 \\ \hline \mathbf{0} & 1 \end{array}\right) = \left(\begin{array}{c|c} \Lambda_1\Lambda_2 & \Lambda_1 a_2 + a_1 \\ \hline \mathbf{0} & 1 \end{array}\right)$$

$$= P(\Lambda_1\, \Lambda_2;\, \Lambda_1\, a_2 + a_1) \in P \tag{163}$$

Furthermore, it also follows from eq. (163) that the inverse of an arbitrary element $P(\Lambda; a) \in P$ is given by $P(\Lambda^{-1}; -\Lambda^{-1}\, a) \in P$ with the identity element being $P(\mathbf{I}; 0) \equiv \mathbf{I}_{5\times 5} \in P$ confirming the group nature of P.

The Poincare transformations do not preserve the scalar product as well as norms, which is shown as follows:

$$\langle x'|y'\rangle = \langle \Lambda x + a|\Lambda y + a\rangle = \langle \Lambda x|\Lambda y\rangle + \langle \Lambda x|a\rangle + \langle a|\Lambda y\rangle + \langle a|a\rangle$$

$$= \langle x|y\rangle + \langle \Lambda x|a\rangle + \langle a|\Lambda y\rangle + \langle a|a\rangle \neq \langle x|y\rangle \tag{164}$$

$$\langle x'|x'\rangle = \langle \Lambda x + a|\Lambda x + a\rangle = \langle \Lambda x|\Lambda x\rangle + \langle \Lambda x|a\rangle + \langle a|\Lambda x\rangle + \langle a|a\rangle \neq \langle x|x\rangle \tag{165}$$

However, the relative norm is preserved, e.g.,

$$(x' - y')^2 = (x')^2 - 2\,\langle x'|y'\rangle + (y')^2 = (\Lambda x + a)^2 - 2\langle x'|y'\rangle + (\Lambda y + a)^2$$

$$= (\Lambda x)^2 + 2\langle \Lambda x|a\rangle + a^2 - 2\langle \Lambda x + a|\Lambda y + a\rangle + (\Lambda y)^2 + 2\langle \Lambda y|a\rangle + a^2$$

$$= (\Lambda x)^2 + 2\langle \Lambda x|a\rangle + a^2 - 2\langle \Lambda x|\Lambda y\rangle - 2\langle \Lambda x|a\rangle - 2\langle a|\Lambda y\rangle$$

$$- 2a^2 + (\Lambda y)^2 + 2\langle \Lambda y|a\rangle + a^2$$

$$= x^2 - 2\langle x|y\rangle + y^2 = (x - y)^2 \tag{166}$$

The Lorentz group $L \equiv \{P(\Lambda; 0)\} \subset P$ and the group of pure translations $T \equiv \{P(\mathbf{I}; a)\} \subset P$ both constitute subgroups of the Poincare group.

The group of translations also constitutes an abelian invariant subgroup of P and the factor group $\frac{P}{T}$ is isomorphic to the Lorentz group.

The transformation $P(\Lambda; a) \in P$ maps the origin to the point a.

1.7.1 Generators and Lie Algebra of the Poincare Group

Relating to each of the ten parameters that describe the Poincare group, there exist ten generators that constitute the infinitesimal transformations corresponding to each of these parameters.

(a) There are four generators of translations representing translations along each of the four dimensions. They are represented by $\hat{P}_\mu$; $\mu = 0, 1, 2, 3$. Their action is defined by the equation

$$\hat{T}(\delta b) = \mathbf{I} - i\hat{P}_\mu\, \delta b^\mu \tag{167}$$

Finite translations can, as usual, be expressed by exponentiation of the infinitesimal ones, e.g.,

$$\hat{T}(b) = e^{-i\hat{P}_\mu b^\mu} \tag{168}$$

The infinitesimal generators of translations in a particular direction manifest themselves as the components of the linear momentum operator in that direction in the coordinate space dynamics.

As mentioned above, the group of translations forms an invariant subgroup of the Poincare group. We have

$$\Lambda T(b)\Lambda^{-1} = P(\Lambda;0)P(\mathbf{I};b)P^{-1}(\Lambda;0) = P(\Lambda;0)\,P(\mathbf{I};b)\,P(\Lambda^{-1};0)$$

$$= \left(\begin{array}{c|c}\Lambda & 0\\ \hline \mathbf{0} & 1\end{array}\right)\left(\begin{array}{c|c}\mathbf{I} & b\\ \hline \mathbf{0} & 1\end{array}\right)\left(\begin{array}{c|c}\Lambda^{-1} & 0\\ \hline \mathbf{0} & 1\end{array}\right) = \left(\begin{array}{c|c}\Lambda & 0\\ \hline \mathbf{0} & 1\end{array}\right)\left(\begin{array}{c|c}\Lambda^{-1} & b\\ \hline \mathbf{0} & 1\end{array}\right)$$

$$= \left(\begin{array}{c|c}\mathbf{I} & \Lambda b\\ \hline \mathbf{0} & 1\end{array}\right) = T(\Lambda b) \in T \tag{169}$$

and

$$P(\Lambda;a)\;T(b)P^{-1}(\Lambda;a) = P(\Lambda;a)\;P(\mathbf{I};b)P^{-1}(\Lambda;a)$$

$$= P(\Lambda;a)\;P(\mathbf{I};b)\;P(\Lambda^{-1};-\Lambda^{-1}a)$$

$$= \left(\begin{array}{c|c}\Lambda & a\\ \hline \mathbf{0} & 1\end{array}\right)\left(\begin{array}{c|c}\mathbf{I} & b\\ \hline \mathbf{0} & 1\end{array}\right)\left(\begin{array}{c|c}\Lambda^{-1} & -\Lambda^{-1}a\\ \hline \mathbf{0} & 1\end{array}\right)$$

$$= \left(\begin{array}{c|c}\Lambda & a\\ \hline \mathbf{0} & 1\end{array}\right)\left(\begin{array}{c|c}\Lambda^{-1} & -\Lambda^{-1}a+b\\ \hline \mathbf{0} & 1\end{array}\right)$$

$$= \left(\begin{array}{c|c}\mathbf{I} & \Lambda b\\ \hline \mathbf{0} & 1\end{array}\right) = T(\Lambda \mathrm{b}) \in T \tag{170}$$

thereby establishing the invariance of the translation subgroup.

The translation generators transform as 4 coordinate vectors under a Lorentz transformation. The left-hand side of eq. (169) gives

$$\Lambda\hat{T}(b)\,\Lambda^{-1} = \Lambda e^{-i\hat{P}_\mu b^\mu}\;\Lambda^{-1} = \Lambda\left[\sum_i \frac{1}{i!}(-i\hat{P}_\mu\, b^\mu)^i\right]\Lambda^{-1} = \left[\sum_i \frac{1}{i!}(-i\Lambda\hat{P}_\mu\, b^\mu\,\Lambda^{-1})^i\right]$$

$$= \left[\sum_i \frac{1}{i!}(-i\Lambda\hat{P}_\mu\,\Lambda^{-1}.\Lambda b^\mu\,\Lambda^{-1})^i\right] = \left[\sum_i \frac{1}{i!}(-i\Lambda\hat{P}_\mu\,\Lambda^{-1}.b^\mu)^i\right]$$

and the right-hand side gives

$$\hat{T}(\Lambda b) = e^{-i\hat{P}_\rho(\Lambda b)^\rho} = \sum_i \frac{1}{i!}\left(-i\hat{P}_\rho\,(\Lambda b)^\rho\right)^i = \sum_i \frac{1}{i!}\left(-i\hat{P}_\rho\,\Lambda^\rho_\sigma\, b^\sigma\right)^i$$

Comparing the two sides, we get the transformation law for the translation generators under Lorentz transformations as

$$\Lambda\hat{P}_\mu\ \Lambda^{-1} = \hat{P}_\nu\ \Lambda^\nu_\mu \tag{171}$$

(b) There are six generators of homogeneous Lorentz transformations that have already been alluded to in the preceding paragraphs, viz. $\hat{J}_i$ $\hat{K}_i$, $i = 1, 2, 3$. These generators can be construed as constituting infinitesimal rotations by trigonometric or hyperbolic angles in various (μ, ν), μ, $\nu = 0$, 1, 2, 3 planes in the four-dimensional spacetime, e.g., $\hat{J}_1$ may be construed as the generator of rotations in the $x^2 - x^3$ plane by the infinitesimal trigonometric angle $\delta\theta$. Similarly, $\hat{K}_1$ corresponds to a "rotation" by an infinitesimal "hyperbolic" angle in the $x^1 - x^0$ plane, etc. Thus, the generators $\hat{J}_i$, $\hat{K}_i$ of homogeneous Lorentz transformations can also be labelled in terms of the plane in which the rotation (by real or imaginary angles) is induced by the generator. For spatial rotations, i.e., μ, $\nu \equiv m$, n = 1, 2, 3 we may write, for example, $2\hat{J}_k = \varepsilon^{kmn}\ \hat{J}_{mn}$ or equivalently $\hat{J}_{mn} = \varepsilon^{mnk}\ \hat{J}_k$ with $k = 1, 2, 3$ or $\hat{J}^{kl} = \varepsilon_{klm}\ \hat{J}^m$, whence $\varepsilon^{nkl}\ \hat{J}_{kl} = \varepsilon^{nkl}\ \varepsilon^{klm}\ \hat{J}_m = 2\delta_{mn}\hat{J}_m = 2\hat{J}_n$.

In the case of generators of "boosts" i.e., rotations by hyperbolic angles, that involve the time axis and one of the spatial axis, we shall use the labelling $\hat{K}_m = \hat{J}_{m0}$ with $m = 1, 2, 3$.

This labelling of $\hat{J}_i$, $\hat{K}_i$ in terms of the plane of rotation is especially useful because in four-dimensional space and also in higher dimensions there is no unique axis associated with a set of one parameter rotations. This is so because the subspace that is perpendicular to a particular plane itself is multidimensional.

In terms of the above labelling scheme, these generators $\hat{J}_{\mu\nu}$, μ, $\nu = 0, 1, 2, 3$ of infinitesimal homogeneous Lorentz transformations may be defined as anti-symmetric tensors of rank two defined in terms of the equation

$$\Lambda(\delta\omega) = \mathbf{I} - \frac{i}{2}\ \hat{J}_{\mu\nu}\ \delta\omega^{\mu\nu} \tag{172}$$

where $\delta\omega^{\mu\nu}$ are anti-symmetric infinitesimal parameters representing infinitesimal angular displacements in the (μ, ν) plane.

Finite transformations can, as usual, be expressed by exponentiation of the infinitesimal ones, e.g.,

$$\Lambda(\omega) = e^{-\frac{i}{2}\hat{J}_{\mu\nu}\omega^{\mu\nu}} \tag{173}$$

The transformation of $\Lambda(\omega)$ under the effect of an arbitrary Lorentz transformation is investigated as

$$\Omega\Lambda(\omega)\Omega^{-1} = \Omega e^{-\frac{i}{2}\hat{J}_{\mu\nu}\omega^{\mu\nu}}\ \Omega^{-1} = \Omega\left[\sum_k \frac{1}{k!}\left(-\frac{i}{2}\hat{J}_{\mu\nu}\ \omega^{\mu\nu}\right)^i\right]\Omega^{-1} = \left[\sum_k \frac{1}{k!}\left(-\frac{i}{2}\Omega\hat{J}_{\mu\nu}\ \omega^{\mu\nu}\ \Omega^{-1}\right)^i\right]$$

$$= \left[\sum_k \frac{1}{k!} \left(-\frac{i}{2} \Omega \hat{J}_{\mu\nu} \, \Omega^{-1} \Omega \omega^{\mu\nu} \, \Omega^{-1} \right)^i \right] = \left[\sum_k \frac{1}{k!} \left(-\frac{i}{2} \hat{J}_{\mu\nu} \, \omega'^{\mu\nu} \right)^i \right] = e^{-\frac{i}{2} \hat{J}'_{\mu\nu} \omega'^{\mu}} \tag{174}$$

where $\omega'^{\mu\nu} = \Omega^{\mu}_{\lambda} \Omega^{\nu}_{\sigma} \, \omega^{\lambda\sigma}$ and $\hat{J}'_{\mu\nu} = \Omega \hat{J}_{\mu\nu} \, \Omega^{-1} = \hat{J}_{\lambda\sigma} \, \Omega^{\lambda}_{\mu} \, \Omega^{\sigma}_{\nu}$.

The complete Lie algebra of the Poincare group is given by

$$\left[\hat{P}_{\mu}, \hat{P}_{\lambda}\right] = 0 \tag{175}$$

$$\left[\hat{P}_{\mu}, \hat{J}_{\lambda\sigma}\right] = i\left(\hat{P}_{\sigma} \, g_{\mu\lambda} - \hat{P}_{\lambda} \, g_{\mu\sigma}\right) \tag{176}$$

$$\left[\hat{J}_{\mu\nu}, \hat{J}_{\lambda\sigma}\right] = i\left(\hat{J}_{\sigma\nu} \, g_{\mu\lambda} - \hat{J}_{\lambda\nu} \, g_{\mu\sigma} - \hat{J}_{\mu\lambda} \, g_{\nu\sigma} + \hat{J}_{\mu\sigma} \, g_{\nu\lambda}\right) \tag{177}$$

in the metric with $(g_{\mu\nu})$ = diag (1, –1, –1, –1). Equation (175) directly follows from the commutativity of the coordinates and the consequential abelian nature of the translational subgroup.

To prove eq. (176), we make use of eq. (171). Since this equation holds for any arbitrary Lorentz transformation Λ, it holds for infinitesimal homogeneous Lorentz transformations as well. Hence, we have, from the left-hand side of eq. (171)

$$\Lambda(\delta\omega) \; \hat{P}_{\mu} \; \Lambda^{-1} \; \delta\omega = \left(\mathbf{I} - \frac{i}{2} \hat{J}_{\lambda\sigma} \, \delta\omega^{\lambda\sigma} \right) \hat{P}_{\mu} \left(\mathbf{I} + \frac{i}{2} \hat{J}_{\lambda\sigma} \, \delta\omega^{\lambda\sigma} \right)$$

$$= \hat{P}_{\mu} + \frac{i}{2} \left[\hat{P}_{\mu}, \hat{J}_{\lambda\sigma}\right] \delta\omega^{\lambda\sigma} + 0(\delta\omega^2) \tag{178}$$

and from the right-hand side

$$\hat{P}_{\nu} \Lambda \, (\delta\omega)^{\nu}_{\mu} = \hat{P}_{\nu} \left(I - \frac{i}{2} \hat{J}_{\lambda\sigma} \, \delta\omega^{\lambda\sigma} \right)^{\nu}_{\mu} = \hat{P}_{\mu} - \frac{i}{2} \hat{P}_{\nu} \left(\hat{J}_{\lambda\sigma}\right)^{\nu}_{\mu} \; \delta\omega^{\lambda\sigma}$$

$$= \hat{P}_{\mu} + \frac{1}{2} (\delta^{\nu}_{\lambda} \, g_{\sigma\mu} - \delta^{\nu}_{\sigma} \, g_{\lambda\mu}) \; \hat{P}_{\nu} \, \delta\omega^{\lambda\sigma} = \hat{P}_{\mu} + \frac{1}{2} \left(\hat{P}_{\lambda} \, g_{\sigma\mu} - \hat{P}_{\sigma} \, g_{\lambda\mu} \right) \delta\omega^{\lambda\sigma} \tag{179}$$

Comparison of the right-hand sides of eqs. (178) and (179) gives eq. (176).

The proof of eq. (177) runs on parallel lines. We have, from the left-hand side of the equation $\Omega \hat{J}_{\mu\nu} \; \Omega^{-1} = \hat{J}_{\lambda\sigma} \; \Omega^{\lambda}_{\mu} \, \Omega^{\sigma}_{\nu}$.

$$\left(\mathbf{I} - \frac{i}{2} \hat{J}_{\alpha\beta} \, \delta\omega^{\alpha\beta} \right) \hat{J}_{\mu\nu} \left(\mathbf{I} + \frac{i}{2} \hat{J}_{\alpha\beta} \, \delta\omega^{\alpha\beta} \right) = \left(\hat{J}_{\mu\nu} - \frac{i}{2} \hat{J}_{\alpha\beta} \, \hat{J}_{\mu\nu} \, \delta\omega^{\alpha\beta} \right) \left(\mathbf{I} + \frac{i}{2} \hat{J}_{\alpha\beta} \, \delta\omega^{\alpha\beta} \right)$$

$$= \hat{J}_{\mu\nu} + \frac{i}{2} \hat{J}_{\mu\nu} \, \hat{J}_{\alpha\beta} \; \delta\omega^{\alpha\beta} - \frac{i}{2} \hat{J}_{\alpha\beta} \; \hat{J}_{\mu\nu} \; \delta\omega^{\alpha\beta} + 0 \; (\delta\omega)^2$$

$$= \hat{J}_{\mu\nu} + \frac{i}{2} \left[\hat{J}_{\mu\nu}, \hat{J}_{\alpha\beta}\right] \delta\omega^{\alpha\beta} + 0 \; (\delta\omega)^2 \tag{180}$$

while the right-hand side gives

$$\hat{J}_{\lambda\sigma}\left[\mathbf{I}-\frac{i}{2}\hat{J}_{\alpha\beta}\,\delta\omega^{\alpha\beta}\right]^{\lambda}_{\mu}\left[\mathbf{I}-\frac{i}{2}\hat{J}_{\alpha\beta}\,\delta\omega^{\alpha\beta}\right]^{\sigma}_{\nu}$$

$$=\left[\hat{J}_{\mu\sigma}-\frac{i}{2}\left(\hat{J}_{\alpha\beta}\right)^{\lambda}_{\mu}\hat{J}_{\lambda\sigma}\,\delta\omega^{\alpha\beta}\right]\left[\mathbf{I}-\frac{i}{2}\hat{J}_{\alpha\beta}\,\delta\omega^{\alpha\beta}\right]^{\sigma}_{\nu}$$

$$=\left[\hat{J}_{\mu\sigma}+\frac{1}{2}(\delta^{\lambda}_{\alpha}\,g_{\beta\mu}-\delta^{\lambda}_{\beta}\,g_{\alpha\mu})\hat{J}_{\lambda\sigma}\,\delta\omega^{\alpha\beta}\right]\left[\mathbf{I}-\frac{i}{2}\hat{J}_{\alpha\beta}\,\delta\omega^{\alpha\beta}\right]^{\sigma}_{\nu}$$

$$=\left[\hat{J}_{\mu\sigma}+\frac{1}{2}(\hat{J}_{\alpha\sigma}\,g_{\beta\mu}-\hat{J}_{\beta\sigma}\,g_{\alpha\mu})\delta\omega^{\alpha\beta}\right]\left[\mathbf{I}-\frac{i}{2}\hat{J}_{\alpha\beta}\,\delta\omega^{\alpha\beta}\right]^{\sigma}_{\nu}$$

$$=\hat{J}_{\mu\nu}-\frac{i}{2}\left(\hat{J}_{\alpha\beta}\right)^{\sigma}_{\nu}\hat{J}_{\mu\sigma}\,\delta\omega^{\alpha\beta}+\frac{1}{2}\left(\hat{J}_{\alpha\nu}\,g_{\beta\mu}-\hat{J}_{\beta\nu}\,g_{\alpha\mu}\right)\delta\omega^{\alpha\beta}+0(\delta\omega)^2$$

$$=\hat{J}_{\mu\nu}+\frac{1}{2}(\delta^{\sigma}_{\alpha}\,g_{\beta\nu}-\delta^{\sigma}_{\beta}\,g_{\alpha\nu})\,\hat{J}_{\mu\sigma}\,\delta\omega^{\alpha\beta}+\frac{1}{2}\left(\hat{J}_{\alpha\nu}\,g_{\beta\mu}-\hat{J}_{\beta\nu}\,g_{\alpha\mu}\right)\delta\omega^{\alpha\beta}+0(\delta\omega)^2$$

$$=\hat{J}_{\mu\nu}+\frac{1}{2}\left(\hat{J}_{\mu\alpha}\,g_{\beta\nu}-\hat{J}_{\mu\beta}\,g_{\alpha\nu}\right)\delta\omega^{\alpha\beta}+\frac{1}{2}\left(\hat{J}_{\alpha\nu}\,g_{\beta\mu}-\hat{J}_{\beta\nu}\,g_{\alpha\mu}\right)\delta\omega^{\alpha\beta}+0(\delta\omega)^2 \tag{181}$$

Comparing eqs. (180) and (181) and making use of the symmetric property of the metric tensor, i.e., $g_{\alpha\beta}=g_{\beta\alpha}$ and renaming the dummy indices we obtain eq. (177).

Now we establish the equivalence of eq. (176) to the following set of equations:

$$\left[\hat{P}^0,\hat{J}_n\right]=0 \tag{182}$$

$$\left[\hat{P}_{m,}\,\hat{J}_n\right]=i\varepsilon^{mnl}\,\hat{P}_l \tag{183}$$

$$\left[\hat{P}_m,\hat{K}_n\right]=-i\delta_{mn}\,\hat{P}_0 \tag{184}$$

$$\left[\hat{P}_0\,\hat{K}_n\right]=-i\hat{P}_n \tag{185}$$

We have, from eq. (176), $\left[\hat{P}_{\mu},\hat{J}_{\lambda\sigma}\right]=i\left(\hat{P}_{\sigma}\,g_{\mu\lambda}-\hat{P}_{\lambda}\,g_{\mu\sigma}\right)$whence

$\left[\hat{P}_{m,}\,\hat{J}_{n0}\right]=i\left(\hat{P}_0\,g_{mn}-\hat{P}_n\,g_{m0}\right)$. Now, $\hat{J}_{n0}\equiv\hat{K}_n$, $g_{m0}=0$, $g_{mn}=-\delta_{mn}$, $\hat{P}^0=g^{0\nu}\,\hat{P}_{\nu}=\hat{P}_0$. Making all these substitutions, we get eq. (184).

Similarly, from eq. (176), we have $\left[\hat{P}_0,\hat{J}_{n0}\right]=-\,i\left(\hat{P}_n\,g_{00}-\hat{P}_0\,g_{0n}\right)$

Now, $\hat{J}_{n0}\equiv\hat{K}_n$, $g_{00}=1$, $g_{0n}=0$, $\hat{P}^0=g^{0\nu}\,\hat{P}_{\nu}=\hat{P}_0$.

Making all these substitutions, we get eq. (185). Equations (182)-(183) may be proved similarly.

We can also establish the equivalence of eq. (177), viz.

$\left[\hat{J}_{\mu\nu},\hat{J}_{\lambda\sigma}\right] = i\left(\hat{J}_{\sigma\nu}\, g_{\mu\lambda} - \hat{J}_{\lambda\nu}\, g_{\mu\sigma} + \hat{J}_{\mu\sigma}\, g_{\nu\lambda} - \hat{J}_{\mu\lambda}\, g_{\nu\sigma}\right)$ to the following:

$$\left[\hat{J}_m,\hat{J}_n\right] = i\varepsilon^{nml}\,\hat{J}_l \tag{186}$$

$$\left[\hat{K}_m,\hat{J}_n\right] = i\varepsilon^{mnl}\,\hat{K}_l \tag{187}$$

$$\left[\hat{K}_m,\hat{K}_n\right] = -i\varepsilon^{mnl}\,\hat{J}_l \tag{188}$$

From eq. (177), we get

$$\left[\hat{J}_{l0},\hat{J}_{mn}\right] = -i\left(\hat{J}_{m0}\, g_{ln} - \hat{J}_{n0}\, g_{lm} + \hat{J}_{lm}\, g_{0n} - \hat{J}_{ln}\, g_{0m}\right)$$

Now, $\hat{J}_{l0} \equiv \hat{K}_l$, $\hat{J}_{mn} = \varepsilon_{mnp}\,\hat{J}^{\,p}$, $\hat{J}_{n0} \equiv \hat{K}_n$, $g_{ln} = -\delta_{ln}$, $g_{lm} = -\delta_{lm}$, $g_{0n} = 0$ so that $\varepsilon_{mnp}\left[\hat{K}_l,\hat{J}^{\,p}\right]$ $= -\, i\left(\hat{K}_n\,\delta_{lm} - \hat{K}_m\,\delta_{ln}\right)$.

Contracting both sides with ε^{mnq}, we get

$$\varepsilon^{mnq}\,\varepsilon_{mnp}\left[\hat{K}_l,\hat{J}^{\,p}\right] = i\varepsilon^{mnq}\left(\hat{K}_m\,\delta_{ln} - \hat{K}_n\,\delta_{lm}\right) \text{ or } 2\left[\hat{K}_l,\hat{J}_m\right] = 2i\varepsilon^{lmn}\,\hat{K}_n$$

which is the required result. The remaining equations can be proved in a similar manner.

For ease of reference, we produce some important properties of the fully anti-symmetric tensor in three dimensions as follows:

$$\varepsilon_{\alpha\beta\gamma}\,\varepsilon^{\alpha\beta\gamma} = 6 \tag{189}$$

$$\varepsilon_{\alpha\beta\gamma}\,\varepsilon^{\mu\beta\gamma} = 2\,\delta^{\mu}_{\alpha} \tag{190}$$

$$\varepsilon_{\alpha\beta\gamma}\,\varepsilon^{\mu\nu\gamma} = \delta^{\mu}_{\alpha}\,\delta^{\nu}_{\beta} - \delta^{\nu}_{\alpha}\,\delta^{\mu}_{\beta} \tag{191}$$

$$\varepsilon_{\alpha\beta\gamma}\,\varepsilon^{\lambda\mu\nu} = \begin{vmatrix} \delta^{\lambda}_{\alpha} & \delta^{\mu}_{\alpha} & \delta^{\nu}_{\alpha} \\ \delta^{\lambda}_{\beta} & \delta^{\mu}_{\beta} & \delta^{\nu}_{\beta} \\ \delta^{\lambda}_{\gamma} & \delta^{\mu}_{\gamma} & \delta^{n}_{\gamma} \end{vmatrix} \tag{192}$$

CHAPTER 2

Representations of the Lorentz and Poincare Groups and Transformation Properties of Physical Quantities

At the very outset, it is pertinent to mention that finite dimensional irreducible representations (IRRs) of both these groups are not unitary and all unitary representations are infinite dimensional since the groups are non-compact.

2.1 UNITARY REPRESENTATIONS OF THE LORENTZ GROUP (WIGNER, 1939; GEL'FAND, 1963; NAIMARK, 1964; OHNUKI, 1976; TUNG, 1984)

2.1.1 Reduction of Representations and Relabelling of Basis

The Lie algebra for the proper Lorentz group is given by

$$\left[\hat{J}_{\mu\nu}, \hat{J}_{\lambda\sigma}\right] = i\left(\hat{J}_{\sigma\nu}\, g_{\mu\lambda} - \hat{J}_{\lambda\nu}\, g_{\mu\sigma} - \hat{J}_{\mu\lambda}\, g_{\nu\sigma} + \hat{J}_{\mu\sigma}\, g_{\nu\lambda}\right) \tag{1}$$

or equivalently by

$$\left[\hat{J}_{m}, \hat{J}_{n}\right] = i\varepsilon^{mnl}\, \hat{J}_{l} \tag{2}$$

$$\left[\hat{K}_{m}, \hat{J}_{n}\right] = i\varepsilon^{mnl}\, \hat{K}_{l} \tag{3}$$

$$\left[\hat{K}_{m}, \hat{K}_{n}\right] = -\, i\varepsilon^{mnl}\, \hat{J}_{l} \tag{4}$$

This Lie algebra may also be written as

$$\left[\hat{M}_m, \hat{M}_n\right] = i\varepsilon^{mnk}\,\hat{M}_k \tag{5}$$

$$\left[\hat{N}_m, \hat{N}_n\right] = i\varepsilon^{mnk}\,\hat{N}_k \tag{6}$$

$$\left[\hat{M}_m, \hat{N}_n\right] = 0 \tag{7}$$

where $\hat{M}_m = \frac{1}{2}\left(\hat{J}_m + i\hat{K}_m\right)$ and $\hat{N}_m = \frac{1}{2}\left(\hat{J}_m - i\hat{K}_m\right)$ with $m, n = 1, 2, 3$. Eqs. (5)-(7) constitute the Lie algebra of the group $SU(2)_M \otimes SU(2)_N$ with the generators of the group transformations being $\hat{M}$, $\hat{N}$ respectively. The Casimir operators for the groups $SU(2)_M$ and $SU(2)_N$ are $\hat{M}^2$ and $\hat{N}^2$ respectively, since $\left[\hat{M}^2, \hat{M}_n\right] = \left[\hat{N}^2, \hat{N}_n\right] = 0$, $n = 1, 2, 3$. The irreducible representations (IRRs) of the two subgroups $SU(2)_M$ and $SU(2)_N$ can be labeled in terms of the eigenvalues of the corresponding Casimir operators. The Casimir operator commutes with all the generators of the group transformations. As a result, by Schur's Lemma, the Casimir operator gets mapped into an integral multiple of the identity matrix in any irreducible representation. Thus, all vectors constituting a basis of an irreducible representation are eigenvectors of the Casimir operator with the same eigenvalue, i.e., eigenvalue of the Casimir operator is invariant for a given IRR. This eigenvalue can, therefore, be used to label an IRR of a group.

Within an IRR, the basis vectors of the representation space can be chosen as the eigenvectors of a maximal set of mutually commuting operators (since only commuting operators can have complete set of simultaneous eigenvectors). By convention, we choose the maximal set of commuting operators to consist of the Casimir operators $\hat{M}^2$ and $\hat{N}^2$ together with one of the generators, usually taken as $\hat{M}_3$ and $\hat{N}_3$ respectively.

We shall, first, establish the construction of IRRs of $SU(2)_M$ using $\left(\hat{M}^2, \hat{M}_3\right)$ as the maximal set of commuting operators. We also define the "raising" and "lowering" operators $\hat{M}_\pm = \hat{M}_1 \pm i\hat{M}_2$ so that the following relations hold:

$$\left[\hat{M}_3, \hat{M}_\pm\right] = \pm\,\hat{M}_\pm \tag{8}$$

$$\left[\hat{M}_+, \hat{M}_-\right] = 2\hat{M}_3 \tag{9}$$

$$\hat{M}^2 = \hat{M}_+\hat{M}_- + \hat{M}_3^2 - \hat{M}_3 = \hat{M}_-\hat{M}_+ + \hat{M}_3^2 + \hat{M}_3 \tag{10}$$

$$\hat{M}_\pm^\dagger = \hat{M}_\mp \tag{11}$$

If $|k\rangle$ is any normalized eigenvector of $\hat{M}_3$ with eigenvalue k, then we have

$$\hat{M}_3|k\rangle = k|k\rangle \tag{12}$$

Since the basis vectors are assumed to be simultaneous eigenvectors of $\hat{M}_3$ and the Casimir operator $\hat{M}^2$, we may label them by using the labels k, which is the eigenvalue of $\hat{M}_3$ and u, that is related to the eigenvalue of the Casimir operator $\hat{M}^2$ and write eq. (12) as

$$\hat{M}_3|uk\rangle = k|uk\rangle \tag{13}$$

and, further

$$\hat{M}^2|uk\rangle = u(u+1)|uk\rangle \tag{14}$$

Of these, the index u corresponding to the eigenvalue of the Casimir operator $\hat{M}^2$ labels out the IRR, since it is left invariant by group operations and hence, is constant across a given IRR, whereas the other index k, which is the eigenvalue of $\hat{M}_3$ labels out the basis vectors within an IRR. We have

$$\hat{M}_3\hat{M}_\pm|uk\rangle = \hat{M}_\pm\hat{M}_3|uk\rangle + \left[\hat{M}_3, \hat{M}_\pm\right]\left|uk\right\rangle$$

$$= \hat{M}_\pm\hat{M}_3|uk\rangle \pm \hat{M}_\pm|uk\rangle = (\mathrm{k} \pm 1)\ \hat{M}_\pm|uk\rangle \tag{15}$$

Thus, $\hat{M}_\pm|uk\rangle$ is an eigenstate of $\hat{M}_3$ with eigenvalue $k \pm 1$. This means that, in line with our labelling scheme for eigenstates, we must have $\hat{M}_\pm|uk\rangle \infty |u, k \pm 1\rangle$ or $\hat{M}_\pm|uk\rangle = A_m^\pm|u, k \pm 1\rangle$. Similar calculations apply to $SU(2)_N$.

Hence, we can label the IRRs of the direct product algebra $SU(2)_M \otimes SU(2)_N$ that provides the IRRs for the Lorentz group by two numbers u, v corresponding to the two Casimir operators $\hat{M}^2$ and $\hat{N}^2$ of the subalgebras $SU(2)_M$ and $SU(2)_N$ respectively. The basis vectors in the representation space are, then, chosen as $\{|uv; kl\rangle\} \equiv \{|kl\rangle\}$ that consist of the direct products of the basis vectors from the canonical basis $\{|uk\rangle\}$ and $\{|vl\rangle\}$ of the two subalgebras $SU(2)_M$ and $SU(2)_N$ respectively. On this basis, the action of the generators of the Lorentz group is given by the following equations:

$$\hat{J}_3|kl\rangle = \left(\hat{M}_3 + \hat{N}_3\right)|kl\rangle = (k+l)|kl\rangle \tag{16}$$

$$\hat{J}_\pm\left|kl\right\rangle = \left(\hat{M}_\pm + \hat{N}_\pm\right)|kl\rangle = A^\pm(u, k)|k \pm 1, l\rangle + B^\pm(v, l)|k, l \pm 1\rangle \tag{17}$$

$$\hat{K}_3\left|kl\right\rangle = i\left(\hat{N}_3 - \hat{M}_3\right)|kl\rangle = i(l-k|kl\rangle \tag{18}$$

$$\hat{K}_\pm\left|kl\right\rangle = i\left(\hat{N}_\pm - \hat{M}_\pm\right)|kl\rangle = -iA^\pm(u, k)|k \pm 1, l\rangle + iB^\pm(v, l)|k, l \pm 1\rangle \tag{19}$$

It is seen from Eqs. (16) and (17) that the restriction of the representation space to the subgroup of three-dimensional rotations is equivalent to that of a direct product of representations of the $SO(3)$ group labelled by the indices u and v respectively.

We can reduce this direct product of representations to a direct sum of invariant subspaces with respect to rotations. For this purpose, we identify these constituents of the direct product representations in terms of u, v corresponding the eigenvalues of the respective Casimir operators and denote them by $D^u[R]$ and $D^v[R]$ respectively. We denote the direct product of these two representations by $D^{u\times v}[R]$. We designate, as above, the basis vectors in the product representations by $\{|uv;kl\rangle\} \equiv \{|kl\rangle\}$ so that $|kl\rangle = |uk\rangle \otimes |vl\rangle$. The action of $D^{u\times v}[R]$ on the direct product space $W = V_M \otimes V_N$ is obtained by its action on the direct product basis $|kl\rangle = |uk\rangle \otimes |vl\rangle$, e.g.,

$$D^{u\times v}[R]|kl\rangle = D^u[R]^m_k\, D^v[R]^n_l|mn\rangle \tag{20}$$

We show that $D^{u\times v}[R]$ defines a representation of $SO(3)$. This can be easily done by computing the character of the direct product representation. We have $\chi^{u+v} = TrD^{u\times v}[R] = D^{u\times v}[R]^k_k = D^u[R]^i_i\; D^v[R]^j_j = \chi^u\,\chi^v$ establishing our proposition.

We examine the reducibility of $D^{u\times v}[R]$. For this purpose, we write

$$D^{u\times v}[R] \sim \sum_{\lambda\oplus} a_\lambda\, D^\lambda[R] \tag{21}$$

i.e., we attempt to reduce the direct product space W into a direct sum of invariant subspaces W^λ_α, where λ labels out the IRR and $\alpha = 1, 2, \ldots \alpha_\lambda$, distinguishes the spaces that correspond to the same λ. We can, then, choose a basis in W corresponding to any IRR, i.e., any λ, such that the first n^λ_1 basis vectors are in W^λ_1, the next n^λ_1 in W^λ_2, and so on. Therefore, the representation matrices are in block diagonal form in the new basis.

To examine the reducibility of our direct product representation, we consider an infinitesimal rotation about an arbitrary axis $\hat{\mathbf{n}}$. Then, we must have

$$D^u[R_n(d\theta)]\; D^v[R_n(d\theta)] = D^{u\times v}[R_n(d\theta)] \tag{22}$$

whence, to first order in $d\theta$

$$\left(\hat{\mathbf{I}}^u - i\hat{J}^u_n\, d\theta\right).\left(\hat{\mathbf{I}}^v - i\hat{J}^v_n\, d\theta\right) = \hat{\mathbf{I}}^u \otimes \hat{\mathbf{I}}^v - i\left(\hat{J}^u_n \otimes \hat{\mathbf{I}}^v + \hat{\mathbf{I}}^u \otimes \hat{J}^v_n\right) d\theta = \left(\hat{\mathbf{I}}^{u\times v} - i\hat{J}^{u\times v}_n\, d\theta\right) \tag{23}$$

which gives $\hat{J}^{u\times v}_n = \hat{J}^u_n \otimes \hat{\mathbf{I}}^v \times \hat{\mathbf{I}}^u \otimes \hat{J}^v_n$. Thus, we find that the generators of a direct product representation are the sums of the corresponding generators of its constituent representations.

The generators defined by $\hat{J}^{u\times v}_n = \hat{J}^u_n \otimes \hat{\mathbf{I}}^v + \hat{\mathbf{I}}^u \otimes \hat{J}^v_n$ satisfy the same Lie algebra as the original one, i.e., $\left[\hat{J}^{u\times v}_k, \hat{J}^{u\times v}_l\right] = i\varepsilon^m_{kl}\,\hat{J}^{u\times v}_m$. We have

$$\begin{aligned}
\left[\hat{J}^{u\times v}_k, \hat{J}^{u\times v}_l\right] &= \left[\left(\hat{J}^u_k \otimes \hat{\mathbf{I}}^v + \hat{\mathbf{I}}^u \otimes \hat{J}^v_k\right), \left(\hat{J}^u_l \otimes \hat{\mathbf{I}}^v + \hat{\mathbf{I}}^u \otimes \hat{J}^v_l\right)\right] \\
&= \left(\hat{J}^u_{\hat{k}} \otimes \hat{\mathbf{I}}^v + \hat{\mathbf{I}}^u \otimes \hat{J}^v_k\right)\left(\hat{J}^u_l \otimes \hat{\mathbf{I}}^v + \hat{\mathbf{I}}^u \otimes \hat{J}^v_l\right) \\
&\quad - \left(\hat{J}^u_l \otimes \hat{\mathbf{I}}^v + \hat{\mathbf{I}}^u \otimes \hat{J}^v_l\right)\left(\hat{J}^u_k \otimes \hat{\mathbf{I}}^v + \hat{\mathbf{I}}^u \otimes \hat{J}^v_k\right) \\
&= \left(\hat{J}^u_k \otimes \hat{\mathbf{I}}^v\right)\left(\hat{J}^u_l \otimes \hat{\mathbf{I}}^v\right) + \left(\hat{\mathbf{I}}^u \otimes \hat{J}^v_k\right)\left(\hat{\mathbf{I}}^u \otimes \hat{J}^v_l\right) \\
&\quad - \left(\hat{J}^u_l \otimes \hat{\mathbf{I}}^v\right)\left(\hat{J}^u_k \otimes \hat{\mathbf{I}}^v\right) - \left(\hat{\mathbf{I}}^u \otimes \hat{J}^v_l\right)\left(\hat{\mathbf{I}}^u \otimes \hat{J}^v_k\right)
\end{aligned}$$

$$= \left(\hat{J}_k^u \hat{J}_l^u\right) \otimes \hat{\mathbf{I}}^v + \hat{\mathbf{I}}^u \otimes \left(\hat{J}_k^v, \hat{J}_l^v\right) - \left(\hat{J}_l^u, \hat{J}_k^u\right) \otimes \hat{\mathbf{I}}^v - \hat{\mathbf{I}}^u \otimes \left(\hat{J}_l^v, \hat{J}_k^v\right)$$

$$= \left[\hat{J}_k^u, \hat{J}_l^u\right] \otimes \hat{\mathbf{I}}^v + \hat{\mathbf{I}}^u \otimes \left[\hat{J}_k^v, \hat{J}_l^v\right] = i\varepsilon_{kl}^m \left(\hat{J}_m^u \otimes \hat{\mathbf{I}}^v + \hat{\mathbf{I}}^u \otimes \hat{J}_m^v\right) = i\varepsilon_{kl}^m J_m^{u\times v}$$

Our next step is to regroup the basis vectors of the product representation $\{|kl\rangle\}$ to form invariant subspaces thereby leading to the reduction of the direct product representation. We also have

$$\hat{J}_3^{u\times v}\left|kl\right\rangle = \left(\hat{J}_3^u \times \hat{\mathbf{I}}^v + \hat{\mathbf{I}}^u \times \hat{J}_3^v\right)(|uk\rangle \times |vl\rangle) = (k+l)(|uk\rangle \times |vl\rangle) = (k+l)|kl\rangle \tag{24}$$

so that $|kl\rangle$ is an eigenvector of $\hat{J}_3^{u\times v}$ with eigenvalue $(k + l)$.

Further, $(\hat{J}^{u\times v})^2 = \left(\hat{J}_3^{u\times v}\right)^2 + \hat{J}_3^{u\times v} + \hat{J}_-^{u\times v}\hat{J}_+^{u\times v}$ whence

$$\left(\hat{J}^{u\times v}\right)^2\left|uk\right\rangle \otimes|vl\rangle = \left(\hat{J}_3^u \otimes \hat{\mathbf{I}}^v + \hat{\mathbf{I}}^u \otimes \hat{J}_3^v\right)\left(\hat{J}_3^u \otimes \hat{\mathbf{I}}^v + \hat{\mathbf{I}}^u \otimes \hat{J}_3^v\right)\left|uk\right\rangle \otimes|vl\rangle$$

$$+ \left(\hat{J}_3^u \otimes \hat{\mathbf{I}}^v + \hat{\mathbf{I}}^u \otimes \hat{J}_3^v\right)\left|uk\right\rangle \otimes|vl\rangle + \left(\hat{J}_-^u \otimes \hat{\mathbf{I}}^v + \hat{\mathbf{I}}^u \otimes \hat{J}_-^v\right)\left(\hat{J}_+^u \otimes \hat{\mathbf{I}}^v + \hat{\mathbf{I}}^u \otimes \hat{J}_+^v\right)|uk\rangle \otimes|vl\rangle$$

$$= (k+l)^2|uk\rangle \otimes|vl\rangle + (k+l)|uk\rangle \otimes |vl\rangle + \left(\hat{J}_-^u \otimes \hat{\mathbf{I}}^v + \hat{\mathbf{I}}^u \otimes \hat{J}_-^v\right)$$

$$[A^+(u, k)|u, k+1\rangle \otimes|v, l\rangle + B^+(v, l)|uk \otimes|v, l+1\rangle]$$

$$= (k+l)^2|uk\rangle \otimes|vl\rangle + (k+l)|uk\rangle \otimes |vl\rangle + A^-(u, k+1)$$

$$A^+(u, k)|u, k\rangle \otimes|v, l\rangle + B^-(v, l+1)B^+(v, l)|u, k\rangle \otimes|v, l\rangle$$

so that $|uk\rangle \otimes |vl\rangle$ is also an eigenvector of $\left(\hat{J}^{u\times v}\right)^2$.

We also know from the elementary theory of angular momentum in three dimensions, of which the group $SO(3)$ provides a representation, that the range of the eigenvalues of the generator $\hat{M} \equiv \hat{J}_3^u$ extends from $-u \le k \le +u$, where u relates to the eigenvalue of the Casimir operator $\hat{M}^2$. Similarly, the range of l extends from $-v \le l \le +v$. Therefore, the highest eigenvalue of $\hat{J}_3^{u\times v}$ is $u+v$ and there is only one eigenvector corresponding to this eigenvalue given by $|uv\rangle$. There will be two eigenvectors corresponding to the eigenvalue $u+v-1$ of $\hat{J}_3^{u\times v}$ given by $|u-1, v\rangle$ and $|u, v-1\rangle$ and so on. The lowest eigenvalue of $\hat{J}_3^{u\times v}$ is $-u-v$ and there is only one eigenvector corresponding to this eigenvalue given by $|-u,-v\rangle$. Since the eigenvalues of $\hat{J}_3^{u\times v}$ change in integral steps, the total number of eigenvalues of $\hat{J}_3^{u\times v}$ will be $2(u+v)+1$ and the corresponding eigenvectors can be obtained by applying the "lowering operator $\hat{J}_-^{u\times v}$" to the highest state $|uv\rangle$.

Since there is only one state with the eigenvalue $u + v$ and there exists no state with a higher eigenvalue of $\hat{J}_3^{u\times v}$, this state i.e. $|uv\rangle$ is the highest member of the basis of the IRR labelled by $u + v$. The remaining basis vectors can be generated by successive application of the lowering operator $\hat{J}_-^{u\times v}$, as mentioned above. The total number of basis vectors in this representation will be $2(u + v) + 1$ (because the range of eigenvalues of $\hat{J}_3^{u\times v}$ is from $u + v$ through 0 to $-(u + v)$) and they constitute the basis of an invariant subspace corresponding to the IRR labelled by $u + v$ relating it to the eigenvalue of $(\hat{J}^{u\times v})^2$.

As mentioned above, there are two linearly independent states corresponding to the eigenvalue $u + v - 1$ of $\hat{J}_3^{u\times v}$ given by $|u - 1, v\rangle$ and $|u, v - 1\rangle$. One of these states is included as the second element of the basis in the IRR labelled by $u + v$. The second state with eigenvalue $u + v - 1$ (orthogonal to the first with the same eigenvalue), will constitute the highest basis vector in another IRR that is labelled by the index $u + v - 1$. This basis will consist of $2(u + v - 1) + 1 = 2(u + v) - 1$ vectors which can be obtained by using the "lowering operator $\hat{J}_-^{u\times v}$."

For the eigenvalue $u + v - 2$ of $\hat{J}_3^{u\times v}$, there will be three linearly independent vectors viz. $|u, v - 2\rangle$, $|u - 1, v - 1\rangle$, $|u - 2, v\rangle$, one of which is included as the third element in the basis of the invariant subspace labelled by $u + v$, the second being included as the second element in the basis of the invariant subspace labelled by $u + v - 1$ and the third will be the highest element of the basis of the invariant subspace labelled by $u + v - 2$.

Now, if $u > v$, the above series will terminate when the eigenvalue of $\hat{J}_3^{u\times v}$ is $u - v$ (because the lowest allowed value of l is $-v$) with the last set of linearly independent vectors being $|u, -v\rangle$, $|u - 1, -v + 1\rangle,\ldots, |u - 2v, v\rangle$. Further, the space labelled by $u - v$ will have a basis consisting of $2(u - v) + 1$ linearly independent vectors (because the range of eigenvalues of $\hat{J}_3^{u\times v}$ in this case will be from $u - v$ through 0 to $-(u - v)$). A similar analysis holds if $v > u$ whence the series will terminate when the eigenvalue of $\hat{J}_3^{u\times v}$ is $v - u$. Hence, we conclude that this iterative process will yield an invariant subspace in each step with the index of the IRR diminishing by one until the index reaches $|u - v|$. The number of vectors that span the invariant subspace in this case will be $2|u - v| + 1$.

The aggregate number of independent basis states of all the IRRs taken together are:

$$[2(u + v) + 1] + [2(u + v - 1) + 1] + \ldots + [2|u - v| + 1] = (2u + 1)(2v + 1) \tag{25}$$

as it should be.

If we represent the new basis vectors by $\{|jm\rangle\}$, where $m = -j,\ldots, +j$ and $j = |u - v|,\ldots, u + v$, then the transformation matrix between the two sets is given by

$$|jm\rangle = \langle kl(uv)\ jm\rangle|kl\rangle \tag{26}$$

$$|kl\rangle = \langle jm(uv)\ kl\rangle|jm\rangle \tag{27}$$

where the coefficients of the basis vectors on the right hand side are called the Clebsch Gordon coefficients. The above notation for the Clebsch Gordon coefficients involves summing over the repeated indices, viz. kl and jm in eqs. (26) and (27) respectively. The indices in parenthesis (uv) are not to be summed over. Further $\langle jm(uv)kl\rangle = \langle kl(uv)jm\rangle^*$ so that the unitarity condition becomes $\langle jm(uv)kl\langle = \langle kl(uv)jm\rangle$.

The above discussion enables us to formulate another scheme for the labelling of IRRs. We can use the labels (j_0, j_1) to identify the IRR, where $j_0 = |u - v|$ and $j_1 = u + v$ are the lowest and highest values of the angular momentum in the representation.

2.1.2 Effect of the Group Generators on the Basis

In view of the foregoing, the representation space carries the label (j_0, j_1) and the vectors within a representation carry the labels (j, m). We also know, from the theory of angular momentum in three dimensions that j can be an integer or a half integer and $-j \le m \le +j$. The action of the generators of rotations $\hat{J}_3^{u\times v}$ on a basis vector is given by

$$\hat{J}_3^{u\times v}|jm\rangle = m|jm\rangle \tag{28}$$

$$\hat{J}_\pm^{u\times v}|jm\rangle = A_m^\pm(j, m)|jm \pm 1\rangle \tag{29}$$

The value of j, being related to the Casimir operator $(\hat{J}^{u\times v})^2$ does not change by the group operations i.e., rotations.

The effect of the generators of the Lorentz boosts $\hat{K}_i$ on the basis vectors is slightly more complicated. In view of eq. (3), the three components of $\hat{K}_i$ constitute the Cartesian components of a 3-vector under rotations. Applying the Wigner Eckart theorem to obtain the matrix elements between irreducible physical states represented in an appropriate basis, we obtain

$$\langle j'm'|\hat{K}_3|jm\rangle = A_j^{j'}\langle j'm'(1, j)0m\rangle \tag{30}$$

and

$$\langle j'm'|\hat{K}_\pm|jm\rangle = \mp\sqrt{2A_j^{j'}}\langle j'm'(1, j) \pm 1m\rangle \tag{31}$$

The coefficients $A_j^{j'}$ represent the reduced matrix elements given by $\langle j'|\hat{K}_3|j\rangle$. $\langle j'm'(1, j)0m\rangle$ and $\langle j'm'(1, j) \pm 1m\rangle$ are the Clebsch Gordon coefficients. The factor of $\mp\sqrt{2}$ arises in Eq. (31) because of the fact that $\hat{K}_3 \mp (2)^{-1/2}\hat{K}_\pm$ constitute a normalized set of irreducible spherical tensor operators. Taking the values of the Clebsch Gordon coefficients from published tables, noting that the coefficients $\langle j'm'(1, j)nm\rangle$ vanish unless $j' = j - 1, j, j + 1$ and $m' = m + n$, we define

$$A_j^+ = A_j^{j+1}[(j + 1)(2j + 1)]^{-1/2} \tag{32}$$

$$A_j = A_j^j[j(j + 1)]^{-1/2} \tag{33}$$

$$A_j^- = A_j^{j-1}[j(2j + 1)]^{-1/2} \tag{34}$$

and write eqs. (30) and (31) as

$$\hat{K}_3|jm\rangle = |j - 1m\rangle[(j + m)(j - m)]^{1/2}A_j^- + |jm\rangle mA_j + |j + 1m\rangle[(j + m + 1)(j - m + 1)]^{1/2}A_j^+ \tag{35}$$

and

$$\hat{K}_{\pm}|jm\rangle = \mp|j-1m\pm1\rangle\,[(j\mp m)(j\mp m-1)]^{1/2}A_j^- + |jm\pm1\rangle\,[(j\mp m)(j\pm m+1)]^{1/2}A_j \pm |j+1m\pm1\rangle\,[(j\pm m+1)(j\pm m+2)]^{1/2}A_j^+ \tag{36}$$

2.1.3 Computing the Coefficients A_j^+, A_j and A_j^-

To obtain constraints on the coefficients A_j^+, A_j and A_j^- and hence, compute their admissible values, we mandate that the group Lie algebra must be satisfied. We start by requiring that $\left[\hat{K}_+,\hat{K}_-\right] = -2\hat{J}_3$ or equivalently

$$\left[\hat{K}_+,\hat{K}_-\right]|jm\rangle = -2\hat{J}_3|jm\rangle \tag{37}$$

Using eq. (36), the left-hand side of eq. (37) becomes

$$\begin{aligned}\left[\hat{K}_+,\hat{K}_-\right]|jm\rangle &= |j-1,m\rangle\,2\sqrt{(j+m)(j-m)}\,A_j^-[(j-1)A_{j-1}-(j+1)A_j]\\ &\quad -|jm\rangle\,(2m)\,[(2j+3)A_{j+1}^-A_j^+ + A_j^2 - (2j-1)A_j^-A_{j-1}^+]\\ &\quad +|j+1,m\rangle\,2\sqrt{(j+m+1)(j-m+1)}\,A_j^+[jA_j+(j+2)A_{j+1}]\end{aligned} \tag{38}$$

and the right hand side gives

$$-2\hat{J}_3|jm\rangle = -|jm\rangle\,(2m) \tag{39}$$

Equating eqs. (38) and (39), we get

$$[(j-1)A_{j-1}-(j+1)A_j]A_j^- = 0 \tag{40}$$

$$A_j^+[(j+2)A_{j+1}-jA_j] = 0 \tag{41}$$

$$(2j-1)A_j^-A_{j-1}^+ - A_j^2 - (2j+3)A_{j+1}^-A_j^+ = 1 \tag{42}$$

Equations (40) and (41) lead to the same constraint. Since the allowed values for j are $0, \frac{1}{2}, 1, \ldots$, they are bounded from below. Therefore, it must have a least value, say j_0. From eq. (35) and (36), it then follows that $A_{j0}^- = 0$. However, when $A_j^{\pm} \neq 0$, the expressions within the square brackets of eqs. (40) and (41) must vanish. For $A_j^- \neq 0$, eq. (41) gives

$$A_{j+1} = \frac{j}{j+2}A_j = A_{j0}\prod_{k=j_0}^{j}\frac{k}{k+2} = A_{j_0}\frac{j_0(j_0+1)}{(j+1)(j+2)} \tag{43}$$

The second step is simply the expansion of the recursion relation and the final step involves cancelling of common factors in the numerator and denominator. Writing $t = -i(j_0+1)A_{j_0}$, we have, from Eq. (43)

$$A_j = \text{it}\,\frac{j_0}{j(j+1)} \tag{44}$$

Using Eq. (44) and writing $B_j^2 = -A_j^-A_{j-1}^+$, eq. (42) becomes

$$(2j+3)B_{j+1}^2 = (2j-1)\,B_j^2 + 1 - \left[\frac{tj_0}{j(j+1)}\right]^2 \tag{45}$$

To solve eq. (45) consider

$$(2j+3)B_{j+1}^2 = (2j-1)B_j^2 + 1 - \left[\frac{tj_0}{j(j+1)}\right]^2$$

$$(2j-1)B_j^2 = \frac{(2j-1)(2j-3)}{2j+1}\,B_{j-1}^2 + \frac{2j-1}{2j+1}\left[1 - \left(\frac{tj_0}{(j-1)j}\right)^2\right]$$

$$\frac{(2j-1)(2j-3)}{2j+1}\,B_{j-1}^2 = \frac{(2j-3)(2j-5)}{2j+1}\,B_{j-2}^2 + \frac{2j-3}{2j+1}\left[1 - \left(\frac{tj_0}{(j-1)(j-2)}\right)^2\right]$$

and so on. These equations are obtained by using the recursion relation (45) such that the LHS of any equation exactly cancels the B term of RHS of the previous equation.

For example, putting $j \equiv j-1$ in Eq. (45), we get

$$(2j+1)B_j^2 = (2j-3)\,B_{j-1}^2 + 1 - \left[\frac{tj_0}{(j-1)\,j}\right]^2$$

so that

$$(2j-1)\,B_j^2 = \frac{(2j-1)(2j-3)}{(2j+1)}\,B_{j-1}^2 + \frac{(2j-1)}{(2j+1)} - \frac{(2j-1)}{(2j+1)}\left[\frac{tj_0}{(j-1)\,j}\right]^2$$

which is the second equation in the series. Similarly, the remaining equations can be manipulated. The last equation in the series will have B_{j_0+1} because $B_{j_0} = 0$ since $A_{j0} = 0$. Adding up all the equations in this series, we get

$$(2j+3)\,B_{j+1}^2\,\frac{1}{2j+1}\sum_{k=j_0+1}^{j+1}(2k-1) - \frac{t^2\,j_0^2}{2j+1}\sum_{k'=j_0+1}^{j+1}\frac{2k'-1}{k'^2\,(k-1)^2} \tag{46}$$

Writing $j \equiv j-1$, we get

$$B_j^2 = \frac{1}{4j^2-1}\sum_{k=j_0+1}^{j}(2k-1) - \frac{t^2\,j_0^2}{4j^2-1}\sum_{k'=j_0+1}^{j}\frac{2k'-1}{k'^2\,(k'-1)^2} \tag{47}$$

Now, the expression $\sum_{k=j_0+1}^{j}(2k-1)$ is simply an arithmetic series $2j_0+1, 2j_0+3, \ldots, 2j-1$, i.e., $j-j_0$ terms. Its summation is $j^2 - j_0^2$. Furthermore

$$-\sum_{k'=j_0+1}^{j}\frac{2k'-1}{k'^2(k'-1)^2}=\sum_{k'=j_0+1}^{j}\frac{k'^2-2k'+1-k'^2}{k'^2(k'-1)^2}$$

$$=\sum_{k'=j_0+1}^{j}\left[\frac{1}{k'^2}-\frac{1}{(k'-1)^2}\right]=\frac{1}{j^2}-\frac{1}{j_0^2} \tag{48}$$

Combining these results, we get

$$B_j^2=\frac{(j^2-j_0^2)(j^2-t^2)}{j^2(4j^2-1)} \tag{49}$$

whence, using $B_j^2=-A_j^-A_{j-1}^+$, we can write

$$A_j^-=B_j\eta_j \text{ and } A_{j-1}^+=-B_j\eta_j^{-1} \tag{50}$$

with η_j being arbitrary.

2.1.4 Computing *t* and η_j

The admissible values of t and η_j are subject to the constraints of unitarity of the representation and the orthonormality of the canonical basis. The requirement of unitarity mandates that $\hat{K}_3^\dagger=\hat{K}_3$ and $\hat{K}_\pm^\dagger=\hat{K}_\mp$ whence, from eqs. (35) and (36), we get

$$A_j=A_j^* \text{ and } A_j^-=-(A_{j-1}^+)^* \tag{51}$$

Substituting eqs. (44), (49) and (50) in eq. (51), we get

$$j_0(t+t^*)=0 \text{ and } |B_j|\,(|\eta_j|^2-e^{-2i\beta_j})=0 \tag{52}$$

where we have obtained the second equation by writing $B_j=|B_j|e^{i\beta_j}$ so that $B_j^*=|B_j|e^{-i\beta_j}$ which gives $A_j^-=B_j\eta_j=-(A_{j-1}^+)^*=-(-B_j^*\eta_j^{-1*})=B_j^*\eta_j^{-1*}$ whence $|B_j|e^{i\beta_j}|\eta_j|^2-|B_j|e^{-i\beta_j}=0=|B_j|\,(|\eta_j|^2-e^{-2i\beta_j})$. If $|B_j|\neq 0$, then $|\eta_j|^2-e^{-2i\beta_j}=0$ whence $\beta_j=0$ and $|\eta_j|^2=1$. Thus, η_j is an arbitrary phase factor associated with the basis vectors. Conventionally, we choose $\eta_j=1$ for all values of j.

The first part of eq. (52) mandates that either (i) $t^*=-t$, i.e., t is imaginary or (ii) $j_0=0$. IRRs corresponding to: (i) are known as the Principal Series while those corresponding to (ii) are classified as Complementary Series. It is pertinent to mention here that the parameter t gets identified with j_1 in the usual (j_0, j_1) representation.

2.1.5 Principal Series and Complementary Series

(i) Writing $t=-i\omega$ with $\omega\in\mathbf{R}$, we have $A_j=\dfrac{\omega j_0}{j(j+1)}$ and $B_j^2=\dfrac{(j^2-j_0^2)(j^2+\omega^2)}{j^2(4j^2-1)}>0$ is automatically satisfied so that $\omega \in \mathbf{R}$ is unconstrained. This set of conditions yields the principal series of unitary representations of the Lorentz group.

(ii) $j_0 = 0$ implies that $A_j = 0$ and $B_j^2 = \dfrac{(j^2 - t^2)}{(4j^2 - 1)}$. Since we require B_j^2 to be real and positive for $j =$ 1, 2, 3,... we must have $1 - t^2 \geq 0$ or $-1 \leq t \leq +1$. These constitute the complementary series.

2.2 FINITE DIMENSIONAL REPRESENTATIONS OF THE LORENTZ GROUP (GEL' FAND, 1963, TUNG, 1984)

The results of the preceding section carry over in entirety to determine the finite dimensional representations of the Lorentz group upto eq. (50) since we have nowhere used the prescription of unitarity upto that stage (eq. (50)). To obtain the finite dimensional representations, we can, therefore, assume that results upto eq. (50) are already established and shall proceed therefrom. The local isomorphism of the proper Lorentz group to the direct product group $SU(2)_M \otimes SU(2)_N$ restricts the permissible values of j of the canonical basis $\{|jm|\}$to $|u - v| \leq j \leq u + v$. We now determine the values of j_0 and $t \equiv j_1$ for the case of finite dimensional representations. As in the foregoing paragraphs j_0 is defined by $j_0 = | u - v |$ and it is the lowest value for j (which is bounded from below). Let us further assume that $j_{\max} = j_m = u + v$.

Now, from eqs. (35) and (36), we see that for this maximum value of $j \equiv j_m$, A_j^+ must necessarily vanish, i.e., $A_{jm}^+ = 0$. Using $B_j^2 = -A_j^- A_{j-1}^+$, this implies that $B_{j_m+1}^2 = -A_{j_m+1}^- A_{jm}^+ = 0$ whence from eq. (49), $(j_m + 1)^2 = t^2$ or $t = \pm (j_m + 1)$. Using $\hat{M}_m = \frac{1}{2}(\hat{J}_m + i\hat{K}_m)$ and $\hat{N}_m = \frac{1}{2}$ $(\hat{J}_m + i\hat{K}_m)$ with $m, n = 1, 2, 3$, we may write

$$\hat{M}^2 = \frac{1}{4}[\hat{J}_3^2 - \hat{K}_3^2 + 2i\hat{J}_3\hat{K}_3 + 2(\hat{J}_3 + i\hat{K}_3) + \hat{J}_-\hat{J}_+ - \hat{K}_-\hat{K}_+ + i(\hat{K}_-\hat{J}_+ + \hat{J}_-\hat{K}_+)] \tag{53}$$

$$\hat{N}^2 = \frac{1}{4}[\hat{J}_3^2 - \hat{K}_3^2 - 2i\hat{J}_3\hat{K}_3 + 2(\hat{J}_3 - i\hat{K}_3) + \hat{J}_-\hat{J}_+ - \hat{K}_-\hat{K}_+ - i(\hat{K}_-\hat{J}_+ + \hat{J}_-\hat{K}_+)] \tag{54}$$

Since these operators are Casimir operators, they commute with all the group generators and, therefore, by Schur's Lemma, get mapped into a constant multiple of the identity operator. We proceed to determine their eigenvalues from the above expansion by operating on our canonical basis vectors. It is emphasized here that the choice of the vector is irrelevant since the eigenvalues of Casimir operators remain unchanged for the basis vectors in a given IRR. Operating $\hat{M}^2$ and $\hat{N}^2$ on $|j_m\, j_m\rangle$ and using eqs. (53) and (54); (35) and (36), we have

$$\hat{M}^2|j_m\, j_m\rangle = \frac{1}{4}[\hat{J}_m^2 - j_m^2 A_{jm}^2 + 2ij_m^2 A_{jm} + 2j_m(1 + iA_{jm})]|j_m\, j_m\rangle \tag{55}$$

$$\hat{N}^2|j_m\, j_m\rangle = \frac{1}{4}[\hat{J}_m^2 - j_m^2 A_{jm}^2 - 2ij_m^2 A_{jm} + 2j_m(1 - iA_{jm})]|j_m\, j_m\rangle \tag{56}$$

where $\hat{J}_+|j_m j_m\rangle = \hat{K}_+|j_m j_m\rangle = 0$ since $j_{max} = j_m = u + v$ represents the highest eigenvalue in the representation. Eq. (44) gives

$$A_{j_m} = i\,\frac{tj_0}{j_m(j_m+1)} \tag{57}$$

which together with $t = \pm(j_m + 1)$ gives $A_{j_m} = \pm i\frac{j_0}{j_m}$ whence from eqs. (55) and (56), we get

$$\hat{M}^2|j_m j_m\rangle = \frac{1}{4}[j_m^2 + j_0^2 \mp 2j_m j_0 + 2j_m \mp 2j_0]|j_m j_m\rangle \tag{58}$$

$$\hat{N}^2|j_m j_m\rangle = \frac{1}{4}[j_m^2 + j_0^2 \pm 2j_m j_0 + 2j_m \pm 2j_0]|j_m j_m\rangle \tag{59}$$

However, we also have, as per the (u, v) labelling of the IRR

$$\hat{M}^2|j_m j_m\rangle = u(u+1)|j_m j_m\rangle \tag{60}$$

$$\hat{N}^2|j_m j_m\rangle = v(v+1)|j_m j_m\rangle \tag{61}$$

From eqs. (58)-(61), we obtain

$$u(u+1) = \frac{1}{4}(j_m \mp j_0)(j_m \mp j_0 + 2) \tag{62}$$

$$v(v+1) = \frac{1}{4}(j_m \pm j_0)(j_m \pm j_0 + 2) \tag{63}$$

Subtracting eq. (63) from (62), we get

$$(u-v)(u+v+1) = \mp j_0(j_m+1) \tag{64}$$

Two situations arise as follows:

(i) $(u - v) > 0$, whence the LHS of eq. (64) is positive. If we are to identify $u = \frac{(j_m - j_0)}{2}$ and $v = \frac{(j_m + j_0)}{2}$ so that $0 < u - v = -j_0$ then the RHS will be positive only if the upper sign '–' holds in the RHS of eq. (64). Correspondingly, the upper sign '+' will hold in $t = \pm(j_m+1)$ and we have $t = (j_m + 1)$.

(ii) $(u - v) < 0$, whence the lower sign '+' holds in the RHS of eq. (64). Correspondingly, the lower sign '–' will hold in $t = \pm(j_m + 1)$ and we have $t = -(j_m + 1)$.

A modification of this scheme of labelling of IRRs and basis vectors is to set $j_0 = u - v$, allowing it to take negative values as well, whence $t = u + v + 1$ irrespective of the sign of $(u - v)$.

The proper Lorentz group is non-compact so that it has no unitary finite dimensional representations. The non-compactness is a manifestation of the existence of unbounded group elements, in particular, the Lorentz boosts that can be parametrized in terms of the rapidity

$\xi = \tanh^{-1}(v/c)$, $-\infty \le \xi \le +\infty$. In fact, Lorentz boosts can be construed as spatiotemporal rotations that mix the spatial and time components of the four-dimensional spacetime by hyperbolic (unbounded) angles.

The non-compact nature of the proper Lorentz group may also be attributed, equivalently, to the non-positive definite nature of the Minkowski metric, which shows up again as the negative sign in the RHS of eq. (4).

As mentioned earlier, the Lie algebra of the Lorentz group consisting of eqs. (5–7) is identical to that of $SU(2)_M \otimes SU(2)_N$ and there exists a local isomorphism between the two so that every representation of the latter provides a finite dimensional representation of the former. The important point, however, is that whereas $SU(2)$ and hence, the direct product $SU(2)_M \otimes SU(2)_N$ is compact, the proper Lorentz group is not. The reason is that while $SU(2)_M \otimes SU(2)_N$ has $\hat{M}$, $\hat{N}$ as the generators and, therefore, group elements can be obtained by exponentiation of $i\hat{M}$, $i\hat{N}$ with a real parameter, the proper Lorentz group has $\hat{J}$, $\hat{K}$ as the generators so that group elements are obtained by exponentiation of $i\hat{J}$, $i\hat{K}$. Now, because of the interelationship $\hat{M}_m = \frac{1}{2}(\hat{J}_m + i\hat{K}_m)$ and $\hat{N}_m = \frac{1}{2}(\hat{J}_m + i\hat{K}_m)$, the two sets of generators $\hat{M}$, $\hat{N}$ and $\hat{J}$, $\hat{K}$ cannot be Hermitian simultaneously. Hence, the representations generated by both these sets cannot simultaneously be unitary.

An important aspect of representations of the Lorentz group is the physical relevance of different classes of representations. Although the non-unitary representations cannot be realized as physical particle states (because all symmetry operations must be represented by unitary operators on the space of physical states), physical variables like position, momentum, angular momentum, energy momentum tensor, particle wavefunctions and fields all do transform as finite dimensional representations. However, physical particle states themselves transform as the unitary (infinite dimensional) representations.

2.3 UNITARY REPRESENTATIONS OF THE POINCARE GROUP (WIGNER, 1939; GEL'FAND, 1963; NAIMARK, 1964; OHNUKI, 1976; TUNG, 1984; FUSHCHICH, 1994)

Since the Poincare group is non-compact, all its unitary representations are infinite dimensional. They may conveniently be derived by the method of induced representation. Before applying this method to the actual construction of the IRRs of the Poincare group, we introduce the steps involved in the procedure and illustrate them with the example of the two-and three-dimensional Euclidean groups.

(a) We define the two-dimensional Euclidean group E_2 as consisting of all continuous linear transformations on the two-dimensional Euclidean space $\mathbf{R}_2$ that leave the length of the 2-vectors invariant. Representing an arbitrary vector $\mathbf{x}$ in $\mathbf{R}_2$ in terms of its components (x^1, x^2), we can write a general linear transformation of this vector as $x'^i = R^i_j x^j + a^i$, $i, j = 1, 2$. The condition of "length preservation", then, mandates that the matrix R should be an orthogonal matrix for, writing the transformation equation in matrix form as $\mathbf{x}' = R.\mathbf{x} + A$, where $A \equiv (a^1\ a^2)^T$ whence $|\mathbf{x}|^2 = \mathbf{x}^T\mathbf{x} = |\mathbf{x}' - A|^2 = (R.\mathbf{x})^T.(R.\mathbf{x}) = \mathbf{x}^T R^T R.\mathbf{x}$ whence $R^T = R^{-1}$. Thus, the homogeneous part of the transformation corresponds to "rotations" in two-dimensional space while the inhomogeneous part corresponds to uniform translations of all points. Hence, to study the representations of E_2, we need to examine the nuances of the translational group and the two-dimensional rotation group.

(b) We first consider the representations of the one-dimensional translational group designated T_1. An arbitrary element of this group corresponds to translation by a distance a and is denoted by $\hat{T}(a)$. The action of $\hat{T}(a)$ on a state $|x\rangle$ localized at the point x is given by $\hat{T}(a)|x\rangle = |a + x\rangle$. The operator $\hat{T}(a)$, then possesses the properties

$$\text{(i)}\ \hat{T}(a)\,\hat{T}(b) = \hat{T}\,(a+b) \tag{65}$$

$$\text{(ii)}\ \hat{T}(0) = \hat{\mathbf{I}} \tag{66}$$

$$\text{(iii)}\ \left[\hat{T}(a)\right]^{-1} = \hat{T}(-a) \tag{67}$$

so that the elements $\hat{T}(a)$ form a one parameter group designated as above by T_1. If the infinitesimal generator of this one parameter group is designated by $\hat{P}$, then $\hat{T}(da) = \hat{\mathbf{I}} - i\hat{P}\,da$, whence $\hat{T}(a + da) = \hat{T}(a)\,\hat{T}(da) = \hat{T}(a)\,[\hat{\mathbf{I}} - i\hat{P}\,da]$. Expanding $\hat{T}(a + da)$ as a Taylor series around $\hat{T}(a)$ to first order, we obtain $\hat{T}(a+da) = \hat{T}(a) + \dfrac{d\hat{T}(a)}{da}da$, whence $\dfrac{d\hat{T}(a)}{da} = -i\hat{P}\,\hat{T}(a)$ which has the solution $\hat{T}(a) = e^{-i\hat{P}a}$ in view of the boundary condition $\hat{T}(0) = \hat{\mathbf{I}}$. All IRRs of T_1 are one dimensional. For unitary representations, the generator $\hat{P}$ must be a Hermitian operator whose action on a state vector $|p\rangle$ is defined by $\hat{P}|p\rangle = p|p\rangle$ where p is the eigenvalue of the operator $\hat{P}$ corresponding to the eigenstate $|p\rangle$. If $\hat{T}(a) \to \hat{T}^p(a) \equiv e^{-i\hat{P}a}$ then $\hat{T}^p(a)|p\rangle \equiv e^{-i\hat{P}a}|p\rangle = e^{-ipa}|p\rangle$ with p being unrestricted.

(c) We now discuss the irreducible representations of the two-dimensional rotation group. For this, we consider a system that is symmetric under rotations in a plane around a fixed point. Let e_1, e_2 be a Cartesian orthonormal basis in the two-dimensional space so that an arbitrary vector can be represented as $x^ie_i \equiv x^1e_1 + x^2e_2$. Let this vector be rotated through an angle θ in the counter-clockwise direction. Then, if the transformed vector is given by $x'^ie_i \equiv x'^1e_1 + x'^2e_2$, by elementary geometry, we have $x'^1 = x^1\cos\theta - x^2\sin\theta$, $x'^2 = x^1\sin\theta + x^2\cos\theta$ or, in matrix notation $\mathbf{x}' = R(\theta)\mathbf{x}$ wherein $\mathbf{x} = (x^1\ x^2)^T$, $\mathbf{x}' = (x'^1\ x'^2)^T$ and $R(\theta) = \begin{pmatrix}\cos\theta & -\sin\theta\\ \sin\theta & \cos\theta\end{pmatrix}$. The transformation rule for components, therefore, is $x'^j = R^j_i\,x^i$. Since rotation leaves the length of the vector invariant, the rotation matrix must be orthogonal, e.g., $|\mathbf{x}|^2 = \mathbf{x}^T\mathbf{x} = |\mathbf{x}'|^2 = (R.\mathbf{x})^T.(R.\mathbf{x}) = \mathbf{x}^TR^TR.\mathbf{x}$, whence $R^T = R^{-1}$. This can also be explicitly verified for the above general form of the two-dimensional rotation matrix. Furthermore, $R^T = R^{-1}$ implies that $(\det R)^2 = \det R\,\det R^T = \det R\,\det R^{-1} = \det RR^{-1} = \det\mathbf{I} = 1$, whence $\det R = \pm 1$. Orthogonal matrices satisfying the condition $\det R = +1$ represent rotations in two-dimensional space. They constitute a group by themselves, called the special orthogonal group designated $SO(2)$.

Orthogonal matrices with $\det R = -1$ correspond to rotations coupled with spatial inversions. This set of matrices is not connected to the identity transformation by a continuous change of parameters.

The law of composition of rotations is given by $R(\alpha)R(\beta) = R(\alpha + \beta)$ with the convention that if $\alpha + \beta > 2\pi$ then $R(\theta) = R(\theta \pm 2\pi)$. Further, since two rotations commute, the group $SO(2)$ is abelian.

Now, consider a rotation by an infinitesimal angle $d\phi$. Since $R(\phi)$ is differentiable with respect to ϕ in the neighboruhood of the identity, it follows that $R(d\phi)$ differs from $R(0) = I$ by an infinitesimal quantity linear in $(d\phi)$ so that we can write $R(d\phi) = R(0) - iJd\phi = \mathbf{I} - iJd\phi$, where J is called the generator of rotations given by $J = i \left.\frac{dR(\phi)}{d\phi}\right|_{\phi=0}$ and is independent of the rotation angle.

Now, the rotation $R(\phi + d\phi)$ may be written as

$$R(\phi + d\phi) = R(\phi)\, R(d\phi) = R(\phi)\, [\mathbf{I} - iJ(d\phi)] = R(\phi) + \frac{dR}{d\phi}\, d\phi \tag{68}$$

where in the last expansion we have used Taylor's series expansion to first order about $R(\phi)$. From this, we have $\frac{dR(\phi)}{d\phi} = -iJR(\phi)$ whence $R(\phi) = e^{-iJ\phi}$ since for $\phi = 0$, $R(\phi) = \mathbf{I}$. The explicit expression for the generator can be obtained from

$$J = i\left.\frac{dR(\phi)}{d\phi}\right|_{\phi=0} = i\left.\frac{d}{d\phi}\begin{pmatrix}\cos\phi & -\sin\phi\\ \sin\phi & \cos\phi\end{pmatrix}\right|_{\phi=0} = i\begin{pmatrix}0 & -1\\ 1 & 0\end{pmatrix} \tag{69}$$

so that

$$R(\phi) + e^{-iJ\phi} = \mathbf{I} - iJ\phi - \mathbf{I}\frac{\phi^2}{2!} + iJ\frac{\phi^3}{3!} + \ldots = \mathbf{I}\cos\phi - iJ\sin\phi = \begin{pmatrix}\cos\phi & -\sin\phi\\ \sin\phi & \cos\phi\end{pmatrix} \text{ since } J^2 = \mathbf{I},$$

thereby reproducing the earlier result.

Now consider the issue of representations of $SO(2)$ on a finite dimensional vector space V. Let, under the representation $R(\phi) \to \hat{U}(\phi)$, $J \to \hat{J}$ where $\hat{J}$, now, is an operator acting on the representation space V. Then, we must have

$$\hat{U}(\alpha)\,\hat{U}(\beta) = \hat{U}(\alpha + \beta) = \hat{U}(\beta)\,\hat{U}(\alpha) \tag{70}$$

$$\hat{U}(\theta) = \hat{U}(\theta \pm 2\,\pi) \tag{71}$$

$$\hat{U}(d\phi) = \hat{\mathbf{I}} - i\hat{J}(d\phi) \tag{72}$$

$$\hat{U}(\phi) = e^{-i\hat{J}\phi} \tag{73}$$

The last equation is an operator equation on V. For unitary representations of $SO(2)$, i.e., unitary $\hat{U}(\phi)$, $\hat{J}$ must be Hermitian. Further, since $SO(2)$ is abelian, all its IRRs are unidimensional and for an arbitrary vector $|j\rangle \in V$ we must have $\hat{J}|j\rangle = j|j\rangle$ and $\hat{U}(\phi)|j\rangle = e^{-i\hat{J}\phi}|j\rangle = e^{-ij\phi}|j\rangle$, where j is the eigenvalue of the operator $\hat{J}$ corresponding to the eigenstate $|j\rangle \in V$. Obviously, then, $\hat{U}(\alpha)\,\hat{U}(\beta)|j\rangle = e^{-ij\alpha}e^{-ij\beta}|j\rangle = e^{-ij(\alpha+\beta)}|j\rangle = \hat{U}(\alpha+\beta)|j\rangle = e^{-ij(\beta+\alpha)}|j\rangle = \hat{U}(\beta)\,\hat{U}(\alpha)|j\rangle$ so that the group composition law is satisfied by the representation operator $\hat{U}$. Further, the condition $\hat{U}(\theta) = \hat{U}(\theta \pm 2\pi)$ implies that we must also have $e^{i2\pi j} = 1$ so that j must be an integer. Denoting these integral values of j by m we label the corresponding representation by this integer m and write.

$$\hat{J}\,|m\rangle = m|m\rangle \tag{74}$$

$$\hat{U}^{m}(\phi)|m\rangle = e^{-im\phi}|m\rangle \tag{75}$$

The defining equation for $R(\theta)$, viz. $R(\theta) = \begin{pmatrix} \cos\theta & -\sin\theta \\ \sin\theta & \cos\theta \end{pmatrix}$ is a two-dimensional representation of the group $SO(2)$. It must, therefore, be reducible. It is, actually, equivalent to a direct sum of the $m = \pm 1$ representations. This can be established by diagonalizing the matrix corresponding to the operator $\hat{J}$, i.e., $\hat{J} = \begin{pmatrix} 0 & -1 \\ i & 0 \end{pmatrix}$ which can be easily done. The eigenvalues of $\hat{J}$ are found to be ± 1 with the corresponding normalized eigenvectors in the orthonormal Cartesian basis $\{e_1, e_2\}$ being $e_\pm = \frac{1}{\sqrt{2}}(\mp e_1 - ie_2)$. With respect to the new eigenbasis we, have $\hat{J}e_\pm = \pm e_\pm$ and $R(\phi)\, e_\pm = e^{\mp i\phi}\, e_\pm$.

(d) We, now, revert our attention to the two-dimensional Euclidean group E_2 that consists of translations and rotations in two-dimensional space. A general transformation in this space can, therefore, be written in terms of the components of the state vectors as $x'^1 = x^1\cos\theta - x^2\sin\theta + b^1$, $x'^2 = x^1\sin\theta + x^2\cos\theta + b^2$. We denote this transformation by $g(b, \theta)$. Using the above explicit representation of an arbitrary transformation in E_2, we can write the group composition law as $g(b_2, \theta_2)\, g(b_1, \theta_1) = g(b_3, \theta_3)$, where $\theta_3 = \theta_1 + \theta_2$ and $b_3 = R(\theta_2)b_1 + b_2$. This also enables determination of the group inverse. e.g., $g(b', \theta')\, g(b, \theta) = g(0, 0)$, whence $\theta' = -\theta$ and $b' = -R(-\theta)b$ giving $g^{-1}(b, \theta) = g(-R(-\theta)\, b, -\theta)$. The subset of E_2 consisting of elements $\{g(0, \theta) = R(\theta)\}$ constitutes the subgroup of rotations in two-dimensional space $SO(2)$ with the representation $\hat{R}(\theta) = e^{-i\hat{J}\theta}$. On the other hand, the subgroup $\{g(b,0) = T(b)\}$ constitutes the subgroup of translations T_2 in two dimensions with two independent one parameter subgroups generated by $\hat{P}_1$and $\hat{P}_2$ respectively. Since the translations in x^1 and x^2 directions are independent of each other, the corresponding generators $\hat{P}_1$ and $\hat{P}_2$ commute and we can write a general translation as $\hat{T}(b) = e^{-i\hat{\mathbf{p}}.\mathbf{b}} = e^{-i\hat{P}_1 b^1}\, e^{-i\hat{P}_2 b^2}$. We can write the general group element of E_2, $g(b, \theta)$ as $g(b, \theta) = T(b)\, R(\theta)$ for, the right-hand side is $T(b)\, R(\theta) = g(b, 0)\, g(0, \theta) = g(b, \theta)$.

The general group element of E_2, $g(b, \theta)$ can be conveniently represented by 3×3 matrices as $g(b, \theta) = \begin{pmatrix} \cos\theta & -\sin\theta & b^1 \\ \sin\theta & \cos\theta & b^2 \\ 0 & 0 & 1 \end{pmatrix}$ with the generators of rotations being represented by $\hat{J} = \begin{pmatrix} 0 & -i & 0 \\ i & 0 & 0 \\ 0 & 0 & 0 \end{pmatrix}$ and the two translation generators in the x^1 and x^2 directions by $\hat{P}_1 = \begin{pmatrix} 0 & 0 & i \\ 0 & 0 & 0 \\ 0 & 0 & 0 \end{pmatrix}$ and $\hat{P}_2 = \begin{pmatrix} 0 & 0 & 0 \\ 0 & 0 & i \\ 0 & 0 & 0 \end{pmatrix}$ respectively.

Using the explicit representations of $\hat{J}$, $\hat{P}_1$ $\hat{P}_2$ given above, we can easily verify that $[\hat{P}_1\ \hat{P}_2] = 0$, $[\hat{J}\ \hat{P}_k] = i\varepsilon^{km}\ \hat{P}_m$, where e^{km} is the two dimensional unit antisymmetric tensor with $e^{kk} = 0$, $e^{km} = 1$, k, m cyclic. Using the Campbell Baker Hausdorff formula, viz. $e^A B e^{-A} = \sum_{n=0}^{\infty} \frac{1}{n!}[A, B]_n$, where $[A, B]_n = [A, [A, B]_{n-1}]$ and $[A, B]_0 = 1$, we have $e^{-i\hat{J}\theta}\ \hat{P}_k\, e^{i\hat{J}\theta} =$

$\sum_{n=0}^{\infty}\frac{1}{n!}[-i\hat{J}\theta,\ \hat{P}_k]_n = \sum_{n=0}^{\infty}\frac{(-i\theta)^n}{n!}\ [\hat{J},\ \hat{P}_k]_n$. Noting that $[\hat{J}\ \hat{P}_1] = i\hat{P}_2$, $[\hat{J}\ \hat{P}_2] = -\ i\hat{P}_1$ so that $[\hat{J}, [\hat{J}\ \hat{P}_1]] = i[\hat{J}, \hat{P}_2] = -\,i^2\ \hat{P}_1$ etc., we have

$$e^{-i\hat{J}\theta}\ \hat{P}_1 e^{i\hat{J}\theta} = \left[\mathbf{I} - \frac{i^2}{2!}(-i\theta)^2 + \frac{i^4}{4!}(-i\theta)^4 - \ldots\right]\hat{P}_1 + \left[\frac{i}{1!}(-i\theta) - \frac{i^3}{3!}(-i\theta)^3 + \frac{i^5}{5!}(-i\theta)^5 - \ldots\right]\hat{P}_2$$

$$= \left[\mathbf{I} - \frac{1}{2!}\theta^2 + \frac{1}{4!}\theta^4 - \ldots\right]\hat{P}_1 + \left[\frac{1}{1!}\theta - \frac{1}{3!}\theta^3 + \frac{1}{5!}\theta^5 - \ldots\right]\hat{P}_2$$

$$= \cos\theta\ \hat{P}_1 + \sin\theta\ \hat{P}_2 \tag{76}$$

with a similar expression for $\hat{P}_2$ so that we may write as a general expression $e^{-i\hat{J}\theta}\ \hat{P}_k e^{i\hat{J}\theta} = \hat{P}_m R(\theta)^m_k$. Proceeding similarly, we also obtain

$$e^{-i\hat{J}\theta}\ \hat{\mathbf{P}}.\mathbf{b}\ e^{i\hat{J}\theta} = e^{-i\hat{J}\theta}\ \hat{P}_1 b^1\ e^{i\hat{J}\theta} + e^{-i\hat{J}\theta}\ \hat{P}_2 b^2\ e^{i\hat{J}\theta}$$

$$= \hat{P}_1\ R(\theta)^1_1\ b^1 + \hat{P}_2\ R(\theta)^2_1\ b^1 + \hat{P}_1\ R(\theta)^1_2\ b^2 + \hat{P}_2\ R(\theta)^2_2\ b^2$$

$$= \hat{P}_1\ [R(\theta)^1_1\ b^1 + R(\theta)^1_2\ b^2] + \hat{P}_2\ [R(\theta)^2_1\ b^1 + R(\theta)^2_2\ b^2]$$

$$= \hat{P}_1\ b'^1 + \hat{P}_2\ b'^2 = \hat{\mathbf{P}}.\mathbf{b}' = \hat{P}_m\ R(\theta)^m_n\ b^n \tag{77}$$

where $b'^m = R(\theta)^m{}_1\ b^1 + R(\theta)^m{}_2\ b^2 = R(\theta)^m_k\ b^k$ so that $\mathbf{b}' = R(\theta).\mathbf{b}$ which leads us to

$$e^{-i\hat{J}\theta}\ \hat{T}(\mathbf{b})\ e^{i\hat{J}\theta} = e^{-i\hat{J}\theta}\ e^{-i\hat{\mathbf{P}}.\mathbf{b}}\ e^{i\hat{J}\theta} = e^{-i\hat{J}\theta}\left(\sum_{n=0}^{\infty}\frac{1}{n!}(-i\,\hat{\mathbf{P}}.\mathbf{b})^n\right)e^{+i\hat{J}\theta}$$

$$= e^{-i\hat{\mathbf{P}}\,R(\theta)\mathbf{b}} = \hat{T}\,[R(\theta)\mathbf{b}] \tag{78}$$

This leads us to another approach to the composition rule for this group as

$$g(a, \theta)\ g(b, \phi) = T(a)\ R(\theta)\ T(b)\ R(\phi) = T(a)\ [R(\theta)T(b)\ R(-\theta)]\ R(\theta)\ R(\phi)$$

$$= T(a)\ T[R(\theta)\,b]\ R(\theta + \phi) = T[R(\theta)b + a]\ R(\theta + \phi)$$

$$= g[R(\theta)b + a,\ \theta + \phi] \tag{79}$$

We define a subgroup H of a group G to be an invariant subgroup of G if the similarity transformation of H by a fixed element $x \in G$ always gives H, i.e., $xHx^{-1} = H$ for every element $x \in G$. It is easy to establish that the translation group $\{T(b)\}$ forms an invariant subgroup T_2 of E_2 for given $T(a) \in T_2$, we have

$$g(b, \theta)\ T(a)\ g(b, \theta)^{-1} = T(b)\ R(\theta)\ T(a)\ R(-\theta)\ T(-b)$$
$$= T(b)\ T\,[R(\theta)a]\ T(-b) = T\,[R(\theta)a] \in T_2$$

The factor group E_2/T_2 defined as the collection of right cosets of T_2 in E_2, i.e., collection of sets of the form $T_2 g\,(b, \theta) = \{T(a)\ g(b, \theta)\colon T(a) \in T_2\}$ is isomorphic to the group of rotations $SO(2)$ for $T(a)\ g\,(b, \theta) = T(a)\ T(b)\ R(\theta) = T(a + b)\ R(\theta) = T(a + b)\ g\,(0, \theta) = T_2\, g(0, \theta)$ where the last step follows from the fact that the set of translations forms a group so that $T(a + b) = T(a)\ T(b)$ with $T(a)$ varying over all possible values in the group T_2 would again reproduce T_2, i.e., $T(a)\ T(b)\ \forall\ T(a) \in T_2 = T(c)\ \forall T(c) \in T_2 = T_2$ so that $T_2 g\,(b, \theta) = T_2 g\,(0, \theta)$. Since, distinct cosets $T_2\, g(0, \theta) = \mathrm{T}_2\, R(\theta)$ are labelled by the one parameter, viz. the angle of rotation θ, there exists a one-one correspondence between these cosets and the group of two dimensional rotations $SO(2)$.

Our next step is to construct the unitary IRRs of the two-dimensional Euclidean group by the induced representation method. This method can be conveniently adopted for constructing group representations of continuous groups that contain an abelian (commutative) subgroup. For the two dimensional Euclidean group E_2, the two-dimensional translation subgroup T_2 constitutes an invariant abelian subgroup. The group E_2 has three generators $\hat{P}_1$, $\hat{P}_2$, $\hat{J}$ of which $\hat{P}_1$, $\hat{P}_2$, are generators of the group T_2 corresponding to translations in the x^1 and x^2 directions, respectively, and $\hat{J}$ is the generator of rotations in the $x^1 - x^2$ plane.

Consider a standard momentum vector $\mathbf{p}_0 \equiv (p, 0)$. The subspace corresponding to this momentum vector consists of all translations along the x^1 direction. As proved earlier, the factor group E_2/T_2 consists of the space of rotations in two-dimensional space and is isomorphic to $SO(2)$. We examine whether the operations in this factor group, i.e., rotations leave the eigenvector $\mathbf{p}_0$ invariant. This will be so if the corresponding generators commute. In this case, however, $[\hat{J}, \hat{P}_1] \neq 0$. It, therefore, follows that $\mathbf{p}_0$ is not left invariant by rotations and hence, there is only one independent eigenstate of $\hat{\mathbf{P}}$ corresponding to the eigenvector $\mathbf{p}_0$. We also have $\hat{P}_1|\mathbf{p}_0\rangle = p|\mathbf{p}_0\rangle$, $\hat{P}_2|\mathbf{p}_0\rangle = 0$, $\hat{\mathbf{P}}^2|\mathbf{p}_0\rangle = p^2|\mathbf{p}_0\rangle$.

To generate the full irreducible invariant subspace, we conduct group operations that generate new eigenvalues of $\hat{\mathbf{P}}$. This can only be done with the help of those generators that do not commute with $\hat{\mathbf{P}}$. In the case of group E_2, we have $[\hat{P}_1, \hat{P}_1] = [\hat{P}_1, \hat{P}_2] = [\hat{P}_2, \hat{P}_1] = [\hat{P}_2, \hat{P}_2] = 0$ so that $[\hat{\mathbf{P}}, \hat{P}_1] = [\hat{\mathbf{P}}, \hat{P}_2] = 0$ but $[\hat{J}, \hat{P}_k] \neq 0$ so that $[\hat{J}\ \hat{\mathbf{P}}] \neq 0$. Hence, new eigenvalues of $\hat{\mathbf{P}}$ can be generated only by the rotation operation.

Now, consider

$$\hat{P}_k\, R(\theta)|\mathbf{p}_0\rangle = R(\theta)[R(\theta)^{-1}\hat{P}_k\, R(\theta)]|\mathbf{p}_0\rangle = R(\theta)\,\hat{P}_l\, R(-\theta)^l_k|\mathbf{p}_0\rangle$$
$$= R(\theta)\ \hat{P}_1\ R(-\theta)^1_k|\mathbf{p}_0\rangle + R(\theta)\ \hat{P}_2\ R(-\theta)^2_k|\mathbf{p}_0\rangle$$

$$= R(\theta)\ \hat{P}_1 R(\theta)^k_1 |\mathbf{p}_0\rangle + R(\theta)\hat{P}_2\, R(\theta)^k_2 |\mathbf{p}_0\rangle$$

$$= pR(\theta)^k_{\ 1}\ R(\theta)|\mathbf{p}_0\rangle = p'^k\ R(\theta)|\mathbf{p}_0\rangle \tag{80}$$

where, in the second step, we have used $e^{-i\hat{J}\theta}\ \hat{P}_k\ e^{i\hat{J}\theta} = \hat{P}_m\ R(\theta)^m_k$, in the third step, the orthogonality of the rotation matrix $R(-\theta)^i_j = R(\theta)^j_i$ and, in the final step $\hat{P}_l|p\rangle_= p|\mathbf{p}_0\rangle\ \hat{P}_2|\mathbf{p}_0\rangle = 0$. The above equation implies that corresponding to the eigenvector $|\mathbf{p}_0\rangle$ with eigenvalue p, the operator $\hat{P}_k$, on rotation, generates the eigenvector $R(\theta)|\mathbf{p}_0\rangle$ with eigenvalue $p'^k = pR(\theta)^k_1$. Further

$$\hat{\mathbf{P}}.\mathbf{b}\ R(\theta)|\mathbf{p}_0\rangle = R(\theta)\ [R(\theta)^{-1}\ \hat{\mathbf{P}}.\mathbf{b}\ R(\theta)]|\mathbf{p}_0\rangle$$

$$= R(\theta)\ [R(\theta)^{-1}\ (\hat{P}_1\, b^1 + \hat{P}_2\, b^2)\ R(\theta)]|\mathbf{p}_0\rangle$$

$$= R(\theta)\ \hat{P}_l\, R(-\theta)^l_k\ b^k|\mathbf{p}_0\rangle = pR(\theta)\ R(-\theta)^l_k\ b^k|\mathbf{p}_0\rangle$$

$$= p''\ R(\theta)|\mathbf{p}_0\rangle \tag{81}$$

where $p'' = [pR(-\theta)^1_1\, b^1 + pR(-\theta)^1_2\ b^2]$ showing that $R(\theta)|\mathbf{p}_0\rangle$ is an eigenvector of $\hat{\mathbf{P}}$.

Let $|\mathbf{p}\rangle = R(\theta)|\mathbf{p}_0\rangle$. Since $R(\theta) = e^{-i\hat{J}\theta}$ is unitary, the normalization of both $|\mathbf{p}\rangle$ and $|\mathbf{p}_0\rangle$ will be the same.

The set of vectors $\{|\mathbf{p}\rangle\}$ constitutes a basis of an irreducible vector space that is invariant under E_2 for

(i) $\{|\mathbf{p}\rangle\}$ is closed under all group operations in E_2, viz.

$$\hat{T}(b)|\mathbf{p}\rangle = e^{-i\hat{\mathbf{P}}b}|\mathbf{p}\rangle = e^{-i\mathbf{p}b}|\mathbf{p}\rangle = e^{-ip''}|\mathbf{p}\rangle \tag{82}$$

since $|\mathbf{p}\rangle$ is an eigenstate of $\hat{\mathbf{p}}b$ with eigenvalue p''.

$$R(\phi)|\mathbf{p}\rangle = R(\phi)\ R(\theta)|\mathbf{p}_0\rangle = R(\phi + \theta)|\mathbf{p}_0\rangle = |\mathbf{p}'\rangle \tag{83}$$

(ii) If $\mathbf{p} \neq \mathbf{p}'$, then the corresponding vectors $|\mathbf{p}\rangle$ and $|\mathbf{p}'\rangle$ are orthogonal being eigenvectors of a Hermitian operator $\hat{\mathbf{P}}$ corresponding to different eigenvalues so that $\langle\mathbf{p}'|\mathbf{p}\rangle = 0$.

We finally examine the normalization of the basis vectors. For $\mathbf{p} \neq \mathbf{p}'$ we know that $\langle\mathbf{p}'|\mathbf{p}\rangle = 0$. For $\mathbf{p} = \mathbf{p}'$, we proceed by noting that $\hat{\mathbf{P}}^2$ commutes with all the three generators of the group for

$$[\hat{J},\ \hat{P}_k^2] = [\hat{J},\ \hat{P}_k]\ \hat{P}_k + \hat{P}_k\ [\hat{J},\ \hat{P}_k] = i\varepsilon^{kl}\ \hat{P}_l\ \hat{P}_k + i\varepsilon^{kl}\hat{P}_k\ \hat{P}_l \tag{84}$$

which gives

$$[\hat{J},\ \hat{\mathbf{P}}^2] = \sum_{k=1}^{2}\left[\hat{J}, \hat{P}_k^2\right] = i\sum_{k=1}^{2}\left(\varepsilon^{kl}\ \hat{P}_l\,\hat{P}_k + \varepsilon^{kl}\ \hat{P}_k\,\hat{P}_l\right) = i(\hat{P}_2\hat{P}_1 - \hat{P}_2\ \hat{P}_1) = 0 \tag{85}$$

Thus, $\hat{\mathbf{p}}^2$ is a Casimir operator and, therefore, has a unique eigenvalue for each irreducible representation. Its eigenvalues p^2 are positive definite. Since these eigenvalues are invariant under all group operations, the labelling of irreducible representations can be done in terms of these eigenvalues. The eigenvectors (basis vectors) in a particular irreducible representation (that is identified by the eigenvalue of the Casimir operator $\hat{\mathbf{P}}^2$) are then labelled by the continuous label θ and we write $|\mathbf{p}\rangle \equiv |p, \theta\rangle$. Further, from $|\mathbf{p}\rangle = R(\theta)|\mathbf{p}_0\rangle$, we find that there exists a one-one correspondence between these basis vectors and the subgroup of rotations $\{R(\theta)\} \equiv SO(2)$. We, therefore, use the invariant measure $\frac{d\theta}{2\pi}$ of the subgroup of rotations for measuring these basis vectors and adopt the normalization scheme $\langle \mathbf{p}'|\mathbf{p}\rangle = \langle p, \theta'|p, \theta\rangle = 2\pi\delta(\theta' - \theta)$.

(e) As another illustration of the induced representation method, we derive the induced representations of the three dimensional Euclidean group E_3. The group operations include the subgroup consisting of translations $\{T_3\text{: } T(b)\}$ in the three spatial directions with infinitesimal generators $\hat{P}_1$, $\hat{P}_2$, $\hat{P}_3$ and rotations in three-dimensional space $\{SO(3)\text{: } R(\alpha, \beta, \gamma)\}$ with corresponding generators $\hat{J}_1$, $\hat{J}_2$, $\hat{J}_3$. The group elements are generated, as usual, by exponentiation of the infinitesimal generators so that $\hat{T}(\mathbf{b}) = e^{-i\hat{\mathbf{P}}.\mathbf{b}} = e^{-i\hat{P}_i b^i}$ and $R(\alpha, \beta, \gamma) = e^{-i(\hat{J}_1\alpha + \hat{J}_2\beta + \hat{J}_3\gamma)}$. The Lie algebra consisting of the six generators $\hat{P}_i$, $\hat{P}_i$ satisfies the following:

$$[\hat{P}_i, \hat{P}_j] = 0 \tag{86}$$

$$[\hat{P}_k, \hat{J}_l] = i\varepsilon_{klm}\, \hat{P}^m \tag{87}$$

$$[\hat{J}_k, \hat{J}_l] = i\varepsilon_{klm}\, \hat{J}^m \tag{88}$$

where ε_{klm} is the three-dimensional totally antisymmetric Levi Civita tensor.

The two operators $\hat{\mathbf{P}}^2$ and $\hat{\mathbf{J}}.\hat{\mathbf{P}}$ commute with all the group elements and therefore, constitute the Casimir operators of the group E_3. This is shown as follows:

$$[\hat{\mathbf{P}}^2, \hat{P}_k] = \left[\sum_l \hat{P}_l^2, \hat{P}_k\right] = \sum_l \left[\hat{P}_l^2, \hat{P}_k\right] = \sum_l \{\hat{P}_l[\hat{P}_l, \hat{P}_k] + [\hat{P}_l, \hat{P}_k]\hat{P}_l\} = 0$$

$$[\hat{\mathbf{P}}^2, \hat{J}_l] = \left[\sum_k \hat{P}_k^2, \hat{J}_l\right] = \sum_k \left[\hat{P}_k^2, \hat{J}_l\right] = \sum_k \{\hat{P}_k[\hat{P}_k, \hat{J}_l] + [\hat{P}_k, \hat{J}_l]\hat{P}_k\}$$

$$= \mathrm{i}\sum_k (e_{klm}\hat{P}_k \hat{P}^m + \varepsilon_{klm}\hat{P}^m \hat{P}_k) = 0 \tag{89}$$

because e_{klm} is antisymmetric in the indices k, m. Similarly,

$$[\hat{\mathbf{J}}.\hat{\mathbf{P}}, \hat{P}_l] = \left[\sum_k \hat{J}_k \hat{P}_k, \hat{P}_l\right] = \sum_l \left[\hat{J}_k \hat{P}_k, \hat{P}_l\right] = \sum_k \{\hat{J}_k[\hat{P}_k, \hat{P}_l] + [\hat{J}_k, \hat{P}_l]\hat{P}_k\}$$

$$= \sum_k \{[\hat{J}_k, \hat{P}_l]\hat{P}_k\} = -i\sum_k \varepsilon_{lkm}\hat{P}^m \hat{P}_k = 0 \tag{90}$$

$$[\hat{\mathbf{J}}.\hat{\mathbf{P}}, \hat{J}_l] = \left[\sum_k \hat{J}_k \hat{P}_k, \hat{J}_l\right] = \sum_k \left[\hat{J}_k \hat{P}_k, \hat{J}_l\right] = \sum_k \{\hat{J}_k[\hat{P}_k, \hat{J}_l] + [\hat{J}_k, \hat{J}_l]\hat{P}_k\}$$

$$= i\sum_k \left(\varepsilon_{klm} \hat{J}_k \hat{P}^m + \varepsilon_{klm} \hat{J}^m \hat{P}_k\right) = 0 \tag{91}$$

In E_2, the group of translations T_2 constitutes an invariant abelian subgroup thereof. Similarly, in E_3, the group of translations in three-dimensional space T_3 constitutes an invariant abelian subgroup of E_3. Furthermore, as in E_2, the following relations also hold in E_3:

$$R(\alpha, \beta, \gamma)\ \hat{P}_i\ [R(\alpha, \beta, \gamma)]^{-1} = e^{-i(\hat{J}_1\alpha + \hat{J}_2\beta + \hat{J}_3\gamma)}\ \hat{P}_i e^{i(\hat{J}_1\alpha + \hat{J}_2\beta + \hat{J}_3\gamma)} = R^j_i \hat{P}_j \tag{92}$$

$$R(\alpha, \beta, \gamma)\ \hat{T}(b)\ [R(\alpha, \beta, \gamma)]^{-1} = \hat{T}(\mathbf{b}'), \quad \text{where } b'^i = R^i_j\, b^j \tag{93}$$

A general element of E_3 can be decomposed as $g = T(\mathbf{b})\ R(\alpha, \beta, \gamma)$. This can be seen by writing $g = R_1\, T_1\, R_2\, T_2 ... R_n\, T_n$ which can always be done because products of translations yield translations and products of rotations yield rotations and

$$g = R_1\, T_1\, R_2\, T_2 ... R_n\, T_n = (R_1\, T_1\, R_1^{-1})\ (R_1\, R_2\, T_2\, R_2^{-1} R_1^{-1})\ (R_1\, R_2\, R_3\, T_3\, R_3^{-1} R_2^{-1} R_1^{-1}) \times .....$$

$$\times\ (R_1 ... R_{n-1}\ T_{n-1}\ R_{n-1}^{-1} ... R_1^{-1})\ (R_1 ... R_n\ T_n\ R_n^{-1} ... R_1^{-1}) R_1 ... R_n = T'_1\ T'_2 ...\ T'_n R = TR$$

where $T'_1 = R_1\, T_1\, R_1^{-1}$, $T'_2 = R_1\, R_2\, T_1\, R_2^{-1}\, R_{n-1}^{-1}$ etc., and $R = R_1 ... R_n$ thereby establishing the desired result.

Just as the two-dimensional translation group T_2 is an invariant subgroup of E_2 and the factor group $\frac{E_2}{T_2}$ is isomorphic to the two-dimensional rotation group $SO(2)$, the three-dimensional translation group T_3 is an invariant subgroup of E_3 and the factor group $\frac{E_3}{T_3}$ is isomorphic to the three-dimensional rotation group $SO(3)$. The generators of this invariant subgroup T_3 are $\hat{P}_1, \hat{P}_2, \hat{P}_3$, i.e., the components of the vector $\hat{\mathbf{P}}$. Additionally, the group E_3 has two Casimir operators, viz. $\hat{\mathbf{P}}^2$ and $\hat{\mathbf{J}}.\hat{\mathbf{P}}$. The eigenvalues of these two Casimir operators can be used for labelling of the IRRs. Within an IRR, identified as above by the eigenvalues of $\hat{\mathbf{P}}^2$ -and $\hat{\mathbf{J}}.\hat{\mathbf{P}}$, we generate the basis for the irreducible vector space to consist of the eigenvectors of the maximal linear momentum set of commuting operators, viz. $\{\hat{\mathbf{P}}^2,\ \hat{\mathbf{J}}.\hat{\mathbf{P}};\hat{\mathbf{P}}\}$ (We can also use the maximal angular momentum set $\{\hat{\mathbf{P}}^2,\ \hat{\mathbf{J}}.\hat{\mathbf{P}};\hat{\mathbf{J}}^2\ \hat{J}_3\}$ or the maximal mixed set $\{\hat{\mathbf{P}}^2,\ \hat{\mathbf{J}}.\hat{\mathbf{P}};\hat{P}_3\ \hat{J}_3\}$ of commuting operators). Accordingly, we will label the eigenvalues by $\{p^2, \lambda p;\ \mathbf{p}\}$. We call $\mathbf{p}$ the momentum vector and λ, the helicity of the eigenstate. We may represent the corresponding eigenvectors, equivalently, by $\{p, \lambda;\ \hat{\mathbf{P}}\}$, where $\hat{\mathbf{P}} = \mathbf{p}|\mathbf{p}|^{-1}$is a unit vector in the direction of $\mathbf{p}$ characterized by the angles (θ, ϕ), in polar coordinates. From the above labelling scheme, the following relations are self-evident:

$$\hat{\mathbf{P}}^2\, |p, \lambda; \hat{\mathbf{P}}\rangle = p^2\, |p, \lambda;\ \hat{\mathbf{P}}\rangle \tag{94}$$

$$\hat{\mathbf{J}}.\hat{\mathbf{P}}\, |p, \lambda;\ \hat{\mathbf{P}}| = \lambda p |p, \lambda; \hat{\mathbf{P}}\rangle \tag{95}$$

$$\hat{\mathbf{P}}|\mathrm{p}, \lambda; \hat{\mathbf{P}}| = \mathbf{p}|p, \lambda; \hat{\mathbf{P}}\rangle \tag{96}$$

Since the factor group $\frac{E_3}{T_3} \cong SO(3)$, a trivial representation of E_3 can be directly induced from the representations of $SO(3)$ by the mapping $g = TR \rightarrow D^J(R)$ in which all pure translations are mapped into the identity element. In this representation all generators of translations are mapped into the null operator and both the Casimir operators $\hat{\mathbf{P}}^2$ and $\hat{\mathbf{J}}.\hat{\mathbf{P}}$ have zero eigenvalues.

Let us now consider the subspace of the given vector space that is characterized by a standard vector $\hat{\mathbf{P}} = \hat{\mathbf{P}}_0 = (0, 0, 1) = e_3$. Now, all the group elements in the factor group $\frac{E_3}{T_3} \cong SO(3)$, i.e., the group of rotations that leave the subspace corresponding to the standard vector $\hat{\mathbf{P}}_0$ invariant constitute a subgroup of the factor group $\frac{E_3}{T_3} \cong SO(3)$. This is proved as follows.

Let us assume that V_0 is the subspace corresponding to the standard vector $\hat{\mathbf{P}}_0$. Further, let G_0 be the set of the elements of the factor group $\frac{E_3}{T_3}$ that leave the space V_0 invariant. (We define V_{inv} to be an A invariant subspace of a given linear transformation A: $V \rightarrow V$ in a vector space V, if and only if $Av \in V_{inv}\ \forall\ v \in V_{inv}$, i.e., $AV_{inv} \subset V_{inv}$). Then, if $g \in G_0$ and $\hat{p} \in V_0$, we must have $g\,\hat{p} \in V_0$. This implies that $g_2 g_1 \hat{p} \in V_0$ for $g_1, g_2 \in G_0$ so that $g_2 g_1 \in G_0$, whence G_0 is closed under the group composition and, therefore, a subgroup of $\frac{E_3}{T_3}$. This subgroup is called the little group of $\hat{\mathbf{P}}_0$.

Since $\hat{\mathbf{P}}_0$ is aligned along the x^3 axis, its value remains unchanged by rotations that are carried out about this (x^3) axis. These rotations are generated by $\hat{J}_3$ by exponentiation, i.e., $R_3(\phi) = e^{-i\hat{J}_3\phi}$. The little group of $\hat{\mathbf{P}}_0$, therefore, consists of rotations in the $x^1 - x^2$ plane and is isomorphic to the group $SO(2)$.

We know from the preceding analysis of the $SO(2)$ group that all its irreducible representations are one dimensional and can be labelled by a single index λ that is the eigenvalue of the generator $\hat{J}_3$. Furthermore, these eigenvalues, being eigenvalues of $\hat{J}_3$ can take all integral values as is well known from the standard theory of angular momentum. Since we have defined our standard vector $\hat{\mathbf{P}} = \hat{\mathbf{P}}_0 = (0, 0, 1) = e_3$, it follows that $\hat{P}_1|\mathbf{p}_0\rangle = \hat{P}_2|\mathbf{p}_0\rangle = 0$ and $\hat{P}_3|\mathbf{p}_0\rangle = \mathbf{p}_0|\mathbf{p}_0\rangle$ so that $\hat{\mathbf{P}}|\mathbf{p}_0\rangle = \mathbf{p}_0|\mathbf{p}_0\rangle$. Also, $\hat{\mathbf{J}}.\hat{\mathbf{P}}|\mathbf{p}_0\rangle = \mathbf{p}_0\hat{\mathbf{J}}|\mathbf{p}_0\rangle = (0, 0, p).(\hat{J}_1, \hat{J}_2, \hat{J}_3)|\mathbf{p}_0\rangle = p\hat{J}_3|\mathbf{p}_0\rangle = p\lambda|\mathbf{p}_0\rangle$ where $p = |\mathbf{p}_0|$.

The basis vectors of the subspace corresponding to the standard vector $\hat{\mathbf{p}}_0$, under transformations of its little group (rotations in two-dimensional space, viz. $SO(2)$) will satisfy

$$R_3(\phi)|p, \lambda; \hat{\mathbf{P}}_0\rangle = e^{-i\hat{J}_3\phi}|p, \lambda; \hat{\mathbf{P}}_0\rangle = e^{-i\lambda\phi}|p, \lambda; \hat{\mathbf{P}}_0\rangle \tag{97}$$

because $\hat{J}_3|\mathbf{P}_0\rangle = \lambda|\mathbf{p}_0\rangle$. Similarly, under translations, we have

$$\hat{T}(\mathbf{b})|p, \lambda; \mathbf{p}_0\rangle = e^{-i\hat{\mathbf{P}}.\mathbf{b}}|p, \lambda; \mathbf{p}_0\rangle = e^{-i\mathbf{p}_0\mathbf{b}}|p, \lambda; \mathbf{p}_0\rangle \tag{98}$$

which follows from $\hat{\mathbf{P}}|\mathbf{p}_0\rangle = \mathbf{p}_0|\mathbf{p}_0\rangle$.

Since the group has two Casimir operators $\hat{\mathbf{P}}^2$ and $\hat{\mathbf{J}}.\hat{\mathbf{P}}$, the labelling of the IRRs of E_3 has to be done using two indices. We use p, λ for this purpose. Since p, λ correspond, respectively, to the Casimir operators $\hat{\mathbf{P}}^2$ and $\hat{\mathbf{J}}.\hat{\mathbf{P}}$ of E_3 their values are invariant for a particular IRR of E_3 which enables labelling of the IRRs using these two indices. Within an IRR, the vectors are identified by the unit vector index $\hat{\mathbf{P}}$.

To construct the full vector space of a particular IRR of E_3 that is labelled by p, λ, we generate the new basis vectors from $|p, \lambda; \mathbf{p}_0\rangle$ by operating thereon by rotations that do not belong to the little group of $\hat{\mathbf{p}}_0$. We have

$$|p, \lambda; \hat{\mathbf{p}}\rangle = R(\omega, \theta, 0)|\, p, \lambda; \hat{\mathbf{p}}_0\rangle \tag{99}$$

where $\hat{\mathbf{p}} = R(\omega, \theta, 0)\hat{\mathbf{p}}_0$. The remaining labels p, λ will remain unchanged by these rotations as such rotations do not result in a new IRR but generate the remaining basis vectors of the same IRR.

It remains to establish that the basis vectors generated by $|p, \lambda; \hat{\mathbf{p}}\rangle = R(\omega, \theta, 0)|p, \lambda; \hat{\mathbf{p}}_0\rangle$ (where in $\hat{\mathbf{p}} = (\omega, \theta)$ in polar coordinates) have the necessary group composition properties of E_3. For this, we have $\hat{T}(\mathbf{b})|p, \lambda; \hat{\mathbf{p}}\rangle = e^{-i\hat{\mathbf{p}}.\mathbf{b}}|p, \lambda; \hat{\mathbf{p}}\rangle = e^{-ipb}|p, \lambda; \hat{\mathbf{p}}\rangle$. Further,

$$\begin{aligned} R(\alpha, \beta, \gamma)|p, \lambda; \hat{\mathbf{p}}\rangle &= R(\alpha, \beta, \gamma)\, R(\omega, \theta, 0)|p, \lambda; \hat{\mathbf{p}}_0\rangle \\ &= R(\omega', \theta', 0)\, [R^{-1}(\omega', \theta', 0)\, R(\alpha, \beta, \gamma)\, R(\omega, \theta, 0)]|p, \lambda; \hat{\mathbf{p}}_0\rangle \\ &= R(\omega', \theta', 0)\, R(0, 0, \psi)|p, \lambda; \hat{\mathbf{p}}_0\rangle \\ &= R(\omega', \theta', 0)\, e^{-i\lambda\psi}|p, \lambda; \hat{\mathbf{p}}_0\rangle = e^{-i\lambda\psi}|p, \lambda; \hat{\mathbf{p}}'\rangle \end{aligned} \tag{100}$$

where $\hat{\mathbf{p}}' = (\omega', \theta')$ polar coordinates.

The fact that a ψ exists satisfying $R^{-1}(\omega', \theta', 0)\, R(\alpha, \beta, \gamma)\, R(\omega, \theta, 0) = R(0, 0, \psi)$ follows from the fact that the three rotations $R(\omega', \theta', 0)^{-1}\, R(\alpha, \beta, \gamma)\, R(\omega, \theta, 0)$ when acting on the state $|p, \lambda; \hat{\mathbf{p}}_0\rangle$ produce

$$\begin{aligned} R(\omega', \theta, 0)^{-1}\, &R(\alpha, \beta, \gamma)\, R(\omega, \theta, 0)|p, \lambda; \hat{\mathbf{p}}_0\rangle \\ &= R(\omega', \theta', 0)^{-1}\, R(\alpha, \beta, \gamma)|p, \lambda; \hat{\mathbf{p}}\rangle \quad \hat{\mathbf{p}} = (\omega, \theta) \\ &= R(\omega', \theta', 0)^{-1}\, R_3(\alpha)\, R_2(\beta)\, R_3(\gamma)|p, \lambda; \hat{\mathbf{p}}\rangle \\ &= R(\omega', \theta'\, 0)^{-1}|p, \lambda; \hat{\mathbf{p}}'\rangle\, e^{-i\lambda\psi} = |p, \lambda; \hat{\mathbf{p}}_0\rangle\, e^{-i\lambda\psi} \end{aligned} \tag{101}$$

(since $R(\alpha, \beta, \gamma)|p, \lambda; \hat{\mathbf{p}}'\rangle = |p, \lambda; \hat{\mathbf{p}}'\rangle\, e^{-i\lambda\psi}$ with $\hat{\mathbf{p}}' = (\omega', \theta')$ so that $R(\omega', \theta', 0)^{-1}|p, \lambda; \hat{\mathbf{p}}'(= \omega', \theta')\rangle = R(\omega', \theta', 0)^{-1}\, [R(\omega', \theta', 0)|p, \lambda; \hat{\mathbf{p}}_0\rangle] = |p, \lambda; \hat{\mathbf{p}}_0\rangle$ thereby leaving the standard vector $\hat{\mathbf{p}}_0$ invariant. Hence, the product of the three rotations $R(\omega', \theta', 0)^{-1}\, R(\alpha, \beta, \gamma)\, R(\omega, \theta, 0)$ constitutes a rotation about the x^3 axis and can, therefore, be represented as $R(0, 0, \psi)$ and, is also a member of the little group of $\hat{\mathbf{p}}_0$.

Thus, the vector space with the basis $\{|p, \lambda;\, \hat{\mathbf{p}}\rangle\}$ is invariant under the E_3 group with the basis vectors satisfying $\hat{T}(\mathbf{b})|p, \lambda;\, \hat{\mathbf{p}}\rangle = e^{i\mathbf{pb}}|p, \lambda;\, \hat{\mathbf{p}}\rangle$ and $R(\alpha, \beta, \gamma)|p, \lambda;\, \hat{\mathbf{p}}\rangle = e^{-i\lambda\psi}|p, \lambda;\, \hat{\mathbf{p}}'\rangle$ with $|p, \lambda;\, \hat{\mathbf{p}}'\rangle = R(\omega', \theta', 0)|p, \lambda;\, \hat{\mathbf{p}}_0\rangle$ under translations and rotations respectively, thereby defining a unitary representation of E_3.

The irreducibility of this representation follows from the fact that no smaller invariant subspace exists and all the basis vectors are generated from a single vector $\mathbf{p}_0$ by group operations.

Since the basis vectors $\{|p, \lambda;\, \hat{\mathbf{p}}\rangle\}$ are generated from $|p, \lambda;\, \hat{\mathbf{p}}_0\rangle$ by rotations and the δ function associated with the invariant measure of the rotation group is defined by $\sin\theta\, d\theta\, d\omega/4\pi$ it is the appropriate normalization measure. Under this, the normalization condition takes the form $\langle \hat{\mathbf{p}}'|\hat{\mathbf{p}}\rangle|\rangle \equiv 4\pi\delta(\cos\theta' - \cos\theta)\, \delta(\omega' - \omega)$.

(f) We now compute the unitary IRRs of the Poincare group using the method of induced representation which has been illustrated in the preceding paragraphs. The method can be used in this case as well because of the existence of an abelian invariant subgroup comprising translations in four-dimensional Minkowski space, viz. $\{T_4\!: T(b)\}$ as existed in the Euclidean spaces E_2, E_3. Group operations of the Poincare group are defined by

$$\hat{L}(\Lambda)\, \hat{L}(\Lambda') = \hat{L}(\Lambda\Lambda') \tag{102}$$

$$\hat{T}(a)\, \hat{T}(b) = \hat{T}(a + b) \tag{103}$$

$$\hat{L}(\Lambda)\, \hat{T}(a) = \hat{T}(\Lambda a)\, \hat{L}(\Lambda) \tag{104}$$

where $\hat{L}(\Lambda)$ and $\hat{T}(a)$ are unitary representations of a Lorentz transformation Λ^{μ}_{ν} and a translation a^{μ} respectively.

The Poincare group is a ten parameter group consisting of translations in the four spacetime directions, spatial rotations along the three spatial axes and spatiotemporal boosts in the direction of the three spatial axes. The corresponding Lie algebra has ten generators viz. $\hat{P}_{\mu}$, $\mu = 0, 1, 2, 3$ and $\hat{J}_{\mu\nu}$, $\mu, \nu = 0, 1, 2, 3$ which are also the basis elements of the algebra. These generators are defined in terms of the commutators

$$[\hat{P}_{\mu}, \hat{P}_{\lambda}] = 0 \tag{105}$$

$$[\hat{P}_{\mu}, \hat{J}_{\lambda\sigma}] = i(\hat{P}_{\sigma}\, g_{\mu\lambda} - \hat{P}_{\lambda}\, g_{\mu\sigma}) \tag{106}$$

$$[\hat{P}_{\mu\varpi}, \hat{J}_{\lambda\sigma}] = i(\hat{J}_{\sigma\varpi}\, g_{\mu\lambda} - \hat{J}_{\lambda\varpi}\, g_{\mu\sigma} - \hat{J}_{\mu\lambda}\, g_{\varpi\sigma} + \hat{J}_{\mu\sigma}\, g_{\varpi\lambda}) \tag{107}$$

Our first step in developing the representation theory of the Poincare algebra is to identify the Casimir operators of the algebra. These Casimir operators need to commute with all the ten generators viz. $\hat{P}_{\mu}$, $\mu = 0, 1, 2, 3$ and $\hat{J}_{\mu\nu}$, $\mu\nu = 0, 1, 2, 3$, $\mu \neq \nu$. Since, they must commute with the generators of rotations $\hat{J}_{ij}$, $i, j = 1, 2, 3$, they must be left invariant by such rotations and hence, must be scalars. However, we do not have any scalar among the basis elements of the Poincare algebra. It follows that we need to look for the Casimir operators in the enveloping algebra, i.e., in the set of operators of the type $\hat{Q}_A\, \hat{Q}_B \ldots \hat{Q}_N$ where $\hat{Q}_i \in \{\hat{P}_{\mu}, \hat{J}_{\mu\nu}\}$. In other words, we need to construct scalars as products of the vector ($\hat{P}_\mu$) and tensor ($\hat{J}_{\mu\nu}$) quantities. In the context, we define vectors and scalars as:

(i) An operator $\hat{A}_\mu$, $\mu = 0, 1, 2, 3$ is classified as a vector if it transforms as $[\hat{J}_{\mu\nu}, \hat{A}_\lambda] = i(g_{\nu\lambda}\hat{A}_\mu - g_{\mu\lambda}\hat{A}_\nu)$;

(ii) An operator $\hat{A}_{\mu\nu}$, $\mu, \nu = 0, 1, 2, 3$ is classified as a tensor of second rank if it transforms as a product of vector components $\hat{A}_\mu \hat{A}_\nu$ under the action of the generators $\hat{J}_{\mu\nu}$. Similarly, higher rank tensors can be defined.

It is well known that we can obtain scalars from vectors and tensors by the operation of contraction (index convolution). In our case, we have to generate scalars from the two sets of generators $\{\hat{P}_\mu\}$and $\{\hat{J}_{\mu\nu}\}$. Let us define the vectors $\hat{W}_\mu = \frac{1}{2}\varepsilon_{\mu\nu\rho\sigma}\hat{J}^{\nu\rho}\hat{P}^\sigma$ and $\hat{\Gamma}_\lambda = \hat{J}_{\lambda\mu}\hat{P}^\mu$. Then, we have $\hat{W}_\mu \hat{P}^\mu = 0$, $\hat{\Gamma}_\mu \hat{P}^\mu = 0$, $[\hat{P}_\mu, \hat{W}_\nu] = 0$, $[\hat{W}_\mu, \hat{W}_\nu] = i\varepsilon_{\mu\nu\rho\sigma}\hat{P}^\rho \hat{W}^\sigma$, $[\hat{W}_\mu, \hat{\Gamma}_\sigma] = -i\hat{P}_\mu \hat{W}_\sigma$ $[\hat{\Gamma}_\mu, \hat{P}_\nu] = i(\delta_{\mu\nu}\hat{P}_\lambda \hat{P}^\lambda - \hat{P}_\mu \hat{P}_\nu)$, $[\hat{\Gamma}_\mu, \hat{\Gamma}_\nu] = -i\hat{J}_{\mu\nu}\hat{P}_\lambda \hat{P}^\lambda$, $\hat{P}^\lambda \hat{J}_{\lambda\mu}\hat{J}^\mu_\sigma = \hat{P}_\sigma\left(6 - \frac{1}{2}\hat{j}^2\right) - \frac{1}{2}\varepsilon_{\sigma\mu\nu\rho}$ $\hat{J}^{\mu\nu}\hat{W}^\rho + 2i\hat{\Gamma}_\sigma$, $= \hat{J}^2 = \hat{J}_{\lambda\rho}\hat{J}^{\lambda\rho}$,

$$\hat{J}_{\mu\lambda}\hat{J}^{\lambda\sigma}\hat{J}_{\sigma\nu} = \frac{1}{2}(ig_{\mu\nu} - \hat{J}_{\mu\nu})\hat{J}^2 + \frac{1}{2}\varepsilon_{\mu\nu\rho\sigma}\hat{J}^{\rho\sigma}\left(\frac{1}{8}\varepsilon_{\alpha\beta\lambda\gamma}\hat{J}^{\alpha\beta}\hat{J}^{\lambda\gamma}\right) + 3i\hat{J}_{\mu\lambda}\hat{J}^\lambda_\nu$$

It is seen from the above that all scalars that can be constituted by the convolution of $\{\hat{P}_\mu\}$ and $\{\hat{J}_{\mu\nu}\}$ can be expressed as the linear combinations of the entities $\hat{J}^2$, $\hat{P}_\mu \hat{P}^\mu$ $\hat{W}_\mu \hat{W}^\mu$, $\hat{\Gamma}_\lambda \hat{\Gamma}^\lambda$, $\hat{W}_\lambda \hat{\Gamma}^\lambda$ and $\varepsilon_{\alpha\beta\lambda\gamma}\hat{J}^{\alpha\beta}\hat{J}^{\lambda\gamma}$. These six scalars exhaust the set of all independent scalars that can be constructed from $\{\hat{P}_\mu\}$ and $\{\hat{J}_{\mu\nu}\}$. Furthermore, out of these six scalars only $\hat{P}_\mu \hat{P}^\mu$, $\hat{W}_\mu \hat{W}^\mu$ commute with all the generators of the Poincare algebra and hence, constitute the two main Casimir operators of the group. This is established below:

$$[\hat{P}^\mu \hat{P}_\mu, \hat{P}_\lambda] = \hat{P}^\mu[\hat{P}_\mu \hat{P}_\lambda] + [\hat{P}^\mu \hat{P}_\lambda]\hat{P}^\mu = 0 \tag{108}$$

$$\begin{aligned}[\hat{P}_0^2 - \hat{\mathbf{P}}^2, \hat{J}_n] &= [\hat{P}_0^2, \hat{J}_n] - [\hat{\mathbf{P}}^2, \hat{J}_n] \\ &= \hat{P}_0[\hat{P}_0, \hat{J}_n] + [\hat{P}_0, \hat{J}_n]\hat{P}_0 - [\hat{P}_1^2 + \hat{P}_2^2 + \hat{P}_3^2, \hat{J}_n] \\ &= -(\hat{P}_1[\hat{P}_1, \hat{J}_n] + [\hat{P}_1, \hat{J}_n]\hat{P}_1) - (\hat{P}_2[\hat{P}_2, \hat{J}_n] + [\hat{P}_2, \hat{J}_n]\hat{P}_2) \\ &\quad - (\hat{P}_3[\hat{P}_3, \hat{J}_n] + [\hat{P}_3, \hat{J}_n]\hat{P}_3) = 0\end{aligned} \tag{109}$$

$$\begin{aligned}[\hat{P}_0^2 - \hat{\mathbf{P}}^2, \hat{K}_n] &= [\hat{P}_0^2, \hat{K}_n] - [\hat{\mathbf{P}}^2, \hat{K}_n] \\ &= \hat{P}_0[\hat{P}_0, \hat{K}_n] + [\hat{P}_0, \hat{K}_n]\hat{P}_0 - [\hat{P}_1^2 + \hat{P}_2^2 + \hat{P}_3^2, \hat{K}_n] \\ &= -i\hat{P}_0\hat{P}_n - i\hat{P}_n\hat{P}_0 - (\hat{P}_1[\hat{P}_1, \hat{K}_n] + [\hat{P}_1, \hat{K}_n]\hat{P}_1)\end{aligned}$$

$$- (\hat{P}_2[\hat{P}_2, \hat{K}_n] + [\hat{P}_2, \hat{K}_n]\hat{P}_2) - (\hat{P}_3[\hat{P}_3, \hat{K}_n] + [\hat{P}_3, \hat{K}_n]\hat{P}_3)$$

$$= -i\hat{P}_0 \hat{P}_n - i\hat{P}_n \hat{P}_0 + i\delta_{1n} (\hat{P}_1\hat{P}_0 + \hat{P}_0\hat{P}_1)$$

$$+ i\delta_{2n} (\hat{P}_2\hat{P}_0 + \hat{P}_0\hat{P}_2) + i\delta_{3n} (\hat{P}_3\hat{P}_0 + \hat{P}_0\hat{P}_3) = 0 \qquad (110)$$

The commutativity of $\hat{C}_1 = \hat{P}_\mu \hat{P}^\mu$ with the generators of the Poincare group may also be directly inferred from the fact that:

(i) Being the scalar product of a 4-vector with itself, it is invariant under homogeneous Lorentz transformations and hence, commutes with the generators of such Lorentz transformations.

(ii) It is also invariant under translations because of the commutativity of the generators of translations, i.e., the abelian nature of the translations subgroup.

We shall now establish that $\hat{C}_2 = \hat{W}_\lambda \hat{W}^\lambda$ (called the Pauli Lubanski vector) constitutes the second Casimir operator of the Poincare group. To establish this, we obtain the following properties of this vector:

$$\text{(i) } \hat{W}^\lambda \hat{P}_\lambda = 0 \qquad (111)$$

$$\text{(ii) } [\hat{W}^\lambda, \hat{P}^\mu] = 0 \qquad (112)$$

$$\text{(iii) } [\hat{W}^\lambda, \hat{J}^{\mu\nu}] = i(\hat{W}^\nu g^{\mu\lambda} - \hat{W}^\mu g^{\lambda\nu}) \qquad (113)$$

$$\text{(iv) } [\hat{W}^\lambda, \hat{W}^\sigma] = i\varepsilon^{\lambda\sigma\mu\nu} \hat{W}_\mu, \hat{P}_\nu \qquad (114)$$

Property (i) follows from the fact that in $\hat{W}^\lambda \hat{P}_\lambda = \frac{1}{2} \varepsilon^{\lambda\mu\nu\sigma} \hat{J}_{\mu\nu} \hat{P}_\sigma \hat{P}_\lambda$, the four dimensional Levi Civita tensor is antisymmetric in the summation indices λ, σ whereas the product of 4-momenta $\hat{P}_\sigma \hat{P}_\lambda$ is symmetric in these indices. This property implies the orthogonality of the vectors $\hat{W}$ and $\hat{P}$.

To prove (ii), we note that

$$[\hat{W}^\lambda, \hat{P}^\mu] = \frac{1}{2}[\varepsilon^{\lambda\alpha\beta\sigma} \hat{J}_{\alpha\beta} \hat{P}_\sigma \hat{P}^\mu] = \frac{1}{2} \varepsilon^{\lambda\alpha\beta\sigma} \{\hat{J}_{\alpha\beta} [\hat{P}_\sigma, \hat{P}^\mu] + [\hat{J}_{\alpha\beta}, \hat{P}^\mu]\hat{P}_\sigma\}$$

$$= \frac{1}{2}\varepsilon^{\lambda\alpha\beta\sigma} [\hat{J}_{\alpha\beta}, \hat{P}^\mu]\hat{P}_\sigma = \frac{i}{2}\varepsilon^{\lambda\alpha\beta\sigma} (\delta^\mu_\alpha \hat{P}_\beta - \delta^\mu_\beta \hat{P}_\alpha)\hat{P}_\sigma$$

$$= \frac{i}{2}\varepsilon^{\lambda\mu\beta\sigma} \hat{P}_\beta \hat{P}_\sigma - \frac{i}{2}\varepsilon^{\lambda\alpha\mu\sigma} \hat{P}_\alpha \hat{P}_\sigma = 0$$

where the last step follows from the antisymmetricity of the four-dimensional Levi Civita tensor in the summation indices whereas the product of 4-momenta is symmetric in these indices. This property of the Pauli Lubanski vector expresses the translation invariance of $\hat{W}$.

Property (iii) implies that the Pauli Lubanski vector $\hat{W}$ transforms as the components of a 4-vector under homogeneous Lorentz transformations. This may be proved by using the commutation relations

$$[\hat{P}_\mu, \hat{J}_{\lambda\sigma}] = i(\hat{P}_\sigma g_{\mu\lambda} - \hat{P}_\lambda g_{\mu\sigma}) = \frac{i}{2}\varepsilon_{\lambda\sigma\rho\nu}\varepsilon^{\rho\nu}_{\mu\alpha}\hat{P}^\alpha$$

and

$$[\hat{J}_{\mu\nu}, \hat{J}_{\lambda\sigma}] = i(\hat{J}_{\sigma\nu} g_{\mu\lambda} - \hat{J}_{\lambda\nu} g_{\mu\sigma} - \hat{J}_{\mu\lambda} g_{\nu\sigma} + \hat{J}_{\mu\sigma} g_{\nu\lambda}) = -i\varepsilon_{\mu\nu\alpha\rho}\varepsilon^{\rho}_{\lambda\sigma\beta}\hat{J}^{\alpha\beta}$$

where the last step follows from the properties of the totally antisymmetric tensor in 4 dimensions. Then we have

$$[\hat{W}^\lambda, \hat{J}^{\mu\nu}] = \frac{1}{2}\varepsilon^{\lambda\rho\sigma\tau}[\hat{J}_{\rho\sigma}, \hat{P}_\tau, \hat{J}^{\mu\nu}] = \frac{1}{2}\varepsilon^{\lambda\rho\sigma\tau}(\hat{J}_{\rho\sigma}, \hat{P}_\tau, \hat{J}^{\mu\nu} - \hat{J}^{\mu\nu}\hat{J}_{\rho\sigma}\hat{P}_\tau)$$

$$= \frac{1}{2}\varepsilon^{\lambda\rho\sigma\tau}([\hat{J}_{\rho\sigma}, \hat{J}^{\mu\nu}]\hat{P}_\tau + \hat{J}_{\rho\sigma}[\hat{P}_\tau + \hat{J}^{\mu\nu}])$$

$$= \frac{1}{2}\varepsilon^{\lambda\rho\sigma\tau}\left(i\varepsilon_{\rho\sigma\alpha\gamma}\varepsilon^{\mu\nu\gamma}_{\beta}\hat{J}^{\alpha\beta}\hat{P}_\tau - \frac{i}{2}\varepsilon^{\mu\nu}_{\alpha\beta}\varepsilon^{\tau\kappa}\hat{J}_{\rho\kappa}\hat{P}_\kappa\right)$$

$$= -i(\varepsilon^{\mu\rho\sigma\tau}\hat{J}_{\rho\sigma}\hat{P}_\tau g^{\lambda\nu} - \varepsilon^{\nu\alpha\beta\kappa}\hat{J}_{\alpha\beta}\hat{P}_\kappa g^{\mu\lambda}) = -i(\hat{W}^\mu g^{\lambda\nu} - \hat{W}^\nu g^{\mu\lambda})$$

Property (iv) is proved as:

$$[\hat{W}^\mu, \hat{W}^\nu] = \frac{1}{4}\varepsilon^{\lambda\alpha\beta\gamma}\varepsilon^{\mu\nu\rho\sigma}[\hat{J}_{\alpha\beta}, \hat{P}_\gamma, \hat{J}_{\nu\rho}, \hat{P}_\sigma]$$

$$= \frac{1}{4}\varepsilon^{\lambda\alpha\beta\gamma}\varepsilon^{\mu\nu\rho\sigma}[\hat{J}_{\alpha\beta}, \hat{P}_\gamma, \hat{J}_{\nu\rho}, \hat{P}_\sigma - \hat{J}_{\nu\rho}\hat{P}_\sigma\hat{J}_{\alpha\beta}\hat{P}_\gamma]$$

$$= \frac{1}{4}\varepsilon^{\lambda\alpha\beta\gamma}\varepsilon^{\mu\nu\rho\sigma}([\hat{J}_{\alpha\beta}, \hat{J}_{\nu\rho}]\hat{P}_\gamma\hat{P}_\sigma + \hat{J}_{\alpha\beta}[\hat{P}_\gamma, \hat{J}_{\nu\rho}]\hat{P}_\sigma - \hat{J}_{\nu\rho}[\hat{P}_\sigma, \hat{J}_{\alpha\beta},]\hat{P}_\gamma)$$

$$= \frac{1}{4}\varepsilon^{\lambda\alpha\beta\gamma}\varepsilon^{\mu\nu\rho\sigma}[\hat{J}_{\alpha\beta}, \hat{J}_{\nu\rho}]\hat{P}_\gamma\hat{P}_\sigma = \frac{i}{4}\varepsilon^{\lambda\alpha\beta\gamma}\varepsilon^{\mu\nu\rho\sigma}\varepsilon_{\alpha\beta\delta\tau}\varepsilon^{\nu\rho\kappa\tau}\hat{J}^{\delta\kappa}\hat{P}_\gamma\hat{P}_\sigma$$

$$= i\varepsilon^{\lambda\alpha\mu\nu}\varepsilon_{\alpha\beta\delta\tau}\varepsilon_{\mu\delta\kappa\tau}\hat{J}^{\delta\kappa}\hat{P}^\tau = i\varepsilon^{\lambda\sigma\mu\nu}\hat{W}_\mu\hat{P}_\nu$$

on the basis of the properties of the antisymmetric tensor and the antisymmetry of $\hat{J}^{\mu\nu}$.

This property implies that the independent components of $\{\hat{W}^\mu\}$ form a Lie algebra on the subspace corresponding to a given 4-vector $\{p^\nu\}$, that is the eigenvalue of the 4-momentum $\hat{P}^\nu$. This Lie algebra constitutes the Lie algebra of the little group of p^ν and the form of the algebra is determined by the nature of the vector $\{\hat{P}^\nu\}$. We shall discuss this issue further in the sequel.

The Casimir nature of $\hat{C}_2 = \hat{W}_\mu\hat{W}^\mu$, i.e., that it commutes with all the generators of the group then follows from the fact that:

(i) It has been proved above that each component of $\hat{W}$ is translationally invariant for $[\hat{W}^\lambda, p^\mu] = 0$. Hence, $\hat{C}_2 = \hat{W}_\mu\hat{W}^\mu$ is also translationally invariant.

(ii) Further, $\hat{C}_2 = \hat{W}_\mu \hat{W}^\mu$ is the scalar product of a 4-vector with itself. Hence, it is the square of the length of the 4-vector which makes it invariant under homogeneous Lorentz transformations.

We shall, now, obtain an explicit expression for the Pauli Lubanski vector in the representation space spanned by the basis which is the orthonormal set of eigenfunctions of the two Casimir operators $\hat{C}_1\, \hat{C}_2$, the four components of the momentum operator $\hat{P}_\mu$ and $\hat{W}_0$. The basis vectors are, accordingly, labelled by the corresponding eigenvalues, so that

$$\hat{C}_{1(2)}|c, p, \omega\rangle = c_{1(2)}|c, p, \omega\rangle \tag{115}$$

$$\hat{P}_\mu|c, p, \omega\rangle = P_\mu|c, p, \omega\rangle \tag{116}$$

$$\hat{W}_0|c, p, \omega\rangle = \omega|\mathbf{p}||c, p, \omega\rangle \tag{117}$$

In this basis, the commutator $[\hat{W}^\lambda, \hat{W}^\sigma] = i\varepsilon^{\lambda\sigma\mu\nu}\, \hat{W}_\mu \hat{P}_\nu$ takes the form $[\hat{W}_a, \hat{W}_b] = i\varepsilon_{abc}$ $(p_0 \hat{W}_c - \hat{W}_0 pc]$, $[\hat{W}_0, \hat{W}_a] = i\varepsilon_{abc} p_b \hat{W}_c$. These commutators determine a Lie algebra for every fixed value of p. In fact, as we shall see later, the character of this Lie algebra is determined by the nature of p, being timelike, null or spacelike. Let us introduce the transformation $\hat{R}_{\mu\nu}$ by the eqs. $\hat{W}_\mu \to \hat{W}'_\mu = \hat{R}_{\mu\nu} \hat{W}^\nu$ and $\hat{P}_\mu \to \hat{P}'_\mu = \hat{R}_{\mu\nu}\, \hat{P}^\nu$. The objective of introducing this transformation is to achieve maximal simplification of the commutators of the components of $\hat{W}_\lambda$. For the purpose we set $\hat{R}_{0a} = \hat{R}_{a0} = 0$, $\hat{R}_{00} = 1$ and $\hat{R}_{ab} = -R_{ab}$ where $R_{ab} = \hat{\mathbf{P}}.\hat{\mathbf{n}}\,\delta_{ab} - \varepsilon_{abc}\,\theta_c + \theta_a\,\theta_b\,(1 + \hat{\mathbf{P}}.\hat{\mathbf{n}})^{-1}$, $\theta_a = \varepsilon_{abc}\, \hat{p}_b \hat{n}_c$, $\hat{p}_a \dfrac{p_a}{|\mathbf{p}|}$ and $\mathbf{n} \equiv (n_1, n_2, n_3)$ is an arbitrary unit vector. Written explicitly,

$$R_{\mu\nu} = \begin{pmatrix} 1 & 0 & 0 & 0 \\ 0 & \hat{\mathbf{p}}.\hat{\mathbf{n}} + \dfrac{(\hat{p}_2\hat{n}_3 - \hat{p}_3\hat{n}_2)^2}{(1+\hat{\mathbf{p}}.\hat{\mathbf{n}})} & \begin{matrix} -(\hat{p}_1\hat{n}_2 - \hat{p}_2\hat{n}_1) \\ +\dfrac{(\hat{p}_2\hat{n}_3 - \hat{p}_3\hat{n}_2)(\hat{p}_3\hat{n}_1 - \hat{p}_1\hat{n}_3)}{(1+\hat{\mathbf{p}}.\hat{\mathbf{n}})} \end{matrix} & \begin{matrix} (\hat{p}_3\hat{n}_1 - \hat{p}_1\hat{n}_3) \\ +\dfrac{(\hat{p}_2\hat{n}_3 - \hat{p}_3\hat{n}_2)(\hat{p}_1\hat{n}_2 - \hat{p}_2\hat{n}_1)}{(1+\hat{\mathbf{p}}.\hat{\mathbf{n}})} \end{matrix} \\ 0 & \begin{matrix} (\hat{p}_1\hat{n}_2 - \hat{p}_2\hat{n}_1) \\ +\dfrac{(\hat{p}_2\hat{n}_3 - \hat{p}_3\hat{n}_2)(\hat{p}_3\hat{n}_1 - \hat{p}_1\hat{n}_3)}{(1+\hat{\mathbf{p}}.\hat{\mathbf{n}})} \end{matrix} & \hat{\mathbf{p}}.\hat{\mathbf{n}} + \dfrac{(\hat{p}_3\hat{n}_1 - \hat{p}_1\hat{n}_3)^2}{(1+\hat{\mathbf{p}}.\hat{\mathbf{n}})} & \begin{matrix} -(\hat{p}_2\hat{n}_3 - \hat{p}_3\hat{n}_2) \\ +\dfrac{(\hat{p}_3\hat{n}_1 - \hat{p}_1\hat{n}_3)(\hat{p}_1\hat{n}_2 - \hat{p}_2\hat{n}_1)}{(1+\hat{\mathbf{p}}.\hat{\mathbf{n}})} \end{matrix} \\ 0 & \begin{matrix} -(\hat{p}_3\hat{n}_1 - \hat{p}_1\hat{n}_3) \\ +\dfrac{(\hat{p}_2\hat{n}_3 - \hat{p}_3\hat{n}_2)(\hat{p}_1\hat{n}_2 - \hat{p}_2\hat{n}_1)}{(1+\hat{\mathbf{p}}.\hat{\mathbf{n}})} \end{matrix} & \begin{matrix} (\hat{p}_2\hat{n}_3 - \hat{p}_3\hat{n}_2) \\ +\dfrac{(\hat{p}_3\hat{n}_1 - \hat{p}_1\hat{n}_3)(\hat{p}_1\hat{n}_2 - \hat{p}_2\hat{n}_1)}{(1+\hat{\mathbf{p}}.\hat{\mathbf{n}})} \end{matrix} & \hat{\mathbf{p}}.\hat{\mathbf{n}} + \dfrac{(\hat{p}_1\hat{n}_2 - \hat{p}_2\hat{n}_1)^2}{(1+\hat{\mathbf{p}}.\hat{\mathbf{n}})} \end{pmatrix}$$

Under this transformation, the four vector p transforms as $p'_a = n_a|\mathbf{p}|$ (which confirms that the transformation is a spatial rotation for $\mathbf{p}'^2 = \sum_{a=1}^{3} p'^2_a = \sum_{a=1}^{3} n_a^2\, \mathbf{p}^2 = \mathbf{p}^2$) and the transformed components of the Pauli Lubanski vector obey the commutation relations $[\hat{W}'_a, \hat{W}'_b] = i\varepsilon_{abc}(p_0 \hat{W}'_c - \hat{n}_c|\mathbf{p}|\hat{W}'_0)$ and $[\hat{W}'_0, \hat{W}'_a] = i|\mathbf{p}|\varepsilon_{abc}\, \hat{n}_b\, \hat{W}'_c$. Writing $\hat{W}'_0 = |\mathbf{p}|\hat{\Omega}_0$ and $\hat{W}'_a = \hat{n}_a\, \hat{\Omega}_0 p_0 + \hat{\Omega}_a$, we obtain

$$[\hat{\Omega}_0, \hat{\Omega}_a] = i\varepsilon_{abc}\, \hat{n}_b \hat{\Omega}_c,\ [\hat{\Omega}_a, \hat{\Omega}_b] = ip_\mu p^\mu\, \varepsilon_{abc}\, \hat{n}_c \hat{\Omega}_0 \tag{118}$$

These sets of commutators determine a Lie algebra whose structure constants depend on the nature of $c_1 \equiv p_\mu p^\mu$ and $\hat{\mathbf{n}}$. Inverting the above set of transformations, we obtain $\hat{W}_0 = \hat{W}'_0 = |\mathbf{p}|\hat{\Omega}_0$ and

$$\hat{W}_a = R^{-1}_{ab}\, \hat{W}'_b = \hat{\Omega}_a + \hat{p}_a\, \hat{\Omega}_a p_0 - \frac{(\hat{p}_a + \hat{n}_a)\hat{\Omega}_b\, \hat{p}_b}{1 + \hat{\mathbf{n}}.\hat{\mathbf{p}}} \tag{119}$$

Now, since the scalar product of three vectors is preserved under spatial rotations, we have $\hat{\mathbf{W}}.\hat{\mathbf{P}} = \hat{\mathbf{W}}'.\hat{\mathbf{P}}'$. whence

$$0 = \hat{W}_\mu\, \hat{P}^\mu = \hat{W}_0\, \hat{P}_0 - \hat{\mathbf{W}}.\hat{\mathbf{P}} = \hat{W}_0\, p_0 - \hat{\mathbf{W}}.\mathbf{p} = \hat{W}'_0 p'_0 - \hat{\mathbf{W}}'.\mathbf{p}'$$

$$= \hat{W}_0 p_0 - \hat{\mathbf{W}}'.\hat{\mathbf{n}}|\mathbf{p}| = |\mathbf{p}|\Omega_0 p_0 - (\hat{\mathbf{n}}\Omega_0 p_0 + \Omega)\hat{\mathbf{n}}|\mathbf{p}| = -\Omega\hat{\mathbf{n}} \tag{120}$$

This condition implies that this Lie algebra has only three linearly independent elements. The Casimir operators for this algebra are

$$\hat{I}_1 = \hat{\Omega}_0^2\, p_\mu p^\mu + \hat{\Omega}_2^1 + \hat{\Omega}_2^2 + \hat{\Omega}_3^2,\ \hat{I}_2 = e^{2i\pi\hat{\Omega}_0} \tag{121}$$

and their eigenvalues can be used for the labelling of the representations. The Casimir nature of $\hat{I}_1$ follows from

$$[\hat{\Omega}_0, \hat{I}_1] = \left[\hat{\Omega}_0, \hat{\Omega}_0^2 c_1 + \sum_{a=1}^{3} \hat{\Omega}_a^2\right] = \sum_{a=1}^{3}\left[\hat{\Omega}_0, \hat{\Omega}_a^2\right]$$

$$= \sum_{a=1}^{3}\left\{\hat{\Omega}_a\left[\hat{\Omega}_0, \hat{\Omega}_a\right] + \left[\hat{\Omega}_0, \hat{\Omega}_a\right]\hat{\Omega}_a\right\} = i\sum_{a=1}^{3}\left\{\varepsilon_{abc}\, \hat{\Omega}_a n_b \hat{\Omega}_c + \varepsilon_{abc}\, n_b \hat{\Omega}_c \hat{\Omega}_a\right\} = 0$$

and $$[\hat{\Omega}_j, \hat{I}_1] = \left[\hat{\Omega}_j, c_1\hat{\Omega}_0^2 + \sum_{a=1}^{3} \hat{\Omega}_a^2\right] = [\hat{\Omega}_j, c_1\hat{\Omega}_0^2] + \left[\hat{\Omega}_j, \sum_{a=1}^{3} \hat{\Omega}_a^2\right]$$

$$= \hat{\Omega}_0[\hat{\Omega}_j, c_1\hat{\Omega}_0] + [\hat{\Omega}_j, c_1\hat{\Omega}_0]\, \hat{\Omega}_0 + \sum_{a=1}^{3}\left\{\hat{\Omega}_a\left[\hat{\Omega}_j, \hat{\Omega}_a\right] + \left[\hat{\Omega}_j, \hat{\Omega}_a\right]\hat{\Omega}_a\right\}$$

$$= \sum_{k,l=1}^{3}\left\{-ic_1\, \varepsilon_{jkl}\, n_k\, \hat{\Omega}_0\, \hat{\Omega}_l - ic_1\, \varepsilon_{jkl}\, n_k\, \hat{\Omega}_l\, \hat{\Omega}_0\right\}$$

$$+ \sum_{a,m=1}^{3}\left\{ic_1\, \varepsilon_{jam}\, n_m\, \hat{\Omega}_a\, \hat{\Omega}_0 + ic_1\, \varepsilon_{jam}\, n_m\, \hat{\Omega}_0\, \hat{\Omega}_a\right\} = 0$$

thereby establishing the Casimir nature of $\hat{I}_1$. The Casimir operator $\hat{I}_2$ has eigenvalues equal to ± 1 with the positive sign corresponding to single valued representations and the negative sign to double valued ones.

To further explore the structure of this Lie algebra and simplify the commutators, we introduce another spatial rotation R' that aligns the unit vector $\hat{\mathbf{n}}$ with the x^3 axis so that $\hat{\mathbf{n}} \to \hat{\mathbf{n}}' \equiv (0, 0, 1) = R'\hat{\mathbf{n}}$. We also have $\hat{\Omega}_a \to \hat{\Omega}'_a = R'_{ab}\,\hat{\Omega}_b$. Written out explicitly, this rotation matrix takes the form

$$R' = \begin{pmatrix} \frac{n_2^2}{1+n_3} + n_3 & -\frac{n_1 n_2}{1+n_3} & -n_1 \\ -\frac{n_1 n_2}{1+n_3} & \frac{n_1^2}{1+n_3} + n_3 & -n_2 \\ n_1 & n_2 & n_3 \end{pmatrix} \tag{122}$$

The commutators then become

$$[\hat{\Omega}_0, \hat{\Omega}'_1] = i\hat{\Omega}'_2,\ [\hat{\Omega}_0, \hat{\Omega}'_2] = -\,i\hat{\Omega}'_1,\ [\hat{\Omega}_1, \hat{\Omega}'_2] = ip_\mu p^\mu\,\hat{\Omega}'_0,\ \hat{\Omega}'_3 = 0 \tag{123}$$

Defining the ladder operators $\hat{\Omega}'_\pm = \hat{\Omega}'_1 \pm i\hat{\Omega}'_2$, we obtain

$$[\hat{\Omega}_0, \hat{\Omega}'_\pm] = \pm\,\hat{\Omega}'_\pm,\ [\hat{\Omega}'_+, \hat{\Omega}'_-] = 2c_1\,\hat{\Omega}_0,\ \hat{\Omega}'_-\,\hat{\Omega}'_+ = \hat{I}_1 - c_1\,\hat{\Omega}_0^2 - c_1\,\hat{\Omega}_0,$$

$$\hat{\Omega}'_+, \hat{\Omega}'_- = \hat{I}_1 - c_1\,\hat{\Omega}_0^2 + c_1\,\hat{\Omega}_0 \tag{124}$$

Using the standard methods of angular momentum, we obtain the complete IRRs of this Lie algebra. For this purpose, we choose, as a basis, a complete set of eigenfunctions of the commuting operators $\hat{C}_1 \equiv \hat{P}_\mu\,\hat{P}^\mu$, $\hat{I}_1 = \hat{\Omega}_0^2 p_\mu p^\mu + \hat{\Omega}_1^2 + \hat{\Omega}_2^2 + \hat{\Omega}_3^2$ and $\hat{\Omega}_0$. The explicit form of the Casimir operators $\hat{I}_1$, $\hat{I}_2$ and $\hat{\Omega}_\mu$ that realize a Hermitian IR of this Lie algebra are given by

$$\hat{P}_\mu\,\hat{P}^\mu|c_1, i_1, \omega\rangle = c_1|c_1, i_1, \omega\rangle \tag{125}$$

$$\hat{\Omega}_0|c_1, i_1, \omega\rangle = \omega|c_1, i_1, \omega\rangle \tag{126}$$

$$\hat{I}_1|c_1, i_1, \omega\rangle = -i_1|c_1, i_1, \omega\rangle \tag{127}$$

and $\hat{I}_1|c_1, i_1, \omega\rangle = e^{2i\pi\hat{\Omega}_0}|c_1, i_1, \omega\rangle = (-1)^{2\omega}|c_1, i_1, \omega\rangle$ because ω can assume only integer or half integer values. We also have

$$\hat{\Omega}_0\,\hat{\Omega}'_\pm|c_1, i_1, \omega\rangle = (\hat{\Omega}'_\pm\,\hat{\Omega}_0 \pm \hat{\Omega}'_\pm)|c_1, i_1, \omega\rangle = (\omega \pm 1)\hat{\Omega}'_\pm|c_1, i_1, \omega\rangle \tag{128}$$

whence $\hat{\Omega}'_\pm|c_1, i_1, \omega\rangle$ are eigenvectors of $\hat{\Omega}_0$ with respective eigenvalues $\omega \pm 1$. It follows that $\hat{\Omega}'_\pm|c_1, i_1, \omega\rangle$ is proportional to $|c_1, i_1, \omega \pm 1\rangle$ respectively so that we can write $\hat{\Omega}'_\pm|c_1, i_1, \omega\rangle = C_\pm|c_1, i_1, \omega \pm 1\rangle$. To determine $C_\pm$, we have $|C_\pm|^2 = |\hat{\Omega}'_\pm|c_1, i_1, \omega\rangle|^2 = \langle c_1, i_1, \omega|\hat{\Omega}'^\dagger_\pm\,\hat{\Omega}'_\pm|c_1, i_1, \omega\rangle$

$$= \langle |c_1, i_1, \omega| \hat{\Omega}'_{\mp} \hat{\Omega}'_{\pm} |c_1, i_1, \omega\rangle$$
$$= \langle c_1, i_1, \omega| \hat{I}_1 - c_1 \hat{\Omega}_0^2 \mp c_1 \hat{\Omega}_0 |c_1, i_1, \omega\rangle$$
$$= -i_1 - c_1\omega(\omega \pm 1) \tag{129}$$

whence

$$\hat{\Omega}'_1 |c_1, i_1, \omega\rangle = \frac{C_+}{2} c_1, i_1, \omega + 1\rangle + \frac{C_-}{2} |c_1, i_1, \omega - 1\rangle \tag{130}$$

$$\hat{\Omega}'_2 |c_1, i_1, \omega\rangle = \frac{C_+}{2i} |c_1, i_1, \omega + 1\rangle - \frac{C_-}{2i} |c_1, i_1, \omega - 1\rangle \tag{131}$$

Now, inverting the transformation between $\hat{\Omega}$ and $\hat{\Omega}'$, we obtain

$$\hat{\Omega}_1 = \left(\frac{n_2^2}{1+n_3} + n_3\right)\hat{\Omega}'_1 - \left(\frac{n_1 n_2}{1+n_3}\right)\hat{\Omega}'_2 \tag{132}$$

$$\hat{\Omega}_2 = \left(\frac{n_1^2}{1+n_3} + n_3\right)\hat{\Omega}'_2 - \left(\frac{n_1 n_2}{1+n_3}\right)\hat{\Omega}'_1 \tag{133}$$

$$\breve{\Omega}_3 = -n_1 \breve{\Omega}'_1 - n_2 \breve{\Omega}'_2 \tag{134}$$

Using these two sets of equations we can obtain the IRs for $\hat{\Omega}_1$, $\hat{\Omega}_2$, $\hat{\Omega}_3$, $\hat{\Omega}_{\pm}$.

As mentioned earlier the structure of this Lie algebra depends on the character of the eigenvalue $p_\mu p^\mu$ of the first Casimir operator $\hat{C}_1$. In fact, it can be shown that this Lie algebra is isomorphic to the group of 3×3 orthogonal matrices $AO(3)$, for, $p_\mu p^\mu > 0$, the two-dimensional Euclidean group E_2 for $p_\mu p^\mu = 0$ and the group of pseudo-orthogonal matrices in $(1 + 2)$ dimensional Minkowski space $AO(1, 2)$ for $p_\mu p^\mu < 0$. This is established by setting

(i) $\hat{\Omega}_0 = \hat{S}_3$, $\hat{\Omega}'_1 = m\hat{S}_1$, $\hat{\Omega}'_2 = m\hat{S}_2$ for $p_\mu p^\mu = M^2 < 0$ whence the commutators become $[\hat{S}_a, \hat{S}_b] = i\varepsilon_{abc} \hat{S}_c$ that is characteristic of $AO(3)$;

(ii) $\hat{\Omega}_0, = \hat{T}_0$, $\hat{\Omega}'_1 = \hat{T}_1$, $\hat{\Omega}'_2 = \hat{T}_2$ for $p_\mu p^\mu = 0$ whence the commutators take the form $[\hat{T}_0, \hat{T}_1] = i\hat{T}_2$, $[\hat{T}_0, \hat{T}_2] = -i\hat{T}_1$ and $[\hat{T}_1, \hat{T}_2] = 0$ which characterize the group algebra E_2;

(iii) $\hat{\Omega}_0 = \hat{S}_{12}$, $\hat{\Omega}'_1 = \eta\hat{S}_{01}$, $\hat{\Omega}'_2 = \eta\hat{S}_{02}$ with $p_\mu p^\mu = -\eta^2 < 0$ whence the commutators take the form $[\hat{S}_{01}, \hat{S}_{02}] = -i\hat{S}_{12}$, $[\hat{S}_{01}, \hat{S}_{12}] = -i\hat{S}_{02}$, and $[\hat{S}_{02}, \hat{S}_{12}] = i\hat{S}_{01}$which are characteristic of the algebra $AO(1, 2)$.

In view of the above discussion, it is natural to classify the irreducible representations of the Poincare group into the following sets on the basis of the character of the eigenvalue $c_1 = p_\mu p^\mu$ of the first Casimir operator:

(a) $[M]$, Timelike vector case corresponding $c_1 > 0$

(b) $[L]$, Null vector case corresponding to $c_1 = 0$, $p^0 = \mathbf{p} = 0$

(c) $[0]$, Lightlike case corresponding to $c_1 = 0$, $\mathbf{p} \neq 0$

(d) $[T]$, Spacelike case corresponding to $c_1 < 0$.

In addition to the two main Casimir operators viz. $\hat{P}_\mu \hat{P}^\mu$ and $\hat{W}_\mu \hat{W}^\mu$ and the operator $e^{2i\pi \hat{J}12}$ $= e^{2i\pi \hat{J}23} = e^{2i\pi \hat{J}31}$ that has the eigenvalues ± 1 and determines the single/double valuedness of the representation, we also have other Casimir operators that are specific to a particular class of representations. For the moment we denote these class specific Casimir operators by $\hat{C}_\alpha$.

To completely comprehend the relevance of the Pauli Lubanski vector in the representation theory of the Poincare group, we need to study the theory underlying the decomposition of the Poincare group into the relevant factor group by the group of translations in four spacetime T_4, that constitutes the abelian invariant subgroup of the Poincare group. For the purpose, we need to address the issue of the labelling of basis vectors within a given IRR. As mentioned above, the four-dimensional translation group T_4 constitutes an invariant subgroup of the Poincare group. We also know that the basis vectors in a representation of the translation group can be labelled by the eigenvalues of the generators of the translation group. However, the specification of a basis vector using a single index corresponding to the generators of translations is incomplete. For a complete specification of the basis vectors, we introduce a second index m that encompasses all the other degrees of freedom of the group (that are necessary to be considered to ensure the completeness of the basis) but is independent of the translation generators. A complete specification of a basis vector in the representation space (c_1, c_2, c_α) would, thus, consist of five indices, two of them related to the eigenvalues of the main Casimir operators $\hat{C}_1 = \hat{P}_\mu \hat{P}^\mu$ and $\hat{C}_2 = \hat{W}_\mu \hat{W}^\mu$ of the Poincare group and the third to the class specific set of Casimir operators $\hat{C}_\alpha$ for identifying the representation itself and the other two (p, m), where p is the eigenvalue of the momentum operator $\hat{P}_\mu$ and m represents the residual set of degrees of freedom (that are necessary for ensuring the completeness of the basis but are not covered by p), for specifying a basis vector within a representation. Now, all the basis vectors in a given representation space correspond to the same eigenvalue $c_1 \equiv M^2$ of the first Casimir operator. We also have the relativistic mass-energy relation $p_0 = \pm \sqrt{\mathbf{p}^2 + M^2}$. It follows that we can as well use the eigenvalues $\mathbf{p}$ of 3-momentum in lieu of the eigenvalues $p \equiv (p_0, \mathbf{p})$ of 4-momentum for specification of the basis vectors.

Now, $exp\{-i[\hat{L}(\Lambda)\hat{P}_\nu \hat{L}(\Lambda)^{-1}]a^\nu\} = \hat{L}(\Lambda)\, e^{-i\hat{P}_\nu a^\nu}\, \hat{L}(\Lambda)^{-1} = \hat{L}(\Lambda)\, \hat{T}(a)\, \hat{L}(\Lambda)^{-1} = \hat{T}(\Lambda a) =$ $e^{-i\hat{P}_\nu \Lambda^\nu{}_\mu a^\mu}$ whence we get the transformation rules for the covariant generators of translations as $\hat{L}(\Lambda)\hat{P}_\mu\, \hat{L}(\Lambda)^{-1} = \hat{P}_\nu \Lambda^\nu{}_\mu$ or equivalently

$$\hat{L}(\Lambda)^{-1}\, \hat{P}_\mu\, \hat{L}(\Lambda) = \hat{P}_\nu (\Lambda^{-1})^\nu{}_\mu \tag{135}$$

Given a basis vector $|\mathbf{p}, m\rangle$, we have $\hat{P}_\mu\, \hat{L}(\Lambda)|\mathbf{p}, m\rangle = \hat{P}_\nu(\Lambda^{-1})^\nu{}_\mu\, \hat{L}(\Lambda)|\mathbf{p}, m\rangle = p_\nu(\Lambda^{-1})^\nu{}_\mu$ $\hat{L}(\Lambda)|\mathbf{p}, m\rangle$ thereby showing that $\hat{L}(\Lambda)|\mathbf{p}, m\rangle$ is also an eigenvector of $\hat{P}_\mu$ with the eigenvalue $p_\nu(\Lambda^{-1})^\nu{}_\mu$.

Let us define the operator $\hat{S}(\Lambda)$ by

$$\hat{S}(\Lambda)|\mathbf{p}, m\rangle = |\Lambda\mathbf{p}, m\rangle \tag{136}$$

$\hat{S}(\Lambda)$ is unitary because it leaves the inner product

$$\langle \mathbf{p}, m|\mathbf{p}', m'\rangle = \delta_{mm'}\, \omega_p\, \delta(\mathbf{p} - \mathbf{p}') \tag{137}$$

invariant for $\langle \Lambda\mathbf{p}', m'|\Lambda\mathbf{p}, m\rangle = \langle \mathbf{p}', m'|\hat{S}^\dagger \hat{S}|\mathbf{p}, m\rangle = \delta_{mm'}\, \omega_{\Lambda p}\, \delta(\Lambda\mathbf{p} - \Lambda\mathbf{p}') = \delta_{mm'}\, \omega_p\, \delta(\mathbf{p} - \mathbf{p}')$ $= \langle \mathbf{p}', m'|\mathbf{p}, m\rangle$ where the penultimate step follows from the Lorentz invariance of $\omega_p\, \delta(\mathbf{p} - \mathbf{p}')$.

Further $\hat{S}(\Lambda)\, \hat{S}(\Lambda') = \hat{S}(\Lambda\Lambda')$ implying that $\hat{S}(\Lambda)$ is a unitary representation of the Lorentz group. Also being independent of m, it is diagonal with respect to the corresponding generators.

We also have $\hat{S}(\Lambda)\, \hat{T}(\Lambda^{-1}a)|\mathbf{p}, m\rangle = \hat{S}(\Lambda)\, e^{-i\hat{P}_\mu(\Lambda^{-1}a)^\mu}|\mathbf{p}, m\rangle$

$= \hat{S}(\Lambda)\, e^{-i\hat{p}_\mu(\Lambda^{-1}a)^\mu}|\mathbf{p}, m\rangle = e^{-i\hat{p}_\mu(\Lambda^{-1}a)^\mu}|\Lambda\mathbf{p}, m\rangle = e^{-i(\Lambda^{-1}\hat{p})_\mu(\Lambda^{-1}a)^\mu}|\Lambda\mathbf{p}, m\rangle$

$= \hat{T}(a)|\Lambda\mathbf{p}, m\rangle = \hat{T}(a)\, \hat{S}(\Lambda)|\mathbf{p}, m\rangle$ whence, because of the completeness of the set of basis vectors, we can infer that

$$\hat{T}(\Lambda^{-1} a)\, \hat{S}(\Lambda)^{-1} = \hat{S}(\Lambda)^{-1}\, \hat{T}(a) \tag{138}$$

From eq. (104), we also infer that $\hat{L}(\Lambda)\hat{T}(\Lambda^{-1} a) = \hat{T}(a)\, \hat{L}(\Lambda)$ which, together with eq. (138) gives

$$[\hat{T}(\Lambda)\, \hat{S}(\Lambda)^{-1}, \hat{T}(a)] = 0 \tag{139}$$

Since this equation holds for arbitrary a, it follows that $\hat{L}(\Lambda)\, \hat{S}(\Lambda)^{-1}$ commutes with the translation operator. This implies that it is diagonal with respect to the generators of translation $\hat{P}_\mu$. Writing it as $\hat{Q}(\Lambda, \hat{p})$, we obtain

$$\hat{L}(\Lambda) = \hat{Q}(\Lambda, \hat{P})\, \hat{S}(\Lambda) \tag{140}$$

Further, since both $\hat{L}(\Lambda)$ and $\hat{S}(\Lambda)$, are unitary, it follows that $\hat{Q}(\Lambda, \hat{P})$ is also unitary.

Now, $\quad \hat{Q}(\Lambda\Lambda', \hat{P})\, \hat{S}(\Lambda\Lambda')|\mathbf{p}, m\rangle = \hat{L}(\Lambda\Lambda')|\mathbf{p}, m\rangle = \hat{L}(\Lambda)\, \hat{L}(\Lambda')|\mathbf{p}, m\rangle$

$= \hat{Q}(\Lambda, \hat{P})\, \hat{S}(\Lambda)\, \hat{Q}(\Lambda', \hat{P})\, \hat{S}(\Lambda')|\mathbf{p}, m\rangle = \hat{Q}(\Lambda, \hat{P})\hat{S}(\Lambda)\, \hat{Q}(\Lambda', \hat{P})|\Lambda'\,\mathbf{p}, m\rangle$

$= \hat{Q}(\Lambda, \hat{P})\, \hat{S}(\Lambda)\, \hat{Q}(\Lambda', \Lambda'\hat{P})|\Lambda'\,\mathbf{p}, m\rangle = \hat{Q}(\Lambda, \hat{P})\, \hat{Q}(\Lambda', \Lambda'\hat{P})\, \hat{S}(\Lambda)|\Lambda'\,\mathbf{p}, m\rangle$

$= \hat{Q}(\Lambda, \hat{P})\, \hat{Q}(\Lambda', \Lambda'\hat{P})|\Lambda\Lambda'\,\mathbf{p}, m\rangle = \hat{Q}(\Lambda, \hat{P})\, \hat{Q}(\Lambda', \Lambda^{-1}\hat{P})|\Lambda\Lambda'\,\mathbf{p}, m\rangle$

$= \hat{Q}(\Lambda, \hat{P})\, \hat{Q}(\Lambda', \Lambda^{-1}\hat{P})\, \hat{S}(\Lambda\Lambda')|\mathbf{p}, m\rangle$ whence

$$\hat{Q}(\Lambda\Lambda', \hat{P}) = \hat{Q}(\Lambda, \hat{P})\, \hat{Q}(\Lambda', \Lambda^{-1}\hat{P}) \tag{141}$$

Using eq. (141), we obtain

$|\mathbf{p}, m\rangle = \hat{L}(\mathbf{I})|\mathbf{p}, m\rangle = \hat{L}(\Lambda\Lambda^{-1})|\mathbf{p}, m\rangle = \hat{Q}(\Lambda, \hat{P})\, \hat{Q}(\Lambda^{-1}, \Lambda^{-1}\hat{P})\, \hat{S}(\Lambda\Lambda^{-1})|\mathbf{p}, m\rangle$

$= \hat{Q}(\Lambda, \hat{P})\, \hat{Q}(\Lambda^{-1}, \Lambda^{-1}\hat{P})|\mathbf{p}, m\rangle = \hat{Q}(\Lambda, \hat{P})\, \hat{Q}(\Lambda, \hat{P})^{-1}|\mathbf{p}, m\rangle$ so that

$$\hat{Q}(\Lambda, \hat{P})^{-1} = \hat{Q}(\Lambda^{-1}, \Lambda^{-1}\hat{P}) \tag{142}$$

Let us assume that there exists an operator $\hat{U}(\hat{P})$ in the irreducible representation space that satisfies (i) unitarity, (ii) is diagonal with respect to $\hat{P}_\mu$ and hence, commutes with $\hat{T}(a)$, (iii) is single valued in $\hat{P}_\mu$ (so that it preserves the single valuedness of the state operator $|\mathbf{p}, m\rangle$ with respect to $\hat{P}_\mu$). Under the action of $\hat{U}(\hat{P})$ (i) the translation operator $\hat{T}(a)$ is not affected since $[\hat{T}(a), \hat{U}(\hat{P})] = 0$, (ii) the Lorentz transformation $\hat{L}(\Lambda)$ is transformed as $\hat{U}(\hat{P})\,\hat{L}(\Lambda)\,\hat{U}(\hat{P})^{-1}$. We have

$$\hat{U}(\hat{P})\ \hat{L}(\Lambda)\ \hat{U}(\hat{P})^{-1}|\mathbf{p}, m\rangle = \hat{U}(\hat{P})\ \hat{Q}(\Lambda, \hat{P})\ \hat{S}(\Lambda)\hat{U}(\hat{P})^{-1}|\mathbf{p}, m\rangle$$
$$\hat{U}(\hat{P})\ \hat{Q}(\Lambda, \hat{P})\ \hat{U}(p)^{-1}\ \hat{S}(\Lambda)|\mathbf{p}, m\rangle = \hat{U}(\hat{P})\ \hat{Q}(\Lambda, \hat{P})\ \hat{U}(p)^{-1}|\Lambda\mathbf{p}, m\rangle$$
$$\hat{U}(\hat{P})\ \hat{Q}(\Lambda, \hat{P})\ \hat{U}(\Lambda^{-1}, \hat{P})^{-1}|\Lambda\mathbf{p}, m\rangle = \hat{U}(\hat{P})\ \hat{Q}(\Lambda, \hat{P})\ \hat{U}(\Lambda^{-1}, \hat{P})^{-1}\ \hat{S}(\Lambda)|\mathbf{p}, m\rangle \quad (143)$$

Thus, $\hat{Q}(\Lambda, \hat{P})$ and $\hat{U}(\hat{P})\ \hat{Q}(\Lambda, \hat{P})\ \hat{U}(\Lambda^{-1}, \hat{P})^{-1}$ provide equivalent representations for given representations $\hat{T}(a)$ of and $\hat{S}(\Lambda)$.

To proceed further with the representation theory of the Poincare group, it is necessary at this point to introduce the concept of "little group". As mentioned above, the IRRs of the Poincare group are classified into timelike [*M*] for $c_1 = p_\mu p^\mu = M^2 > 0$ null [*L*] for $c_1 = p_0 = \mathbf{p} = 0$, lightlike [0] for $c_1 = 0 \neq \mathbf{p}$ and spacelike [*T*] for $c_1 = p_\mu p^\mu = M^2 < 0$ on the basis of the nature of the eigenvalues of the Casimir operator $\hat{C}_1$. Now, for the classes [*M*] and [0], the sign of the energy $\frac{p_0}{|p_0|}$ commutes with all the generators of group operations $\hat{L}(\Lambda)$ and $\hat{T}(a)$ and hence, constitutes another Casimir operator (that is specific to the classes [*M*] and [0] of the representations). As a result of this, there exist invariant subspaces of these two classes of representations that are determined by the energy sign viz. $[M_\pm]$. and $[0_\pm]$. Thus, given an IRR, the corresponding identifying eigenvalue of $\hat{C}_1$ would be classified into one of the classes on the basis of being timelike, null, lightlike or spacelike with a subclassification within this on the basis of the energy sign. In other words, for a given IRR, the operator $\hat{p}^2$ and its eigenvalue $c_1 \equiv M^2$ would fall into one of [*M*], [0], [*L*] or [*T*] whereas the operator $\hat{p}_\mu$ and its associated eigenvalue p_μ would be classified into one of $[M_+]$, $[M_-]$, $[0_+]$, $[0_-]$, [*L*] or [*T*]. In view of the fact that $\hat{p}_\mu \hat{L}(\Lambda)|\mathbf{p}, m\rangle = p_\nu \Lambda^\nu_\mu \hat{L}(\Lambda)|\mathbf{p}, m\rangle$, the eigenvalues within a class are connected with each other by a suitable Lorentz transformation. It is to be noted that in the case of class [*T*] of representations we cannot subclassify the representations on the basis of the energy sign because in this case the energy eigenvalue p_0 can change its character (real/imaginary) by an appropriate Lorentz transformation within the same representation space. Thus, the energy sign is not unique to a particular IRR in this case. Physically, the class [*T*] corresponds to imaginary mass $p^2 \equiv M^2 < 0$ which implies that $p_0^2 = \mathbf{p}^2 + M^2$ corresponding to such a representation could be made either positive or negative by a Lorentz transformation thereby implying that p_0 can be real or imaginary. Additionally, this representation also does not have a well defined and reasonable non-relativistic limit. The class [*L*] corresponds to the case of zero momentum and energy.

Let us denote by $\{P\}$, the set of eigenvalues p_μ of a particular class. Let us identify an element q_μ arbitrarily from the set $[p]$. As mentioned above, the various eigenvalues in $\{p\}$ are related to each other through suitable Lorentz transformations. Let $\{\eta\}$ be the set of Lorentz transformations that leave q_μ invariant, i.e.,

$$\eta^\mu_\nu q^\nu = q^\mu \text{ or } \eta q = q \tag{144}$$

$\{\eta\}$ is a subset of the Lorentz group and is called a "little group" of the Lorentz group. Now, since the elements of $\{p\}$ are connected inter se by Lorentz transformation, given a $p_\mu \in \{p\}$, there would exist a Lorentz transformation κ_p such that

$$p = \kappa_p q \tag{145}$$

Corresponding to an arbitrary element $q_\mu \in \{p\}$, the Lorentz transformation $\eta_p = \kappa^{-1}_{\Lambda p} \Lambda\kappa_p$ or equivalently $\eta_p = \kappa^{-1}_p \Lambda\kappa_{\Lambda^{-1}p}$ is an element of the little group $\{\eta\}$ for, we have

$$\eta_p q = \kappa^{-1}_{\Lambda p} \Lambda\kappa_p q = \kappa^{-1}_{\Lambda p} \Lambda p = q \text{ or}$$

$$\eta_p q = \kappa^{-1}_p \Lambda\kappa^{-1}_{\Lambda p} q = \kappa^{-1}_p \Lambda\Lambda^{-1} p = q \tag{146}$$

where we have used eq. (145). Using eqs. (141) and (146), we obtain

$$\hat{Q}(\Lambda, \hat{P}) = \hat{Q}(\kappa_{\Lambda p}, \eta_p \kappa^{-1}_{p}, \hat{P}) = \hat{Q}(\kappa_{\Lambda p}, \hat{P})\, \hat{Q}(\eta_p\, \kappa^{-1}_{p}, \kappa^{-1}_{\Lambda p}\, \hat{P})$$

$$\hat{Q}(\kappa_{\Lambda p}, \hat{P})\, \hat{Q}(\eta_p, \kappa^{-1}_{\Lambda p}\, \hat{P})\, \hat{Q}(\kappa^{-1}_{p}, \eta^{-1}_{p}, \kappa^{-1}_{\Lambda p}\, \hat{P}) \tag{147}$$

Now, operating the right hand side of eq. (147) on the state $|\Lambda\mathbf{p}, m\rangle$, we obtain

$$\hat{Q}(\kappa_{\Lambda p}, \hat{P})\, \hat{Q}(\eta_p, \kappa^{-1}_{\Lambda p}\, \hat{P})\, \hat{Q}(\kappa^{-1}_{p}, \eta^{-1}_{p}, \kappa^{-1}_{\Lambda p}\, \hat{P})|\Lambda\mathbf{p}, m\rangle$$

$$\hat{Q}(\kappa_{\Lambda p}, \hat{P})\, \hat{Q}(\eta_p, \kappa^{-1}_{\Lambda p}\, \hat{P})\, \hat{Q}(\kappa^{-1}_{p}, \eta^{-1}_{p}, \kappa^{-1}_{\Lambda p}\, \Lambda p)|\Lambda\mathbf{p}, m\rangle$$

$$\hat{Q}(\kappa_{\Lambda p}, \hat{P})\, \hat{Q}(\kappa^{-1}_{p}, \eta^{-1}_{p}, \kappa^{-1}_{\Lambda p}\, \Lambda p)\, \hat{Q}(\eta_p, \kappa^{-1}_{\Lambda p}\, \hat{P})|\Lambda\mathbf{p}, m\rangle$$

$$\hat{Q}(\kappa_{\Lambda p}, \hat{P})\, \hat{Q}(\kappa^{-1}_{p}, \eta^{-1}_{p}, \kappa^{-1}_{\Lambda p}\, \Lambda p)\, \hat{Q}(\eta_p, \kappa^{-1}_{\Lambda p}\, \Lambda p)|\Lambda\mathbf{p}, m\rangle$$

$$\hat{Q}(\kappa_{\Lambda p}, \hat{P})\, \hat{Q}(\eta_p, q)\, \hat{Q}(\kappa^{-1}_{p}, \eta^{-1}_{p}, q)|\Lambda\mathbf{p}, m\rangle$$

$$\hat{Q}(\kappa_{\Lambda p}, \hat{P})\, \hat{Q}(\eta_p, q)\, \hat{Q}(\kappa^{-1}_{p}, q)|\Lambda\mathbf{p}, m\rangle$$

$$= \hat{Q}(\kappa_{\Lambda p}, \Lambda_p)\, \hat{Q}(\eta_p, q)\, \hat{Q}(\kappa^{-1}_{p}, q)|\Lambda\mathbf{p}, m\rangle \quad \text{whence}$$

$$\hat{Q}(\Lambda, \hat{P}) = \hat{Q}(\kappa_{\Lambda p}, \Lambda\hat{P})\, \hat{Q}(\eta_p, q)\, \hat{Q}(\kappa^{-1}_{p}, q) \tag{148}$$

Eq. (148) is identical to the expression $\hat{U}(\Lambda p)^{-1}\, \hat{Q}(\eta_p, q)\, \hat{U}(p)$ if we identify $\hat{U}(\Lambda p)^{-1}$ $= \hat{Q}(\kappa_{\Lambda p}, \Lambda p)$ or $\hat{U}(\Lambda p) = \hat{Q}(\kappa_{\Lambda p}, \Lambda p)^{-1} = \hat{Q}(\kappa^{-1}_{\Lambda p}, \kappa^{-1}_{\Lambda p}\, \Lambda p) = \hat{Q}(\kappa^{-1}_{\Lambda p}, q)$ and $\hat{U}(p) = \hat{Q}(\kappa^{-1}_{p}, q)$.

Since $\hat{U}(p)^{-1}$ is unitary, it follows that $\hat{U}(p)$ is unitary as well. It, therefore, follows from eq. (148) that $\hat{Q}(\eta_p, q)$ generates a representation of the Poincare group that is unitarily equivalent to $\hat{Q}(\Lambda, \hat{P})$. Hence, in lieu of eq. (140), we can write the equivalent expression $\hat{L}(\Lambda) = \hat{Q}(\eta_p, q)$ $\hat{S}(\Lambda)$ without any loss of generality. Further, using eq. (142), we get

$$\hat{Q}(\kappa_{p,}\ \hat{P})^{-1} = \hat{Q}(\kappa_{p'}^{-1}\ \kappa^{-1}_{\ p}\ \hat{P}) = \hat{Q}(\kappa_{p'}^{-1}\ q) \tag{149}$$

Making use of eq. (141), we can also write $\hat{Q}(\eta_p\ \eta'_p,\ \hat{P}) = \hat{Q}(\eta_p,\ \hat{P})\ \hat{Q}(\eta'_p\ \eta^{-1}_p,\ \hat{P})$ and $\hat{Q}(\eta_p\ \eta'_p,\ q) = \hat{Q}(\eta_p,\ q)\ \hat{Q}(\eta'_p\ \eta^{-1}_p,\ q)\ \hat{Q}(\eta_p,\ q)\ \hat{Q}(\eta'_p,\ q)$ where $\eta_p = \kappa^{-1}_{\Lambda p}\ \Lambda\kappa_p$ and $\eta'_{\ p} = \kappa^{-1}_{\Lambda p}\ \Lambda'\kappa_p$.

This shows that $\hat{Q}(\eta_p,\ q)$ is a unitary representation of the relevant little group $\hat{Q}(\Lambda,\ \hat{P})$. The operator $\hat{Q}(\eta_p,\ q)$ is called "Wigner Rotation". It is simple to see that $\hat{Q}(\eta_p,\ q)\ \hat{S}(\Lambda)$ is a unitary representation of the Lorentz group. Unitarity follows from the unitarity of $\hat{Q}(\eta_p,\ q)$ and $\hat{S}(\Lambda)$. Further, $\hat{Q}(\eta_p,\ q)\ \hat{S}(\Lambda)\ \hat{Q}(\eta'_{\ p},\ q)\ \hat{S}(\Lambda')$.

$\hat{Q}(\eta_p,\ q)\ \hat{Q}(\eta'_{\Lambda^{-1}p},\ q)\ \hat{S}(\Lambda)\ \hat{S}(\Lambda') = \hat{Q}(\eta_p\ \eta'_{\Lambda^{-1}p},\ q)\ \hat{S}(\Lambda\Lambda')$ and $\eta_p\ \eta'_{\Lambda^{-1}p} = \kappa^{-1}_{\Lambda p}\ \Lambda\kappa_p\ (\kappa_p)^{-1}\ \Lambda'\kappa_{\Lambda^{-1}p} = \kappa^{-1}_{\Lambda p}\ \Lambda\Lambda'\ \kappa_{\Lambda^{-1}p} = \kappa^{-1}_{\Lambda\Lambda'\cdot\Lambda^{-1}p}\ \Lambda\Lambda'\ \kappa_{\Lambda^{-1}p}$, thereby establishing that $\hat{Q}(\eta_p,\ q)\ \hat{S}(\Lambda)$ is a representation of the Lorentz group. There is, however, a rider in the sense that since the representations of the Lorentz group are at most double valued, only those $\hat{Q}(\eta_p,\ q)$ contribute to the representations of the Lorentz group that have at most double valued representations.

It remains to be proved that:

(i) In a unitary representation of the Poincare group $\hat{Q}(\eta_p,\ q)$, constitutes a unitary irreducible representation of a "little group"; and

(ii) If $\hat{Q}(\eta_p,\ q)$ is a unitary irreducible representation of a "little group", there exists a corresponding unitary representation of the Poincare group.

To prove (i), let us assume that $\hat{Q}(\eta_p,\ q)$ is reducible. Now, consider eq. (140). In this eq.$\hat{S}(\Lambda)$ is diagonal in generators relating to the set of eigenvalues collectively represented by m. It follows that the entire dependence of $\hat{L}(\Lambda)$ on 'm' related generators is encompassed in $\hat{Q}(\eta_p,\ q)$ (which, incidentally, is diagonal in $\hat{p}_\mu$ and hence, independent of the translation operator). Therefore, the reducibility, if any, of $\hat{Q}(\eta_p,\ q)$ would emanate from the degrees of freedom in 'm'. In other words, if $\hat{Q}(\eta_p,\ q)$ is to be reducible, such reducibility comes from the freedom associated with 'm'. Since, $\hat{L}(\Lambda)$ is 'm' dependent and the entire such dependence comes from $\hat{Q}(\eta_p,\ q)$, reducibility of $\hat{Q}(\eta_p,\ q)$ would manifest itself as reducibility of $\hat{L}(\Lambda)$. Again, since $\hat{T}(a)$ is diagonal with respect to both $\hat{p}_\mu$ and 'm' related degrees of freedom, reducibility of $\hat{L}(\Lambda)$ would imply reducibility of the Poincare group. Consequently, we cannot have a situation wherein the given representation of the Poincare group is irreducible yet the corresponding representation of $\hat{Q}(\eta_p,\ q)$ is reducible.

To prove (ii), let us assume that $\hat{Q}(\eta_p,\ q)$ has, at most, double valued representations. Let φ be an element of the Poincare group and let $D(\varphi)$ be its unitary representation consisting of $\hat{L}(\Lambda)$ expressed as $\hat{Q}(\eta_p,\ q)\ \hat{S}(\Lambda)$ together with the representation of the translation operator $\hat{T}(a)$. Let us assume that this representation is reducible. If the representation is reducible, there will exist state vectors $|\Psi\rangle$ and $|\Phi\rangle$ such that $\langle\Phi|D(\varphi)|\Psi\rangle = 0$. Now, because $D(\varphi)\hat{T}(a)$ is also a representation of the Poincare group, we will have $\langle\Phi|D(\varphi)\hat{T}(a)|\Psi\rangle = 0$. Writing the two state vectors $|\psi\rangle$ and $|\Phi\rangle$ as superpositions of the basis vectors, we have $|\Psi\rangle = \sum_m \int_\Sigma dp\, C_1(\mathbf{p},\, m)|\mathbf{p},\, m\rangle$ and $|\Phi\rangle = \sum_m \int_{\Sigma'} dp'\, C_2\, (\mathbf{p}',\, m')|\mathbf{p}',\, m\rangle$ where the regions of integration are the area formed by all the p_μ belonging to a given class. Substituting these values, we get

$$\sum_{m,m'} \int\int_{\Sigma,\Sigma'} dpdp'\, C_1(\mathbf{p}, m)\, C_2^*(\mathbf{p}',\, m')\langle\mathbf{p}',\, m'|D(\varphi)|\mathbf{p},\, \mathrm{m}\rangle\, e^{-ip_\mu a^\mu} = 0$$

Since a^μ is arbitrary, we have

$$\sum_{m',m''} C_1\,(\mathbf{p}',\, m')\; C_2^*\,(\mathbf{p}'',\, m'')\;\langle\mathbf{p}'',\, m''|D(\varphi)|\mathbf{p}',\, m'\rangle = 0 \tag{150}$$

Let us assume that $D(\varphi) = \hat{L}(\kappa_{p''})\;\hat{L}(\eta)\;\hat{L}(\kappa_{p'}^{-1})$ for an arbitrary element η of the little group. Then, we have the following results:

$$\hat{L}(\kappa_{p'}^{-1}) \sum_{m'} C_1\,(\mathbf{p}',\, m')|\mathbf{p}',\, m'\rangle = \hat{Q}(\kappa_p^{-1}\;\kappa_{p'}^{-1}\;\kappa_{\kappa p.p},\, q)\;\hat{S}\,(\kappa_p^{-1})\sum_{m'} C_1\,(\mathbf{p}',\, m')|\mathbf{p}',\, m'\rangle$$

$$= \hat{Q}(\kappa_p^{-1}\;\kappa_{p'}^{-1}\;\kappa_{\kappa p}p,\, q)\sum_{m'} C_1\,(\mathbf{p}',\, m')|\kappa_q^{-1}\;\mathbf{p}',\, m'\rangle$$

$$= \hat{Q}(\kappa_p^{-1},\, q)\sum_{m'} C_1\,(\mathbf{p}',\, m')|\kappa_p^{-1}\;\mathbf{p}',\, m'\rangle\,\hat{Q}(\kappa_q^{-1},\, q)\sum_{m'} C_1\,(\mathbf{p}',\, m')|q,\, m'\rangle$$

Similarly,

$$\hat{L}(\kappa_{p''})^\dagger \sum_{m''} C_2\,(\mathbf{p}'',\, m'')|\mathbf{p}'',\, m''\rangle = \hat{L}(\kappa_{p''}^{-1})\sum_{m''} C_2\,(\mathbf{p}'',\, m'')|\mathbf{p}'',\, m''\rangle$$

$$= \hat{Q}(\kappa_q^{-1},\, q)\sum_{m''} C_2\,(\mathbf{p}'',\, m'')|\kappa_{p''}^{-1}\;\mathbf{p}'',\, m''\rangle = \hat{Q}(\kappa_q^{-1},\, q)\sum_{m''} C_2\,(\mathbf{p}'',\, m'')|q,\, m''\rangle$$

It may be noted that both $\sum_{m'} C_1\,(\mathbf{p}',\, m')|q,\, m'\rangle$, $\sum_{m'} C_2\,(\mathbf{p}'',\, m'')|q,\, m''\rangle$ are non-zero because $\hat{S}$ is a unitary operator. We also have

$$\hat{P}_\mu \sum_{m'} C_1\,(\mathbf{p}',\, m')|\kappa_{p'}^{-1}\;\mathbf{p}',\, m'\rangle = q_\mu \sum_{m'} C_1\,(\mathbf{p}',\, m')|\kappa_{p'}^{-1}\;\mathbf{p}',\, m'\rangle$$

and

$$\hat{P}_\mu \sum_{m''} C_2\,(\mathbf{p}'',\, m'')|\kappa_{p''}^{-1}\;\mathbf{p}'',\, m''\rangle = q_\mu \sum_{m''} C_2\,(\mathbf{p}'',\, m'')|\kappa_{p''}^{-1}\;\mathbf{p}'',\, m''\rangle$$

where we have used the fact that $\kappa_{p'}\, q = p'$ and $\kappa_{p''}\, q = p''$.

Again, using $\hat{L}(\eta) = \hat{Q}(\kappa_p^{-1}\eta\kappa_{\eta^{-1}p}, q)$, $\hat{S}(\eta)$ and noting that is η an element of the little group so that $\eta q = q$, whence we can write $\hat{L}(\eta) = \hat{Q}(\kappa_p^{-1}\eta\kappa_q, q)$, $\hat{S}(\eta)$ and hence obtain

$$\sum_{m',m''} C_1(\mathbf{p}', m')\, C_2^*(\mathbf{p}'', m'')\, \langle \mathbf{p}'', m''|\hat{L}(\kappa_{p''})\, \hat{L}(\eta)\, \hat{L}(\kappa_{p'}^{-1})|\mathbf{p}', m'\rangle$$

$$= \sum_{m',m''} C_1(\mathbf{p}', m')\, C_2^*(\mathbf{p}'', m'')\, \langle q, m''|\hat{Q}(\kappa_q, q)\, \hat{Q}(\kappa_q^{-1}\eta\, k_q, q)\, \hat{S}(\eta)\, \hat{Q}(\kappa_q^{-1}, q)|q, m'\rangle$$

$$= \sum_{m',m''} C_1(\mathbf{p}', m')\, C_2^*(\mathbf{p}'', m'')\, \langle q, m''|\hat{Q}(\kappa_q, q)\, \hat{Q}(\kappa_q^{-1}\eta\, k_q, q)\, \hat{Q}(\kappa_{\eta^{-1}q}^{-1}, q)\, \hat{S}(\eta)|q, m'\rangle$$

$$= \sum_{m',m''} C_1(\mathbf{p}', m')\, C_2^*(\mathbf{p}'', m'')\, \langle q, m''|\hat{Q}(\kappa_q, q)\, \hat{Q}(\kappa_q^{-1}\eta\, k_q, q)\, \hat{Q}(\kappa_{\eta^{-1}q}^{-1}, q)|\eta q, m'\rangle$$

$$= \sum_{m',m''} C_1(\mathbf{p}', m')\, C_2^*(\mathbf{p}'', m'')\, \langle q, m''|\hat{Q}(\kappa_q, q)\, \hat{Q}(\kappa_q^{-1}\eta\, k_q, q)\, \hat{Q}(\kappa_{\eta^{-1}q}^{-1}, q)|q, m'\rangle$$

$$= \sum_{m',m''} C_1(\mathbf{p}', m')\, C_2^*(\mathbf{p}'', m'')\, \langle q, m''|\hat{Q}(\eta, q)|q, m'\rangle \tag{151}$$

The left hand side of eq. (151) vanishes because of eq. (150). Since η is an arbitrary element of the little group, eq. (151) implies that $\hat{Q}(p, q)$ is a reducible representation of the little group thereby contradicting our assumption. In other words, reducibility of the Poincare group implies reducibility of $\hat{Q}(p, q)$. Conversely, if $\hat{Q}(p, q)$ is irreducible, then the corresponding representation of the Poincare group must necessarily be irreducible. It is reiterated here that the index $\mathbf{p}$ used for identifying the basis vectors is to be construed as equivalent to the four vector $\mathbf{p} \equiv \left(\sqrt{\mathbf{p}^2 + c_1}, \mathbf{p}\right)$.

It is in the context of the "little group" decomposition of the Poincare group that the Pauli Lubanski vector assumes its physical significance. Given a unit four vector $\{p_\mu\}$, the independent components of $\hat{W}^\mu$ form a Lie algebra on the subspace of $\{p_\mu\}$. This Lie algebra is the Lie algebra of the "little group" of p_μ. Stated otherwise, the generators of the "little group" of p_μ are the independent components of $\hat{W}^\mu$ corresponding to p_μ.

2.4 IRRs CORRESPONDING TO MASSIVE PARTICLES (WIGNER, 1939; GEL'FAND, 1963; NAIMARK, 1964; OHNUKI, 1976; TUNG, 1984; FUSHCHICH, 1994; SOO, 2003; STEFANOVICH, 2007)

Now, for the study of the representations corresponding to the massive $[M_\pm]$ class and the identification of the corresponding "little group" let us define the "standard vector" as

$q = (\mathrm{sgn}(p_0)M, 0, 0, 0)$. We have identified the standard vector in the "rest frame" so that all the spatial components are zero. The factor group of the Poincare group with respect to the subgroup of translations in four dimensions, T_4, (which is abelian) is the homogeneous Lorentz group. The maximal subgroup of the homogeneous Lorentz group that leaves q invariant is the group of three-dimensional rotations, identifiable with $SO(3)$ which, thus, constitutes the "little group" for this class of representations.

This may also be seen by invoking the fact that the generators of the little group of the standard vector, e.g., q are the independent components of the Pauli Lubanski vector $\hat{W} \equiv (\hat{W}^0, \hat{\mathbf{W}}) = \frac{1}{2}\varepsilon^{\lambda\mu\nu\sigma}\hat{J}_{\mu\nu}\hat{P}_{\sigma} = (\hat{\mathbf{P}}.\hat{\mathbf{J}},\ \hat{P}^0\ \hat{\mathbf{P}} \times \hat{\mathbf{K}})$ corresponding to the standard vector q. The second Casimir operator for massive particles corresponding to q is, therefore, $\hat{W}^{\mu}\hat{W}_{\mu} = M\hat{J}^{\mu}M\hat{J}_{\mu} = M^2$ $\hat{\mathbf{J}}^2$ (as $\mathbf{p} = 0$, $p_0 = M$ for the standard vector). The corresponding Lie algebra takes the form $[\hat{W}^i, \hat{W}^j] = i\varepsilon^{ijk0}\,\hat{W}_k\,\hat{P}_0$ or equivalently $[\hat{J}^i, \hat{J}^j] = i\varepsilon^{ijk}\,\hat{J}_k$ which is the Lie algebra of the group of rotations in three-dimensional space thereby confirming that the little group corresponding to massive representations is $SO(3)$.

As is common in the literature on relativity, we can segregate Lorentz transformation in two types with distinctly identifiable features viz. (i) the θ transformations that constitute spatial rotations about the three spatial axes respectively and do not involve any mixing of spatial and temporal coordinates and (ii) the τ transformations that involve Lorentz boosts along the three spatial axes that are in the nature of spatiotemporal rotations. The coordinate transformation equations under these transformations can be summarized as

$$\left.\begin{aligned} \mathbf{x} &\rightarrow \mathbf{x} + \mathbf{x} \times \theta \\ x^0 &\rightarrow x^0 \end{aligned}\right\} \quad \theta \textit{ tranformation} \tag{152}$$

$$\left.\begin{aligned} \mathbf{x} &\rightarrow \mathbf{x} - \tau x^0 \\ x^0 &\rightarrow x^0 - \tau \mathbf{x} \end{aligned}\right\} \quad \tau \textit{ tranformation} \tag{153}$$

We shall, now, obtain explicit expressions for the infinitesimaWigner rotation for each of these transformations. Corresponding to $q \equiv (\pm M, 0, 0, 0)$, the Lorentz transformation that satisfies the equation $p = \kappa_p\, q$ takes the explicit form

$$\kappa_p = \begin{pmatrix} \omega_p / M & \pm p_1 / M & \pm p_2 / M & \pm p_3 / M \\ \pm p_1 / M & 1 + \varpi_p\, p_1^2 & \varpi_p\, p_1\, p_2 & \varpi_p\, p_1\, p_3 \\ \pm p_2 / M & \varpi_p\, p_2\, p_1 & 1 + \varpi_p\, p_2^2 & \varpi_p\, p_2\, p_3 \\ \pm p_3 / M & \varpi_p\, p_3\, p_1 & \varpi_p\, p_3\, p_2 & 1 + \varpi_p\, p_3^2 \end{pmatrix} \tag{154}$$

where $\varpi_p = \dfrac{1}{p^2}\left(\dfrac{\omega_p}{M} - 1\right)$. In terms of the rapidity, $\varsigma = \cosh^{-1}\dfrac{\omega_p}{M} = \sinh^{-1}\dfrac{|\mathbf{p}|}{M}$, eq. (154) may be written as $(\kappa_p)^i_j = \delta^i_j + \dfrac{p_i\, p_j}{\mathbf{p}^2}(\cosh\varsigma - 1)$, $(\kappa_p)^i_0 = (\kappa_p)^0_i = \dfrac{p_i}{|\mathbf{p}|}\sinh\zeta$ and $(\kappa_p)^0_0 = \cosh\zeta$.

Using this expression we can calculate the Wigner rotation $\eta_p = \kappa_p^{-1}\Lambda\kappa_{\Lambda^{-1}p}$ or $\eta_p = \kappa_{\Lambda p}^{-1}\Lambda\kappa_p$. Now, it is obvious that η_p is a member of the "little group" which in the case of $[M_\pm]$ constitutes the group of three dimensional rotations. The action of η_p on a four vector will, therefore, leave the time component invariant and we can write for $x \equiv (x^0, \mathbf{x})$, $\eta_p x = \eta_p(x^0, \mathbf{x}) = (x^0, \mathbf{x} + \mathbf{x}\times\zeta)$ where ζ is the rotation induced by the element $\eta_p \in SO(3)$. For an infinitesimal θ transformation $\boldsymbol{\zeta} = \boldsymbol{\theta}$, and for a τ transformation $\zeta = \pm\dfrac{\mathbf{p}\times\tau}{M+\omega_p}$ if we consider only first order terms in the angles. This can be established by invoking a Taylor's expansion to the first order of the Wigner rotation. For convenience, we rewrite the expression for the Wigner rotation in terms of the unitary representations of the respective operators as $\hat{Q}(\Lambda, p) = \hat{L}^{-1}(\Lambda p)\,\Lambda\hat{L}(p)$ with the identification $\hat{L}^{-1}(\Lambda p) = \hat{L}^{-1}(\kappa_{\Lambda p})$ and $\hat{L}(p) = \hat{L}(\kappa_p)$. We, then, have for infinitesimal Wigner rotations

$$\hat{Q}(\Lambda, p) = \hat{Q}(\Lambda, p)\Big|_{\substack{\omega=0\\(\Lambda=\mathbf{I})}} + \frac{\omega_{\alpha\beta}}{2}\frac{d\hat{Q}(\Lambda, p)}{d\omega_{\alpha\beta}}\Bigg|_{\substack{\omega=0\\(\Lambda=\mathbf{I})}} + \ldots$$

$$= \mathbf{I} + \frac{\omega_{\alpha\beta}}{2}\left[\frac{d\hat{L}^{-1}(\Lambda p)}{d\omega_{\alpha\beta}}\right]\Lambda\hat{L}(p)\Big|_{\substack{\omega=0\\\Lambda=\mathbf{I}}} + \frac{\omega_{\alpha\beta}}{2}\left[\hat{L}^{-1}(\Lambda p)\left(\frac{d\Lambda(\omega)}{d\omega_{\alpha\beta}}\right)\hat{L}(p)\right]\Bigg|_{\substack{\omega=0\\\Lambda=\mathbf{I}}} + \ldots$$

$$= \mathrm{I} - \frac{\omega_{\alpha\beta}}{2}\left[\hat{L}^{-1}(\Lambda p)\frac{d\hat{L}(\Lambda p)}{d\omega_{\alpha\beta}}\right]\Bigg|_{\substack{\omega=0\\\Lambda=\mathbf{I}}} + \frac{i\omega_{\alpha\beta}}{2}\,[\hat{L}^{-1}(p)\,\hat{M}^{\alpha\beta}\,\hat{L}(p)] + \ldots \tag{155}$$

with, $\Lambda(\omega) = \mathbf{I} + \dfrac{i}{2}\omega_{\mu\nu}\hat{M}^{\mu\nu}$, $\Lambda^\mu_\nu = p^\mu + \dfrac{i}{2}(\omega_{\alpha\beta}\hat{M}^{\alpha\beta})^\mu_\nu\, p^\nu = \delta^\mu_\nu + \omega^\mu_\nu$.

The matrix elements of the generators $\hat{K}^i = \hat{M}^{0i}$ and $\hat{J}^i = \dfrac{1}{2}e_{ijk}\hat{M}^{jk}$ are given by

$$(\hat{K}^i)_{ab} = i(\delta^i_a\,\delta_{0b} + \delta_{0a}\,\delta^i_b) \tag{156}$$

$$(\hat{J}^i)_{ab} = -i\varepsilon_{iab} \tag{157}$$

Writing the Lorentz boost $\hat{L}(p)$ in terms of the rapidity as $(\hat{L}(p))^i_j = \delta^i_j + \dfrac{p_i\, p_j}{\mathbf{p}^2}(\cosh\varsigma - 1)$, $(\hat{L}(p))^i_0 = (\hat{L}(p))^0_i = \dfrac{p_i}{|\mathbf{p}|}\sinh\varsigma$, and $(\hat{L}(p))^0_0 = \cosh\varsigma$, we obtain the expression for the Wigner rotation as

$$\hat{Q}(\Lambda, p) = \mathbf{I} + \frac{i}{2}\left[\omega_{ij} - \frac{1}{p^0 + M}\left(p_i\,\omega_{j0} - p_j\,\omega_{i0}\right)\right]\hat{M}^{ij} \tag{158}$$

with $\quad p^0 \equiv \sqrt{|\mathbf{p}|^2 + M^2} = \omega_p.$

We can compute the finite Wigner rotations for general Lorentz transformations by decomposing such transformation as $\Lambda = \hat{R}(\psi).\hat{L}(\alpha)$ where the former factor is a pure spatial rotation and the latter is a pure Lorentz boost. Using eq. (141), we then have $\hat{Q}(\Lambda, p) = \hat{Q}(\hat{R}(\psi), \hat{L}(\alpha)p).\hat{Q}(\hat{L}(\alpha).p)$. Now, irrespective of the momenta, Wigner angles are degenerate with ordinary rotation angles when the boost parameters are zero. Therefore, the first factor is just $(i\psi.\hat{\mathbf{J}})$, while the second factor is the Wigner rotation for an arbitrary pure boost which can be written in the form $\hat{Q}(\hat{L}(\tau), p) \equiv \exp(i\phi(\tau).\hat{\mathbf{J}})$.

$$\cos\phi = \frac{[\cosh\tau + \cosh\zeta + \sinh\tau\sinh\zeta(\hat{\tau}.\hat{\mathbf{p}}) + (\cosh\tau - 1)(\cosh\zeta - 1)(\hat{\tau}.\hat{\mathbf{p}})^2]}{[1 + \cosh\tau\cosh\zeta + \sinh\tau\sinh\zeta(\hat{\tau}.\hat{\mathbf{p}})}$$

$$= \frac{[M\cosh\tau + p^0 + \sinh\tau(\hat{\tau}.\hat{\mathbf{p}}) + (\cosh\tau - 1)(p'^0 - M)(\hat{\tau}.\hat{\mathbf{p}})^2]}{[M + p^0\cosh\tau + \sinh\tau(\hat{\tau}.\hat{\mathbf{p}})]}$$

$$(\sin\phi)\,\hat{\phi} = \frac{[\sinh\tau\sinh\zeta + (\cosh\tau - 1)(\cosh\zeta - 1)(\hat{\tau}.\hat{\mathbf{p}})]}{[1 + \cosh\tau\cosh\zeta + \sinh\tau\sinh\zeta(\hat{\tau}.\hat{\mathbf{p}})]}(\hat{\tau}\times\hat{\mathbf{p}})$$

$$= \frac{[|\mathbf{p}|\sinh\tau + (p^0 - M)(\cosh\tau - 1)(\hat{\tau}.\hat{\mathbf{p}})]}{[M + p^0\cosh\tau + p\sinh\tau(\hat{\tau}.\hat{\mathbf{p}})]}(\hat{\tau}\times\hat{\mathbf{p}})$$

The complete expression for the Wigner rotation may, therefore, be written as

$$\hat{Q}(\Lambda, p) = \exp(i\boldsymbol{\theta}_w.\hat{\mathbf{J}}) = \exp(i\psi.\hat{\mathbf{J}}).\exp(i\phi(\boldsymbol{\alpha}).\hat{\mathbf{J}}) \tag{159}$$

with

$$\cos\left(\frac{\theta_W}{2}\right) = \left(\cos\frac{\psi}{2}\right)\left(\cos\frac{\phi}{2}\right) - \left(\sin\frac{\psi}{2}\right)\left(\sin\frac{\phi}{2}\right)(\hat{\psi}.\hat{\phi})$$

$$\sin\left(\frac{\theta_W}{2}\right)\hat{\theta}_W = \left(\cos\frac{\psi}{2}\right)\left(\sin\frac{\phi}{2}\right)\hat{\phi} + \left(\sin\frac{\theta}{2}\right)\left(\cos\frac{\phi}{2}\right)\hat{\psi} + \left(\sin\frac{\psi}{2}\right)\left(\sin\frac{\phi}{2}\right)(\hat{\phi}\times\hat{\psi})$$

Full derivation using the SL(2,C) representation of the Lorentz group is given in a subsequent section.

2.4.1 Construction of the Basis Vectors for $[M_{\pm}]$

The Hilbert space Hof a massive particle can be represented as a direct sum of the eigenspaces H_P of the momentum operator, i.e., $\mathbf{H} = \otimes_{p\varepsilon\mathbb{C}^3} H_p$. In the sequel, we consider the "positive energy" subspace of the massive, $[M_{\pm}]$, representations i.e. the representation space corresponding to the representation class $[M_+]$.

To construct the basis we select the "standard vector" as $p_t^{\mu} = (p^0, \mathbf{p}) = (M, \mathbf{0})$. Since $\mathbf{p} = 0$, this scenario corresponds to the particle being in a state of rest with mass (rest energy) equal to M. The vector p_t^{μ} is invariant under all rotations in the three spatial dimensions. Hence, the "little group" of p_t is the subgroup consisting of rotations in the three dimensional space which is isomorphic to $SO(3)$. The Pauli Lubanski vector that corresponds to the second Casimir operator, in this case, becomes proportional to $\hat{\mathbf{J}}^2$, as we have already shown. Each irreducible unitary representation, labeled by the eigenvalue of $\hat{\mathbf{J}}^2$, of the rotation group $SO(3)$, being the little group of p_t, induces an irreducible unitary representation of the Poincare group. The IRRs are labeled by the eigenvalue $c_1 > 0$ of the Casimir operator $\hat{C}_1 = \hat{P}_{\mu}\hat{P}^{\mu}$ (or equivalently, by M) alongwith the labeling index $s(s + 1)$ corresponding to the operator $\hat{\mathbf{J}}^2$ of the unitary representations of the rotation group in the three spatial dimensions.

Within a representation labeled by (M, s) we shall label the basis vectors of the subspace H_0 corresponding to the rest frame eigenvalues p_t of $\hat{P}^{\mu}$ by $|\mathbf{0}\lambda\rangle$ since $p^0 = M$, $\mathbf{p} = 0$, and λ is the eigenvalue of $\hat{J}_3$. For a particular IRR, M, $s(s+1)$ remain invariant and hence, are used for labeling the IRRs. In designating the basis vectors of a particular IRR, they may be omitted in the labeling scheme of the basis vectors and we may abbreviate the basis vectors as $|\mathbf{0}\lambda\rangle$. These basis vectors are defined by the following relations:

$$\hat{P}^{\mu}|\mathbf{0}\lambda\rangle = \hat{P}_t^{\mu}|\mathbf{0}\lambda\rangle, \quad \text{where } p_t^{\mu} = (M, \mathbf{0}) \tag{160}$$

$$\hat{\mathbf{J}}^2|\mathbf{0}\lambda\rangle = s(s+1)|\mathbf{0}\lambda\rangle \tag{161}$$

$$\hat{J}_3|\mathbf{0}\lambda\rangle = \lambda|\mathbf{0}\lambda\rangle \tag{162}$$

As mentioned above, this subspace, H_0, is invariant under the action of the little group being the group of rotations in the three spatial dimensions. It has the following properties:

(i) It is invariant with respect to rotations for

$$\hat{\mathbf{P}}(\hat{\mathbf{I}} - i\hat{\mathbf{J}}\theta|\mathbf{0}\lambda\rangle = \hat{\mathbf{P}}|\mathbf{0}\lambda\rangle - i\hat{\mathbf{P}}\hat{\mathbf{J}}\theta|\mathbf{0}\lambda\rangle = 0 \tag{163}$$

showing that the rotated vector has also zero momentum and hence belongs to H_0. In arriving at this result, we have used the commutation relation $[\hat{p}_m, \hat{J}_n] = i\varepsilon^{mnl}\hat{P}_l$;

(ii) If we define the relativistic spin operator for positive energy massive particles as

$$\hat{S} = \frac{1}{M}\hat{P}^0\hat{\mathbf{J}} - \frac{1}{M}\hat{\mathbf{P}} \times \hat{\mathbf{K}} - \frac{1}{(\hat{P}_0 + M)M}\hat{\mathbf{P}}(\hat{\mathbf{P}}.\hat{\mathbf{J}}), \text{ then}$$

$$\hat{S}_3|\mathbf{0}\lambda\rangle = \left[\frac{1}{M}\hat{P}^0\hat{J}_3 - \frac{1}{M}(\hat{P}_1\hat{K}_2 - \hat{P}_2\hat{K}_1) - \frac{1}{(\hat{P}_0 + M)M}\hat{P}_3(\hat{\mathbf{P}}.\hat{\mathbf{J}})\right]|\mathbf{0}\lambda\rangle$$

$$= \left[\frac{1}{M}\hat{J}_3\hat{P}^0 - \frac{1}{M}(\hat{K}_2\hat{P}_1 - \hat{K}_1\hat{P}_2) - \frac{1}{(\hat{P}_0 + M)M}(\hat{\mathbf{P}}.\hat{\mathbf{J}})\hat{P}_3\right]|\mathbf{0}\lambda\rangle = \frac{1}{M}\hat{J}_3\hat{P}^0|\mathbf{0}\lambda\rangle = \hat{J}_3|\mathbf{0}\lambda\rangle \tag{164}$$

thereby showing that for the subspace H_0 relating to the rest frame of the massive particle, we have. Here we have used the commutators $[\hat{P}_0, \hat{J}_n] = 0$, $[\hat{P}_m, \hat{K}_n] = i\delta_{mn}\hat{P}^0$.

(iii) A basis can be constructed in H_0 consisting of the eigenvectors of $\hat{S}_3$ and labeled by the eigenvalue of $\hat{S}_3 \equiv \hat{J}_3$, e.g., $|0\lambda$ where $\hat{J}_3|0\lambda\rangle = \hat{S}_3|0\lambda\rangle = \lambda|0\lambda\rangle$. The action of a finite rotation operator on these basis vectors is

$$e^{-i\hat{\mathbf{J}}\varphi}|\mathbf{0}\lambda\rangle = e^{-i\hat{S}\varphi}|0\lambda\rangle = \sum_{\lambda'=-s}^{s} D^s_{\lambda',\lambda}(\varphi)|0\lambda\rangle \tag{165}$$

where D^s are the $(2s + 1) \times (2s + 1)$ rotation matrices.

In order to generate a complete basis of this irreducible representation consisting of general eigenvectors of $\hat{P}^\mu$, we shall operate on the basis vector $|\mathbf{0}\lambda\rangle$ by the remaining transformations of the factor group. We note that the factor group of the Poincare group with respect to the full translation group is the homogeneous Lorentz group. We also know that the homogeneous Lorentz group comprises rotations in the three spatial coordinates and Lorentz "boosts" that involve the time axis. Since the standard vector p_t^μ is invariant under rotations, operating on p_t^μ by rotations will not lead to new basis elements (the subgroup consisting of rotations in the three-dimensional space which is isomorphic to $SO(3)$ constitutes the "little group" of p_t^μ). Rotations simply transform a basis vector to another existing basis vector, i.e., they do not generate new basis vectors. Therefore, for constructing our basis, we need now to implement the effect of Lorentz "boosts" on the basis vector $|0\lambda\rangle$. The basis vectors $|\mathbf{0}\lambda\rangle$ of H_0 can be transformed by pure boost transformations to generate basis vectors $|\mathbf{p}\lambda\rangle$ of H_P. The appropriate boost transformation that does the trick is $\lambda_p \equiv e^{-i\hat{K}\theta_p} \approx (\hat{\mathbf{I}} - i\hat{\mathbf{K}}\theta_\mathbf{p})$ where $\theta_\mathbf{p} = \frac{\mathbf{p}}{|\mathbf{p}|}\sinh^{-1}\frac{|\mathbf{p}|}{M}$ so that, as a first order approximation $\theta_\mathbf{p} \approx \frac{\mathbf{p}}{M}$. Hence, $|\mathbf{p}\lambda\rangle = N(\mathbf{p})\lambda_\mathbf{p}|\mathbf{0}\lambda\rangle = \mathrm{N}(\mathbf{p})\, e^{-i\hat{\mathbf{k}}\theta_\mathbf{p}}|0\lambda\rangle \approx N(\mathbf{p})\,(\hat{\mathbf{I}} - i\hat{\mathbf{K}}\theta_\mathbf{p})|0\lambda\rangle$ (with $N(p)$ being a normalization factor. We, then have,

$$\begin{aligned}
\hat{\mathbf{P}}|\mathbf{p}\lambda\rangle &= N(\mathbf{p})\hat{\mathbf{P}}e^{-i\hat{\mathbf{K}}\theta_\mathbf{p}}|\mathbf{0}\lambda\rangle = N(\mathbf{p})\, e^{-i\hat{\mathbf{k}}\theta_\mathbf{p}}\, e^{+i\hat{\mathbf{k}}\theta_\mathbf{p}}\, \hat{\mathbf{P}}e^{-i\hat{\mathbf{k}}\theta_\mathbf{p}}|0\lambda\rangle \\
&\approx N(\mathbf{p})\,(\hat{\mathbf{I}} - i\hat{\mathbf{K}}\theta_\mathbf{p})\,[(\hat{\mathbf{I}} + i\hat{\mathbf{K}}\theta_\mathbf{p})\,\hat{\mathbf{P}}(\hat{\mathbf{I}} - i\hat{\mathbf{K}}\theta_\mathbf{p})]|\mathbf{0}\lambda\rangle \\
&\approx N(\mathbf{p})\,(\hat{\mathbf{I}} - i\hat{\mathbf{K}}\theta_\mathbf{p})\,[\hat{\mathbf{P}} + i[\hat{\mathbf{K}}\theta_\mathbf{p}, \hat{\mathbf{P}}]]|\mathbf{0}\lambda\rangle = N(\mathbf{p})\,(\hat{\mathbf{I}} - i\hat{\mathbf{K}}\boldsymbol{\theta}_\mathbf{p})[\hat{\mathbf{P}} + \hat{P}^0\boldsymbol{\theta}_\mathbf{p}]|\mathbf{0}\lambda\rangle \\
&= N(\mathbf{p})\,(\hat{\mathbf{I}} - i\hat{\mathbf{K}}\theta_\mathbf{p})\, M\theta_p|\mathbf{0}\lambda\rangle \approx N(\mathbf{p})\,(\hat{\mathbf{I}} - i\hat{\mathbf{K}}\theta_\mathbf{p})\mathbf{p}|\mathbf{0}\lambda\rangle \\
&= \mathbf{p}N(\mathbf{p})\,(\hat{\mathbf{I}} - i\hat{\mathbf{K}}\theta_\mathbf{p})|\mathbf{0}\lambda\rangle = \mathbf{p}|\mathbf{p}\lambda\rangle
\end{aligned} \tag{166}$$

confirming that $|\mathbf{p}\lambda\rangle$ is an eigenstate of the momentum operator with momentum $\mathbf{p}$. Now, consider

$$\hat{S}_3|\mathbf{p}\lambda\rangle = N(\mathbf{p})\,\hat{S}_3\,e^{-i\hat{\mathbf{k}}\,\theta_{\mathbf{p}}}|\mathbf{0}\lambda\rangle = N(\mathbf{p})\,e^{-i\hat{\mathbf{k}}\,\theta_{\mathbf{p}}}\,e^{-i\hat{\mathbf{k}}\,\theta_{\mathbf{p}}}\,\hat{S}_3\,e^{-i\hat{\mathbf{k}}\,\theta_{\mathbf{p}}}|\mathbf{0}\lambda\rangle$$
$$\approx N(\mathbf{p})\,(\hat{\mathbf{I}} - i\hat{\mathbf{K}}\theta_{\mathbf{p}})\,[(\hat{\mathbf{I}} + i\hat{\mathbf{K}}\theta_{\mathbf{p}})\,\hat{S}_3(\hat{\mathbf{I}} - i\hat{\mathbf{K}}\theta_{\mathbf{p}})]|\mathbf{0}\lambda\rangle$$
$$\approx N(\mathbf{p})\,(\hat{\mathbf{I}} - i\hat{\mathbf{K}}\theta_{\mathbf{p}})\,(\hat{S}_3 + i[\hat{\mathbf{K}}\theta_{\mathbf{p}},\,\hat{S}_3])|\mathbf{0}\lambda\rangle \tag{167}$$

We have $\hat{S}_3|0\lambda\rangle = \lambda|\mathbf{0}\lambda\rangle$ by definition of the basis vector, since it is labelled by the eigenvalue of $\hat{S}_3$. To determine $[\hat{\mathbf{K}}\theta_p,\,\hat{S}_3]|\mathbf{0}\lambda\rangle$, we make use of the explicit representation of the spin operator as

$$\hat{S}_3 = \frac{1}{M}\hat{P}^0\hat{J}_3 - \frac{1}{M}(\hat{P}_1\,\hat{K}_2 - \hat{P}_2\,\hat{K}_1) - \frac{1}{(\hat{P}_0 + M)M}\,\hat{P}_3(\hat{\mathbf{P}}.\hat{\mathbf{J}}) \tag{168}$$

We, then, have $[\hat{K}_1\,\theta^1_{\mathbf{p}},\,\hat{P}_0\,\hat{J}_3]|\mathbf{0}\lambda\rangle = \{\hat{P}_0[\hat{K}_1\theta^1_{\mathbf{p}},\,\hat{J}_3] + [\hat{K}_1\theta^1_{\mathbf{p}},\,\hat{P}_0]\hat{J}_3\}|\mathbf{0}\lambda\rangle$

$$= (-\,i\hat{P}_0\,\hat{K}_2\,\theta^1_{\mathbf{p}} - i\hat{P}_1\theta^1_{\mathbf{p}}\hat{J}_3)|\mathbf{0}\lambda\rangle = (-i\hat{P}_0\,\hat{K}_2\,\theta^1_{\mathbf{p}}|\mathbf{0}\lambda\rangle = -\,i(i\hat{P}_2 + \hat{K}_2\,\hat{P}_0)\theta^1_{\mathbf{p}}|\mathbf{0}\lambda\rangle$$
$$= -iM\hat{K}_2\,\theta^1_{\mathbf{p}}|\mathbf{0}\lambda\rangle$$

$$[\hat{K}_2\,\theta^2_{\mathbf{p}},\,\hat{P}_0\,\hat{J}_3]|\mathbf{0}\lambda\rangle = \{\hat{P}_0[\hat{K}_2\,\theta^2_{\mathbf{p}},\,\hat{J}_3] + [\hat{K}_2\,\theta^2_{\mathbf{p}},\,\hat{P}_0]\hat{J}_3\}|\mathbf{0}\lambda\rangle$$
$$= (i\hat{P}_0\,\hat{K}_1\,\theta^2_{\mathbf{p}} - i\hat{P}_2\,\theta^2_{\mathbf{p}}\,\hat{J}_3)|\mathbf{0}\lambda\rangle = (i\hat{P}_0\,\hat{K}_1\,\theta^2_{\mathbf{p}})|\mathbf{0}\lambda\rangle = i(i\hat{P}_1 + \hat{K}_1\,\hat{P}_0)\theta^2_{\mathbf{p}}|\mathbf{0}\lambda\rangle$$
$$= iM\hat{K}_1\,\theta^2_{\mathbf{p}}|\mathbf{0}\lambda\rangle$$

$$[\hat{K}_3\,\theta^3_{\mathbf{p}},\,\hat{P}_0\,\hat{J}_3]|\mathbf{0}\lambda\rangle = \{\hat{P}_0[\hat{K}_3\,\theta^3_{\mathbf{p}},\,\hat{J}_3] + [\hat{K}_3\,\theta^3_{\mathbf{p}},\,\hat{P}_0]\hat{J}_3\}|\mathbf{0}\lambda\rangle$$
$$= (-i\hat{P}_3\,\theta^3_{\mathbf{p}}\,\hat{J}_3|\mathbf{0}\lambda\rangle = 0 \quad \text{so that } [\hat{K}\theta_{\mathbf{p}},\,\hat{P}_0\hat{J}_3]|\mathbf{0}\lambda\rangle = iM\,[\hat{K}_1\,\theta^2_{\mathbf{p}} - \hat{K}_2\,\theta^1_{\mathbf{p}}]|\mathbf{0}\lambda\rangle.$$

Similarly, $[\hat{K}_1\,\theta^1_{\mathbf{p}},\,\hat{P}_1\,\hat{K}_2]|\mathbf{0}\lambda\rangle = \{\hat{P}_1[\hat{K}_1\,\theta^1_{\mathbf{p}},\,\hat{K}_2] + [\hat{K}_1\,\theta^1_{\mathbf{p}},\,\hat{P}_1]\hat{K}_2\}|\mathbf{0}\lambda\rangle$

$$= (-i\hat{P}_1\,\hat{J}_3\theta^1_{\mathbf{p}} - i\hat{P}_0\,\theta^1_{\mathbf{p}}\,\hat{K}_2)|\mathbf{0}\lambda\rangle = (-i\hat{P}_0\hat{K}_2\theta^1_{\mathbf{p}})|\mathbf{0}\lambda\rangle$$
$$= -i(i\hat{P}_2 + \hat{K}_2\,\hat{P}_0)\theta^1_{\mathbf{p}}|\mathbf{0}\lambda\rangle = -iM\hat{K}_2\,\theta^1_{\mathbf{p}}|\mathbf{0}\lambda\rangle$$

$$[\hat{K}_2\,\theta^2_{\mathbf{p}},\,\hat{P}_1\hat{K}_2]|\mathbf{0}\lambda\rangle = \{\hat{P}_1[\hat{K}_2\,\theta^2_{\mathbf{p}},\,\hat{K}_2] + [\hat{K}_2\,\theta^2_{\mathbf{p}},\,\hat{P}_1]\hat{K}_2\}|\mathbf{0}\lambda\rangle = 0$$
$$[\hat{K}_3\,\theta^3_{\mathbf{p}},\,\hat{P}_1\hat{K}_2]|\mathbf{0}\lambda\rangle = \{\hat{P}_1[\hat{K}_3\,\theta^3_{\mathbf{p}},\,\hat{K}_2] + [\hat{K}_3\,\theta^3_{\mathbf{p}},\,\hat{P}_1]\hat{K}_2\}|\mathbf{0}\lambda\rangle$$
$$= (i\hat{P}_1\hat{J}_1\,\theta^3_{\mathbf{p}})|\mathbf{0}\lambda\rangle = 0$$

so that $\quad [\hat{K}\theta_{\mathbf{p}},\,\hat{P}_1\hat{K}_2]|\mathbf{0}\lambda\rangle = -iM\hat{K}_2\,\theta^1_{\mathbf{p}}|\mathbf{0}\lambda\rangle.$

Similarly, $[\hat{K}\theta_{\mathbf{p}},\,\hat{P}_2\hat{K}_1]|\mathbf{0}\lambda\rangle = -iM\hat{K}_1\,\theta^2_{\mathbf{p}}|\mathbf{0}\lambda\rangle$ and

$$[\hat{K}\theta_{\mathbf{p}},\,\hat{P}_3(\hat{\mathbf{P}}.\hat{\mathbf{J}})]|\mathbf{0}\lambda\rangle = \{\hat{P}_3[\hat{K}\boldsymbol{\theta}_{\mathbf{p}},\,(\hat{\mathbf{P}}.\hat{\mathbf{J}})] + [\hat{K}\theta_{\mathbf{p}},\,\hat{P}_3](\hat{\mathbf{P}}.\hat{\mathbf{J}})\}|\mathbf{0}\lambda\rangle = 0$$

Putting all these pieces together, we get

$$\hat{S}_3|\mathbf{p}\lambda\rangle = \lambda N(\mathbf{p})\,(\hat{\mathbf{I}} - i\hat{\mathbf{K}}\theta_{\mathbf{p}})|\mathbf{0}\lambda\rangle = \lambda|\mathbf{p}\lambda\rangle \tag{169}$$

showing that $|\mathbf{p}\lambda\rangle$ is an eigenstate of $\hat{S}_3$ with eigenvalue λ. The effect of various transformations constituting the Poincare group on the basis vectors is summarized thus:

(a) Translations: We have

$$e^{-i\mathbf{P}a}|\mathbf{p}\lambda\rangle = e^{-i\mathbf{p}a}|\mathbf{p}\lambda\rangle,\ e^{i\hat{H}x^0}|\mathbf{p}\lambda\rangle = e^{i\omega_{\mathbf{p}}x^0}|\mathbf{p}\lambda\rangle \tag{170}$$

with $\omega_{\mathbf{p}}^2 = \mathbf{p}^2 + M^2$;

(b) Spatial Rotations: We have $e^{-i\hat{\mathbf{J}}\varphi}|\mathbf{p}\lambda\rangle = N(\mathbf{p})e^{-i\hat{\mathbf{J}}\varphi}\, e^{-i\hat{\mathbf{K}}\theta_{\mathbf{p}}}|\mathbf{0}\lambda\rangle$

$$= N(\mathbf{p})e^{-i\hat{\mathbf{J}}\varphi}\, e^{-i\hat{\mathbf{K}}\theta_{\mathbf{p}}} e^{i\hat{\mathbf{J}}\varphi} e^{-i\hat{\mathbf{J}}\varphi}|\mathbf{0}\lambda\rangle = N(\mathbf{p})e^{-i(R(\varphi)^{-1}\hat{\mathbf{K}})\theta_{\mathbf{p}}} \sum_{\lambda'=-s}^{s} D_{\lambda'\lambda}(\varphi)|\mathbf{0}\lambda'\rangle$$

$$= N(\mathbf{p})e^{-i\hat{\mathbf{K}}R(\varphi)\theta_{\mathbf{p}}} \sum_{\lambda'=-s}^{s} D_{\lambda'\lambda}(\varphi)|0\lambda'\rangle = \sum_{\lambda'=-s}^{s} D_{\lambda'\lambda}(\varphi)|R(\varphi)\mathbf{p},\lambda'\rangle \tag{171}$$

where we have used

$$\hat{U}(R)\,\hat{P}_i|\mathbf{p}\rangle = \hat{U}(R)\,\hat{P}_i\hat{U}(R)^{-1}\,\hat{U}(R)|\mathbf{p}\rangle = \hat{U}(R)\hat{P}_i\,\hat{U}(R)^{-1}|\mathbf{p}'\rangle$$

$$= \hat{U}(R)\,p_i|\mathbf{p}\rangle = p_i|\mathbf{p}'\rangle = \sum_j (R^{-1})_i^j\, p_j'|\mathbf{p}'\rangle = \sum_j (R^{-1})_i^j\, \hat{P}_j|\mathbf{p}'\rangle \text{ whence}$$

$$\hat{U}(R)\hat{P}_i\hat{U}(\mathrm{R})^{-1} = \sum_j (R^{-1})_i^j\, \hat{P}_j \tag{172}$$

(c) Lorentz Boosts: Let us apply a Lorentz boost Λ to a basis vector $|\mathbf{p}\lambda\rangle$ to obtain $\Lambda|\mathbf{p}\lambda\rangle = \Lambda N(\mathbf{p})\tau_{\mathbf{p}}|0\lambda\rangle$ where $\tau_{\mathbf{p}}$ transforms a vector of zero momentum to a vector of momentum $\mathbf{p}$. Now, the right hand side, being the product of two boosts, is also a Lorentz transformation and hence, can be represented by the product of a spatial rotation followed by a boost, i.e., $\Lambda|\mathbf{p}\lambda\rangle = N(\mathbf{p})\,\Lambda\tau_{\mathbf{p}}|0\lambda\rangle = N(\mathbf{p})\tau_{\mathbf{p}}.R(\varphi_W(\mathbf{p},\Lambda))|0\lambda\rangle$ or equivalently $\tau^{-1}_{\mathbf{p}}.\Lambda\tau_{\mathbf{p}}|0\lambda\rangle = R(\varphi_W(\mathbf{p},\Lambda)|0\lambda\rangle$. Now, we have shown above that a rotation keeps invariant the subspace of zero momentum. It follows that the sequence of boosts on the left hand side must return each vector of zero momentum to the subspace of zero momentum. Now, $\tau_{\mathbf{p}}$ transforms a vector of zero momentum to a vector of momentum $\mathbf{p}$. Subsequent application of the boost Λ would transform this vector's momentum to $\Lambda\mathbf{p}$. It follows that the boost $\tau_{\mathbf{p}'}$ will transform the vector with momentum $\Lambda\mathbf{p}$ back to zero, i.e, $\tau_{\mathbf{p}'} = \tau_{\Lambda\mathbf{p}}$. Therefore,

$$e^{-i\hat{\mathbf{K}}\theta}|\mathbf{p}\lambda\rangle = N(\mathbf{p})\,\Lambda\tau_{\mathbf{p}}|0\lambda\rangle = N(\mathbf{p})\,\tau_{\Lambda\mathbf{p}}\,R(\varphi_W(\mathbf{p}.\Lambda))|0\lambda\rangle$$

$$= N(\mathbf{p})\,\tau_{\Lambda\mathbf{p}} \sum_{\lambda'=-s}^{s} D^s_{\lambda'\lambda}(\varphi_W(\mathbf{p},\Lambda))|0\lambda'\rangle$$

$$= \frac{N(\mathbf{p})}{N(\Lambda\mathbf{p})} \sum_{\lambda'=-s}^{s} D^{s}_{\lambda'\lambda} \, (\varphi_W(\mathbf{p}, \Lambda))|\Lambda\mathbf{p}, 0\lambda'\rangle \tag{173}$$

where $D^{s}_{\lambda'\lambda}(\varphi_W(\mathbf{p}, \Lambda))$ are the $(2s + 1) \times (2s + 1)$ unitary rotation matrix representations of the Wigner rotations.

(d) Arbitrary Poincare Group Element-Consider the action of an arbitrary Poincare group element $\hat{U}(\Lambda, x)$ on the basis vector $|\mathbf{p}\lambda\rangle$ (where we assume that the translation is performed first followed by rotations/boosts), We, then, have

$$\hat{U}(\Lambda, x)|\mathbf{p}\lambda\rangle = \hat{U}(\Lambda, 0)\, e^{-i\hat{P}_\mu x^\mu}|\mathbf{p}\lambda\rangle = \frac{N(\mathbf{p})}{N(\Lambda\mathbf{p})} e^{-ip_\mu x^\mu} \sum_{\lambda'=-s}^{s} D_{\lambda'\lambda} \, (\varphi_W(\mathbf{p}, \Lambda))|\Lambda\mathbf{p}, \lambda'\rangle$$

To determine the normalization factor $N(\mathbf{p})$ we invoke the requirement of invariance of identity under Lorentz boosts, i.e., $\mathbf{I} = e^{-i\hat{\mathbf{K}}\theta}\, \mathbf{I} e^{i\hat{\mathbf{K}}\theta}$. Introducing the spectral resolution of the identity in the momentum representation $\mathbf{I} = \int d\mathbf{p}|\mathbf{p}\rangle\langle\mathbf{p}|$ and omitting the spin indices for brevity, we have $\mathbf{I} = e^{-i\hat{\mathbf{K}}\theta} \left(\int d\mathbf{p}|\mathbf{p}\rangle\langle\mathbf{p}|\right) e^{i\hat{\mathbf{K}}\theta}$

$$= \int d\mathbf{p} \left|\frac{N(\mathbf{p})}{N(\Lambda\mathbf{p})}\right|^2 |\Lambda\mathbf{p}\rangle\langle\Lambda\mathbf{p}| = \int d(\Lambda\mathbf{p}) \det\left|\frac{d(\mathbf{p})}{d(\Lambda\mathbf{p})}\right| \left|\frac{N(\mathbf{p})}{N(\Lambda\mathbf{p})}\right|^2 |\Lambda\mathbf{p}\rangle\langle\Lambda\mathbf{p}|$$

Since the Jacobian of the transformation $\mathbf{p} \to \Lambda\mathbf{p}$ viz. $\det\left|\frac{d(\mathbf{p})}{d(\Lambda\mathbf{p})}\right|$ should not depend on the direction of the boost, we choose a boost along the direction of x^3 axis so that

$$p_1 = (\Lambda p)_1,\; p_2 = (\Lambda p)_2,\; p_3 = (\Lambda p)_3 \cosh\theta - \omega_{\Lambda p} \sinh\theta$$

$$\omega_p = \sqrt{M^2 + (\Lambda p)_1^2 + (\Lambda p)_2^2 + [(\Lambda p)_3 \cosh\theta - \omega_{\Lambda p}\sinh\theta]^2}$$

$\omega_{\Lambda p} = \cosh\theta - (\Lambda_p)_3 \sinh\theta$. This gives

$$\det\left|\frac{d(\mathbf{p})}{d(\Lambda\mathbf{p})}\right| = \det\begin{pmatrix} 1 & 0 & 0 \\ 0 & 1 & 0 \\ \frac{(\Lambda p)_1 \sinh\theta}{\omega_{(\Lambda p)}} & \frac{(\Lambda p)_2 \sinh\theta}{\omega_{(\Lambda p)}} & \cosh\theta - \frac{(\Lambda p)_3 \sinh\theta}{\omega_{(\Lambda p)}} \end{pmatrix}$$

$$= \cosh\theta - \frac{(\Lambda p)_3 \sinh\theta}{\omega_{\Lambda p}} = \frac{\omega_p}{\omega_{\Lambda p}}$$ whence we have $N(\mathbf{p}) = \omega_p^{-1/2}$.

2.5 TRANSFORMATION RULES FOR THE WAVEFUNCTIONS (MOMENTUM REPRESENTATION)

Two different eigenstates of the momentum operator, being different eigenstates of a Hermitian operator must necessarily be orthogonal, i.e., $\langle \mathbf{p}|\mathbf{p}'\rangle = 0$ if $\mathbf{p} \neq \mathbf{p}'$. Had the spectrum of eigenvalues $Spec\{\mathbf{p}\}$ been discrete, we would have the simple normalization $\langle \mathbf{p}|\mathbf{p}'\rangle = \delta_{\mathbf{pp}'}$. However, such a normalization is obviously not possible for continuous spectra. To obviate this problem and obtain normalizable vectors, we introduce momentum space wavefunctions $\psi(\mathbf{p})$ and write an arbitrary state vector as $|\Psi\rangle = \int d\mathbf{p}\psi(\mathbf{p})|\mathbf{p}\rangle$ with the normalization $\langle \mathbf{p}|\mathbf{p}'\rangle = \delta(\mathbf{p} - \mathbf{p}')$ so $\langle \mathbf{p}|\psi\rangle = \langle \mathbf{p}|\int d\mathbf{p}'\, \psi(\mathbf{p}')|\mathbf{p}'\rangle = \int d\mathbf{p}'\, \psi(\mathbf{p}')\, \langle \mathbf{p}'|\mathbf{p}\rangle = \int d\mathbf{p}'\, \psi(\mathbf{p}')\, \delta(\mathbf{p} - \mathbf{p}') = \psi(\mathbf{p})$ where we have, for the time being omitted the spin index and identified the eigenstates only by the momentum eigenvalue.

Transformation properties of the momentum space wavefunction $\psi(\mathbf{p})$ are summarized below:

$$\text{(i)}\ \hat{P}_\mu\, \psi(\mathbf{p}) = \langle \mathbf{p}|\hat{P}_\mu|\psi\rangle = \langle \mathbf{p}|\hat{P}_\mu^\dagger|\psi\rangle = p_\mu\langle \mathbf{p}|\psi\rangle = p_\mu\psi(\mathbf{p}) \tag{174}$$

$$\text{(ii)}\ \hat{J}_i\, \psi(\mathbf{p}) = (\hat{x}^2\hat{P}_3 - \hat{x}^3\hat{P}_2)\, \psi(p) = i\left(p_3\frac{d}{dp_2} - p_2\frac{d}{dp_3}\right)\psi(\mathbf{p}) \tag{175}$$

$$\text{(iii)}\ \hat{K}_1\, \psi(\mathbf{p}) = i\text{Lim}_{\theta\to 0}\frac{d}{d\theta}[e^{-i\hat{K}_1\theta}\, \psi(\mathbf{p})] = i\text{Lim}_{\theta\to 0}\frac{d}{d\theta}\, [\langle \mathbf{p}|e^{-i\hat{K}_1\theta}|\Psi\rangle]$$

$$= i\text{Lim}_{\theta\to 0}\frac{d}{d\theta}[\langle \mathbf{p}|e^{(i\hat{K}_1\theta)\dagger}|\Psi\rangle] = i\text{Lim}_{\theta\to 0}\frac{d}{d\theta}\, [\langle \mathbf{p}|(e^{i\hat{K}_1\theta})^\dagger|\Psi\rangle]$$

$$= i\text{Lim}_{\theta\to 0}\frac{d}{d\theta}\left[\sqrt{\frac{\omega_{\Lambda^{-1}p}}{\omega_p}}\left\langle \Lambda^{-1}\mathbf{p}, \lambda\,|\,\Psi\right\rangle\right] = i\text{Lim}_{\theta\to 0}\frac{d}{d\theta}\left[\sqrt{\frac{\omega_{\Lambda^{-1}p}}{\omega_p}}\,\psi(\Lambda^{-1}\mathbf{p})\right]$$

$$= i\text{Lim}_{\theta\to 0}\frac{d}{d\theta}\left[\sqrt{\frac{\sqrt{\mathbf{p}^2 + M^2}\cosh\theta - p_1\sinh\theta}{\sqrt{\mathbf{p}^2 + M^2}}}\,\psi(p_1\cosh\theta - \sqrt{\mathbf{p}^2 + M^2}\sinh\theta, p_2, p_3)\right]$$

$$= -i\left[\omega_p\frac{d}{dp_1} + \frac{p_1}{2\omega_p}\right]\psi(\mathbf{p}) \tag{176}$$

With explicit spin dependence, the basis vectors $|\mathbf{p}\lambda\rangle$ transform under a Lorentz transformation as $U[\Lambda]|\mathbf{p}\lambda\rangle = \sum_\sigma D_{\sigma\lambda}\, [W(\Lambda, p)]|\Lambda\mathbf{p}, \sigma\rangle$ where $D_{\sigma\lambda}$ are the matrix elements of the unitary operator D that corresponds to the Wigner rotation $W(\Lambda, p)$. Let $|\Psi\rangle = \sum_\lambda \int |\,p\lambda\rangle\psi^\lambda\, [\mathbf{p}]\, dp$ be a quantum state, so that the wavefunction $\psi^\lambda(\mathbf{p}) = \langle \mathbf{p}\lambda|\Psi\rangle$ whence

$$|\Psi'\rangle = \hat{U}[\Lambda]\Psi\rangle = \sum_{\sigma,\nu}\int |\Lambda\mathbf{p},\sigma\rangle\, D^s[W(\Lambda,p)]^\sigma_\nu\, \psi^\nu(\mathbf{p})dp$$

$$\sum_{\sigma,\nu}\int |\mathbf{p},\sigma\rangle\, D^s\,[W(\Lambda,\Lambda^{-1}\,p)]^\sigma_\nu\, \psi^\nu(\Lambda^{-1}\,\mathbf{p})\,dp = \sum_{\lambda}\int |\mathbf{p}\lambda\rangle\, \psi'^\lambda(\mathbf{p})\,dp$$

whence we get

$$\Psi'^\lambda(\mathbf{p}) = \sum_{\lambda'} D^s\,[W[\Lambda,\Lambda^{-1}\,p)]^\lambda_{\lambda'}\, \psi^{\lambda'}(\Lambda^{-1}\,\mathbf{p}) \tag{177}$$

which elucidates the rule for the transformation of multicomponent wavefunctions under arbitrary Lorentz transformations.

2.6 TRANSFORMATION OF VECTORS

Consider an operator $\hat{P}$. We have

$$\begin{aligned}\hat{U}(\Lambda)\,\hat{P}^\mu|\mathbf{p}\rangle &= \hat{U}(\Lambda)\hat{P}^\mu\,\hat{U}(\Lambda)^{-1}\,\hat{U}(\Lambda)|\mathbf{p}\rangle = \hat{U}(\Lambda)\,\hat{P}^\mu\hat{U}(\Lambda)^{-1}|\mathbf{p}'\rangle \\ &= \hat{U}(\Lambda)\,p^\mu|\mathbf{p}\rangle = p^\mu|\mathbf{p}'\rangle = (\Lambda^{-1})^\mu_\nu\,p'^\nu|\mathbf{p}'\rangle = (\Lambda^{-1})^\mu_\nu\,\hat{P}^\nu|\mathbf{p}'\rangle\end{aligned}$$

whence

$$\hat{U}(\Lambda)\,\hat{P}^\mu\,\hat{U}(\Lambda)^{-1} = (\Lambda^{-1})^\mu_\nu\,\hat{P}^\nu \tag{178}$$

For a covariant vector, we have $\hat{U}(\Lambda)\,\hat{p}^\mu\,\hat{U}(\Lambda)^{-1} = \hat{p}_\nu\,\Lambda^\nu_\mu$. These vectors transform according to the $(u, v) \equiv \left(\frac{1}{2},\frac{1}{2}\right)$ representation of the proper Lorentz group because under three-dimensional rotations, the time component is unchanged (corresponding to $j_0 = 0$) and the spatial components transform as a 3-vector (corresponding to $j_1 = 1$) whence $u = \frac{j_1 + j_0}{2} = \frac{1}{2}$ and $v = \frac{j_1 - j_0}{2} = \frac{1}{2}$. The representation matrices corresponding to the $(u, v) \equiv \left(\frac{1}{2},\frac{1}{2}\right)$ representation are the $SL(2)$ matrices.

2.7 TRANSFORMATION OF SECOND RANK ANTISYMMETRIC TENSORS

The generators of the proper Lorentz group $\hat{J}_{\mu\nu}$ are second rank antisymmetric tensors. It is, therefore, important to study their transformation properties.

Let us construct a basis for the space of the second rank tensors in Minkowski space. Let $\{e_\mu, \mu = 0, 1, 2, 3\}$ constitute a basis for the Minkowski space. Second rank tensors over this space form a 16-dimensional space that is spanned by $e_\mu \otimes e_\nu$. Using these basis vectors, we constitute the six antisymmetric tensors $e_{[\mu\nu]} := e_\mu \otimes e_\nu - e_\nu \otimes e_\mu$. The set of these six antisymmetric tensors constitutes a basis set for the space of second rank antisymmetric tensors and we can write an arbitrary second rank antisymmetric tensor in terms of these basis vectors as $\hat{T} = T^{\mu\nu} e_{[\mu\nu]}$. Let us designate these six basis vectors as $\{f_i, i = 1, 2, \ldots 6\}$.

Our first step is to prove that the space of rank two antisymmetric tensors is invariant under the action of Lorentz transformations. For the purpose we write $\hat{T}^{\mu}_{\nu} = \frac{T^{\mu}_{\nu} - T^{\nu}_{\mu}}{2}$ so that $\hat{T}'^{\mu}_{\nu} = [\hat{U}(\Lambda)\hat{T}\hat{U}(\Lambda)^{-1}]^{\mu}_{\nu}$

$$= \frac{1}{2}\{\hat{U}(\Lambda)^{\mu}_{\lambda}\, T^{\lambda}_{\kappa}\,[\hat{U}(\Lambda)^{-1}]^{\kappa}_{\nu} - \hat{U}(\Lambda)^{\nu}_{\rho}\, T^{\rho}_{\sigma}\,[\hat{U}(\Lambda)^{-1}]^{\sigma}_{\mu}\} = \frac{1}{2}(T'^{\mu}_{\nu} - T'^{\nu}_{\mu})$$

which is antisymmetric so that the subspace of antisymmetric second rank tensors is invariant under Lorentz transformations. Having established the invariance of the relevant subspace we shall now show that it admits the representation $(1, 0) \otimes (0, 1)$ by explicitly decomposing this space into a direct sum of minimal invariant subspaces.

We, now, examine the behaviour of the basis tensors $e_{[\mu\nu]}$ under various Lorentz transformations.

We first consider pure rotations. Let us examine the purely spatial components of $e_{[\mu\nu]}$ viz. $e_{[12]}, e_{[23]}, e_{[31]}$. Let us write $e_i = \frac{1}{2}\varepsilon^{ijk} e_{jk}$. Its image under an $SO(3)$ transformation will be $e'_k = \frac{1}{2}\varepsilon^{ij}_k e'_{ij} = \frac{1}{2}\varepsilon^{ij}_k R^l_i e_{lm} R^m_j$. Multiplying the right hand side by $\delta^n_k = R^n_a R^a_k$ (which follows from the orthogonality of the rotation matrices), renaming $k \to n$ and then summing over n, we get $e'_k = \frac{1}{2}\varepsilon^{ij}_n R^l_i e_{lm} R^m_j \delta^n_k = \frac{1}{2}\varepsilon^{ij}_n R^l_i e_{lm} R^m_j R^n_a R^a_k$. Noting that for a rotation matrix we have $R^i_l R^j_m R^k_n \varepsilon^{lmn} \varepsilon^{ijk}$ whence

$R^l_i R^m_j R^n_a \varepsilon^{ij}_n = R^l_i R^m_j R^n_a \varepsilon^{ija} \delta_{an} = \varepsilon^{lmn} \delta_{an}$ so that $e'_k = \frac{1}{2}\varepsilon^{lm}_a e_{lm} R^a_k = e_a R^a_k$ which is the transformation law for 3-vectors with spin one. The other three basis vectors viz. $e_{[01]} = e_{[02]}, e_{[03]}$ also transform among each other as vectors since $e_{[0i]} = e_0 e_i - e_i e_0$ and e_0, being the time component, is invariant under rotations and, therefore, a scalar. The above analysis implies that we can decompose the space of second rank antisymmetric tensors into a direct sum of two invariant spin one subspaces insofar as rotations is concerned.

Let us, now, consider Lorentz boosts. To study the effect of boosts on these basis vectors, we need to ascertain the eigenvectors and eigenvalues of the operators $\hat{M}_3 = \frac{(\hat{J}_3 + i\hat{K}_3)}{2}$ and $\hat{N}_3 = \frac{(\hat{J}_3 - i\hat{K}_3)}{2}$. Let us start with boosts along the x^3 direction. We have

$$L^3\ (\xi) = \begin{pmatrix} \cosh\xi & 0 & 0 & \sinh\xi \\ 0 & 1 & 0 & 0 \\ 0 & 0 & 1 & 0 \\ \sinh\xi & 0 & 0 & \cosh\xi \end{pmatrix}$$

whence $L_3(\xi)\, e_{[01]} = e_{\rho\sigma}(L_3)^{\rho}{}_0\, (L_3)^{\sigma}_1 = e_{\rho 1}(L_3)^{\rho}_0 = \cosh\xi e_{[01]} + \sinh\xi\, e_{[31]}$ so that $e^{i\xi\hat{K}_3}\, e_{[01]} = \cosh\xi\, e_{[01]} + \sinh\xi\, e_{[31]}$ whence, on performing an expansion of both sides in powers of ξ and retaining only first order terms, we get $i\hat{K}_3\,\xi\, e_{[01]} = -e_{[31]}$. Similarly, we get $i\hat{K}_3\, e_{[02]} = e_{[23]}$, $i\hat{K}_3\, e_{[03]} = 0$, $i\hat{K}_3\, e_{[12]} = 0$, $i\hat{K}_3\, e_{[23]} = e_{[02]}$ and $i\hat{K}_3\, e_{[31]} = -e_{[01]}$. We also have $\hat{J}_3\, e_{[01]}, ie_{[02]}, \hat{J}_3\, e_{[02]} = -ie_{[01]}$, $\hat{J}_3\, e_{[03]} = 0$, $\hat{J}_3\, e_{[12]} = 0$, $\hat{J}_3\, e_{[23]} = ie_{[31]}$ and $\hat{J}_3\, e_{[31]} = -\, ie_{[23]}$.

Our objective here is to establish the decomposition of the second rank antisymmetric tensor space according to the $(1, 0) \otimes (0, 1)$ representation. Now, if this decomposition does exist, then the space of such second rank antisymmetric tensors should contain three eigenstates of $\hat{M}_3$ with eigenvalues $-1, 0, +1$ that are annihilated by $\hat{N}_3$ and also vice versa. Explicit calculations yield the following eigenvectors of $\hat{M}_3$ corresponding to the respective eigenvalues $-1,0$ and 1. We give an illustrative calculation. For the purpose, we write an arbitrary vector in the six dimensional space of the second rank antisymmetric tensors as $|\Phi\rangle = \sum_{i=1}^{6} a^i f_i$ and solve the eigenvalue equation $\hat{M}_3|\Phi\rangle$.

$\left(\dfrac{\hat{J}_3 + i\hat{K}_3}{2}\right)|\Phi\rangle = \left(\dfrac{\hat{J}_3 + i\hat{K}_3}{2}\right)\sum_{i=1}^{6} a^i f_i = m\sum_{i=1}^{6} a^i f_i$. Making use of the transformation properties of the basis vectors under the operators $\hat{J}_3\ \hat{K}_3$ and then comparing the coefficients of the basis vectors on both sides, we obtain the equations $ma_1 = -\frac{1}{2}(a_6 + ia_2)$, $ma_2 = \frac{1}{2}(ia_1 + a_5)$, $ma_5 = \frac{1}{2}\ (a_2 - ia_6)$, $ma_6 = \frac{1}{2}(ia_5 - a_1)$. Solving this set of equations yields $m = 0, \pm 1$, e.g., $m^2 a_1 = -\frac{1}{2}ma_6 - \frac{1}{2}mia_2$
$= -\frac{1}{2}\left[\frac{1}{2}(ia_5 - a_1)\right] - \frac{1}{2}\left[\frac{i}{2}(ia_1 - a_5)\right] = -\frac{1}{2}(ia_5 - a_1) = -\frac{1}{2}\left[\frac{i}{2m}(a_2 - ia_6) - a_1\right] = a_1$. For $m = -1$, we get $a_1 = a_6$, $a_2 = -ia_6$, $a_5 = ia_6$. Writing $a_6 = 1$, we get (i) below. Calculation of the other eigenvectors is on similar lines.

(i) $e_{[01]} - ie_{[02]} + ie_{[23]} + e_{[31]}$ (Eigenvalue -1)

(ii) $-ie_{[01]} + e_{[02]} + ie_{[23]} + e_{[31]}$ (Eigenvalue $+1$)

(iii) $e_{[03]} + e_{[12]}$ (Eigenvalue 0)

Similarly, the various eigenvectors of $\hat{N}_3$ are

(iv) $ie_{[01]} + e_{[02]} + e_{[23]} - ie_{[31]}$ (Eigenvalue −1)

(v) $e_{[01]} + ie_{[02]} - ie_{[23]} + e_{[31]}$ (Eigenvalue +1)

(vi) $e_{[03]} + e_{[12]}$ (Eigenvalue 0)

It can easily be seen by explicit calculations that for each of the above eigenvectors (*i*) – (*vi*) (designated $|\phi\rangle$), we have $\hat{N}_3\ \hat{M}_3|\phi\rangle = \hat{M}_3\ \hat{N}_3|\phi\rangle = (\hat{J}_3^2 + \hat{K}_3^2)|\phi\rangle = 0$ thereby showing that the eigenvectors of $\hat{M}_3$ are annihilated by $\hat{N}_3$ and vice versa. This establishes our result that the space of second rank antisymmetric tensors can be written as the direct sum of two subspaces that form the (1, 0) ⊗ (0, 1) representation of the proper Lorentz group.

The generators $\hat{J}_{\mu\nu}$, $\mu, \nu = 0, 1, 2, 3$ of infinitesimal homogeneous Lorentz transformations are antisymmetric tensors of rank two defined in terms of the equation

$$\Lambda(\delta\omega) = \mathbf{I} - \frac{i}{2}\hat{J}_{\mu\nu}\ \delta\omega^{\mu\nu} \tag{179}$$

where $\delta\omega^{\mu\nu}$ are antisymmetric infinitesimal parameters representing infinitesimal angular displacements in the (μ, ν) plane.

The transformation of $\Lambda(\omega)$ under the effect of an arbitrary Lorentz transformation is investigated as

$$\Omega\Lambda(\omega)\Omega^{-1} = \Omega e^{-i\hat{J}_{\mu\nu}\omega^{\mu\nu}}\ \Omega^{-1} = \Omega\left[\sum_i \frac{1}{i!}(-i\hat{J}_{\mu\nu}\ \omega^{\mu\nu})^i\right]\Omega^{-1} = \left[\sum_i \frac{1}{i!}(-i\Omega\hat{J}_{\mu\nu}\omega^{\mu\nu}\ \Omega^{-1})^i\right]$$

$$= \left[\sum_i \frac{1}{i!}(-i\Omega\hat{J}_{\mu\nu}\ \Omega^{-1}\ \Omega\omega^{\mu\nu}\ \Omega^{-1})^i\right] = \left[\sum_i \frac{1}{i!}(-i\hat{J}'_{\mu\nu}\ \omega'^{\mu\nu})^i\right] = e^{-i\hat{J}_{\mu\nu}\omega^{\mu\nu}} \tag{180}$$

where $\omega'^{\mu\nu} = \Omega^{\mu}_{\lambda}\ \omega^{\lambda\sigma}\ \Omega^{\nu}_{\sigma}$ and $\hat{J}'_{\mu\nu} = \Omega\hat{J}_{\mu\nu}\ \Omega^{-1} = \Omega^{\lambda}_{\mu}\ \hat{J}_{\lambda\sigma}\ \Omega^{\sigma}_{\nu}.$

2.8 TRANSFORMATION OF TENSOR TRACE

Consider a second rank tensor **T**. Given a unitary representation of the proper Lorentz group $\hat{L}(\Lambda)$, **T** would transform as $\mathbf{T}' = \hat{L}(\Lambda)\ \mathbf{T}\hat{L}(\Lambda)^{-1}$. Then, $Tr\ \mathbf{T}' = Tr\ [\hat{L}(\Lambda)\ \mathbf{T}\hat{L}(\Lambda)^{-1}] = Tr\ \mathbf{T}$ showing that the tensor trace is invariant under proper Lorentz transformation i.e. it is a Lorentz scalar and transforms as the (0, 0) representation of the Lorentz group.

2.9 TRANSFORMATION OF THE SYMMETRIC SECOND RANK TENSOR

We have so far established the following:

(i) The Lorentz vectors transform as the $\left(\frac{1}{2}, \frac{1}{2}\right)$ representation;

(ii) Therefore, second rank tensors in Minkowski space transform under the reducible $\left(\frac{1}{2}\otimes\frac{1}{2}\right)\otimes\left(\frac{1}{2}\otimes\frac{1}{2}\right)$ representation.

(iii) The $\left(\frac{1}{2}\otimes\frac{1}{2}\right)\otimes\left(\frac{1}{2}\otimes\frac{1}{2}\right)$ representation can be decomposed as: $\left(\frac{1}{2}\otimes\frac{1}{2}\right)\otimes\left(\frac{1}{2}\otimes\frac{1}{2}\right)$ $= (0\oplus 1)\otimes(0\oplus 1) = (0\otimes 0)\oplus(1\otimes 0\oplus 0\otimes 1)\oplus(1\otimes 1)$.

(iv) The trace of the tensor, being a Lorentz scalar, transforms as the (0, 0) representation.

(v) The antisymmetric part transforms as the $(1\otimes 0)\otimes(0\otimes 1)$ representation.

It, follows, therefore, that the traceless symmetric part transforms as the $(1\otimes 1)$ representation of the proper Lorentz group.

2.10 THE POINCARE GROUP & RELATIVISTIC WAVE EQUATION (WIGNER, 1939; BARGMANN, 1948; FOLDY, 1956; SHIROKOV, 1958; OHNUKI, 1976; TUNG, 1984; FUSHCHICH, 1994)

As is well known, quantum mechanical dynamics are described by wave equations whose solutions represent physical states of definite mass and spin. A generic wave equation can be represented in the form

$$\hat{\Pi}(M, i\partial)^{\alpha}_{\beta}\,\hat{\Psi}^{\beta}(x) = 0 \tag{181}$$

where $\hat{\Pi}$ is a linear differential operator and $\hat{\Psi}(x)$ is an operator that depends on the space variables as well. It is emphasized that single particle quantum theory can be developed either in terms of the c number wavefunctions $\psi(x)$ (that are single component or multicomponent single column vectors) or the corresponding space dependent operators $\hat{\Psi}(x)$. For many body systems, e.g., quantum field theory, these c number wavefunctions $\psi(x)$ take the form of operators $\hat{\Psi}(x)$ that operate on the vector space of physical states. For the common wave equations $\hat{\Pi}$ is of the first or second degree in ∂_{μ}. A solution of this operator valued equation can be obtained by taking the Fourier transform to get

$$\hat{\Psi}^{\alpha}(x) = \frac{1}{(2\pi)^{3}}\int dp\,\hat{\Phi}^{\alpha}(p)\,e^{-ipx} \tag{182}$$

where $\hat{\Phi}^{\alpha}(p)$ satisfies

$$\hat{\Pi}(M, p)^{\alpha}_{\beta}\,\hat{\Phi}^{\beta}(p) = 0 \tag{183}$$

which is easily seen by substituting the latter eq. in the former. Since, e^{-ipx} is a Lorentz scalar it follows that the transformation properties of $\hat{\Psi}(x)$ and $\hat{\Phi}(p)$ under Lorentz transformations would be identical.

We can easily obtain the transformation properties of $\hat{\Psi}(x)$ by exploiting its relationship with the c number wavefunction $\psi(x)$. Given an arbitrary single particle state $|\psi\rangle$, the c number wavefunction corresponding to the operator $\hat{\Psi}(x)$ is obtained as $\langle 0|\hat{\Psi}^{\alpha}(x)|\psi\rangle = \psi^{\alpha}(x)$ where $|0\rangle$ is the vacuum state. Consider an arbitrary Lorentz transformation with the representation $\hat{U}(\Lambda)$. We, then have $\hat{U}(\Lambda)|\psi\rangle = |\psi'\rangle$ and, as proved earlier, $\psi'^{\alpha}(x) = D^{s}[\Lambda]^{\alpha}_{\beta}\,\psi^{\beta}(\Lambda^{-1}x)$. Making use of the fact that the vacuum state is invariant under any Lorentz transformation, we can write

$$\langle 0|\hat{U}(\Lambda)\hat{\Psi}'^{\alpha}(x)\hat{U}(\Lambda)^{-1}|\psi'\rangle = \langle 0|\hat{\Psi}^{\alpha}(x)|\psi\rangle = \psi^{\alpha}(x) = D^{s}[\Lambda^{-1}]^{\alpha}_{\lambda}\,\psi'^{\lambda}(\Lambda x)$$

Now, multiplying the equation $\langle 0|\hat{\Psi}^{\alpha}(x)|\psi\rangle = \psi^{\alpha}(x)$ on the left by $D^{s}[\Lambda^{-1}]$, substituting ψ' for ψ and ψ for $x' = \Lambda x$, for x we obtain $0 = \langle 0|D^{s}[\Lambda^{-1}]^{\alpha}_{\lambda}\,\hat{\Psi}^{\lambda}(\Lambda x)|\psi'\rangle = D^{s}[\Lambda^{-1}]^{\alpha}_{\lambda}\,\psi'^{\lambda}(\Lambda x)$ Comparing the left hand sides of the two equations, we get the transformation law

$$\hat{U}(\Lambda)\;\hat{\Psi}^{\alpha}(x)\;\hat{U}(\Lambda)^{-1} = D^{s}[\Lambda^{-1}]^{\alpha}_{\lambda}\,\hat{\Psi}^{\lambda}(\Lambda x) \tag{184}$$

In order that our proposed wave equation be relativistically acceptable:

(a) It must be covariant, i.e., $D(\Lambda)\hat{\Pi}(M,p)\,D(\Lambda^{-1}) = \hat{\Pi}(M,\Lambda p)$ must lead to $\hat{\Pi}(M,\Lambda p)\hat{\Phi}(\Lambda p) = 0$. This follows from $\hat{\Pi}(M,p)\hat{\Phi}(p) = 0$

$$\Rightarrow \qquad \hat{\Pi}(M,p)\;(\hat{U}(\Lambda^{-1})\;\hat{U}(\Lambda))\;\hat{\Phi}(p) = 0$$

$$\Rightarrow \qquad \hat{U}(\Lambda)\;\hat{\Pi}(M,p)\;(\hat{U}(\Lambda^{-1})\;\hat{U}(\Lambda))\;\hat{\Phi}(p)\;\hat{U}(\Lambda^{-1}) = 0$$

$$\Rightarrow \qquad \hat{\Pi}(M,\Lambda p)\;(D[\Lambda^{-1}]^{\alpha}_{\beta}\;\hat{\Phi}^{\beta}(\Lambda p)) = 0$$

$$\Rightarrow \qquad D[\Lambda^{-1}]^{\alpha}_{\beta}\;(\hat{\Pi}\;(M,\Lambda p)\;\hat{\Phi}^{\beta}(\Lambda p) = 0$$

$$\Rightarrow \qquad \hat{\Pi}(M,\Lambda p)\hat{\Phi}^{\beta}(\Lambda p) = 0 \tag{185}$$

where the penultimate step follows from the fact that the $D[\Lambda^{-1}]^{\alpha}_{\beta}$ are scalars and in the final step we note that these scalars are not all zero for any Λ.

(b) We must also have the mass shell condition $(p^2 - M^2)\,\hat{\Phi}(p) = 0$ or $\hat{\Phi}(p) = \delta(p^2 - M^2)\,\tilde{\Phi}(p)$ in the metric characterized by $g \equiv diag\,(+1, -1, -1\,-1)$. The mass shell condition implies that $\hat{\Phi}(p)$ is non-vanishing only when its argument satisfies $p^0 = \pm\,(\mathbf{p}^2 + M^2)^{1/2}$ corresponding to the positive energy and negative energy solutions respectively. Using, then, the Lorentz invariant integration measure on the mass shell viz. $\tilde{d}p = \dfrac{d\mathbf{p}}{(2\pi)^3\,2p^0}$, we have

$$\hat{\psi}(x) = \int \tilde{d}p\,[\hat{\Phi}_{+}(p)\,e^{-ipx} + \hat{\Phi}_{-}(p)e^{ipx}] \tag{186}$$

where

$$\hat{\Phi}_{\pm}(p) = \tilde{\Phi}[p^0 = \pm\,(\mathbf{p}^2 + M^2)^{1/2}, \pm\,\mathbf{p}] \tag{187}$$

and

$$\hat{\Pi}(M, \pm\,p)\;\hat{\Phi}_{+}(p) = 0 \tag{188}$$

Since any occurrence of p^0 in $\hat{\Pi}$ can be replaced by $(\mathbf{p}^2 + M^2)^{1/2}$, if we focus only on the positive energy solutions, as is the case hereafter, it follows then that $\hat{\Pi}$ can be considered a function of the 3-vector $\mathbf{p}$ in lieu of the 4-vector p.

(c) Let us assume that the field operator $\hat{\Phi}(p)$ transforms according to the (u, v) representation of the Lorentz group. Then, we have $|u - v| \le j \le (u + v)$. Now, if $u \ne 0$, $v \ne 0$ then j will have multiple values implying that the $\hat{\Phi}(p)$ has multiple spin states. However, in order that our relativistic equation $\hat{\Pi}(M, p)^{\alpha}_{\beta}\, \hat{\Phi}^{\beta}(p) = 0$ be physically acceptable and the solutions thereof correspond to particles with unique spin, $\hat{\Pi}(M. p)$ must necessarily act as a projection operator, which acting on the field operator $\hat{\Phi}(p)$ would project out the appropriate spin components corresponding to the spin characteristic of the particle being described.

Let us now assume that we are describing a particle of spin s. Since λ can take $(2s + 1)$ viz., $\lambda = -s, -s + 1, \ldots 0, \ldots s - 1, s$ values, the matrix equation $\hat{\Pi}(M, \pm p)\hat{\Phi}\pm(p) = 0$ must necessarily have $(2s + 1)$ independent solutions. Considering the particle in the rest frame $p \equiv pt = (p^0 = M, \mathbf{p} = 0)$, we write these independent solutions as $u(\mathbf{0}\lambda)$ so that $\hat{\Pi}(M, p_t)^{\alpha}_{\beta}\, u^{\beta}(\mathbf{0}\lambda) = 0$ for $\lambda = -s, \ldots s$. The solutions corresponding to an arbitrary momentum vector $\mathbf{p}$ can be obtained as

$$u^{\alpha}(\mathbf{p}\lambda) = D[\varphi_W(p, \Lambda)]^{\alpha}_{\beta}\, u^{\beta}(\mathbf{0}\lambda) \tag{189}$$

where $\varphi_W(p, \Lambda)$ is a Wigner rotation that brings the rest frame momentum vector $p_t \equiv (M, \mathbf{0})$ to the desired momentum $p \equiv (p^0, \mathbf{p})$. We also have

$$\begin{aligned}\hat{\Pi}(M, p)\, u(\mathbf{p}\lambda) &= D[\varphi_W(p, \Lambda)]\; \hat{\Pi}(M, p_t)\, D[\varphi_W(p, \Lambda)]^{-1}\; u(\mathbf{p}\lambda)\\ &= D[\varphi_W(p, \Lambda)]\; \hat{\Pi}(M, p_t)\, u(\mathbf{0}\lambda) = 0\end{aligned} \tag{190}$$

thereby confirming that $u(\mathbf{p}\lambda)$ is a solution of the wave equation. $u(\mathbf{p}\lambda)$ constitute concrete realizations of the linear momentum basis states $\{|\mathbf{p}\lambda\rangle\}$ of the time like irreducible representations of the Poincare group. These realizations are identified by (i) the IRR representation labels (M, s) (ii) the representation labels (u, v) for identifying the representation of the proper Lorentz group (iii) the set of Poincare labels $(\mathbf{p}, \lambda)$ for identifying the basis vectors where the range of λ is determined by the label s, i.e., $\lambda = -s, \ldots, 0, \ldots, +s$ and (iv) the set of Lorentz labels collectively indicated by α whose range is determined by (u, v). To summarize, then, we have the following situation:

(i) If the field operator $\hat{\Phi}(p)$ has multiple spin content characterized by $|u - v| \le j \le (u + v)$, then the differential operator $\hat{\Pi}(M, p)$ acting on $\hat{\Phi}(p)$ projects out the desired spin components corresponding to the particle states being described.

(ii) If the spin of the particle being described is designated s (which is projected out by the action of $\hat{\Pi}(M, p)$ on $\hat{\Phi}(p)$), then corresponding to this s, there would exist $(2s + 1)$ independent solutions to the wave equation $\hat{\Pi}(M, \pm p)\, \hat{\Phi}_{\pm}(p) = 0$ identified with $\lambda = -s, \ldots 0, \ldots +s$.

(iii) The general solution to the wave equation can be written as

$$\hat{\Psi}^a(x) = \sum_\lambda \int \tilde{d}p[b(\mathbf{p}, \lambda)u^a(\mathbf{p}, \lambda)e^{-ipx}] + \text{negative energy term.} \tag{191}$$

The coefficients $b(\mathbf{p}\lambda)$ carry the operator property on the right hand side of the above equation and constitute the annihilation operators for the particle states. The adjoints $b^\dagger(\mathbf{p}\lambda)$ are the creation operators and can be used to write the basis vectors $\{|\mathbf{p}\lambda\rangle\}$ as $|\mathbf{p}\lambda\rangle = b^\dagger(\mathbf{p}, \lambda)|0\rangle$. Since the basis vectors $\{|\mathbf{p}\lambda\rangle\}$ transform under a proper Lorentz transformation as $\Lambda|\mathbf{p}\lambda\rangle = D^s[\varphi_W(\Lambda, p)]^{\lambda'}_{\lambda}|\Lambda\mathbf{p}, \lambda'\rangle$, we have

$$\hat{U}(\Lambda)|\mathbf{p}\lambda\rangle = \hat{U}(\Lambda)b^\dagger(\mathbf{p}, \lambda)\,\hat{U}(\Lambda)^{-1}\,\hat{U}(\Lambda)|0\rangle$$

$= \hat{U}(\Lambda)b^\dagger(\mathbf{p}, \lambda)\,\hat{U}(\Lambda)^{-1}|0\rangle = D^s[\varphi_W(\Lambda, p)]^{\lambda'}{}_{\lambda}\,b^\dagger(\Lambda\mathbf{p}, \lambda')|0\rangle$ whence we get the transformation law for $b^\dagger(\mathbf{p}\lambda)$ as

$$\hat{U}(\Lambda)b^\dagger(\mathbf{p}, \lambda)\,\hat{U}(\Lambda)^{-1} = D^s[\varphi_W(\Lambda, p)]^{\lambda'}_{\lambda}b^\dagger(\Lambda\mathbf{p}, \lambda') \tag{192}$$

The Lorentz transformation law for the annihilation operators follows by conjugation as

$$\hat{U}(\Lambda)b(\mathbf{p}, \lambda)\,\hat{U}(\Lambda)^{-1} = D^s[\varphi_W(\Lambda, p)^{-1}]^{\lambda}_{\lambda'}\,b(\Lambda p, \lambda') \tag{193}$$

Making use of the orthonormality of the momentum spin basis states viz. $\langle\mathbf{p}'\,\lambda'|\mathbf{p}\lambda\rangle = \delta^{\lambda}_{\lambda'}\,\delta(\mathbf{p}-\mathbf{p}')$, we obtain $u^\alpha(\mathbf{p}\lambda)\,e^{-ipx} = \langle 0|\hat{\Psi}^\alpha(x)|\mathbf{p}\lambda\rangle$. Noting that $\langle 0|b(\mathbf{p}', \lambda') = \langle\mathbf{p}'\,\lambda'|$, $b^\dagger(\mathbf{p}, \lambda)|0\rangle = |\mathbf{p}\lambda\rangle$, we get $\langle 0|b(\mathbf{p}', \lambda')b^\dagger(\mathbf{p}, \lambda)|0\rangle = \langle\mathbf{p}', \lambda'|\mathbf{p}\lambda\rangle = \delta^{\lambda'}_{\lambda}\,\delta(\mathbf{p}-\mathbf{p}')$. Let us, now, obtain the transformation law for the c number wavefunctions $u^\alpha(\mathbf{p}, \lambda)$. For the purpose, we have $\hat{U}(\Lambda)\hat{\Psi}^\alpha(x)\,\hat{U}(\Lambda^{-1}) = D[\Lambda^{-1}]^\alpha_\beta\,\hat{\Psi}^\beta(\Lambda x)$

$$= D[\Lambda^{-1}]^\alpha_\beta \sum_\lambda \int \tilde{d}q\,[b(\mathbf{q}, \lambda)u^\beta(\mathbf{q}, \lambda)e^{-iq\Lambda x} + \text{negative energy term}]$$

$$= D[\Lambda^{-1}]^\alpha_\beta \sum_\lambda \int \tilde{d}p\,[b(\Lambda\mathbf{p}, \lambda)u^\beta(\Lambda\mathbf{p}, \lambda)e^{-iqx} + \text{negative energy term}]$$

where in the last step we have changed the integration variable from q to $p = \Lambda^{-1}q$. The infinitesimal integration volume element is invariant under Lorentz transformations. We also have

$$\hat{U}(\Lambda)\sum_{\lambda,}\int \hat{d}p\,[b(\mathbf{p}, \lambda)\,u^\alpha(\mathbf{p}, \lambda)e^{-ipx}] + \text{negative energy term}\}\,\hat{U}(\Lambda^{-1})$$

$$= \sum_{\lambda,\lambda'}\int \hat{d}p\;\{D^s[\varphi_W(\Lambda, p)^{-1}]^{\lambda}_{\lambda'}\,b(\Lambda\mathbf{p}, \lambda')\,u^\alpha(\mathbf{p}, \lambda)e^{-ipx} + \text{negative energy term}\}$$

where we have used $\hat{U}(\Lambda)\,b(\mathbf{p}, \lambda)\,\hat{U}(\Lambda)^{-1} = D^s[\varphi_W(\Lambda, p)^{-1}]^{\lambda}_{\lambda'}\,b(\Lambda\mathbf{p}, \lambda')$. Interchanging $\lambda \leftrightarrow \lambda'$ and then comparing the two equations, we get

$$D[\Lambda^{-1}]^\alpha_\beta\,u^\beta(\Lambda\mathbf{p}, \lambda) = D^{(s)}[\varphi_W(\Lambda, p)^{-1}]^{\lambda'}_{\lambda}\,u^\alpha(\mathbf{p}, \lambda') \tag{194}$$

or equivalently

$$D[\Lambda]^\alpha_\beta\,u^\beta(\mathbf{p}, \lambda) = u^\alpha(\Lambda\mathbf{p}, \lambda')\,D^{(s)}[\varphi_W(\Lambda, p)]^{\lambda'}_{\lambda} \tag{195}$$

Finally, we have $D[\Lambda]\, u(\mathbf{p}, \lambda) = D[\Lambda]\, D[\kappa_p]\, u(\mathbf{0}, \lambda)$

$$= D[\kappa_{\Lambda p}]\, D[\kappa_{\Lambda p}]^{-1}\, D[\Lambda]\, D[\kappa_p]\, u(\mathbf{0}, \lambda) = D[\kappa_{\Lambda p}]\, D[\varphi_W(\Lambda, p)]\, u(\mathbf{0}, \lambda)$$
$$= D[\kappa_{\Lambda p}]\, u(0, \lambda')\, D^{(s)}\, [\varphi_W(\Lambda, p)]^{\lambda}_{\lambda'} = u(\Lambda\mathbf{p}, \lambda')\, D^{(s)}\, [\varphi_W(\Lambda, p)]^{\lambda}_{\lambda'} \tag{196}$$

thereby explicitly confirming the above transformation law.

The above analysis sets out the relationship between the transformation properties of the various physical quantities encountered in quantum mechanics. We have:

(a) The particle states are physical states and hence, must necessarily transform according the unitary representation (which happens to be infinite dimensional) of the Poincare group. As a corollary, the creation and annihilation operators also have identical transformation properties;

(b) The field operators are not necessarily physical observables so that their transformation need not necessarily be unitary. They may transform according to some non-unitary and, therefore, finite dimensional representation of the Poincare group.

(c) The c number wavefunctions $u(\mathbf{p}, \lambda)e^{-ipx}$ are the coefficient functions that connect the set of operators $\{b(\mathbf{p}, \lambda)\}$ (that transform according to the infinite dimensional unitary representations of the Poincare group and the set of field operators $\{\hat{\Psi}(x)\}$ that transform according to the non-unitary finite dimensional representations of the proper Poincare group.

2.11 EXPLICIT REALIZATIONS OF THE GENERATORS FOR THE MASSIVE CASE (SHIROKOV, 1958; RITUS, 1961; OHNUKI, 1976; CHAKRABARTI, 1988; FUSHCHICH, 1994)

As has been mentioned earlier, each of the irreducible unitary representations of the Poincare group is characterized by ten generators, four corresponding to translations in the three spatial and one time dimensions, three relating to spatial rotations along the three spatial axes and the remaining three generating spatiotemporal rotations (boosts) along the three spatial axes. Our objective in this section is to obtain the explicit form of each of these generators on the basis of the commutation relations satisfied by them. It is important to mention here that although the Lie algebra of the Poincare group consisting of the commutators between the ten generators retains its structure in any basis set, explicit forms of the generators would be different in different bases. For example, explicit forms of these generators have been obtained in a linear momentum basis (in which the momentum and energy operators are diagonal), a Lorentz basis (in which the two Casimir operators of the homogeneous Lorentz group are diagonal), a basis of a subgroup of the homogeneous Lorentz group, a basis consisting of the IRRs of the group O(3,1) etc.

In the sequel, we shall focus on obtaining the realizations of the Poincare generators in the linear momentum basis, i.e., a basis in which the four momentum operators are diagonal. For the sake of easy reference, we reproduce below the Lie algebra of the Poincare group:

$$[\hat{P}^{\mu}, \hat{P}^{\nu}] = 0 \tag{197}$$
$$[\hat{P}^{\rho}, \hat{J}^{\mu\nu}] = i(g^{\rho\mu}\, \hat{P}^{\nu} - g^{\rho\nu}\, \hat{P}^{\mu}) \tag{198}$$
$$[\hat{J}^{\mu\nu}, \hat{J}^{\rho\sigma}] = i(g^{\nu\rho}\, \hat{J}^{\mu\sigma} + g^{\mu\sigma}\, \hat{J}^{\nu\rho} - g^{\mu\rho}\, \hat{J}^{\nu\sigma} - g^{\nu\sigma}\, \hat{J}^{\mu\rho}) \tag{199}$$

The Pauli Lubanski vector $\hat{W}^{\mu} = \frac{1}{2}\,\varepsilon^{\mu}_{\nu\rho\sigma}\,\hat{J}^{\nu\rho}\,\hat{P}^{\sigma}$ ($\hat{W}^0 = \hat{\mathbf{J}}.\hat{\mathbf{P}}$, $\hat{\mathbf{W}} = \hat{\mathbf{J}}.\hat{P}^0 - \hat{\mathbf{K}} \times \hat{\mathbf{P}}$)(satisfies the following commutators:

$$[\hat{W}^{\lambda}, \hat{W}^{\sigma}] = i\varepsilon^{\lambda\sigma\mu\nu}\,\hat{W}_{\mu}\,\hat{P}_{\nu} \tag{200}$$

$$[\hat{W}^{\lambda}, \hat{J}^{\mu\nu}] = i(\hat{W}^{\nu}\,g^{\mu\lambda} - \hat{W}^{\mu}\,g^{\lambda\nu}] \tag{201}$$

$$[\hat{W}^{\mu}, \hat{P}^{\sigma}] = 0 \tag{202}$$

For the massive case under consideration, the little group is the three-dimensional spatial rotation group $SO(3)$ so that the independent components of $\hat{W}^{\mu}$ coincide with the generators of spatial three-dimensional rotations that satisfy $[\hat{S}^i, \hat{S}^j] = i\varepsilon^{ijk}\,\hat{S}^k$. We follow the method of Shirokov to obtain the Foldy-Shirokov realizations of the Poincare group. Introducing the operator $\hat{g}$ by

$$\hat{g}^{\mu} = \hat{J}^{\mu}_{\nu}\,\hat{P}^{\nu} \tag{203}$$

so that

$$\hat{g}^0 = \hat{J}^0_{\nu}\,\hat{P}^{\nu} = \hat{\mathbf{K}}.\hat{\mathbf{P}},\; \hat{\mathbf{g}} = \hat{\mathbf{K}}\hat{P}^0 + \hat{\mathbf{J}} \times \hat{\mathbf{P}} \tag{204}$$

Since we are working in a linear momentum basis in which the momentum operators are diagonal, we have

$$\hat{g}^{\mu}\hat{P}_{\mu} = 0 \tag{205}$$

Using eqs. (200)-(202) and the definition (203), we arrive at the following commutators:

$$[\hat{g}^{\mu}, \hat{W}^{\sigma}] = i\hat{W}^{\mu}\,\hat{P}^{\sigma} \tag{206}$$

$$[\hat{g}^{\mu}, \hat{P}^{\sigma}] = i(\hat{P}^{\mu}\,\hat{P}^{\nu} - g^{\mu\nu}\,\hat{P}^{\sigma}\,\hat{P}_{\sigma}) \tag{207}$$

$$[\hat{g}^{\mu}, \hat{g}^{\nu}] = i(\hat{g}^{\mu}\,\hat{P}^{\nu} - \hat{g}^{\nu}\,\hat{P}^{\mu} - \varepsilon^{\mu\nu}_{\rho\sigma}\,\hat{W}^{\rho}\,\hat{P}^{\sigma}) \tag{208}$$

Using the definition of the Pauli Lubanski vector and eqs. (202), (204), we obtain:

$$M^2\,\hat{\mathbf{J}} = \hat{\mathbf{g}} \times \hat{\mathbf{P}} + \hat{\mathbf{W}}\hat{P}^0 - \hat{W}^0\,\hat{\mathbf{P}} \tag{209}$$

$$M^2\,\hat{\mathbf{K}} = -\,\hat{\mathbf{W}} \times \hat{\mathbf{P}} + \hat{\mathbf{g}}\,\hat{P}^0 - \hat{g}^0\,\hat{\mathbf{P}} \tag{210}$$

or equivalently, in a linear momentum basis,

$$\hat{J}^{\mu\nu} = \frac{\hat{g}^{\mu}\,\hat{P}^{\nu} - \hat{g}^{\nu}\,\hat{P}^{\mu} - \varepsilon^{\mu\nu}_{\rho\sigma}\,\hat{W}^{\rho}\,\hat{P}^{\sigma}}{\hat{P}^{\xi}\,\hat{P}_{\xi}} \tag{211}$$

In the linear momentum basis, $\hat{W}(p)$ is the restriction of the Pauli Lubanski vector ($\hat{W}^0 = \hat{\mathbf{J}}.\hat{\mathbf{P}}$, $\hat{\mathbf{W}} = \hat{\mathbf{J}}.\hat{P}^0 - \hat{\mathbf{K}} \times \hat{\mathbf{P}}$) to the eigenspace of $\hat{P}^{\mu}$ corresponding to the eigenvalue p. In the rest frame $p_R \equiv (M, 0)$, we have the corresponding restriction ($\hat{W}^0_R = 0, \hat{\mathbf{W}}_R = M\hat{\mathbf{J}}_R = M\hat{\mathbf{S}}_R$). Using eq. (200), it is seen that the components of $\hat{\mathbf{S}}_R$ satisfy $[\hat{S}^i, \hat{S}^j] = i\varepsilon^{ijk}\,\hat{S}^k$ and hence, constitute a three-dimensional spatial rotation structure. To obtain the restriction of $\hat{W}$ corresponding to an arbitrary p, we impart a pure Lorentz boost L_p to the system that transforms the particle of mass M from rest to a state of momentum $\mathbf{p}$. The matrix representation of L_p is $(L_p)^0_0, \frac{p^0}{M}, p^0 = (|\mathbf{p}|^2 + M^2)^{1/2}$, $(L_p)^i_0 = \frac{p^i}{M}, (L_p)^0_i = -\frac{p_i}{M}, (L_p)^i_j = \delta^i_j - \frac{p^i p_j}{M(p^0 + M)}$. We know from eqs. (201) & (202) that the components of $\hat{W}$ transform as a vector under Lorentz transformation so that $\hat{W}^0 = \hat{\mathbf{S}}.\mathbf{p}$, $\hat{\mathbf{W}} = M\hat{\mathbf{S}} + \frac{\hat{\mathbf{S}}.\mathbf{p}}{p^0 + M}\mathbf{p}$ where we have replaced the momentum operators with the corresponding eigenvalues which is permitted since we are working in a basis in which the momentum operators are diagonal. Our next step is to obtain explicit expressions for the components of $\hat{g}$.

From eq. (207), we obtain $[\hat{g}^0, \hat{p}^i] = i\hat{P}^0\hat{P}^i$ whence we can write $\hat{g}^0 = i\hat{P}^0\hat{O}$ where the operator $\hat{O}$ satisfies $[\hat{O}, \hat{P}^i] = \hat{P}^i$ which has the solution

$$\hat{O} = \frac{\partial}{\partial \mathbf{p}}.\mathbf{p} = \mathbf{p}.\frac{\partial}{\partial \mathbf{p}} + 3 \tag{212}$$

whence we get

$$\hat{g}^0 = ip^0\left(\mathbf{p}.\frac{\partial}{\partial \mathbf{p}}\right) + 3ip^0 \tag{213}$$

where we have replaced the momentum operators by their respective eigenvalues. Since $\hat{g}^\mu p_\mu = 0$, we have

$$\hat{\mathbf{g}}.\mathbf{p} = \hat{g}^0 p^0 = ip^0\left(\mathbf{p}.\frac{\partial}{\partial \mathbf{p}}\right)p^0 + 3i(p^0)^2 = ip^0\left[p^0\left(\mathbf{p}.\frac{\partial}{\partial \mathbf{p}}\right) + \frac{|\mathbf{p}|^2}{p^0}\right]3i(p^0)^2$$

$$= i|\mathbf{p}|^2\left(\mathbf{p}.\frac{\partial}{\partial \mathbf{p}}\right) + iM^2\left(\mathbf{p}.\frac{\partial}{\partial \mathbf{p}}\right) + i|\mathbf{p}|^2 + 3i(p^0)^2$$

$$= i\mathbf{p}.\left[\mathbf{p}\left(\mathbf{p}.\frac{\partial}{\partial \mathbf{p}}\right) + \mathbf{p}\right] + iM^2\left[\left(\frac{\partial}{\partial \mathbf{p}}.\mathbf{p}\right) - 3\right] + 3i(p^0)^2$$

$$= i\mathbf{p}.\left(\mathbf{p}.\frac{\partial}{\partial \mathbf{p}}\right)\mathbf{p} + iM^2\left(\frac{\partial}{\partial \mathbf{p}}.\mathbf{p}\right) + 3i\mathbf{p}.\mathbf{p} = i\left[\mathbf{p}\left(\mathbf{p}.\frac{\partial}{\partial \mathbf{p}}\right) + M^2\frac{\partial}{\partial \mathbf{p}} + 3\mathbf{p}\right].\mathbf{p}$$

whence

$$\hat{\mathbf{g}} = i\left[\mathbf{p}\left(\mathbf{p}.\frac{\partial}{\partial \mathbf{p}}\right) + M^2\frac{\partial}{\partial \mathbf{p}} + 3\mathbf{p}\right] + (\hat{O} \times \mathbf{p}) \tag{214}$$

Using eqs. (206) together with $\hat{W}^0 = \hat{\mathbf{S}}.\mathbf{p}$ we obtain $[g, \hat{W}^0] = i\mathbf{W}p^0$ or

$$\left[\mathbf{p}\left(\mathbf{p}.\frac{\partial}{\partial \mathbf{p}}\right)+M^2\frac{\partial}{\partial \mathbf{p}}+3\mathbf{p}-i(\hat{O}\times p),\hat{\mathbf{S}}.\mathbf{p}\right]=p^0\left(M\hat{\mathbf{S}}+\frac{\hat{\mathbf{S}}.\mathbf{p}}{p^0+M}\mathbf{p}\right)\text{ or}$$

$$\mathbf{p}(\hat{\mathbf{S}}.\mathbf{p})+M^2\hat{\mathbf{S}}-i[\hat{O}\times\mathbf{p},\hat{\mathbf{S}}.\mathbf{p}]=p^0\left(M\hat{\mathbf{S}}+\frac{\hat{\mathbf{S}}.\mathbf{p}}{p^0+M}\mathbf{p}\right)\text{ or}$$

$$[\hat{O}\times\mathbf{p},\hat{\mathbf{S}}.\mathbf{p}]=\frac{iM}{p^0+M}[|\mathbf{p}|^2\hat{\mathbf{S}}-(\hat{\mathbf{S}}.\mathbf{p})\mathbf{p}]\text{ whence }\hat{O}=\frac{M}{p^0+M}\hat{\mathbf{S}}\text{ giving}$$

$$\hat{\mathbf{g}}=i\mathbf{p}\left(\mathbf{p}.\frac{\partial}{\partial \mathbf{p}}\right)+\mathrm{i}M^2\frac{\partial}{\partial \mathbf{p}}+3i\mathbf{p}+\frac{M}{p^0+M}(\hat{\mathbf{S}}\times\mathbf{p}) \tag{215}$$

Using eqs. (209), (210), (213), (214) and $\hat{\mathbf{W}}=M\hat{S}+\frac{\hat{\mathbf{S}}.\mathbf{p}}{P^0+M}\mathbf{p}$, $\hat{W}^0=\hat{\mathbf{S}}.\mathbf{p}$, we, finally, obtain

$$\hat{\mathbf{J}}=-i\mathbf{p}\times\frac{\partial}{\partial \mathbf{p}}+\hat{\mathbf{S}} \tag{216}$$

$$\hat{\mathbf{K}}=ip^0\frac{\partial}{\partial \mathbf{p}}-\frac{\hat{\mathbf{S}}\times\mathbf{p}}{P^0+M} \tag{217}$$

which constitute the Shirokov form of the realizations of the Poincare group generators.

2.12 PARTICLE SPIN IN RELATIVISTIC QUANTUM MECHANICS

A particle is said to possess "spin s" if the squared angular momentum operator $\mathbf{J}^2$ acting on the states of the particle in its rest frame returns eigenvalues of the form $s(s+1)$. We note from the above that $J^2|\mathbf{0}\lambda\rangle=s(s+1)|\mathbf{0}\lambda\rangle$ implying that the index s can be identified with the spin of the particle. We have already shown above that M corresponds to particle's mass and $\mathbf{p}$ to its 3-momentum. To obtain an explicit physical interpretation of λ, we consider an arbitrary state and consider its motion in a chosen frame of reference such that the x^3 axis coincides with the direction of motion of the particle. We designate the state with reference to the chosen frame as $|p\hat{\mathbf{z}},\lambda\rangle$ so that $P_1|p\hat{\mathbf{z}},\lambda\rangle=P_2|p\hat{\mathbf{z}},\lambda\rangle=0$, $\mathbf{P}|p\hat{\mathbf{z}},\lambda\rangle=P_3|p\hat{\mathbf{z}},\lambda\rangle=p|p\hat{\mathbf{z}},\lambda\rangle$. Since the operators $\mathbf{P}$ and $\mathbf{J}.\mathbf{P}$ commute with each other, they possess simultaneous eigenvectors—let us assume that $|p\hat{\mathbf{z}},\lambda\rangle$ is a simultaneous eigenvector of $\mathbf{P}$ and $\mathbf{J}.\mathbf{P}$. Then $\mathbf{J}.\mathbf{P}|p\hat{\mathbf{z}},\lambda\rangle\, J_3P_3|p\hat{\mathbf{z}},\lambda\rangle$, $\lambda p|p\hat{\mathbf{z}},\lambda\rangle$ whence

$$\frac{\mathbf{J}.\mathbf{P}}{p}|p\hat{\mathbf{z}},\lambda\rangle=J_3|p\hat{\mathbf{z}},\lambda\rangle=\lambda|p\hat{\mathbf{z}},\lambda\rangle \tag{218}$$

As there can be no orbital angular momentum along the direction of motion, we can interpret the index λ as the eigenvalue of the total angular momentum $\mathbf{J}$ along the direction of linear motion of the particle.

2.13 SUMMARY OF THE REPRESENTATION THEORY FOR MASSIVE PARTICLES

Important features of this class are summarized below:

(i) The little group of the timelike class of representations is the group of three-dimensional rotations;

(ii) In order to generate a complete basis of this irreducible representation consisting of general eigenvectors of $\hat{P}^{\mu}$, we shall operate on the basis vector $|\mathbf{0}\lambda\rangle$ by the remaining transformations of the factor group. We note that the factor group of the Poincare group with respect to the full translation group is the homogeneous Lorentz group. We also know that the homogeneous Lorentz group comprises rotations in the three spatial coordinates and Lorentz "boosts" that involve the time axis. Since we have already used the "rotations" (the subgroup consisting of rotations in the three-dimensional space which is isomorphic to $SO(3)$ constitutes the "little group" of p_t^{μ}) for constructing our basis, we need now to implement the effect of Lorentz "boosts" on the basis vector $|\mathbf{0}\lambda\rangle$.

In this context, we note that a general homogeneous Lorentz transformation can be factorized as $\Lambda = R(\alpha, \beta, 0)\, L_3(\xi)\, R(\phi, \theta, \psi)^{-1}$. Now, if we operate with this transformation on the vector $|\mathbf{0}\lambda\rangle$, the first rotation on the right, being a member of the "little group" $SO(3)$, shall leave the subspace corresponding to the "standard vector" invariant i.e. the effect of this rotation will be merely to move into another of the basis vectors that have already been generated so that no new basis vectors will be generated. Hence, we need consider here only the impact of a Lorentz boost $L_3(\xi)$ followed by a rotation $R(\alpha, \beta, 0)$. Let us symbolize

$$|\mathbf{p}\lambda\rangle \equiv R(\alpha, \beta, 0)|p\hat{\mathbf{z}}\ \lambda\rangle \equiv R(\alpha, \beta, 0)\, L_3(\xi)|\mathbf{0}\lambda\rangle \equiv H(p)|\mathbf{0}\lambda\rangle \tag{219}$$

p being the magnitude of the 3 momentum and α, β, the polar and azimuthal angles of the momentum vector $\mathbf{p}$. The Lorentz transformation $R(\alpha, \beta, 0)\, L_3(\xi) \equiv H(p)$ transforms our "standard vector" corresponding to $\mathbf{p} = 0$, $p^0 = M$, to a general vector p^{μ}, thereby completing the basis.

(iii) $|\mathbf{p}\lambda\rangle \equiv R(\alpha, \beta, 0)|p\hat{\mathbf{z}}\lambda\rangle \equiv R(\alpha, \beta, 0)\, L_3(\xi)|\mathbf{0}\lambda\rangle \equiv H(p)|\mathbf{0}\lambda\rangle$ is an eigenstate of P^{μ}, with eigenvalues $(p^0, \mathbf{p})$, where $p^0 = M\cosh\xi$ and $|\mathbf{p}| = M\sinh\xi$. This implies that $p_{\mu}p^{\mu} = p_0^2 - \mathbf{p}^2 = M^2$. The proof follows from the fact that T_4, the group of translations with generators $\{P^{\mu}\}$ is an invariant abelian subgroup of the Poincare group. We have

$$\begin{aligned} P^{\mu}|\mathbf{p}\lambda\rangle &= P^{\mu}\, H(p)|\mathbf{0}\lambda\rangle = H(p)\, H(p)^{-1}\, P^{\mu}H(p)|\mathbf{0}\lambda\rangle \\ &= H(p)\, H(p)^{\mu}_{\ \nu}\, P^{\nu}|\mathbf{0}\lambda\rangle = H(p)\, H(p)^{\mu}_{\ \nu}\, p_t^{\nu}|\mathbf{0}\lambda\rangle = P^{\mu}|\mathbf{p}\lambda\rangle \end{aligned} \tag{220}$$

where $p_t^{\nu} \equiv (p^{\circ}, 0, 0, 0)$ are the momentum eigenvalues of the standard vector $|0\lambda\rangle$. Also

$$p^{\mu} = H(p)^{\mu}_{\ \nu}\, p_t^{\nu} = H(p)^{\mu}_{\ 0}\, p_t^0 + H(p)^{\mu}_{\ 1}\, p_t^1 + H(p)^{\mu}_{\ 2}\, p_t^2 + H(p)^{\mu}_{\ 3}\, p_t^3 = H(p)^{\mu}_{\ 0}\, p_t^0 \tag{221}$$

The fact that $p^0 = M\cosh\xi$ and $|\mathbf{p}| = M\sinh\xi$ can be established by computing the explicit representation of $H(p)$. We have

$$H(p) \equiv R(\alpha, \beta, 0)\, L_3(\xi)$$

$$\begin{pmatrix} 1 & 0 & 0 & 0 \\ 0 & \cos\alpha\cos\beta & -\sin\alpha & \cos\alpha\sin\beta \\ 0 & \sin\alpha\cos\beta & \cos\alpha & \sin\alpha\sin\beta \\ 0 & -\sin\beta & 0 & \cos\beta \end{pmatrix} \begin{pmatrix} \cosh\xi & 0 & 0 & \sinh\xi \\ 0 & 1 & 0 & 0 \\ 0 & 0 & 1 & 0 \\ \sinh\xi & 0 & 0 & \cosh\xi \end{pmatrix}$$

$$= \begin{pmatrix} \cosh\xi & 0 & 0 & \sinh\xi \\ \cos\alpha\sin\beta\sinh\xi & \cos\alpha\cos\beta & -\sin\alpha & \cos\alpha\sin\beta\cosh\xi \\ \sin\alpha\sin\beta\sinh\xi & \sin\alpha\cos\beta & \cos\alpha & \sin\alpha\sin\beta\cosh\xi \\ \cos\beta\sinh\xi & -\sin\beta & 0 & \cos\beta\cosh\xi \end{pmatrix} \tag{222}$$

so that

$$p^0 = H(p)^0_0\, p^0_t = M\cosh\xi,\; p^1 = H(p)^1_0\, p^0 = M\cos\alpha\sin\beta\sinh\xi$$
$$p^2 = H(p)^2_0\, p^0 = M\sin\alpha\sin\beta\sinh\xi \text{ and } p^3 = H(P)^3_0\, p^0 = M\cos\beta\sinh\xi \tag{223}$$

Obviously, $\mathbf{p}^2 = p_1^2 + p_2^2 + p_3^2 = M^2\sinh^2\xi$ so that $|\mathbf{p}| = M\sinh\xi$.

(iv) As an immediate corollary to (1), we have

$$T(b)|\mathbf{p}\lambda\rangle = e^{-ib^{\mu}p_{\mu}}|\mathbf{p}\lambda\rangle \tag{224}$$

(v) We have

$$\Lambda|\mathbf{p}\lambda\rangle = \Lambda H(p)|\mathbf{0}\lambda\rangle = H(p')\,[H(p')^{-1}\,\Lambda H(p)]|\mathbf{0}\lambda\rangle$$
$$= D^s[R(\Lambda, p)]^{\lambda'}_{\lambda}|p'\,\lambda'\rangle \tag{225}$$

where $p'^{\mu} = \Lambda^{\mu}_{\ \nu}\, p^{\nu}$ and $R(\Lambda, p) \equiv H(p')^{-1}\,\Lambda H(p)$, and $D^s\,[R]$ is the representation matrix of the group of three-dimensional rotations, $SO(3)$ corresponding to the angular momentum s.

(vi) To confirm that $R(\Lambda, p) \equiv H(p')^{-1}\,\Lambda H(p)$ is a rotation in three dimensions, we examine the sequence-wise effect of each of the three operations on $|\mathbf{0}\lambda\rangle$. We have $H\,(p)|\mathbf{0}\lambda\rangle = |\mathbf{p}\,\lambda\rangle$, $\Lambda|\mathbf{p}\lambda\rangle = D^s\,[R]|\mathbf{p}'\,\lambda'\,\rangle$ and $H(p')^{-1}D^s[R]|\mathbf{p}'\,\lambda'\rangle = D^s[R]|\mathbf{0}\lambda\rangle$, and whence the effect of the combined Lorentz transformation embodied in $R(\mathrm{L}, p) \equiv H(p')^{-1}\,\Lambda H(p)$ is to leave the standard vector invariant. This implies that $R(\Lambda, p)$ is a member of the little group and hence, a rotation in the three-dimensional space.

(vii) In view of (ii) and (iii), we conclude that the vector space spanned by the vectors $\{\mathbf{p}\lambda\}$ is invariant under the transformations of the Poincare group.

(viii) The irreducibility of the representation follows from the fact that all the basis vectors are generated from the action of group transformation generators $\hat{J}$ and group transformations $H(p)$ on a single "standard vector".

(ix) The unitarity of the representation follows from: (i) generators of the group transformations being realized as Hermitian operators and (ii) the representation matrices of the translation operations $e^{-ib^{\mu}p_{\mu}}$ and the $D^s[R]$ matrices representing rotations are both unitary.

2.14 NULL VECTOR BASED REPRESENTATIONS OF POINCARE GROUP (WIGNER, 1939; OHNUKI, 1976; TUNG, 1984; FUSHCHICH, 1994)

Consider the null vector designated as $p_n^\mu = 0$. This vector is invariant under all homogeneous Lorentz transformations for $\Lambda^\mu_\nu p_n^\nu = 0$ for all homogeneous Lorentz transformations Λ. Hence, the "little group" of p_n is the homogeneous Lorentz group $\tilde{L}_+$. Each irreducible unitary representation of the homogeneous Lorentz group, therefore, induces an irreducible unitary representation of the Poincare group. These IRRs are labelled by the eigenvalue $c_1 = 0$ of the Casimir operator C_1 alongwith the labelling indices (j_0, v) of the unitary representations of the homogeneous Lorentz group, viz. $\tilde{L}_+$.

The basis vectors in a particular representation are taken as the simultaneous eigenvectors of the three commuting operators $(P^\mu, \mathbf{J}^2, J_3)$ with the corresponding eigenvalues $(p_n^\mu = 0, j, m)$, respectively. The basis vectors satisfy the equations

$$T(b)|0\, jm\rangle = |0\, jm\rangle \tag{226}$$

$$\Lambda|0\, jm\rangle = D^{j_0 v}\,[\Lambda]^{j'm'}_{jm}|\mathbf{0}\, j'm'\rangle \tag{227}$$

where $D^{j_0 v}\,[\Lambda]$ are identified as the unitary representation matrices of the homogeneous Lorentz group.

In addition to the usual Casimir operator $\hat{C}_1 = \hat{P}^\mu \hat{P}_\mu$ mentioned above, this class of representations also admits the Casimir operators $\hat{C}_{n1} = \hat{J}_{\mu\nu}\hat{J}^{\mu\nu}$ and $\hat{C}_{n2} = \frac{1}{8}\,\varepsilon_{\mu\nu\rho\sigma}\,\hat{J}^{\nu\rho}\,\hat{J}^{\mu\sigma}$, being the Casimir operators of the homogeneous Lorentz group (which is the little group for this class of representations). Besides, we also have the universal Casimir operator $\hat{C}_3 = \exp(2i\pi\hat{J}_{12}) = \exp(2i\pi\hat{J}_{23})\exp(2i\pi\hat{J}_{31})$ that has eigenvalues of ± 1 corresponding to the representation being single valued or double valued respectively.

2.15 LIGHT LIKE REPRESENTATIONS OF POINCARE GROUP (WIGNER, 1939; OHNUKI, 1976; TUNG, 1984; FUSHCHICH, 1994)

The light like case is characterized by $c_1 = 0$, $\mathbf{p} \neq 0$, the former condition $(c_1 = 0)$ implies that $p^\mu p_\mu = 0$, whence $|p^0| = |\mathbf{p}|$. Hence, in this case, a "standard vector" can be picked to be of the form $p_l^\mu = (\omega_0, 0, 0, \omega_0)$. A general state of momentum $p^\mu = (\omega, \mathbf{p}) \equiv (\omega, \omega, \hat{\mathbf{p}})$, where $\omega = p^0 = |\mathbf{p}| = \frac{\mathbf{p}}{\hat{\mathbf{p}}}$, can be generated from the standard vector by

(i) Applying a Lorentz boost $L_3(\xi)$ to the standard vector to transform the energy from ω_0 to ω.

(ii) Thereafter applying a rotation $R(\phi, \theta, 0)$ (where the unit vector $\hat{\mathbf{p}}$ is characterized by the angles (θ, ϕ) to bring the x^3 axis in the direction of unit vector $\hat{\mathbf{p}}$.

We denote this composite transformation from p_l^μ to p^μ by $H(p)$ so that, we have $p^\mu = H(p)^\mu_\nu\, p_l^\nu = [R(\phi, \theta, 0)L_3(\xi)]^\mu_\nu\, p_l^\nu$.

As mentioned above, the independent components of the Pauli Lubanski vector $\hat{W}^{\lambda} = \frac{1}{2}\varepsilon^{\lambda\mu\nu\sigma}$ $\hat{J}_{\mu\nu} p_{\sigma}$ constitute the generatorsc of the little group of the standard vector p_l^{μ}. Noting that $p_l^{\mu} = (\omega_0, 0, 0, \omega_0)$, we get

$$\hat{W}^0 = \hat{W}^3 = \omega_0 \hat{J}_{12} = \omega_0 \hat{J}_3,\ \hat{W}^1 = \omega_0 (\hat{J}_{23} + \hat{J}_{20}) = \omega_0 (\hat{J}_1 + \hat{K}_2) \tag{228}$$

and

$$\hat{W}^2 = \omega_0(\hat{J}_{31} - \hat{J}_{10}) = \omega_0 (\hat{J}_2 - \hat{K}_1) \tag{229}$$

Hence, the second Casimir operator for the lightlike case is given by

$$\hat{C}_2 = \hat{W}^{\mu} \hat{W}_{\mu} = (\hat{W}^0)^2 - \hat{\mathbf{W}}^2 = -(\hat{W}^1)^2 - (\hat{W}^2)^2 \tag{230}$$

and the general Lie algebra for the Pauli Lubanski vector becomes, in the lightlike case, $[\hat{W}^1, \hat{W}^2] = 0$, $[\hat{W}^2, \hat{J}_3] = i\hat{W}^1$, $[\hat{W}^1, \hat{J}_3] = -i\hat{W}^2$.

We find that this reduced Lie algebra for the lightlike case is same as that of the two dimensional Euclidean group E_2 with the nonvanishing components of the Pauli Lubanski vector $\hat{W}^1$, $\hat{W}^2$ replacing the components of the translation generators. The representation theory for E_2, therefore, holds. Accordingly, we infer that.

(i) The degenerate irreducible unitary representations are one dimensional and correspond to $w = 0$, where w is the eigenvalue of the Pauli Lubanski vector and, in this case, the basis vectors are labelled by λ, the eigenvalue of $\hat{J}_3$.

(ii) The non-degenerate unitary IRRs are all infinite dimensional, they correspond to $w > 0$ and, in this case, the basis vectors are identified by two indices w, λ, i.e., $\{|w\lambda\rangle, \lambda \in \mathbb{C}\}$, where w labels the IRR and λ labels the basis vectors within the IRR.

A basis for a representation of the full Poincare group can be constructed by starting from any of these representations, labelled by $M = 0$ and w, by applying the homogeneous Lorentz transformation $H(p)$ to the standard basis vector of the chosen representation.

The non-degenerate representations corresponding to $M = 0$, $w > 0$ have not been found to be realized as elementary particles so far. The degenerate representations corresponding to $M = 0$, $w = 0$, however, manifest themselves as photons for $\lambda = \pm 1$, neutrino for $\lambda = -1/2$ and antineutrino $\lambda = 1/2$ states.

As mentioned above, in the degenerate case, the representations and the subspace corresponding to the standard vector p_l^{μ} are one dimensional. The basis vectors are defined by

$$\hat{P}^{\mu}|\mathbf{p}_l \lambda\rangle = p_l^{\mu}|\mathbf{p}_l \lambda\rangle,\ p_l^{\mu} = (\omega_0, 0, 0, \omega_0) \tag{231}$$

$$\hat{J}_3|\mathbf{p}_l \lambda\rangle = \lambda|\mathbf{p}_l \lambda\rangle \tag{232}$$

$$\hat{W}^i|\mathbf{p}_l \lambda\rangle = 0,\ i = 1, 2 \tag{233}$$

As in the case of timelike representations, we can obtain the general basis vector by a combination of a Lorentz boost followed by a rotation on the standard vector for

(i) By applying a Lorentz boost $L_3(\xi)$ to the standard vector we transform the energy from ω_0 to ω.

(ii) Thereafter by applying a rotation $R(\phi, \theta, 0)$ (where the unit vector $\hat{\mathbf{p}}$ is characterized by the angles (θ, ϕ)), we bring the x^3 axis in the direction of unit vector $\hat{\mathbf{p}}$.

The above also follows from the fact that in order to generate a complete basis of this irreducible representation consisting of general eigenvectors of $\hat{P}^\mu$, we shall operate on the basis vector $|\mathbf{p}_l\, \lambda\rangle$ by the remaining transformations of the factor group. We note that:

(i) The factor group of the Poincare group with respect to the full translation group is the homogeneous Lorentz group.

(ii) The homogeneous Lorentz group comprises rotations in the three spatial coordinates and Lorentz "boosts" that involve the time axis.

(iii) The little group of the standard vector i.e. the group comprising Lorentz transformations that leave the subspace of the standard vector invariant is the two-dimensional Euclidean group with the generators $\hat{W}^1$, $\hat{W}^2$, $\hat{J}_3$. In other words, we can say that the transformations induced by these generators shall simply move one basis vector into another and will not result in the generation of new basis vectors. As a corollary, it follows to generate new basis vectors, we need to operate on the standard vector by the residual transformations of the factor group. Now, in terms of the components of the Pauli Lubanski vector for the standard vector, $(\omega_0, 0, 0, \omega_0)$, we can write the generators of the homogeneous Lorentz group as $\hat{W}^1 = \omega_0\ (\hat{J}_1 + \hat{K}_2)$, $\hat{W}^2 = \omega_0(\hat{J}_2 - \hat{K}_1)$, $\hat{J}_3$, $\hat{K}_3$, $\hat{Z}^1$. $= \omega_0\ (\hat{J}_1 - \hat{K}_2)$, $\hat{Z}^2 = \omega_0(\hat{J}_2 + \hat{K}_1)$.

Out of these, the first three generators relate to the little group of the standard vector, $(\omega_0, 0, 0, \omega_0)$, as mentioned earlier. Hence, operation with these generators will not generate any new basis vectors. The other three generators when operating on the standard vector will generate new basis vectors.

Hence, we need consider here only the impact of a Lorentz boost $L_3(\xi)$ followed by a rotation $R(\phi, \theta, 0)$. Let us symbolize

$$|\mathbf{p}\lambda\rangle = H(p)|\mathbf{p}_l\, \lambda\rangle = R(\phi, \theta, 0)|p\hat{\mathbf{z}}\lambda\rangle = R(\phi, \theta, 0)\ L_3(\xi)|\mathbf{p}_l\, \lambda\rangle \tag{234}$$

$p = |\mathbf{p}| = \omega_0\, e^{\xi}$ being the magnitude of the 3 momentum and θ, ϕ, the polar and azimuthal angles of the momentum vector $\mathbf{p}$. The Lorentz transformation $R(\theta, \phi, 0)\ L_3(\xi) \equiv H(p)$ transforms our "standard vector" corresponding to $(\omega_0, 0, 0, \omega_0)$, to a general vector p^μ, thereby completing the basis.

These basis vectors $\{|\mathbf{p}\lambda\rangle\}$ transform as follows under the group transformation operations of the Poincare group:

$$T(b)|\mathbf{p}\lambda\rangle = e^{-ib^\mu p_\mu}|\mathbf{p}\lambda\rangle \tag{235}$$

$$\Lambda|\mathbf{p}\lambda\rangle = e^{-i\lambda\theta(\Lambda,\, p)}|\mathbf{\Lambda p}\lambda\rangle \tag{236}$$

where $\theta(\Lambda, p)$ is determined from $e^{-i\lambda\theta(\Lambda,\, p)} = \langle \mathbf{p}_l \lambda|\, H^{-1}(\mathbf{\Lambda p})\ \Lambda H(p)\,|\mathbf{p}_l \lambda\rangle$. The first equation follows exactly as in the case of timelike representations. For the second equation, we have

$$\Lambda|\mathbf{p}\lambda\rangle = \Lambda H(p)|\mathbf{p}_l\, \lambda\rangle = H(\Lambda p)\ [H(\Lambda p)^{-1}\ \Lambda H(p)]|\mathbf{p}_l\, \lambda\rangle$$

$$= H(\Lambda p)|\mathbf{p}_l\ \lambda\rangle\ \langle \mathbf{p}_l\ \lambda\rangle|[H(\Lambda p)^{-1}\ \Lambda H(p)]|\mathbf{p}_l\ \lambda\rangle$$
$$= |\Lambda \mathbf{p}\lambda\rangle\ \langle \mathbf{p}_l\ \lambda|[H(\Lambda p)^{-1}\ \Lambda H(p)]|\mathbf{p}_l\ \lambda\rangle$$
$$= e^{-i\lambda\theta(\Lambda,\ p)}|\Lambda \mathbf{p}\lambda\rangle$$

Following are some important features that characterize $M = 0$ or light like representations:

(i) For integral $\lambda \in \mathbb{C}$, the representations are single valued whereas for λ odd half integer, the representations are double valued. This is due to the fact that on taking the limit $M \to 0$ of $c_1^2 = M^2 > 0$, the little group of the $c_1^2 = M^2 > 0$ representation, viz., $SO(3)$ picks the single and double valued representations of the restricted little group $SO(2)$ (that is the relevant little group for the $M = 0$ case) from its multivalued representations.

(ii) The basis vectors in lightlike case $\{|\mathbf{p}\lambda\rangle\}$ are labelled by the momentum $\mathbf{p}$ and the helicity λ. This helicity state can take $(2s + 1)$ values. For timelike cases this index can transform among all the possible $(2s + 1)$ values under Lorentz transformation as is seen from the transformation equation for the basis states for massive particles viz. $\Lambda|\mathbf{p}\lambda\rangle = |\mathbf{p}'\ \lambda'\rangle\ D^s\,[R(\Lambda, p)]^{\lambda}_{\lambda'}$ which mixes the helicity states. However, this helicity index is invariant under Lorentz transformations for massless states as is seen from the transformation equation $\Lambda|\mathbf{p}\lambda\rangle = e^{-i\lambda\theta(\Lambda,\ p)}|\Lambda\mathbf{p}\lambda\rangle$.

2.16 SPACE LIKE REPRESENTATIONS OF POINCARE GROUP (WIGNER, 1939; OHNUKI, 1976; TUNG, 1984; FUSHCHICH, 1994)

The space like representations are characterized by $c_1 = p_\mu p^\mu = M^2 < 0$. We write $Q^2 = -M^2 = -c_1 > 0$ and represent the "standard vector" by $p_s^\mu = (0, 0, 0, Q)$. The components of Pauli Lubanski vector $\hat{W}^\lambda = \frac{1}{2}\ \varepsilon_{\lambda\mu\nu\sigma}\ \hat{J}^{\mu\nu}\ p^\sigma$ in the case of subspace corresponding to the standard vector $p_s^\mu = (0, 0, 0, Q)$ become $\hat{W}^0 = Q\hat{J}_3$, $\hat{W}^1 = -Q\hat{K}_2$, $\hat{W}^3 = Q\hat{K}_1$, $\hat{W}^2 = 0$. Using these values, the second Casimir operator is calculated as $\hat{C}_2 = Q^2\ (\hat{J}_3^2 - \hat{K}_1^2 - \hat{K}_2^2)$ and the general Lie algebra for the Pauli Lubanski vector becomes, in the spacelike case, $[\hat{K}_2, \hat{J}_3] = i\hat{K}_1$, $[\hat{K}_1, \hat{J}_3] = -i\hat{K}_2$, and $[\hat{K}_1, \hat{K}_2] = -i\hat{J}_3$.

This Lie algebra is similar to that of $SO(3)$ with $\hat{K}_1$ and $\hat{K}_2$ replacing $\hat{J}_1$ and $\hat{J}_2$, with a negative sign in the last commutator. The presence of this negative sign implies that the little group is not $SO(3)$ but $SO(2, 1)$. Now, $SO(2, 1)$ is a non-compact group and so all its unitary irreducible representations are infinite dimensional.

We obtain the unitary IRRs of the $SO(2, 1)$ corresponding to the Lie algebra $[\hat{K}_2, \hat{J}_3] = i\hat{K}_1$, $[\hat{K}_1, \hat{J}_3] = -i\hat{K}_2$ and $[\hat{K}_1, \hat{K}_2] = i\hat{J}_3$. For this purpose, we define the raising and lowering operators in the usual manner as $\hat{K}_\pm = \hat{K}_1 \pm i\hat{K}_2$ in terms of which the Lie algebra takes the form $[\hat{J}_3, \hat{K}_\pm] = \pm\ \hat{K}_\pm$ and $[\hat{K}_+, \hat{K}_-] = -2\hat{J}_3$. We write the second Casimir operator as

$$\hat{C}_2 = \frac{1}{p_s^2}\ \hat{W}^2 = \frac{1}{-Q^2}\ Q^2\ (\hat{J}_2^3 - \hat{K}_1^2 - \hat{K}_2^2) = (\hat{K}_1^2 + \hat{K}_2^2 - \hat{J}_3^2) = (\hat{K}_+\hat{K}_- + \hat{J}_3 - \hat{J}_3^2) \quad (237)$$

instead of $\hat{C}_2 = Q^2\ (\hat{J}_3^2 - \hat{K}_1^2 - \hat{K}_2^2)$ because $Q^2 = -c_1$ is the eigenvalue of the first Casimir operator that remains invariant for a particular IRR. As in the case of $SO(3)$, we choose the basis vectors to be simultaneous eigenvectors of $\hat{J}_3$ and $\hat{C}_2$ so that we may label them by the eigenvalues of these operators m and c_2, respectively, and write $\hat{J}_3|c_2\, m\rangle$ and $m|c_2\, m\rangle$ and $\hat{C}_2|c_2\, m\rangle = c_2|c_2\, m\rangle$. Then, we have

$$\hat{J}_3\, \hat{K}_\pm|c_2\, m\rangle = \hat{K}_\pm\, \hat{J}_3|c_2\, m\rangle + [\hat{J}_3,\, \hat{K}_\pm]|c_2\, m\rangle$$
$$= \hat{K}_\pm\, \hat{J}_3|c_2\, m\rangle \pm \hat{K}_\pm|c_2\, m\rangle = (m \pm 1)\, \hat{K}_\pm|c_2\, m\rangle \tag{238}$$

Thus, $\hat{K}_\pm|c_2\, m\rangle$ is an eigenstate of $\hat{J}_3$ with eigenvalue $m \pm 1$, i.e, in line with our labelling scheme for eigenstates, we must have $\hat{K}_\pm|c_2\, m\rangle \propto |c_2,\, m \pm 1\rangle$ or $\hat{K}_\pm|c_2\, m\rangle = A_m^\pm|c_2,\, m \pm 1\rangle$.

Now, unitarity of the representations mandates that the generators are Hermitian or that $\hat{K}_\pm^\dagger = \hat{K}_\mp$ whence, assuming normalization of states

$$\langle c_2\, m|\hat{K}_\pm^\dagger\, \hat{K}_\pm|c_2\, m\rangle = \langle c_2\, m|\hat{K}_\mp\, \hat{K}_\pm|c_2\, m\rangle = \langle c_2\, m \pm 1|c_2\, m \pm 1\rangle|A_m^\pm|^2 = |A_m^\pm|^2$$

We also have, using the expression for the Casimir operator

$$\langle c_2\, m|\hat{K}_\pm^\dagger\, \hat{K}_\pm|c_2\, m\rangle = \langle c_2\, m|\hat{C}_2\, \hat{J}_3^2\, \hat{J}_3|c_2\, m\rangle = c_2 + m(m \pm 1)$$

whence $|A_m^\pm|^2 = c_2 + m(m \pm 1) > 0$ which represents the unitarity constraint. Two cases arise:

(a) $c_2 > 0$. Since, except for this positivity condition, c_2 is unconstrained, it may take infinitesimal positive values. Therefore, $m(m \pm 1)$ must itself also take only non-negative values for the above constraint to be satisfied. We also know that m being the eigenvalue of the angular momentum operator $\hat{J}_3$ can increase or decrease only in integral steps. Thirdly, from $m(m \pm 1) \geq 0$ we also find that while $m = 0$ and $m = 1$ are permitted values but no real number between them is permitted. A combination of all the above considerations implies that the allowed values are $m = 0, \pm 1, \pm 2, \ldots$.

(b) $c_2 < 0$. We then have $m(m \pm 1) > |c_2|$. This implies that c_2 can be written as $c_2 = -j(j + 1)$ with $j = 0, \pm 1, \pm 2, \ldots$, whence m can take values in integral steps outside the range $[-j, j]$.

In conclusion, we find that c_2 can either be positive and take continuous values in the range $0 < c_2 < \infty$ or be negative and take discrete values $c_2 = -j(j + 1)$. These eigenvalues of the second Casimir operator can be used to label the IRRs of the little group $SO(2, 1)$.

Any of these unitary IRRs can be used to construct the unitary IRRs of the full Poincare group by applying, on a vector from the subspace of the selected representation (of the little group), homogeneous Lorentz transformations from outside the relevant little group. The procedure to transform the "standard vector" $p_s^\mu = (0, 0, 0, Q)$ to a general vector p^μ is illustrated as follows:

$L_3(\xi)\, p_s^\mu(0, 0, 0, Q) \to p_s'^\mu\,(Q \sinh \xi, 0, 0, Q \cosh \xi)$, for example,

$$p_s'^\mu = \begin{pmatrix} \cosh\xi & 0 & 0 & \sinh\xi \\ 0 & 1 & 0 & 0 \\ 0 & 0 & 1 & 0 \\ \sinh\xi & 0 & 0 & \cosh\xi \end{pmatrix} \begin{pmatrix} 0 \\ 0 \\ 0 \\ Q \end{pmatrix} = \begin{pmatrix} Q\sinh\xi \\ 0 \\ 0 \\ Q\cosh\xi \end{pmatrix}$$

$L_1(\zeta) p_s'^\mu\,(Q \sinh \xi, 0, 0, Q \cosh \xi)$

$$\to p_s''^{\mu}\,(Q \sinh \xi \cosh \zeta, Q \sinh \xi \sinh \zeta, 0, Q \cosh \xi)$$

as seen in

$$p_s''^{\mu} = \begin{pmatrix} \cosh\zeta & \sinh\zeta & 0 & 0 \\ \sinh\zeta & \cosh\zeta & 0 & 0 \\ 0 & 0 & 1 & 0 \\ 0 & 0 & 0 & 1 \end{pmatrix} \begin{pmatrix} Q\sinh\xi \\ 0 \\ 0 \\ Q\cosh\xi \end{pmatrix} = \begin{pmatrix} Q\sinh\xi\cosh\zeta \\ Q\sinh\xi\sinh\zeta \\ 0 \\ Q\cosh\xi \end{pmatrix}$$

$R_3(\phi)p_s''^{\mu}\,(Q \sinh \xi \cosh \zeta, Q \sinh \xi \sinh \zeta, 0, Q \cosh \xi) \to p^{\mu}(Q \sinh \xi \cosh \zeta, Q \sinh \xi \sinh \zeta \cos \phi, Q \sinh \xi \sinh \zeta \sin \phi, Q \cosh \xi)$ as follows:

$$p^{\mu} = \begin{pmatrix} 1 & 0 & 0 & 0 \\ 0 & \cos\phi & -\sin\phi & 0 \\ 0 & \sin\phi & \cos\phi & 1 \\ 0 & 0 & 0 & 1 \end{pmatrix} \begin{pmatrix} Q\sinh\xi\cosh\zeta \\ Q\sinh\xi\sinh\zeta \\ 0 \\ Q\cosh\xi \end{pmatrix} = \begin{pmatrix} Q\sinh\xi\cosh\zeta \\ Q\sinh\xi\sinh\zeta\cos\phi \\ Q\sinh\xi\sinh\zeta\sin\phi \\ Q\cosh\xi \end{pmatrix}$$

The parameters (ξ, ζ, ϕ) of the aforesaid sequence of transformations will be determined with the identification $p^{\mu}(Q \sinh \xi \cosh \zeta, Q \sinh \xi \sinh \zeta \cos \phi, Q \sinh \xi \sinh \zeta \sin \phi, Q \cosh \xi) \equiv (p^0, p^1, p^2, p^3)$

The general vector may be expressed in the usual notation as $|\mathbf{p}\lambda\rangle$ so that

$$|\mathbf{p}\lambda\rangle = H(p)|p_s\,\lambda\rangle \tag{239}$$

where $H(p) = R_3(\phi)\, L_1(\zeta)\, L_3(\xi)$.

The effect of the group operations on the general vector may be expressed, as in the preceding cases, by

$$T(b)|\mathbf{p}\lambda\rangle = e^{-ib_{\mu}p_{\mu}}|\mathbf{p}\lambda\rangle,\ \Lambda|\mathbf{p}\lambda\rangle = |\Lambda\mathbf{p}\lambda'\rangle = D^{c_2}\,[H^{-1}(\Lambda p)\,\Lambda H(p)]^{\lambda}_{\lambda'} \tag{240}$$

where D^{c2} symbolizes the representation matrix for the little group $SO(2,1)$ corresponding to the eigenvalue c_2 of the Casimir operator $\hat{C}_2$.

2.17 UNITARITY & NORMALIZATION OF REPRESENTATIONS OF POINCARE GROUP(WIGNER, 1939; OHNUKI, 1976; TUNG, 1984; FUSHCHICH, 1994)

In constructing unitary representations by the induced representation method, all generators are mandated to be Hermitian operators and the eigenvectors of a maximal set of commuting operators are designed to constitute the set of basis vectors. Since, the eigenvectors of Hermitian operators are invariably orthogonal, orthogonality of our basis is achieved by the very process of their construction.

We now consider the issue of their normalization. In constructing representations by the induced representation method, all the basis vectors are obtained from a "standard vector" say $\tilde{p} \equiv p_n, p_t, p_l, p_s$ by a unitary transformation, i.e., $|\mathbf{p}\lambda\rangle = H(p)|\tilde{p}\lambda\rangle$. Now, if we desire that the scalar product $\langle \mathbf{p}'\lambda'|\mathbf{p}\lambda\rangle$ is to be Lorentz invariant, then, we must have $\langle p'\lambda'|\mathbf{p}\lambda\rangle = \langle \mathbf{p}'\lambda'|\Lambda^{\dagger}\,\Lambda|\mathbf{p}\lambda\rangle$ which, in the case of timelike representations, becomes

$$\langle \mathbf{p}'\,\lambda'|\Lambda^{\dagger}\Lambda|\mathbf{p}\lambda\rangle = D^{s\dagger}\,[R(\Lambda, p')]^{\lambda'}_{\sigma'}\,\langle \Lambda\mathbf{p}'\,\sigma'|\Lambda\mathbf{p}\sigma\rangle\, D^{s}\,[R(\Lambda, p)]^{\sigma}_{\lambda} \tag{241}$$

with similar expressions for the other cases. Now, as mentioned above, the basis vectors that are eigenvectors of Hermitian operators with different eigenvalues, must necessarily be orthogonal so that $\langle \mathbf{p}'\,\lambda'|\mathbf{p}\lambda\rangle = N(p)\delta^3\,(\mathbf{p} - \mathbf{p}')\delta^{\lambda'}_{\lambda}$, where $N(p)$ is a normalization constant.

Combining these two equations and using the unitarity of the D^s matrices, we obtain

$$\begin{aligned} N(p)\delta^3\,(\mathbf{p} - \mathbf{p}')\,\delta^{\lambda'}_{\lambda} &= \langle \mathbf{p}'\,\lambda'|\mathbf{p}\lambda\rangle = \langle \mathbf{p}'\,\lambda'|\Lambda^{\dagger}\Lambda|\mathbf{p}\lambda\rangle \\ &= D^{s\dagger}\,[R(\Lambda, p')]^{\lambda'}_{\sigma'}\,\langle \Lambda\mathbf{p}'\,\sigma'|\Lambda\mathbf{p}\sigma\rangle\, D^{s}\,[R(\Lambda, p)]^{\sigma}_{\lambda} \\ &= N(p)\delta^3\,(\Lambda\mathbf{p} - \Lambda\mathbf{p}')\,\delta^{\lambda'}_{\lambda} \end{aligned} \tag{242}$$

i.e. $N(p)\delta^3\,(\mathbf{p} - \mathbf{p}') = N(\Lambda p)\delta^3\,(\Lambda\mathbf{p} - \Lambda\mathbf{p}')$ as the normalization condition.

Since, in determining the normalization constant, we need to sum over continuous index $\{p^i\}$, we must define a suitable integration measure. For this purpose, we expand an arbitrary state vector $|\psi\rangle$ in the basis $\{|\mathbf{p}\lambda\rangle\}$. We have, $|\psi\rangle = \sum_{\lambda}\int |\mathbf{p}\lambda\rangle\, \psi^{\lambda}(p)\tilde{d}p$. Since our basis is orthogonal, we also have $\psi^{\lambda'}(p') = \langle \mathbf{p}'\,\lambda'|\psi\rangle$. We, therefore, have

$$\psi^{\lambda'}(p') = \sum_{\lambda}\int \langle \mathbf{p}'\lambda'|\mathbf{p}\lambda\rangle\, \psi^{\lambda}(p)\tilde{d}p$$

$$= \int N(p)\delta^3\,(\mathbf{p} - \mathbf{p}')\delta^{\lambda'}_{\lambda}\psi^{\lambda}\,(p)\tilde{d}p = \int \psi^{\lambda'}\,(p)\; N(p)\delta^3\,(\mathbf{p} - \mathbf{p}')\;\tilde{d}p \tag{243}$$

Since $|\psi\rangle$ is an arbitrary state, we must have $\tilde{d}p = \dfrac{1}{N(p)}\,d^3p = \dfrac{1}{N(p)}\,dp^1\,dp^2\,dp^3$. Now, if $N(p)$ is covariant with respect to any homogeneous Lorentz transformation, i.e., $N(p)$ satisfies $N(p)\delta^3(\mathbf{p} - \mathbf{p}') = N(\Lambda p)\delta^3\,(\Lambda\mathbf{p} - \Lambda\mathbf{p}')$, then $\tilde{d}p = \tilde{d}\Lambda p$ so that the measure $\tilde{d}p$ is Lorentz invariant. To determine $N(p)$ explicitly, we note that $d^4\,(\Lambda p) = (\det \Lambda)d^4\,p = d^4\,p$ since the Lorentz transformation has unit determinant and so the 4-volume element is invariant under a Lorentz transformation. Since, the first Casimir operator $\hat{C}_1 = \hat{P}_{\mu}\,\hat{P}^{\mu}$, we have, in terms of eigenvalues $c_1 = p^2 = (p^0)^2 - \mathbf{p}^2$, whence $p^0 = \pm\,(\mathbf{p}^2 + c_1)^{1/2}$. The positive root is relevant to the timelike and lightlike cases which have physical applications whence an appropriate expression for $\tilde{d}p$ is $\tilde{d}p = \dfrac{1}{N}\theta\,(p^0)\,\delta(c_1 - p^2)d^4\,p = \dfrac{1}{N}\dfrac{d^3\,p}{2p^0} = \dfrac{1}{N}\,\dfrac{d^3\,p}{2(c_1 + \mathbf{p}^2)^{1/2}}$. The presence of $\theta(p^0)$ mandates that we consider only positive values of $p^0 = (\mathbf{p}^2 + c_1)^{1/2}$ whereas the δ term ensures that $c_1{=}p^2{=}\,(p^0)^2 - \mathbf{p}^2$. Further, $p^0 = +\,(\mathbf{p}^2 + c_1)^{1/2}$ implies $d^4\,p \equiv dp^0\,d^3p = \dfrac{1}{2p^0}\,d^3p$. Conventionally, N is chosen as $N \equiv (2\pi)^3$ so that $\tilde{d}p = \dfrac{1}{(2\pi)^3}\dfrac{d^3\,p}{2\,p^0}$ and the normalization condition reads

$$\langle \mathbf{p}'\,\lambda'|\mathbf{p}\lambda\rangle = (2\pi)^3\,2p^0\,\delta^3(\mathbf{p} - \mathbf{p}')\delta^{\lambda'}_{\lambda} \equiv \tilde{\delta}(p - p')\delta^{\lambda'}_{\lambda} \tag{244}$$

2.18 SL(2,C) (SPINOR) REPRESENTATION OF THE LORENTZ GROUP & EXPLICIT CALCULATION OF WIGNER ANGLE (HALPERN, 1968; OHNUKI, 1976)

The three Pauli matrices $\{\sigma_i\ i = 1, 2, 3\}$ constitute the basic instruments for obtaining the spinor representation. To this set, we append the matrix $\sigma_0 = \mathbf{I}_{2\times 2}$. Together, this set of four matrices constitutes a basis for the four-dimensional space of 2×2 matrices so that any such matrix can be expressed as a linear combination of these four matrices i.e. given a 2×2 matrix $\mathbf{M}$, we can write it as $\mathrm{M} = \sum_{\mu=0}^{3} c^{\mu}\sigma_{\mu}$ where $c^{\mu} = \frac{1}{2} Tr\,(\sigma_{\mu}{}^{\dagger}\,\mathbf{M}) = \frac{1}{2} Tr\,(\sigma_{\mu}\,\mathbf{M})$. It can be explicitly verified that

$$\mathbf{M}\sigma_2\,\mathbf{M}^T\,\sigma_2 = \det(\mathbf{M}) \tag{245}$$

Further, we can write a 4-vector with components x^{μ}, $\mu = 0, 1, 2, 3$ in the above basis as a 2×2 matrix as $x \equiv \mathbf{X} = x^{\mu}\,\sigma_{\mu} = \begin{pmatrix} x^0 + x^3 & x^1 - ix^2 \\ x^1 + ix^2 & x^0 - x^3 \end{pmatrix}$. The matrix $\mathbf{X}$ is Hermitian if x is real. Also $\det(\mathbf{X}) = (x^0)^2 - \mathbf{x}^2 = x^2$. Let us assume that $\mathbf{X}$ is transformed by a matrix of the $SL(2, \mathbb{C})$ group as $\mathbf{X}' = \mathbf{AXA}^{\dagger}$ where $\det(\mathbf{A}) = 1$. Then, $\det(\mathbf{X}') = \det(\mathbf{X})$ and the components of x are given by $x^{\mu} = \frac{1}{2} Tr(\sigma_{\mu}\,\mathbf{X}')$. Writing $x' = x'^{\mu}\,\sigma_{\mu}$, we obtain $x'^{\mu} = \frac{1}{2} Tr(\sigma_{\mu}\mathbf{X}') = \frac{1}{2} Tr(\sigma_{\mu}\,\mathbf{AXA}^{\dagger})$ $= \frac{1}{2} Tr(\sigma_{\mu}\,\mathbf{A}\sigma_{\nu}x^{\nu}\,\mathbf{A}^{\dagger}) = \frac{1}{2} Tr(\sigma_{\mu}\,\mathbf{A}\sigma_{\nu}\,\mathbf{A}^{\dagger})x^{\nu} = \Lambda^{\mu}_{\nu}(\mathbf{A})x^{\nu}$ where $\Lambda^{\mu}_{\nu}(\mathbf{A}) = \frac{1}{2} Tr(\sigma_{\mu}\,\mathbf{A}\sigma_{\nu}\,\mathbf{A}^{\dagger})$ is a Lorentz transformation because $x^2 = \det(\mathbf{X}) = \det(\mathbf{X}') = x'^2$. Since $\Lambda^{\mu}_{\nu}(\mathbf{A}) = \frac{1}{2} Tr(\sigma_{\mu}\,\mathbf{A}\sigma_{\nu}\,\mathbf{A}^{\dagger})$, it follows that we can write $\mathbf{A}\sigma_{\mu}\,\mathbf{A}^{\dagger} = \sum_{\nu} \Lambda^{\nu}_{\mu}\,\sigma_{\mu}$. We also have the following properties for Λ^{ν}_{μ}:

(i) We have $\Lambda^{\lambda}_{\nu}(\mathbf{AB}) = \frac{1}{2} Tr[\sigma_{\lambda}\,\mathbf{AB}\sigma_{\nu}\,(\mathbf{AB})^{\dagger}] = \frac{1}{2} Tr[\sigma_{\lambda}\,\mathbf{AB}\sigma_{\nu}\,\mathbf{B}^{\dagger}\mathbf{A}^{\dagger}]$

$$= \frac{1}{2} Tr[\mathbf{A}^{\dagger}\,\sigma_{\lambda}\,\mathbf{AB}\sigma_{\nu}\,\mathbf{B}^{\dagger}] = \frac{1}{2} Tr[\mathbf{A}^{\dagger}\,\sigma_{\lambda}\,\mathbf{A}\sigma_{\mu}\,\sigma_{\mu}\,\mathbf{B}\sigma_{\nu}{}^{\dagger}\,\mathbf{B}^{\dagger}]$$

$$= \sum_{\mu} \frac{1}{4} Tr\,[\mathbf{A}^{\dagger}\sigma_{\lambda}\mathbf{A}\sigma_{\mu}]\ Tr\,[\sigma_{\mu}\mathbf{B}\sigma_{\nu}\mathbf{B}^{\dagger}]$$

$$= \sum_{\mu} \frac{1}{4} Tr\,[\sigma_{\lambda}A\sigma_{\mu}\mathbf{A}^{\dagger}]\ Tr\,[\sigma_{\mu}\mathbf{B}\sigma_{\nu}\mathbf{B}^{\dagger}] = \sum_{\mu} \Lambda^{\lambda}_{\mu}\,(\mathbf{A})\Lambda^{\mu}{}_{\nu}(\mathbf{B})$$

$$= [\Lambda(\mathbf{A})\Lambda(\mathbf{B})]^{\lambda}{}_{\nu} \quad \text{whence } \Lambda(\mathbf{AB}) = \Lambda(\mathbf{A})\ \Lambda(\mathbf{B})$$

(ii) $\Lambda^{\lambda *}_{\ \nu} = \frac{1}{2}[Tr(\sigma_\nu \mathbf{A}\sigma_\mu \mathbf{A}^\dagger)]^* = \frac{1}{2}\ Tr\,(\sigma_\nu \mathbf{A}\sigma_\nu \mathbf{A}^\dagger)^\dagger = \frac{1}{2}\ Tr(\mathbf{A}\sigma_\mu \mathbf{A}^\dagger \sigma_\nu) = \ \frac{1}{2}Tr(\sigma_\nu \mathbf{A}\sigma_\mu \mathbf{A}^\dagger\ ) = \Lambda^\nu_{\ \mu}$

thereby confirming that all elements of the matrix are real. In the above derivation, we have used the Hermitian nature of the Pauli matrices.

$$\text{(iii) } \sigma_2\mathbf{A}^{\dagger T}\sigma_\mu^T\mathbf{A}^T\sigma_2 = \sigma_2\mathbf{A}^{\dagger T}\sigma_\mu^T\sigma_2\sigma_2\mathbf{A}^T\sigma_2 = \det(\mathbf{A})\sigma_2\mathbf{A}^{\dagger T}\sigma_\mu^T\sigma_2\mathbf{A}^{-1} = \sigma_2\mathbf{A}^{\dagger T}\sigma_\mu^T\sigma_2\mathbf{A}^{-1}$$

$$= \sigma_2\mathbf{A}^{\dagger T}\sigma_2\sigma_2\sigma_\mu^T\sigma_2\mathbf{A}^{-1} = \det(\mathbf{A}^\dagger)(\mathbf{A}^\dagger)\ ^{-1}\sigma_2\sigma_\mu^T\sigma_2\mathbf{A}^{-1}$$

$$= \det(\sigma_\mu)(\mathbf{A}^\dagger)^{-1} = \sigma^{-1}_{\ \mu}\mathbf{A}^{-1} = \det(\sigma_\mu)(\mathbf{A}^\dagger)^{-1}\ \sigma_\mu\mathbf{A}^{-1}$$

$$= \sigma_2\left[\sum_\nu (\Lambda^\nu_\mu\sigma_\nu)^T\right]\sigma_2 = \sum_\nu [(\Lambda_\mu{}^\nu)\sigma_2(\sigma_\nu)^T\sigma_2]$$

$$= \sum_\nu \{(\Lambda_\mu{}^\nu)[\det(\sigma_\nu)](\sigma_\nu^{-1})\} = \sum_\nu [\det(\sigma_\nu)(\Lambda_\mu{}^\nu)(\sigma_\nu)]$$

whence we get $\det\,(\sigma_\mu)\ (\mathbf{A}^\dagger)^{-1}\ \sigma_\mu\mathbf{A}^{-1} = \sum_\nu [\det(\sigma_\nu)\ (\Lambda_\mu{}^\nu)\ (\sigma_\nu)]$ (246)

$$\text{(iv) } \mathbf{A}\sigma_\mu\sigma_\nu\mathbf{A}^{-1} = [\mathbf{A}\sigma_\mu\mathbf{A}^\dagger]\ [(\mathbf{A}^\dagger)^{-1}\sigma_\nu\mathbf{A}^{-1}] = \frac{1}{\det(\sigma_\mu)\det(\sigma_\nu)} \times$$

$$\left\{\sum_\lambda [\det(\sigma_\lambda)(\Lambda^\lambda_\mu\sigma_\lambda)]\right\}^{-1} \left\{\sum_\rho [\det(\sigma_\rho)(\Lambda^\rho_\nu\sigma_\rho)]\right\} = \frac{1}{\det(\sigma_\mu)\det(\sigma_\nu)} \times$$

$$\left\{\sum_\lambda [(\Lambda^\lambda_\mu\sigma_\lambda)]\right\} \left\{\sum_\rho [\det(\sigma_\rho)(\Lambda^\rho_\nu\sigma_\rho)]\right\}$$

Taking trace of both sides, we get $2\delta^\mu_\nu = Tr(\sigma_\mu\sigma_\nu) = Tr\ (\mathbf{A}\sigma_\mu\sigma_\nu\mathbf{A}^{-1})$

$$= \frac{1}{\det(\sigma_\mu)\det(\sigma_\nu)}\ Tr\left\{\left\{\sum_\lambda [(\Lambda^\lambda_\mu\sigma_\lambda)]\right\}\left\{\sum_\rho [\det(\sigma_\rho)(\Lambda^\rho_\nu\sigma_\rho)]\right\}\right\}$$

$$= \frac{1}{\det(\sigma_\mu)\det(\sigma_\nu)}\ Tr\left\{\sum_{\lambda,\rho} [\det(\sigma_\rho)(\Lambda^\lambda_\mu)(\Lambda^\rho_\nu)(\sigma_\lambda)(\sigma_\rho)]\right\}$$

$$= \frac{2}{\det(\sigma_\mu)\det(\sigma_\nu)}\left\{\sum_{\lambda,\rho} [\det(\sigma_\rho)(\Lambda^\lambda_\mu)(\Lambda^\rho_\nu)\delta^{\lambda\rho}]\right\}$$

$$= \frac{2}{\det(\sigma_\mu)\det(\sigma_\nu)}\left\{\sum_\lambda [\det(\sigma_\lambda)(\Lambda_\mu^\lambda)(\Lambda_\nu^\lambda)]\right\}$$

$$= \frac{2}{\det(\sigma_\mu)\det(\sigma_\nu)}\left\{\sum_\lambda [\det(\sigma_\lambda)(\Lambda_\lambda^\mu)(\Lambda_\nu^\lambda)]\right\}$$

whence $\delta_{\mu\nu} = \left\{\sum_\lambda [\det(\sigma_\lambda)(\Lambda_\mu^\lambda)(\Lambda_\nu^\lambda)]\right\}$. This equation also gives $(\Lambda_0^0)^2 \geq 1$ which reads

with $\Lambda_0^0\,(\mathbf{A}) = \frac{1}{2} Tr\,(\sigma_0 \mathbf{A} \sigma_0 \mathbf{A}^\dagger) = \frac{1}{2} Tr\,(\mathbf{A}\mathbf{A}^\dagger) \geq 0$ yields $\Lambda_0^0 \geq 1$.

(v) To obtain the determinant of Λ, we introduce a transformation of the basis vector $\mathbf{I} = \sigma_0 \to i\sigma_0 = \sigma_4$. In terms of this new set of basis vectors $\{\sigma_1, \sigma_2, \sigma_3, \sigma_4\} = i\sigma_0$, we have the following relations:

(a) Any 2×2 matrix $\mathbf{M}$ may be expressed as $\mathbf{M} = \sum_{\mu=1}^{4} b^\mu \sigma_\mu$ where $b^\mu = \frac{1}{2} Tr\,(\sigma_\mu^\dagger \mathbf{M})$;

(b) $\det(\mathbf{M}) = \mathbf{M}\sigma_2 \mathbf{M}^T \sigma_2$;

(c) For any $\mathbf{B} \in SL(2, \mathbb{C})$, we can write $\mathbf{B}\sigma_\mu \mathbf{B}^\dagger = \sum_{\nu=1}^{4} \Lambda_\mu^\nu \sigma_\nu$ where $\Lambda_\mu^\nu \frac{1}{2}\; Tr\,(\sigma_\nu^\dagger \mathbf{B} \sigma_\mu \mathbf{B}^\dagger)$;

(d) Λ_j^i, Λ_4^4 are all real and Λ_4^i, Λ_j^4 are purely imaginary;

(e) $\mathbf{B}^{-1\dagger}\, \sigma_\mu^\dagger \mathbf{B}^{-1} = \Lambda_\mu^\nu \sigma_\nu^\dagger$;

(f) $\delta_{\mu\nu} = \sum_\lambda [(\Lambda_\mu^\lambda)(\Lambda_\nu^\lambda)]$

(g) To compute the determinant of Λ, we have, in the new basis

$$2 = Tr\,(\sigma_1 \sigma_2^\dagger \sigma_3 \sigma_4^\dagger) = Tr\,(\mathbf{A}\sigma_1 \sigma_2^\dagger \sigma_3 \sigma_4^\dagger \mathbf{A}^{-1})$$

$$= Tr\,[(\mathbf{A}\sigma_1 \mathbf{A}^\dagger)\,(\mathbf{A}^{\dagger-1}\sigma_2^\dagger \mathbf{A}^{-1})(\mathbf{A}\sigma_3 \mathbf{A}^\dagger])\,(\mathbf{A}^{\dagger-1}\sigma_4^\dagger \mathbf{A}^{-1})]$$

$$= Tr\left(\sum_{\mu,\nu,\lambda,\rho=1}^{4} \Lambda_1^\mu \Lambda_2^\nu \Lambda_3^\lambda \Lambda_4^\rho \sigma_\mu \sigma_\nu^\dagger \sigma_\lambda \sigma_\rho^\dagger\right)$$

$$= \sum_{\mu,\nu,\lambda,\rho=1}^{4} Tr(\Lambda_1^\mu \Lambda_2^\nu \Lambda_3^\lambda \Lambda_4^\rho \sigma_\mu \sigma_\nu^\dagger \sigma_\lambda \sigma_\rho^\dagger)$$

$$= \sum_{\mu,\nu,\lambda,\rho=1}^{4} \Lambda_1^\mu \Lambda_2^\nu \Lambda_3^\lambda \Lambda_4^\rho\; Tr\,(\sigma_\mu \sigma_\nu^\dagger \sigma_\lambda \sigma^\dagger \rho)$$

$$= 2 \sum_{\mu,\nu,\lambda,\rho=1}^{4} (\Lambda_1^\mu \Lambda_2^\nu \Lambda_3^\lambda \Lambda_4^\rho) \times (\delta_\nu^\mu \delta_\rho^\lambda - \delta_\lambda^\mu \delta_\rho^\nu + \delta_\rho^\mu \delta_\lambda^\nu + \varepsilon_{\mu\nu\lambda\rho})$$

$$= 2 \sum_{\mu,\nu,\lambda,\rho=1}^{4} \varepsilon_{\mu\nu\lambda\rho}(\Lambda_1^\mu \Lambda_2^\nu \Lambda_3^\lambda \Lambda_4^\rho) = 2\det(\Lambda)$$

whence det $(\Lambda) = 1$. The above properties sindicate the fact that Λ is a Lorentz matrix.

It follows from all the above properties that there exists a homomorphism from the $SL(2,\mathbb{C})$ group to the Lorentz group. Corresponding to each element **A** of the $SL(2,\mathbb{C})$ group, there exists an element $\Lambda(\mathbf{A})$ of the Lorentz group and this correspondence is preserved under the group operation of multiplication, i.e., $\Lambda(\mathbf{AB}) = \Lambda(\mathbf{A})\Lambda(\mathbf{B})$. Now, since the equation $\Lambda_\nu^\mu(\mathbf{A}) = \frac{1}{2} Tr(\sigma_\mu \mathbf{A} \sigma_\nu \mathbf{A}^\dagger)$ is quadratic in the elements of **A** so that $\Lambda(\mathbf{A})$ and $\Lambda(-\mathbf{A})$ are equal.

We now examine whether any elements of $SL(2,\mathbb{C})$ other than **A** and $-$**A** correspond to $\Lambda(\mathbf{A})$. For this purpose, we identify the elements of $SL(2,\mathbb{C})$ that correspond to the identity of the Lorentz group. We revert to our original basis $\{\sigma_0 = \mathbf{I}, \sigma_i, i = 1, 2, 3\}$ and let $\mathbf{A}_0 = \begin{pmatrix} a & b \\ c & d \end{pmatrix}$ correspond to the identity element of the Lorentz group. Then, we must have $\delta_{\mu\nu} = \frac{1}{2} Tr(\sigma_\mu \mathbf{A}_0 \sigma_\nu \mathbf{A}_0^\dagger)$. This represents a set of 16 equations for the four elements of $\mathbf{A}_0$. We consider the equations corresponding to $\mu = \nu = 0, 1, 3$ that are $a^*d + cb^* + c^*b + d^*a = 2$, $aa^* - bb^* - cc^* + dd^* = 2$ and $aa^* + bb^* + cc^* + dd^* = 2$. From the last pair of equations we obtain $aa^* + dd^* = 2$ and $b = c = 0$. Hence, from the first equation, we get $a^*d + ad^* = 2$ whence $(a^* - d^*)(a - d) = 0$ or $|a - d|^2 = 0$ giving $a = d$. It, thus, follows that the matrix $\mathbf{A}_0$ is a constant multiple of the identity matrix. Further, since $\mathbf{A}_0 \in SL(2,\mathbb{C})$, we must have $\det(\mathbf{A}_0) = 1$, whence the constant multiple can only take the values ± 1. Now, if the matrix $\mathbf{A} \in SL(2,\mathbb{C})$ corresponds to the Lorentz transformation $\Lambda(\mathbf{A})$ and $\mathbf{A}^{-1}$ to $\Lambda(\mathbf{A}^{-1})$, then $\Lambda(\mathbf{A})\Lambda(\mathbf{A}^{-1}) = \Lambda(\mathbf{AA}^{-1}) = \Lambda(\mathbf{I}) = \mathbf{I}$ implying that $\Lambda(\mathbf{A}^{-1}) = \Lambda^{-1}(\mathbf{A})$. Again, if $\Lambda(\mathbf{A}) = \Lambda(\mathbf{B})$ then $\Lambda(\mathbf{AB}^{-1}) = \Lambda(\mathbf{A})\Lambda(\mathbf{B}^{-1}) = \Lambda(\mathbf{A})\Lambda^{-1}(\mathbf{B}) = \Lambda(\mathbf{B})\Lambda^{-1}(\mathbf{B}) = \mathbf{I}$ whence, in line with the above argument regarding $\mathbf{A}_0$, we must have $\mathbf{AB}^{-1} = \pm\mathbf{I}$, i.e., $\mathbf{A} = \pm\mathbf{B}$. This analysis, thus, establishes two things (i) each element $\mathbf{A} \in SL(2,\mathbb{C})$ corresponds to an element $\Lambda(\mathbf{A})$ of the Lorentz group and (ii) $\Lambda(\mathbf{A}) \neq \Lambda(\mathbf{B})$ unless $\mathbf{B} = \pm\mathbf{A}$.

What remains to be shown in completing the correspondence between the $SL(2,\mathbb{C})$ group and the Lorentz group is that given a Lorentz transformation, Λ, does there exist an $\mathbf{A} \in SL(2,\mathbb{C})$ such that $\Lambda = \Lambda(\mathbf{A})$. To establish this correspondence, we make use of the fact that any Lorentz transformation can be factored into a rotation followed by a boost and again another rotation. It would therefore, suffice to prove here that there exist $SL(2,\mathbb{C})$ matrices that represent pure rotations and pure boosts. We, shall establish this in the following steps.

Let $\mathbf{A} = \begin{pmatrix} a & b \\ c & d \end{pmatrix} \in SL(2,\mathbb{C})$, $a, b, c, d \in \mathbb{C}$ be arbitrary. We can, then, write

$$\mathbf{A} = \begin{pmatrix} a & b \\ c & d \end{pmatrix} = \frac{b+c}{2}\begin{pmatrix} 0 & 1 \\ 1 & 0 \end{pmatrix} + \frac{i(b-c)}{2}\begin{pmatrix} 0 & -i \\ i & 0 \end{pmatrix}$$

$$+ \frac{a-d}{2}\begin{pmatrix} 1 & 0 \\ 0 & -1 \end{pmatrix} + \frac{a+d}{2}\begin{pmatrix} 1 & 0 \\ 0 & 1 \end{pmatrix}.$$

Writing
$$\theta = \cos^{-1}\left(\frac{a+d}{2}\right) = \sin^{-1}\left\{\frac{[4-(a+d)^2]^{1/2}}{2}\right\},$$

we obtain
$$\mathbf{A} = \cos\theta\begin{pmatrix} 1 & 0 \\ 0 & 1 \end{pmatrix} + i\sin\theta\,\frac{b+c}{i[4-(a+d)^2]^{1/2}}\begin{pmatrix} 0 & 1 \\ 1 & 0 \end{pmatrix}$$

$$+ i\sin\theta\,\frac{b-c}{[4-(a+d)^2]^{1/2}}\begin{pmatrix} 0 & -i \\ i & 0 \end{pmatrix}$$

$$+ i\sin\theta\,\frac{a-d}{i[4-(a+d)^2]^{1/2}}\begin{pmatrix} 1 & 0 \\ 0 & -1 \end{pmatrix}.$$

The sum of the squares of the coefficients of $i\sin\theta$. σ_j, $j = 1, 2, 3$ is

$$\frac{-(b+c)^2 + (b-c)^2 - (a-d)^2}{4-(a+d)^2}$$

$$= \frac{4(ad-bc) - (a+d)^2}{4-(a+d)^2} = 1$$

where, in the last step we have used the fact that $\mathbf{A} \in SL(2, \mathbb{C})$ implying $\det(\mathbf{A}) = ad - bc = 1$. The above analysis enables us to parametrize an arbitrary $\mathbf{A} \in SL(2, \mathbb{C})$ as $\mathbf{A} = \cos\theta + i\sin\theta\boldsymbol{\sigma}.\mathbf{n}$ with $\mathbf{n}^2 = 1$. This representation explicitly shows the existence of six parameters, viz., the real and imaginary components of θ, $\mathbf{n} \equiv (n_1, n_2, n_3)$ reduced by two due to the constraints $\cos^2\theta + \sin^2\theta = 1$, and $\mathbf{n}^2 = 1$. We can express the unit vector $\mathbf{n} \equiv (n_1, n_2, n_3)$ in terms of the radial angles by writing $n_3 = \cos\alpha$, $n_1 = \sin\alpha\cos\beta$ and $n_2 = \sin\alpha\sin\beta$ with $0 \le \mathrm{Re}(\alpha) \le \pi$, $\sin\beta = \dfrac{b-c}{2\sin\theta\sin\alpha}$, $\cos\beta = \dfrac{b+c}{2i\sin\theta\sin\alpha}$, and $0 \le \mathrm{Re}(\beta) \le 2\pi$.

Having established a convenient parametrization for the $SL(2, \mathbb{C})$ matrices, we examine under what circumstances such matrices represent rotations and boosts respectively. We take up the issue of rotations first. It is well known that rotation matrices must necessarily be unitary so that $\mathbf{A}^\dagger = \mathbf{A}^{-1}$ whence $\cos^*\theta - i\sin^*\theta(\boldsymbol{\sigma}.\mathbf{n})^\dagger = \cos\theta - i\sin\theta(\boldsymbol{\sigma}.\mathbf{n})$ thereby mandating that both θ, $\mathbf{n}$ are real. To confirm that we indeed have a rotation matrix for θ, $\mathbf{n}$ real, let us construct the transformed matrix $\mathbf{M}' = \mathbf{AMA}^\dagger = \mathbf{AMA}^{-1}$ corresponding to an arbitrary $\mathbf{M} = x^\mu\sigma_\mu = x^0 + \boldsymbol{\sigma}\mathbf{x}$. We have

$$\mathbf{M}' = \left[\cos\frac{\theta}{2} + i\sin\frac{\theta}{2}(\boldsymbol{\sigma}.\mathbf{n})\right][x^0 + \boldsymbol{\sigma}.\mathbf{x}]\left[\cos\frac{\theta}{2} - i\sin\frac{\theta}{2}(\boldsymbol{\sigma}.\mathbf{n})\right]$$

$$= x^0 + \cos^2\frac{\theta}{2}(\boldsymbol{\sigma}.\mathbf{x}) + i\sin\frac{\theta}{2}\cos\frac{\theta}{2}[(\boldsymbol{\sigma}.\mathbf{n})(\boldsymbol{\sigma}.\mathbf{x}) - (\boldsymbol{\sigma}.\mathbf{x})(\boldsymbol{\sigma}.\mathbf{n})]$$

$$+ \sin^2\frac{\theta}{2}(\boldsymbol{\sigma}.\mathbf{n})(\boldsymbol{\sigma}.\mathbf{x})(\boldsymbol{\sigma}.\mathbf{n})$$

$$= x^0 + \sigma.\{(\mathbf{n.x})\mathbf{n} + \cos\theta[\mathbf{x} - (\mathbf{n.x}).\mathbf{n}] + \sin\theta(\mathbf{x}\times\mathbf{n})\}.$$

Since, $\mathbf{M}' = x'^{\mu}\sigma_{\mu} = x'^0 + \boldsymbol{\sigma}\mathbf{x}'$, we have $x'^0 = x^0$ and $\mathbf{x}' = (\mathbf{n.x})\mathbf{n} + \cos\theta\,[\mathbf{x} - (\mathbf{n.x}).\mathbf{n}] + \sin\theta\,(\mathbf{x}\times\mathbf{n})$. The fact that the above transformation represents a rotation is clearly evident from (i) $x'^0 = x^0$ and (ii) $\mathbf{x}'.\mathbf{x}' = \mathbf{xx}$. Actually, the above transformation is a rotation with the vector $\mathbf{n}$ as the axis through an angle θ in the clockwise direction when we look towards the origin along $\mathbf{n}$. In the expression for the transformed vector $\mathbf{x}'$, the component $(\mathbf{n.x})\mathbf{n}$ represents the unaltered component of $\mathbf{x}$ in the direction of $\mathbf{n}$; the component $[\mathbf{x} - (\mathbf{n.x}).\mathbf{n}]$ is the component perpendicular to $\mathbf{n}$ - it is scaled by $\cos\theta$ as a result of the rotation and the third component $(\mathbf{x}\times\mathbf{n})$ is perpendicular to both $\mathbf{n}$ and $\mathbf{x}$ and its coefficient gets rescaled from 0 to $\sin\theta$ due to the rotation. In conclusion, we note that any rotation is specified by the axis of rotation $\mathbf{n}$ and the angle of rotation θ and the corresponding pair of unitary $SL(2,\mathbb{C})$ matrices is given by $\pm\mathbf{A} = \cos\theta + i\sin\theta\boldsymbol{\sigma}.\mathbf{n}$ with $\mathbf{n}$ & θ both real.

We, now, address the issue of representation of Lorentz boosts by $SL(2,\mathbb{C})$ matrices. For the purpose, we consider Hermitian $SL(2,\mathbb{C})$ matrices. The condition of Hermiticity manifests itself as $\mathbf{A}^{\dagger} = \mathbf{A}$ or $\cos^*\theta - i\sin^*\theta(\boldsymbol{\sigma}.\mathbf{n})^{\dagger} = \cos\theta - i\sin\theta(\boldsymbol{\sigma}.\mathbf{n})$ thereby implying that $\mathbf{n}$ must necessarily be real and θ purely imaginary so that we can write such a matrix as $\cosh\theta + \sinh\theta(\boldsymbol{\sigma}.\mathbf{n})$. The transformation of a 4-vector by such a matrix is evaluated as in the case of rotations and we have

$$\mathbf{M}' = \left[\cosh\frac{\alpha}{2} + \sinh\frac{\alpha}{2}(\boldsymbol{\sigma}.\mathbf{n})\right][x^0 + \boldsymbol{\sigma}.\mathbf{x}]\left[\cosh\frac{\alpha}{2} + \sinh\frac{\alpha}{2}(\boldsymbol{\sigma}.\mathbf{n})\right]$$

$$= \left(\cosh^2\frac{\alpha}{2} + \sinh^2\frac{\alpha}{2}\right)x^0 + \sinh\frac{\alpha}{2}\cosh\frac{\alpha}{2}[(\boldsymbol{\sigma}.\mathbf{n})(\boldsymbol{\sigma}.\mathbf{x}) + (\boldsymbol{\sigma}.\mathbf{x})(\boldsymbol{\sigma}.\mathbf{n})]$$

$$+ 2\sinh\frac{\alpha}{2}\cosh\frac{\alpha}{2}(\boldsymbol{\sigma}.\mathbf{n})\,x^0 + \cosh^2\frac{\alpha}{2}(\boldsymbol{\sigma}.\mathbf{x}) + \sinh^2\frac{\alpha}{2}(\boldsymbol{\sigma}.\mathbf{n})(\boldsymbol{\sigma}.\mathbf{x})(\boldsymbol{\sigma}.\mathbf{n})$$

$$= \cosh\alpha x^0 + \sinh\alpha(\mathbf{n.x}) + \boldsymbol{\sigma}.[\sinh\alpha x^0\mathbf{n} + \mathbf{x} + (\cosh\alpha - 1)(\mathbf{n.x})\mathbf{n}]$$

whence $x'^0 = \cosh\alpha x^0 + \sin\alpha(\mathbf{n.x})$ and $\mathbf{x}' = [\mathbf{x} - (\mathbf{n.x})\mathbf{n}] + \mathbf{n}[\cosh\alpha(\mathbf{n.x}) + \sinh\alpha x^0]$. It can easily be verified that $(x'^0)^2 - (\mathbf{x}')^2 = (x^0)^2 - (\mathbf{x})^2$ thereby confirming that $x \to x'$ is indeed a Lorentz transformation. It is instructive to resolve $\mathbf{x}'$ into components parallel and perpendicular to $\mathbf{n}$. We have $x'^0 = \cosh\alpha x'^0 + \sinh\alpha(\mathbf{n.x})$, $(\mathbf{n.x}') = \sinh\alpha x^0 + \cosh\alpha(\mathbf{n.x})$ and $x' - (\mathbf{n.x}')\mathbf{n} = \mathbf{x} - (\mathbf{n.x})\mathbf{n}$. Thus, the component of $\mathbf{x}$ perpendicular to $\mathbf{n}$ is left unchanged by such a transformation while the time coordinate and the component of $\mathbf{x}$ parallel to $\mathbf{n}$ undergo the usual two-dimensional Lorentz boost. We have, thus, established that any Lorentz boost is specified by the axis of the boost **n** and the hyperbolic angle α (called rapidity) and the corresponding pair of Hermitian $SL(2, \mathbb{C})$ matrices is given by $\pm\mathbf{A} = \cosh\alpha + \sinh\alpha\boldsymbol{\sigma}.\mathbf{n}$ with **n** & α both real (which may also be written in the usual form as $\pm\mathbf{A} = \cos\theta + i\sin\theta\boldsymbol{\sigma}.\mathbf{n}$ with **n** real and θ purely imaginary).

Since every Lorentz transformation with $\Lambda_0^0 \geq +1$ can be written as a composition of rotations and boosts and, as shown above, there exist a pair of $SL(2, \mathbb{C})$ matrices corresponding to any rotation or boost and further, that the set of all $SL(2,\mathbb{C})$ matrices forms a group, it immediately follows that given any Lorentz transformation, we can associate a pair of $SL(2,\mathbb{C})$ matrices with such transformation.

Having established the existence of homomorphism between the Lorentz group and the $SL(2,\mathbb{C})$ group, we make use of this connection to obtain the explicit expression for the Wigner rotation $R(\varphi_W, \mathbf{p}) = L^{-1}(p')\,\Lambda L(p)$ with $\Lambda p = p'$. For the purpose, we write

$$\Lambda = \cosh\frac{\alpha}{2} + \sin\frac{\alpha}{2}(\boldsymbol{\sigma}.\mathbf{e}) \tag{247}$$

$$L(p) = \cosh\frac{\beta}{2} + \sinh\frac{\beta}{2}(\boldsymbol{\sigma}.\mathbf{f})$$

$$= \left(\frac{\omega_p + M}{2M}\right)^{1/2} + \left(\frac{\omega_p + M}{2M}\right)^{1/2}\frac{\sigma.\mathbf{p}}{|\mathbf{p}|} \tag{248}$$

where $\omega_p = M\cosh\beta$ and $\mathbf{p} = M\sinh\beta\mathbf{f}$ and

$$L(p') = \cosh\frac{\gamma}{2} + \sinh\frac{\gamma}{2}(\boldsymbol{\sigma}.\mathbf{g}) = \left(\frac{\omega_{p'} + M}{2M}\right)^{1/2} + \left(\frac{\omega_{p'} - M}{2M}\right)^{1/2}\frac{\sigma.\mathbf{p}'}{|\mathbf{p}'|} \tag{249}$$

with $\omega_{p'} = M\cosh\gamma$, $\mathbf{p}' = M\sinh\gamma\mathbf{g}$, $\omega_{p'} = \omega_p\cosh\alpha + \mathbf{p.e}\sinh\alpha$ and $\mathbf{p}' = [\mathbf{p} - (\mathbf{p.e})\mathbf{e}] + [\omega_p\sinh\alpha + \mathbf{p.e}\cosh\alpha]\mathbf{e}$. We, then, have

$$2MR(\varphi_W, \mathbf{p}) = [(\omega_p + M)(\omega_{p'} + M)]^{1/2}\left[\cosh\frac{\alpha}{2} + \sinh\frac{\alpha}{2}(\sigma.\mathbf{e})\right]$$

$$-\frac{1}{|\mathbf{p}'|}[(\omega_p + M)(\omega_{p'} - M)]^{1/2}\left\{\cosh\frac{\alpha}{2}(\sigma.\mathbf{p}') + \sinh\frac{\alpha}{2}[\mathbf{e}.\mathbf{p}' + i\sigma.(\mathbf{p}'\times e)]\right\}$$

$$+\frac{1}{|\mathbf{p}|}[(\omega_p - M)(\omega_{p'} + M)]^{1/2}\left\{\cosh\frac{\alpha}{2}(\sigma.\mathbf{p}) + \sinh\frac{\alpha}{2}[\mathbf{e.p} + i\sigma.(e\times\mathbf{p})]\right\}$$

$$-\frac{1}{|\mathbf{p}||\mathbf{p}'|}[(\omega_p - M)(\omega_{p'} - M)]^{1/2}\left\{\begin{matrix}\cosh\frac{\alpha}{2}[\mathbf{p}.\mathbf{p}' + i\sigma.(\mathbf{p}'\times\mathbf{p})] + \sinh\frac{\alpha}{2} \\ \times\begin{bmatrix} i(\mathbf{p}'\times \mathrm{e}).\mathbf{p} + (\sigma.\mathrm{p})(\mathbf{p}'.\mathrm{e}) \\ -(\sigma.\mathrm{e})(\mathbf{p}.\mathbf{p}') + (\sigma.\mathbf{p}')(\mathbf{p}.\mathrm{e})\end{bmatrix}\end{matrix}\right\}$$

We write the above as $2MR\,(\varphi_W, \mathbf{p}) = A + B\sigma.\mathbf{p} + C\sigma.\mathbf{e} + iD\sigma.(\mathbf{p}\times\mathbf{e})$ by expressing $\mathbf{p}'$ in terms of $\mathbf{p}$ and $\mathbf{e}$. We now proceed to simplify these coefficients A, B, C, D. We have

$$A = [(\omega_p + M)\,(\omega_{p'} + M)]^{1/2}\cosh\frac{\alpha}{2} - \frac{1}{|\mathbf{p}'|}[(\omega_p + M)\,(\omega_{p'} - M)]^{1/2}$$

$$\times\sinh\frac{\alpha}{2}(\mathbf{e}.\mathbf{p}') + \frac{1}{|\mathbf{p}|}[(\omega_p + M)\,(\omega_{p'} - M)]^{1/2}\sinh\frac{\alpha}{2}\,(\mathbf{e}.\mathbf{p})$$

$$-\frac{1}{|\mathbf{p}||\mathbf{p}'|}\,[(\omega_p - M)\,(\omega_{p'} - M)]^{1/2}\cosh\frac{\alpha}{2}\,(\mathbf{p}.\mathbf{p}')$$

$$= [(\omega_p + M)\,(\omega_{p'} + M)]^{1/2}\cosh\frac{\alpha}{2} - \left(\frac{\omega_p + M}{\omega_{p'} + M}\right)^{1/2}\sinh\frac{\alpha}{2}$$

$$\times[\omega_p\sinh\alpha + (\mathbf{p}.\mathbf{e})\cosh\alpha] + \left(\frac{\omega_{p'} + M}{\omega_p + M}\right)^{1/2}\sinh\frac{\alpha}{2}\,(\mathbf{e}.\mathbf{p})$$

$$-\frac{1}{[(\omega_p + M)(\omega_{p'} + M)]^{1/2}}\cosh\frac{\alpha}{2}\begin{bmatrix}\mathbf{p}^2 + (\mathbf{p}.\mathbf{e})^2(\cosh\alpha - 1)\\ + \omega_p(\mathbf{p}.\mathbf{e})\sinh\alpha\end{bmatrix}$$

$$= \frac{(\omega_p + M)^{-1/2}}{(\omega_{p'} + M)^{1/2}}\left\{\begin{matrix}\cosh\dfrac{\alpha}{2}\begin{bmatrix}(\omega_p + M)(\omega_p\cosh\alpha + (\mathbf{p}.\mathbf{e})\sinh\alpha + M)\\ -\mathbf{p}^2 - (\mathbf{p}.\mathbf{e})^2(\cosh\alpha - 1) - \omega_p(\mathbf{p}.\mathbf{e})\sinh\alpha\end{bmatrix}\\ +\sinh\dfrac{\alpha}{2}\begin{bmatrix}-(\omega_p + M)(\omega_p\sinh\alpha + (\mathbf{p}.\mathbf{e})\cosh\alpha)\\ +(\omega_p\cosh\alpha + (\mathbf{p}.\mathbf{e})\sinh\alpha + M)(\mathbf{e}.\mathbf{p})\end{bmatrix}\end{matrix}\right\}$$

$$= \frac{(\omega_p + M)^{-1/2}}{(\omega_{p'} + M)^{1/2}}\left\{\begin{matrix}\cosh\dfrac{\alpha}{2}\begin{bmatrix}(\omega_p^2\cosh\alpha + \omega_p\,\mathrm{M}\,(\cosh\alpha + 1) + \mathrm{M}\,(\mathrm{p.e})\\ \sinh\alpha + \mathrm{M}^2 - \mathrm{p}^2 - (\mathrm{p.e})^2\,(\cosh\alpha - 1)\end{bmatrix}\\ +\sin\dfrac{\alpha}{2}\begin{bmatrix}(\mathrm{p.e})^2\sinh\alpha + \mathrm{M}(\mathrm{p.e})(1 - \cosh\alpha)\\ -(\omega_p^2\sinh\alpha - \mathrm{M}\omega_p\sinh\alpha\end{bmatrix}\end{matrix}\right\}$$

$$= \frac{1}{[(\omega_p + M)(\omega_{p'} + M)]^{1/2}}\begin{bmatrix}\cosh\dfrac{\alpha}{2}(\omega_p^2 + 2M\omega_p + M^2 - \mathrm{p}^2)\\ +2M(\mathrm{p.e})\sinh\dfrac{\alpha}{2}\end{bmatrix}$$

$$= \frac{1}{[(\omega_p + M)(\omega_{p'} + M)]^{1/2}} \left[\cosh\frac{\alpha}{2}(2M^2 + 2M\omega_p) + 2M(\mathbf{p.e})\sinh\frac{\alpha}{2}\right]$$

$$= \frac{2M}{[(\omega_p + M)(\omega_{p'} + M)]^{1/2}} \left[\cosh\frac{\alpha}{2}(\omega_p + M) + \sinh\frac{\alpha}{2}(\mathrm{p.e})\right] \tag{250}$$

$$B = -\frac{1}{|\mathbf{p}'|}[(\omega_p + M)(\omega_{p'} - M)]^{1/2}\cosh\frac{\alpha}{2} + \frac{1}{|\mathbf{p}|}[(\omega_p - M)(\omega_{p'} + M)]^{1/2}$$

$$\times \cosh\frac{\alpha}{2} - \frac{1}{|\mathbf{p}||\mathbf{p}'|}[(\omega_p + M)(\omega_{p'} - M)]^{1/2}\sinh\frac{\alpha}{2}$$

$$\times [\omega_p \sinh\alpha + (\mathbf{p.e})\cosh\alpha + (\mathbf{p.e})]$$

$$= \frac{(\omega_p + M)^{-1/2}}{(\omega_{p'} + M)^{1/2}} \left\{ \begin{array}{l} -(\omega_p + M)\cosh\frac{\alpha}{2} + [\omega_p\cosh\alpha + (\mathbf{p.e})\sinh\alpha + M] \\ \times\cosh\frac{\alpha}{2} - \sinh\frac{\alpha}{2}[\omega_p\sinh\alpha + (\mathbf{p.e})\cosh\alpha + (\mathbf{p.e})] \end{array} \right\}$$

$$= \frac{(\omega_p + M)^{-1/2}}{(\omega_{p'} + M)^{1/2}} \left\{ -(\omega_p + M)\cosh\frac{\alpha}{2} + \omega_p\cosh\frac{\alpha}{2} + M\cosh\frac{\alpha}{2} \right\} = 0 \tag{251}$$

$$C = [\omega_p + M)(\omega_{p'} + M)]^{1/2}\sinh\frac{\alpha}{2} - \frac{1}{|\mathbf{p}'|}[(\omega_p + M)(\omega_{p'} - M)]^{1/2}$$

$$\times \cosh\frac{\alpha}{2}[(\mathbf{p.e})(\cosh\alpha - 1) + \omega_p\sinh\alpha] - \frac{1}{|\mathbf{p}||\mathbf{p}'|}[(\omega_p - M)(\omega_{p'} - M)]^{1/2}$$

$$\times \sinh\frac{\alpha}{2} \left[\begin{array}{l} -\mathbf{p}^2 + (\mathbf{p.e})^2 - \omega_p(\mathbf{p.e})\sinh\alpha - (\mathbf{p.e})^2\cosh\alpha \\ -(\mathbf{p.e})^2 + \omega_p(\mathbf{p.e})\sinh\alpha + (\mathbf{p.e})^2\cosh\alpha \end{array} \right]$$

$$= \frac{(\omega_p + M)^{-1/2}}{(\omega_{p'} + M)^{1/2}} \left\{ \begin{array}{l} (\omega_p + M)[\omega_p\cosh\alpha + (\mathbf{p.e})\sinh\alpha + M]\sinh\frac{\alpha}{2} \\ -(\omega_p + M)\cosh\frac{\alpha}{2}[(\mathbf{p.e})(\cosh\alpha - 1) + \omega_p\sinh\alpha] \\ +\sinh\frac{\alpha}{2}\mathbf{p}^2 \end{array} \right\}$$

$$= \frac{(\omega_p + M)^{-1/2}}{(\omega_{p'} + M)^{1/2}} \begin{Bmatrix} -(\omega_p + M)\omega_p \sinh\frac{a}{2} + (\mathbf{p.e})(\omega_p + M) \\ \times\left(\cosh\frac{\alpha}{2} - \cosh\frac{\alpha}{2}\right) + [(\omega_p + M)M + \mathbf{p}^2]\sinh\frac{\alpha}{2} \end{Bmatrix}$$

$$= \frac{(\omega_p + M)^{-1/2}}{(\omega_{p'} + M)^{1/2}} \sinh\frac{\alpha}{2} \begin{Bmatrix} -\mathbf{p}^2 - M^2 - M\omega_p \\ + M\omega_p + M^2 + \mathbf{p}^2 \end{Bmatrix} = 0 \tag{252}$$

$$D = -\frac{1}{|\mathbf{p}'|} [(\omega_p + M)(\omega_{p'} - M)]^{1/2} \sinh\frac{\alpha}{2} - \frac{1}{|\mathbf{p}|} [(\omega_p - M)(\omega_{p'} + M)]^{1/2}$$

$$\times \sinh\frac{\alpha}{2} + \frac{1}{|\mathbf{p}||\mathbf{p}'|} [(\omega_p - M)(\omega_{p'} - M)]^{1/2}$$

$$\times \cosh\frac{\alpha}{2} [\omega_p \sinh\alpha + (\mathbf{p.e})(\cosh\alpha - 1)]$$

$$= \frac{(\omega_p + M)^{-1/2}}{(\omega_{p'} + M)^{1/2}} \begin{Bmatrix} -[\omega_p + M + \omega_p \cosh\alpha + (\mathbf{p.e})\sinh\alpha + M]\sinh\frac{\alpha}{2} \\ +[\omega_p \sinh\alpha + (\mathbf{p.e})(\cosh\alpha - 1)]\cosh\frac{\alpha}{2} \end{Bmatrix}$$

$$= \frac{(\omega_p + M)^{-1/2}}{(\omega_{p'} + M)^{1/2}} \begin{Bmatrix} -\omega_p\left(\sinh\frac{\alpha}{2} - \sinh\frac{\alpha}{2}\right) - 2M\sinh\frac{\alpha}{2} \\ -(\mathbf{p.e})\left(\cosh\frac{\alpha}{2} - \cosh\frac{\alpha}{2}\right) \end{Bmatrix}$$

$$= -2M\frac{(\omega_p + M)^{-1/2}}{(\omega_{p'} + M)^{1/2}} \sinh\frac{\alpha}{2} \tag{253}$$

Substituting the values of A, B, C, D on the expression for the Wigner rotation $R(\varphi_W, p)$, we obtain $R(\varphi_W, p) = \dfrac{(\omega_p + M)}{(\omega_{p'} + M)^{1/2}}$

$$\times \left\{\cosh\frac{\alpha}{2}(\omega_p + M) + \sinh\frac{\alpha}{2}(\mathbf{p.e}) - i\sinh\frac{\alpha}{2}\sigma.(\mathbf{p}\times\mathbf{e})\right\}. \tag{254}$$

We can also express the Wigner rotation in terms of the rapidity angles α, β, γ as follows. We have $\omega_p = M\cosh\beta$, $\mathbf{p} = M\sinh\beta\mathbf{f}$, $\omega_{p'} = M\cosh\gamma$ and $\mathbf{p}' = M\sinh\boldsymbol{\gamma}\mathbf{g}$ so that

$$\cosh\,\gamma = \frac{\omega_{p'}}{M}\ \frac{\omega_p \cosh\alpha + (\mathbf{p.e})\sinh\alpha}{M}$$

$$= \cosh\alpha\,\cosh\beta + \sinh\alpha\,\sinh\beta(\mathbf{e.f})$$

$$\sinh\,\boldsymbol{\gamma g} = \frac{\mathbf{p}'}{M} = \frac{\mathbf{p} - (\mathbf{p.e})\mathbf{e} + [\omega_p \sinh\alpha + (\mathbf{p.e})\cosh\alpha]\mathbf{e}}{M}$$

$$= \sinh\,\beta[\mathbf{f} - (\mathbf{e.f})\mathbf{e}] + [\sinh\alpha\,\cosh\beta + \cosh\alpha\,\sinh\beta(\mathbf{e.f})]\mathbf{e}$$

$$= \cosh\,\frac{\gamma}{2} = \frac{[1+\cosh\alpha\,\cosh\beta + \sinh\alpha\,\sinh\beta(\mathbf{e.f})]^{1/2}}{2},\ \sin\,\frac{\gamma}{2}\,\mathbf{g} = \frac{\sinh\gamma g}{2\cosh\frac{\gamma}{2}}$$

$$R(\varphi_W, p) = \frac{\cosh\frac{\alpha}{2}\cosh\frac{\beta}{2} + \sinh\frac{\alpha}{2}\sinh\frac{\beta}{2}(\mathbf{e.f}) + i\sinh\frac{\alpha}{2}\,\sinh\frac{\beta}{2}\sigma.(\mathbf{e}\times\mathbf{f})}{\left[\frac{1}{2}+\frac{1}{2}\cosh\alpha\,\cosh\beta + \frac{1}{2}\sinh\alpha\,\sinh\beta(\mathbf{e.f})\right]^{1/2}}$$

If we write $R(\boldsymbol{\varphi}_W, p) = \cos\,\frac{\xi}{2} + i\,\sin\,\frac{\xi}{2}(\boldsymbol{\sigma}.\mathbf{n})$, then the angle ξ is given by

$$\cos\,\frac{\xi}{2} = \frac{\cosh\frac{\alpha}{2}\,\cosh\frac{\beta}{2} + \sinh\frac{\alpha}{2}\,\sinh\frac{\beta}{2}(\mathrm{e.f})}{\left[\frac{1}{2}+\frac{1}{2}\cosh\alpha\,\cosh\beta + \frac{1}{2}\sinh\alpha\,\sinh\beta(\mathrm{e.f})\right]^{1/2}} \tag{255}$$

$$\sin\,\frac{\xi}{2}\,\mathbf{n} = \frac{\sinh\frac{\alpha}{2}\sinh\frac{\beta}{2}(\mathbf{e}\times\mathbf{f})}{\left[\frac{1}{2}+\frac{1}{2}\cosh\alpha\,\cosh\beta + \frac{1}{2}\sinh\alpha\,\sinh\beta(\mathbf{e.f})\right]^{1/2}} \tag{256}$$

2.19 THE RELATIVISTIC SPIN & POSITION OPERATORS FOR MASSIVE PARTICLES (PRYCE, 1948; NEWTON, 1949; BERG, 1965; JORDAN, 1980; STEFANOVICH, 2007)

2.19.1 The Relativistic Spin Operator

To identify an appropriate spin operator in the relativistic quantum theory, we postulate that a suitable relativistic spin operator, **S**, must necessarily possess the following properties:

(i) It must commute with the orbital angular momentum, **L**, whence, its commutator with the total angular momentum, $\mathbf{J} = \mathbf{L} + \mathbf{S}$, is

$$[J_j, S_k] = i \sum_l \varepsilon_{jkl} S_l \tag{257}$$

(ii) Since, **S** is an angular momentum, it must satisfy the same commutators among its components as the total and orbital angular momenta, i.e.,

$$[S_j, S_k] = i \sum_l \varepsilon_{jkl} S_l \tag{258}$$

(iii) It should be possible to measure spin and linear momentum simultaneously so that the corresponding operators must commute, i.e.,

$$[P_i, S_j] - 0 \tag{259}$$

(iv) It should be possible to measure spin and position simultaneously so that the corresponding operators must commute, i.e.,

$$[R_i, S_j] - 0 \tag{260}$$

(v) It must be expressible as some combination of the complete set of generators of the Poincare group, i.e., P^0, **P, J, K**

(vi) It must behave as a 3-vector under spatial rotations whence it must be linear in the generators **J, K** so that we can express it as

$\mathbf{S} = f_1\,\mathbf{J} + f_2\,\mathbf{K} + f_3\,\mathbf{P} + f_4(\mathbf{P.J})\mathbf{P} + f_5(\mathbf{P.K})\mathbf{P} + f_6(\mathbf{P}\times\mathbf{J}) + f_7(\mathbf{P}\times\mathbf{K})$ where each $f_i \equiv f_i(\mathbf{P}^2, P^0)$ (because each of these coefficients must be three-dimensional scalars). In view of (iii), we get $f_2 = -f_5\mathbf{P}^2$ and $0 = [S^i, P^i] = i(f_1 + f_7P^0)\varepsilon^{ijk}P^k + i(f_2P^0 + f_6\mathbf{P}^2)\delta^{ij} + i(f_5P^0 - f_6)P^iP_J$ whence $f_1 + f_7P^0 = 0$, $f_2P^0 + f_6\mathbf{P}^2 = 0$ and $f_5P^0 = f_6 = -f_2P^0\mathbf{P}^{-2}$. Again, using (ii), we get

$$f_1^2\left\{\left[\mathrm{J} - \frac{1}{P^0}(\mathrm{P}\times\mathrm{K})\right]^k - \frac{1}{(P^0)^2}P^k(\mathrm{P.J})\right\}$$

$$+ f_1f_2\left\{-\left[\mathrm{K} - \frac{1}{\mathbf{P}^2}(\mathbf{P.K})\mathbf{P} - \frac{P^0}{\mathrm{P}^2}(\mathbf{P}\times\mathbf{J})\right]^k - \frac{1}{P^0}(\mathbf{P}\times\mathbf{J})^k\right\}$$

$$+ f_2f_4\,\mathbf{P}^2\left[\mathbf{K} - \frac{1}{\mathbf{P}^2}(\mathbf{P.K})\mathbf{P} - \frac{P^0}{\mathbf{P}^2}(\mathbf{P}\times\mathbf{J})\right]^k$$

$$+ f_1f_4\left\{\mathbf{P}^2\left[\mathbf{J} - \frac{1}{P^0}(\mathbf{P}\times\mathbf{K})\right]^k - \left[\mathbf{J.P} + \frac{1}{P^0}(\mathbf{P}\times\mathbf{K}).\mathbf{P}\right]P^k\right\}$$

$$+ f_2^2\left[2(\mathbf{K}-\mathbf{J})\left(1 + \frac{M^2}{\mathbf{P}^2}\right) + 2(\mathbf{P}\times\mathbf{J})\frac{P^0}{\mathbf{P}^2} + \mathbf{P}(\mathbf{P.J})\frac{M^2}{\mathbf{P}^4}\right]^k$$

$$= f_1 \left[\mathbf{J} - \frac{1}{P^0} (\mathbf{P} \times \mathbf{K}) \right]^k + f_2 \left[\mathbf{K} - \frac{1}{\mathbf{P}^2} (\mathbf{P}.\mathbf{K})\mathbf{P} - \frac{P^0}{\mathbf{P}^2} (\mathbf{P} \times \mathbf{J}) \right]^k$$

$+ f_3\, P^k + f_4\, (\mathbf{P.J})P^k$ whence taking product with P^k and summing over k, we get

$$f_1^2 \frac{M^2}{(P^0)^2} (\mathbf{P.J}) + f_2^2 \left[2\left(1 + \frac{M^2}{\mathbf{P}^2}\right)(\mathbf{P}.\mathbf{K}) - \left(2 + \frac{M^2}{\mathbf{P}^2}\right)(\mathbf{P}.\mathbf{J}) \right]$$

$= f_1(\mathbf{P.J}) + f_3\, \mathbf{P}^2 + f_4 (\mathbf{P.J})\mathbf{P}^2$, Since $f_i \equiv f_i(\mathbf{P}^2, P^0)$ the above expression mandates $f_2 = f_3 = 0$ and $f_1^2 \dfrac{M^2}{(P^0)^2} = f_1 + f_4\, \mathbf{P}^2$. Substituting these values, we get $S^k = f_1 \left[\mathbf{J} - \dfrac{1}{P^0} (\mathbf{P} \times \mathbf{K}) \right]^k + f_4(\mathbf{P.J})P^k$. Setting this value of S^k and comparing the left and right hand sides, we also obtain $f_1^2 + f_1 f_4\, \mathbf{P}^2 = f_1$ and $f_1^2 + f_1 f_4 (P^0)^2 = -f_4 (P^0)^2$ so that $f_1 = \pm \dfrac{P^0}{M}$ and $f_4 = -\dfrac{1}{M(M \pm P^0)}$ thereby providing a unique expression (upto time reversal) for the relativistic spin operator as:

$$\mathbf{S} = \frac{\mathbf{W}}{M} - \frac{W_0\, \mathbf{P}}{M(M + P_0)}$$

$$= \frac{P^0\, \mathbf{J}}{M} - \frac{\mathbf{P} \times \mathbf{K}}{M} - \frac{\mathbf{P}(\mathbf{P}.\mathbf{J})}{M(M + P_0)} \tag{261}$$

This operator meets all the above requirements of a relativistic spin operator for positive massive particles. We have

(i) Making use of the properties of the Pauli Lubanski vector viz. $[\mathbf{J}, W_0] = 0$,

$$[J_j, W_k] = i \sum_{l=1}^{3} \varepsilon_{jkl}\, W_l\, ,\ [J_j, P_k] = i \sum_{l=1}^{3} \varepsilon_{jkl}\, P_l\, , \text{ we obtain}$$

$$[J_j, S_k] = \frac{1}{M} [J_j, W_k] - \frac{1}{M(M + P_0)} [J_j, W_0\, P_k]$$

$$= \frac{1}{M} [J_j, W_k] - \frac{1}{M(M + P_0)} \{W_0 [J_j, P_k] + [J_j, W_0] P_k\}$$

$$= \frac{i}{M} \sum_{l=1}^{3} \varepsilon_{jkl}\, W_l - \frac{i W_0}{M(M + P_0)} \sum_{l=1}^{3} \varepsilon_{jkl}\, P_l = i \sum_{l=1}^{3} \varepsilon_{jkl}\, S_l \tag{262}$$

(ii) Similarly, to prove (258), we make use of the definition of the Pauli Lubanski vector as $\mathbf{W}^0 = (\mathbf{P.\ J})$, $\mathbf{W} = P^0.\mathbf{J} - (\mathbf{P} \times \mathbf{K})$ so that $\mathbf{P.W} = P^0(\mathbf{P.J}) = P^0\, W^0$. We also write $F = -\dfrac{1}{M(M + P_0)}$

and use the vector identity $\mathbf{A} \times (\mathbf{B} \times \mathbf{C}) = \mathbf{B}(\mathbf{A}.\mathbf{C}) - \mathbf{C}(\mathbf{A}.\mathbf{B})$ to get

$$[S_1, S_2] = \left[FW_0 P_1 + \frac{W_1}{M}, FW_0 P_2 + \frac{W_2}{M}\right]$$

$$= i\left[-\frac{FP_1 (\mathbf{P} \times \mathbf{W})_2}{M} + \frac{FP_2 (\mathbf{P} \times \mathbf{W})_1}{M} + \frac{P_0 W_3 - W_0 P_3}{M^2}\right]$$

$$= i\left[-\frac{F(\mathbf{p} \times (\mathbf{P} \times \mathbf{W}))_3}{M} + \frac{P_0 W_3 - W_0 P_3}{M^2}\right]$$

$$= i\left[-\frac{F(P_3(\mathbf{P}.\mathbf{W}) - W_3 \mathbf{P}^2)}{M} + \frac{P_0 W_3 - W_0 P_3}{M^2}\right]$$

$$= i\left[-\frac{F(P_3 P^0 W^0 - W_3 \mathbf{P}^2)}{M} + \frac{P_0 W_3 - W_0 P_3}{M^2}\right]$$

$$= iW_3 \left(\frac{\mathbf{P}^2 F}{M} + \frac{P_0}{M^2}\right) + iP_3 W^0 \left(-\frac{P_0 F}{M} - \frac{1}{M^2}\right). \tag{263}$$

Now, $$\frac{\mathbf{P}^2 F}{M} + \frac{P_0}{M^2} = -\frac{\mathbf{P}^2}{M^2 (M + P^0)} + \frac{P_0}{M^2}$$

$$= \frac{P_0 (M + P_0) - \mathbf{P}^2}{M^2 (M + P_0)} = \frac{P_0 (M + P_0) - (M + p_0)(P_0 - M)}{M^2 (M + P_0)} = \frac{1}{M}$$

and

$$-\frac{P_0 F}{M} - \frac{1}{M^2} = \frac{P_0}{M^2 (M + P_0)} - \frac{1}{M^2} = \frac{P_0 - (M + P_0)}{M^2 (M + P_0)} = -\frac{1}{M(M + P_0)} = F$$

whence from eq. (263) we get

$$[S_1, S_2] = i\left(\frac{W_3}{M} FW^0 P_3\right) = iS_3 \tag{264}$$

We can, similarly prove commutation relations among other combinations of the spin operator components.

(iii) We also have

$$[P_\mu, S_k] = \frac{1}{M} [P_\mu, W_k] - \frac{1}{M(M + P_0)} [P_\mu, W_0, P_k]$$

$$= \frac{1}{M}[P_\mu, W_k] - \frac{1}{M(M+P_0)}\{W_0[P_\mu, P_k] + [P_\mu, W_0]P_k\} = 0 \tag{265}$$

(iv) We also have

$$\mathbf{S}^2 = \left(\frac{\mathbf{W}}{M} + W_0\,\mathbf{P}F\right)^2 = \frac{\mathbf{W}^2}{M^2} + \frac{2W_0\,F\mathbf{P}.\mathbf{W}}{M} + W_2^0\,\mathbf{P}^2 F^2$$

$$= \frac{\mathbf{W}^2}{M^2} + \frac{2W_0\,FP_0\,\mathbf{P}.\mathbf{J}}{M}\;W_2^0\,\mathbf{P}^2 F^2 = \frac{\mathbf{W}^2}{M^2}\;W_2^0\,F\left(\frac{2P_0}{M} + \mathbf{P}^2 F\right)$$

$$= \frac{\mathbf{W}^2}{M^2} + W_2^0\,F\,\frac{2P_0(M+P_0) - \mathbf{P}^2}{M(M+H)} = \frac{\mathbf{W}^2}{M^2} - W_0^2\,\frac{P_0^2 + 2P_0M + M^2}{M^2(M+P_0)^2}$$

$$\frac{\mathbf{W}^2 - W_0^2}{M^2} = -\frac{W^2}{M^2} \tag{266}$$

2.19.2 The Relativistic Position Operator, R, for Massive Particles

To identify an appropriate position operator in the relativistic quantum theory, we postulate that a suitable relativistic position operator, **R**, must necessarily possess the following properties:

(i) It should be possible to measure spin and position simultaneously so that the corresponding operators must commute, i.e.,

$$[R_i, S_j] - 0 \tag{266a}$$

(ii) Space translations of the observer should simply result in corresponding changes in the value of the position, e.g., $e^{-iP_1a}R_1e^{iP_1a} = R_1 - a$, $e^{-iP_1a}R_{2,3}e^{iP_1a} = R_{2,3}$ which implies the commutation relation

$$[R_j, P_k] = i\delta_{jk} \tag{267}$$

(iii) We also require the position operator to transform as a vector so that its commutators with the angular momentum operator would be

$$[J_j, R_k] = i\sum_l \varepsilon_{jkl} R_l \tag{268}$$

We postulate the position operator (Newton Wigner form) to be given by

$$R = -\frac{1}{2}[(P_0)^{-1}\,\mathbf{K} + \mathbf{K}(P_0)^{-1}] - \frac{\mathbf{P}\times\mathbf{S}}{P_0(M+P_0)}$$

$$= -(P_0)^{-1}\,\mathbf{K} - \frac{i}{2}\frac{\mathbf{P}}{(P_0)^2} - \frac{\mathbf{P}\times\mathbf{W}}{MP_0(M+P_0)} \tag{269}$$

(i) We, then, have

$$[J_1, R_2] = -\frac{1}{2}[J_1, P_0^{-1} K_2] - \frac{1}{2}[J_1, K_2 P_0^{-1}] - \left[J_1, \frac{(P_3 S_1 - P_3 S_1)}{P_0 (P_0 + M)}\right]$$

$$= -\frac{i}{2}(P_0^{-1} K_3 + K_3 P_0^{-1}) + \frac{i}{P_0 (P_0 + M)} (P_2 S_1 - P_1 S_2) = iR_3 \tag{270}$$

with similar commutators for the other components.

(ii) We also have

$$[R_1, P_1] = -\frac{1}{2}[(P_0^{-1} K_1 + K_1 P_0^{-1}), P_1] = \frac{i}{2} (P_0^{-1} P_0 + K_0 P_0^{-1}) = i \tag{271}$$

$$[R_1, P_2] = -\frac{1}{2}[(P_0^{-1} K_1 + K_1 P_0^{-1}), P_2] = 0 \tag{272}$$

as desired.

(iii) We now prove that $\mathbf{J} = \mathbf{R} \times \mathbf{P} + \mathbf{S}$. To do so, we make use of the fact that $[K_1 P_2 - K_2 P_1, P_0] = -i(P_1 P_2 - P_2 P_1) = 0$ so that $\mathbf{K} \times \mathbf{P}$ commutes with P_0 and the operator $P_0^{-1} \mathbf{K} \times \mathbf{P}$ is Hermitian.

Furthermore, we also use $\mathbf{P.S} = \frac{\mathbf{P}.P_0 \mathbf{J}}{M} - \frac{\mathbf{P}^2 \mathbf{P.J}(P_0 - M)}{\mathbf{P}^2 M} = \mathbf{P.J}$. We then have

$$\mathbf{J} - \mathbf{R} \times \mathbf{P} = \mathbf{J} + P_0^{-1} \mathbf{K} \times \mathbf{P} + \frac{(\mathbf{P} \times \mathbf{S}) \times \mathbf{P}}{P_0 (M + P_0)}$$

$$= \mathbf{J} + P_0^{-1} \mathbf{K} \times \mathbf{P} - \frac{1}{P_0 (M + P_0)} (\mathbf{P}(\mathbf{P.S}) - \mathbf{S}\mathbf{P}^2)$$

$$= \mathbf{J} + P_0^{-1} \mathbf{K} \times \mathbf{P} - \frac{1}{P_0 (M + P_0)} [(\mathbf{P}(\mathbf{P.S}) - \mathbf{S}(P_0 - M)(P_0 + M)]$$

$$= \mathbf{J} + P_0^{-1} \mathbf{K} \times \mathbf{P} - \frac{\mathbf{P}(\mathbf{P.S})}{P_0 (M + P_0)} - \frac{M}{P_0}\mathbf{S} + \mathbf{S} = \mathbf{S} \tag{273}$$

thereby showing that the total angular momentum is the aggregate of the orbital angular momentum and the spin angular momentum just as in classical physics.

(iv) We also have

$$[S_1, R_1] = [J_1 - (\mathbf{R} \times \mathbf{P})_1, R_1] = [J_1, R_1] - [(\mathbf{R} \times \mathbf{P})_1, R_1]$$

$$= \left[J_1, -\frac{1}{2}(P_0^{-1} K_1 + K_1 P_0^{-1}) - \frac{(\mathbf{P} \times \mathbf{S})_1}{P_0 (P_0 + M)}\right] - [(R_2 P_3 - R_3 P_2), R_1]$$

$$= \left[J_1, -\frac{1}{2}(P_0^{-1} K_1 + K_1 P_0^{-1}) - \frac{(P_2 S_3 - P_3 S_2)}{P_0 (P_0 + M)} \right] - [(R_2 P_3 - R_3 P_2), R_1]$$

$$= \left[J_1, -\frac{(P_2 S_3 - P_3 S_2)}{P_0 (P_0 + M)} \right] = -\frac{1}{P_0 (P_0 + M)} \{[J_1, P_2 S_3] - [J_1, P_3 S_2]\}$$

$$= -\frac{1}{P_0 (P_0 + M)} \{P_2[J_1, S_3] + [J_1, P_2] S_3 - P_3[J_1, S_2] - [J_1, P_3]S_2\}$$

$$= -\frac{1}{P_0 (P_0 + M)} \{- iP_2 S_2 + iP_3 S_3 - iP_3 S_3 + iP_2 S_2\} = 0 \qquad (274)$$

and

$$[S_1, R_2] = [J_1 - (\boldsymbol{R} \times \mathbf{P})_1, R_2] = [J_1, R_2] - [(\mathbf{R} \times \mathbf{P})_1, R_2]$$
$$= iR_3 - [(R_2 P_3 - R_3 P_2), R_2] = iR_3 - [R_2, P_3, R_2] + [R_3 P_2, R_2]$$
$$= iR_3 - R_2[P_3, R_2] - [R_2, R_2]P_3 + R_3[P_2, R_2] + [R_3, R_2]P_2 = iR_3 - iR_3 = 0$$

where we have assumed the commutativity of the various components of the position operator, which is proved below.

(v) We now prove that all components of the position operator commute with each other, i.e., $[R_i, R_j] = 0$. We have

$$[P_0 R_1, P_0 R_2] = [P_0 R_1, P_0]R_2 + P_0\{P_0 R_1, R_2]$$
$$= P_0[R_1, P_0]R_2 + P_0[P_0, R_2]R_1 + P_0^2[R_1, R_2]$$
$$= i(P_1 R_2 - P_2 R_1) + P_0^2 [R_1, R_2]$$
$$= i(\mathbf{P} \times \mathbf{R})_3 + P_0^2 [R_1, R_2] = -iJ_3 + iS_3 + P_0^2[R_1, R_2] \qquad (275)$$

We also have, on using eq. (269)

$$[P_0 R_1, P_0 R_2] = \left[-K_1 - \frac{i}{2}\frac{P_1}{P_0} + F(\mathbf{P} \times \mathbf{W})_1, -K_2 - \frac{i}{2}\frac{P_2}{P_0} + F(\mathbf{P} \times \mathbf{W})_2 \right] \qquad (276)$$

wherein we have

$$[-K_1, - K_2] = [K_1, K_2] = -iJ_3 \qquad (277)$$

$$\left[-\frac{i}{2}\frac{P_1}{P_0}, - K_2 \right] = -\frac{i}{2}\left[K_2, \frac{P_1}{P_0} \right] = -\frac{i}{2}\left[K_2, \frac{P_0 P_1}{P_0^2} \right]$$

$$= -\frac{i}{2}\left\{ P_0\left[K_2, \frac{P_1}{P_0^2} \right] + \left[K_2, \frac{P_0}{P_0^2} \right] P_1 \right\} = -\frac{i}{2}\left(-i\frac{P_2 P_1}{P_0^2} \right) \qquad (278)$$

$$\left[-K_1, -\frac{i}{2}\frac{P_2}{P_0}\right] = \frac{i}{2}\left[K_1, \frac{P_2}{P_0}\right] = \frac{i}{2}\left[K_1, \frac{P_0 P_2}{P_0^2}\right] = \frac{i}{2}\left(-i\frac{P_1 P_2}{P_0^2}\right) \tag{279}$$

$$[-K_1, F(\mathbf{P}\times\mathbf{W})_2] = \frac{1}{M}\left[K_1, \frac{P_3 W_1 - P_1 W_3}{P_0 + M}\right]$$

$$= \frac{1}{M}\left\{-\frac{P_3 W_1 - P_1 W_3}{(P_0 + M)^2}[K_1, P_0] + \frac{P_3[K_1, W_1]}{P_0 + M} - \frac{[K_1, P_1]W_3}{P_0 + M}\right\}$$

$$= \mathrm{i}\{MF^2 (P_3W_1 - P_1W_3)P_1 + FP_3W_0 - FP_0 W_3\} \tag{280}$$

$$[F(\mathbf{P}\times\mathbf{W})_1, -K_2] = F\{MF(\mathbf{P}\times\mathbf{W})_1 [K_2, P_0] + P_3[K_2, W_2] - [K_2, P_2]W_3\}$$

$$= i[MF^2(P_2 W_3 - P_3 W_2)P_2 + FP_3 W_0 - FP_0 W_3] \tag{281}$$

$$[F(\mathbf{P}\times\mathbf{W})_1, F(\mathbf{P}\times\mathbf{W})_2] = F^2[P_2 W_3 - P_3 W_2, P_3 W_1 - P_1 W_3]$$

$$= F^2\{P_2 P_3[W_3, W_1] - P_3^2 [W_2, W_1] + P_3 P_1[W_2, W_3]\}$$

$$= iF^2\{P_2 P_3[P_0 W_2 - W_0 P_2) + P_3^2(P_0 W_3 - W_0 P_3) + P_3 P_1(P_0 W_1 - W_0 P_1)\}$$

$$= iF\{-W_0 P_3(P_1^2 + P_2^2 + P_3^2) + P_0 P_3(P_1 W_1 + P_2 W_2 + P_3 W_3)\}$$

$$= iF^2\{-W_0P_3\mathbf{P}^2 + P_0P_3(\mathbf{P.W})\} = iF^2(-W_0P_3\mathbf{P}^2 + P_0^2P_3 W_0)$$

$$= iF^2 W_0\{-P_3(P_2^0 - M^2) + P_0^2 P_3\} = \frac{iW_0 P_3}{(P_0 + M)^2} \tag{282}$$

Adding eqs. (280) & (281), we get

$$[-K_1, F(\mathbf{P}\times\mathbf{W})_2] + [F(\mathbf{P}\times\mathbf{W})_1, -K_2]$$

$$= i\{MF^2(\mathbf{P}\times(\mathbf{P}\times\mathbf{W}))_3 + 2FP_3W_0 - 2FP_0 W_3\}$$

$$= i\{MF^2[P_3(\mathbf{P.W}) - W_3 \mathbf{P}^2] + 2FP_3 W_0 - 2FP_0W_3\}$$

$$= i\{MF^2[P_3(P_0.W_0) - W_3\mathbf{P}^2] + 2FP_3W_0 - 2FP_0W_3\}$$

$$= iMF^2\{P_3W_0[P_0 - 2(P_0 + M)] + W_3[-(P_0 - M)(P_0 + M) + 2P_0(P_0 + M)]\}$$

$$= iP_3 W_0(F - M^2 F^2) + \frac{iW_3}{M} \tag{283}$$

Now, adding eqs. (277), (278), (279), (282) & (283), we get

$$[P_0 R_1, P_0 R_2] = -iJ_3 + iP_3W_0(F - M^2 F^2) + \frac{iW_3}{M} + iM^2 F^2 W_0 P_3$$

$$= -iJ_3 + i\left(-\frac{P_3 W_0}{M(P_0 + M)} + \frac{W_3}{M}\right) = -iJ_3 + iS_3 \tag{284}$$

whence, we have, on using eq. (275)

$$P_0^2[R_1, R_2] = 0 \quad \Rightarrow [R_1, R_2] = 0 \tag{285}$$

(vi) From the commutation relation $[R_i, P_j] = i\delta_{ij}$, we have

$$[R_i, P_i^n] = inP_i^{n-1} \tag{286}$$

and, in general, given any function $f(P_1)$,

$$[R_i, f(P_i)] = i\,\frac{\partial f(P_i)}{\partial P_i} \tag{287}$$

Further,

$$[\mathbf{R}, P_0] = \left[\mathbf{R}, \sqrt{\mathbf{P}^2 + M^2}\right] = i\frac{\partial\sqrt{\mathbf{P}^2 + M^2}}{\partial \mathbf{P}} = i\frac{\mathbf{P}}{p_0} = i\mathbf{V} \tag{288}$$

(vi) Time evolution of the Position Operator is given by;

$$\mathbf{R}(x^0) = \exp(ip_0 x^0)\,\mathbf{R}\,\exp(-ip_0 x^0)$$
$$= \mathbf{R} + i[P_0, \mathbf{R}]x^0 = \mathbf{R} + \mathbf{V}x^0 \tag{289}$$

(vii) The transformation of the position operator under a "boost" transformation is derived as follows:

For simplicity, we consider the case of a particle without spin, whence the position operator takes the form

$$\mathbf{R} = -\frac{1}{2}(\mathbf{K}P_0^{-1} + P_0^{-1}\mathbf{K}) \tag{290}$$

Now, we have

$$K_2(\theta) = e^{-iK_1\theta}\,K_2\,e^{iK_1\theta}$$

$$= K_2 - i[K_1, K_2] + \frac{\theta^2}{2!}[K_1, [K_1, K_2]] - \frac{i}{3!}[K_1, [K_1,[K_1, K_2]]] + \ldots$$

$$= K_1 - \theta J_3 + \frac{\theta^2}{2!}K_1 - \frac{\theta^3}{3!}J_3 + \ldots$$
$$= K_1 \cosh\theta - J_3 \sinh\theta$$

Hence, the various components of the position operator in the frame O' that is boosted with a velocity $v = \tanh\theta$ along the x^1 axis are given by

$$R_2(\theta) = e^{-iK_1\theta}\,R_2\,e^{iK_1\theta} = -\frac{1}{2}e^{-iK_1\theta}\,(K_2 P_0^{-1} + P_0^{-1} K_2)e^{iK_1\theta}$$

$$= -\frac{1}{2}(K_2 \cosh\theta - J_3 \sinh\theta)(P_0 \cosh\theta - P_1 \sinh\theta)^{-1}$$

$$-\frac{1}{2}(P_0 \cosh\theta - P_1 \sinh\theta)^{-1}(K_2 \cosh\theta - J_3 \sinh\theta)$$

$$= -\frac{1}{2}K_2(1 - K_2^{-1} J_3 \tanh\theta)\, P_0^{-1}(1 - P_0^{-1} P_1 \tanh\theta)^{-1}$$

$$-\frac{1}{2}\, P_0^{-1}(1 - P_0^{-1} P_1 \tanh\theta)^{-1} K_2(1 - K_2^{-1} J_3 \tanh\theta)$$

$$= -\frac{1}{2}K_2(1 - K_2^{-1} J_3 v)\, P_0^{-1}(1 - V_1 v)^{-1} - \frac{1}{2}P_0^{-1}(1 - V_1 v)^{-1} K_2(1 - K_2^{-1} J_3 v)$$

$$= -\frac{1}{2}(1 - V_1 v)^{-1}[(K_2 P_0^{-1} + P_0^{-1} K_2) + (J_3 P_0^{-1} + P_0^{-1} J_3)v] = \beta(R_1 + J_3 P_0^{-1} v) \qquad (291)$$

$$R_1(\theta) = -\frac{1}{2}K_1(P_0 \cosh\theta - P_1 \sinh\theta)^{-1} - \frac{1}{2}(P_0 \cosh\theta - P_1 \sinh\theta)^{-1} K_1 = \beta\frac{R_1}{\cosh\theta} \qquad (292)$$

$$R_3(\theta) = -\frac{1}{2}(K_3 \cosh\theta + J_2 \sinh\theta)(P_0 \cosh\theta - P_1 \sinh\theta)^{-1}$$

$$-\frac{1}{2}(P_0 \cosh\theta - P_1 \sinh\theta)^{-1}(K_3 \cosh\theta + J_y \sinh\theta) = \beta(R_3 - J_2 P_0^{-1} v) \qquad (293)$$

Eqs. (291) – (293) give the components of the position operator in the boosted frame as functions of the relative velocity between the two frames. In other words, $R_i(\theta)$ will give us the position of the particle with reference to the frame O' at an instant of time (as measured in the rest frame O) when the coordinates of the particle in O are R_i. The time dependence of the position operator in the boosted frame O' is obtained as follows:

We have

$$P_0' = e^{-iK_1\theta} P_0\, e^{iK_1\theta} \text{ so that}$$

$$\mathbf{R}(\theta, x'^0) = e^{iP'_0 x'^0}\, \mathbf{R}(\theta)\, e^{-iP_0 x'^0} = e^{iP'_0 x'^0}\, e^{-iK_1\theta}\, \mathbf{R} e^{iK_1\theta}\, e^{-iP'_0 x'^0}$$

$$= (e^{-iK_1\theta} e^{iP_0 x'^0} e^{iK_1\theta})\, e^{-iK_1\theta} \mathbf{R} e^{iK_1\theta} (e^{-iK_1\theta} e^{-iP_0 x'^0} e^{iK_1\theta})$$

$$= e^{-iK_1\theta} e^{iP_0 x'^0} \mathbf{R} e^{-iP_0 x'^0} e^{iK_1\theta} = e^{-iK_1\theta}(\mathbf{R} + \mathbf{V}x'^0) e^{iK_1\theta} = -\mathbf{R}(\theta) + \mathbf{V}(\theta) x'^0 \qquad (294)$$

where we have used eq. (289). The components of $\mathbf{R}(\theta)$ are given by eqs. (291) – (293). $\mathbf{V} = \mathbf{P}P_0^{-1}$ is the velocity operator with $\mathbf{V}(\theta)$ being the corresponding operator in the moving frame that may be expressed as

$$V_1(\theta) = e^{-iK_1\theta}\frac{P_1}{P_0}\, e^{-iK_1\theta} = \frac{P_1 \cosh\theta - P_0 \sinh\theta}{P_0 \cosh\theta - P_1 \sinh\theta} = \frac{P_1 P_0^{-1} - \tanh\theta}{1 - P_1 P_0^{-1} \tanh\theta} = \beta(V_1 - v) \qquad (295)$$

$$V_2(\theta) = e^{-iK_1\theta}\frac{P_2}{P_0} e^{iK_1\theta} = \frac{P_2}{P_0 \cosh\theta - P_1 \sinh\theta} = \frac{P_2 P_0^{-1}}{(1 - P_1 P_0^{-1} \tanh\theta)} = \frac{\beta V_2}{\cosh\theta} \qquad (296)$$

$$V_3(\theta) = \frac{P_3}{P_0 \cosh\theta - P_1 \sinh\theta} = \frac{P_3 P_0^{-1}}{(1 - P_1 P_0^{-1} \tanh\theta)} = \frac{\beta V_3}{\cosh\theta} = \beta V_3 \sqrt{1 - v^2} \tag{297}$$

In deriving eqs. (295)-(297), we have used the transformation rules from the translation generators under boosts viz.

$$\mathbf{P}(\boldsymbol{\theta}) = e^{-i\mathbf{K}\theta}\, \mathbf{P} e^{i\mathbf{K}\theta} = \mathbf{P} + \frac{\theta}{\theta}\left[\left(\mathbf{P}.\frac{\theta}{\theta}\right)(\cosh\theta - 1) - P_0 \sinh\theta\right] \tag{298}$$

$$P_0(\theta) = P_0 \cosh\theta - \left(\mathbf{P}.\frac{\theta}{\theta}\right) \sinh\theta \tag{299}$$

These relations are easily obtained as

$$\frac{\partial^2 P_1(\theta)}{\partial\theta^2} = \frac{\partial}{\partial\theta}(-ie^{-iK_1\theta} K_1 P_1\, e^{iK_1\theta} + e^{-iK_1\theta} P_1 iK_1\, e^{iK_1\theta}) = -i\frac{\partial}{\partial\theta}\{e^{-iK_1\theta}[K_1, P_1]e^{iK_1\theta}\}$$

$$= -\frac{\partial}{\partial\theta}(e^{-iK_1\theta} P_0\, e^{iK_1\theta}) = -(-ie^{-iK_1\theta} K_1 P_0\, e^{iK_1\theta} + e^{-iK_1\theta} P_0\, iK_1 e^{iK_1\theta})$$

$$= ie^{-iK_1\theta}[K_1, P_0]e^{iK_1\theta} = e^{-iK_1\theta} P_1\, e^{iK_1\theta} = P_1(\theta)$$

which has the solution $P_1(\theta) = P_1 \cosh\theta - P_0 \sinh\theta$ where we have used the boundary conditions $P_1(0) = P_1$ & $\left.\frac{\partial P_1(\theta)}{\partial\theta}\right|_{\theta=0} = -P_0$. Similarly,

$$\frac{\partial P_2(\theta)}{\partial\theta} = (-ie^{-iK_1\theta} K_1 P_2\, e^{iK_1\theta} + e^{-iK_1\theta} P_2\, iK_1 e^{iK_1\theta})$$

$$= -i\{e^{-iK_1\theta}[K_1, P_2]e^{iK_1\theta}\} = 0$$

whence $P_2(\theta) = P_2(0) = P_2$ and so on.

Combining the set of eqs. (291)–(293), 294 & (295)–(297), we obtain

$$R_1(\theta, x'^0) = \beta\left[\frac{R_1}{\cosh\theta} + (V_1 - v)x'^0\right] \tag{300}$$

$$R_2(\theta, x'^0) = \beta\left(R_2 + \frac{J_3 v}{P_0} + \frac{V_2 x'^0}{\cosh\theta}\right) \tag{301}$$

$$R_3(\theta, x'^0) = \beta\left(R_3 - \frac{J_2 v}{P_0} + \frac{V_3 x'^0}{\cosh\theta}\right) \tag{302}$$

2.19.3 Consistency of the Position Operator with Lorentz Transformations of Special Relativity

For simplicity, we consider a quantum particle moving along the x^1 axis. We also assume that the relative motion between the unprimed and the primed frame is also along the same axis so that the other spatial coordinates can be ignored in our analysis. In such a situation, the spacetime Lorentz transformation equations are given by

$$x'^1 = x^1 \cosh\theta - x^0 \sinh\theta,\ x'^0 = x^0 \cosh\theta - x^1 \sinh\theta \tag{303}$$

so that, we have

$$R_1(\theta, x'^0) = R_1(0, x^0) \cosh\theta - x^0 \sinh\theta = (R_1 + V_1 x^0) \cosh\theta - x^0 \sinh\theta \tag{304}$$

where

$$x'^0 = x^0 \cosh\theta - R_1(0 + x^0) \sinh\theta = x^0 \cosh\theta - (R_1 + V_1 x^0) \sinh\theta \tag{305}$$

Taking the difference between eqs. (300) & (304) and using eq. (305)

$$\frac{\beta R_1}{\cosh\theta} + (V_1 - v)\,\beta[x^0 \cosh\theta - (R_1 + V_1 x^0)\sinh\theta]$$

$$- (R_1 + V_1 x^0) \cosh\theta + x^0 \sinh\theta = \beta R_1 \left(\frac{1}{\cosh\theta} - \sinh\theta (V_1 - v) - \beta^{-1} \cosh\theta\right)$$

$$+ \beta x^0 [(V_1 - v) \cosh\theta - (V_1 - v) V_1 \sinh\theta - \beta^{-1} V_1 \cosh\theta + \beta^{-1} \sinh\theta] = 0$$

where we have used $\beta = (1 - V_1 v)^{-1}$, $\cosh\theta = (1 - v^2)^{-1/2}$, $\sinh\theta = v\,(1 - v^2)^{-1/2}$. This establishes the consistency of the position operator with the Lorentz transformations.

2.19.4 Uniqueness of the Spin Operator

We now establish that the relativistic spin and position operators as defined in the preceding sections are unique. However, to do so, we need to introduce the concepts of "canonical form" and "power" of operators.

It is obvious that any operator constituted by the generators of the Poincare group viz. P_μ, J_i, K_l with, $\mu = 0, 1, 2, 3$, $i, l = 1, 2, 3$ can be expressed in several equivalent forms using the commutators between these generators. We define the canonical form as the form in which the constituent generators appear in the sequence $C(P_0, P_1, P_2, P_3)$, $J_1, J_2, J_3, K_1, K_2, K_3$ from left to right where $C(P_0, P_1, P_2, P_3)$ are the functions exclusively of the translation generators. Thus, every such operator admits a unique representation in the canonical form as

$$O = C^{00} + \sum_{i=1}^{3} C_i^{10}\, J_i + \sum_{i=1}^{3} C_i^{10}\, K_i + \sum_{i,j=1}^{3} C_i^{11}\, J_i K_j + \sum_{i,j=1}^{3} C_i^{20} J_i J_j + \sum_{i,j=1}^{3} C_i^{02} K_i K_j + \ldots \tag{306}$$

where $C(P_0, P_1, P_2, P_3)$ are functions of the translation generators.

Having set up the operator in the canonical form, we define the power of the constituent terms of the canonical representation as the number of factors J_i, K_l that appear in the term. The power of the operator is defined as the highest power among the terms that constitute the canonical form of the operator. Obviously then $Pow(W_0) = Pow(\mathbf{W}) = Pow(\mathbf{S}) = Pow(\mathbf{R}) = 1$ and $Pow(P_0) = Pow(\mathbf{P}) = 0$. The power of operators has the following properties:

(i) $Pow([L, R]) = Pow(L) + Pow(R) - 1$ if $[L, R] \neq 0$ and $L, R \in C(P_0, P_1, P_2, P_3), J_1, J_2, J_3, K_1, K_2, K_3$. The proof is simple. There can be only two situations in which $L, R \in C(P_0, P_1, P_2, P_3), J_1, J_2, J_3, K_1, K_2, K_3$ and $[L, R] \neq 0$. Firstly, when (i) $Pow(L) = 1$ & $Pow(R) = 0$ or (ii) Pow(L) = 0 & $Pow(R) = 1$. It is seen from the commutation relations between the various Lorentz generators and translation operators that the only nonvanishing commutators are functions of the translation generators and hence, have zero power. The same holds if either of L, R are functions of translation generators. The second case of nonvanishing $[L, R]$ is when $Pow(L) = Pow(R) = 1$, i.e., when both L, R contain: (i) at least one term in the generators J_i, K_l and (ii) no term containing more than one of these generators. In this case again, the commutators between the above generators imply that $[L, R]$ shall contain at least one term in J_i, K_l and no term having more than one of these generators so that $Pow([L, R]) = 1$.

Hence, we always have $Pow([L, R]) = Pow(L) + Pow(R) - 1$.

(ii) Let us consider a term in an operator that is not in the canonical form i.e. that contains some of the generators $C(P_\mu)$, J_i, K_l but not in the desired sequence viz. $C(P_0, P_1, P_2, P_3), J_1, J_2, J_3, K_1, K_2, K_3$. To transform this term to the canonical form, we make use of the commutators among these generators, i.e., write the product of two generators L, R (that is not in the desired order) as $LR = RL + [L, R]$. In such a case, we call the term containing RL in the revised form, the primary term and the term containing $[L, R]$, the secondary term. Now, it is obvious that the primary term contains exactly the same number of generators (albeit in a different order) as the original term so that the power of the primary term is the same as the original term. Extending the same logic and using property (i) above, the power of the secondary term shall be either zero or one less than the power of the primary term depending on whether L, R commute or not. The power of tertiary or higher order terms would be less than the power of the secondary term. Hence, for any two terms $Pow(AB) = Pow(BA) = Pow(A) + Pow(B)$.

(iii) For any two terms A, B, we have $Pow([A, B]) = Pow(A) + Pow(B) - 1$ if $[A, B] \neq 0$. This is so because the primary term in both AB & BA is identical and hence, cancels out in $[A, B]$. However, the secondary terms will be different because the operators are non-commuting so that $[A, B] \neq [B, A]$. Hence, these terms will not cancel. Their power as per (ii) is $Pow(A) + Pow(B) - 1$ which will also be the power of $[A, B]$.

Let us now assume that the relativistic spin operator defined by us is not unique so that there exists another spin operator $\mathbf{S}'$ satisfying the requisite physical properties (outlined earlier). Let $Pow(S'_i) = p$ so that

$$2p - 1 = Pow([S'_1, S'_2] = Pow(S'_3) = p \Rightarrow p = 1 \tag{307}$$

Hence, the components of $\mathbf{S}'$ must have power of unity. Now, the most general pseudovector that can be constructed of power unity is

$$\mathbf{S}' = \alpha(M, P^2)\mathbf{S} + \beta(M, P^2)\,\mathbf{P} \times \mathbf{R} + \chi(M, P^2)\,(\mathbf{P.S})\mathbf{S} \tag{308}$$

Now, $[\mathbf{S}', P] = 0 \Rightarrow \beta(M, P^2) = 0$. Further,

$$[S_1', S_2'] = [\alpha S_1 + \chi(\mathbf{S.P})P_1, \alpha S_2 + \chi(\mathbf{S.P})P_2]$$
$$= \alpha^2[S_1, S_2] - i\alpha\chi\, P_1(\mathbf{S}\times\mathbf{P})_2 + i\alpha\chi\, P_2\,(\mathbf{S}\times\mathbf{P})_1$$
$$= i\alpha^2\, S_3 - i\alpha\chi\, P_1(\mathbf{P}\times(\mathbf{S}\times\mathbf{P}))_3$$
$$= i\alpha^2\, S_3 - i\alpha\chi\mathbf{P}^2\, S_3 + i\alpha\chi(\mathbf{S.P})P = iS'_3$$
$$= i\alpha S_3 + i\chi(\mathbf{S.P})P_3$$

whence

$$\alpha^2 - \alpha\chi\,\mathbf{P}^2 = \alpha \;\&\; \alpha\chi = \chi \text{ giving } \alpha = 1 \;\&\; \chi = 0 \text{ so that } \mathbf{S}' = \mathbf{S}.$$

2.19.5 Uniqueness of the Position Operator

Let us assume that, in addition to the Newton-Wigner position operator, there exists another position operator having the requisite physical properties. In view of the commutator $[R_i, P_j] = i\delta_{ij}$ and the property that $Pow\,([A, B]) = Pow\,(A) + Pow\,(B) - 1$ with $Pow\,(P_j) = 0$, we have $Pow\,(R_j) = 1$. Additionally, the position operator must be a vector. The most general vector of power unity is

$$\mathbf{R}' = a(P^2, M)\mathbf{R} + b(P^2, M)\;\mathbf{S}\times\mathbf{P} + c(P^2, M)\mathbf{P} \tag{309}$$

Now,

$$0 = [R_1', S_2] = b[S_2\,P_3 - S_3\,P_2, S_2] = ibS_1S_2 \Rightarrow b = 0 \tag{310}$$

Also,

$$i = [R_1', P_1] = a[R_1, P_1] = ia \Rightarrow a = 1 \tag{311}$$

whence

$$\mathbf{R}' = \mathbf{R} + c\,(P^2, M)\mathbf{p} \tag{312}$$

To establish that $c\,(P^2, M) = 0$, we need to examine the consistency of eq. (310) with the transformations of special relativity by retracing the steps of eqs. (290)-(305). We, shall, however, for simplicity, consider a quantum particle moving along the x^1 axis. We also assume that the relative motion between the unprimed and the primed frames is also along the same axis so that the other spatial coordinates can be ignored in our analysis. In such a situation, the spacetime Lorentz transformation equations are given by eqs. (303)-(305). We also have, for the x^1 component of $\mathbf{R}'$

$$R_1'\,(\theta) = R_1(\theta) + ce^{-iK_1\theta}\,P_1e^{-iK_1\theta} = \beta\frac{R_1}{\cosh\theta} + (P_1\,\cosh\theta - P_0\,\sinh\theta) \tag{313}$$

$$R_1'(\theta, x'^0) = e^{ip'_0x'^0}\,R_1'(\theta)e^{-iP'_0x'^0} = e^{iP'_0x'^0}\,[R_1(\theta) + ce^{-iK_1\theta}\,P_1e^{-iK_1\theta}]\,e^{-iP'_0x'^0}$$
$$= (e^{-iK_1\theta}\,e^{-iP_0x'^0}\,e^{iK_1\theta})\,e^{-iK_1\theta}\,(R_1 + cP_1)\,e^{iK_1\theta}\,(e^{iK_1\theta}\,e^{-iP_0x'^0}\,e^{iK_1\theta})$$
$$= e^{-iK_1\theta}e^{iP_0x'^0}R_1\,e^{-iP_0x'^0}\,e^{iK_1\theta} + ce^{-iK_1\theta}e^{iP_0x'^0}\,P_1\,e^{-iP_0x'^0}\,e^{iK_1\theta}$$
$$= e^{-iK_1\theta}\,(R_1 + V_1x'^0)e^{iK_1\theta} + ce^{-iK_1\theta}\,P_1\,e^{iK_1\theta}$$
$$= R_1(\theta) + V_1(\theta)x'^0 + c(P_1\,\cosh\theta - P_0\,\sinh\theta)$$

$$= \beta\left[\frac{R_1}{\cosh\theta} + (V_1 - v)x'^0\right] + c(P_1 \cosh\theta - P_0 \sinh\theta) \tag{314}$$

Taking the difference between eqs. (314) & (304) and using eq. (305), we conclude that must necessarily equal zero in order that this difference vanishes and hence, implies the consistency of the position operator with special relativity.

CHAPTER 3

The Klein Gordon Equation

3.1 THE SCHRÖDINGER EQUATION

The time evolution of a quantum mechanical state is described by

$$i\partial_0|\psi, x^0\rangle = \hat{H}|\psi, x^0\rangle \tag{1}$$

where $\hat{H}$ is the Hamiltonian operator representing total energy and $|\psi, x^0\rangle$ the time dependent quantum mechanical state of the quantum system in the Schrödinger representation.

The non-relativistic momentum-energy relationship for a spinless free particle is given by

$$E = \frac{\mathbf{p}^2}{2m} \tag{2}$$

where E is the energy of the particle of mass m corresponding to the momentum $\mathbf{p}$.

Substituting the operator representations for momentum and energy, as suggested by the correspondence principle, viz. $E \to i\partial_0$, $\mathbf{p} \to -i\nabla$, we obtain the Schrödinger equation for the free particle as

$$i\partial_0\psi(\mathbf{x}, x^0) = -\frac{1}{2m}\nabla^2\psi(\mathbf{x}, x^0) \tag{3}$$

where $\psi(\mathbf{x}. x^0)$ is the particle wavefunction in position space and is given by $\psi(\mathbf{x}, x^0) = \langle\mathbf{x}|\psi, x^0\rangle$. As seen from eqs. (1) and (3), the quantum mechanical Hamiltonian for the non-relativistic evolution of a free particle is given by $\hat{H} = -\frac{1}{2m}\nabla^2$.

3.2 INVARIANCE PROPERTIES OF THE SCHRÖDINGER EQUATION

We examine the invariance properties of the Schrödinger equation under 3-dimensional rotations and Lorentz boosts separately.

(A) Rotations

Consider the following rotation by an infinitesimal angle ε about the x^3 axis

$$x^1 \to x'^1 = x^1 + \varepsilon x^2,\ x^2 \to x'^2 = x^2 - \varepsilon x^1,\ x^3 \to x'^3 = x^3,\ x^0 \to x'^0 = x^0$$

If $\psi(x)$ is the free particle wavefunction in the unprimed coordinate system that takes the form $\psi'(x)$ in the primed system and $R(\varepsilon)$ is the operator representing the infinitesimal rotation ε, then we have

$$R(\varepsilon):\ \psi(x) \to \psi'(x) \equiv \hat{U}\ (R(\varepsilon))\psi(x) \equiv \psi(R^{-1}(\varepsilon)(x)) \equiv \psi(x^1 - \varepsilon x^2, x^2 + \varepsilon x^1, x^3, x^0) \qquad (4)$$

where $\hat{U}\ (R(\varepsilon))$ is the unitary operator representing the effect of rotations on the wavefunction $\psi(x)$. Invariance of the Schrödinger equation under rotations shall be established if the transformed wavefunction $\psi'(x)$ satisfies eq. (3).

Expanding $\psi'(x) \equiv \psi(x^1 - \varepsilon x^2, x^2 + \varepsilon x^1, x^3, x^0)$ about x as a Taylor's series, we obtain

$$\psi'(x) \cong \psi(x) + (-\varepsilon x^2\partial_1 + \varepsilon x^1\partial_2)\ \psi(x) \qquad (5)$$

Hence, we have invariance if

$$\left(i\partial_0 + \frac{1}{2m}\nabla^2\right)(-\varepsilon x^2\partial_1 + \varepsilon x^1\partial_2)\psi(x) = 0$$

or $$\left[-ix^2\partial_0\partial_1 + ix^1\partial_0\partial_2 - \frac{1}{2m}(x^2\nabla^2\partial_1 + 2\partial_2\partial_1) + \frac{1}{2m}(x^1\nabla^2\partial_2 + 2\partial_1\partial_2)\right]\psi(x) = 0$$

Noting that $[\partial_0, \partial_i] = [\nabla^2, \partial_i] = [\partial_i, \partial_j] = 0$ for $i, j = 1, 2, 3$, we can write the above equation as

$$\left[-x^2\partial_1\left(i\partial_0 + \frac{1}{2m}\nabla^2\right) + x^1\partial_2\left(i\partial_0 + \frac{1}{2m}\nabla^2\right)\right]\psi(x) = 0$$

which is, indeed true since $\psi(x)$ obeys eq. (3).

(B) Lorentz Boosts

Consider an infinitesimal pure Lorentz boost ξ directed along the x^3 axis

$$x^1 \to x'^1 = x^1,\ x^2 \to x'^2 = x^2,\ x^3 \to x'^3 = x^3 - \xi x^0,\ x^0 \to x'^0 = x^0 - \xi x^3$$

Let the free particle wavefunction in the unprimed coordinate system $\psi(x)$ take the form $\psi'(x)$ in the primed system and $L(\xi)$ be the operator representing the infinitesimal boost ξ, then we have

$$L(\xi){:}\psi(x) \to \psi'(x) \equiv U(L(\xi))\ \psi(x) \equiv \psi(L^{-1}(\xi)(x)) \equiv \psi(x^1, x^2, x^3 + \xi x^0, x^0 + \xi x^3) \qquad (6)$$

Schrödinger equation shall be invariant under Lorentz boosts if the transformed wavefunction $\psi'(x)$ satisfies eq. (3).

Expanding $\psi'(x) \equiv \psi(x^1, x^2, x^3 + \xi x^0, x^0 + \xi x^3)$ about x as a Taylor's series, we obtain

$$\psi'(x) \cong \psi(x) + (\xi x^0 \partial_3 + \xi x^3 \partial_0)\psi(x) \tag{7}$$

Hence, we have invariance if

$$\left(i\partial_0 + \frac{1}{2m}\nabla^2\right)(x^0\partial_3 + x^3\partial_0)\psi(x) = 0$$

or

$$\left[i\partial_3 + ix^0\partial_0\partial_3 + ix^3\partial_0^2 + \frac{1}{2m}(x^0\nabla^2\partial_3 + x^3\nabla^2\partial_0 + 2\partial_3\partial_0)\right]\psi(x) = 0$$

Noting that $[\partial_0, \partial_i] = [\nabla^2, \partial_i] = [\nabla^2, \partial_0] = [\partial_i, \partial_j] = 0$ for $i, j = 1, 2, 3$, we can write the above equation as

$$\left[i\partial_3 + x^0\partial_3\left(i\partial_0 + \frac{1}{2m}\nabla^2\right) + x^3\partial_0\left(i\partial_0 + \frac{1}{2m}\nabla^2\right) + \frac{1}{m}\partial_3\partial_0)\right]\psi(x) = 0$$

which, in view of eq. (3), is equal to $i\partial_3\left(1 - \frac{i}{m}\partial_0\right)\psi(x) = 0$. This equation does not hold for arbitrary $\psi(x)$.

This shows that the Schrödinger equation is not invariant under Lorentz boosts. The origin of this anomaly is traced to the fact that the Schrödinger equation does not epitomize the correct relativistic momentum-energy relationship which requires that time and energy be treated on equal footing.

3.3 THE KLEIN GORDON EQUATION

We have seen above that the Schrödinger equation is not invariant under Lorentz boosts (although it is invariant under rotations). We can, however, obtain a relativistically acceptable wave equation for the quantum dynamics of a particle by noting that the length of the 4-vector $(E, \mathbf{p})$ is invariant under Lorentz transformations. We have

$$p = \left(p^\mu p_\mu\right)^{\frac{1}{2}} = \left(E^2 - \mathbf{p}^2\right)^{\frac{1}{2}} = m \tag{8}$$

where m is the particle mass which is relativistically invariant. Furthermore, eq. (8) encapsulates the correct relativistic mass-energy relationship $E = (\mathbf{p}^2 + m^2)^{1/2}$.

Substituting the operator representations $E \to i\partial_0$, $\mathbf{p} \to i\nabla$ for the energy E and momentum $\mathbf{p}$ and operating on the free particle wavefunction $\psi(x)$, we obtain

$$i\partial_0\psi(x) = \sqrt{-\nabla^2 + m^2\psi(x)} \tag{9}$$

However, eq. (9), although originating from the correct relativistic mass-energy relationship, presents several problems, as it stands, e.g.,

(a) First, contrary to the basic tenets of relativity, it does not provide equivalent treatment of time and space coordinates – while the time derivative appears outside the square root, space derivatives appear within it. Relativity mandates total symmetry among space and time coordinates.

(b) Furthermore, the treatment and interpretation of the ∇^2 operator within the square root is troublesome. If the square root term is expanded as an infinite series in powers of ∇^2, we find that eq. (9) is not localized in space because there would be an infinite number of spatial derivatives acting on $\psi(x)$.

The above issues can, however, be resolved by squaring the operators on both sides of eq. (9) before operating on the wavefunction $\psi(x)$. We then obtain

$$(\partial_0^2 - \nabla^2 + m^2)\psi(x) = 0 = (\Box + m^2)\psi(x) \tag{10}$$

where $\Box = \partial_0^2 - \nabla^2$ is the d'Alembertian operator. Equation (10) is called the Klein Gordon equation and is symmetric in both space and time derivatives.

3.4 INVARIANCE PROPERTIES OF THE KLEIN GORDON EQUATION

The concept of "invariance" essentially means that the "physics must look the same in all inertial frames" or, in other words, the predicted results of all physical experiments should be the same irrespective of them being observed by observers that are placed in different inertial frames. Specifically, we consider an observer A located in an inertial frame I predicting the value of the wavefunction ψ of a free particle at a spacetime point that has coordinates x with respect to frame I. Similarly, A' is an observer in another inertial frame I' that predicts the value of the wavefunction ψ' representing the same free particle at the same spacetime point that has coordinates x' with reference to the frame I'. To establish invariance, we need to show that $\psi'(x') = \psi(x)$ throughout spacetime, i.e., $\psi(x)$ and $\psi'(x')$ must necessarily obey the same equation of motion. Let us assume that the inertial frames I and I' are related through a generalized Lorentz transformation of the form $x'^{\mu} = \Lambda^{\mu}{}_{\nu} x^{\nu} + a^{\mu}$, where a^{μ} represents a translation vector and Λ^{μ}_{ν} is a matrix representing a Lorentz transformation. Both a^{μ} and Λ^{μ}_{ν} are independent of x^{μ}. Now

$$\begin{aligned} &= (\partial'_{\sigma})^2 - (\nabla')^2 = (\partial')^{\mu}(\partial')_{\mu} = (\Lambda^{\mu}_{\rho}\partial^{\rho})(\Lambda_{\mu}{}^{\sigma}\partial_{\sigma}) = (\Lambda^{\mu}_{\rho}\Lambda^{\sigma}_{\mu})(\partial^{\rho}\partial_{\sigma}) \\ &= g^{\mu\sigma}g_{\rho\mu}(\partial^{\rho}\partial_{\sigma}) = \delta^{\sigma}_{\rho}(\partial^{\rho}\partial_{\sigma}) = \partial^{\rho}\partial_{\rho} \end{aligned} \tag{11}$$

so that

$$0 = (' + m^2)\psi'(x') = (+ m^2)\psi'(x') = (+ m^2)\psi(x) \tag{12}$$

which implies that $\psi'(x')$ is a solution of the Klein Gordon equation with reference to frame I and hence, is a wavefunction representing the free particle with reference to inertial frame I. Since, $\psi(x)$ also represents the same particle with reference to frame I, the values of $\psi(x)$ and $\psi'(x')$ must necessarily agree on all spacetime points and we have $\psi(x) \equiv \psi'(x')$.

The invariance of the Klein Gordon equation with regard to Lorentz transformations can also be established as for the Schrödinger equation.

For rotations, we have

$$(\partial_0^2 - \nabla^2 + m^2)(-\varepsilon x^2\partial_1 + \varepsilon x^1\partial_2)\psi(x)$$

$$= \varepsilon\,[-x^2\,\partial_0^2\,\partial_1 + x^1\,\partial_0^2\,\partial_2 + (x^2\nabla^2\partial_1 + 2\partial_2\,\partial_1) - (x^1\,\nabla^2\,\partial_2 + 2\partial_1\,\partial_2)$$
$$-m^2(x^2\partial_1 - x^1\partial_2)]\psi(x)$$
$$= \varepsilon[-x^2\,\partial_1(\partial_0^2 - \nabla^2 + m^2) + x^1\,\partial_2(\partial_0^2 - \nabla^2 + m^2)]\,\psi(x) = 0$$

Similarly, for boosts, we have

$$(\partial_0^2 - \nabla^2 + m^2)(\xi x^0\partial_3 + \xi x^3\partial_0)\psi(x)$$
$$= \xi[x^0\partial_0^2\partial_3 + 2\partial_0\partial_3 + x^3\partial_0^3 - (x^0\,\nabla^2\partial_3 + x^3\nabla^2\partial_0 + 2\partial_3\partial_0)$$
$$+ m^2(x^0\partial_3 + x^3\partial_0)]\psi(x)$$
$$= \xi[x^0\partial_3(\partial_0^2 - \nabla^2 + m^2) + x^3\partial_0\,(\partial_0^2 - \nabla^2 + m^2)]\psi(x) = 0$$

thereby confirming Lorentz invariance.

The wavefunction $\psi(x)$ associated with the Klein Gordon eq. (10) is a single component Lorentz scalar. It represents spinless particles, for, if s were to be the spin of this wavefunction, then it should necessarily possess $2s + 1$ components. Hence, we have, $2s + 1 = 1$ or $s = 0$ showing that the Klein Gordon wavefunction describes spin 0 particles.

3.5 CONSTRUCTION OF THE 4-CURRENT j_μ

The Klein Gordon equation is given by eq. (10) with the complex conjugate

$$(\partial_0^2 - \nabla^2 + m^2)\psi^* = 0 \tag{13}$$

Multiplying eq. (10) from the left by ψ^* and eq. (13) by ψ (also from the left) and subtracting, we obtain

$$(\psi^*\partial_0^2\psi - \psi\partial_0^2\psi^*) - (\psi^*\nabla^2\psi - \psi\nabla^2\psi^*) = 0$$
$$= (\psi^*\partial_0^2\psi + \partial_0\psi^*\partial_0\psi - \partial_0\psi^*\partial_0\psi - \psi\partial_0^2\,\psi^*)$$
$$- (\psi^*\nabla^2\psi + \nabla\psi^*\nabla\psi - \Delta\psi^*\nabla\psi - \psi\nabla^2\,\psi^*)$$
$$= (\psi^*\partial_0^2\psi + \partial_0\psi^*\partial_0\psi - \partial_0\psi\partial_0\psi^* - \psi\partial_0^2\psi^*)$$
$$- (\psi^*\nabla^2\,\psi + \nabla\psi^*\nabla\psi - \nabla\psi\nabla\psi^* - \psi\nabla^2\psi^*)$$
$$= \partial_0\,(\psi^*\partial_0\psi) - \partial_0(\psi\partial_0\psi^*) - \nabla(\psi^*\nabla\psi) + \nabla(\psi\nabla\psi^*)$$
$$= \partial_0\,(\psi^*\partial_0\psi - \psi\,\partial_0\psi^*) + \nabla[\psi^*(-\nabla)\psi - \psi\,(-\nabla)\psi^*]$$
$$= \partial_0\,(\psi^*\partial_0\psi - \psi\partial_0\psi^*) + \nabla[\psi^*(-\nabla)\psi - \psi\,(-\nabla)\psi^*]$$
$$= \partial_\mu\,(\psi^*\partial^\mu\psi - \psi\partial^\mu\psi^*) = \partial_\mu\,j^\mu \tag{14}$$

where $\nabla_\mu = (\partial_0, \nabla)$ and $\nabla^\mu = (\partial^0, -\nabla)$ are, respectively, the covariant and contravariant gradient operators in 4-dimensional space-time. Equation (14) is similar in form to the continuity equation

$$\partial_\mu j^\mu = 0 \tag{15}$$

so that we can identify

$$j^\mu = (\psi^*\partial^\mu\psi - \psi\partial^\mu\psi^*) \tag{16}$$

as the 4-current density in the Klein Gordon relativistic framework.

It is, however, usual to express j^μ as

$$j^\mu = \frac{i\hbar}{2m}(\psi^*\partial^\mu\psi - \psi\partial^\mu\psi^*) \tag{17}$$

so that j^0 has the dimensions of probability density. In natural units, we can write

$$j^\mu = \frac{i}{2m}(\psi^*\partial^\mu\psi - \psi\partial^\mu\psi^*) \tag{18}$$

This premultiplication also ensures that we obtain compatible non-relativistic expressions for j^μ. Written explicitly (in natural units) the components of j^μ take the form

$$\rho \equiv j_0 = j^0 = \frac{i}{2m}(\psi^*\partial^0\psi - \psi\partial^0\psi^*),\ \mathbf{j} \equiv -j_i$$

$$= j^i = -\frac{i}{2m}(\psi^*\nabla\psi - \psi\nabla\psi^*),\ i = 1, 2, 3 \tag{19}$$

3.6 INTERPRETATION OF j_0

Integrating the continuity eq. (15) over the entire configuration space, we obtain

$$\partial_0 \int_V \rho dx = \int_V \partial_0\rho d\mathbf{x} = -\int_V \nabla.\mathbf{j}d\mathbf{x} = -\int_S \mathbf{j}.d\mathbf{S} = 0$$

showing that $\int_V \rho d\mathbf{x}$ is constant in time. In the case of the non-relativistic Schrödinger equation, we had interpreted $\rho \equiv j_0$ as a probability density since, in the non-relativistic scenario, we had $\rho = \psi^*\psi$ which is, obviously, positive definite and hence, conforms to the axioms of a probability density.

However, in the Klein Gordon case, such an interpretation is not physically acceptable. In this case, ρ is a function of ψ, $\partial_0\psi$ and the complex conjugates thereof. At a given point in time $t \equiv x^0$, ψ, $\partial_0\psi$, ψ^*, $\partial_0\psi^*$, may take arbitrary values and so ρ may also attain negative values. In other words, there is no mathematical constraint on ρ that would mandate that it attains only positive values. Hence, we cannot physically identify ρ as a probability density in the relativistic Klein Gordon framework.

As mentioned above, the 4-current density in the Klein Gordon framework is given by $j^\mu = \frac{i}{2m}(\psi^*\partial^\mu\psi - \psi\partial^\mu\psi^*)$. Multiplication of this expression by the particle charge e yields the current density of charge as

$$j_\mu^c = \frac{ie}{2m}(\psi^*\partial_\mu\psi - \psi\partial_\mu\psi^*) = (\rho', -\mathbf{j}'), \quad \text{where}$$

$$\rho' = \frac{ie}{2m}\left(\psi^* \frac{\partial\psi}{\partial x^0} - \psi\frac{\partial\psi^*}{\partial x^0}\right) \tag{20}$$

is called the charge density and

$$\mathbf{j}' = -\frac{ie}{2m}(\psi^*\nabla\psi - \psi\nabla\psi^*) \tag{21}$$

the charge current density.

3.7 FREE PARTICLE SOLUTIONS OF THE KLEIN GORDON EQUATION

The Klein Gordon eq. (10) has a solution for the free particle given by

$$\begin{aligned}\psi &= A\exp(-ip_\mu x^\mu)\\ &= A\exp[-i(p_0 x^0 - \mathbf{px})]\\ &= A\exp[i(\mathbf{px} - Ex^0)]\end{aligned} \tag{22}$$

Substituting this value of ψ into the Klein Gordon eq. (10), we obtain $-E^2 + \mathbf{p}^2 + m^2 = 0$ or $E = \pm\sqrt{\mathbf{p}^2 + m^2}$. For every value of the momentum $\mathbf{p}$, there are, therefore, two possible solutions of the Klein Gordon equation – one with positive energy and the other with negative energy, viz., $\psi^{(\pm)} = A^{(\pm)}\exp[i(\mathbf{px} \mp E_p x^0)]$, $E = \pm E_p = \pm\sqrt{\mathbf{p}^2 + m^2}$ with $A^{(\pm)}$ being normalization constants.

Substituting these values of the wavefunction $\psi^{(\pm)}$ in expression (20) for charge density, we get $\rho'^{(\pm)} = \pm\frac{eE_p}{m}\psi^{*(\pm)}\psi^{(\pm)}$. We can, therefore, identify $\rho'^{(\pm)}$ with particles of positive and negative charge, respectively.

The general solution of the wave equation will, obviously, be a linear combination of both the positive and negative energy wavefunctions.

3.8 BOX NORMALIZATION WITH PERIODIC BOUNDARY CONDITIONS

In order to perform box normalization of the Klein Gordon wavefunction, we confine the waves to a large cubic box (normalization box) with edge length L. We demand periodic boundary conditions at the walls of the box.

We attempt a solution of the Klein Gordon equation by the separation of variables and write

$$\psi(x) = X(x^1)Y(x^2)Z(x^3)T(x^0) \tag{23}$$

and substitute this expression in eq. (10) to obtain

$$\frac{1}{T}\frac{\partial^2 T}{\partial(x^0)^2}-\frac{1}{X}\frac{\partial^2 X}{\partial(x^1)^2}-\frac{1}{Y}\frac{\partial^2 Y}{\partial(x^2)^2}-\frac{1}{Z}\frac{\partial^2 Z}{\partial(x^3)^2}+m^2=0 \tag{24}$$

or

$$\frac{1}{X}\frac{\partial^2 X}{\partial(x^1)^2}=-\frac{1}{Y}\frac{\partial^2 Y}{\partial(x^2)^2}-\frac{1}{Z}\frac{\partial^2 Z}{\partial(x^3)^2}+\frac{1}{T}\frac{\partial^2 T}{\partial(x^0)^2}+m^2 \tag{25}$$

The LHS of this equation is a function of x^1 only while the RHS is independent of x^1. Hence, each side must be equal to a constant. We write $\frac{1}{X}\frac{\partial^2 X}{\partial x^2}=-k_1^2$ which has the solutions $X(x^1)=e^{ik_1x^1}$ and $X(x^1)=e^{-ik_1x^1}$. Since $\exp(ik_1x^1)$ and $\exp(-ik_1x^1)$ are linearly independent, they can be taken to constitute a basis and the general solution can be expressed as linear combinations of these two basis vectors, e.g., $X(x^1)=Ae^{ik_1x^1}+Be^{-ik_1x^1}$.

As we are assuming periodic boundary conditions at the walls of the box, $\exp[\pm ik_1(x^1+L)]$ must also be solutions of $\frac{1}{X}\frac{\partial^2 X}{\partial x^2}=-k_1^2$ and hence can be expressed as a linear combination of our chosen basis. We may, therefore, write $e^{ik_1(x^1+L)}=C_1e^{ik_1x^1}+C_2e^{-ik_1x^1}$ and $e^{-ik_1(x^1+L)}=D_1e^{ik_1x^1}+D_2e^{-ik_1x^1}$. Now $X(x^1)=Ae^{ik_1x^1}+Be^{-ik_1x^1}$ also implies

$$\begin{aligned}X(x^1+L)&=Ae^{ik_1(x^1+L)}+Be^{-ik_1(x^1+L)}\\&=A[C_1e^{ik_1x^1}+C_2e^{-ik_1x^1}]+\mathrm{B}[D_1e^{ik_1x^1}+D_2e^{-ik_1x^1}]\\&=(AC_1+BD_1)e^{ik_1x^1}+(AC_2+BD_2)e^{-ik_1x^1}\end{aligned}$$

Since we require $X(x^1+nL)$, $\forall n\in N$ also to be solutions of $\frac{1}{X}\frac{\partial^2 X}{\partial x^2}=-k_1^2$, we look for solutions with the property $X(x^1+L)=\alpha X(x^1)$ for constant α. This condition will be met if $AC_1+BD_1=\alpha A$ and $AC_2+BD_2=\alpha B$ or, equivalently, $(C_1-\alpha)A+BD_1=0$ and $AC_2+(D_2-\alpha)B=0$.

Solving this homogeneous set of equations, we find that they can have a non-trivial solution only if $\begin{vmatrix}C_1-\alpha & D_1\\ C_2 & D_2-\alpha\end{vmatrix}=0$, otherwise the only solution is $A=B=0$.

This condition leads to a quadratic in α with two solutions α_1 and α_2 leading to two functions $X_1(x^1)$ and $X_2(x^1)$ satisfying, respectively, $X_1(x^1+L)=\alpha_1X_1(x^1)$ and $X_2(x^1+L)=\alpha_2X_2(x^1)$.

The Wronskian determinant also satisfies $D(x^1+L)=\alpha_1\alpha_2D(x^1)$, where $D=X_1X'_2-X_2X'_1$ because

$$\begin{aligned}D(x^1+L)&=X_1(x^1+L)X'_2(x^1+L)-X_2(x^1+L)X'_1(x^1+L)\\&=[\alpha_1X_1(x^1)][\alpha_2X'_2(x^1)]-[\alpha_2X_2(x^1)][\alpha_1X'_1(x^1)]\\&=\alpha_1\alpha_2[X_1(x^1)X'_2(x^1)-X_2(x^1)X'_1(x^1)]=\alpha_1\alpha_2D(x^1)\end{aligned}$$

Now, according to Green's theorem D must be a constant, independent of x^1. It, therefore, follows that $\alpha_1\alpha_2 = 1$. Further, if $|\alpha_{1,2}| > 1$, then the amplitudes of $X(x^1)$ will grow and fall indefinitely for $x^1 \to \pm\infty$. The opposite will hold for $|\alpha_{1,2}| < 1$. Such solutions cannot be normalized, so physically acceptable solutions exist only if $|\alpha_{1,2}| = 1$. Hence, we can write $\alpha_{1,2} = \exp(\pm ilL)$ with real l. Now, writing $l = \dfrac{2n\pi}{L} + m, -\dfrac{\pi}{L} \le m \le \dfrac{\pi}{L}, n \in \mathbb{Z}$, we have $\alpha_{1,2} = \exp(\pm imL), -\dfrac{\pi}{L} \le m \le \dfrac{\pi}{L}$ which gives the complete set of possible wavefunctions.

The two sets of wavefunctions that satisfy $X(x^1 + L) = \alpha X(x^1)$ are, therefore, given by $X_1(x^1 + L) = e^{imL}X_1(x^1)$ and $X_2(x^1 + L) = e^{-imL}X_2(x^1)$ which can hold if and only if $X_{1,2}$ are of the form $X_1(x^1) = N_1e^{imx^1}$ and $X_2(x^1) = N_2e^{-imx^1}$, respectively. Comparing this with the general solution $X(x^1) = Ae^{ik_1x^1} + Be^{ik_1x^1}$ we finally obtain two sets of solutions that satisfy the second order differential equation $\dfrac{1}{X}\dfrac{\partial^2 X}{\partial x^2} = -k_1^2$ and also meet the periodic boundary conditions $X(x^1 + L) = \alpha X(x^1)$, viz. $X_1(x^1) = Ae^{ik_1x^1}$ and $X_2(x^1) = Be^{-ik_1x^1}$. Since the two solutions differ merely in respect of the direction of propagation of the wave, we can take either one as a representative solution, say $X(x^1) = Ae^{ik_1x^1}$.

Substituting $\dfrac{1}{X}\dfrac{\partial^2 X}{\partial x^2} = -k_1^2$ in eq.(25) we get $\dfrac{1}{Y}\dfrac{\partial^2 Y}{\partial (x^2)^2} = -\dfrac{1}{Z}\dfrac{\partial^2 Z}{\partial (x^3)^2} + \dfrac{1}{T}\dfrac{\partial^2 T}{\partial (x^0)^2} + m^2 + k_1^2$ (26)

Here, again the LHS is a function of x^2 only and the RHS is independent of x^2. Hence, we may write $\dfrac{1}{Y}\dfrac{\partial^2 Y}{\partial (x^2)^2} = -k_2^2$. Proceeding as earlier, we conclude that $Y(x^2) = A'e^{ik_2x^2}$

Again substituting $\dfrac{1}{Y}\dfrac{\partial^2 Y}{\partial (x^2)^2} = -k_2^2$ in eq., (26), we obtain $\dfrac{1}{Z}\dfrac{\partial^2 Z}{\partial (x^3)^2} = \dfrac{1}{T}\dfrac{\partial^2 T}{\partial (x^0)^2} + m^2 + k_1^2 + k_2^2$ (27)

Again setting $\dfrac{1}{Z}\dfrac{\partial^2 Z}{\partial (x^3)^2} = -k_3^2$, we get $Z(x^3) = A''e^{ik_3x^3}$.

Substituting $\dfrac{1}{Z}\dfrac{\partial^2 Z}{\partial (x^3)^2} = -k_3^2$ in eq., (27) we get $\dfrac{1}{T}\dfrac{\partial^2 T}{\partial (x^0)^2} = -m^2 - k_1^2 - k_2^2 - k_3^2$ (28)

which has the solution $T(x^0) = A'''e^{k_0x^0}$,

where $k_0^2 = k_1^2 + k_2^2 + k_3^2 + m^2$

Drawing analogy with the relativistic energy-momentum relation $E^2 = \mathbf{p}^2 + m^2 = p_1^2 + p_2^2 + p_3^2 + m^2$, we can identify $k_{1,2,3} \equiv p_{1,2,3}$ and $k_0 \equiv \mp E$ so that our solution becomes, finally,

$$\psi = X(x^1)Y(x^2)Z(x^3)T(x^0) = Ne^{i(p_1x^1 + p_2x^2 + p_3x^3 \mp Ex^0)} = Ne^{i(\mathbf{px} \mp Ex^0)}$$

To accommodate the periodic boundary conditions, we must have $\psi(\boldsymbol{x} + L) = \psi(\boldsymbol{x})$ which necessitates $e^{i\{\mathbf{p}(\mathbf{x}+L) - Ex^0\}} = e^{i(\mathbf{px} - Ex^0)}$. This is possible if $\mathbf{p}\cdot L = 2\pi\mathbf{n}$ or $\mathbf{p} = \dfrac{2\pi}{L}\mathbf{n}$ where $\mathbf{n} = (n_1, n_2, n_3)$, $n_i \in \mathrm{N}$ is a discrete vector in the lattice with the components along the three axes.

Let us define $E_n \equiv \sqrt{\mathbf{p}_n^2 + m^2}$ and $\mathbf{p}_n = \dfrac{2\pi}{L}\mathbf{n}$, then we can write the discrete version of the wavefunction in the lattice space as $\psi_{\mathrm{n}}^{[\pm]} = N_{n(\pm)} \exp[i(\mathbf{p}_n\mathbf{x} \mp E_n x^0)]$.

The normalization factor can be determined by using the expression for the charge density

$$\rho'^{(\pm)} = \pm \frac{eE_n}{m}\psi^{(\pm)*}\psi^{(\pm)}$$

We have
$$\pm e = \int_{L^3} d\mathbf{x}\rho'^{(\pm)}(\mathbf{x}) = \pm \frac{eE_n}{m}|N_{n(\pm)}|^2 L^3$$

so that
$$N_{n(\pm)} = \sqrt{\frac{m}{E_n L^3}}.$$

Hence
$$\psi_n^{(\pm)} = \sqrt{\frac{m}{E_n L^3}} \exp[i(\mathbf{p}_n\mathbf{x} \mp E_n x^0)] \tag{29}$$

Either type of solution (corresponding to positive and negative charge) has the same normalization. The only difference is due to the time factor exp $(\pm E_n x^0)$.

The most general solution of the Klein Gordon equation for positive and negative spin zero particles is a superposition of the above solution over n, i.e.,

$$\psi^{(+)} = \sum_n a_n \psi_n^{(+)}$$
$$= \sum_n a_n \sqrt{\frac{m}{E_n L^3}} \exp[i(\mathbf{p}_n\mathbf{x} - E_n x^0)] \tag{30}$$

and

$$\psi^{(-)} = \sum_n b_n \psi_n^{(-)}$$
$$= \sum_n b_n \sqrt{\frac{m}{E_n L^3}} \exp[i(\mathbf{p}_n\mathbf{x} - E_n x^0)] \tag{31}$$

respectively.

We can also construct solutions for spin 0 particles with zero charge. In the case of charge zero particles, the charge density vanishes, i.e., $\rho' = \frac{ie}{2m}\left(\psi^* \frac{\partial\psi}{\partial x^0} - \psi\frac{\partial\psi^*}{\partial x^0}\right) = 0$ which implies that $\psi^* \frac{\partial\psi}{\partial x^0} = \psi\frac{\partial\psi^*}{\partial x^0}$ or that $\psi^* = \psi$, i.e., that the Klein Gordon wavefunction ψ has to be real for neutral particles.

For neutral, particles we can construct the Klein Gordon wavefunction as a superposition of the positive and negative functions, viz.,

$$\psi_n^{(0)} = \frac{1}{\sqrt{2}}\left(\psi_n^{(+)}(\mathbf{p}_n) + \psi_n^{(-)}(-\mathbf{p}_n)\right)$$

$$= \sqrt{\frac{m}{2E_nL^3}}\{\exp[i(\mathbf{p}_n\mathbf{x} - E_nx^0)] + \exp[i(-\mathbf{p}_n\mathbf{x} + E_nx^0)]\}$$

$$= \sqrt{\frac{m}{2E_nL^3}}\{\exp[i(\mathbf{p}_n\mathbf{x} - E_nx^0)] + \exp[-i(\mathbf{p}_n\mathbf{x} - E_nx^0)]\}$$

$$= \sqrt{\frac{2m}{E_nL^3}}\cos(\mathbf{p}_n\mathbf{x} - E_nx^0) \tag{32}$$

The current charge density $\mathbf{j}'(\mathbf{x}, x^0)$ given by $\mathbf{j}' = -\frac{ie}{2m}(\psi^*\nabla\psi - \psi\nabla\psi^*)$ of neutral particles also vanishes. Therefore, in this case, there is no conservation law.

From the above analysis, it is seen that in the relativistic quantum theory, the Klein Gordon equation for a free spin 0 particle has three solutions corresponding to the electric charge (+, –, 0) of the particle for every well-defined momentum **p**. Hence, in the relativistic theory additional degrees of freedom come into play corresponding to the electric charge of the particle. This is in contrast to the non-relativistic quantum mechanics in which the Schrödinger equation for free, spin 0 particles has a unique solution.

3.9 THE NON-RELATIVISTIC LIMIT OF THE KLEIN GORDON EQUATION

The solution of the free particle Klein Gordon equation for positive energies is given by eq. (22), viz.

$$\psi = A\exp(-ipx) = A\exp(-ip_\mu x^\mu)$$
$$= A\exp[-i(p_0x^0 - \mathbf{px}] = A\exp[i(\mathbf{px} - Ex^0)]$$

where $\quad p_0 = E = +(m^2 + \mathbf{p}^2)^{1/2}$

$$= + m\left(1+\frac{1}{m^2}\mathbf{p}^2\right)^{1/2} = m + \frac{1}{2m}\mathbf{p}^2 + ...$$

so that the solution (22) can be written as

$$\psi = \varphi(x)\exp(-imx^0) \tag{33}$$

where
$$\varphi(x) = \exp\left[i\left(\mathbf{px} - \frac{\mathbf{p}^2}{2m}x^0 - ......\right)\right].$$

In the non-relativistic limit the difference between the total energy E of the particle and its rest mass energy m is small, i.e., the kinematic energy $E' = E - m$ is negligible compared to the rest mass energy m or $E' \ll m$. Using eq. (33), we have

$$E\psi = i\partial_0\psi = i(\partial_0\varphi - im\varphi)\exp(-imx^0) \tag{34}$$

so that

$$|E\psi| = |i\partial_0\varphi - im\varphi)\exp(-imx^0)|$$
$$= |i\partial_0\varphi + m\varphi||\exp(-imx^0)| = |i\partial_0\varphi + m\varphi|$$
$$= [(\partial_0\varphi)^2 + m^2\varphi^2]^{1/2} \approx m\varphi$$

because

$$m\varphi \gg |E'\varphi| = |(E - m)\varphi| = |E\varphi| - |m\varphi|$$
$$= |i\partial_0\varphi + m\varphi| - |m\varphi| \approx |i\partial_0\varphi|$$

where the third step follows from the fact that both E and m are positive and $E > m$.

Therefore,

$$\partial_0\psi = (\partial_0\varphi - im\varphi)\exp(-imx^0)$$
$$\approx - im\varphi\exp(-imx^0) \tag{35}$$

and

$$\partial_0^2\psi = \partial_0(\partial_0\varphi - im\varphi)\exp(-imx^0)$$
$$= (\partial_0^2\varphi - 2im\partial_0\varphi - m^2\varphi)\exp(-imx^0)$$
$$\approx -(2im\partial_0\varphi + m^2\varphi)\exp(-imx^0) \tag{36}$$

Substituting these expressions in the Klein Gordon eq. (10), we obtain

$$-(2im\partial_0\varphi + m^2\varphi)\exp(-imx^0) - \nabla^2\varphi\exp(-imx^0) + m^2\varphi\exp(-imx^0) = 0$$

or

$$i\partial_0\varphi + \frac{1}{2m}\nabla^2\varphi = 0 \tag{37}$$

which is the free particle Schrödinger equation.

Thus, we find that in the non-relativistic limit, i.e., when, $\frac{|\mathbf{p}|}{m} \to 0$, if $\psi = \varphi(x)\exp(-imx^0)$ is a, solution of the Klein Gordon eq. (10), then $\varphi(x)$ becomes a solution of the corresponding

Schrödinger equation. In other words, the relativistic mass energy relationship manifests itself as an additional term exp $(-imx^0)$ in the expression for the wavefunction solution of the corresponding wave equation. Thus, there occurs a displacement of energies by a factor of m due to the relativistic correction to first order.

This is also seen as follows:

$$\hat{H}_R = (m^2 - \nabla^2)^{1/2} = m\left(1 - \frac{1}{m^2}\nabla^2\right)^{\frac{1}{2}} \approx m\left(1 - \frac{1}{2m^2}\nabla^2\right) = m - \frac{1}{2m}\nabla^2 = m + \hat{H}_{NR} \tag{38}$$

where $\hat{H}_R$, $\hat{H}_{NR}$ are the relativistic and the corresponding non-relativistic Hamiltonian operators, respectively.

3.10 SCALAR PRODUCT FOR THE KLEIN GORDON EQUATION

The solution of the free particle Klein Gordon equation for positive energies is given by

$$\begin{aligned}\psi(x) &= A\exp(-ip_\mu x^\mu)\\ &= \varphi(\mathbf{x})\exp(-ip_0x^0)\\ &= \varphi(\mathbf{x})\exp(-iEx^0)\end{aligned} \tag{39}$$

where $\quad p_0 = E = +(m^2 + \mathbf{p}^2)^{1/2}.$

For solutions with well-defined momentum $\mathbf{p}$, solution (39) must additionally satisfy

$$\hat{\mathbf{p}}\,\psi^{\mathbf{p}}(x) = -i\nabla\psi^{\mathbf{p}}(x) = \mathbf{p}\psi^{\mathbf{p}}(x) \tag{40}$$

Using eqs. (22) and (39) the solution of eq. (40) is

$$\varphi^{\mathbf{p}}(\mathbf{x}) = \xi(\mathbf{p})\exp(i\mathbf{px}) \tag{41}$$

so that

$$\begin{aligned}\psi^{\mathbf{p}}(x) &= \xi(\mathbf{p})\exp(-ipx)\\ &= \xi(\mathbf{p})\exp(-ip_\mu x^\mu)\\ &= \xi(\mathbf{p})\exp[i(\mathbf{p.x} - p_0x^0)]\end{aligned} \tag{42}$$

A complete solution of the Klein Gordon eq. (10) is, then, a superposition of the plane waves represented by eq. (42) in space, viz.

$$\Psi(x) = \frac{1}{(2\pi)^{3/2}}\int d\mathbf{p}\,\xi(\mathbf{p})\exp(-ipx) \tag{43}$$

with its Fourier transform in $\mathbf{p}$ (momentum) space being

$$\tilde{\Psi}(p) = \frac{1}{(2\pi)^{3/2}}\int d\mathbf{x}\,\tilde{\xi}(\mathbf{x})\exp(-ipx) \tag{44}$$

We now construct the invariant scalar product of two Klein Gordon wavefunctions. We approach the problem in $\mathbf{p}$ space wherein the product takes a particularly simple form.

We postulate the ansatz

$$\left\langle \tilde{\Psi}_1 | \tilde{\Psi}_2 \right\rangle = \int d\mathbf{p}\phi(\mathbf{p})\tilde{\Psi}_1^*(p)\tilde{\Psi}_2(p) \tag{45}$$

We fix $\phi(\mathbf{p})$ by demanding Lorentz invariance of $\langle|\rangle$. For rotational invariance, we need $\phi(\mathbf{p}) = \phi(|\,\mathbf{p}\,|)$.

We now consider Lorentz boosts along the x^3 axis represented by

$$x^1 \to x'^1 = x^1, x^2 \to x'^2 = x^2, x^3 \to x'^3 = \beta(x^3 - vx^0), x^0 \to x'^0$$

$$= \beta(x^0 - vx^3), \beta = \left(1 - \frac{v^2}{c^2}\right)^{-1/2} \tag{46}$$

We then have

$$p'_1 = p_1, p'_2 = p_2, p'_3 = \beta^{-1} p_3 + (v\beta)^{-1} p_0\, p'_0 = \beta^{-1} p_0 + (v\beta)^{-1} p_3 \tag{47}$$

so that

$$dp'_1 dp'_2 dp'_3 = dp_1 dp_2 dp_3 \left(\frac{1}{\beta} + \frac{1}{(v\beta)}\frac{dp_0}{dp_3}\right)$$

$$= dp_1 dp_2 dp_3 \left(\frac{1}{\beta} + \frac{1}{(v\beta)}\frac{p_3}{p_0}\right) = dp_1 dp_2 dp_3 \left(\frac{p'_0}{p_0}\right)$$

since $$p_0 = (m^2 + \mathbf{p}^2)^{1/2} = (m^2 + p_1^2 + p_2^2 + p_3^2)^{1/2}, \frac{dp_0}{dp_3} = \frac{p_3}{p_0}$$

or $$\frac{dp'_1 dp'_2 dp'_3}{p'_0} = \frac{dp_1 dp_2 dp_3}{p_0} \tag{48}$$

Hence, $\phi(\mathbf{p})$ would be invariant under Lorentz transformations if $\phi(\mathbf{p}) = \phi(|\mathbf{p}|) = \dfrac{A}{p_0}$ so that the relativistically invariant scalar product in $\mathbf{p}$ space takes the form

$$\left\langle \tilde{\Psi}_1 | \tilde{\Psi}_2 \right\rangle = A \int \frac{d\mathbf{p}}{p_0} \tilde{\Psi}_1^*(p)\tilde{\Psi}_2(p) \tag{49}$$

where A is a number which is usually taken as $\dfrac{1}{2}$ for massive particles, in natural units.

To define the invariant scalar product in position space, we write eq. (43) in the form

$$\Psi(\mathbf{x}, x^0) = \frac{1}{(2\pi)^{3/2}} \int \frac{d\mathbf{p}}{2p_0} \eta(\mathbf{p}, x^0) \exp(i\mathbf{px}) \tag{50}$$

where $\eta(\mathbf{p}, x^0) = 2p_0\xi(\mathbf{p}) \exp(-ip_0 x^0)$, with the Fourier inverse

$$\tilde{\Psi}(\mathbf{p}, x^0) = \frac{2p_0}{(2\pi)^{3/2}} \int d\mathbf{x}\tilde{\eta}(\mathbf{x}, x^0) \exp(i\mathbf{px}) \tag{51}$$

Substituting these expressions in eq. (49), we obtain, for equal times

$$\langle\Psi_1|\Psi_2\rangle = \frac{2}{(2\pi)^3}\int\frac{d\mathbf{p}}{p_0}(p_0)^2\int d\mathbf{x}_1\int d\mathbf{x}_2 e^{i\mathbf{p}.(\mathbf{x}_1-\mathbf{x}_2)}\tilde{\eta}_1^*(\mathbf{x}_1,x^0)\tilde{\eta}_2(\mathbf{x}_2,x^0)$$

$$= \frac{2}{(2\pi)^3}\int d\mathbf{p}\int d\mathbf{x}_1\int d\mathbf{x}_2 p_0 e^{i\mathbf{p}.(\mathbf{x}_1-\mathbf{x}_2)}\tilde{\eta}_1^*(\mathbf{x}_1,x^0)\tilde{\eta}_2(\mathbf{x}_2,x^0)$$

$$= \frac{2}{(2\pi)^3}\int d\mathbf{p}\int d\mathbf{x}_1\int d\mathbf{x}_2\; e^{i\mathbf{p}.(\mathbf{x}_1-\mathbf{x}_2)}p_0[\tilde{\eta}_1^*(\mathbf{x}_1,x^0)\tilde{\eta}_2(\mathbf{x}_2,x^0)]$$

$$= \frac{1}{(2\pi)^3}\int d\mathbf{p}\int d\mathbf{x}_1\int d\mathbf{x}_2\; e^{i\mathbf{p}.(\mathbf{x}_1-\mathbf{x}_2)}$$

$$[p_0\tilde{\eta}_1^*(\mathbf{x}_1,x^0)\tilde{\eta}_2(\mathbf{x}_2,x^0) + \tilde{\eta}_1^*(\mathbf{x}_1.x^0).p_0\tilde{\eta}_2(\mathbf{x}_2,x^0)] \tag{52}$$

Now, p_0 is the eigenvalue of the operator $i\partial_0$ acting upon a well-defined quantum state ψ, i.e. $i\partial_0\psi = p_0\psi$. Furthermore, since p_0 is real, we also have $-i\partial_0\psi^* = p_0\psi^*$. Making use of these results, we can write eq. (52) as

$$\langle\Psi_1|\Psi_2\rangle = \frac{1}{(2\pi)^3}\int d\mathbf{p}\int d\mathbf{x}_1\int d\mathbf{x}_2\, e^{i\mathbf{p}.(x_1-x_2)}$$

$$\{[-i\partial_0\tilde{\eta}_1^*(\mathbf{x}_1,x^0)]\tilde{\eta}_2(\mathbf{x}_2,x^0) + \tilde{\eta}_1^*(\mathbf{x}_1,x^0)\;[i\partial_0\tilde{\eta}_2(\mathbf{x}_2,x^0)]\}$$

Performing the $d\mathbf{p}$ integration, using

$$\frac{1}{(2\pi)^3}\int e^{i\mathbf{p}.(\mathbf{x}_1-\mathbf{x}_2)}\,d\mathbf{p} = \delta(\mathbf{x}_1-\mathbf{x}_2) \tag{53}$$

we obtain

$$\langle\Psi_1|\Psi_2\rangle = -i\int d\mathbf{x}\{[\partial_0\tilde{\eta}_1^*(\mathbf{x},x^0)]\;\tilde{\eta}_2(\mathbf{x},x^0) - \tilde{\eta}_1^*(\mathbf{x},x^0)[\partial_0\tilde{\eta}_2(\mathbf{x},x^0)]\}$$

$$= i\int d\mathbf{x}\tilde{\eta}_1^*(\mathbf{x},x^0)\overleftrightarrow{\partial}^0\;\tilde{\eta}_2(\mathbf{x},x^0) \tag{54}$$

3.10.1 Properties of the Scalar Product

1. The scalar product defined by eq. (54) possesses all the standard properties of a scalar product, viz.

$$\langle\phi|\psi\rangle = \overrightarrow{\langle\psi|\phi\rangle}$$

$$\langle\phi_1+\phi_2|\psi\rangle = \langle\phi_1|\psi\rangle + \langle\phi_2|\psi\rangle$$

$$\langle\phi|\psi_1+\psi_2\rangle = \langle\phi|\psi_1\rangle + \langle\phi|\psi_2\rangle$$

$$\langle\phi|\phi\rangle \geq 0$$

2. If $\phi(x)$ and $\psi(x)$ are solutions of the equation

$$i\partial_0\psi(x) = \sqrt{m^2-\nabla^2}\;\psi(x)$$

$$-i\partial_0\phi^*(x) = \sqrt{m^2 - \nabla^2}\,\phi^*(x)$$

We can write the scalar product as

$$\langle\phi|\psi\rangle = i\int d\mathbf{x}[\phi^*(x)\partial_0\,\psi(x) - \partial_0\phi^*(x)\psi(x)]$$

$$= \int d\mathbf{x}[\phi^*(x)\sqrt{m^2 - \nabla^2}\,\psi(x) + \sqrt{m^2 - \nabla^2}\,\phi^*(x)\psi(x)]$$

so that the time derivatives are eliminated from the scalar product and, therefore, quantities involved in the scalar product can be evaluated with reference to a specific instant of time x^0 as is the case in non-relativistic quantum mechanics.

3. The relativistic invariance of the scalar product in momentum (**p**) space has been established in defining the product in **p** space. Its invariance can also be shown directly in configuration space as follows:

As defined in eq. (54) the integral in the scalar product is taken over all space at a given time say $x^0 = x_0^0$. We can generalize this to an integral over a space like surface σ.

For this purpose, we define the spacelike surface, σ, as being such that for any two points x_1 and x_2 on it, $(x_1 - x_2)^2$ is always spacelike, i.e., $(x_1 - x_2)^2 < 0$ always so that no two points on σ can be connected by a light signal.

Let $\Omega(\sigma)$ be a function of the spacelike surface σ. We define the functional derivative $\dfrac{\delta}{\delta\sigma(x)}$ of $\Omega(\sigma)$ at a space like point x as

$$\frac{\delta\Omega(\sigma)}{\delta\sigma(x)} = \lim_{\omega\to 0}\frac{\Omega(\sigma') - \Omega(\sigma)}{\omega(x)} \tag{55}$$

where σ' constitutes an infinitesimal deformation of the space like surface σ in the neighbourhood of space-time point x and $\omega(x)$ is the volume enclosed between $\sigma'(x)$ and $\sigma(x)$. Obviously, $\omega(x) \to 0$ as $\sigma' \to \sigma$.

Taking

$$\Omega(\sigma) = i\int d\sigma^\mu(x)[\phi^*(x)\partial_\mu\chi(x) - \partial_\mu\phi^*(x)\chi(x)]$$

$$= i\int d\sigma^\mu\phi^*(x)\overleftrightarrow{\partial}_\mu\chi(x) \tag{56}$$

We obtain

$$\frac{\delta\Omega(\sigma)}{\delta\sigma(x)} = i\lim_{\omega\to 0}\frac{\int_{\sigma'} d\sigma^\mu(x)\phi^*(x)\overleftrightarrow{\partial}_\mu\chi(x) - \int_{\sigma} d\sigma^\mu(x)\phi^*(x)\overleftrightarrow{\partial}_\mu\chi(x)}{\omega(x)} \tag{57}$$

Making use of Gauss's theorem, which, in 4-space takes the form

$$\int_V d^4x\partial^\mu\Lambda_\mu(x) = \int_S d\sigma^\mu(x)\Lambda_\mu(x) \tag{58}$$

We further obtain

$$\frac{\delta\Omega(\sigma)}{\delta\sigma(x)} = i \lim_{\omega\to 0} \frac{\int_{\omega(x)} d^4x\partial^\mu[\phi^*(x)\overleftrightarrow{\partial}_\mu\chi(x)]}{\omega(x)} = i\partial^\mu[\phi^*(x)\overleftrightarrow{\partial}_\mu \chi(x)]$$

$$= i\partial^\mu[\phi^*(x)\partial_\mu\chi(x) - i\partial_\mu\phi^*(x)\,\chi(x)]$$

$$= i[\phi^*(x)\partial^\mu\partial_\mu\chi(x) - \partial^\mu\partial_\mu\phi^*(x)\chi(x)]$$

$$= i\{\phi^*(x)[-m^2\chi(x)] - [-m^2\,\phi^*(x)]\,\chi(x)\} = 0 \tag{59}$$

where the second last step follows from the fact that both $\phi^*(x)$, $\chi(x)$ satisfy the Klein Gordon equation.

This shows that the scalar product defined by $i\int d\sigma^\mu\phi^*(x)\overleftrightarrow{\partial}_\mu\chi(x)$ is independent of the space like surface σ on which it is evaluated provided, of course, that ϕ, χ obey the Klein Gordon equation.

Thus, if we choose the surface $x^0 = k$ (constant), we get the definition of scalar product in eq. (54). This is explained by noting that if $n^\mu(x)$ is the unit normal to the surface element σ at a point x then in the case of a spacelike surface, we have $n_\mu(\mathrm{x})n^\mu(x) = 1$ for all x on σ. Now, a plane $x^0 = k$ constant is a special case of such a spacelike surface for which $n^\mu(x) \equiv (1,0,0,0)$. Now, the pseudo vector surface element $d\sigma^\mu(x) = n^\mu(x)d\sigma$ for an arbitrary three-dimensional surface S in space-time has the components $d\sigma^\mu = \{dx^1\,dx^2\,dx^3,\ dx^0\,dx^2\,dx^3,\ dx^0\,dx^1\,dx^3,\ dx^0\,dx^1\,dx^2\}$ and the Gauss's theorem in four space is $\int_V d^4x\partial^\mu F_\mu(x) = \int_S d\sigma^\mu(x)F_\mu(x)$. For the special case of the plane $x^0 = k$(constant) for which $n^\mu(x) \equiv (1,0,0,0)$ so that $d\sigma^\mu(x) = n^\mu(x)\,d\sigma = (d\sigma, 0, 0, 0)$ whence the right-hand side integral of Gauss's dstheorem becomes $\int_{S(x^0=k)} d\sigma(x)F_0(x)$. As a corollary to the above, the scalar product of eq. (54) is also independent of the spacelike surface $\sigma(x)$ and is conserved in time, i.e., $\langle\phi|\chi\rangle_{x_0^0} = \langle\phi|\chi\rangle_{x_1^0}$.

4. The set of positive energy solutions of the Klein Gordon equation constitutes a linear vector space with the scalar product defined by eq. (54).

3.11 TWO COMPONENT REPRESENTATION OF THE KLEIN GORDON EQUATION (THE Ψ REPRESENTATION)

The Klein Gordon equation is a second order equation in time. Hence, in order to have a unique solution, we need initial values of the wavefunction ψ as well as of the derivative $\partial_0\psi$. We can, however, transform this second order equation into a set of a pair of coupled first order differential equations in time by introducing another degree of freedom as the time derivative of the wavefunction ψ, i.e., by substituting $\frac{i}{m}\,\partial_0\psi = \xi$. The Klein Gordon equation then takes the form

$$i\partial_0 \xi = -\frac{1}{m} \nabla^2\psi + m\psi \tag{60}$$

This can be further simplified and put in a more compact form by substituting $\psi = \phi + \chi$ and $\xi = \phi - \chi$. Using eq. (60) and $\frac{i}{m}\partial_0\psi = \xi$, we obtain

$$i\partial_0\phi = -\frac{1}{2m}\nabla^2(\phi + \chi) + m\phi \tag{61}$$

and

$$i\partial_0\chi = \frac{1}{2m}\nabla^2(\phi + \chi) - m\chi \tag{62}$$

The fact that the above pair of equations is equivalent to the Klein Gordon eq. (10) is easily established. We have, on subtracting (62) from (61)

$$i\partial_0(\phi - \chi) = -\frac{1}{m}\nabla^2(\phi + \chi) + m(\phi + \chi) \tag{63}$$

or $(\partial_0^2 - \nabla^2 + m^2)\psi = 0$, which is the Klein Gordon equation. Here we have used $\psi = \phi + \chi$, $\xi = \phi - \chi$, i and $\frac{i}{m}\partial_0\psi = \xi$.

The pair of eqs. (61)-(62) may be further put in a more compact form as

$$i\partial_0\Psi = \frac{1}{2m}\nabla^2\begin{pmatrix} -1 & -1 \\ 1 & 1 \end{pmatrix}\Psi + m\begin{pmatrix} 1 & 0 \\ 0 & -1 \end{pmatrix}\Psi$$

$$= -\frac{1}{2m}\nabla^2(\sigma_3 + i\sigma_2)\Psi + m\sigma_3\Psi \tag{64}$$

where $\Psi = \begin{pmatrix} \phi \\ \chi \end{pmatrix}$ and σ_1, σ_2, σ_3 are the Pauli matrices

$$\sigma_1 = \begin{pmatrix} 0 & 1 \\ 1 & 0 \end{pmatrix}, \sigma_2 = \begin{pmatrix} 0 & -i \\ i & 0 \end{pmatrix}, \sigma_3 = \begin{pmatrix} 1 & 0 \\ 0 & -1 \end{pmatrix} \tag{65}$$

that satisfy the relations $\sigma_i^2 = \mathbf{I}$, $\sigma_i\sigma_j = i\varepsilon_{ijk}\sigma_k$ and i, j, k cyclic.

Both the components of Ψ, viz. ϕ and χ individually satisfy the Klein Gordon eq. (10). We have from eq. (64)

$$0 = \left\{i\partial_0 + \left[\frac{1}{2m}\nabla^2(\sigma_3 + i\sigma_2) - m\sigma_3\right]\right\}\Psi = \left\{i\partial_0 - \left[\frac{1}{2m}\nabla^2(\sigma_3 + i\sigma_2) - m\sigma_3\right]\right\}$$

$$\times\left\{i\partial_0 + \left[\frac{1}{2m}\nabla^2(\sigma_3 + i\sigma_2) - m\sigma_3\right]\right\}\Psi$$

$$= \left\{-\partial_0^2 - \left[\frac{1}{2m}\nabla^2(\sigma_3 + i\sigma_2) - m\sigma_3\right]^2\right\}\Psi \text{ (since } [\partial_0, \nabla^2] = 0)$$

Now, $$\left[\frac{1}{2m}\nabla^2(\sigma_3 + i\sigma_2) - m\sigma_3\right]^2$$

$$= \frac{1}{4m^2}\nabla^4(\sigma_3 + i\sigma_2)^2 + m^2\,\sigma_3^2$$

$$-\frac{1}{2}\nabla^2[(\sigma_3 + i\sigma_2)\sigma_3 + \sigma_3(\sigma_3 + i\sigma_2)]$$

$$= \frac{1}{4m^2}\nabla^4[\sigma_3^2 - \sigma_2^2 + i(\sigma_3\sigma_2 + \sigma_2\sigma_3)] + m^2 - \nabla^2 = m^2 - \nabla^2$$

so that

$$(-\partial_0^2 + \nabla^2 - m^2)\Psi = 0 \tag{66}$$

which shows that both ϕ, χ satisfy the Klein Gordon eq. (10).

The charge density in this representation is obtained by substituting $\psi = \phi + \chi$ and $\xi = \phi - \chi$ in eq. (20) to obtain

$$\rho' = \frac{ie}{2m}(\psi^*\partial_0\psi - \psi\partial_0\psi^*)$$

$$= \frac{e}{2}[\phi^* + \chi^*](\phi - \chi) + (\phi + \chi)(\phi^* - \chi^*)]$$

$$= e(\phi^*\phi - \chi^*\chi) \tag{67}$$

since $$[\phi, \phi^*] = [\chi, \chi^*] = [\phi, \chi^*] = [\chi, \phi^*] = 0.$$

Equation (67) can be written in the matrix form as $\rho' = e\Psi^\dagger\sigma_3\Psi$. Further, from the condition of charge normalization, we also have

$$\pm\, e = \int \rho'(x)d\mathbf{x} = e\int \Psi^\dagger\sigma_3\Psi d\mathbf{x}$$

$$= e\int (\phi^*\phi - \chi^*\chi)d\mathbf{x} \tag{68}$$

Similarly, the expression for the current vector in the Ψ representation is given by

$$\mathbf{j} = -\frac{ie}{2m}(\psi^*\nabla\psi - \psi\nabla\psi^*) = \frac{ie}{2m}[(\phi^* + \chi^*)\,\nabla(\phi + \chi) - (\phi + \chi)\nabla(\phi^* + \chi^*)]$$

$$= -\frac{ie}{2m}(\phi^*\nabla\phi + \phi^*\nabla\chi + \chi^*\nabla\phi + \chi^*\nabla\chi - \phi\nabla\phi^* - \phi\nabla\chi^* - \chi\nabla\phi^* - \chi\nabla\chi^*)$$

$$= -\frac{ie}{2m}\left[(\phi^*\chi^*)\,\nabla\begin{pmatrix}1 & 1\\ 1 & 1\end{pmatrix}\begin{pmatrix}\phi\\ \chi\end{pmatrix} - \nabla(\phi^*\,\chi^*)\begin{pmatrix}1 & 1\\ 1 & 1\end{pmatrix}\begin{pmatrix}\phi\\ \chi\end{pmatrix}\right]$$

$$= -\frac{ie}{2m}\,[\Psi^\dagger\nabla\sigma_3(\sigma_3 + i\sigma_2)\Psi - \nabla\psi^\dagger\sigma_3(\sigma_3 + i\sigma_2)\Psi] \tag{69}$$

3.12 FREE PARTICLE SOLUTION OF THE KLEIN GORDON EQUATION IN THE Ψ REPRESENTATION

For well defined momentum states, we can substitute the energy and momentum operators by their respective eigenvalues $i\partial_0 \equiv E, -i\Delta \equiv \mathbf{p}$ in eq. (64) to get

$$E\Psi = \left[\frac{1}{2m}\mathbf{p}^2\begin{pmatrix}1 & 1\\ -1 & -1\end{pmatrix} + m\begin{pmatrix}1 & 0\\ 0 & -1\end{pmatrix}\right]\Psi = 0 \tag{70}$$

We postulate a solution of the operator eq. (64) of the same form as the one component solution for the free particle Klein Gordon equation and write, in terms of the eigenvalues of the respective operators

$$\Psi = \Psi_0 \exp[i\ \mathbf{px} - p_0\mathbf{x}^0)] \tag{71}$$

Substituting eq. (71) in (70) and using $\Psi = (\phi, \chi)^T$ and $\Psi_0 = (\phi_0\ \chi_0)^T$, we obtain

$$E\begin{pmatrix}\phi_0\\ \chi_0\end{pmatrix} = \begin{pmatrix}\frac{1}{2m}\mathbf{p}^2(\phi_0 + \chi_0) + m\phi_0\\ -\frac{1}{2m}\mathbf{p}^2(\phi_0 + \chi_0) - m\chi_0\end{pmatrix} \tag{72}$$

which is equivalent to the following set of coupled equations:

$$\left(E - \frac{1}{2m}\mathbf{p}^2 - m\right)\phi_0 - \frac{1}{2m}\mathbf{p}^2\chi_0 = 0 \tag{73}$$

and

$$\left(E + \frac{1}{2m}\mathbf{p}^2 + m\right)\chi_0 + \frac{1}{2m}\mathbf{p}^2\phi_0 = 0 \tag{74}$$

with the determinant $\Delta = \begin{vmatrix} E - \frac{1}{2m}\mathbf{p}^2 - m & -\frac{1}{2m}\mathbf{p}^2\\ \frac{1}{2m}\mathbf{p}^2 & E + \frac{1}{2m}\mathbf{p}^2 + m\end{vmatrix}$ which must necessarily vanish for the existence of a non-trivial solution of eqs. (73)-(74).

Hence, $E^2 - \left(\frac{1}{2m}\mathbf{p}^2 + m\right)^2 + \left(\frac{1}{2m}\mathbf{p}^2\right)^2 = 0$ or $E^2 = m^2 + \mathbf{p}^2$ which is the correct relativistic mass-energy relationship.

We, therefore, again end up with positive and negative energy solutions for $E = +(m^2 + \mathbf{p}^2)^{1/2}$ and $E = -(m^2 + \mathbf{p}^2)^{1/2}$, respectively.

(A) Positive Energy Solutions

Let $E = E_p$, where $E_p = (\mathbf{p}^2 + m^2)^{1/2} > 0$

From eq. (72), we obtain

$$\phi_0 = \left(\frac{m + E_p}{m - E_p}\right)\chi_0 \tag{75}$$

Since the set of eqs. (72) are not linearly independent, the choice of one of the variables ϕ_0 or χ_0 is left arbitrary. For convenience, we choose $\chi_0 = (m - E)\, k$ so that $\phi_0 = (m + E)\, k$, where k is to be determined from the normalization condition given by eq. (68), we have

$$\begin{aligned}\pm 1 &= \int_{L^3} d\mathbf{x}(\phi\phi^* - \chi\chi^*) \\ &= \int_{L^3} d\mathbf{x}(\phi_0\phi_0^* - \chi_0\chi_0^*) \\ &= k^2L^3\lfloor(m + E_p)^2 - (m - E_p)^2\rfloor \\ &= 4k^2mE_pL^3\end{aligned}$$

or $k^2 = \pm\dfrac{1}{4mE_pL^3}$. Since, we demand that k must necessarily be real, being the normalization factor, we have

$$k = \left(\frac{1}{4mE_pL^3}\right)^{1/2} \tag{76}$$

Hence, the complete two-component wavefunction for the positive energy state is

$$\begin{aligned}\Psi^{(+)} &= \begin{pmatrix}\phi^{(+)} \\ \chi^{(+)}\end{pmatrix} = \begin{pmatrix}\phi_0^{(+)} \\ \chi_0^{(+)}\end{pmatrix}\exp[i(\mathbf{px} - E_px^0)] \\ &= \left(\frac{1}{4mE_pL^3}\right)^{1/2}\begin{pmatrix}m + E_p \\ m - E_p\end{pmatrix}\exp[i(\mathbf{px} - E_px^0)]\end{aligned} \tag{77}$$

(B) Negative Energy Solutions

Writing $E = -E_p$ with $E_p = +\sqrt{\mathbf{p}^2 + m^2} > 0$ and proceeding as in case (A), we arrive at the following two-component solution of the Klein Gordon eq. (70) for negative energies:

$$\Psi^{(-)} = \begin{pmatrix}\phi^{(-)} \\ \chi^{(-)}\end{pmatrix} = \left(\frac{1}{4mE_pL^3}\right)^{1/2}\begin{pmatrix}m - E_p \\ m + E_p\end{pmatrix}\exp[i(\mathbf{px} + E_px^0)] \tag{78}$$

3.13 NON-RELATIVISTIC LIMITS OF TWO COMPONENT SOLUTIONS

For positive energy solutions with $E \equiv E_p = \sqrt{\mathbf{p}^2 + m^2}$, we have

$$\begin{pmatrix}\phi_0^{(+)} \\ \chi_0^{(+)}\end{pmatrix} = \left(\frac{1}{4mE_pL^3}\right)^{1/2}\begin{pmatrix}m + E_p \\ m - E_p\end{pmatrix} = \frac{1}{L^{3/2}}\begin{pmatrix}\dfrac{m + E_p}{\sqrt{4mE_p}} \\ \dfrac{m - E_p}{\sqrt{4mE_p}}\end{pmatrix} \tag{79}$$

Now

$$E_p = \sqrt{m^2 + \mathbf{p}^2} = m\sqrt{\left(1 + \frac{1}{m^2}\mathbf{p}^2\right)} \approx m\left(1 + \frac{1}{2m^2}\mathbf{p}^2\right)$$

so that

$$\frac{m + E_p}{\sqrt{4mE_p}} = \frac{2m + \frac{1}{2m}\mathbf{p}^2}{\sqrt{4m^2 + 2\mathbf{p}^2}} \approx 1$$

in the non-relativistic limit $\frac{\mathbf{p}}{m} \to 0$ and

$$\frac{m - E_p}{\sqrt{4mE_p}} = \frac{-\frac{1}{2m}\mathbf{p}^2}{\sqrt{4m^2 + 2\mathbf{p}^2}} \approx 0$$

Hence, in the non-relativistic limit, we have

$$\Psi_0^{(+)} = \begin{pmatrix} \phi_0^{(+)} \\ \chi_0^{(+)} \end{pmatrix} = \frac{1}{L^{3/2}}\begin{pmatrix} 1 \\ 0 \end{pmatrix} \tag{80}$$

for positive energy states.

In an analogous manner, we can show that

$$\Psi_0^{(-)} = \begin{pmatrix} \phi_0^{(-)} \\ \chi_0^{(-)} \end{pmatrix} = \frac{1}{L^{3/2}}\begin{pmatrix} 0 \\ 1 \end{pmatrix} \tag{81}$$

for negative energy states.

3.14 FESHBACH VILLARS REPRESENTATION (Φ REPRESENTATION)

In the Feshbach Villars representation, the two-component wavefunction solution of the Klein Gordon eq. (70) takes the form

$$\Phi^{(+)} = \begin{pmatrix} \phi_{FV}^{(+)} \\ \chi_{FV}^{(+)} \end{pmatrix} = \Phi_0^{(+)} \exp[i(\mathbf{px} - E_p x^0)]$$

$$= \begin{pmatrix} \phi_{0FV}^{(+)} \\ \chi_{0FV}^{(+)} \end{pmatrix} \exp[i(\mathbf{px} - E_p x^0)]$$

$$= \frac{1}{L^{3/2}}\begin{pmatrix} 1 \\ 0 \end{pmatrix} \exp[i(\mathbf{px} - E_p x^0)] \tag{82}$$

for positive energy states and

$$\Phi^{(-)} = \begin{pmatrix} \phi_{FV}^{(-)} \\ \chi_{FV}^{(-)} \end{pmatrix} = \Phi_0^{(-)} \exp[i(\mathbf{px} + E_p x^0)]$$

$$= \begin{pmatrix} \phi_{0FV}^{(-)} \\ \chi_{0FV}^{(-)} \end{pmatrix} \exp[i(\mathbf{px} + E_p x^0)]$$

$$= \frac{1}{L^{3/2}} \begin{pmatrix} 0 \\ 1 \end{pmatrix} \exp[i(\mathbf{px} + E_p x^0)] \tag{83}$$

for negative energy states with $E_p = +\sqrt{\mathbf{p}^2 + m^2}$.

To obtain the representation of the operator $\hat{U}$ that transforms the general two component representation (Ψ representation) $\Psi = \begin{pmatrix} \phi \\ \chi \end{pmatrix}$ to the Feshbach Villars representation (Φ representation) $\Phi = \begin{pmatrix} \phi_{\text{FV}} \\ \chi_{\text{FV}} \end{pmatrix}$ we assume $\hat{U} = \begin{pmatrix} a & c \\ b & d \end{pmatrix}$ and note that for positive energy states

$$\Phi^{(+)} = \hat{U}\Psi^{(+)} \text{ or } \begin{pmatrix} \phi_{FV}^{(+)} \\ \chi_{FV}^{(+)} \end{pmatrix}$$

$$= \begin{pmatrix} a & c \\ b & d \end{pmatrix} \begin{pmatrix} \phi^{(+)} \\ \chi^{(+)} \end{pmatrix} \text{ or, using eqs. (77) and (82)}$$

$$\frac{1}{L^{3/2}} \begin{pmatrix} 1 \\ 0 \end{pmatrix} = \begin{pmatrix} \phi_{0FV}^{(+)} \\ \chi_{0FV}^{(+)} \end{pmatrix} = \begin{pmatrix} a & c \\ b & d \end{pmatrix} \begin{pmatrix} \phi_0^{(+)} \\ \chi_0^{(+)} \end{pmatrix}$$

$$= \frac{1}{\sqrt{4mE_p L^3}} \begin{pmatrix} a & c \\ b & d \end{pmatrix} \begin{pmatrix} m + E_p \\ m - E_p \end{pmatrix}$$

which gives

$$\frac{1}{\sqrt{4mE_p}} [a(m + E_p) + c(m - E_p)] = 1, \; b(m + E_p) + d(m - E_p) = 0$$

Similarly, for negative energy states, we have

$$a(m - E_p) + c(m + E_p) = 0, \; \frac{1}{\sqrt{4mE_p}} [b(m - E_p) + d\,(m + E_p)] = 1$$

Solving these sets of equations, we obtain

$$\hat{U} = \frac{1}{\sqrt{4mE_p}}\begin{pmatrix} E_p+m & E_p-m \\ E_p-m & E_p+m \end{pmatrix} \tag{84}$$

with the inverse

$$\hat{U}^{-1} = \frac{1}{\sqrt{4mE_p}}\begin{pmatrix} E_p+m & -(E_p-m) \\ -(E_p-m) & E_p+m \end{pmatrix} \tag{85}$$

and the adjoint $U^{\dagger} = \hat{U}$ and hence $\hat{U}$ is not unitary.

In terms of Pauli matrices, we can write $\hat{U}$, $\hat{U}^{-1}$ and $\hat{U}^{\dagger}$ as

$$\hat{U} = \frac{(E_p+m)\mathbf{I}+(E_p-m)\sigma_1}{\sqrt{4mE_p}} = \hat{U}^{\dagger},$$

$$\hat{U}^{-1} = \frac{(E_p+m)\mathbf{I}-(E_p-m)\sigma_1}{\sqrt{4mE_p}} \tag{86}$$

3.15 NORMALIZATION OF THE FESHBACH VILLARS WAVEFUNCTION

The normalization of the Feshbach Villars wavefunction is achieved using eq. (68) as follows:

$$\pm 1 = \int d\mathbf{x}(\phi\phi^* - \chi\chi^*) = \int d\mathbf{x}(\phi_0\phi_0^* - \chi_0\chi_0^*)$$

$$= \int d\mathbf{x}(\phi_0^* - \chi_0^*)\begin{pmatrix} 1 & 0 \\ 0 & -1 \end{pmatrix}\begin{pmatrix} \phi_0 \\ \chi_0 \end{pmatrix}$$

$$= \int d\mathbf{x}\begin{pmatrix} \phi_0 \\ \chi_0 \end{pmatrix}^{\dagger}\begin{pmatrix} 1 & 0 \\ 0 & -1 \end{pmatrix}\begin{pmatrix} \phi_0 \\ \chi_0 \end{pmatrix} = \int d\mathbf{x}\Psi_0^{\dagger}\sigma_3\Psi_0 \tag{87}$$

Now, $\hat{U}\Psi_0 = \Phi_0$ so that $\Psi_0 = \hat{U}^{-1}\Phi_0$ and $(\hat{U}_0\Psi_0)^{\dagger} = \Phi_0^{\dagger}$ or $\Psi_0^{\dagger}\hat{U}^{\dagger} = \Phi_0^{\dagger}$ or $\Psi_0^{\dagger}\hat{U} = \Phi_0^{\dagger}$ or $\Psi_0^{\dagger} = \Phi_0^{\dagger}\hat{U}^{-1}$. Hence, using eq. (87), we have

$$\pm 1 = \int d\mathbf{x}\Phi_0^{\dagger}\hat{U}^{-1}\sigma_3\hat{U}^{-1}\Phi_0 = \int d\mathbf{x}\Phi_0^{\dagger}\sigma_3\Phi_0 \tag{88}$$

since

$$\hat{U}^{-1}\sigma_3 = \frac{1}{\sqrt{4mE_p}}[(E_p+m)\mathbf{I} - (E_p-m)\sigma_1]\sigma_3$$

$$= \frac{1}{\sqrt{4mE_p}}[(E_p+m)\sigma_3 + (E_p-m)\sigma_3\sigma_1]$$

$$= \sigma_3 \frac{1}{\sqrt{4mE_p}} [(E_p + m)\mathbf{I} + (E_p - m)\sigma_1] = \sigma_3 \hat{U}$$

Hence, we find that the Φ scalar product defined as $\langle \Psi_0 | \Xi_0 \rangle_\Phi = \int \Psi_0^\dagger \sigma_3 \Xi_0 d\mathbf{x}$ is invariant under the transformation $\Psi_0 \to \hat{U}\Psi_0$, $\Xi_0 \to \hat{U}\Xi_0$ from the Ψ representation to the Φ representation.

In the Feshbach Villars representation, we define the expectation of an operator $\hat{O}$ as

$$\langle \hat{O} \rangle_\Phi = \langle \Phi, \hat{O}\Phi \rangle = \int \Phi^\dagger \sigma_3 \hat{O} \Phi d\mathbf{x} \tag{89}$$

If we desire that $\langle \hat{O} \rangle_\Phi$ be real, then we must have

$$\begin{aligned} \langle \Phi, \hat{O}\Phi \rangle_\Phi &= \int \Phi^\dagger \sigma_3 \hat{O} \Phi d\mathbf{x} \\ &= \langle \hat{O} \rangle_\Phi = \langle \hat{O}^\dagger \rangle_\Phi = \langle \Phi, \hat{O}\Phi \rangle_\Phi \\ &= \langle \hat{O}\Phi, \Phi \rangle_\Phi = \int (\hat{O}\Phi)^\dagger \sigma_3 \Phi d\mathbf{x} \\ &= \int \Phi^\dagger \hat{O}^\dagger \sigma_3 \Phi d\mathbf{x} \end{aligned} \tag{90}$$

which gives $\qquad \hat{O}^\dagger \sigma_3 = \sigma_3 \hat{O}$ or $\hat{O}^\dagger = \sigma_3 \hat{O} \sigma_3$

or

$$\hat{O} = \sigma_3 \hat{O}^\dagger \sigma_3 \tag{91}$$

since $\sigma_3^2 = \mathbf{I}$ which defines Hermiticity in the Feshbach Villars representation as

$$\hat{O}^\dagger{}_{FV} = \sigma_3 \hat{O}^\dagger \sigma_3 = \hat{O} \tag{92}$$

Hence, we can define 'unitarity' in the Feshbach Villars representation as

$$\hat{O}^\dagger{}_{FV} = \sigma_3 \hat{O}^\dagger \sigma_3 = \hat{O}^{-1} \tag{93}$$

We then have, in the Φ product, for an operator $\hat{O}$, that is FV unitary

$$\begin{aligned} \langle \Phi, \Phi \rangle_\Phi &= \langle \Phi, \hat{O}^{-1} \hat{O} \Phi \rangle_\Phi = \langle (\hat{O}^{-1})^\dagger_{FV} \Phi, \hat{O}\Phi \rangle_\Phi \\ &= \langle (\hat{O}^\dagger_{FV})^\dagger_{FV} \Phi, \hat{O}\Phi \rangle_\Phi = \langle \hat{O}\Phi, \hat{O}\Phi \rangle_\Phi \end{aligned} \tag{94}$$

since

$$\begin{aligned} (\hat{O}^\dagger{}_{FV})^\dagger{}_{FV} &= \sigma_3 (\hat{O}^\dagger{}_{FV}) \sigma_3 \\ &= \sigma_3 (\sigma_3 \hat{O}^\dagger \sigma_3)^\dagger \sigma_3 \\ &= \hat{O} \text{ and } \sigma_3^\dagger = \sigma_3,\ \sigma_3^2 = \mathbf{I} \end{aligned}$$

The Hamiltonian in the Ψ representation given by $\hat{H} = -(\sigma_3 + i\sigma_2)\,\dfrac{1}{2m}\,\nabla^2 + \sigma_3 m$ is Φ Hermitian since

$$\begin{aligned}\hat{H}_{FV}^{\dagger} &= \sigma_3 \hat{H}^{\dagger}\, \sigma_3 = \sigma_3 \hat{H} \sigma_3 \\ &= -\,\sigma_3\left[(\sigma_3 + i\sigma_2)\frac{1}{2m}\nabla^2 - \sigma_3 m\right]\sigma_3 \\ &= -\,(\sigma_3 + i\sigma_2)\,\frac{1}{2m}\nabla^2 + \sigma_3 m = \hat{H} \qquad (95)\end{aligned}$$

(as $\hat{H}^{\dagger} = \hat{H}$).

3.16 TRANSFORMATION RULES FOR OPERATORS TO THE Φ REPRESENTATION

The transformation rules for operators from the Ψ representation to the Φ representation depend on the scalar product with respect to which the transformation is to be made.

If the scalar product is defined by $\langle\Psi|\Psi'\rangle = \int \Psi^{\dagger}\Psi'\, d\mathbf{x}$, then the matrix elements of any operator $\hat{O}$ will be given by

$$\begin{aligned}\langle\Psi|\hat{O}|\Psi'\rangle &= \int \Psi^{\dagger}\hat{O}\Psi' d\mathbf{x} \\ &= \int \Phi^{\dagger}(\hat{U}^{-1})^{+}\, \hat{O}\hat{U}^{-1}\Phi' d\mathbf{x} \\ &= \int \Phi^{\dagger}\hat{O}_{\Psi}\Phi' d\mathbf{x} \qquad (96)\end{aligned}$$

where $\hat{O}_{\Psi} = (\hat{U}^{-1})^{\dagger}\hat{O}\hat{U}^{-1} = \hat{U}^{-1}\hat{O}\hat{U}^{-1}$.

However, if the scalar product is defined as $\langle\Psi|\Psi'\rangle_{\Phi} = \int\Psi^{\dagger}\sigma_3\Psi' d\mathbf{x}$, then the matrix elements of an operator $\hat{O}$ are given by

$$\begin{aligned}\langle\Psi|\hat{O}|\Psi'\rangle_{\Phi} &= \int \Psi^{\dagger}\sigma_3\hat{O}\Psi' d\mathbf{x} \\ &= \int \Phi^{\dagger}\, (\hat{U}^{-1})^{\dagger}\sigma_3\hat{O}\hat{U}^{-1}\Phi' d\mathbf{x} \\ &= \int \Phi^{\dagger}\hat{U}^{-1}\sigma_3\hat{O}\hat{U}^{-1}\Phi' d\mathbf{x} \\ &= \int \Phi^{\dagger}\sigma_3\hat{U}\,\hat{O}\hat{U}^{-1}\Phi' d\mathbf{x} \\ &= \int \Phi^{\dagger}\sigma_3\hat{O}_{\Phi}\Phi' d\mathbf{x} \qquad (97)\end{aligned}$$

where $\hat{O}_{\Phi} = \hat{U}\,\hat{O}\hat{U}^{-1}$ and since

$$(\hat{U}^{-1})^{\dagger}\, \sigma_3 = (\hat{U}^{-1})^{\dagger}\, \sigma_3^{\dagger} = (\sigma_3\hat{U}^{-1})^{\dagger} = (\hat{U}\sigma_3)^{\dagger}\, \sigma_3^{\dagger} = \hat{U}^{\dagger} = \sigma_3\hat{U}^{\dagger} = \sigma_3\hat{U}$$

3.17 THE HAMILTONIAN IN THE FESHBACH VILLARS REPRESENTATION

The Hamiltonian $\hat{H}$ in the Feshbach Villars representation (Φ representation) is given by

$$\hat{H}_\Phi = \hat{U}\,\hat{H}\,\hat{U}^{-1}$$

where

$$\hat{U} = \frac{1}{\sqrt{4mE_p}}\begin{pmatrix} m+E_p & E_p-m \\ E_p-m & m+E_p \end{pmatrix},$$

$$\hat{U}^{-1} = \frac{1}{\sqrt{4mE_p}}\begin{pmatrix} m+E_p & m-E_p \\ m-E_p & m+E_p \end{pmatrix}$$

and

$$\hat{H} = \begin{pmatrix} -1 & -1 \\ 1 & 1 \end{pmatrix}\frac{1}{2m}\nabla^2 + \begin{pmatrix} 1 & 0 \\ 0 & -1 \end{pmatrix} m.$$

Performing explicit calculations with the above expressions, we obtain

$$\hat{H}_\Phi = \frac{1}{2E_p}\left[\begin{pmatrix} -1 & -1 \\ 1 & 1 \end{pmatrix}\nabla^2 + \begin{pmatrix} m^2+E_p^2 & m^2-E_p^2 \\ E_p^2-m^2 & -(m^2+E_p^2) \end{pmatrix}\right] \tag{98}$$

Noting that $E_p^2 = m^2 + \mathbf{p}^2$ and making the replacement $-i\nabla \equiv \hat{\mathbf{p}}$ by the corresponding momentum eigenvalue $\mathbf{p}$ for well-defined momentum states, we obtain

$$\hat{H}_\Phi = \begin{pmatrix} 1 & 0 \\ 0 & -1 \end{pmatrix} E_p = \sigma_3 E_p \tag{99}$$

3.18 KLEIN GORDON EQUATION AND ITS SOLUTION FOR A FREE PARTICLE IN THE FESHBACH VILLARS REPRESENTATION

The Klein Gordon equation in the Ψ representation is given by eq. (64), viz.

$$i\partial_0\Psi = -\left[(\sigma_3 + i\sigma_2)\frac{1}{2\mathrm{m}}\nabla^2 - \sigma_3 m\right]\Psi = \hat{H}\Psi$$

Multiplying this equation by the transformation operator $\hat{U}$, we obtain

$$\hat{U}(i\partial_0\Psi) = i\partial_0(\hat{U}\Psi) = \hat{U}\hat{H}\Psi = \hat{U}\hat{H}\hat{U}^{-1}(\hat{U}\Psi)$$

or,

$$i\partial_0\Phi = \hat{H}_\Phi\Phi = \sigma_3 E_p\Phi \tag{100}$$

since $\hat{U}\hat{H}\hat{U}^{-1} = \hat{H}_\Phi = \sigma_3 E_p$ for well-defined states of the momentum operator.

In fact, we can obtain positive and negative energy solutions for the free particles in the Φ representation directly from eq. (100).

We write $\Phi = \begin{pmatrix} \varepsilon_1 \\ \varepsilon_2 \end{pmatrix} \exp(i\mathbf{px})$ (where Φ's satisfy $-i\nabla\Phi = \mathbf{p}\Phi$) so that $i\partial_0 \begin{pmatrix} \varepsilon_1 \\ \varepsilon_2 \end{pmatrix} = E_p \begin{pmatrix} \varepsilon_1 \\ -\varepsilon_2 \end{pmatrix}$ with the solutions $\varepsilon_1 = A_1 \exp(-iE_p x^0)$ and $\varepsilon_2 = A_2 \exp(iE_p x^0)$, where A_1 and A_2 are normalization constants to be determined from

$$\pm 1 = \int_{L^3} \Phi^\dagger \sigma_3 \Phi \, d\mathbf{x} \int_{L^3} \begin{pmatrix} \varepsilon_1 \\ \varepsilon_2 \end{pmatrix}^\dagger \sigma_3 \begin{pmatrix} \varepsilon_1 \\ \varepsilon_2 \end{pmatrix} d\mathbf{x} = (|A_1|^2 - |A_2|^2)L^3 \quad \text{so that}$$

$$|A_1|^2 - |A_2|^2 = \pm \frac{1}{L^3}.$$

Hence, we have

$$\Phi^{(+)} = \frac{1}{L^{3/2}} \begin{pmatrix} 1 \\ 0 \end{pmatrix} \exp\left[i(\mathbf{px} - E_p x^0)\right] \tag{101}$$

$$\Phi^{(-)} = \frac{1}{L^{3/2}} \begin{pmatrix} 0 \\ 1 \end{pmatrix} \exp\left[i(\mathbf{px} + E_p x^0)\right] \tag{102}$$

3.19 CALCULATION OF CURRENT DENSITY IN THE Φ REPRESENTATION

The current density for the Klein Gordon equation is given by eq. (19), viz.

$$\mathbf{j} = -\frac{ie}{2m}(\psi^* \nabla \psi - \psi \nabla \psi^*)$$

Now

$$\begin{aligned}(\psi^* \nabla \psi - \psi \nabla \psi^*) &= (\phi^* + \chi^*)\nabla(\phi + \chi) - [\nabla(\phi^* + \chi^*)](\phi + \chi) \\ &= (\phi^* \;\; \chi^*) \begin{pmatrix} 1 & 1 \\ 1 & 1 \end{pmatrix} \nabla \begin{pmatrix} \phi \\ \chi \end{pmatrix} - [\nabla(\phi^* \; \chi^*)] \begin{pmatrix} 1 & 1 \\ 1 & 1 \end{pmatrix} \begin{pmatrix} \phi \\ \chi \end{pmatrix} \\ &= \Psi^\dagger(\mathbf{I} + \sigma_1)\nabla\Psi - \nabla\Psi^\dagger(\mathbf{I} + \sigma_1)\Psi \end{aligned} \tag{103}$$

To transform the above expression to the Φ representation, we note that

$$\Psi = \hat{U}^{-1}\Phi, \; \Psi^\dagger = \Phi^\dagger(\hat{U}^{-1})^\dagger = \Phi^\dagger \hat{U}^{-1}. \text{ Further}$$

$$\hat{U}^{-1}(\mathbf{I} + \sigma_1)\hat{U}^{-1} = \frac{1}{4mE_p} \begin{pmatrix} m + E_p & m - E_p \\ m - E_p & m + E_p \end{pmatrix} \begin{pmatrix} 1 & 1 \\ 1 & 1 \end{pmatrix}$$

$$\begin{pmatrix} m + E_p & m - E_p \\ m - E_p & m + E_p \end{pmatrix} = \frac{m}{E_p}(\mathrm{I} + \sigma_1) \tag{104}$$

Hence, we have

$$\begin{aligned}\psi^* \nabla \psi - \psi \nabla \psi^* &= \Psi^\dagger(\mathbf{I} + \sigma_1)\nabla\Psi - \nabla\Psi^\dagger(\mathbf{I} + \sigma_1)\Psi \\ &= \Phi^\dagger \hat{U}^{-1}(\mathbf{I} + \sigma_1)\nabla \hat{U}^{-1}\Phi - \nabla(\Phi^\dagger \hat{U}^{-1})(\mathbf{I} + \sigma_1)\hat{U}^{-1}\Phi\end{aligned}$$

$$= \frac{m}{E_p}[\Phi^\dagger(\mathbf{I} + \sigma_1)\nabla\Phi - \nabla\Phi^\dagger(\mathbf{I} + \sigma_1)\Phi] \tag{105}$$

Therefore,

$$\mathbf{j} = -\frac{ie}{2E_p}[\Phi^\dagger(\mathbf{I} + \sigma_1)\nabla\Phi - \nabla\Phi^\dagger(\mathbf{I} + \sigma_1)\Phi] \tag{106}$$

3.20 KLEIN GORDON EQUATION IN THE Φ – p REPRESENTATION

The Klein Gordon Hamiltonian in the $\Phi - \mathbf{p}$ representation is given by $\sigma_3 E_p = \begin{pmatrix} 1 & 0 \\ 0 & -1 \end{pmatrix} E_p$.

Hence, the Klein Gordon equation will take the form

$$i\partial_0\tilde{\Phi}(p) = \sigma_3 E_p \tilde{\Phi}(p) \tag{107}$$

where $\tilde{\Phi}(p)$ now satisfies $\hat{\mathbf{P}}\tilde{\Phi}(p) = \mathbf{p}\tilde{\Phi}(p)$ and $i\nabla\mathbf{p}\tilde{\Phi}(p) = \mathbf{x}\tilde{\Phi}(p)$ so that $\tilde{\Phi}(p) = (\varepsilon_1(p_0)\ \varepsilon_2(p_0))^T \exp(-i\mathbf{px})$ or,

$$i\partial_0\begin{pmatrix} \varepsilon_1 \\ \varepsilon_2 \end{pmatrix}\exp(-i\mathbf{px}) = \begin{pmatrix} 1 & 0 \\ 0 & -1 \end{pmatrix} E_p \begin{pmatrix} \varepsilon_1 \\ \varepsilon_2 \end{pmatrix}\exp(-i\mathbf{px}) = E_p\begin{pmatrix} \varepsilon_1 \\ -\varepsilon_2 \end{pmatrix}\exp(-i\mathbf{px}) \tag{108}$$

which leads to the solutions

$$\tilde{\Phi}^{(+)}(p) = N\begin{pmatrix} 1 \\ 0 \end{pmatrix}\exp[-i(\mathbf{px} + E_p x^0)]$$

and

$$\tilde{\Phi}^{(-)}(p) = N\begin{pmatrix} 0 \\ 1 \end{pmatrix}\exp[-i(\mathbf{px} - E_p x^0)] \text{ respectively.}$$

The fact that the Hamiltonian in the $\Phi - \mathbf{p}$ representation is also given by $\sigma_3 E_p$ as in the $\Phi - \mathbf{x}$ representation is established as follows:

$$\langle \mathbf{p}|\hat{H}|\mathbf{p}'\rangle = \iint \langle \mathbf{p}|\mathbf{x}'\rangle d\mathbf{x}'\langle \mathbf{x}'|\hat{H}|\mathbf{x}''\rangle\, d\mathbf{x}''\langle \mathbf{x}''|\mathbf{p}'\rangle$$

Now,

$$\langle \mathbf{x}'|\hat{H}|\mathbf{x}''\rangle = \sigma_3 E_p \delta(\mathbf{x}' - \mathbf{x}'')$$

and

$$\langle \mathbf{p}|\mathbf{x}'\rangle = \frac{1}{(2\pi)^{3/2}}\exp(-i\mathbf{px}').$$

Hence,

$$\langle \mathbf{p}|\hat{H}|\mathbf{p}'\rangle = \sigma_3 E_p \frac{1}{(2\pi)^3}\iint e^{-i\mathbf{px}' + i\mathbf{p}'\mathbf{x}''}\delta(\mathbf{x}' - \mathbf{x}'')d\mathbf{x}'d\mathbf{x}''$$

$$= \sigma_3 E_p \frac{1}{(2\pi)^3}\int e^{-i(\mathbf{p} - \mathbf{p}')\mathbf{x}'}d\mathbf{x}' = \sigma_3 E_p \delta(\mathbf{p} - \mathbf{p}')$$

3.21 FREE PARTICLE SOLUTIONS IN THE Φ – p AND Ψ – p REPRESENTATIONS

The free particle solutions of the Klein Gordon equation in the $\Phi - \mathbf{x}$ representation are given by eqs. (101)-(102), viz.

$$\Phi^{(+)} = \frac{1}{L^{3/2}} \begin{pmatrix} 1 \\ 0 \end{pmatrix} \exp[i(\mathbf{px} - E_p x^0)]$$

and

$$\Phi^{(-)} = \frac{1}{L^{3/2}} \begin{pmatrix} 0 \\ 1 \end{pmatrix} \exp[i(\mathbf{px} + E_p x^0)]$$

for positive and negative energy particles respectively, with $E_p = (m^2 + \mathbf{p}^2)^{1/2}$. In the momentum space they are given by:

$$\widetilde{\Phi}^{(+)}(p) = \frac{1}{(2\pi)^{3/2}} \begin{pmatrix} 1 \\ 0 \end{pmatrix} \exp[-i(\mathbf{px} + E_p x^0)] \tag{109}$$

and

$$\widetilde{\Phi}^{(-)}(p) = \frac{1}{(2\pi)^{3/2}} \begin{pmatrix} 0 \\ 1 \end{pmatrix} \exp[-i(\mathbf{px} - E_p x^0)] \tag{110}$$

for positive and negative energies respectively. We have normalized these states to δ functions.

To obtain the solutions in the $\Psi - \mathbf{p}$ representation, we need to apply $\hat{U}^{-1}$ to $\widetilde{\Phi}(p)$ to get

$$\widetilde{\Psi}^{(+)}(p) = \hat{U}^{-1}\widetilde{\Phi}^{(+)}(p)$$

$$= \frac{1}{\sqrt{4mE_p}} \begin{pmatrix} m+E_p & m-E_p \\ m-E_p & m+E_p \end{pmatrix} \frac{1}{(2\pi)^{3/2}} \begin{pmatrix} 1 \\ 0 \end{pmatrix}$$

$$\exp[-i(\mathbf{px} + E_p x^0)] = \frac{1}{(2\pi)^{2/3}\sqrt{4mE_p}} \begin{pmatrix} m+E_p \\ m-E_p \end{pmatrix} [-i(\mathbf{px} + E_p x^0)] \tag{111}$$

and similarly

$$= \widetilde{\Psi}^{(-)}(p) = \hat{U}^{-1}\widetilde{\Phi}^{(-)}(p) \frac{1}{\sqrt{4mE_p}} \begin{pmatrix} m+E_p & m-E_p \\ m-E_p & m+E_p \end{pmatrix} \frac{1}{(2\pi)^{3/2}} \begin{pmatrix} 0 \\ 1 \end{pmatrix}$$

$$\exp[-i(\mathbf{px} - E_p x^0)] = \frac{1}{(2\pi)^{3/2}\sqrt{4mE_p}} \begin{pmatrix} m+E_p \\ m-E_p \end{pmatrix} [-i(\mathbf{px} - E_p x^0)] \tag{112}$$

3.22 ONE PARTICLE OPERATORS

The Hamiltonian in the Ψ representation for a free spin 0 particle is given by eq. (64) which leads to the equation of motion

$$\hat{H}\Psi = \left[-(\sigma_3 + i\sigma_2)\frac{1}{2m}\nabla^2 + m\sigma_3\right]\begin{pmatrix}\phi\\ \chi\end{pmatrix} = E\begin{pmatrix}\phi\\ \chi\end{pmatrix}. \tag{113}$$

On simplification, this leads to the following pair of differential equations:

$$-\frac{1}{2m}\nabla^2(\phi + \chi) + m\phi = E\phi \tag{114}$$

and

$$\frac{1}{2m}\nabla^2(\phi + \chi) - m\chi = E\chi \tag{115}$$

Noting that $-\nabla^2\phi = \mathbf{p}^2\phi$ and $-\nabla^2\chi = \mathbf{p}^2\chi$ for well-defined momentum states, we obtain $m(\phi - \chi) = E(\phi + \chi)$ and $\frac{\mathbf{p}^2}{m}(\phi + \chi) + m(\phi + \chi) = E(\phi - \chi)$ and, therefore, $E = \pm\sqrt{\mathbf{p}^2 + m^2}$, i.e., the issue of negative energy as eigenvalues of the Hamiltonian still remains unresolved.

It can, however, be shown that the average energy computed in the Feshbach Villars representation is always positive. We have

$$E_{av} = \langle \hat{H}\rangle_\Phi = \int \Psi^\dagger\sigma_3\hat{H}\Psi d\mathbf{x}$$

$$= \pm E_p\int \Psi^\dagger\sigma_3\ \Psi d\mathbf{x} = E_p \tag{116}$$

since $\hat{H}\Psi = \pm E_p\Psi$ and $\int \Psi^\dagger\sigma_3\Psi\ d\mathbf{x} = \pm 1$ and thereby showing that the average energy is always positive.

3.23 POSITION OPERATOR IN Φ – p REPRESENTATION

In the Schrödinger picture, the position operator in the momentum space is given by $\hat{\mathbf{x}} = i\nabla_p = i\left(\frac{\partial}{\partial p_x}, \frac{\partial}{\partial p_y}, \frac{\partial}{\partial p_z}\right)$. Transformation to the Feshbach Villars representation (Φ representation) is achieved through the transformation operator $\hat{U}$, i.e., $\hat{\mathbf{x}}_\Phi = \hat{U}(i\nabla_p)\hat{U}^{-1}$, where

$$\hat{U} = \frac{1}{\sqrt{4mE_p}}\begin{pmatrix} m + E_p & E_p - m\\ E_p - m & m + E_p\end{pmatrix}$$

and

$$\hat{U}^{-1} = \frac{1}{\sqrt{4mE_p}}\begin{pmatrix} m + E_p & m - E_p\\ m - E_p & m + E_p\end{pmatrix}.$$

Furthermore

$$\nabla_p E_p = \nabla_p(m^2 + \mathbf{p}^2)^{1/2}\ \frac{1}{2(m^2 + \mathbf{p}^2)^{1/2}}\cdot 2\mathbf{p} = \frac{\mathbf{p}}{E_p}$$

Hence,

$$\nabla_p \hat{U}^{-1} = \nabla_p \left[\frac{1}{\sqrt{4mE_p}} \begin{pmatrix} m+E_p & m-E_p \\ m-E_p & m+E_p \end{pmatrix} \right]$$

$$= \nabla_p \begin{pmatrix} \sqrt{\dfrac{m}{4E_p}} + \sqrt{\dfrac{E_p}{4m}} & \sqrt{\dfrac{m}{4E_p}} - \sqrt{\dfrac{2E_p}{4m}} \\ \sqrt{\dfrac{m}{4E_p}} - \sqrt{\dfrac{E_p}{4m}} & \sqrt{\dfrac{m}{4E_p}} + \sqrt{\dfrac{E_p}{4m}} \end{pmatrix}$$

$$= \left(\begin{matrix} -\sqrt{\dfrac{m}{4}} \cdot \dfrac{\mathbf{p}}{2E_p^{5/2}} + \sqrt{\dfrac{1}{4m}} \cdot \dfrac{\mathbf{p}}{2E_p^{3/2}} + \left(\sqrt{\dfrac{m}{4E_p}} + \sqrt{\dfrac{E_p}{4m}} \right) \nabla_p \\ -\sqrt{\dfrac{m}{4}} \cdot \dfrac{\mathbf{p}}{2E_p^{5/2}} - \sqrt{\dfrac{1}{4m}} \cdot \dfrac{\mathbf{p}}{2E_p^{3/2}} + \left(\sqrt{\dfrac{m}{4E_p}} - \sqrt{\dfrac{E_p}{4m}} \right) \nabla_p \end{matrix} \right.$$

$$\left. \begin{matrix} -\sqrt{\dfrac{m}{4}} \cdot \dfrac{\mathbf{p}}{2E_p^{\frac{5}{2}}} - \sqrt{\dfrac{1}{4m}} \cdot \dfrac{\mathbf{p}}{2E_p^{\frac{3}{2}}} + \left(\sqrt{\dfrac{m}{4E_p}} - \sqrt{\dfrac{E_p}{4m}} \right) \nabla_p \\ -\sqrt{\dfrac{m}{4}} \cdot \dfrac{\mathbf{p}}{2E_p^{\frac{5}{2}}} + \sqrt{\dfrac{1}{4m}} \cdot \dfrac{\mathbf{p}}{2E_p^{\frac{3}{2}}} + \left(\sqrt{\dfrac{m}{4E_p}} + \sqrt{\dfrac{E_p}{4m}} \right) \nabla_p \end{matrix} \right)$$

$$= \frac{\mathbf{p}}{2E_p^2 \sqrt{4mE_p}} \begin{pmatrix} E_p - m & -(E_p+m) \\ -(E_p+m) & E_p - m \end{pmatrix}$$

$$+ \frac{1}{\sqrt{4mE_p}} \begin{pmatrix} E_p + m & m - E_p \\ m - E_p & E_p + m \end{pmatrix} \nabla_p$$

Thus, $$\hat{\mathbf{x}}_\Phi = \hat{U}(i\nabla_p)\hat{U}^{-1}$$

$$= \frac{1}{\sqrt{4mE_p}} \begin{pmatrix} E_p + m & E_p - m \\ E_p - m & E_p + m \end{pmatrix} \left[\frac{i\mathbf{p}}{2E_p^2 \sqrt{4mE_p}} \right.$$

$$\begin{pmatrix} E_p - m & -(E_p+m) \\ -(E_p+m) & E_p - m \end{pmatrix}$$

$$+ \frac{i}{\sqrt{4mE_p}} \begin{pmatrix} E_p + m & m - E_p \\ m - E_p & E_p + m \end{pmatrix} \nabla^p \Bigg]$$

$$= \frac{i\mathbf{p}}{2E_p^2(4mE_p)} \begin{pmatrix} 0 & -4mE_p \\ -4mE_p & 0 \end{pmatrix} + \frac{i}{4mE_p} \begin{pmatrix} 4mE_p & 0 \\ 0 & 4mE_p \end{pmatrix} \nabla_p$$

$$= -i\sigma_1 \frac{\mathbf{p}}{2(\mathbf{p}^2 + m^2)} + i\mathbf{I}\nabla_p \tag{117}$$

Since, the position operator $\hat{\mathbf{x}}_\Phi$ in the Φ representation must also be diagonal and σ_1 is a non-diagonal matrix, it follows that the appropriate position operator in the Φ representation is given by $i\mathbf{I}\nabla p$.

The velocity operator is now computed as

$$\frac{d}{dx^0} \hat{\mathbf{x}}_\Phi = \frac{1}{i} [\hat{\mathbf{x}}_\Phi, \hat{H}_\Phi] = \frac{1}{i} [i\mathbf{I}\nabla_p, \sigma_3 E_p]$$

$$= \left[\mathbf{I}\nabla_p, \sigma_3 \sqrt{\mathbf{p}^2 + m^2}\right]$$

$$= \sigma_3 \left(\frac{\mathbf{p}}{\sqrt{\mathbf{p}^2 + m^2}} + \sqrt{\mathbf{p}^2 + m^2}\nabla_p - \sqrt{\mathbf{p}^2 + m^2}\nabla_p \right)$$

$$= \sigma_3 \left(\frac{\mathbf{p}}{\sqrt{p^2 + m^2}} \right) = \sigma_3 \frac{\mathbf{p}}{E_p} \tag{118}$$

Thus, the velocity is given by $\sigma_3 \frac{\mathbf{p}}{E_p}$ with the eigenvalues $\pm \frac{\mathbf{p}}{E_p}$. The corresponding eigenfunctions are obtained by solving

$$\frac{d\hat{\mathbf{x}}_\Phi}{dt} \Phi^{(\pm)}(p) = \sigma_3 \frac{\mathbf{p}}{E_p} \Phi^{(\pm)}(p)$$

$$= \begin{pmatrix} 1 & 0 \\ 0 & -1 \end{pmatrix} \frac{\mathbf{p}}{E_p} \Phi^{(\pm)}(p)$$

$$= \pm \frac{\mathbf{p}}{E_p} \Phi^{(\pm)}(p) \tag{119}$$

3.24 THEORY OF CHARGED SPIN 0 PARTICLES

We consider a charged spinless particle of charge e in an electromagnetic field described by the 4-vector

$$A^{\mu} = \{A^0, \mathbf{A}\} = \{A_0, A_x, A_y, A_z\}$$

or
$$A_{\mu} = \{A^0, -\mathbf{A}\} = \{A_0, -A_x, -A_y, -A_z\} \tag{120}$$

As in the non-relativistic quantum mechanics of a charged particle in an electromagnetic field, we make the operator substitutions $\hat{\mathbf{p}} \equiv -i\nabla \rightarrow -i\nabla - e\mathbf{A}$, $\hat{E} \equiv i\partial_0 \rightarrow i\partial_0 - eA_0$ in the Klein Gordon equation $(\partial_0{}^2 - \nabla^2 + m^2)\psi = 0$ or $\{-(i\partial_0)^2 + (-i\nabla)^2 + m^2\}\psi = 0$ to obtain

$$(i\partial_0 - eA_0)^2\ \psi = [(i\nabla + e\mathbf{A})^2 + m^2]\psi \tag{121}$$

3.25 CHARGE AND CURRENT DENSITIES

To obtain charge and current densities, we proceed as usual. Multiplying eq. (121) by ψ^* and subtracting therefrom the complex conjugate of the equation so obtained, we have

$$\psi^*[(i\partial_0 - eA_0)^2 - (i\nabla + e\mathbf{A})^2]\ \psi = m^2\psi^*\psi \tag{122}$$

$$\psi[(-i\partial_0 - eA_0)^2 - (-i\nabla + e\mathbf{A})^2]\psi^* = m^2\psi^*\psi \tag{123}$$

On subtracting, we get

$$\begin{aligned}&[\psi^*\ (i\partial_0 - eA_0)^2\ \psi - \psi(-i\partial_0 - e\mathbf{A}_0)^2\psi^*]\\ &- [\psi^*(i\nabla + e\mathbf{A})^2\ \psi - \psi\ (-i\Delta + e\mathbf{A})^2\ \psi^*] = 0\end{aligned} \tag{124}$$

Simplification yields

$$\begin{aligned}0 &= \psi^*\ (-\partial_0^2 + \nabla^2)\ \psi - \psi\ (-\partial_0^2 + \nabla^2)\psi^* - ie\\ &\quad [\psi^*\partial_0 A_0\psi + \psi^*\nabla\mathbf{A}\psi + \psi A_0\partial_0\psi^* + \psi\mathbf{A}\nabla\psi^*]\\ &\quad - ie\ [\psi\partial_0 \mathrm{A}_0\psi^* + \psi\nabla\mathbf{A}\psi^* + \psi^* A_0\partial_0\psi + \psi^*\mathbf{A}\nabla\psi]\\ &= -\psi^*\partial_\mu\partial^\mu\psi + \psi\partial_\mu\partial^\mu\psi^* - ie\partial_\mu\ (\psi^*\ A^\mu\psi) - ie\partial_\mu\ (\psi A^\mu\psi^*)\\ &= -\ \psi^*\partial_\mu\partial^\mu\psi + \psi\partial_\mu\partial^\mu\psi^* - 2ie\partial_\mu(\psi A^\mu\psi^*)\\ &= \partial_\mu[(-\psi^*\partial^\mu\psi + \psi\ \partial^\mu\psi^*)\ -2ie\psi A^\mu\psi^*\end{aligned} \tag{125}$$

We now define the 4. current density j^μ as $j^\mu = \dfrac{ie}{2m}\ [\psi^*\partial^\mu\psi - \psi\partial^\mu\psi^*] - \dfrac{e^2}{m}\ \Psi A^\mu\ \Psi^*$ so that we can write $\partial_\mu j^\mu = 0$ which is a continuity equation.

The charge and current densities are respectively obtained as

$$\rho' = j_0 = \frac{ie}{2m}\ [\psi^*\partial_0\psi - \psi\partial_0\psi^*] + \frac{e^2}{m}\ \psi A_0\psi^* \tag{126}$$

and

$$\mathbf{j} = -\ \frac{ie}{2m}[\psi^*\nabla\psi - \psi\nabla\psi^*] - \frac{e^2}{m}\ \psi\mathbf{A}\psi^* \tag{127}$$

3.26 GAUGE INVARIANCE

Consider the gauge transformation given $A'^{\mu}(x) = A^{\mu}(x) + \partial^{\mu}\phi(x)$ or equivalently

$$A'_0 = A'^0 = A^0 + \partial^0\phi(x) = A_0 + \partial_0\phi(x),\ \mathbf{A}' = \mathbf{A} - \nabla\phi(x) \tag{128}$$

We postulate that the gauge transformation results in a phase transformation of the wavefunction ψ to $\psi' = \psi \exp[if(x)]$ where $f(x)$ is to be determined by

$$\begin{aligned} 0 &= (i\partial_0 - eA'_0)^2\,\psi - (i\nabla + e\mathbf{A}')^2\,\psi - m^2\psi \\ &= (i\partial_0 - eA_0)^2\,\psi' - (i\nabla + e\mathbf{A})^2\,\psi' - m^2\psi' \end{aligned} \tag{129}$$

We then obtain, using the latter part of the above equation and eqs. (121) and (128)

$$\begin{aligned} 0 &= \{i\partial_0 - e[A_0 + \partial_0\phi(x)]\}^2\,\psi - (i\nabla + e[\mathbf{A} - \nabla\phi(x)]\}^2\,\psi - m^2\psi \\ &= (i\partial_0 - eA_0)^2\,\psi' - (i\nabla + e\mathbf{A})^2\,\psi' - m^2\psi' \end{aligned} \tag{130}$$

which on simplification yields

$$f(x) = e\phi(x) \tag{131}$$

so that the transformed wavefunction as a consequence of the gauge transformation (128) is $\psi' = \psi \exp[ie\phi(x)]$. Hence, the gauge transformation causes a change in the phase of the wavefunction ψ by an amount $e\phi(x)$.

3.27 NON-RELATIVISTIC LIMIT OF THE KLEIN GORDON EQUATION WITH FIELDS

As in our analysis of the non-relativistic limit of the Klein Gordon equation for the free particle, we split the Klein Gordon wavefunction $\psi(x)$ as

$$\psi(x) = \phi(x) \exp(-imx^0) \tag{132}$$

where $\phi(x)$ represents the non-relativistic part. Further, as proved earlier, $|\,i\partial_0\phi\,| << m|\,\phi\,|$, i.e., the non-relativistic energy is insignificant in relation to the mass rest energy.

We further assume that $|\,eA_0\phi\,| << m|\,\phi\,|$, i.e., the potentials are also negligible in relation to the mass rest energy. We then have, using eq. (132)

$$\begin{aligned} (i\partial_0 - eA_0)\,\psi &= (i\partial_0 - eA_0)\,[\phi(x)\exp(-imx^0)] \\ &= [i(\partial_0\phi(x)) + m\phi(x) - eA_0\phi(x)]\exp(-imx^0) \end{aligned}$$

Hence,

$$(i\partial_0 - eA)^2\,\psi = (i\partial_0 - eA_0)[i\partial_0\phi(x) + m\phi(x) - eA_0\phi(x)]\exp(-imx^0)$$

$$= \begin{bmatrix} -(\partial_0^2\phi(x)) + im(\partial_0\phi(x)) + im(\partial_0\phi(x)) + m^2\phi(x) - ie(\partial_0 A_0)\phi(x) \\ -ieA_0(\partial_0\phi(x)) - emA_0\phi(x) - ieA_0(\partial_0\phi(x)) - emA_0\phi(x) + e^2A_0^2\,\phi(x) \end{bmatrix}\exp(-imx^0) \tag{133}$$

In view of the limiting assumption of $|\,i\partial_0\phi\,| << m|\,\phi\,|$ and $|\,eA_0\phi\,| << m\,|\,\phi\,|$, we can neglect the quadratic terms in $\partial_0\phi$ and $eA_0\phi$ to obtain.

$$\begin{aligned} (i\partial_0 - eA_0)^2\,\psi &= [2im\partial_0\phi(x) + m^2\phi(x) - ie(\partial_0 A_0)\phi(x) \\ &\quad - 2emA_0\phi(x)]\exp[--imx^0] \end{aligned} \tag{134}$$

Substituting this expression in eq. (121), we obtain

$$[2im\partial_0 + m^2 - ie(\partial_0 A_0) - 2emA_0]\phi(x) = [(i\nabla + e\mathbf{A})^2 + m^2]\phi(x)$$

or

$$i\partial_0\phi(x) = \frac{1}{2m}[(i\nabla + e\mathbf{A})^2] + 2emA_0 + ie(\partial_0 A_0)]\phi(x)$$

$$= \left[\frac{1}{2m}(i\nabla + e\mathbf{A})^2 + eA_0 + \frac{ie}{2m}(\partial_0 A_0)\right]\phi(x)$$

$$\approx \left[\frac{\mathbf{p}^2}{2m} - \frac{e}{2m}\mathbf{A}.\mathbf{p} + eA_0\frac{ie}{2m}(\nabla.\mathbf{A}) + \frac{ie}{2m}(\partial_0 A_0)\right]\phi(x)$$

(ignoring $\mathbf{A}^2$ term in the last step)

$$= \left[\frac{\mathbf{p}^2}{2m} - \frac{e}{2m}\mathbf{A}.\mathbf{p} + eA_0\right]\phi(x) \tag{135}$$

in the Lorentz gauge where $\nabla.\mathbf{A} + \partial_0 A_0 = 0$.

Equation (135) is the Schrödinger's equation with electromagnetic fields.

3.28 SYMMETRIES OF THE KLEIN GORDON EQUATION (FUSHCHICH, 1995)

To explore the symmetries of the Klein Gordon equation, we write it as

$$\hat{L}\psi = (\hat{P}^\mu \hat{P}_\mu - m^2)\,\psi = 0 \tag{136}$$

We define a symmetry operator (SO) as an operator $\hat{Q}$ (may be linear, curvilinear, differential, integral, etc.) that transforms solutions of eq. (136) into solutions, i.e. that satisfies

$$\hat{L}(\hat{Q}\psi) = 0 \tag{137}$$

We study, here, the SO that belong to the class of the first order linear differential operators — such operators can be identified as the Lie derivatives or generators of continuous group transformations.

In this section, we consider solutions that are defined on an open set D of the four-dimensional manifold M. An arbitrary point on this manifold is identified by the coordinates $x \equiv (x^\mu, \mu = 0, 1, 2, 3)$. The coordinate functions are assumed analytic in these real variables $x \equiv (x^\mu, \mu = 0, 1, 2, 3)$. It can be easily verified that the set of solutions $\{\psi\}$ of eq. (136) constitutes a complex vector space designated V_0 for, if $\psi_1, \psi_2 \in V_0$, then we have $\hat{L}(\alpha_1\psi_1 + \alpha_2\psi_2) = \alpha_1\hat{L}(\psi_1) + \alpha_2\hat{L}(\psi_2) = 0$ implying that $(\alpha_1\psi_1 + \alpha_2\psi_2) \in V_0$ for $\alpha_1, \alpha_2 \in \boldsymbol{C}$. Let V be a vector space of complex valued functions defined on the open set D and real analytic. Let $\hat{L} \equiv \hat{P}^\mu\hat{P}_\mu - m^2$ be a linear differential operator defined on V. Now, if $\psi \in V$ then $\hat{L}\psi = (\hat{P}^\mu\hat{P}_\mu - m^2)\psi \in V$. Further, V_0 (which is the solution space of $\hat{L}\psi = (\hat{P}^\mu\hat{P}_\mu - m^2)\psi = 0$ is a subspace of V and coincides with the null space or kernel of the operator $\hat{L}$. Let M_1 be the set of the first order differential operators defined on V. Then, a linear operator of the first order

$$\hat{Q} = A^{\mu}\hat{P}_{\mu} + B \tag{138}$$

with A^{μ}, $B \in V$ is a symmetry operator of the Klein Gordon equation in class M_1 if $\left[\hat{Q},\hat{L}\right] = \alpha_Q \hat{L}, \alpha_Q \in V$, because, given $\psi \in V_0$, we have $\hat{L}\hat{Q}\psi = (\hat{Q} - \alpha_Q)\hat{L}\psi = 0$ so that if ψ is a solution of $\hat{L}\psi = 0$ then so is $\hat{Q}\psi$. Conversely, if $\hat{L}\hat{Q}\psi = 0$ for $\psi \in V_0$, then $\left[\hat{Q},\hat{L}\right]\psi = (\hat{Q}\hat{L} - \hat{L}\hat{Q})\psi = \hat{Q}(\hat{L}\psi) = \alpha_Q \hat{L}\psi$ for some $\alpha_Q \in V$.

It is also easily seen that the SO of the Klein Gordon equation in the class M_1, i.e., of the first order differential operators form a complete Lie algebra, i.e., if $\hat{Q}_1$, $\hat{Q}_2$ are symmetry operators then SO are $a_1\hat{Q}_1 + a_2\hat{Q}_2$ and also $[\hat{Q}_1, \hat{Q}_2]$. The proof immediately follows from the fact that the operators $\hat{Q}_1$, $\hat{Q}_2$, being SOs satisfy $[\hat{Q}_1, \hat{L}] = \alpha_{Q_1}\hat{L}$ and $[\hat{Q}_2, \hat{L}] = \alpha_{Q_2}\hat{L}$ whence $\hat{Q}_3 = a_1\hat{Q}_1 + a_2\hat{Q}_2$ and $\hat{Q}_4 = [\hat{Q}_1, \hat{Q}_2]$ both belong to M_1. Further,

$$\begin{aligned}
[\hat{Q}_3, \hat{L}] &= [a_1\hat{Q}_1 + a_2\hat{Q}_2, \hat{L}] = a_1[\hat{Q}_1, \hat{L}] + a_2[\hat{Q}_2, \hat{L}] \\
&= a_1\alpha_{Q_1}\hat{L} + a_2\alpha_{Q_2}\hat{L} = (a_1\alpha_{Q_1} + a_2\alpha_{Q_2})\hat{L} = \alpha_{Q_3}\hat{L} \\
[\hat{Q}_4, \hat{L}] &= [[\hat{Q}_1, \hat{Q}_2], \hat{L}] = [(\hat{Q}_1\hat{Q}_2 - \hat{Q}_2\hat{Q}_1), \hat{L}] = [\hat{Q}_1\hat{Q}_2, \hat{L}] - [\hat{Q}_2\hat{Q}_1, \hat{L}] \\
&= \hat{Q}_1[\hat{Q}_2, \hat{L}] + [\hat{Q}_1\hat{L}]\hat{Q}_2 - \hat{Q}_2[\hat{Q}_1, \hat{L}] - [\hat{Q}_2, \hat{L}]\hat{Q}_1 \\
&= \hat{Q}_1\alpha_{Q_2}\hat{L} + \alpha_{Q_1}\hat{L}\hat{Q}_2 - \hat{Q}_2\alpha_{Q_1}\hat{L} - \alpha_{Q_2}\hat{L}\hat{Q}_1 \\
&= \hat{Q}_1\alpha_{Q_2}\hat{L} - \alpha_{Q_2}\hat{Q}_1\hat{L} + \alpha_{Q_2}\hat{Q}_1\hat{L} + \alpha_{Q_1}\hat{L}\hat{Q}_2 - \hat{Q}_2\alpha_{Q_1}\hat{L} + \\
&\quad \alpha_{Q_1}\hat{Q}_2\hat{L} - \alpha_{Q_1}\hat{Q}_2\hat{L} - \alpha_{Q_2}\hat{L}\hat{Q}_1 \\
&= [\hat{Q}_1, \alpha_{Q_2}]\hat{L} + \alpha_{Q_2}[\hat{Q}_1, \hat{L}] - \alpha_{Q_1}[\hat{Q}_2, \hat{L}] - [\hat{Q}_2, \alpha_{Q_1}]\hat{L} \\
&= [\hat{Q}_1, \alpha_{Q_2}]\hat{L} + \alpha_{Q_2}\alpha_{Q_1}\hat{L} - \alpha_{Q_1}\alpha_{Q_2}\hat{L} - [\hat{Q}_2, \alpha_{Q_1}]\hat{L} \\
&= ([\hat{Q}_1, \alpha_{Q_2}] - [\hat{Q}_2, \alpha_{Q_1}])\hat{L} = \alpha_{Q_4}\hat{L}
\end{aligned} \tag{139}$$

It follows that the symmetry operators of the Klein Gordon equation of class M_1 constitute a Lie algebra, usually called the Invariance Algebra (IA). This algebra may be finite or infinite dimensional.

Explicitly, we call the Lie algebra G the Invariant Algebra of the Klein Gordon equation if (a) the set of linear differential operators $\{\hat{Q}_A, A = 1, 2, 3, ...\}$ forms a basis of G and (b) any $\hat{Q}_A \in \{\hat{Q}_A\}$ satisfies $[\hat{Q}_A, \hat{L}] = \alpha_Q \hat{L}$.

We shall, now, establish that the Lie algebra generated by the operators $\hat{P}^{\mu} \equiv i(\partial_0, -\nabla)$, $\hat{J}_{\mu\nu} = \hat{x}^{\mu}, \hat{P}_{\nu}, - \hat{x}^{\nu}, \hat{P}_{\mu}$ $\mu\nu = 0, 1, 2, 3$ (which we recognize as the Poincare group algebra) constitutes the maximally extensive IA of the Klein Gordon equation in class M_1. To prove this identity, we write the (for the moment unknown) SO, $\hat{Q}_A$, as

$$\hat{Q} = \frac{1}{2}\ \{\hat{K}^{\mu},\ \hat{P}_{\mu}\} + \hat{C} \tag{140}$$

whence

$$\hat{Q} = \frac{1}{2}\ \{\hat{K}^{\mu}, \hat{P}_{\mu}\} + \hat{C} = \frac{1}{2}\ (\hat{K}^{\mu}\hat{P}_{\mu} + \hat{P}_{\mu}\hat{K}^{\mu}) + \hat{C}$$

$$= \hat{K}^{\mu}\hat{P}_{\mu} + \frac{1}{2}\ (\hat{P}_{\mu}\hat{K}^{\mu} - \hat{K}^{\mu}\hat{P}_{\mu}) + \hat{C} = \hat{K}^{\mu}\hat{P}_{\mu} + \frac{1}{2}\ [\hat{P}_{\mu}, \hat{K}^{\mu}] + \hat{C} \tag{141}$$

leading to the identification $\hat{K}^{\mu} = A^{\mu}$, $\hat{C} = B + \frac{1}{2}[\hat{K}^{\mu},\ \hat{P}_{\mu}]$. We also have

$$\alpha_Q m^2 = \alpha_Q \hat{P}^{\mu}\ \hat{P}_{\mu} - [\hat{Q},\ \hat{L}] = \alpha_Q \hat{P}^{\mu}\hat{P}_{\mu} - \left[\frac{1}{2}\{\hat{K}^{\nu}, \hat{P}_{\nu}\}, + \hat{C}, \hat{P}^{\mu}\hat{P}_{\mu} - m^2\right]$$

$$= \alpha_Q\ \hat{P}^{\mu}\hat{P}_{\mu} - \left[\frac{1}{2}\{\hat{K}^{\nu}, \hat{P}_{\nu}\}, \hat{P}^{\mu}\hat{P}_{\mu}\right] - [\hat{C},\ \hat{P}^{\mu}\hat{P}_{\mu}] \tag{142}$$

Also

(i) $[\hat{C},\ \hat{P}^{\mu}\hat{P}_{\mu}] = \hat{C}\,\hat{P}^{\mu}\hat{P}_{\mu} - \hat{P}^{\mu}\hat{P}_{\mu}\hat{C} = \hat{C}\,\hat{P}^{\mu}\hat{P}_{\mu} - \hat{P}^{\mu}(\hat{P}_{\mu}\hat{C}) - \hat{P}^{\mu}(\hat{C}\,\hat{P}_{\mu})$

$$= -\,(\hat{P}^{\mu}\hat{P}_{\mu}\hat{C}) - (\hat{P}_{\mu}\hat{C})\hat{P}^{\mu} - (\hat{P}^{\mu}\hat{C})\ \hat{P}_{\mu}$$

$$= -\,(\hat{P}_{\mu}\hat{C})\hat{P}^{\mu} - \hat{P}^{\mu}(\hat{P}_{\mu}\hat{C}) = -\ \{(\hat{P}_{\mu}\hat{C}), \hat{P}^{\mu}\}$$

(ii)

$$\left(\frac{1}{4}\Big\{\{\alpha_Q,\ \hat{P}^{\mu}\}, \hat{P}_{\mu}\Big\} - \frac{1}{2}\{(\hat{P}^{\mu}\alpha_Q), \hat{P}_{\mu}\} + \frac{1}{4}(\hat{P}_{\mu}\hat{P}_{\mu}\alpha_Q)\right)\psi$$

$$= \begin{pmatrix} \frac{1}{4}(\{a_Q, \hat{P}^{\mu}\}\hat{P}_{\mu} + \hat{P}_{\mu}\{a_Q, \hat{P}^{\mu}\}) \\ -\frac{1}{2}\Big[(\hat{P}^{\mu}\alpha_Q)\hat{P}_{\mu} + \hat{P}_{\mu}(\hat{P}^{\mu}\alpha_Q)\Big] + \frac{1}{4}(\hat{P}^{\mu}\hat{P}_{\mu}\alpha_Q) \end{pmatrix}\psi$$

$$= \frac{1}{4}(\alpha_Q\hat{P}^{\mu}\hat{P}_{\mu}\,\psi + \hat{P}^{\mu}(\alpha_Q\hat{P}_{\mu}\ \psi) + \hat{P}_{\mu}(\alpha_Q\hat{P}^{\mu}\ \psi) + \hat{P}_{\mu}\hat{P}^{\mu}\ (\alpha_Q\psi))$$

$$-\frac{1}{2}\ ((\hat{P}^{\mu}\alpha_Q)\ \hat{P}_{\mu}\,\psi + \hat{P}_{\mu}\ [(\hat{P}^{\mu}\alpha_Q)\psi]) + \frac{1}{4}\ (\hat{P}^{\mu}\hat{P}_{\mu}\,\alpha_Q)\psi$$

$$= \frac{1}{4}\begin{pmatrix} \alpha_Q\hat{P}^{\mu}\hat{P}_{\mu}\psi + (\hat{P}^{\mu}\alpha_Q)(\hat{P}_{\mu}\psi) + \alpha_Q(\hat{P}^{\mu}\hat{P}_{\mu}\psi) + \\ (\hat{P}_{\mu}\,\alpha_Q)(\hat{P}^{\mu}\psi) + \alpha_Q(\hat{P}_{\mu}\hat{P}^{\mu}\psi) + (\hat{P}_{\mu}\hat{P}^{\mu}\,\alpha_Q)\psi \\ +(\hat{P}^{\mu}\alpha_Q)(\hat{P}_{\mu}\ \psi) + (\hat{P}_{\mu}\alpha_Q)(\hat{P}^{\mu}\psi) + \alpha_Q(\hat{P}_{\mu}\hat{P}^{\mu}\psi) \end{pmatrix}$$

$$-\frac{1}{2}\ ((\hat{P}^{\mu}\ \alpha_Q)\ (\hat{P}_{\mu}\psi) + (\hat{P}_{\mu}\hat{P}^{\mu}\alpha_Q)\psi + (\hat{P}^{\mu}\alpha_Q)\ (\hat{P}_{\mu}\psi)) + \frac{1}{4}(\hat{P}^{\mu}\hat{P}_{\mu}\alpha_Q)\psi$$

$$= \alpha_Q\hat{P}^{\mu}\hat{P}_{\mu}\,\psi$$

(iii) $$\left[\frac{1}{2}\{\hat{K}^{\mu},\hat{P}_{\mu}\},\hat{P}_{\nu}\hat{P}^{\nu}\right] = \frac{1}{2}\hat{P}_{\nu}\,[\{\hat{K}^{\mu},\,\hat{P}_{\mu}\},\hat{P}^{\nu}] + \frac{1}{2}[\{\hat{K}^{\mu},\,\hat{P}_{\mu}\},\hat{P}_{\nu}]\hat{P}^{\nu}$$

$$= \frac{1}{2}\hat{P}_{\nu}\{\hat{K}^{\mu},\,\hat{P}_{\mu}\}\hat{P}^{\nu} - \frac{1}{2}\hat{P}_{\nu}\hat{P}^{\nu}\{\hat{K}^{\mu},\,\hat{P}_{\mu}\} + \frac{1}{2}\,\{\hat{K}^{\mu},\,\hat{P}_{\mu}\}\hat{P}_{\nu}\hat{P}^{\nu} - \frac{1}{2}\,\hat{P}_{\nu}\,\{\hat{K}^{\mu},\,\hat{P}_{\mu}\}\hat{P}^{\nu}$$

$$= -\frac{1}{2}\hat{P}_{\nu}\hat{P}^{\nu}\{\hat{K}^{\mu},\,\hat{P}_{\mu}\} + \frac{1}{2}\,\{\hat{K}^{\mu},\,\hat{P}_{\mu}\}\,\hat{P}_{\nu}\hat{P}^{\nu} = -\frac{1}{2}\,A + \frac{1}{2}\,B$$

$$B\psi = (\hat{K}^{\mu}\hat{P}_{\mu} + \hat{P}_{\mu}\hat{K}^{\mu})\hat{P}_{\nu}\hat{P}^{\nu}\,\psi = 2\hat{K}^{\mu}\hat{P}_{\mu}\hat{P}_{\nu}\hat{P}^{\nu}\,\psi + (\hat{P}_{\mu}\hat{K}^{\mu})\,(\hat{P}_{\nu}\hat{K}^{\nu}\psi)$$

$$A\psi = \hat{P}_{\nu}\hat{P}^{\nu}\,(\hat{K}^{\mu}\hat{P}_{\mu} + \hat{P}_{\mu}\hat{K}^{\mu})\,\psi = \hat{P}_{\nu}\hat{P}^{\nu}\,[\hat{K}^{\mu}(\hat{P}_{\mu}\psi)] + \hat{P}_{\nu}\hat{P}^{\nu}\hat{P}_{\mu}\hat{K}^{\mu}\psi$$

$$= \hat{P}_{\nu}\,[(\hat{P}^{\nu}\hat{K}^{\mu})\,(\hat{P}_{\mu}\psi) + \hat{K}^{\mu}\,(\hat{P}^{\nu}\hat{P}_{\nu}\,\psi)] + \hat{P}_{\nu}\hat{P}^{\nu}\,[(\hat{P}_{\mu}\hat{K}^{\mu})\,\psi + \hat{K}^{\mu}(\hat{P}_{\mu}\psi)]$$

$$= (\hat{P}_{\nu}\,\hat{P}^{\nu}\hat{K}^{\mu})\,(\hat{P}_{\mu}\psi) + (\hat{P}^{\nu}\hat{K}^{\mu})\,(\hat{P}_{\nu}\,\hat{P}_{\mu}\,\psi) + (\hat{P}_{\nu}\,\hat{K}^{\mu})\,(\hat{P}^{\nu}\hat{P}_{\mu}\,\psi) + \hat{K}^{\mu}(\hat{P}_{\nu}\,\hat{P}^{\mu}\hat{P}_{\mu}\,\psi)$$

$$+ \hat{P}_{\nu}\,[(\hat{P}^{\nu}\,\hat{P}_{\mu}\hat{K}^{\mu})\psi + (\hat{P}_{\mu}\,\hat{K}^{\mu})\,(\hat{P}^{\nu}\,\psi) + (\hat{P}^{\nu}\hat{K}^{\mu})\,(\hat{P}_{\mu}\,\psi) + \hat{K}^{\mu}\,(\hat{P}^{\nu}\,\hat{P}_{\mu}\,\psi)]$$

$$= (\hat{P}_{\nu}\hat{P}^{\nu}\hat{K}^{\mu})\,(\hat{P}_{\mu}\psi) + (\hat{P}^{\nu}\hat{K}^{\mu})\,(\hat{P}_{\nu}\,\hat{P}_{\mu}\psi) + (\hat{P}_{\nu}\hat{K}^{\mu})\,(\hat{P}^{\nu}\hat{P}_{\mu}\psi) + \hat{K}^{\mu}\,(\hat{P}^{\nu}\,\hat{P}_{\mu}\psi)$$

$$+ (\hat{P}_{\nu}\hat{P}^{\nu}\hat{P}_{\mu}\hat{K}^{\mu}).\,\psi + (\hat{P}^{\nu}\hat{P}_{\mu}\hat{K}^{\mu})\,(\hat{P}_{\nu}\psi) + (\hat{P}_{\nu}\,\hat{P}_{\mu}\hat{K}^{\mu})\,(\hat{P}^{\nu}\psi) + (\hat{P}_{\mu}\hat{K}^{\mu})\,(\hat{P}_{\nu}\hat{P}^{\nu}\psi)$$

$$+ (\hat{P}_{\nu}\hat{P}^{\nu}\hat{K}^{\mu})\,(\hat{P}_{\mu}\,\psi) + (\hat{P}^{\nu}\hat{K}^{\mu})\,(\hat{P}_{\nu}\hat{P}_{\mu}\,\psi) + (\hat{P}_{\nu}\hat{P}^{\mu})\,(\hat{P}^{\nu}\hat{P}_{\mu}\,\psi)\;\hat{K}^{\mu}\,(\hat{P}_{\nu}\,\hat{P}^{\nu}\hat{P}_{\mu}\psi)$$

$$= 2(\hat{P}_{\nu}\hat{P}^{\nu}\hat{K}^{\mu})\,(\hat{P}_{\mu}\psi) + 2\,(\hat{P}^{\nu}\hat{K}^{\mu})\,(\hat{P}_{\nu}\hat{P}_{\mu}\psi) + 2\,(\hat{P}_{\nu}\hat{K}^{\mu})\,(\hat{P}^{\nu}\hat{P}_{\mu}\psi) + 2\hat{K}^{\mu}(\hat{P}_{\nu}\,\hat{P}^{\nu}\hat{P}_{\mu}\psi)$$

$$+ 2\,(\hat{P}^{\nu}\hat{P}_{\mu}\hat{K}^{\mu})\,(\hat{P}_{\nu}\psi) + (\hat{P}_{\nu}\hat{P}^{\nu}\hat{P}_{\mu}\hat{K}^{\mu})\,\psi + (\hat{P}_{\mu}\hat{K}^{\mu})\,(\hat{P}_{\nu}\,\hat{P}^{\nu}\psi)$$

so that $$\left[\frac{1}{2}\{\hat{K}^{\mu},\hat{P}_{\mu}\},\hat{P}_{\nu}\hat{P}^{\nu}\right]\psi = -(\hat{P}_{\nu}\hat{P}^{\nu}\hat{K}^{\mu})\,(\hat{P}_{\mu}\psi) - (\hat{P}^{\nu}\hat{K}^{\mu})\,(\hat{P}_{\nu}\,\hat{P}_{\mu}\psi)$$

$$- (\hat{P}_{\nu}\hat{K}^{\mu})\,(\hat{P}^{\nu}\hat{P}_{\mu}\psi) - (\hat{P}^{\nu}\hat{P}_{\mu}\hat{K}^{\mu})\,(\hat{P}_{\nu}\psi) - \frac{1}{2}\,(\hat{P}_{\nu}\hat{P}^{\nu}\,\hat{P}_{\mu}\,\hat{K}^{\mu})\psi \tag{143}$$

(iv) $$\{\{(\hat{P}^{\nu}\hat{K}^{\mu}),\,\hat{P}_{\mu}\},\,\hat{P}_{\nu}\}\psi = \{(\hat{P}^{\nu}\hat{K}^{\mu}),\,\hat{P}_{\mu}\}\hat{P}_{\nu}\psi + \hat{P}_{\nu}\{(\hat{P}^{\nu}\hat{K}^{\mu}),\,\hat{P}_{\mu}\}\psi$$

$$= [(\hat{P}^{\nu}\hat{K}^{\mu})\hat{P}_{\mu} + \hat{P}_{\mu}(\hat{P}^{\nu}\hat{K}^{\mu})]\,(\hat{P}_{\nu}\psi) + \hat{P}_{\nu}\,[(\hat{P}^{\nu}\hat{K}^{\mu})\hat{P}_{\mu} + \hat{P}_{\mu}\,(\hat{P}^{\nu}\hat{K}^{\mu})]\psi$$

$$= (\hat{P}^{\nu}\hat{K}^{\mu})\,(\hat{P}_{\mu}\hat{P}_{\nu}\psi) + (\hat{P}_{\mu}\hat{P}^{\nu}\hat{K}^{\mu})\,(\hat{P}_{\nu}\psi) + (\hat{P}^{\nu}\hat{K}^{\mu})\,(\hat{P}_{\mu}\hat{P}_{\nu}\psi)$$

$$+ \hat{P}_{\nu}\,[(\hat{P}^{\nu}\hat{K}^{\mu})(\hat{P}_{\mu}\psi) + (\hat{P}_{\mu}\hat{P}^{\nu}\hat{K}^{\mu})\,\psi + (\hat{P}^{\nu}\hat{K}^{\mu})\,(\hat{P}_{\mu}\psi)]$$

$$= (\hat{P}^{\nu}\hat{K}^{\mu})\,(\hat{P}_{\mu}\hat{P}_{\nu}\psi) + (\hat{P}_{\mu}\hat{P}^{\nu}\hat{K}^{\mu})\,(\hat{P}_{\nu}\psi) + (\hat{P}^{\nu}\hat{K}^{\mu})\,(\hat{P}_{\mu}\hat{P}_{\nu}\psi)$$

$$+ (\hat{P}_{\nu}\hat{P}^{\nu}\hat{K}^{\mu})\,(\hat{P}_{\mu}\psi) + (\hat{P}^{\nu}\hat{K}^{\mu})\,(\hat{P}_{\nu}\hat{P}_{\mu}\psi) + (\hat{P}_{\nu}\hat{P}_{\mu}\hat{P}^{\nu}\hat{K}^{\mu})\psi$$

$$+ (\hat{P}_{\mu}\hat{P}^{\nu}\hat{K}^{\mu})\,(\hat{P}_{\nu}\psi) + (\hat{P}_{\nu}\hat{P}^{\nu}\hat{K}^{\mu})\,(\hat{P}_{\mu}\psi) + (\hat{P}^{\nu}\hat{K}^{\mu})\,(\hat{P}_{\nu}\hat{P}_{\mu}\psi)$$

$$= 4\,(\hat{P}^{\nu}\hat{K}^{\mu})\,(\hat{P}_{\mu}\hat{P}_{\nu}\psi) + 2\,(\hat{P}_{\mu}\hat{P}^{\nu}\hat{K}^{\mu})\,(\hat{P}_{\nu}\psi)$$

$$+ 2\,(\hat{P}_{\nu}\hat{P}^{\nu}\hat{K}^{\mu})\,(\hat{P}_{\mu}\psi) + (\hat{P}_{\nu}\hat{P}_{\mu}\hat{P}^{\nu}\hat{K}^{\mu})\psi \tag{144}$$

(v) From (iii) & (iv), we have

$$\left[\frac{1}{2}\{\hat{K}^{\mu}, \hat{P}_{\mu}\}, \hat{P}_{\nu}\hat{P}^{\nu}\right] = -\frac{1}{2}\{\{(\hat{P}^{\nu}\hat{K}^{\mu}),\ \hat{P}_{\mu}\},\ \hat{P}_{\nu}\} \tag{145}$$

Using the results of (i)-(v) in eq. (142), we obtain

$$\frac{1}{2}\ \{\{(\hat{P}^{\nu}\hat{K}^{\mu}),\ \hat{P}_{\mu}\},\ \hat{P}_{\nu}\} + \{(\hat{P}_{\mu}\hat{C}),\ \hat{P}^{\mu}\} = -\frac{1}{4}\{\{\alpha_Q,\ \hat{P}^{\mu}\},\ \hat{P}_{\mu}\}$$

$$+\frac{1}{2}\ \{(\hat{P}^{\mu}\alpha_Q),\ \hat{P}_{\mu}\} - \frac{1}{4}\ (\hat{P}^{\mu}\hat{P}_{\mu}\alpha_Q) + \alpha_Q m^2 \tag{146}$$

which leads us to the following identification:

$$\hat{P}^{\nu}\hat{K}^{\mu} + \hat{P}^{\mu}\hat{K}^{\nu} = -\frac{1}{2}\ g^{\mu\nu}\alpha_Q \tag{147}$$

$$\hat{P}^{\mu}\hat{C} = \hat{P}^{\mu}\alpha_Q \tag{148}$$

$$m^2\alpha_Q = 0 \tag{149}$$

For non-zero m, we have $\alpha_Q = 0$ whence $\hat{P}^{\nu}\hat{K}^{\mu} + \hat{P}^{\mu}\hat{K}^{\nu} = 0$ and $\hat{P}^{\mu}\hat{C} = 0$, the first of which can be easily integrated to yield

$$\hat{K}^{\mu} = a^{(\mu\sigma)}\ x_{\sigma} + b^{\mu} \tag{150}$$

with $a^{(\mu\sigma)} = -\,a^{(\sigma\mu)}$ and b^{μ} are arbitrary numbers. Further, $\hat{P}^{\mu}\hat{C} = 0$ implies that $\hat{C}$ is independent of the coordinates. Using eq. (140), we obtain the general expression of the symmetry operator as

$$\hat{Q} = a^{(\mu\sigma)}\ \hat{x}^{\mu}\ \hat{P}_{\sigma} + b^{\mu}\hat{P}_{\mu} + C \tag{151}$$

which is a linear combination of the generators of the Poincare group, viz.

$$\hat{P}_0 = i\ \frac{\partial}{\partial x_0},\ \hat{P}_j = -\ \frac{\partial}{\partial x_j},\ \hat{J}_{\mu\nu} = \hat{x}_{\mu}\ \hat{P}_{\nu} - \hat{x}_{\nu}\ \hat{P}_{\mu} \tag{152}$$

and the unit operator. It can easily be verified that $\hat{P}_0 = i\dfrac{\partial}{\partial x_0}$, $\hat{P}_j = -\dfrac{\partial}{\partial x_j}$, $\hat{J}_{\mu\nu} = \hat{x}_{\mu}\ \hat{P}_{\nu} - \hat{x}_{\nu}\ \hat{P}_{\mu}$ satisfy the commutation relations of the generators of the Poincare algebra, viz.,

$$[\hat{P}_{\mu},\ \hat{P}_{\nu}] = 0\ ,\ [\hat{P}_{\mu},\ \hat{J}_{\nu\lambda}] = i\ (g_{\mu\nu}\hat{P}_{\lambda} - g_{\mu\lambda}\hat{P}_{\nu})$$

and

$$[\ \hat{J}_{\mu\nu},\ \hat{J}_{\lambda\sigma}] = i(g_{\mu\sigma}\ \hat{J}_{\nu\lambda} + g_{\nu\lambda}\hat{J}_{\mu\sigma} - g_{\mu\lambda}\ \hat{J}_{\nu\sigma} - g_{\nu\sigma}\ \hat{J}_{\mu\lambda}).$$

To investigate further, the Invariance Algebra of the Klein Gordon equation, we make use of two results from the theory of Lie groups insofar as they relate to symmetries of differential equations, viz. (i) if a differential equation possesses an Invariance Algebra of class M_1, then it is locally invariant under the continuous transformation group acting on the dependent and independent variables and (ii) any basis element of the Invariance Algebra corresponds to a one parameter transformation group.

Using (ii), we can write

$$x \to x' = g_\theta (x),\ \psi (x) \to \psi'(x') = \hat{T}_{g_\theta} (\psi(x)) = \hat{D}(\theta, x)\psi(x) \tag{153}$$

where θ is a transformation parameter, g_θ and $\hat{D}$ are analytic functions of θ and x and $\hat{T}_{g_\theta}$ are linear operators defined on the space V. To obtain explicit expressions for g_θ and $\hat{D}$, we integrate the Lie equations:

$$\frac{dx'^\mu}{d\theta} = K^\mu (x') = A^\mu(x'),\ x'^\mu|_{\theta = 0} = x^\mu \tag{154}$$

$$\frac{d\psi'}{d\theta} = iB (x')\psi',\ \psi'|_{\theta = 0} = \psi \tag{155}$$

Now, comparing eqs. (138) and (152), we find that in the case of any of the generators $\hat{P}_0 = i\dfrac{\partial}{\partial x_0}$, $\hat{P}_j = -\dfrac{\partial}{\partial x_j}$, $\hat{J}_{\mu\nu} = \hat{x}_\mu \hat{P}_\mu - \hat{x}_\nu \hat{P}_\mu$, $B = 0$ always, so that the solution of eq. (155) takes the form

$$\psi' (x') = \psi(x),\ \psi'(x) = \psi(g_\theta^{-1} (x)) \tag{156}$$

Again, comparing eqs. (138) and (152), we obtain $A^\mu = K^\mu = 1$ if $Q = P_\mu$ and $K^\mu = A^\mu = x_\sigma g^\mu_\lambda - x_\lambda g^\mu_\sigma$ if $Q = J_{\mu\sigma}$. Writing $\theta = b^\mu$ for the case $Q = P_\mu$, we have $\dfrac{dx'^\mu}{db^\mu} = 1$ with $x'^\mu|_{\theta = 0} = x^\mu$ which has the solution

$$x'^\mu = x^\mu + b^\mu \tag{157}$$

Proceeding similarly, we have, in the case of $Q = J_{\mu\sigma}$

$$\begin{aligned}
x'^i &= x^i \cos \theta_{ij} + x^j \sin \theta_{ij}, \\
x'^j &= x^j \cos \theta_{ij} - x^i \sin \theta_{ij}, \\
x'^\mu &= x^\mu,\ \mu \neq i, j,\ i, j \neq 0, \\
x'^i &= x^i \cosh \theta_{0i} - x^0 \sinh \theta_{0i}, \\
x'^0 &= x^0 \cosh \theta_{0i} - x^i \sinh \theta_{0i}, \\
x'^\mu &= x^\mu,\ \mu \neq 0,\ i
\end{aligned} \tag{158}$$

where θ_{ij}, θ_{0i} are transformation parameters. It is easily seen that the set of transformations (157) & (158) constitutes the Poincare group. These transformations conserve the space-time interval, i.e., $S (x^{(1)} - x^{(2)}) = S(x'^{(1)} - x'^{(2)})$ where $S(x) = x_0^2 - x_1^2 - x_2^2 - x_3^2$.

As we know, the Poincare group is the group of inhomogeneous linear transformations of coordinates $x_\mu \rightarrow x'_\mu = a_{\mu\sigma}x^\sigma + b_\mu$ (where $a_{\mu\sigma}$, b_μ are real parameters satisfying $a_{\mu\sigma}\, a^{\lambda\mu} = g_\sigma^\lambda$), that conserve the interval $S(x)$. The subgroup of the full Poincare group determined by $\det\|a_{\mu\sigma}\| = +1$ and $a_{00} \geq 1$ constitutes a Lie group (sometimes called the proper Poincare group), although the full Poincare group is not a Lie group because, in this case, the determinant of the transformation matrix $(a_{\mu\sigma})$ is not a continuous function of the group parameters. A general set of solutions of the proper Poincare group conditions $\det\|a_{\mu\sigma}\| = +1$ and $a_{00} \geq 1$ is

$$a_{00} = \cos\gamma\cos^2\varphi + \cosh\beta\sin^2\varphi - \frac{\lambda^2}{R^2}(\cos\gamma - \cosh\beta) \tag{159}$$

$$a_{0j} = \frac{1}{R}\begin{bmatrix}\sin\gamma\cos\beta(\lambda_j\cos\varphi - \theta_j\cos\varphi) \\ -\cos\gamma\sinh\beta(\lambda_j\sin\varphi + \theta_j\cos\varphi)\end{bmatrix} + \frac{1}{R^2}\varepsilon_{jkl}\lambda_k\theta_l(\cos\gamma - \cosh\beta) \tag{160}$$

$$a_{j0} = a_{0j} - \frac{2}{R^2}\varepsilon_{jkl}\lambda_k\theta_l(\cos\gamma - \cosh\beta)$$

$$a_{jk} = \frac{1}{R}\varepsilon_{ijk}\theta_i(\sin\gamma\cosh\beta\cos\varphi - \cos\gamma\sinh\beta\sin\varphi) -$$

$$(\lambda_j\lambda_k + \theta_j\theta_k - \theta^2\delta_{jk})(\cos\gamma - \cosh\beta) + \delta_{jk}(\cos\gamma\cos^2\varphi + \cosh\beta\sin^2\varphi) \tag{161}$$

where $\theta = (\theta_1^2 + \theta_2^2 + \theta_3^2)^{1/2}$, $\lambda = (\lambda_1^2 + \lambda_2^2 + \lambda_3^2)^{1/2}$, $\gamma = R\cos\varphi$, $\beta = R\sin\varphi$, $R = [(\theta^2 - \lambda^2)^2 + 4(\lambda_i\theta_i)^2]^{1/2}$ and $\varphi = \arctan\dfrac{\lambda_i\theta_i}{\theta^2 - \lambda^2}$. This explicitly establishes the dependence of the Poincare group on ten real parameters, viz. b^μ,. λ_i, θ_i. Explicit expressions for the basis elements can also be obtained from $\hat{P}_\mu = i\left.\dfrac{\partial A(a,b)}{\partial b_\mu}\right|_{b_\mu=\lambda_i=\theta_i=0}$, $\hat{J}_{lm} = i\varepsilon_{lmn}\left.\dfrac{\partial A(a,b)}{\partial\theta_n}\right|_{b_\mu=\lambda_i=\theta_i=0}$, $\hat{J}_{0k} = \hat{J}_{0k}$ $\left.\dfrac{\partial A(a,b)}{\partial\lambda_k}\right|_{b_\mu=\lambda_i=\theta_i=0}$ where $A(a, b)$ is the matrix $A(a,b) = \begin{pmatrix}(a_{\mu\nu}) & (b_\mu) \\ \mathbf{0} & 1\end{pmatrix}$ with $\mu, \nu = 0, 1, 2, 3$ and $\mathbf{0} = (0, 0, 0, 0)$. The matrix $A(a, b)$ can be reconstructed from the above basis elements by the usual process of exponentiation $A(a,b) = \exp\left(\dfrac{i}{2}\theta_{\mu\sigma}\hat{J}^{\mu\sigma}\right)\exp(i\hat{P}_\mu b^\mu)$.

To explore the higher order symmetries of the Klein Gordon equation, we write it in (1 + 3) dimensional Minkowski space as

$$\hat{L}\varphi = \left(g^{\mu\nu}\frac{\partial}{\partial x_\mu}\frac{\partial}{\partial x_\nu} - m^2\right)\varphi = 0 \tag{162}$$

where m is real and $g^{\mu\nu} = \begin{cases}1 \text{ for } \mu = \nu = 1 \\ -1 \text{ for } \mu = \nu = 2,3,4 \\ 0 \text{ for } \mu \neq \nu, \mu, \nu = 1,2,3,4\end{cases}$ and $\varphi \equiv \varphi(x^\mu, \mu = 1, 2, 3, 4)$ is a function of the space-time coordinates. As in the case of first order symmetry operators, we consider solutions of eq. (162) defined on an open set D of the four-dimensional manifold $\mathbf{R}^4$ consisting of points with coordinates (x^μ, $\mu = 1, 2, 3, 4$). The solutions of eq. (162) are assumed to be analytic with respect to the real space-time coordinates (x^μ, $\mu = 1, 2, 3, 4$). We have already proved that the solution set of all such solutions forms a complex vector space V_0.

We define V as a vector space of all complex valued functions defined on D that are real analytic. The linear operator $\hat{L} \equiv \left(g^{\mu\nu}\dfrac{\partial}{\partial x_\mu}\dfrac{\partial}{\partial x_\nu} - m^2\right)$ is defined on the space V so that $\hat{L}\varphi \in V \;\forall \varphi \in V$. V_0 is, obviously, a subspace of the vector space V that coincides with the kernel of the operator $\hat{L}$.

Let M_n be the set of differential operators of order n defined on the vector space V.

We define a symmetry operator of order n or a symmetry operator in the class M_n as a linear differential operator of order n that satisfies

$$\left[\hat{Q}, \hat{L}\right] = \hat{\alpha}_Q\, \hat{L} \tag{163}$$

with $\hat{\alpha}_Q \in M_{n-1}$. Now, an arbitrary operator in class M_n can be expressed as the sum

$$\hat{Q} = \sum_{j=0}^{n} \hat{Q}_j \tag{164}$$

with

$$\hat{Q}_j = \Xi^{\lambda_1\lambda_2\lambda_3\ldots\lambda_j}\,\frac{\partial^j}{\partial x_{\lambda_1}\partial x_{\lambda_2}\partial x_{\lambda_3}\ldots\partial x_{\lambda_j}} \tag{165}$$

and $\Xi^{\lambda_1\lambda_2\lambda_3\ldots\lambda_j} \in V$, $\lambda_i = 1, 2, 3, 4$. As in the case of first order symmetry operators, we express the constituents $\hat{Q}_j$ of the symmetry operator $\hat{Q}$ in terms of anticommutators as

$$\hat{Q}_j = \left\{\ldots\left\{\left\{\hat{K}^{(\lambda_1\lambda_2\lambda_j)}, \frac{\partial}{\partial x_{\lambda_1}}\right\}, \frac{\partial}{\partial x_{\lambda_2}}\right\}, \ldots, \frac{\partial}{\partial x_{\lambda_j}}\right\} \tag{166}$$

Similarly, we express $\hat{\alpha}_Q \in M_{n-1}$ in terms of anticommutators as

$$\hat{\alpha}_Q = \sum_{j=0}^{n-1} \hat{\alpha}_j \tag{167}$$

with

$$\hat{\alpha}_j = \left\{\ldots\left\{\left\{\alpha^{\lambda_1\lambda_2\ldots\lambda_j}, \frac{\partial}{\partial x_{\lambda_1}}\right\}, \frac{\partial}{\partial x_{\lambda_2}}\right\}, \ldots, \frac{\partial}{\partial x_{\lambda_j}}\right\} \tag{168}$$

Using eq. (168), we can write

$$\alpha_Q \hat{L} \equiv \frac{1}{4}\left\{\left\{\hat{\alpha}_Q, \frac{\partial}{\partial x_\mu}\right\}, \frac{\partial}{\partial x^\mu}\right\} - \frac{1}{2}\left\{\frac{\partial \alpha_Q}{\partial x_\mu}, \frac{\partial}{\partial x_\mu}\right\} + \frac{1}{4}\left(\frac{\partial^2 \alpha_Q}{\partial x_\mu \partial x^\mu}\right) - \alpha_Q m^2 \tag{169}$$

where the $\hat{K}^{\lambda_1\lambda_2\ldots\lambda_j}$ and $\hat{\alpha}^{\lambda_1\lambda_2\ldots\lambda_j}$ are symmetric tensors of rank j. We can also write

$$\left[\hat{Q}, \hat{L}\right] \equiv -\left\{\frac{\partial \hat{Q}}{\partial x_\mu}, \frac{\partial}{\partial x^\mu}\right\} \tag{170}$$

We can reduce the operator eq. (153) by substituting from eqs. (164)-(170) into a system of equations for the coefficients $\hat{K}^{(\lambda_1\lambda_2\ldots\lambda_j)}$and $\alpha^{\lambda_1\lambda_2\ldots\lambda_j}$ to obtain

$$\partial^{(\lambda_{j+1}}\hat{K}^{\lambda_1\lambda_2\ldots\lambda_j)} = 0 \tag{171}$$

$$\alpha^{(\lambda_1\lambda_2\ldots\lambda_j)} \equiv 0 \tag{172}$$

with $\partial^{\lambda_{j+1}} = \dfrac{\partial}{\partial x_{\lambda_{j+1}}}$. The parentheses indicate that the tensors are symmetric with respect to the indices enclosed in the parentheses so that

$$\frac{1}{j!}\partial^{(\lambda_{j+1}}\hat{K}^{\lambda_1\,\lambda_2\ldots\lambda_j)} = \partial^{\lambda_{j+1}}\hat{K}^{\lambda_1\lambda_2\ldots\lambda_j} + \partial^{\lambda_1}\hat{K}^{\lambda_{j+1}\lambda_2\ldots\lambda_j}$$
$$+ \partial^{\lambda_2}\hat{K}^{\lambda_1\,\lambda_{j+1}\,\ldots\,\lambda_j} + \ldots + \partial^{\lambda_j}\hat{K}^{\lambda_1\,\lambda_2\ldots\lambda_{j+1}} \tag{173}$$

The term $\partial^{(\lambda_{j+1}}\hat{K}^{\lambda_1\lambda_2\ldots\lambda_j)}$ is a symmetric tensor of rank $j + 1$ in the four-dimensional space. We write it as $\partial^{(\lambda_{j+1}}\hat{K}^{\lambda_1\lambda_2\ldots\lambda_j)} \equiv \mathbf{K}^{\lambda_1\,\lambda_2\lambda_j\ldots\,\lambda_{j+1})}$. The $\hat{K}^{\lambda_1\lambda_2\,\ldots\,\lambda_j}$ are symmetric tensors of rank j in the same space. We need to determine the components of these tensors. The system of equations (171) constitutes a set of $\binom{j+4}{j+1}$ equations for $\binom{j+3}{j}$ unknowns and is, therefore, an overdetermined system of partial differential equations.

Using the standard approach to solving such systems of overdetermined partial differential equations, we differentiate each term of $\partial^{(\lambda_{j+1}}\hat{K}^{\lambda_1\lambda_2\ldots\lambda_j)}$, k times by x_{β_i} $(i = 1, 2, 3,\ldots,k)$. As a result, for each value of k, we have a system of linear algebraic homogeneous equations, e.g.,

$$\partial^{\beta_1}\partial^{\beta_2}\ldots\partial^{\beta_k}\partial^{\lambda_{j+1}}\hat{K}^{\lambda_1\lambda_2\ldots\lambda_j} \equiv \hat{K}^{(\lambda_1\lambda_2\ldots\lambda_j,\,\lambda_{j+1})\beta_1\,\beta_2\ldots\beta_k} = 0 \tag{174}$$

The left-hand side of eq. (174) is a tensor of rank $j + k + 1$ that is symmetric with respect to the $j + 1$ indices $\lambda_1, \lambda_2, \lambda_3,\ldots, \lambda_{j+1}$ and also with respect to the k indices $\beta_1, \beta_2,\ldots, \beta_k$. Of these, it is reiterated that the unknown functions are components of the symmetric tensor of rank j that is symmetric with respect to the indices $\lambda_1, \lambda_2, \lambda_3,\ldots, \lambda_j$. It follows that in this new system of equations given by eq. (174) the number of equations are given by $n_{eq} = \binom{j+4}{j+1} \times \binom{k+3}{k}$ while the number of unknowns is $n_{un} = \binom{j+3}{j} \times \binom{k+4}{k+1}$. Now, for $k = j$, the number of equations equals the number of unknowns whereas for $k < j$, the number of equations is less than the number of unknowns. As mentioned above, for $k = j$, the number of equations equals the number of unknowns and the system of equations (174) has the solution

$$\hat{K}^{\lambda_1\lambda_2...\lambda_j,\lambda_{j+1}\beta_1\beta_2...\beta_k} = 0 \tag{175}$$

which implies that the coefficients of $\hat{K}^{\lambda_1\lambda_2...\lambda_j}$ are polynomials on (x^μ, μ = 1, 2, 3, 4) of order j. The number of arbitrary parameters in the solution will be

$$n_j = \sum_{k=0}^{4}(n_{eq}^k - n_{un}^k) = \frac{1}{4}\binom{j+4}{3}\binom{j+4}{3} \tag{176}$$

For the case $k < j$, eq. (171) has n_j linearly independent solutions forming a complete system. We obtain below the explicit form of the solutions for this case. For the purpose, we note that if $\hat{K}^{\lambda_1\lambda_2...\lambda_j,\lambda_{j0}}$ is a solution of eq. (171) for $j = j_0$ and $\hat{K}^{\lambda}$ is a solution of the said equation for $j = 1$, then $\hat{K}^{\lambda_1\lambda_2...\lambda_j,\lambda_{j0+1}} = \hat{K}^{(\lambda_1\lambda_2...\lambda_j\,\lambda_{j0},}\hat{K}^{\lambda_{j0+1})}$ is a solution of eq. (171) for $j = j_0 + 1$. Using this algorithm iteratively, we can obtain solution for any desired value of j. We have the following recursive relation:

$$\hat{K}^{\lambda_1\lambda_2...\lambda_j} = g^{(\lambda_{j-1},\lambda_j}\hat{K}^{\lambda_1\lambda_2...\lambda_{j-2})} + k^{\lambda_1\lambda_2...\lambda_j} \tag{177}$$

where $\hat{K}^{\lambda_1\lambda_2...\lambda_{j-2}}$ is the solution of eq. (171) for $j \to j-2$ involving n_{j-2} arbitrary parameters and $k^{\lambda_1\lambda_2...\lambda_j}$ is a solution of eq. (171) with $n_j - n_{j-2}$ arbitrary parameters. The number of independent solutions are given by using eq. (176) as $n = n_j - n_{j-2} = \frac{1}{4!}(j+1)(j+2)(2j+3)(j^2+3j+4)$. The solutions are identified by using a set of three integers $\{c_1, c_2, c_3\}$ that satisfy $0 \le c_1 \le 2\left\{\frac{j}{2}\right\}$, $0 \le c_2 \le j - i\left\{\frac{c_1+1}{2}\right\}$, $0 \le c_3 \le j - 2\left\{\frac{c_1+1}{2}\right\} - 2c_2$. In addition, for each value of c, the solutions include the number of arbitrary parameters given by:

$$n_c = \begin{cases} (c_1 + 2c_3 + 1)^2 & c_1 = c_2 + 2c_3 \\ 2(c_2 + 2c_3 + 1)(2c_1 - c_2 - 2c_3 + 1) & c_1 \ne c_2 + 2c_3 \end{cases} \tag{178}$$

The explicit expression for $k^{\lambda_1\lambda_2...\lambda_j}$ is given by:

$$k_c^{\lambda_1\lambda_2...\lambda_j} = \sum_\mu M_\mu \kappa^{B_\mu,(A_\mu,D_c}\left(\prod_{i=A_\mu+D_c+1}^{A_\mu+D_c+L_\mu} x^{\lambda_i}\right)^* \times \left(\prod_{k=\left\{\frac{1}{2}(A_\mu+D_c+L_\mu)\right\}+1}^{\left\{\frac{j}{2}\right\}} g^{\lambda_{2k+1}\lambda_{2k})}\right)^* (x^2)^{F_\mu} \tag{179}$$

where $\mu = (\mu_1, \mu_2, \mu_3, \mu_4, \mu_5)$, $x^2 = x_a x_b g^{ab}$, $a, b = 1, 2, 3, 4$, $D_c = c_3$,

$$M_\mu = (-1)^{\mu_1 + \mu_3 + \mu_5} 2\mu_3, \frac{\left\{\frac{c_2}{2}\right\}!}{\mu_2!\mu_3!\mu_4!}\begin{pmatrix}\frac{c_1}{2}\\ \mu_1\end{pmatrix}, l = (-1)^{n+1}, B_\mu = 2\mu_2 + \mu_3 + \mu_5, \; A_\mu = j - c_1 - B_\mu - D_c,$$

$F_\mu = \mu_1 + \mu_4$, $L_\mu = c_1 + \mu_3 - 2\mu_1 - \mu_5$. Also $\kappa^{B_\mu, A_\mu, D_c} = \kappa^{\beta_1\beta_2 \ldots \beta_{B_\mu}\lambda_1\lambda_2\ldots\lambda_{A\mu}[\lambda_{A\mu+1}d_1]\ldots[\lambda_{A\mu+Dc}d_{Dc}]} \times x_{\beta_1}x_{\beta_2}\ldots x_{\beta_{B_\mu}} x_{d1} x_{d2} \ldots x_{dD_c}$. The summation in eq. (179) is to be done over all possible values of μ that are nonnegative integers satisfying the conditions $0 \le \mu_1 \le \left(\frac{c_1}{2}\right)$, $\mu_2 + \mu_3 + \mu_4 = \left\{\frac{c_2}{2}\right\}$ and $0 \le \mu_5 \le \frac{1}{2}[1 - (-1)^{c_1}]$. Furthermore, symmetrization over the indices $\lambda_1, \lambda_2, \ldots, \lambda_j$ is implied in eq. (179). Finally, we obtain explicit expressions for the symmetry operators. The first step to doing so is to identify all possible permissible combinations for $c \equiv \{c_1, c_2, c_3\}$, i.e., those values that satisfy the constraints $0 \le c \le j$, $0 \le c_1 \le 2\left\{\frac{j}{2}\right\}$, $0 \le c_2 \le j - i\left\{\frac{c_1+1}{2}\right\}$, $0 \le c_3 \le j - 2\left\{\frac{c_1+1}{2}\right\} - 2c_2$. The next step is to construct the various Killing tensors viz. $k_c^{\lambda_1\lambda_2\ldots\lambda_j}$ of rank j using eq. (179). We can, then, obtain the values of $\hat{K}^{\lambda_1\lambda_2\ldots\lambda_j}$ using eq. (177) recursively. Finally, we substitute $\hat{K}^{\lambda_1\lambda_2\ldots\lambda_j}$ in eq. (164) & (166) to obtain the symmetry operators and do the summation over j from 0 to n. The number of linearly independent symmetry operators of order n are given by the sum of the number of linearly independent solutions for $n = j, j-1$ so that we have 11 first order symmetry operators, 60 second order ones, 225 third order ones and so on. The first, second and third order symmetry operators are given below:

(a) First Order Symmetry Operators

$$K^\lambda = \kappa^{\lambda}_{(0,0,0)} + \kappa^{[\lambda d_1]}_{(0,0,1)} x_{d_1}$$

(b) Second Order Symmetry Operators

$$K^{\lambda_1\lambda_2} = g^{\lambda_1\lambda_2}K^{(0)} + \kappa^{\lambda_1\lambda_2}_{(0,0,0)} + \kappa^{(\lambda_1[\lambda_2 d_1])}_{(0,0,1)} x_{d_1} + \kappa^{[\lambda_1 d_1][\lambda_2 d_2]}_{(0,0,2)} x_{d_1}x_{d_2}$$

$$\kappa^{\lambda_1\lambda_2}_{(0,1,0)}x^2 - 2\kappa^{\beta(\lambda_1 \quad x^{\lambda_2})}_{(0,1,0)} x_\beta + \kappa^{\beta_1\beta_2}_{(0,1,0)} x_{\beta_1}x_{\beta_2} g^{\lambda_1\lambda_2} + \kappa^{(\lambda_1 \; x^{\lambda_2})}_{(1,0,0)}$$

$$-\kappa^{\beta}_{(1,0,0)} x_\beta g^{\lambda_1\lambda_2} + \kappa_{(2,0,0)} x^{\lambda_1}x^{\lambda_2} - \kappa_{(2,0,0)} g^{\lambda_1\lambda_2}x^2$$

(c) Third Order Symmetry Operators

$$K^{\lambda_1\lambda_2\lambda_3} = g^{(\lambda_1\lambda_2}K^{\lambda_3)} + \kappa^{\lambda_1\lambda_2\lambda_3}_{(0,0,0)} + \kappa^{(\lambda_1\lambda_2[\lambda_3 d_1])}_{(0,0,1)} x_{d_1} + \kappa^{(\lambda_1[\lambda_2 d_1][\lambda_3 d_2])}_{(0,0,2)} x_{d_1}x_{d_2}$$

$$+\kappa^{[\lambda_1 d_1][\lambda_2 d_2][\lambda_3 d_3]}_{(0,0,3)} x_{d_1}x_{d_2}x_{d_3} - 2\kappa^{\beta_1 \; (\lambda_1\lambda_2}_{(0,1,0)} x^{\lambda_3)} x_{\beta_1} + \kappa^{\lambda_1\lambda_2\lambda_3}_{(0,1,0)} x^2$$

$$+\kappa^{\beta_1\beta_2\,(\lambda_1}_{(0,1,0)} g^{\lambda_2\lambda_3)} x_{\beta_1}x_{\beta_2} + \kappa^{(\lambda_1\lambda_2[\lambda_3 d_1])}_{(0,1,1)} x^2 x_{d_1} - 2\kappa^{\beta_1(\lambda_1[\lambda_2 d_1]}_{(0,1,1)} x^{\lambda_3)} x_{\beta_1}x_{d_1}$$

$$+\kappa_{(0,1,1)}^{\beta_1\beta_2([\lambda_1 d_1]} g^{\lambda_2\lambda_3)} x_{\beta_1} x_{\beta_2} x_{d_1} + \kappa_{(1,0,0)}^{(\lambda_1\lambda_2} x^{\lambda_3)} - \kappa_{(1,0,0)}^{\beta_1(\lambda_1} g^{\lambda_2\lambda_3)} x_{\beta_1} + \kappa_{(1,0,1)}^{(\lambda_1[\lambda_2 d_1]} x^{\lambda_3)} x_{d_1}$$

$$-\kappa_{(1,0,1)}^{\beta_1([\lambda_1 d_1]} g^{\lambda_2\lambda_3)} x_{\beta_1} x_{d_1} + \kappa_{(2,0,0)}^{(\lambda_1} x^{\lambda_2} x^{\lambda_3)} - \kappa_{(2,0,0)}^{(\lambda_1} g^{\lambda_2\lambda_3)} x^2 + \kappa_{(2,0,1)}^{([\lambda_1 d_1] x^{\lambda_2\lambda_3)}} x d_1$$

$$-\kappa_{(2,0,1)}^{([\lambda_1 d_1])} g^{\lambda_2\lambda_3)} x^2 x_{d_1}$$

CHAPTER **4**

The Dirac Equation

4.1 THE DIRAC EQUATION

One of the cardinal shortcomings of the Klein Gordon equation, as perceived by physicists of the 1920s, was the existence of negative probability densities and the emergence of extraneous negative energy states. Although these problems were surmounted and explained with the discovery of "antiparticles" like the positron, anti-proton etc. in naturally occurring physical phenomena and the according of a "charge" based interpretation to the Klein Gordon equation in the quantum field theoretic model of elementary particles, the Klein Gordon equation was discarded by most physicists of that era as being physically unsustainable. On the contrary, efforts were directed towards finding an equation that was first order in the time derivative so that a positive definite probability density could be constructed as in the case of non-relativistic Schrödinger equation. These efforts bore fruit with Paul Dirac propounding his first order equation (hereinafter referred to as the "Dirac" equation) in 1928 to explain the relativistic dynamics of particles at the quantum scale.

The following considerations went into the formulation of the quantum mechanical equation proposed by Dirac:

(a) The equation should be relativistically covariant.

(b) It should be of the form $i\partial_0\psi = \hat{H}\psi$.

(c) The equation should be linear in the time derivative and should lead to a positive definite probability density.

(d) Since the equation is to be linear in the time derivative and relativistic considerations mandate that there be completely symmetric treatment of space and time coordinates in any acceptable relativistically covariant theory, it is necessary that the Hamiltonian be linear in the space derivatives as well.

(e) The equation should also be linear in all the four coordinates $\{x^\mu, \mu = 0, 1, 2, 3\}$ so that the principle of superposition holds. Thus, we are looking for an equation that is linear in x^μ's and their first derivatives.

(f) Finally, the equation must lead to the correct relativistic momentum-energy relationship $E^2 = \mathbf{p}^2 + m^2$ or equivalently, the wavefunction ψ should satisfy the Klein Gordon equation $\partial_0{}^2\psi = (\nabla^2 - m^2)\psi$.

The most general linear first order differential equation that meets the aforesaid specifications is the one that expresses the time derivative of each component of the wavefunction (that is assumed for the present to be an N component vector) as a linear combination of all the other components alongwith their spatial derivatives. In natural units $\hbar = c = 1$, it would be of the form

$$\frac{\partial \psi_l}{\partial x^0} + \sum_{k=1}^{3}\sum_{n=1}^{N} \alpha_{ln}^k \frac{\partial \psi_n}{\partial x^k} = +im\sum_{n=1}^{N} \beta_{\text{ln}}\,\psi_n \;= 0,\; l = 1, 2, ..., N \tag{1}$$

In view of the requirement of homogeneity of space and time mandated by relativistic considerations, α_{ln}^k and β_{ln} must be dimensionless constants, independent of the apace time coordinates x^0, x^1, x^2, x^3. We have mentioned above that ψ is not a single component scalar but an N component vector. Similarly, the α^k's cannot be simple numbers. The reasons are:

(a) If the α^k's were simple numbers and ψ a single component scalar, we would have an equation of the form

$$i\frac{\partial \psi}{\partial x^0} = -i\left(\alpha^1\frac{\partial \psi}{\partial x^1} + \alpha^2\frac{\partial \psi}{\partial x^2} + \alpha^3\frac{\partial \psi}{\partial x^3}\right) + \beta m\psi \tag{2}$$

However, this equation is not form invariant under spatial rotations (if the α^k's are simple numbers) as shown below.

Consider a rotation by an infinitesimal angle ε about the x^3 axis. The rotation operator is given by $R(\varepsilon) = 1 + A_3\varepsilon$, where A_3 is the generator of infinitesimal rotations about the x^3 axis and is given by

$$A_3 = \frac{d}{d\theta}R^{(3)}(\theta)\bigg|_{\theta=0} = \frac{d}{d\theta}\begin{pmatrix}\cos\theta & \sin\theta & 0\\ -\sin\theta & \cos\theta & 0\\ 0 & 0 & 1\end{pmatrix}\Bigg|_{\theta=0} = \begin{pmatrix}0 & 1 & 0\\ -1 & 0 & 0\\ 0 & 0 & 0\end{pmatrix}$$

so that

$$\begin{pmatrix}x'^1\\ x'^2\\ x'^3\end{pmatrix} = R^{(3)}(\varepsilon)\begin{pmatrix}x^1\\ x^2\\ x^3\end{pmatrix} = (1 + A_3\varepsilon)\begin{pmatrix}x^1\\ x^2\\ x^3\end{pmatrix} = \begin{pmatrix}x^1 + \varepsilon x^2\\ x^2 - \varepsilon x^1\\ x^3\end{pmatrix}$$

We, therefore, obtain

$$\begin{aligned}\psi_R(\mathbf{x}, x^0) &\cong \psi(x^1 - \varepsilon x^2, x^2 + \varepsilon x^1, x^3, x^0)\\ &= \psi(\mathbf{x}, x^0) + \left(-\varepsilon x^2\frac{\partial}{\partial x^1} + \varepsilon x^1\frac{\partial}{\partial x^2}\right)\psi(\mathbf{x}, x^0)\end{aligned}$$

Now, we will have invariance if $(i\partial_0 - \hat{H})\,\psi_R(\mathbf{x}, x^0) = 0$ or equivalently,

$$(i\partial_0 - \hat{H})\left(-\varepsilon x^2 \frac{\partial}{\partial x^1} + \varepsilon x^1 \frac{\partial}{\partial x^2}\right)\psi(\mathbf{x}, x^0) = 0 \text{ since } (i\partial_0 - \hat{H})\,\psi(\mathbf{x}, x^0) = 0$$

where

$$\hat{H} = -i\left(\alpha^1 \frac{\partial}{\partial x^1} + \alpha^2 \frac{\partial}{\partial x^2} + \alpha^3 \frac{\partial}{\partial x^3} + i\beta m\right) \tag{3}$$

Hence, invariance requires

$$(i\partial_0 + i\alpha^1\partial_1 + i\alpha^2\partial_2 + i\alpha^3\partial_3 - \beta m)(-\varepsilon x^2\partial_1 + \varepsilon x^1\partial_2)\,\psi = 0$$

or

$$\begin{pmatrix} -ix^2\partial_0\partial_1 + ix^1\partial_0\partial_2 - i\alpha^1 x^2\partial_1^2 + i\alpha^1\partial_2 + i\alpha^1 x^1\partial_1\partial_2 - i\alpha^2\partial_1 - i\alpha^2 x^2\partial_2\partial_1 \\ + i\alpha^2 x^1\partial_2^2 - i\alpha^3 x^2\partial_3\partial_1 + i\alpha^3 x^1\partial_3\partial_2 + \beta m x^2\partial_1 - \beta m x^1\partial_2 \end{pmatrix}\psi = 0$$

or

$$\begin{bmatrix} -x^2\partial_1\,(i\partial_0 + i\alpha^1\partial_1 + i\alpha^2\partial_2 + i\alpha^3\partial_3 - \beta m) \\ +x^1\partial_2\,(i\partial_0 + i\alpha^1\partial_1 + i\alpha^2\partial_2 + i\alpha^3\partial_3 - \beta m) + i(\alpha^1\partial_2 - \alpha^2\partial_1) \end{bmatrix}\psi = 0$$

which on simplification, using (3) and $(i\partial_0 - \hat{H})\,\psi(\mathbf{x}, x^0) = 0$, becomes $(i\alpha^1\partial_2 - i\alpha^2\partial_1)\,\psi = 0$ with the trivial solution $\alpha^1 = \alpha^2 = 0$. Similarly, by considering rotations about the x^2 axis we will get $\alpha^1 = \alpha^3 = 0$. Requirement of rotational symmetry, therefore, mandates the trivial solution $\alpha^1 = \alpha^2 = \alpha^3 = 0$. Hence, we cannot think of the α^k's as simple numbers because that leads to violation of rotational invariance of the Dirac equation.

(b) Secondly, in the relativistic setup ψ cannot be a one component scalar because the probability density $\rho = \psi^*\psi$ has to be the time component of a conserved 4-vector. Only then its integral over all space, at a given instant, can be invariant.

To get over these two difficulties, Dirac proposed that the α^k's and β be $N \times N$ matrices, designated $\boldsymbol{\alpha}^k$'s and $\boldsymbol{\beta}$ and ψ be an N component vector, designated Ψ (where $N = 4$, which shall be proved in the sequel). He, accordingly proposed eq. (1) as a matrix equation

$$\frac{\partial \Psi}{\partial x^0} + \sum_{k=1}^{3} \alpha^k \frac{\partial \psi}{\partial x^k} + im\beta\Psi = 0 \tag{4}$$

with each component of Ψ satisfying the above equation.

4.2 REPRESENTATIONS AND PROPERTIES OF THE DIRAC MATRICES

The Dirac eq. (4) can be written compactly in matrix form as

$$\begin{aligned}\partial_0\Psi + (\alpha\cdot\nabla)\,\Psi + im\beta\Psi &= 0 = i\partial_0\Psi - (\alpha\cdot - i\nabla)\Psi - m\beta\Psi \\ &= i\partial_0\Psi - (\alpha\cdot\hat{\mathbf{p}})\Psi - m\beta\Psi = \hat{\mathbf{p}}_0\Psi - (\alpha\cdot\hat{\mathbf{p}})\Psi - m\beta\Psi \end{aligned} \tag{5}$$

where $\Psi = (\psi_1,..., \psi_N)^T$ is an N column vector and $\alpha = \{\alpha^1, \alpha^2, \alpha^3\}$, β are all $N \times N$ matrices.

The first requirement that the Dirac equation must meet is that it must lead to a continuity equation of the form

$$\partial_\mu j^\mu = 0 = \partial_0\rho + \nabla \cdot \mathbf{j} \tag{6}$$

where $\rho = \psi^\dagger\psi \sum_{n=1}^{N} \psi_n^* \psi_n \sum_{n=1}^{N} |\psi_n|^2$ must be positive definite which happens to be the case since the components of Ψ are assumed real.

Writing the conjugate of eq. (5), we have

$$\partial_0\Psi^\dagger + (\nabla\Psi^\dagger \bullet \alpha^\dagger) - im\Psi^\dagger\beta^\dagger = 0 \tag{7}$$

Multiplying eq. (5) on the left by $\Psi^\dagger$ and eq. (7) on the right by Ψ and then adding the resulting equations, we obtain

$$0 = [\Psi^\dagger \cdot \partial_0\Psi + \partial_0\Psi^\dagger \cdot \Psi] + (\Psi^\dagger(\alpha \cdot \nabla\Psi] + [\nabla\Psi^\dagger \cdot \alpha^\dagger]\Psi) + im[\Psi^\dagger\beta\Psi - \Psi^\dagger\beta^\dagger \Psi] \tag{8}$$

In order that eq. (8) corresponds to a continuity equation of the form of eq. (6), we need to impose the following conditions on α, β:

(i) $\beta = \beta^\dagger$, i.e., β must be a Hermitian matrix so that the last term in eq. (8) vanishes. This is necessary since the term contains no derivatives whereas the continuity equation contains only terms that are time and spatial derivatives.

(ii) We also need to identify the second term as a 3-divergence. This is possible only if $\alpha = \alpha^\dagger$ for then, we have

$$\Psi^\dagger(\alpha \cdot \nabla\Psi) + (\nabla\Psi^\dagger \cdot \alpha)\Psi = (\Psi^\dagger \cdot \alpha)\nabla\Psi + \nabla(\Psi^\dagger \cdot \alpha)\Psi = \nabla(\Psi^\dagger \cdot \alpha.\Psi) \tag{9}$$

Hence, $\alpha = \{\alpha^1, \alpha^2, \alpha^3\}$ must each be a Hermitian matrix.

The Hermiticity requirement of these matrices can also be established by noting that

$$i\partial_0\Psi = \hat{H}\Psi = [\alpha \cdot (-i\nabla)]\Psi + m\beta\Psi = \alpha \cdot \hat{\mathbf{p}}\Psi + m\beta\Psi \tag{10}$$

Since the Hamiltonian operator $\hat{H}$ must necessarily be Hermitian (so that its eigenvalues, that represent measurable energy states, are real), it follows that α^k's and β must necessarily be Hermitian as shown below:

$$\begin{aligned}\hat{H}^\dagger\Psi &= (\alpha \cdot \hat{\mathbf{p}})^\dagger\Psi + m\beta^\dagger\Psi = (\hat{\mathbf{p}}^\dagger \cdot \alpha^\dagger)\Psi + m\beta^\dagger\Psi = (\hat{\mathbf{p}} \cdot \alpha^\dagger)\Psi + m\beta^\dagger\Psi \\ &= (\alpha^\dagger \cdot \hat{\mathbf{p}})\Psi + m\beta^\dagger\Psi = \hat{H}\Psi = (\alpha \cdot \hat{\mathbf{p}})\Psi + m\beta\Psi\end{aligned} \tag{11}$$

If α^k's and β are taken Hermitian, we can write eq. (8) as

$$0 = \partial_0(\Psi^\dagger\Psi) + \nabla(\Psi^\dagger \cdot \alpha \cdot \Psi) \tag{12}$$

which leads to the identification of $\Psi^\dagger\Psi$ (a positive definite quantity) as the probability density ρ (being the fourth component of the 4-current vector) and $\Psi^\dagger.\alpha.\Psi$ as the 3-current density $\mathbf{j}$.

Now we examine under what conditions the Dirac equation implies the correct relativistic momentum energy relationship. We premultiply eq. (5) by the operator $\partial_0 - (\boldsymbol{\alpha} \cdot \nabla) - im\boldsymbol{\beta}$ to get

$$\partial_0^2 \Psi + \partial_0(\boldsymbol{\alpha} \cdot \nabla)\Psi + im\boldsymbol{\beta}\partial_0\Psi - (\boldsymbol{\alpha} \cdot \nabla)\partial_0\Psi - (\boldsymbol{\alpha} \cdot \nabla) \cdot (\boldsymbol{\alpha} \cdot \nabla)\Psi$$
$$- im(\boldsymbol{\alpha} \cdot \nabla)\boldsymbol{\beta}\Psi - im\boldsymbol{\beta}\partial_0\Psi - im\boldsymbol{\beta}(\boldsymbol{\alpha} \cdot \nabla)\Psi + m^2\boldsymbol{\beta} \cdot \boldsymbol{\beta}\Psi = 0$$

or

$$\partial_0^2 \Psi - (\boldsymbol{\alpha} \cdot \nabla) \cdot (\boldsymbol{\alpha} \cdot \nabla)\Psi - im[(\boldsymbol{\alpha} \cdot \nabla)\boldsymbol{\beta} + \boldsymbol{\beta} \cdot (\boldsymbol{\alpha} \cdot \nabla)]\, \Psi + m^2\boldsymbol{\beta} \cdot \boldsymbol{\beta}\Psi = 0$$

In order that this equation becomes equivalent to the Klein Gordon equation $(\partial_0{}^2 - \nabla^2 + m^2)\Psi = 0$, the matrices α^k's and β must meet the following further conditions:

(a) $(\boldsymbol{\alpha}^k)^2 = \boldsymbol{\beta}^2 = \mathbf{I}$ (13)

(b) $\boldsymbol{\alpha}^k\boldsymbol{\alpha}^l + \boldsymbol{\alpha}^l\boldsymbol{\alpha}^k = 2\delta^{kl}$; $k, l = 1, 2, 3$ (14)

(c) $\boldsymbol{\alpha}^k\boldsymbol{\beta} + \boldsymbol{\beta}\boldsymbol{\alpha}^k = 0$; $k = 1,2,3$ (15)

We next establish that the dimension of the matrices $\boldsymbol{\alpha}^k$'s and $\boldsymbol{\beta}$ must necessarily be even for, we have

$$\boldsymbol{\beta}\boldsymbol{\alpha}^k = -\boldsymbol{\alpha}^k\boldsymbol{\beta} = -\mathbf{I}\boldsymbol{\alpha}^k\boldsymbol{\beta} \Rightarrow |\boldsymbol{\beta}\boldsymbol{\alpha}^k| = |\boldsymbol{\beta}||\boldsymbol{\alpha}^k|$$
$$= |\boldsymbol{\alpha}^k||\boldsymbol{\beta}| = |-\mathbf{I}||\boldsymbol{\alpha}^k||\boldsymbol{\beta}| \Rightarrow |-\mathbf{I}| = +1 = (-1)^N \tag{16}$$

showing that N is even. Therefore, the matrices $\boldsymbol{\alpha}^k$'s and $\boldsymbol{\beta}$ must necessarily be even dimensional.

Furthermore, these matrices $\boldsymbol{\alpha}^k$'s and $\boldsymbol{\beta}$ must also be traceless, i.e., with vanishing trace. This is shown as follows:

Since, each of these matrices is Hermitian, they can be diagonalized, although not simultaneously (because they do not commute with each other). We choose a basis in which $\boldsymbol{\beta}$ is diagonal, i.e., $\boldsymbol{\beta} = \text{diag}\,(b_1,..., b_N)$. Since $\beta^2 = \mathbf{I}$, we have, $b_i^2 = 1$ or $b_i = \pm 1$; $\forall i = 1, 2, 3,..., N$. Also, since $|\boldsymbol{\beta}|, |\boldsymbol{\alpha}^k| \neq 0$ as $(\alpha^k)^2 = \beta^2 = \mathbf{I}$, each of these matrices is non-singular and hence, invertible so that

$$\beta\alpha^k = -\alpha^k\beta \Rightarrow (\alpha^k)^{-1}\,\beta\alpha^k = -\beta \Rightarrow Tr\,[(\alpha^k)^{-1}\,\beta\alpha^k]$$
$$= Tr[\alpha^k(\alpha^k)^{-1}\,\beta] = Tr\,(\beta)$$
$$= Tr(-\beta) = -Tr(\beta) \Rightarrow Tr\beta = 0 \tag{17}$$

Similarly, $Tr(\alpha^k) = 0$.

The even dimensionality of these matrices can also be established as a corollary to the traceless property as follows:

In the diagonal representation of $\boldsymbol{\beta}$ let there be m number of $b_i = +1$ and n number of $b_i = -1$.

Then, obviously, $m + n = N$. But since $Tr(\boldsymbol{\beta}) = 0$, we have $m - n = 0$ whence $m = n$, $N = 2m = 2n$ or N is even.

It is emphasized once again that the matrices $\boldsymbol{\alpha}^k$'s and $\boldsymbol{\beta}$ being Hermitian, can be diagonalized in a basis that is specific to that matrix but they cannot be simultaneously diagonalized in a single basis since they do not commute with each other.

To establish the dimensionality of the matrices $\boldsymbol{\alpha}^k$'s and $\boldsymbol{\beta}$, we first make the following transformations:

$$\boldsymbol{\gamma}^k = \boldsymbol{\beta\alpha}^k \tag{18}$$

$$\boldsymbol{\gamma}^0 = \boldsymbol{\beta} \tag{19}$$

Since both $\boldsymbol{\alpha}^k$, $\boldsymbol{\beta}$ are $N \times N$ matrices, it follows that $\boldsymbol{\gamma}^k$, $\boldsymbol{\gamma}^0$ are also $N \times N$ matrices. The $\boldsymbol{\gamma}^k$ matrices are unitary $(\boldsymbol{\gamma}^k)^{-1} = (\boldsymbol{\gamma}^k)^\dagger$ for $(\boldsymbol{\gamma}^k)^{-1} = (\boldsymbol{\beta\alpha}^k)^{-1} = (\boldsymbol{\alpha}^k)^{-1}\boldsymbol{\beta}^{-1} = \boldsymbol{\alpha}^k\boldsymbol{\beta} = -\boldsymbol{\beta\alpha}^k = -\boldsymbol{\beta}^\dagger(\boldsymbol{\alpha}^k)^\dagger = (-\boldsymbol{\alpha}^k\boldsymbol{\beta})^\dagger = (\boldsymbol{\beta\alpha}^k)^\dagger = \boldsymbol{\gamma}^{k\dagger}$ and anti-Hermitian $(\boldsymbol{\gamma}^k)^\dagger = (\boldsymbol{\beta\alpha}^k)^\dagger = (\boldsymbol{\alpha}^k)^\dagger\boldsymbol{\beta}^\dagger = \boldsymbol{\alpha}^k\boldsymbol{\beta} = -\boldsymbol{\beta\alpha}^k = -\boldsymbol{\gamma}^k$, while $\boldsymbol{\gamma}^0$ is unitary and Hermitian $(\boldsymbol{\gamma}^0)^{-1} = (\boldsymbol{\gamma}^0)^\dagger = \boldsymbol{\gamma}^0$. We also have $\boldsymbol{\gamma}_\mu = \mathbf{g}_{\mu\nu}\boldsymbol{\gamma}^\nu$ so that, in the standard metric

$$\boldsymbol{\gamma}_k = -\boldsymbol{\gamma}^k,\ \boldsymbol{\gamma}_0 = \boldsymbol{\gamma}^0 \tag{20}$$

Consider the following 16 entities:

$$\begin{aligned}
&\mathbf{I},\\
&\boldsymbol{\gamma}_0,\ i\boldsymbol{\gamma}_1,\ i\boldsymbol{\gamma}_2,\ i\boldsymbol{\gamma}_3,\\
&i\boldsymbol{\gamma}_1\boldsymbol{\gamma}_2,\ i\boldsymbol{\gamma}_2\boldsymbol{\gamma}_3,\ i\boldsymbol{\gamma}_3\boldsymbol{\gamma}_1,\ \boldsymbol{\gamma}_1\boldsymbol{\gamma}_0,\ \boldsymbol{\gamma}_2\boldsymbol{\gamma}_0,\ \boldsymbol{\gamma}_3\boldsymbol{\gamma}_0,\\
&\boldsymbol{\gamma}_1\boldsymbol{\gamma}_2\boldsymbol{\gamma}_3,\ i\boldsymbol{\gamma}_1\boldsymbol{\gamma}_2\boldsymbol{\gamma}_0,\ i\boldsymbol{\gamma}_3\boldsymbol{\gamma}_1\boldsymbol{\gamma}_0,\ i\boldsymbol{\gamma}_2\boldsymbol{\gamma}_3\boldsymbol{\gamma}_0,\\
&i\boldsymbol{\gamma}_1\boldsymbol{\gamma}_2\boldsymbol{\gamma}_3\boldsymbol{\gamma}_0
\end{aligned} \tag{21}$$

We denote by Γ_i; $i = 1, 2, \ldots\ldots, 16$ the various elements of the above array sequentially. Then we have the following properties of the set of 16 elements comprising the above array, commonly referred to as the Dirac algebra:

(i) $\boldsymbol{\gamma}_\mu = g_{\mu\nu}\boldsymbol{\gamma}^\nu,\ \boldsymbol{\gamma}^\mu = g^{\mu\nu}\boldsymbol{\gamma}_\nu$ (22)

(ii) $\{\boldsymbol{\gamma}_\mu, \boldsymbol{\gamma}_\nu\} = 2g_{\mu\nu}\mathbf{I},\ \{\boldsymbol{\gamma}^\mu, \boldsymbol{\gamma}^\nu\} = 2g^{\mu\nu}\mathbf{I}$ (for example) (23)

$$\{\boldsymbol{\gamma}_0, \boldsymbol{\gamma}_0\} = \boldsymbol{\gamma}_0^2 + \boldsymbol{\gamma}_0^2 = \beta^2 + \beta^2 = 2\beta^2 = 2\mathbf{I} = 2g_{00}\mathbf{I}$$

$$\begin{aligned}
\{\boldsymbol{\gamma}_0, \boldsymbol{\gamma}_k\} &= \boldsymbol{\gamma}_0\boldsymbol{\gamma}_k + \boldsymbol{\gamma}_k\boldsymbol{\gamma}_0 = \boldsymbol{\beta}\cdot\boldsymbol{\beta\alpha}_k + \boldsymbol{\beta\alpha}_k\cdot\boldsymbol{\beta} = \boldsymbol{\beta}\cdot\boldsymbol{\beta\alpha}_k + \boldsymbol{\beta}\cdot\boldsymbol{\alpha}_k\boldsymbol{\beta}\\
&= \boldsymbol{\beta}\cdot\boldsymbol{\beta\alpha}_k - \boldsymbol{\beta}\cdot\boldsymbol{\beta}\cdot\boldsymbol{\alpha}_k = 0 = 2g_{0k}\mathbf{I}
\end{aligned}$$

$$\begin{aligned}
\{\boldsymbol{\gamma}_i, \boldsymbol{\gamma}_i\} &= \boldsymbol{\gamma}_i^2 + \boldsymbol{\gamma}_i^2 = \boldsymbol{\beta\alpha}_i\cdot\boldsymbol{\beta\alpha}_i + \boldsymbol{\beta\alpha}_i\cdot\boldsymbol{\beta\alpha}_i = 2\boldsymbol{\beta\alpha}_i\cdot\boldsymbol{\beta\alpha}_i\\
&= 2\boldsymbol{\beta}\cdot\boldsymbol{\alpha}_i\boldsymbol{\beta\alpha}_i = -2\boldsymbol{\beta\beta\alpha}_i\boldsymbol{\alpha}_i = -2\boldsymbol{\alpha}_i^2 = -2\mathbf{I} = 2g_{ii}\mathbf{I}
\end{aligned}$$

$$\begin{aligned}
\{\boldsymbol{\gamma}_i, \boldsymbol{\gamma}_j\} &= \boldsymbol{\gamma}_i\boldsymbol{\gamma}_j + \boldsymbol{\gamma}_j\boldsymbol{\gamma}_i = \boldsymbol{\beta\alpha}_i\cdot\boldsymbol{\beta\alpha}_j + \boldsymbol{\beta\alpha}_j\cdot\boldsymbol{\beta\alpha}_i = \boldsymbol{\beta}\cdot\boldsymbol{\alpha}_i\boldsymbol{\beta\alpha}_j + \boldsymbol{\beta}\cdot\boldsymbol{\alpha}_j\boldsymbol{\beta\alpha}_i\\
&= -\boldsymbol{\beta\beta\alpha}_i\boldsymbol{\alpha}_j - \boldsymbol{\beta\beta\alpha}_j\boldsymbol{\alpha}_i = -\boldsymbol{\alpha}_i\boldsymbol{\alpha}_j - \boldsymbol{\alpha}_j\boldsymbol{\alpha}_i\\
&= -\{\boldsymbol{\alpha}_i, \boldsymbol{\alpha}_j\} = 0 = 2g_{ij}\mathbf{I}
\end{aligned}$$

(iii) $\Gamma_i^2 = \mathbf{I}\ \forall i = 1, 2, 3, \ldots, 16$ (for example) (24)

$$\boldsymbol{\gamma}_0^2 = \beta^2 = \mathbf{I}$$

$$(i\boldsymbol{\gamma}_i)^2 = -(\boldsymbol{\beta\alpha}_i)\cdot(\boldsymbol{\beta\alpha}_i) = (\boldsymbol{\beta\alpha}_i)\cdot(\boldsymbol{\alpha}_i\boldsymbol{\beta}) = \boldsymbol{\beta\alpha}_i\cdot\boldsymbol{\alpha}_i\boldsymbol{\beta} = \boldsymbol{\beta}^2 = \mathbf{I}$$

$$(\boldsymbol{\gamma}_1\boldsymbol{\gamma}_0)^2 = (\boldsymbol{\gamma}_1\boldsymbol{\gamma}_0)(\boldsymbol{\gamma}_1\boldsymbol{\gamma}_0) = -(\boldsymbol{\gamma}_1\boldsymbol{\gamma}_0)(\boldsymbol{\gamma}_0\boldsymbol{\gamma}_1) = -\boldsymbol{\gamma}_1^2 = \mathbf{I}$$

$$\begin{aligned}
(\boldsymbol{\gamma}_1\boldsymbol{\gamma}_2\boldsymbol{\gamma}_3)^2 &= (\boldsymbol{\gamma}_1\boldsymbol{\gamma}_2\boldsymbol{\gamma}_3)\cdot(\boldsymbol{\gamma}_1\boldsymbol{\gamma}_2\boldsymbol{\gamma}_3) = -(\boldsymbol{\gamma}_1\boldsymbol{\gamma}_2\boldsymbol{\gamma}_3)\cdot(\boldsymbol{\gamma}_1\boldsymbol{\gamma}_3\boldsymbol{\gamma}_2) = (\boldsymbol{\gamma}_1\boldsymbol{\gamma}_2\boldsymbol{\gamma}_3).(\boldsymbol{\gamma}_3\boldsymbol{\gamma}_1\boldsymbol{\gamma}_2)\\
&= -(\boldsymbol{\gamma}_1\boldsymbol{\gamma}_2)\cdot(\boldsymbol{\gamma}_1\boldsymbol{\gamma}_2) = (\boldsymbol{\gamma}_1\boldsymbol{\gamma}_2)\cdot(\boldsymbol{\gamma}_2\boldsymbol{\gamma}_1) = -\boldsymbol{\gamma}_1^2 = \mathbf{I}
\end{aligned}$$

$$(i\boldsymbol{\gamma}_1\boldsymbol{\gamma}_2\boldsymbol{\gamma}_0)^2 = -(\boldsymbol{\gamma}_1\boldsymbol{\gamma}_2\boldsymbol{\gamma}_0)\cdot(\boldsymbol{\gamma}_1\boldsymbol{\gamma}_2\boldsymbol{\gamma}_0) = (\boldsymbol{\gamma}_1\boldsymbol{\gamma}_2\boldsymbol{\gamma}_0)\cdot(\boldsymbol{\gamma}_1\boldsymbol{\gamma}_0\boldsymbol{\gamma}_2)$$

$$= -(\gamma_1\gamma_2\gamma_0)\cdot(\gamma_0\gamma_1\gamma_2)$$
$$= -(\gamma_1\gamma_2)\cdot(\gamma_1\gamma_2) = (\gamma_1\gamma_2)\cdot(\gamma_2\gamma_1) = -\gamma_1^2 = \mathbf{I}$$
$$(i\gamma_1\gamma_2\gamma_3\gamma_0)^2 = -(\gamma_1\gamma_2\gamma_3\gamma_0)\cdot(\gamma_1\gamma_2\gamma_3\gamma_0) = (\gamma_1\gamma_2\gamma_3\gamma_0)\cdot(\gamma_1\gamma_2\gamma_0\gamma_3)$$
$$= -(\gamma_1\gamma_2\gamma_3\gamma_0)\cdot(\gamma_1\gamma_0\gamma_2\gamma_3) = (\gamma_1\gamma_2\gamma_3\gamma_0)\cdot(\gamma_0\gamma_1\gamma_2\gamma_3)$$
$$= (\gamma_1\gamma_2\gamma_3)\cdot(\gamma_1\gamma_2\gamma_3) = \mathbf{I}$$

(iv) $\Gamma_i\Gamma_j = \rho\Gamma_k;\ \rho = \pm 1, \pm i$ (25)

This property follows from the fact that each of the Γ_i consists of one or more of the $\boldsymbol{\gamma}$ matrices $\gamma_0, \gamma_1, \gamma_2, \gamma_3$ prefixed with $\rho_i = 1, i$. Hence, the expression $\Gamma_i\Gamma_j$ will be of the form $(\rho_i\gamma_1^b\gamma_2^c\gamma_3^d\gamma_0^a).(\rho_j\gamma_1^f\gamma_2^g\gamma_3^h\gamma_0^e)$ where the exponents (small alphabets) can take the values 0, 1. Using the properties (23) and (24) this expression can be reduced to the form $\rho_i\rho_j$ $(\gamma_1^q\gamma_2^r\gamma_3^s\gamma_0^p)$ with the exponents again taking values 0,1 or equivalently to $\rho\Gamma_k$ with $\rho = \pm 1$, $\pm i$, as desired.

(v) For each Γ_i, except $\Gamma_i = \mathbf{I}$, $\exists\Gamma_j$ such that

$$\Gamma_j\Gamma_i\Gamma_j = -\Gamma_i \text{ i.e., } \{\Gamma_j, \Gamma_i\} = 0 \quad \text{(for example)} \tag{26}$$

For elements of the second row of the array, we have

$$(\gamma_k\gamma_0)\gamma_0(\gamma_k\gamma_0) = \gamma_k\gamma_k\gamma_0 = -\gamma_0$$
$$(\gamma_k\gamma_0)i\gamma_k(\gamma_k\gamma_0) = -i\gamma_k\gamma_0^2 = -i\gamma_k$$

For elements of the third row of the array, we have

$$i\gamma_i(i\gamma_i\gamma_j)i\gamma_i = i\gamma_j\gamma_i = -i\gamma_i\gamma_j$$
$$\gamma_0(\gamma_k\gamma_0)\gamma_0 = \gamma_0\gamma_k = -\gamma_k\gamma_0$$

For elements of the fourth row of the array, we have

$$\gamma_0(\gamma_1\gamma_2\gamma_3)\gamma_0 = -\gamma_1\gamma_0\gamma_2\gamma_3\gamma_0 = \gamma_1\gamma_2\gamma_0\gamma_3\gamma_0 = -\gamma_1\gamma_2\gamma_3$$
$$\gamma_k\gamma_0(i\gamma_i\gamma_j\gamma_0)\gamma_k\gamma_0 = -i\gamma_k\gamma_0\gamma_i\gamma_j\gamma_k = -i\gamma_0\gamma_i\gamma_j = -i\gamma_i\gamma_j\gamma_0$$

For the only element of the fifth row, we have

$$\gamma_0(i\gamma_1\gamma_2\gamma_3\gamma_0)\gamma_0 = i\gamma_0\gamma_1\gamma_2\gamma_3 = -i\gamma_1\gamma_2\gamma_3\gamma_0$$

so that the property is established.

(vi) For $\Gamma_j \neq \mathbf{I}$, $Tr\,\Gamma_j = 0$ since (27)

$$\Gamma_j\Gamma_i\Gamma_j = -\Gamma_i \Rightarrow Tr(\Gamma_i) = Tr(\Gamma_j^2\Gamma_i) = Tr(\Gamma_j\Gamma_i\Gamma_j)$$
$$= Tr(-\Gamma_i) = -Tr(\Gamma_i) \Rightarrow Tr(\Gamma_i) = 0$$

Further, $Tr\,\mathbf{I} = d$, where d is the dimension of the unit matrix $\mathbf{I}$.

(vii) The set $\{\Gamma_i, i = 1, 2, 3,......, 16\}$ constitutes a basis, i.e., the elements Γ_i, $i = 1, 2, 3,......,$ 16 are linearly independent and the equation

$$\sum_{i=1}^{16} a_i\Gamma_i = 0 \tag{28}$$

holds if and only if $a_i = 0$, $\forall i = 1, 2, 3,......, 16$. This is established as follows:

Taking trace of both sides of eq. (28), we obtain $a_1 = 0$ if we take $\Gamma_1 = \mathbf{I}$. Now multiplying eq. (28) by Γ_h where Γ_h is any element of the above algebra, we obtain

$$\sum_{\substack{i=1\\ i\neq h}}^{16} a_i\Gamma_i\Gamma_h + a_h\mathbf{I} = 0$$

because $\Gamma_h^2 = \mathbf{I}$. On using eq. (25) this can be written in the form.

$$\sum_{k=2}^{16} c_k\Gamma_k + a_h\mathbf{I} = 0$$

Again, on taking trace of both sides we end up with $a_h = 0$. Continuing this process of multiplying eq. (28) by each element of the algebra Γ_h; $h = 1, \ldots\ldots, 16$ and then taking trace of the resulting equation, we obtain $a_i = 0$, $\forall i = 1, 2, 3, \ldots\ldots, 16$ as required.

We also conclude as a corollary to the above, that it is not possible to represent the $\boldsymbol{\gamma}$ matrices by matrices whose dimension is less than 4×4 since we cannot construct 16 linearly independent elements from matrices whose dimension is less than 4×4.

(viii) ***Dimensionality of the γ Matrices***

We have shown above that γ matrices cannot be represented by matrices whose dimension is less than 4×4. We now establish that $\boldsymbol{\gamma}$ matrices must necessarily be 4×4 matrices. Let us assume that $\boldsymbol{\gamma}$ matrices are d dimensional. Then, from above, we have $Tr\Gamma_i\Gamma_j = \delta_{ij} Tr\ \mathbf{I}_{d\times d} = d\delta_{ij};\ i, j = 1, 2, 3, \ldots, 16.$

Hence, $\det(Tr\Gamma_i\Gamma_j) = \det(d\delta_{ij}) = \det[\text{diag}(d, d, d, \ldots, d)] = d^{16} \neq 0$

Here $\det(Tr\ \Gamma_i\Gamma_j)$ is the determinant of the 16×16 matrix whose elements are $Tr\ \Gamma_i\Gamma_j;\ i, j = 1, 2, 3, \ldots, 16$. Hence, we conclude that the Dirac algebra consisting of the 16 Γ_i matrices is semisimple.

By Frobenius theorem for finite semisimple groups, the number of possible inequivalent irreducible representations (IRR) of such an algebra (that contains an identity element $\mathbf{I}$) is equal to the number of elements in the algebra that commute with all the other elements of the algebra. Furthermore, in the Dirac algebra there is only one element $\mathbf{I}$ that commutes with each of the other elements and hence, there exists only one IRR. As far as the dimensionality of the IRRs is concerned, we also have $N = d_1^2 + d_2^2 + \ldots d_r^2$ where N is the number of elements of the algebra, d_i the dimension of the ith IRR and r the number of such IRRs. In the case of Dirac algebra $N = 16$, $r = 1$ so that $d = 4$. Thus, the Dirac algebra has only one IRR and that is four dimensional. We can, therefore, represent all the elements of the Dirac algebra by 4×4 matrices. Additionally, there are exactly 16 linearly independent 4×4 matrices.

(ix) As a corollary, since Γ_i, $i = 1, 2, 3, \ldots, 16$ are linearly independent and constitute a basis, every 4×4 matrix can be expressed as a linear combination of these basis vectors, i.e., for any 4×4 matrix $\mathbf{X}$, we have

$$\mathbf{X} = \sum_{i=1}^{16} x_i\Gamma_i \tag{29}$$

The coefficients x_i can be obtained by multiplying eq. (29) by Γ_i, $i = 1, 2, 3, ..., 16$ successively and then taking the trace. We have

$$Tr(\mathbf{X}\Gamma_k) = Tr\left[x_k\Gamma_k^2 + \sum_{\substack{i=1\\ i\neq k}}^{16} x_i\Gamma_i\Gamma_k\right] = Tr\left[x_k\,\mathbf{I} + \sum_{j=2}^{16} y_j\,\Gamma_j\right] = Tr\,x_k\mathbf{I} = x_k Tr\,\mathbf{I} = 4x_k$$

so that

$$x_k = \frac{1}{4}\,Tr(\mathbf{X}\Gamma_k) \tag{30}$$

(x) Any 4×4 matrix that commutes with each of the $\boldsymbol{\gamma}$ matrices $\gamma_0, \gamma_1, \gamma_2, \gamma_3$ and therefore, with each of the Γ_i, $i = 1, 2, 3, ..., 16$ is a constant multiple of unity. Let us assume that $\mathbf{Y}$ is a matrix that commutes with each of the $\boldsymbol{\gamma}$ matrices $\gamma_0, \gamma_1, \gamma_2, \gamma_3$ and, therefore, with each of the Γ_i, $i = 1, 2, 3, ..., 16$. We write $\mathbf{Y}$ in the form

$$\mathbf{Y} = y_i\,\Gamma_i + \sum_{j\neq 1} y_j\,\Gamma_j;\ (\Gamma_j \neq \mathbf{I}) \tag{31}$$

Let $\mathbf{\Gamma}_k$ be an element of the Dirac algebra such that $\mathbf{\Gamma}_k\mathbf{\Gamma}_i\mathbf{\Gamma}_k = -\mathbf{\Gamma}_i$. Then, multiplying eq. (31) on left and right sides by $\mathbf{\Gamma}_k$, we obtain (since $\mathbf{Y\Gamma}_k = \mathbf{\Gamma}_k\mathbf{Y}$)

$$y_i\Gamma_i + \sum_{j\neq 1} y_j\,\Gamma_j = \mathrm{Y} = \Gamma_k^2\,\mathrm{Y} = \Gamma_k\,\mathrm{Y}\Gamma_k$$

$$= -\,y_i\Gamma_i + \sum_{j\neq 1} y_j\,\Gamma_k\Gamma_j\Gamma_k = -\,y_i\Gamma_i + \sum_{j\neq 1} z_j\Gamma_j \tag{32}$$

Multiplying eq. (32) by $\mathbf{\Gamma}_i$ and taking the trace, we get $y_i = -y_i = 0$. Since, $\mathbf{\Gamma}_i$ is arbitrary except from being the identity matrix, the only non-vanishing y_i is the coefficient of the identity matrix. Hence, we have $\mathbf{Y} = y_i\mathbf{I}$, as desired.

(xi) Given two representations of the Dirac algebra, i.e., two sets of 4×4 matrices γ_μ, γ'_μ; $\mu = 0, 1, 2, 3$ that satisfy the anticommutation relations (23), i.e., $\{\gamma_\mu, \gamma_\nu\} = 2g_{\mu\nu}\mathbf{I}$, $\{\gamma'_\mu, \gamma'_\nu\} = 2g_{\mu\nu}\mathbf{I}$, then we can write

$$\gamma'_\mu = \mathbf{S}\gamma_\mu\mathbf{S}^{-1} \tag{33}$$

where $\mathbf{S}$ is a non-singular matrix given by

$$\mathbf{S} = \sum_{i=1}^{16} \Gamma'_i\ \mathbf{F}\mathbf{\Gamma}_i \tag{34}$$

and $\mathbf{F}$ is an arbitrary 4×4 matrix which can be chosen in a manner that $\mathbf{S}$ is non-singular. We have, from eq. (25) $\Gamma_i\Gamma_j = \rho\Gamma_k$, where $\rho = \pm 1, \pm i$. Hence,

$$\mathbf{\Gamma}_i\mathbf{\Gamma}_j = \mathbf{\Gamma}_i\mathbf{\Gamma}_j\mathbf{\Gamma}_i\mathbf{\Gamma}_j\mathbf{\Gamma}_j\mathbf{\Gamma}_i = (\mathbf{\Gamma}_i\mathbf{\Gamma}_j)(\mathbf{\Gamma}_i\mathbf{\Gamma}_j)\,\mathbf{\Gamma}_j\mathbf{\Gamma}_i$$
$$= (\rho\mathbf{\Gamma}_k)^2\,\mathbf{\Gamma}_j\mathbf{\Gamma}_i = \rho^2\mathbf{\Gamma}_j\mathbf{\Gamma}_i = \rho^{-2}\mathbf{\Gamma}_j\mathbf{\Gamma}_i$$

so that $\mathbf{\Gamma}_j\mathbf{\Gamma}_i = \rho^2\mathbf{\Gamma}_i\mathbf{\Gamma}_j = \rho^3\mathbf{\Gamma}_k$. Similarly, in the primed system, since the commutation rules are the same, we have $\mathbf{\Gamma}'_i\mathbf{\Gamma}'_j = \rho\mathbf{\Gamma}'_k$ and $\mathbf{\Gamma}'_j\mathbf{\Gamma}'_i = \rho^3\mathbf{\Gamma}'_k$.

Now, if $S = \sum_{i=1}^{16} \Gamma_i' \ FT_i$, then

$$\Gamma_j' \mathbf{S}\Gamma_j = \sum_i \Gamma_j' \Gamma_i' \mathbf{F}\Gamma_i \Gamma_j = \sum_k \rho^4 \Gamma_k' \mathbf{F}\Gamma_k = \sum_k \Gamma_k' \mathbf{F}\Gamma_k = \mathbf{S} \tag{35}$$

(The summation over i for fixed j runs over all the 16 elements of the algebra and can, therefore, be replaced by a summation over k which also runs over the entire algebra.)

Since $\boldsymbol{\gamma}$ matrices are irreducible, it follows by Schur's Lemma that **S** is non-singular. We now establish that **S** is uniquely determined upto a factor. To prove this assertion, let us assume that there exist two non-singular matrices $\mathbf{S}_1$ and $\mathbf{S}_2$ such that

$$\begin{aligned}\gamma'_\mu = \mathbf{S}_1\gamma_\mu\mathbf{S}_1^{-1} = \mathbf{S}_2\gamma_\mu\mathbf{S}_2^{-1} &\Rightarrow \gamma_\mu = \mathbf{S}_2^{-1}\mathbf{S}_1\,\gamma_\mu\,\mathbf{S}_1^{-1}\mathbf{S}_2 \\ = (\mathbf{S}_2^{-1}\mathbf{S}_1)\gamma_\mu\,(\mathbf{S}_2^{-1}\mathbf{S}_1)^{-1} &\Rightarrow \gamma_\mu = (\mathbf{S}_2^{-1}\mathbf{S}_1) = (\mathbf{S}_2^{-1}\mathbf{S}_1)\gamma_\mu\end{aligned}$$

Thus, $\mathbf{S}_2^{-1}\mathbf{S}_1$ commutes with all the γ matrices and is, therefore, a constant multiple of the identity matrix, i.e., $\mathbf{S}_2^{-1}\mathbf{S}_1 = c\mathbf{I} \Rightarrow \mathbf{S}_1 = c\mathbf{S}_2$. If we further require that det $\mathbf{S} = 1$, then **S** is completely determined upto a factor $\pm 1, \pm i$.

We now establish that there exists 4×4 matrix **F** which can be chosen in a manner that **S** does not vanish. Let, if possible, $\mathbf{S} = \mathbf{0}\ \forall \mathbf{F}$. Then consider a choice of the matrix **F** such that $\mathbf{F}_{\alpha\beta} = 1$ and $\mathbf{F}_{\mu\nu} = 0\ \forall_\mu,\ \nu = 0, 1, 2, 3;\ \mu \neq \alpha,\ \nu \neq \beta$, i.e., **F** is a 4×4 matrix that has a single $(\alpha, \beta)^{th}$ element as unity and all the other elements vanishing. Then, $\mathbf{S} = \sum_{i=1}^{16} \Gamma_i' \mathbf{F}\Gamma_i = \sum_{i=1}^{16} (\Gamma_i)_{\mu\alpha}(\Gamma_i)_{\beta\sigma} = 0$. This equation must hold for all values of α, β due to the arbitrary choice of the matrix **F**. Setting various values for $\alpha, \beta = 0, 1, 2, 3$, we get the matrix equation $\sum_{i=1}^{16} (\Gamma_i')_{\mu\nu} \boldsymbol{\Gamma}_i = 0$. But $(\boldsymbol{\Gamma}_i')^2 = \mathbf{I}$. Hence, $(\boldsymbol{\Gamma}'_i)_{\mu\nu}$ cannot vanish for all $\mu, \nu = 0, 1, 2, 3$. This contradicts the linear independence of the $\boldsymbol{\Gamma}$ matrices. Hence, $\mathbf{S} \neq \mathbf{0}$.

We finally prove that the matrix **F** can be so chosen that **S** is non-singular. Consider the matrix $\mathbf{T} = \sum_{i=1}^{16} \Gamma_i \mathbf{G}\boldsymbol{\Gamma}'_i$. We then have $\Gamma_j \mathbf{T}\Gamma'_j = \sum_{i=1}^{16} \Gamma_j\, \boldsymbol{\Gamma}'_i \mathbf{G}\boldsymbol{\Gamma}'_i \boldsymbol{\Gamma}'_j$. Now, as proved earlier, $\boldsymbol{\Gamma}_i\boldsymbol{\Gamma}_j = \rho\boldsymbol{\Gamma}_k;\ \rho = \pm 1, \pm i$ and $\boldsymbol{\Gamma}_j\boldsymbol{\Gamma}_i = (\boldsymbol{\Gamma}_j)^{-1}(\boldsymbol{\Gamma}_i)^{-1} = (\rho\boldsymbol{\Gamma}_k)^{-1} = \rho^{-1}\boldsymbol{\Gamma}_k$. We claim that, for a fixed j, k takes all values from 1 to 16 as i runs from 1 to 16. If this is not so, then we must necessarily have $\boldsymbol{\Gamma}_i\boldsymbol{\Gamma}_j = \rho\boldsymbol{\Gamma}_k$ and $\boldsymbol{\Gamma}_m\boldsymbol{\Gamma}_j = \rho'\boldsymbol{\Gamma}_k;\ \boldsymbol{\Gamma}_i \neq \boldsymbol{\Gamma}_m$. This implies that $\boldsymbol{\Gamma}_i = \rho\boldsymbol{\Gamma}_k(\boldsymbol{\Gamma}_j)^{-1} = \rho\boldsymbol{\Gamma}_k\boldsymbol{\Gamma}_j = \rho\rho'^{-1}\boldsymbol{\Gamma}_m$, thereby violating the linear independence of the $\boldsymbol{\Gamma}$ matrices which is a contradiction. The $\boldsymbol{\Gamma}'_i$'s are obtained from $\boldsymbol{\gamma}'^{\mu}$'s in the same manner as the $\boldsymbol{\Gamma}_i$ are obtained from $\boldsymbol{\gamma}^\mu$. Furthermore, $\boldsymbol{\gamma}^\mu$, $\boldsymbol{\gamma}'^\mu$ are subject to the same commutation relations. Therefore, we have $\boldsymbol{\Gamma}'_i\boldsymbol{\Gamma}'_j = \rho\boldsymbol{\Gamma}'_k$. Making use of these relations, we have

$$\boldsymbol{\Gamma}_j\mathbf{T}\boldsymbol{\Gamma}'_j = \sum_{i=1}^{16} \boldsymbol{\Gamma}_j\boldsymbol{\Gamma}_i\mathbf{G}\boldsymbol{\Gamma}'_i\boldsymbol{\Gamma}'_j = \sum_{k=1}^{16}\rho^{-1}\boldsymbol{\Gamma}_k\mathbf{G}(\rho\boldsymbol{\Gamma}'_k)$$

$$= \sum_{i=1}^{16} \boldsymbol{\Gamma}_k\mathbf{G}\boldsymbol{\Gamma}'_k = \mathbf{T} \text{ or } \boldsymbol{\Gamma}_j\,\mathbf{T} = \mathbf{T}(\boldsymbol{\Gamma}'_j)^{-1} = \mathbf{T}\boldsymbol{\Gamma}'_j$$

Furthermore, from eq. (35) $\boldsymbol{\Gamma}'_j\mathbf{S} = \mathbf{S}\boldsymbol{\Gamma}_j$ so that $\boldsymbol{\Gamma}_j\mathbf{T}\mathbf{S} = \mathbf{T}\boldsymbol{\Gamma}'_j\mathbf{S} = \mathbf{T}\mathbf{S}\boldsymbol{\Gamma}_j$. Hence, the matrix **TS** commutes with all the $\boldsymbol{\Gamma}_j$'s. Therefore, according to Schur's Lemma **TS** must be a multiple of the unit matrix $\mathbf{TS} = c\mathbf{I}$. As proved in the immediately preceding paragraph, we can choose **G** such that $\mathbf{T} \neq \mathbf{0}$. We now prove that with this choice of **G** (such that $\mathbf{T} \neq \mathbf{0}$), we can choose **F** such that $c \neq 0$. We assume the contrary, i.e., that $c = 0\ \forall\mathbf{F}$. This assumption gives

$$0 = \mathbf{TS} = \sum_{i=1}^{16} \mathbf{T}\boldsymbol{\Gamma}'_i\,\mathbf{F}\boldsymbol{\Gamma}_i \tag{36}$$

If we choose **F** such that the $(\alpha, \beta)^{\text{th}}$ element of **F** is unity and all the other elements vanish, then it follows from eq. (36) that $0 = \sum_{i=1}^{16} (\mathbf{T}\boldsymbol{\Gamma}'_i)_{\mu\alpha}(\Gamma_i)_{\beta\nu}$. This equation must hold for all values of α, β due to the arbitrary choice of the matrix **F**. Setting various values for $\alpha, \beta = 0, 1, 2, 3$, we get the matrix equation $\sum_{i=1}^{16} (\mathbf{T}\boldsymbol{\Gamma}'_i)_{\mu\nu}\Gamma_i = \mathbf{0}$. This requires that $(\mathbf{T}\Gamma'_i)_{\mu\nu} = 0\ \forall\ \mu, \nu = 0, 1, 2, 3$ since the $\boldsymbol{\Gamma}$ matrices are linearly independent, i.e., $\mathbf{T} = \mathbf{0}$ (because one of the $\boldsymbol{\Gamma}$ matrices is the identity matrix) which contradicts our assumption of $\mathbf{T} \neq \mathbf{0}$. Therefore, **S** must be non-singular and $\boldsymbol{\gamma}'^{\mu} = \mathbf{S}\boldsymbol{\gamma}^{\mu}\mathbf{S}^{-1}$.

(xii) We now establish that if $\boldsymbol{\gamma}^{\mu\dagger} = \mathbf{g}^{\mu\mu}\boldsymbol{\gamma}^{\mu}$, $\boldsymbol{\gamma}'^{\mu\dagger} = \mathbf{g}^{\mu\mu}\boldsymbol{\gamma}'^{\mu}$ then there exists a unitary matrix **U** such that $\boldsymbol{\gamma}'^{\mu} = \mathbf{U}\boldsymbol{\gamma}^{\mu}\mathbf{U}^{-1}$. We have already proved that there exists a non-singular matrix **S** such that $\boldsymbol{\gamma}'^{\mu} = \mathbf{S}\boldsymbol{\gamma}^{\mu}\mathbf{S}^{-1}$. Let us choose $\mathrm{U} = \dfrac{\mathbf{S}}{\det \mathbf{S}}$ so that $\mathbf{U}\boldsymbol{\gamma}^{\mu}\mathbf{U}^{-1} = \mathbf{S}\boldsymbol{\gamma}^{\mu}\mathbf{S}^{-1} = \boldsymbol{\gamma}'^{\mu}$. Also det $\mathbf{U} = \dfrac{\det \mathbf{S}}{\det \mathbf{S}} = \mathbf{1}$. Thus, $\mathrm{U} = \dfrac{\mathbf{S}}{\det \mathbf{S}}$ uniquely determines **U** except for a factor ± 1, $\pm i$ that subsists because $\det(\pm\mathbf{U}) = \det(\pm i\mathbf{U}) = \dfrac{\det \mathbf{S}}{\det \mathbf{S}} = \mathbf{1}$. A contrary assumption implies that there exist two matrices $\mathbf{U}_1$, $\mathbf{U}_2$ such that $\boldsymbol{\gamma}'^{\mu} = \mathbf{U}_1\boldsymbol{\gamma}^{\mu}\mathbf{U}_1^{-1}$, det $\mathbf{U}_1 = 1$ and $\boldsymbol{\gamma}'^{\mu} = \mathbf{U}_2\boldsymbol{\gamma}^{\mu}\mathbf{U}_2^{-1}$, det $\mathbf{U}_2 = 1$. Hence, $\boldsymbol{\gamma}'^{\mu} = \mathbf{U}_1\boldsymbol{\gamma}^{\mu}\mathbf{U}_1^{-1} = \mathbf{U}_2\boldsymbol{\gamma}^{\mu}\mathbf{U}_2^{-1} \Rightarrow \mathbf{U}_2^{-1}\mathbf{U}_1\boldsymbol{\gamma}^{\mu} = \boldsymbol{\gamma}^{\mu}\mathbf{U}_2^{-1}\mathbf{U}_1$, i.e., the matrix $\mathbf{U}_2^{-1}\mathbf{U}_1$ commutes with all the $\boldsymbol{\gamma}^{\mu}$. Therefore, $\mathbf{U}_2^{-1}\mathbf{U}_1$ must be a constant multiple of the identity matrix and we have $\mathbf{U}_1 = k\mathbf{U}_2$ with $k = \pm 1, \pm i$ since det $k\mathbf{U}_2$ = det $\mathbf{U}_1$ = det $\mathbf{U}_2$ =

$1 \Rightarrow k = \pm 1, \pm i$. Taking conjugate of $\gamma'^{\mu} = U\gamma^{\mu}U^{-1}$, we have $(\gamma'^{\mu})^{\dagger} = (U\gamma^{\mu}U^{-1})^{\dagger} = (U^{-1})^{\dagger} \gamma^{\mu\dagger}U^{\dagger}$ or $g^{\mu\mu}\gamma'^{\mu} = (U^{-1})^{\dagger}(g^{\mu\mu}\gamma^{\mu})U^{\dagger}$ or $\gamma'^{\mu} = (U^{\dagger})^{-1}\gamma^{\mu}U^{\dagger}$. Hence, $(U^{\dagger})^{-1}$ also satisfies the same conditions as U. Therefore, $\gamma^{\mu} = U\gamma^{\mu}U^{-1} = (U^{\dagger})^{-1}\gamma^{\mu}U^{\dagger}$ so that $U^{\dagger}U\gamma^{\mu} = \gamma^{\mu}U^{\dagger}U$ or $U^{\dagger}U$ commutes with all the γ^{μ}. Therefore, $U^{\dagger}U$ must be a constant multiple of the identity matrix and we have $U^{\dagger}U = lI$ or $U^{\dagger} = lU^{-1}$ with $l = \pm 1, \pm i$ since det lU^{-1} = det $U^{\dagger}$ = det $U = 1 \Rightarrow l = \pm 1, \pm i$. Now, the diagonal elements of the product of a matrix and its Hermitian conjugate have the form $(U^{\dagger}U)_{\alpha\alpha} = \sum_{\rho} (U^{\dagger})_{\alpha\rho} U_{\rho\alpha} = \sum_{\rho} (U)^{*}_{\rho\alpha} U_{\rho\alpha}$ and $(U)^{*}_{\rho\alpha} U_{\rho\alpha}$ is real and positive definite. Therefore, the factor l must also be positive definite and real, i.e., $l = +1$. Thus, $U^{\dagger}U = I$, as required.

(xiii) The matrix $\gamma^5 \equiv \gamma_5 = i\gamma^0\gamma^1\gamma^2\gamma^3 = -i\gamma_0\gamma_1\gamma_2\gamma_3$ anticommutes with all the matrices γ^{μ}, i.e.,

$$\{\gamma^{\mu}, \gamma^5\} = 0; \mu = 0, 1, 2, 3 \quad \text{(for example)} \tag{37}$$

$$\{\gamma^0, \gamma^5\} = \gamma^0\gamma^5 + \gamma^5\gamma^0 = \gamma^0 \cdot i\gamma^0\gamma^1\gamma^2\gamma^3 + i\gamma^0\gamma^1\gamma^2\gamma^3 \cdot \gamma^0$$
$$= i\gamma^0\gamma^1\gamma^2\gamma^3 - i\gamma^1\gamma^2\gamma^3\gamma^0 \cdot \gamma^0 = 0$$
$$\{\gamma^1, \gamma^5\} = \gamma^1\gamma^5 + \gamma^5\gamma^1 = \gamma^1 \cdot i\gamma^0\gamma^1\gamma^2\gamma^3 + i\gamma^0\gamma^1\gamma^2\gamma^3 \cdot \gamma^1 = i\gamma^0\gamma^2\gamma^3 - i\gamma^0\gamma^2\gamma^3 = 0$$
$$\{\gamma^2, \gamma^5\} = \gamma^2\gamma^5 + \gamma^5\gamma^2 = \gamma^2 \cdot i\gamma^0\gamma^1\gamma^2\gamma^3 + i\gamma^0\gamma^1\gamma^2\gamma^3 \cdot \gamma^2 = -i\gamma^0\gamma^1\gamma^3 + i\gamma^0\gamma^1\gamma^3 = 0$$
$$\{\gamma^3, \gamma^5\} = \gamma^3\gamma^5 + \gamma^5\gamma^3 = \gamma^3 \cdot i\gamma^0\gamma^1\gamma^2\gamma^3 + i\gamma^0\gamma^1\gamma^2\gamma^3 \cdot \gamma^3 = i\gamma^0\gamma^1\gamma^2 - i\gamma^0\gamma^1\gamma^2 = 0$$

Furthermore,

$$\left(\gamma^5\right)^2 = i\gamma^0\gamma^1\gamma^2\gamma^3 \cdot i\gamma^0\gamma^1\gamma^2\gamma^3 = -\gamma^0\gamma^1\gamma^2\gamma^3 \cdot \gamma^0\gamma^1\gamma^2\gamma^3$$
$$= -\gamma^0\gamma^1\gamma^2 \cdot \gamma^0\gamma^1\gamma^2 = \gamma^0\gamma^1 \cdot \gamma^0\gamma^1 = \gamma^0 \cdot \gamma^0 = 1 \tag{38}$$

Also

$$(\gamma^5)^{\dagger} = (i\gamma^0\gamma^1\gamma^2\gamma^3)^{\dagger} = -i\gamma^{3\dagger}\gamma^{2\dagger}\gamma^{1\dagger}\gamma^{0\dagger} = i\gamma^3\gamma^2\gamma^1\gamma^0$$
$$= -i\gamma^2\gamma^1\gamma^0\gamma^3 = -i\gamma^1\gamma^0\gamma^2\gamma^3 = i\gamma^0\gamma^1\gamma^2\gamma^3 = \gamma^5 \tag{39}$$

(xiv) The trace of the product of an odd number of Dirac γ^{μ} matrices vanishes. We have

$$\gamma^5\gamma^{\mu}(\gamma^5)^{-1} = \gamma^5\gamma^{\mu}\gamma^5 = -\left(\gamma^5\right)^2\gamma^{\mu} = (-1)\gamma^{\mu} \tag{40}$$

$$\gamma^5\gamma^{\mu}\gamma^{\nu}(\gamma^5)^{-1} = \gamma^5\gamma^{\mu}\gamma^{\nu}\gamma^5 = (-1)^2\left(\gamma^5\right)^2\gamma^{\mu}\gamma^{\nu} = (-1)^2\gamma^{\mu}\gamma^{\nu} \tag{41}$$

and similarly

$$\gamma^5\gamma^{\mu_1}\gamma^{\mu_2}.....\gamma^{\mu_n}(\gamma^5)^{-1} = \gamma^5\gamma^{\mu_1}\gamma^{\mu_2}.....\gamma^{\mu_n}\gamma^5 = (-1)^n\gamma^{\mu_1}\gamma^{\mu_2}.....\gamma^{\mu_n} \tag{42}$$

Taking trace of both sides of eq. (42), we have

$$(-1)^n Tr(\gamma\gamma^{\mu_1}\gamma^{\mu_2}.....\gamma^{\mu_n}) = Tr\,[\gamma^5\gamma^{\mu_1}\gamma^{\mu_2}.....\gamma^{\mu_n}\left(\gamma^5\right)^{-1}] = Tr(\gamma\gamma^{\mu_1}\gamma\gamma^{\mu_2}.....\gamma^{\mu_n}) \tag{43}$$

Hence, for odd values of n, we have

$$(-1)^n Tr(\gamma^{\mu_1}\gamma^{\mu_2}......\gamma^{\mu_n}) = (-1)\, Tr(\gamma^{\mu_1}\gamma^{\mu_2}......\gamma^{\mu_n})$$
$$= Tr(\gamma^{\mu_1}\gamma^{\mu_2}......\gamma^{\mu_n}) = 0 \tag{44}$$

(xv) The trace of the product of an even number of Dirac γ^μ matrices can always be reduced to one of $(n-2)$ matrices, e.g., for $n = 2$, we have

$$Tr\,(\gamma^\mu\gamma^\nu) = Tr\,(\gamma^\nu\gamma^\mu) = Tr\left[\frac{1}{2}(\gamma^\mu\gamma^\nu + \gamma^\nu\gamma^\mu)\right] = Tr(g^{\mu\nu}\mathbf{I} = 4g^{\mu\nu} \tag{45}$$

Similarly, for $(n-4)$ we have

$$\begin{aligned} Tr\,(\gamma^\mu\gamma^\nu\gamma^\rho\gamma^\sigma) &= Tr\,(\gamma^\sigma\gamma^\mu\gamma^\nu\gamma^\rho) = Tr(\gamma^\rho\gamma^\sigma\gamma^\mu\gamma^\nu) = Tr(\gamma^\nu\gamma^\rho\gamma^\sigma\gamma^\mu) \\ &= \frac{1}{4}\,Tr(\gamma^\mu\gamma^\nu\gamma^\rho\gamma^\sigma + \gamma^\sigma\gamma^\mu\gamma^\nu\gamma^\rho + \gamma^\rho\gamma^\sigma\gamma^\mu\gamma^\nu + \gamma^\nu\gamma^\rho\gamma^\sigma\gamma^\mu) \end{aligned} \tag{46}$$

Now

$$\begin{aligned} \gamma^\nu\gamma^\rho\gamma^\sigma\gamma^\mu &= (2g^{\nu\rho} - \gamma^\rho\gamma^\nu)\,\gamma^\sigma\gamma^\mu = 2g^{\nu\rho}\,\gamma^\sigma\gamma^\mu - \gamma^\rho\gamma^\nu\gamma^\sigma\gamma^\mu \\ &= 2g^{\nu\rho}\,\gamma^\sigma\gamma^\mu - \gamma^\rho = (2g^{\nu\sigma} - \gamma^\sigma\gamma^\nu)\gamma^\mu \\ &= 2g^{\nu\rho}\,\gamma^\sigma\gamma^\mu - 2g^{\nu\sigma}\gamma^\rho\gamma^\mu + \gamma^\rho\gamma^\sigma\gamma^\nu\gamma^\mu \\ &= 2g^{\nu\rho}\,\gamma^\sigma\gamma^\mu - 2g^{\nu\sigma}\gamma^\rho\gamma^\mu + \gamma^\rho\gamma^\sigma(2g^{\mu\nu} - \gamma^\mu\gamma^\nu) \\ &= 2g^{\nu\rho}\,\gamma^\sigma\gamma^\mu - 2g^{\nu\sigma}\gamma^\rho\gamma^\mu + 2g^{\mu\nu}\gamma^\rho\gamma^\sigma - \gamma^\rho\gamma^\sigma\gamma^\mu\gamma^\nu \end{aligned}$$

so that

$$\gamma^\rho\gamma^\sigma\gamma^\mu\gamma^\nu + \gamma^\nu\gamma^\rho\gamma^\sigma\gamma^\mu = 2g^{\nu\rho}\gamma^\sigma\gamma^\mu - 2g^{\nu\sigma}\gamma^\rho\gamma^\mu + 2g^{\mu\nu}\,\gamma^\rho\gamma^\sigma \tag{47}$$

Similarly,

$$\begin{aligned} \gamma^\sigma\gamma^\mu\gamma^\nu\gamma^\rho &= (2g^{\mu\sigma} - \gamma^\mu\gamma^\sigma)\,\gamma^\nu\gamma^\rho = 2g^{\mu\sigma}\,\gamma^\nu\gamma^\rho - \gamma^\mu\gamma^\sigma\gamma^\nu\gamma^\rho \\ &= 2g^{\mu\sigma}\gamma^\nu\gamma^\rho - \gamma^\mu\,(2g^{\nu\sigma} - \gamma^\nu\gamma^\sigma)\gamma^\rho \\ &= 2g^{\mu\sigma}\gamma^\nu\gamma^\rho - 2g^{\nu\sigma}\gamma^\mu\gamma^\rho + \gamma^\mu\gamma^\nu\gamma^\sigma\gamma^\rho \\ &= 2g^{\mu\sigma}\gamma^\nu\gamma^\rho - 2g^{\nu\sigma}\gamma^\mu\gamma^\rho + \gamma^\mu\gamma^\nu\,(2g^{\rho\sigma} - \gamma^\rho\gamma^\sigma) \\ \text{or} \quad &= 2g^{\mu\sigma}\gamma^\nu\gamma^\rho - 2g^{\nu\sigma}\gamma^\mu\gamma^\rho + 2g^{\rho\sigma}\gamma^\mu\gamma^\nu - \gamma^\mu\gamma^\nu\gamma^\rho\gamma^\sigma \end{aligned}$$

$$\gamma^\mu\gamma^\nu\gamma^\rho\gamma^\sigma + \gamma^\sigma\gamma^\mu\gamma^\nu\gamma^\rho = 2g^{\mu\sigma}\gamma^\nu\gamma^\rho - 2g^{\nu\sigma}\gamma^\mu\gamma^\rho + 2g^{\rho\sigma}\gamma^\mu\gamma^\nu \tag{48}$$

Adding eqs. (47) and (48), we have

$$\begin{aligned} \gamma^\mu\gamma^\nu\gamma^\rho\gamma^\sigma + \gamma^\sigma\gamma^\mu\gamma^\nu\gamma^\rho + \gamma^\rho\gamma^\sigma\gamma^\mu\gamma^\nu + \gamma^\nu\gamma^\rho\gamma^\sigma\gamma^\mu &= 2g^{\nu\rho}\gamma^\sigma\gamma^\mu - 2g^{\nu\sigma}\gamma^\rho\gamma^\mu \\ &+ 2g^{\mu\nu}\gamma^\rho\gamma^\sigma + 2g^{\mu\sigma}\gamma^\nu\gamma^\rho - 2g^{\nu\sigma} - \gamma^\mu\gamma^\rho + 2g^{\rho\sigma}\gamma^\mu\gamma^\nu \end{aligned} \tag{49}$$

On taking trace of both sides and using eq. (45), we have

$$\begin{aligned} Tr\,(\gamma^\mu\gamma^\nu\gamma^\rho\gamma^\sigma + \gamma^\sigma\gamma^\mu\gamma^\nu\gamma^\rho + \gamma^\rho\gamma^\sigma\gamma^\mu\gamma^\nu + \gamma^\nu\gamma^\rho\gamma^\sigma\gamma^\mu) \\ = 8g^{\nu\rho}g^{\mu\sigma} - 8g^{\nu\sigma}g^{\mu\rho} + 8g^{\mu\nu}g^{\rho\sigma} + 8g^{\mu\sigma}g^{\nu\rho} - 8g^{\nu\sigma}g^{\mu\rho} + 8g^{\rho\sigma}g^{\mu\nu} \\ = 16g^{\nu\rho}g^{\mu\sigma} - 16g^{\nu\sigma}g^{\mu\rho} + 16g^{\mu\nu}g^{\rho\sigma} \end{aligned} \tag{50}$$

Hence, from eq. (46)

$$
\begin{aligned}
Tr(\gamma^{\mu}\gamma^{\nu}\gamma^{\rho}\gamma^{\sigma}) &= Tr(\gamma^{\sigma}\gamma^{\mu}\gamma^{\nu}\gamma^{\rho}) = Tr(\gamma^{\rho}\gamma^{\sigma}\gamma^{\mu}\gamma^{\nu}) = Tr(\gamma^{\nu}\gamma^{\rho}\gamma^{\sigma}\gamma^{\mu}) \\
&= \frac{1}{4}\, Tr(\gamma^{\mu}\gamma^{\nu}\gamma^{\rho}\gamma^{\sigma} + \gamma^{\sigma}\gamma^{\mu}\gamma^{\nu}\gamma^{\rho} + \gamma^{\rho}\gamma^{\sigma}\gamma^{\mu}\gamma^{\nu} + \gamma^{\nu}\gamma^{\rho}\gamma^{\sigma}\gamma^{\mu}) \\
&= 4g^{\nu\rho}g^{\mu\sigma} - 4g^{\nu\sigma}g^{\mu\rho} + 4g^{\mu\nu}g^{\rho\sigma} \qquad (51)
\end{aligned}
$$

and so on.

4.3 DIRAC EQUATION IN COVARIANT FORM

The Dirac eq. (5) can be put in a more elegant and covariant form as follows:

Multiplying eq. (5) throughout by $\boldsymbol{\beta}$, we have $\boldsymbol{\beta}\partial_0\Psi + (\boldsymbol{\beta\alpha})\cdot\nabla\Psi + im\Psi = 0$. Using the definition of $\boldsymbol{\gamma}$ matrices of eqs. (18) (19), we obtain $\boldsymbol{\gamma}^0\partial_0\Psi + \boldsymbol{\gamma}\cdot\nabla\Psi + im\Psi = 0$ or equivalently

$$
\boldsymbol{\gamma}^0\partial_0\Psi + \boldsymbol{\gamma}^k\,\partial_k\Psi + im\Psi = 0 = \boldsymbol{\gamma}^{\mu}\partial_{\mu}\Psi + im\Psi = \partial\!\!\!/\,\Psi + im\Psi \qquad (52)
$$

Identifying $i\partial_{\mu} \equiv \hat{p}_{\mu}$, the 4-momentum operator, eq. (52) may also be written as

$$
0 = \boldsymbol{\gamma}^{\mu}\,\hat{p}_{\mu}\Psi - m\Psi = (\hat{p}\!\!\!/ - m)\Psi \qquad (53)
$$

The above expressions for the Dirac equation are explicitly covariant since space and time derivatives are treated alike. The dagger notation $\partial\!\!\!/ \equiv \boldsymbol{\gamma}^{\mu}\partial_{\mu}$, $\hat{p}\!\!\!/ \equiv \boldsymbol{\gamma}^{\mu}\,\hat{p}_u$ was introduced by Robert Feynmann to simplify representation.

4.4 EXPLICIT REPRESENTATIONS OF THE DIRAC MATRICES

One of the commonly used explicit representations of the Dirac matrices is

$$
\boldsymbol{\gamma}^1 = \begin{pmatrix} 0 & 0 & 0 & -1 \\ 0 & 0 & -1 & 0 \\ 0 & 1 & 0 & 0 \\ 1 & 0 & 0 & 0 \end{pmatrix}, \boldsymbol{\gamma}^2 = \begin{pmatrix} 0 & 0 & 0 & i \\ 0 & 0 & -i & 0 \\ 0 & -i & 0 & 0 \\ i & 0 & 0 & 0 \end{pmatrix}
$$

$$
\boldsymbol{\gamma}^3 = \begin{pmatrix} 0 & 0 & -1 & 0 \\ 0 & 0 & 0 & 1 \\ 1 & 0 & 0 & 0 \\ 0 & -1 & 0 & 0 \end{pmatrix}, \boldsymbol{\gamma}^0 = \begin{pmatrix} 0 & 0 & 1 & 0 \\ 0 & 0 & 0 & 1 \\ 1 & 0 & 0 & 0 \\ 0 & 1 & 0 & 0 \end{pmatrix} \qquad (54)
$$

Another representation called the Dirac representation, also used in explicit calculations, is given by

$$
\boldsymbol{\gamma}^{k(D)} = \begin{pmatrix} 0 & \sigma_k \\ -\sigma_k & 0 \end{pmatrix};\ k = 1, 2, 3, \text{ and } \boldsymbol{\gamma}^{0(D)} = \begin{pmatrix} I & 0 \\ 0 & -I \end{pmatrix} \qquad (55)
$$

where σ_k ($k = 1, 2, 3$) are the well known Pauli matrices

$$\boldsymbol{\sigma}_1 = \begin{pmatrix} 0 & 1 \\ 1 & 0 \end{pmatrix}, \boldsymbol{\sigma}_2 = \begin{pmatrix} 0 & -i \\ i & 0 \end{pmatrix}, \boldsymbol{\sigma}_3 = \begin{pmatrix} 1 & 0 \\ 0 & -1 \end{pmatrix} \tag{56}$$

The above representations of the Dirac matrices are equivalent and the transformation matrix, for example, from (54) to (55), is given by

$$\mathbf{S} = \frac{\rho}{\sqrt{2}} \begin{pmatrix} 1 & 0 & 1 & 0 \\ 0 & 1 & 0 & 1 \\ 1 & 0 & -1 & 0 \\ 0 & 1 & 0 & -1 \end{pmatrix} \quad \text{with } \rho = \pm 1, \pm i \tag{57}$$

Another representation of the Dirac matrices that finds application in some computations is the Majorana representation

$$\boldsymbol{\gamma}^{(M)1} = -i\,\boldsymbol{\gamma}^{(D)2}, \boldsymbol{\gamma}^{(M)2} = \boldsymbol{\gamma}^{(D)0}, \boldsymbol{\gamma}^{(M)3} = -i\,\gamma_1^{(D)}\gamma_2^{(D)}\gamma_3^{(D)}\gamma_0^{(D)}, \boldsymbol{\gamma}^{(M)0} = -i\,\boldsymbol{\gamma}^{(D)1} \tag{58}$$

In the Dirac representation of γ matrices, the following relations also hold:

$$\boldsymbol{\gamma}^0(\boldsymbol{\gamma}^\mu)^\dagger\,\boldsymbol{\gamma}^0 = \boldsymbol{\gamma}^\mu \tag{59}$$

and, on taking transpose of eq. (59)

$$[\boldsymbol{\gamma}^0(\boldsymbol{\gamma}^\mu)^\dagger\boldsymbol{\gamma}^0]^T = (\boldsymbol{\gamma}^0)^T[(\boldsymbol{\gamma}^\mu)^\dagger]^T(\boldsymbol{\gamma}^0)^T = \boldsymbol{\gamma}^0(\boldsymbol{\gamma}^\mu)^*\,\boldsymbol{\gamma}^0 = (\boldsymbol{\gamma}^\mu)^T \tag{60}$$

Also

$$(\boldsymbol{\gamma}^0)^T = \boldsymbol{\gamma}^0, (\boldsymbol{\gamma}^1)^T = -\boldsymbol{\gamma}^1, (\boldsymbol{\gamma}^2)^T = \boldsymbol{\gamma}^2, (\boldsymbol{\gamma}^3)^T = -\boldsymbol{\gamma}^3 \tag{61}$$

These can be easily verified by explicit calculations.

4.5 NON-RELATIVISTIC LIMIT OF THE DIRAC EQUATION

We now examine whether the free particle Dirac equation corresponds to the free particle Schrödinger equation in the non-relativistic limit. The Dirac equation for a free particle given by eq. (5) can be written as

$$\hat{p}_0\,\Psi = (\alpha\cdot\hat{\mathbf{p}})\Psi + m\beta\Psi \tag{62}$$

where Ψ is a 4-component spinor that can conveniently be written as a pair of 2-component spinors $\Psi \equiv (\phi\ \chi)^T$, where $\phi \equiv (\phi_1\ \phi_2)^T$ constitutes the upper two components and $\chi \equiv (\chi_1\ \chi_2)^T$ makes the lower two components. We can then write eq. (62), using the representation $\boldsymbol{\alpha}^k = \begin{pmatrix} 0 & \sigma_k \\ \sigma_k & 0 \end{pmatrix}$; $k = 1, 2, 3$, $\beta = \begin{pmatrix} \mathbf{I} & 0 \\ 0 & -\mathbf{I} \end{pmatrix}$ as

$$\hat{p}_0\begin{pmatrix} \phi \\ \chi \end{pmatrix} = \begin{pmatrix} \mathbf{0} & \sigma \\ \sigma & \mathbf{0} \end{pmatrix}\hat{\mathbf{p}}\begin{pmatrix} \phi \\ \chi \end{pmatrix} + m\begin{pmatrix} \mathrm{I} & \mathbf{0} \\ \mathbf{0} & -\mathrm{I} \end{pmatrix}\begin{pmatrix} \phi \\ \chi \end{pmatrix}$$

which gives the following pair of coupled first order differential equations:

$$\hat{p}_0 \begin{pmatrix} \phi \\ \chi \end{pmatrix} = \begin{pmatrix} \sigma \hat{\mathbf{p}} \chi \\ \sigma \hat{\mathbf{p}} \phi \end{pmatrix} + m \begin{pmatrix} \phi \\ -\chi \end{pmatrix} \tag{63}$$

As in the case of evaluating the non-relativistic limit of the Klein Gordon equation, we separate out the rest energy related term from the wavefunction Ψ by writing

$$\Psi = \Psi_{nr} \exp(-imx^0) = \begin{pmatrix} \phi \\ \chi \end{pmatrix} = \begin{pmatrix} \phi_{nr} \\ \chi_{nr} \end{pmatrix} \exp(-imx^0) \tag{64}$$

Hence,

$$\hat{p}_0 \Psi = i\partial_0 \Psi = i\partial_0 [\Psi_{nr} \exp(-imx^0)] = [(i\partial_0 \Psi_{nr}) + m\Psi_{nr}]\exp(-imx^0) \tag{65}$$

$$= \left[i\partial_0 \begin{pmatrix} \phi_{nr} \\ \chi_{nr} \end{pmatrix} + m \begin{pmatrix} \phi_{nr} \\ \chi_{nr} \end{pmatrix} \right] \exp(-imx^0)$$

Substituting eqs. (64) and (65) in eq. (63), we get

$$I\partial_0 \begin{pmatrix} \phi_{nr} \\ \chi_{nr} \end{pmatrix} + m \begin{pmatrix} \phi_{nr} \\ \chi_{nr} \end{pmatrix} = \begin{pmatrix} \sigma \hat{\mathbf{p}} \chi_{nr} \\ \sigma \hat{\mathbf{p}} \phi_{nr} \end{pmatrix} + m \begin{pmatrix} \phi_{nr} \\ -\chi_{nr} \end{pmatrix} \Rightarrow i\partial 0 \begin{pmatrix} \phi_{nr} \\ \chi_{nr} \end{pmatrix} = \begin{pmatrix} \sigma \hat{\mathbf{p}} \chi_{nr} \\ \sigma \hat{\mathbf{p}} \phi_{nr} \end{pmatrix} - 2m \begin{pmatrix} 0 \\ \chi_{nr} \end{pmatrix} \tag{66}$$

Now, the non-relativistic condition effectively means that the kinetic energy of the free particle is far less than the rest energy, i.e., $|i\partial_0 \chi| << |m\chi|$. Then, from the second of eqs. (66), we have

$$\chi_{nr} = \frac{\sigma \cdot \hat{\mathbf{p}}}{2m} \phi_{nr} \rightarrow 0 \text{since } |\hat{\mathbf{p}}\phi| << m \tag{67}$$

Thus, χ_{nr} represents the smaller pair of components of the wavefunction Ψ. Inserting the value of χ_{nr} from eq. (67) in the first of eq. (66), we get

$$i\partial_0 \phi_{nr} = \sigma \hat{\mathbf{p}} \chi_{nr} = \frac{(\sigma \hat{\mathbf{p}})^2}{2m} \phi_{nr} = \frac{\hat{\mathbf{p}}^2}{2m} \phi_{nr} \tag{68}$$

which is exactly the free particle Schrödinger equation.

We, therefore, conclude that in the non-relativistic limit, the upper 2-component spinor $\phi \equiv (\phi_1\ \phi_2)^T$ satisfies the Schrödinger equation while the lower 2-component spinor $\chi \equiv (\chi_1\ \chi_2)^T$ vanishes. Hence, in the non-relativistic case, the Dirac equation transforms to the Pauli equation which is the non-relativistic equation for spin 1/2 particles.

Proceeding in an exactly similar manner, we can establish that for negative energy solutions, the lower 2-component spinor $\chi \equiv (\chi_1\ \chi_2)^T$ satisfies the Schrödinger equation, while the upper 2-component spinor $\chi \equiv (\chi_1\ \chi_2)^T$ vanishes.

From eq. (67) we also conclude that in the limit of non-relativistic velocities, the 4-component Dirac spinor will take the following form for various combinations of helicity and energy for a free particle at rest:

Abbreviation	Energy Sign	Helicity	Expression
$\omega^{(1)}(\mathbf{p})$	+	+1	$(1\ 0\ 0\ 0)^T$
$\omega^{(2)}(\mathbf{p})$	+	−1	$(0\ 1\ 0\ 0)^T$
$\omega^{(3)}(\mathbf{p})$	−	+1	$(0\ 0\ 1\ 0)^T$
$\omega^{(4)}(\mathbf{p})$	−	−1	$(0\ 0\ 0\ 1)^T$

CHAPTER 5

Relativistic Invariance of the Dirac Equation

5.1 PROPERTIES OF LORENTZ TRANSFORMATIONS

Before discussing relativistic invariance of the Dirac equation, we summarize below the salient features of Lorentz transformations.

(a) Consider two different observers A and B located in different inertial frames I and I' and observing the same set of physical phenomenon from their respective positions. Further, assume that a point P in four-dimensional space-time is identified by the coordinates x^μ by observer A with respect to the inertial frame I and coordinates $(x')^\mu$ by observer B with reference to frame I'. The two sets of coordinates are then connected through a Lorentz transformation and expressed as

$$(x')^\nu = \sum_{\mu=0}^{3} a^\nu{}_\mu x^\mu \equiv a^\nu{}_\mu x^\mu, \; x = \Lambda x \tag{1}$$

where Λ is a 4×4 matrix and x a 4-component vector.

(b) The Lorentz transformation represented by the 4×4 matrix Λ is a linear homogeneous transformation. The constituents of the transformation matrix depend only on the spatial orientation and relative velocities of the two inertial frames.

(c) The Lorentz transformation leaves the distance between two space-time points x^μ and $(x')^\mu$ invariant, i.e.,

$$\begin{aligned} ds'^2 &= (dx')^\mu (dx')_\mu = g'_{\mu\nu}(dx')^\mu (dx')^\nu \\ &= g_{\mu\nu}(dx')^\mu (dx')^\nu \\ &= g_{\mu\nu} dx^\mu dx^\nu = dx^\mu dx_\mu = ds^2 \end{aligned} \tag{2}$$

(d) The requirement of invariance of space-time distance between two points connected by a Lorentz transformation translates to the following constraints on the transformation matrix:

$$(dx')^\mu (dx')_\mu = a^\mu{}_\nu a_\mu{}^\sigma dx^\nu dx_\sigma = dx^\nu dx_\nu = \delta^\sigma_\nu dx^\nu dx_\sigma \text{ so that } a^\mu{}_\nu a_\mu{}^\sigma = \delta^\sigma_\nu \tag{3}$$

(e) Equation (3) further implies that $\left[\det\left(a^\mu{}_\nu\right)\right]^2 = 1$ or $\det\left(a^\mu{}_\nu\right) = \pm 1$. Lorentz transformations having $\det(a^\mu{}_\nu) = +1$ constitute the group of proper Lorentz transformations. They contain the identity element **I**. Any element of this group can be obtained from the identity by an infinite succession of infinitesimal transformations. Three-dimensional rotations and Lorentz boosts belong to the group of proper Lorentz transformations.

Lorentz transformations with $\det(a^\mu{}_\nu) = -1$ are called improper Lorentz transformations. They cannot be obtained from identity by successive infinitesimal Lorentz transformations and comprise reflections in space and time.

(f) The infinitesimal Lorentz transformations are given by

$$a^\nu{}_\mu = \delta^\nu_\mu + \Delta\omega^\nu{}_\mu \text{ with } \Delta\omega^{\nu\mu} = -\Delta\omega^{\mu\nu} \tag{4}$$

which can be established by ignoring second and higher order terms in $a^\mu{}_\nu a_\mu{}^\rho = \delta^\rho_\nu$, e.g., (to first order in $\Delta\omega$)

$$\left(\delta^\mu_\nu + \Delta\omega^\mu{}_\nu\right)\left(\delta^\rho_\mu + \Delta\omega_\mu{}^\rho\right) = \delta^\mu_\nu \delta^\rho_\mu + \delta^\mu_\nu \Delta\omega_\mu{}^\rho + \delta^\rho_\mu \Delta\omega^\mu{}_\nu$$

$$= \delta^\rho_\nu + \Delta\omega_\nu{}^\rho + \Delta\omega^\rho{}_\nu \Rightarrow \Delta\omega_\nu{}^\rho + \Delta\omega^\rho{}_\nu = 0$$

or

$$\Delta\omega_\nu{}^\rho = -\Delta\omega^\rho{}_\nu \text{ or } g_{\mu\nu}\Delta\omega^{\mu\rho} = -g_{\nu\mu}\Delta\omega^{\rho\mu} \text{ or } \Delta\omega^{\mu\rho} = -\Delta\omega^{\rho\mu}$$

as required.

(g) Of the 16 possible parameters of the Lorentz transformation $\Delta\omega^{\mu\nu}$; $\mu, \nu = 0, 1, 2, 3$, the four diagonal elements vanish since $\Delta\omega^{\mu\mu} = -\Delta\omega^{\mu\mu} = 0$, six of the remaining twelve are dependent on the other six since $\Delta\omega^{\mu\nu} = -\Delta\omega^{\nu\mu}$, so that we have six independent parameters. Each of these independent parameters constitute a rotation angle in 4-dimensional Minkowski space and generates an infinitesimal Lorentz transformation.

5.2 LORENTZ COVARIANCE OF DIRAC EQUATION

In order to establish Lorentz covariance of the Dirac equation, we need to prove that the "physics of Dirac particles is the same in every inertial frame". In other words, if $\psi(x)$ is the wavefunction describing a certain state as observed by A in inertial frame I and it satisfies the Dirac equation

$$(i\gamma^\mu \partial_\mu - m)\,\psi(x) = 0 \tag{5}$$

then the wavefunction $\psi'(x')$ that describes the same state as observed by B in inertial frame I' should satisfy an equation of the same form, e.g.,

$$(i\gamma'^{\mu}\,\partial'_{\mu} - m)\,\psi'(x') = 0 \tag{6}$$

where the γ'^{μ} satisfy anticommutation relations similar to those of γ^{μ}, i.e.,

$$\{\gamma'^{\mu}, \gamma'^{\nu}\} = 2g^{\mu\nu}\mathbf{I} \tag{7}$$

and also meet the Hermiticity requirement

$$\gamma'^{0\dagger} = \gamma'^{0} \text{ and } \gamma'^{k} = -\gamma'^{k};\ k = 1, 2, 3 \tag{8}$$

We write Dirac eq. (6) in the primed frame as

$$i\gamma'^{0}\,\partial'_0\psi'(x') = \left(-i\sum_{k=1}^{3}\gamma'^{k}\partial'_k + m\right)\psi'(x')$$

or equivalently as

$$i\partial'_0\psi'(x') = \left(-i\sum_{k=1}^{3}\gamma'^{0}\,\gamma'^{k}\partial'_k + \gamma'^{0}\,m\right)\psi'(x') \equiv \hat{H}'\psi'(x') \tag{9}$$

where $\hat{H}'$ is the Hamiltonian in the primed frame given by

$$\hat{H}' \equiv -i\sum_{k=1}^{3}\gamma'^{0}\,\gamma'^{k}\,\partial'_k + \gamma'^{0}m \tag{10}$$

In order that this Hamiltonian have real energy eigenvalues, it must be Hermitian, i.e., $(\hat{H}')^{\dagger} \equiv \hat{H}'$ which necessitates $\gamma'^{0\dagger} = \gamma'^{0}$ and $\gamma'^{k\dagger} = -\gamma'^{k}$; $k = 1, 2, 3$, i.e., γ matrices in the primed frame obey properties similar to their counterparts in the unprimed frame. We have already shown in Chapter 4 that if two sets of γ matrices obey the same anticommutation relation (7), then they are unitarily equivalent, i.e. there exists a unitary transformation $\mathbf{U}$ such that $\gamma'^{\mu} = \mathbf{U}^{\dagger}\gamma^{\mu}\mathbf{U}$, $\mathbf{U}^{\dagger}\mathbf{U} = \mathbf{U}\mathbf{U}^{\dagger} = \mathbf{I}$. We can, therefore, replace γ' by γ in eq. (10) for the primed system, i.e. we do not need to differentiate between the two sets of γ matrices, γ' and γ. Hence, we can write the Dirac equation in the primed system as

$$\left(i\gamma^{0}\partial'_0 + i\sum_{k=1}^{3}\gamma^{k}\partial'_k - m\right)\psi'(x') = \left(i\gamma^{\mu}\partial'_{\mu} - m\right)\psi'(x') = 0 \tag{11}$$

Let $\hat{S}(\Lambda)$ represent the transformation operator that transforms the wavefunction in the unprimed system $\psi(x)$, to that in the primed system $\psi'(x')$, i.e.,

$$\psi'(x') = \hat{S}(\Lambda)\psi(x) \tag{12}$$

$\hat{S}(\Lambda)$ must be linear in the space-time coordinates since both, the Lorentz transformation and the Dirac equation, are linear in the space-time coordinates. Furthermore, $x' = \Lambda x$, where the matrix Λ is the Lorentz transformation from the unprimed to the primed frame.

Hence, we have

$$\psi'(x') = \psi'(\Lambda x) = \hat{S}(\Lambda)\psi(x) = \hat{S}(\Lambda)\psi(\Lambda^{-1}x') \tag{13}$$

Obviously, $\hat{S}(\Lambda)$ is a function of the parameters of the Lorentz transformation $a^{\mu}{}_{\nu}$ and depends on the relative velocities and spatial orientations of the two frames.

Now, by the Equivalence Principle of Relativity, the transformation matrix $\hat{S}(\Lambda)$ must necessarily be non-singular, i.e., corresponding to $\hat{S}(\Lambda)$ there must exist $\hat{S}^{-1}(\Lambda)$ that enables a transformation of the wavefunction from the primed to the unprimed frame, i.e.,

$$\begin{aligned}\hat{S}^{-1}(\Lambda)\psi'(\Lambda x) &= \hat{S}^{-1}(\Lambda)\psi'(x') = \psi(x) = \hat{S}(\mathbf{I})\psi(x) = \hat{S}(\Lambda^{-1}\Lambda)\psi(x)\\ &= \hat{S}(\Lambda^{-1})\hat{S}(\Lambda)\psi(x) = \hat{S}(\Lambda^{-1})\psi'(x')\\ &= \hat{S}(\Lambda^{-1})\psi'(\Lambda x) \Rightarrow \hat{S}^{-1}(\Lambda) = \hat{S}(\Lambda^{-1})\end{aligned} \tag{14}$$

This is the fundamental property that must be satisfied by the transformation operator $\hat{S}(\Lambda)$.

To find other properties of this operator, we return to the Dirac equation in the unprimed system, viz. eq. (5). Using eq. (14), we have

$$\begin{aligned}0 &= (i\gamma^{\mu}\partial_{\mu} - m)\hat{S}^{-1}(\mathbf{\Lambda})\psi'(x')\\ &= (i\gamma^{\mu}\hat{S}^{-1}(\mathbf{\Lambda})\partial_{\mu} - \hat{S}^{-1}(\mathbf{\Lambda})m)\psi'(x')\end{aligned} \tag{15}$$

Multiplying eq. (15) from the left by $\hat{S}(\Lambda)$ and noting that

$$\partial_{\mu} = \frac{\partial}{\partial x^{\mu}} = \frac{\partial x'^{\nu}}{\partial x^{\mu}}\frac{\partial}{\partial x'^{\nu}} = a^{\nu}{}_{\mu}\frac{\partial}{\partial x'^{\nu}} = a^{\nu}{}_{\mu}\partial'_{\nu} \text{ since}$$

$$(x')^{\nu} = \sum_{\mu=0}^{3} a^{\nu}_{\mu}x^{\mu} \equiv a^{\nu}{}_{\mu}x^{\mu}, \text{ we obtain}$$

$$\left[i\hat{S}(\Lambda)\gamma^{\mu}\hat{S}^{-1}(\Lambda)a^{\nu}{}_{\mu}\partial'_{\nu} - m\right]\psi'(x') = 0 \tag{16}$$

Since we require form invariance of the Dirac equation, i.e., eq. (16) must be identical to the Dirac equation in the primed system, i.e., eq. (11), we must identify (corresponding to the Lorentz transformation $(x')^{\nu} = a^{\nu}{}_{\mu}x^{\mu}$)

$$\hat{S}(\mathbf{\Lambda})\gamma^{\mu}\hat{S}^{-1}(\mathbf{\Lambda})a^{\nu}{}_{\mu} = \boldsymbol{\gamma}^{\nu}$$

or equivalently

$$\hat{S}(\mathbf{\Lambda})\gamma^{\mu}\hat{S}^{-1}(\mathbf{\Lambda}) = a^{\mu}{}_{\nu}\boldsymbol{\gamma}^{\nu} \text{ or } \hat{S}^{-1}(\mathbf{\Lambda})\gamma^{\nu}\hat{S}(\mathbf{\Lambda}) = a^{\nu}{}_{\mu}\boldsymbol{\gamma}^{\mu} \tag{17}$$

From eq. (17), it can be shown that if we define the γ matrices for the primed frame as $\gamma'^{\nu} = a^{\nu}{}_{\mu}\gamma^{\mu}$, then they also satisfy $\{\gamma'^{\mu}, \gamma'^{\nu}\} = 2g^{\mu\nu}\mathbf{I}$ for $\gamma'^{\nu}\gamma'^{\lambda} = S^{-1}(\Lambda)\gamma^{\nu}\gamma^{\lambda}S(\Lambda) = S^{-1}(\Lambda)(2g^{\nu\lambda}\mathrm{I} - \gamma^{\lambda}\gamma^{\nu})S(\Lambda) = 2g^{\nu\lambda}\mathrm{I} - \gamma'^{\lambda}\gamma'^{\nu}$.

Thus, in order to prove the covariance of the Dirac equation, we need to show explicitly that there exists a solution to eq. (17). In the sequel, we will construct the $\hat{S}$ operator for proper and improper Lorentz transformations.

5.3 $\hat{S}$ OPERATOR FOR INFINITESIMAL PROPER LORENTZ TRANSFORMATIONS

We now construct the $\hat{S}$ operator for infinitesimal proper Lorentz transformations given by eq. (4). In this case, $\hat{S}$ will be a function of the parameters $\Delta\omega^{\mu\nu}$ of the infinitesimal Lorentz transformation. Therefore, expanding $\hat{S}(\Delta\omega^{\mu\nu})$ as a series in powers of $\Delta\omega^{\mu\nu}$ and retaining only terms to first order in $\Delta\omega^{\mu\nu}$, we have

$$\hat{S}(\Delta\omega^{\mu\nu}) = \hat{I} + \hat{\sigma}_{\mu\nu}\Delta\omega^{\mu\nu} \tag{18}$$

and

$$\hat{S}^{-1}(\Delta\omega^{\mu\nu}) = \hat{I} - \hat{\sigma}_{\mu\nu}\Delta\omega^{\mu\nu} \tag{19}$$

The coefficients $\hat{\sigma}_{\nu\mu}$ are operators represented by 4×4 matrices that also obey

$$\hat{\sigma}_{\mu\nu} = -\hat{\sigma}_{\nu\mu} \tag{20}$$

since

$$\begin{aligned}\hat{I} - \hat{\sigma}_{\mu\nu}\Delta\omega^{\mu\nu} &= \hat{S}^{-1}(\Delta\omega^{\mu\nu}) \\ &= \hat{S}(\Delta\omega^{\mu\nu})^{-1} = \hat{S}'(\Delta\omega^{\mu\nu}) \\ &= \hat{I} + \hat{\sigma}_{\nu\mu}\Delta\omega^{\mu\nu} = \hat{I} - \hat{\sigma}_{\nu\mu}\Delta\omega^{\nu\mu}\end{aligned}$$

Substituting eqs. (4), (18), (19) in eq. (17), we obtain

$$(\hat{I} + \hat{\sigma}_{\alpha\beta}\Delta\omega^{\alpha\beta})\,\gamma^{\mu}(\hat{I} - \hat{\sigma}_{\alpha\beta}\Delta\omega^{\alpha\beta}) = (\delta^{\mu}_{\nu} + \Delta\omega^{\mu}{}_{\nu})\gamma^{\nu} \tag{21}$$

which gives, to first order in $\Delta\omega^{\mu\nu}$

$$\hat{\sigma}_{\alpha\beta}\Delta\omega^{\alpha\beta}\gamma^{\mu} - \gamma^{\mu}\hat{\sigma}_{\alpha\beta}\Delta\omega^{\alpha\beta} = \Delta\omega^{\mu}{}_{\nu}\gamma^{\nu} \tag{22}$$

$$\begin{aligned}\Delta\omega^{\alpha\beta}(\hat{\sigma}_{\alpha\beta}\gamma^{\mu} - \gamma^{\mu}\hat{\sigma}_{\alpha\beta}) &= \Delta\omega^{\mu}{}_{\nu}\gamma^{\nu} = g^{\mu}{}_{\sigma}\Delta\omega^{\sigma}{}_{\nu}\gamma^{\nu} = \Delta\omega^{\sigma}{}_{\nu}g^{\mu}{}_{\sigma}\gamma^{\nu} \\ &= \Delta\omega^{\alpha}{}_{\beta}g^{\mu}{}_{\alpha}\gamma^{\beta} = \Delta\omega^{\beta\alpha}g^{\mu}_{\alpha}\gamma_{\beta} = \Delta\omega^{\alpha\beta}g^{\mu}_{\beta}\gamma_{\alpha} = -\Delta\omega^{\beta\alpha}g^{\mu}{}_{\beta}\gamma_{\alpha}\end{aligned} \tag{23}$$

so that

$$\Delta\omega^{\alpha\beta}(\hat{\sigma}_{\alpha\beta}\gamma^{\mu} - \gamma^{\mu}\hat{\sigma}_{\alpha\beta}) = \frac{1}{2}\Delta\omega^{\beta\alpha}(g^{\mu}{}_{\alpha}\gamma_{\beta} - g^{\mu}{}_{\beta}\gamma_{\alpha}) = -\frac{1}{2}\Delta\omega^{\alpha\beta}(g^{\mu}{}_{\alpha}\gamma_{\beta} - g^{\mu}{}_{\beta}\gamma_{\alpha}) \tag{24}$$

or

$$[\hat{\sigma}_{\alpha\beta}, \gamma^{\mu}] = -\frac{1}{2}(g^{\mu}{}_{\alpha}\gamma_{\beta} - g^{\mu}{}_{\beta}\gamma_{\alpha}) \tag{25}$$

We now show that

$$\hat{\sigma}_{\alpha\beta} = \frac{1}{8}[\gamma_{\alpha}, \gamma_{\beta}] \tag{26}$$

satisfies eq. (25). We have, on substituting eq. (26) in (25)

$$\left[\frac{1}{8}[\gamma_{\alpha}, \gamma_{\beta}], \gamma^{\mu}\right] = \frac{1}{8}[(\gamma_{\alpha}\gamma_{\beta} - \gamma_{\beta}\gamma_{\alpha}), \gamma^{\mu}]$$

$$= \frac{1}{8}[(\gamma_{\alpha}\gamma_{\beta} - \gamma_{\beta}\gamma_{\alpha})\gamma^{\mu} - \gamma^{\mu}(\gamma_{\alpha}\gamma_{\beta} - \gamma_{\beta}\gamma_{\alpha})]$$

$$= \frac{1}{4}(\gamma^{\mu}\gamma_{\beta}\gamma_{\alpha} - \gamma_{\beta}\gamma_{\alpha}\gamma^{\mu}) = \frac{1}{4}(\gamma^{\mu}\gamma_{\beta}\gamma_{\alpha} + \gamma_{\beta}\gamma^{\mu}\gamma_{\alpha} - \gamma_{\beta}\gamma_{\alpha}\gamma^{\mu} - \gamma_{\beta}\gamma^{\mu}\gamma_{\alpha})$$

$$= \frac{1}{4}(\{\gamma^{\mu}, \gamma_{\beta}\}\gamma_{\alpha} - \gamma_{\beta}\{\gamma^{\mu}, \gamma_{\alpha}\}) = \frac{1}{2}(g^{\mu}{}_{\beta}\gamma_{\alpha} - g^{\mu}{}_{\alpha}\gamma_{\beta}) \tag{27}$$

Substituting the value of $\hat{\sigma}_{\alpha\beta} = \frac{1}{8}[\gamma_{\alpha}, \gamma_{\beta}]$ in eq. (18), we obtain the expression for the $\hat{S}$ operator for infinitesimal proper Lorentz transformation as

$$\hat{S}(\Delta\omega^{\mu\nu}) = \hat{I} + \frac{1}{8}[\gamma_{\mu}, \gamma_{\nu}]\,\Delta\omega^{\mu\nu} \tag{28}$$

5.4 $\hat{S}$ OPERATOR FOR FINITE PROPER LORENTZ TRANSFORMATIONS

Now we construct the $\hat{S}$ operator for finite proper Lorentz transformations. For a finite rotation angle ω generated by successive rotations by infinitesimal angles $\Delta\omega$, i.e., $\omega = \lim_{N\to\infty}(N\Delta\omega)$ around a generalized axis in the direction of unit vector **n,** we have

$$\psi'(x') = \lim_{N\to\infty}[\hat{I} + \hat{\sigma}_{\mu\nu}(\Delta\omega)^{\mu\nu}]^{N}\,\psi(x)$$

$$= \lim_{N \to \infty} \left[\hat{I} + \frac{\omega}{N} \hat{\sigma}_{\mu\nu} (\hat{I}_n)^{\mu\nu} \right]^N \psi(x)$$

$$= \{\exp[\omega \hat{\sigma}_{\mu\nu} (I_n)^{\mu\nu}]\} \psi(x) \tag{29}$$

where $I_\mathbf{n}$ is the operator represented by a 4×4 matrix corresponding to a unit Lorentz rotation around the **n** axis.

5.5 $\hat{S}^{-1}$ OPERATOR FOR FINITE PROPER LORENTZ TRANSFORMATIONS

We calculate the inverse of the spinor transformation operator $\hat{S}^{-1}$, first, for pure rotations and then for Lorentz boosts. For pure rotations, the components $(\hat{I}_\mathbf{n})^{0\mu}$ of the operator $I_\mathbf{n}$ vanish $\forall \mu = 0, 1, 2, 3$ because, for spatial rotations, the time coordinates remain unchanged. Hence, we have, from eq. (29)

$$\hat{S}(\omega) = \exp[\omega \hat{\sigma}_{ij} (I_n)^{ij}] = \exp(\omega^{ij} \hat{\sigma}_{ij}) \Rightarrow \hat{S}^\dagger(\omega)$$

$$= \exp(\omega^{ij} \hat{\sigma}_{ij}^\dagger) = \exp(-\omega^{ij} \hat{\sigma}_{ij}) = \hat{S}^{-1}(\omega) \tag{30}$$

where the penultimate step follows from

$$\hat{\sigma}_{ij}^\dagger = \frac{1}{8}[\gamma_i, \gamma_j]^\dagger = \frac{1}{8}(\gamma_i\gamma_j - \gamma_j\gamma_i)^\dagger = \frac{1}{8}[(\gamma_i\gamma_j)^\dagger - (\gamma_j\gamma_i)^\dagger]$$

$$= \frac{1}{8}(\gamma_j^\dagger\gamma_i^\dagger - \gamma_i^\dagger\gamma_j^\dagger)$$

$$= \frac{1}{8}[(-\gamma_j)(-\gamma_i) - (-\gamma_i)(-\gamma_j)] = \frac{1}{8}(\gamma_j\gamma_i - \gamma_i\gamma_j)$$

$$= \frac{1}{8}[\gamma_j, \gamma_i] = -\frac{1}{8}[\gamma_i, \gamma_j] = -\hat{\sigma}_{ij} \tag{31}$$

Further,

$$\gamma_0 \hat{\sigma}_{ij} = \frac{1}{8}\gamma_0(\gamma_i\gamma_j - \gamma_j\gamma_i) = \frac{1}{8}(\gamma_i\gamma_j\gamma_0 - \gamma_j\gamma_i\gamma_0)$$

$$= \frac{1}{8}(\gamma_i\gamma_j - \gamma_j\gamma_i)\gamma_0 = \frac{1}{8}[\gamma_i, \gamma_i]\gamma_0 = \hat{\sigma}_{ij}\gamma_0 \tag{32}$$

i.e., $\hat{\sigma}_{ij}$, and therefore, $\hat{S}$ and $\hat{S}^\dagger$ commute with γ_0. Thus,

$$\gamma_0 \hat{S}^\dagger \gamma_0 = \hat{S}^\dagger \gamma_0^2 = \hat{S}^\dagger = \hat{S}^{-1} \tag{33}$$

To compute the $\hat{S}^{-1}$ operator for Lorentz boosts, we rotate the coordinate system so that the boost coincides with the x^1 axis. For this transformation, the spinor transformation operator $\hat{S}(\omega) = \exp[\omega\hat{\sigma}_{\mu\nu}(I_n)^{\mu\nu}]$ will take the form

$$\hat{S}(\omega) = \exp\left\{\omega\left[\hat{\sigma}_{01}(I_{x^1})^{01} + \hat{\sigma}_{10}(I_{x^1})^{10}\right]\right\}$$

$$= \exp\{\omega[\hat{\sigma}_{01}(+1) + \hat{\sigma}_{10}(-1)]\} = \exp(2\omega\hat{\sigma}_{01}) \tag{34}$$

where I_{x^1} is the 4×4 matrix representing a unit Lorentz boost in the direction of the x^1 axis and is given by

$$(I_{x^1})^{\nu}{}_{\mu} = \begin{pmatrix} 0 & -1 & 0 & 0 \\ -1 & 0 & 0 & 0 \\ 0 & 0 & 0 & 0 \\ 0 & 0 & 0 & 0 \end{pmatrix} \tag{35}$$

This is proved as follows:

Infinitesimal Lorentz transformations are given by eq. (4), viz. $a^{\nu}{}_{\mu} = \delta^{\nu}{}_{\mu} + \Delta\omega^{\nu}{}_{\mu}$ with $\Delta\omega^{\nu\mu} = -\Delta\omega^{\mu\nu}$. For a Lorentz boost along the x^1 axis, we have $\Delta\omega^{01} = -\Delta\omega^{10} \equiv -\Delta\beta \neq 0$ and all other $\Delta\omega^{\mu\nu} = 0$ so that $\Delta\omega^{0}{}_{1} = g_{1\sigma}\Delta\omega^{\sigma 0} = g_{11}\Delta\omega^{10} = -\Delta\omega^{10} = -\Delta\beta$, $\Delta\omega^{1}{}_{0} = g_{0\sigma}\Delta\omega^{\sigma 1} = g_{00}\Delta\omega^{01} = \Delta\omega^{01} = -\Delta\beta$ and $\Delta\omega^{j}{}_{i} = g_{ik}\Delta\omega^{kj} = 0$. Therefore,

$$a^{\nu}{}_{\mu} = \delta^{\nu}_{\mu} + \Delta\omega^{1}{}_{0}\delta^{\nu}_{1}\delta^{0}_{\mu} + \Delta\omega^{0}{}_{1}\delta^{\nu}_{0}\delta^{1}_{\mu} = \delta^{\nu}_{\mu} - \Delta\beta(\delta^{\nu}_{1}\delta^{0}_{\mu} + \delta^{\nu}_{0}\delta^{1}_{\mu})$$

Hence, with $\Delta\omega = \Delta\beta$,

$$(I_{x^1})^{\nu}{}_{\mu} = -(\delta^{\nu}_{1}\delta^{\mu}_{0} + \delta^{0}_{\nu}\delta^{1}_{\mu}) = \begin{pmatrix} 0 & -1 & 0 & 0 \\ -1 & 0 & 0 & 0 \\ 0 & 0 & 0 & 0 \\ 0 & 0 & 0 & 0 \end{pmatrix} \text{ and } (I_{x^1})^{\mu\nu} = g^{\mu\rho}(I_{x^1})^{\nu}{}_{\rho}$$

whence

$$(I_{x^1})^{01} = g^{1\rho}(I_{x^1})^{0}{}_{\rho} = g^{11}(I_{x^1})^{0}{}_{1} = -(I_{x^1})^{0}{}_{1} = +1$$

$$(I_{x^1})^{10} = g^{0\rho}(I_{x^1})^{1}{}_{\rho} = g^{00}(I_{x^1})^{1}{}_{0} = +(I_{x^1})^{1}{}_{0} = -1 \tag{36}$$

Therefore, from eq. (34)

$$\hat{S}^{-1}(\omega) = \exp(-2\omega\hat{\sigma}_{01}) = \sum_{n=0}^{\infty}(-2\omega\hat{\sigma}_{01})^n = \sum_{n=0}^{\infty}[2\omega(\gamma_0\hat{\sigma}_{01}\gamma_0)]^n$$

$$= \sum_{n=0}^{\infty} \{[\gamma_0(2\omega\hat{\sigma}_{01})\gamma_0].[\gamma_0(2\omega\hat{\sigma}_{01})\gamma_0]...n\ times\} = \sum_{n=0}^{\infty} [\gamma_0(2\omega\hat{\sigma}_{01})^n \gamma_0]$$

$$= \gamma_0 \sum_{n=0}^{\infty} (2\omega\hat{\sigma}_{01})^n \gamma_0 = \gamma_0[\exp(2\omega\hat{\sigma}_{01})]\gamma_0 = \gamma_0\hat{S}\gamma_0 = \gamma_0\hat{S}^\dagger \gamma_0 \quad (37)$$

where we have used

$$\gamma_0\hat{\sigma}_{01}\gamma_0 = \frac{1}{8}\gamma_0(\gamma_0\gamma_1 - \gamma_1\gamma_0)\gamma_0 = \frac{1}{8}(\gamma_1\gamma_0 - \gamma_0\gamma_1)] = -\hat{\sigma}_{01} \quad (38)$$

and

$$\hat{\sigma}_{01}^\dagger = \frac{1}{8}[\gamma_0, \gamma_1]^\dagger = \frac{1}{8}(\gamma_0\gamma_1 - \gamma_1\gamma_0)^\dagger = \frac{1}{8}[(\gamma_0\gamma_1)^\dagger - (\gamma_1\gamma_0)^\dagger]$$

$$= \frac{1}{8}(\gamma_1^\dagger\gamma_0^\dagger - \gamma_0^\dagger\gamma_1^\dagger)$$

$$= \frac{1}{8}[(-\gamma_1)(\gamma_0) - (\gamma_0)(-\gamma_1)] = \frac{1}{8}(\gamma_0\gamma_1 - \gamma_1\gamma_0) = \frac{1}{8}[\gamma_0, \gamma_1] = \hat{\sigma}_{01}$$

so that

$$\hat{S}^\dagger = \exp(2\omega\hat{\sigma}_{01}^\dagger) = \exp(2\omega\hat{\sigma}_{01}) = \hat{S} \quad (39)$$

5.6 SPINOR TRANSFORMATION OPERATOR FOR IMPROPER LORENTZ TRANSFORMATIONS

We now construct the spinor transformation operator for improper Lorentz transformations. First consider the case of spatial reflections. A spatial reflection transforms a 4-vector as

$$\mathbf{x} \to -\mathbf{x} \equiv \mathbf{x}', x^0 \to x^0 \equiv (x')^0 \quad (40)$$

Hence, the transformation matrix is

$$a^\nu{}_\mu = \begin{pmatrix} 1 & 0 & 0 & 0 \\ 0 & -1 & 0 & 0 \\ 0 & 0 & -1 & 0 \\ 0 & 0 & 0 & -1 \end{pmatrix}. \quad (41)$$

The spinor transformation operator corresponding to spatial reflections is also called the "Parity" operator $\hat{P}$. The parity operator is discrete and cannot be generated by successive infinitesimal transformations acting on the identity. However, eq. (17) still holds and we have

$$\hat{P}(\Lambda)\gamma^{\mu}\hat{P}^{-1}(\Lambda) = a^{\mu}{}_{\nu}\gamma^{\nu} \tag{42}$$

which gives

$$\delta^{\sigma}_{\nu}\gamma^{\nu} = a^{\sigma}_{\mu}a_{\nu}{}^{\mu}\gamma^{\nu} = \hat{p}a^{\sigma}_{\mu}\gamma^{\mu}\hat{p}{-1}$$

$$= \hat{P}\left(\sum_{\mu=0}^{3} g^{\sigma\mu}\gamma^{\mu}\right)\hat{P}{-1} = \hat{P}_{g}{}^{\sigma\sigma}\gamma^{\sigma}\hat{P}^{-1}$$

or, equivalently

$$\hat{P}^{-1}\gamma^{\sigma}\hat{P} = g^{\sigma\sigma}\gamma^{\sigma} \tag{43}$$

where we have used $a^{\sigma}{}_{\mu} = g^{\sigma\mu}$ in the third step and the fact that $g^{\sigma\mu} = 0\ \forall\ \mu \neq \sigma$ since $g^{\sigma\mu}$ is diagonal. A possible solution of eq. (43) is

$$\hat{P} = e^{i\phi}\gamma^{0},\ \hat{P}^{-1} = e^{-i\phi}(\gamma^{0})^{-1} = e^{-i\phi}\gamma^{0} \tag{44}$$

because

$$\hat{P}^{-1}\gamma^{k}\hat{P} = e^{-i\phi}\gamma^{0}\gamma^{k}e^{i\phi}\gamma^{0} = \gamma^{0}\gamma^{k}\gamma^{0} = -\gamma^{k}$$

$$= g^{kk}\gamma^{k},\ \hat{P}^{-1}\gamma^{0}\hat{P} = e^{-i\phi}\gamma^{0}\gamma^{0}e^{i\phi}\gamma^{0} = \gamma^{0} = g^{00}\gamma^{0} \tag{45}$$

ϕ is an arbitrary phase angle that may be so chosen that four consecutive space inversions reproduce the original spinor in analogy with proper Lorentz transformations wherein a rotation of 4π reproduces the original spinor. In that case, $\hat{P}^{4}\psi(x) = e^{4i\phi}(\gamma^{0})^{4}\psi(x) = e^{4i\phi}\psi(x) = \psi(x)$ so that $e^{4i\phi} = 1$ or $e^{i\phi} = \pm 1, \pm i$. The Parity operator, as defined above, is unitary, i.e.,

$$\hat{P}^{-1} = (e^{i\phi}\gamma^{0})^{-1} = (\gamma^{0})^{-1}e^{-i\phi} = \gamma^{0}e^{-i\phi}$$

$$= (\gamma^{0})^{\dagger}(e^{i\phi})^{\dagger} = (e^{i\phi}\gamma^{0})^{\dagger} = \hat{P}^{\dagger} \tag{46}$$

Furthermore,

$$\gamma^{0}\hat{P}^{\dagger}\gamma^{0} = \gamma^{0}(e^{i\phi}\gamma^{0})^{\dagger}\gamma^{0} = \gamma^{0}(\gamma^{0\dagger}e^{-i\phi})\gamma^{0} = \gamma^{0}(\gamma^{0}e^{-i\phi})\gamma^{0}$$

$$= e^{-i\phi}\gamma^{0} = e^{-i\phi}(\gamma^{0})^{-1} = (\gamma^{0}e^{i\phi})^{-1}$$

$$= (e^{i\phi}\gamma^{0})^{-1} = \hat{P}^{-1} \tag{47}$$

We also have

$$\psi'(x') = \psi(\mathbf{x}', x'^{0}) = \psi'(-\mathbf{x}, x^{0}) = \hat{P}\psi(x) = \hat{P}\psi(\mathbf{x}, x^{0})$$

$$= e^{i\phi}\gamma^{0}\psi(\mathbf{x}, x^{0}) = \hat{P}\psi(-\mathbf{x}', x'^{0}) \tag{48}$$

The spinor $\psi'(\mathbf{x}', x^{0})$ is called the spatially reflected spinor. The impact of the Parity operator on various operators of interest is as follows:

$\hat{P}\mathbf{x}\hat{P}^{-1} = -\mathbf{x}$ because

$$\hat{P}\mathbf{x}\hat{P}^{-1}\psi(\mathbf{x}, x^0) = \hat{P}\mathbf{x}\psi(-\mathbf{x}, x^0) = -\mathbf{x}\hat{P}\psi(-\mathbf{x}, x^0) = -\mathbf{x}\psi(\mathbf{x}, x^0) \tag{49}$$

$$\hat{P}x^0\hat{P}^{-1} = x^0 \text{ since}$$

$$\hat{P}x^0\hat{P}^{-1}\psi(\mathbf{x}, x^0) = \hat{P}x^0\psi(-\mathbf{x}, x^0) = x^0\hat{P}\psi(-\mathbf{x}, x^0) = x^0\psi(\mathbf{x}, x^0) \tag{50}$$

$$\hat{P}\hat{\mathbf{p}}\hat{P}^{-1} = -\hat{\mathbf{p}} \text{ because}$$

$$\hat{P}\hat{\mathbf{p}}\hat{P}^{-1}\psi(\mathbf{x}, x^0) = \hat{P}\hat{\mathbf{p}}\,\psi(-\mathbf{x}, x^0) = \hat{P}(-i\nabla_x)\psi(-\mathbf{x}, x^0)$$

$$= \hat{P}(i\nabla_{-x})\psi(-\mathbf{x}, x^0)$$

$$= (i\nabla_x)\hat{P}\psi(-\mathbf{x}, x^0) = (i\nabla_x)\psi(\mathbf{x}, x^0) = -\hat{\mathbf{p}}\psi(\mathbf{x}, x^0) \tag{51}$$

$$\hat{P}\,\hat{p}\,\hat{P}^{-1} = \hat{p}_0 \text{ because}$$

$$\hat{P}\,\hat{p}_0\hat{P}^{-1}\psi(\mathbf{x}, x^0) = \hat{P}\,\hat{p}_0\psi(-\mathbf{x}, x^0) = \hat{P}(i\partial_0)\psi(-\mathbf{x}, x^0)$$

$$= (i\partial_0)\hat{P}\psi(-\mathbf{x}, x^0) = (i\partial_0)\psi(\mathbf{x}, x^0) = \hat{p}_0\psi(\mathbf{x}, x^0) \tag{52}$$

We now examine the impact of the Parity operator on the Dirac equation. For this, we write the Dirac equation in the form

$$(i\partial_\mu\gamma^\mu - m)\psi(\mathbf{x}, x^0) = 0 = (i\partial_\mu\gamma^\mu - m)\hat{P}^{-1}\psi'(\mathbf{x}', x^0)$$

$$= (i\partial_0\gamma^0 + i\partial_k\gamma^k - m)\hat{P}^{-1}\psi'(\mathbf{x}', x^0) \tag{53}$$

Multiplying eq. (53) from the left by $\hat{P}$ and noting that

$$\hat{P}(\partial_0\gamma^0)\hat{P}^{-1} = (e^{i\phi}\gamma^0)(\partial_0\gamma^0)(e^{i\phi}\gamma^0)^{-1} = (e^{i\phi}\gamma^0)(\partial_0\gamma^0)(e^{-i\phi}\gamma^0) = \gamma^0\partial_0 = \partial_0\gamma^0$$

and

$$\hat{P}(\partial_k\gamma^k)\hat{P}^{-1} = (e^{i\phi}\gamma^0)(\partial_k\gamma^k)(e^{i\phi}\gamma^0)^{-1} = (e^{i\phi}\gamma^0)(\partial_k\gamma^k)(e^{-i\phi}\gamma^0) = -\partial_k\gamma^k$$

we have

$$0 = (i\partial_0\gamma^0 - i\partial_k\gamma^k - m)\psi'(\mathbf{x}', x^0) = (i\partial_0'\gamma^0 + i\partial_k'\gamma^k - m)\psi'(\mathbf{x}', x'^0)$$

$$= (i\partial_\mu'\gamma^\mu - m)\psi'(\mathbf{x}', x'^0) \tag{54}$$

where

$$x' \equiv (\mathbf{x}', x'^0) = (-\mathbf{x}, x^0),\ \partial_0' = \frac{\partial}{\partial x'^0} = \frac{\partial}{\partial x^0} = \partial_0,\ \partial_k' = \frac{\partial}{\partial x'^k} = -\frac{\partial}{\partial x^k} = -\partial_k$$

which shows that the transformed wavefunction $\psi'(x')$ obeys the same Dirac equation as the original wavefunction. Hence, the Dirac equation is invariant under Parity transformations.

Now consider the case of Time Reversal transformations. The Time Reversal transformation transforms a 4-vector as

$$\mathbf{x} \to \mathbf{x} \equiv \mathbf{x}', x^0 \to -x^0 \equiv (x')^0 \tag{55}$$

The Time Reversal transformation is also an improper Lorentz transformation and cannot be obtained by operating successively infinitesimally on the identity. To examine further properties of the Time Inversion operator $\hat{T}$, we write the Dirac equation as

$$i\partial_0\psi(\mathbf{x}, x^0) = (-i\alpha.\nabla + \beta m)\psi(\mathbf{x}, x^0) = \hat{H}\psi(\mathbf{x}, x^0) \tag{56}$$

where $\hat{H} \equiv (-i\alpha.\nabla + \beta m)$ is the Dirac Hamiltonian. Operating on eq. (56) by the Time Inversion operator, we can write

$$\hat{T}[i\partial_0\psi(\mathbf{x}, x^0)] = \hat{T}[(-i\alpha.\nabla + \beta m)\psi(\mathbf{x}, x^0)] \tag{57}$$

or

$$(\hat{T}i\hat{T}^{-1})\hat{T}[\partial_0\psi(\mathbf{x}, x^0)] = \hat{T}[(-i\alpha.\nabla + \beta m)]\hat{T}^{-1}[\hat{T}\psi(\mathbf{x}, x^0)] \tag{58}$$

Now

$$\partial_0\psi(\mathbf{x}', x'^0) = \partial_0\psi(\mathbf{x}, -x^0) = \frac{\partial}{\partial x^0}\psi(\mathbf{x}, -x^0) = \frac{\partial(-x^0)}{\partial x^0}.\frac{\partial}{\partial(-x^0)}\psi(\mathbf{x}, -x^0)$$

$$= -\frac{\partial}{\partial(-x^0)}\psi(x, -\mathbf{x}^0) = -\partial_0'\psi(\mathbf{x}', x'^0) \tag{59}$$

so that $\hat{T}\partial_0\psi'(\mathbf{x}', x'^0) = -\partial_0'\psi'(\mathbf{x}', x'^0)$ and eq. (58) becomes

$$-(\hat{T}i\hat{T}^{-1})[\partial_0'\psi'(\mathbf{x}', x'^0)] = \hat{T}[(-i\alpha.\nabla + \beta m)]\hat{T}^{-1}\psi(\mathbf{x}', x'^0) \tag{60}$$

If the Dirac equation is to be invariant under time inversion, eq. (60) must be equivalent to

$$i\partial'_0\psi'(\mathbf{x}', x'^0) = (-i\alpha.\nabla + \beta m)\psi'(\mathbf{x}', x'^0) \tag{61}$$

Thus, comparing eqs. (60) and (61), we obtain two possibilities, viz.

$$\hat{T}i\hat{T}^{-1} = -i \text{ and } \hat{T}\hat{H}\hat{T}^{-1} = \hat{T}(-i\alpha.\nabla + \beta m)\hat{T}^{-1} = -i\alpha.\nabla + \beta m = \hat{H} \tag{62}$$

or

$$\hat{T}i\hat{T}^{-1} = i \text{ and } \hat{T}\hat{H}\hat{T}^{-1} = \hat{T}(-i\alpha.\nabla + \beta m)\hat{T}^{-1} = i\alpha.\nabla - \beta m = -\hat{H} \tag{63}$$

However, the latter option, i.e., eq. (63) is not physically acceptable for, in the case of time independent Hamiltonian, $\hat{T}\hat{H}\hat{T}^{-1} = -\hat{H}$ implies that the energy eigenvalue spectrum is altered by time inversion, a physically untenable situation. To further corroborate this fact, we consider the Hamiltonian of a Dirac particle in an electromagnetic field represented by the 4-vector $A_\mu(x)$, i.e.,

$$\hat{H} = \boldsymbol{\alpha}[-i\nabla - e\mathbf{A}(\mathbf{x}, x^0)] + \beta m + eA_0(\mathbf{x}, x^0) \tag{64}$$

We consider the impact of time inversion on each of the operators involved in the Hamiltonian. We have $\hat{T}\mathbf{x}\hat{T}^{-1} = \mathbf{x}$, $\hat{T}\nabla\hat{T}^{-1} = \nabla$ since $\hat{T}$ does not effect the spatial coordinates $\hat{T}\mathbf{A}(\mathbf{x}, x^0)\hat{T}^{-1} = \mathbf{A}(\mathbf{x}, -x^0) = -\mathbf{A}(\mathbf{x}, x^0)$ because the vector current $\mathbf{j}(x)$, that creates the vector potential $\mathbf{A}$, will change sign under the effect of $\hat{T}$. However, the fourth component of the 4-potential $A_\mu(x)$, viz., $A_0(x)$ will remain unaffected by time inversion because the corresponding charge density $\rho(x)$ will not change sign under operation by $\hat{T}$, i.e., $\hat{T} A_0(\mathbf{x}, x^0)\hat{T}^{-1} = A_0(\mathbf{x}, -x^0) = A_0(\mathbf{x}, x^0)$.

In view of the aforesaid, compliance with $\hat{T}\hat{H}\hat{T}^{-1} = -\hat{H}$ would necessitate the following: $\hat{T}\alpha\hat{T}^{-1} = -\alpha$ because we must have

$$\hat{T}(-i\boldsymbol{\alpha}\nabla)\hat{T}^{-1} = -(\hat{T}i\hat{T}^{-1})\,(\hat{T}\alpha\hat{T}^{-1})\,(\hat{T}\nabla\hat{T}^{-1}) = -i(\hat{T}\boldsymbol{\alpha}\hat{T}^{-1})\nabla = --i(-\boldsymbol{\alpha})\nabla$$

and $\hat{T}\boldsymbol{\alpha}\hat{T}^{-1} = +\boldsymbol{\alpha}$ because

$$\hat{T}(-e\boldsymbol{\alpha}\mathbf{A})\hat{T}^{-1} = -(\hat{T}e\hat{T}^{-1})\,(\hat{T}\boldsymbol{\alpha}\hat{T}^{-1})\,(\hat{T}\mathbf{A}\hat{T}^{-1})$$

$$= -e(\hat{T}\boldsymbol{\alpha}\hat{T}^{-1})\,(-\mathbf{A}) = -e(\boldsymbol{\alpha})\,(-\mathbf{A}) = e\boldsymbol{\alpha}\mathbf{A}$$

thus, creating an inconsistency. Hence, we must accept $\hat{T}i\hat{T}^{-1} = -i$ and

$$\hat{T}\hat{H}\hat{T}^{-1} = \hat{T}(-i\boldsymbol{\alpha}.\nabla + \boldsymbol{\beta}m)\hat{T}^{-1} = -i\boldsymbol{\alpha}.\nabla + \boldsymbol{\beta}m = \hat{H}$$

as the only plausible option.

This solution implies that

$$\hat{T}(-i\boldsymbol{\alpha}\nabla)\hat{T}^{-1} = -(\hat{T}i\hat{T}^{-1})\,(\hat{T}\boldsymbol{\alpha}\hat{T}^{-1})\,(\hat{T}\nabla\hat{T}^{-1}) = i(\hat{T}\boldsymbol{\alpha}T^{-1})\nabla = i(-\boldsymbol{\alpha})\nabla$$

so that

$$\hat{T}\boldsymbol{\alpha}\hat{T}^{-1} = -\boldsymbol{\alpha} \text{ and } \hat{T}\boldsymbol{\beta}\hat{T}^{-1} = \beta \tag{65}$$

because $\hat{T}(\beta m)\hat{T}^{-1} = (\hat{T}\beta\hat{T}^{-1})\,(\hat{T}m\hat{T}^{-1}) = (\hat{T}\beta\hat{T}^{-1})m = \beta m.$

Our next step is to determine an explicit representation of the Time Inversion operator. For this, we write $\hat{T} = \hat{T}_0\hat{K}$ where $\hat{K}$, is the complex conjugation operator, i.e., $\hat{K}i\hat{K}^{-1} = -i$. Substituting $\hat{T} = \hat{T}_0\hat{K}$ in eq. (65), we have the following:

Since $\boldsymbol{\alpha}_1$, $\boldsymbol{\alpha}_3$ and $\boldsymbol{\beta}$ are all real, the operator $\hat{K}$ will not affect these matrices so that $\hat{T}_0\boldsymbol{\alpha}_1\hat{T}_0^{-1} = -\boldsymbol{\alpha}_1$, $\hat{T}_0\boldsymbol{\alpha}_3\hat{T}_0^{-1} = -\boldsymbol{\alpha}_3$, $\hat{T}_0\boldsymbol{\beta}\hat{T}_0^{-1} = \boldsymbol{\beta}$.

However, in the case of $\boldsymbol{\alpha}_2$, which is purely imaginary, the operation $\hat{T}\boldsymbol{\alpha}_2\hat{T}^{-1} = -\boldsymbol{\alpha}_2$ will be performed by the complex conjugation operator $\hat{K}$ so that $\hat{T}_0\boldsymbol{\alpha}_2\hat{T}_0^{-1} = \boldsymbol{\alpha}_2$ for

$$\hat{T}\boldsymbol{\alpha}_2\hat{T}^{-1} = \hat{T}_0\hat{K}\boldsymbol{\alpha}_2(\hat{T}_0\hat{K})^{-1} = \hat{T}_0(\hat{K}\boldsymbol{\alpha}_2\hat{K}^{-1})\hat{T}_0^{-1}$$

$$= \hat{T}_0(-\boldsymbol{\alpha}_2)\hat{T}_0^{-1} = -\hat{T}_0\boldsymbol{\alpha}_2\hat{T}_0^{-1} = -\boldsymbol{\alpha}_2$$

Making use of the properties of the Dirac matrices, it is seen that

$$\hat{T}_0 = -i\boldsymbol{\alpha}_1\boldsymbol{\alpha}_3 \tag{66}$$

meets all the aforesaid requirements. Furthermore, $\hat{T}_0 = -i\boldsymbol{\alpha}_1\boldsymbol{\alpha}_3$ also ensures unitarity of $\hat{T}$ for

$$\hat{T}^\dagger = (\hat{T}_0\hat{K})^\dagger = \hat{K}^\dagger\hat{T}_0^\dagger = \hat{K}^\dagger(-i\boldsymbol{\alpha}_1\boldsymbol{\alpha}_3)^\dagger = \hat{K}^\dagger(-i)^\dagger\boldsymbol{\alpha}_3^\dagger\boldsymbol{\alpha}_1^\dagger = \hat{K}^\dagger i\boldsymbol{\alpha}_3\boldsymbol{\alpha}_1$$

$$= \hat{K}^{-1}i\boldsymbol{\alpha}_3^{-1}\boldsymbol{\alpha}_1^{-1} = \hat{K}^{-1}(-i\boldsymbol{\alpha}_1\boldsymbol{\alpha}_3)^{-1} = \hat{K}^{-1}\hat{T}_0^{-1} = (\hat{T}_0\hat{K})^{-1} = \hat{T}^{-1} \tag{67}$$

where we have used the fact that $\hat{K}$ is unitary being the complex conjugation operator.

We have the following further relations for the Time Inversion operator:

(i) $\boldsymbol{\psi}'(\mathbf{x}', x'^0) = \boldsymbol{\psi}'(\mathbf{x}, -x^0) = \hat{T}\boldsymbol{\psi}(\mathbf{x}, x^0) = \hat{T}_0\boldsymbol{\psi}^*(\mathbf{x}, x^0)$ (68)

(ii) $\hat{T}_0 = -i\boldsymbol{\alpha}_1\boldsymbol{\alpha}_3 = -i\boldsymbol{\beta}^2\boldsymbol{\alpha}_1\boldsymbol{\alpha}_3 = i\boldsymbol{\beta}\boldsymbol{\alpha}_1\boldsymbol{\beta}\boldsymbol{\alpha}_3 = i\gamma_1\gamma_3$ (69)

(iii) $\hat{T}_0^\dagger = (i\boldsymbol{\gamma}_1\boldsymbol{\gamma}_3)^\dagger = -i(\boldsymbol{\gamma}_3^\dagger\boldsymbol{\gamma}_1^\dagger) = -i(-\boldsymbol{\gamma}_3)(-\boldsymbol{\gamma}_1) = -i\boldsymbol{\gamma}_3\boldsymbol{\gamma}_1 = i\boldsymbol{\gamma}_1\boldsymbol{\gamma}_3 = \hat{T}_0$ (70)

(iv) $\boldsymbol{\psi}'^\dagger(\mathbf{x}', x'^0) = \boldsymbol{\psi}'^\dagger(\mathbf{x}, -x^0) = [\hat{T}\boldsymbol{\psi}(\mathbf{x}, x^0)]^\dagger = [\hat{T}_0\boldsymbol{\psi}^*(\mathbf{x}, x^0)]^\dagger$

$$= [\boldsymbol{\psi}^*(\mathbf{x}, x^0)]^\dagger\hat{T}_0^\dagger = [\boldsymbol{\psi}(\mathbf{x}, x^0)]^T\hat{T}_0 \tag{71}$$

(v) $\hat{T}_0^\dagger\hat{T}_0 = (\hat{T}_0)^2 = (i\boldsymbol{\gamma}_1\boldsymbol{\gamma}_3).(i\boldsymbol{\gamma}_1\boldsymbol{\gamma}_3) = -(\boldsymbol{\gamma}_1\boldsymbol{\gamma}_3).(\boldsymbol{\gamma}_1\boldsymbol{\gamma}_3)$

$$= (\boldsymbol{\gamma}_1\boldsymbol{\gamma}_3).(\boldsymbol{\gamma}_3\boldsymbol{\gamma}_1) = -(\boldsymbol{\gamma}_1\boldsymbol{\gamma}_1) = \mathbf{I} \tag{72}$$

(vi) $\hat{T}_0\boldsymbol{\gamma}_\mu\hat{T}_0 = (\boldsymbol{\gamma}^\mu)^*$

for

$$\hat{T}_0\boldsymbol{\gamma}_0\hat{T}_0 = (i\boldsymbol{\gamma}_1\boldsymbol{\gamma}_3)\boldsymbol{\gamma}_0(i\boldsymbol{\gamma}_1\boldsymbol{\gamma}_3) = \boldsymbol{\gamma}_1\boldsymbol{\gamma}_0\boldsymbol{\gamma}_1 = \boldsymbol{\gamma}_0 = (\boldsymbol{\gamma}_0)^* = (\boldsymbol{\gamma}^0)^*$$

$$\hat{T}_0\boldsymbol{\gamma}_1\hat{T}_0 = (i\boldsymbol{\gamma}_1\boldsymbol{\gamma}_3)\boldsymbol{\gamma}_1(i\boldsymbol{\gamma}_1\boldsymbol{\gamma}_3) = \boldsymbol{\gamma}_1\boldsymbol{\gamma}_3\boldsymbol{\gamma}_3 = -\boldsymbol{\gamma}_1 = (-\boldsymbol{\gamma}_1)^* = (\boldsymbol{\gamma}^1)^*$$

$$\hat{T}_0\boldsymbol{\gamma}_2\hat{T}_0 = (i\boldsymbol{\gamma}_1\boldsymbol{\gamma}_3)\boldsymbol{\gamma}_2(i\boldsymbol{\gamma}_1\boldsymbol{\gamma}_3) = \boldsymbol{\gamma}_1\boldsymbol{\gamma}_2\boldsymbol{\gamma}_1 = \boldsymbol{\gamma}_2 = (-\boldsymbol{\gamma}_2)^* = (\boldsymbol{\gamma}^2)^*$$

$$\hat{T}_0\boldsymbol{\gamma}_3\hat{T}_0 = (i\boldsymbol{\gamma}_1\boldsymbol{\gamma}_3)\boldsymbol{\gamma}_3(i\boldsymbol{\gamma}_1\boldsymbol{\gamma}_3) = \boldsymbol{\gamma}_1\boldsymbol{\gamma}_1\boldsymbol{\gamma}_3 = -\boldsymbol{\gamma}_3 = (-\boldsymbol{\gamma}_3)^* = (\boldsymbol{\gamma}^3)^* \tag{73}$$

5.7 COVARIANCE OF 4-CURRENT DENSITY

The 4-current density in the Dirac framework is given by

$$j(x) = \{\boldsymbol{\psi}^\dagger\boldsymbol{\psi}, \boldsymbol{\psi}^\dagger\boldsymbol{\alpha}\boldsymbol{\psi}\} = \boldsymbol{\psi}^\dagger\boldsymbol{\gamma}^0\boldsymbol{\gamma}\boldsymbol{\psi}, \text{ i.e., } j^\mu(x) = \boldsymbol{\psi}^\dagger\boldsymbol{\gamma}^0\boldsymbol{\gamma}^\mu\boldsymbol{\psi} \tag{74}$$

since $(\boldsymbol{\gamma}^0)^2 = 1$, $\boldsymbol{\gamma}^0\boldsymbol{\gamma}^k = \boldsymbol{\beta}(\boldsymbol{\beta}\boldsymbol{\alpha}^k) = \boldsymbol{\alpha}^k$. Hence, for a Lorentz transformation $\boldsymbol{\Lambda}$, the 4-current density transforms as

$$j'^\mu(x') = \boldsymbol{\psi}'^\dagger(x')\boldsymbol{\gamma}^0\boldsymbol{\gamma}^\mu\boldsymbol{\psi}'(x') = \boldsymbol{\psi}^\dagger(x)\hat{S}^\dagger\boldsymbol{\gamma}^0\boldsymbol{\gamma}^\mu\hat{S}\boldsymbol{\psi}(x)$$

$$= \boldsymbol{\psi}^\dagger(x)\,\boldsymbol{\gamma}^0(\boldsymbol{\gamma}^0\hat{S}^\dagger\boldsymbol{\gamma}^0)\boldsymbol{\gamma}^\mu\hat{S}\boldsymbol{\psi}(x)$$

$$= \boldsymbol{\psi}^\dagger(x)\boldsymbol{\gamma}^0\hat{S}^\dagger\boldsymbol{\gamma}^\mu\hat{S}\boldsymbol{\psi}(x) = \boldsymbol{\psi}^\dagger(x)\boldsymbol{\gamma}^0\hat{S}^{-1}\boldsymbol{\gamma}^\mu\hat{S}\boldsymbol{\psi}(x)$$

$$= a^\mu{}_\nu\boldsymbol{\psi}^+(x)\boldsymbol{\gamma}^0\boldsymbol{\gamma}^\nu\boldsymbol{\psi}(x) = a^\mu{}_\nu j^\nu(x) \tag{75}$$

so that the 4-current transforms as a 4-vector under Lorentz transformations.

We now examine explicitly the impact of the 4-current under space and time inversion. For space inversion, we have

$$j'^0(x') = \psi'^\dagger(x')\psi'(x') = \psi^\dagger(x)\hat{P}^\dagger\hat{P}\psi(x) = \psi^\dagger(x)\psi(x) = j^0(x) \tag{76}$$

$$j'^k(x') = \psi'^\dagger(x')\gamma^0\gamma^k\psi'(x') = \psi^\dagger(x)\hat{P}^\dagger\gamma^0\gamma^k\hat{P}\psi(x) = \psi^\dagger(x)\hat{P}^{-1}\gamma^0\gamma^k\hat{P}\psi(x)$$

$$= \psi^\dagger(x)(\hat{P}^{-1}\gamma^0\hat{P})(\hat{P}^{-1}\gamma^k\hat{P})\psi(x) = -\psi^\dagger(x)\gamma^0\gamma^k\psi(x) = -j^k(x) \tag{77}$$

Under time reversal, we have

$$j'^0(x') = \psi'^\dagger(x')\psi'(x') = \psi^\dagger(x)\hat{T}^\dagger\hat{T}\psi(x) = \psi^\dagger(x)\psi(x) = j^0(x) \tag{78}$$

$$j'^k(x') = \psi'^\dagger(x')\gamma^0\gamma^k\psi'(x') = \psi^\dagger(x)\hat{T}^\dagger\gamma^0\gamma^k\hat{T}\psi(x) = \psi^\dagger(x)\hat{T}^{-1}\gamma^0\gamma^k\hat{T}\psi(x)$$

$$= \psi^\dagger(x)(\hat{T}^{-1}\gamma^0\hat{T})(\hat{T}^{-1}\gamma^k\hat{T})\psi(x) = -\psi^\dagger(x)\gamma^0\gamma^k\psi(x) = -j^k(x) \tag{79}$$

5.8 COVARIANCE OF CONTINUITY EQUATION

The continuity equation, in covariant form, is given by $\partial_\mu j^\mu(x) = 0$. But $\partial_\mu . = \frac{\partial}{\partial x^\mu} = \frac{\partial x'^\nu}{\partial x^\mu}\frac{\partial}{\partial x'^\nu} = a^\nu{}_\mu \frac{\partial}{\partial x'^\nu} = a^\nu{}_\mu \partial'_\nu$ since $(x')^\nu = \sum_{\mu=0}^{3} a^\nu{}_\mu x^\mu \equiv a^\nu{}_\mu x^\mu$. Further, from eq. (75), we have $j'^\mu(x') = a^\mu{}_\nu j^\nu(x) \Rightarrow j^\nu(x) = a_\mu{}^\nu j'^\mu(x')$. Therefore,

$$\partial_\mu j^\mu(x) = (a_\mu{}^\nu \partial'_\nu)[a_\sigma{}^\mu j'^\sigma(x')] = \delta^\nu_\sigma \partial'_\nu j'^\sigma(x') = \partial'_\nu j'^\nu(x') \tag{80}$$

thereby establishing the invariance of the continuity equation. The above also confirms that $j^0(x) = \psi^\dagger\psi$ transforms like the time component of a 4-vector. Furthermore, this probability density is conserved under Lorentz transformations for

$$j'^0(x') = \psi'^\dagger(x')\psi'(x') = \psi^\dagger(x)\hat{S}^\dagger\hat{S}\psi(x)$$

$$= \psi^\dagger(x)\hat{S}^{-1}\hat{S}\psi(x) = \psi^\dagger(x)\psi(x) = j^0(x) \tag{81}$$

5.9 BILINEAR COVARIANTS

Bilinear covariants are quantities constructed from the 4-component Dirac wavefunction $\psi(x)$, its adjoint $\overline{\psi}(x)$ and the 16 Γ matrices Γ_i, $i = 1, 2, 3, \ldots, 16$. In this section, we examine the transformation properties of such covariants under Lorentz transformations. Such quantities are useful in constructing interaction functions in Quantum Field Theory.

Before proceeding to explore the properties of these billinear forms, we need to determine the transformation properties of the adjoint spinor

$$\overline{\psi}(x) = \psi(x)^\dagger\gamma^0 \tag{82}$$

We have, from eq. (17), $\hat{S}(\Lambda)\gamma^\mu\hat{S}^{-1}(\Lambda)a^\nu{}_\mu = \gamma^\nu$ or equivalently

$$a^{\nu}{}_{\mu}\boldsymbol{\gamma}^{\mu} = \hat{S}^{-1}(\boldsymbol{\Lambda})\boldsymbol{\gamma}^{\nu}\hat{S}(\boldsymbol{\Lambda}) \tag{83}$$

Since the matrix elements $a^{\nu}{}_{\mu}$ are real, we obtain, on taking the Hermitian adjoint of eq. (83)

$$(a^{\nu}_{\mu}\boldsymbol{\gamma}^{\mu})^{\dagger} = \left(a^{\nu}_{0}\gamma^{0} + \sum_{k=1}^{3} a^{\nu}_{k}\gamma^{k}\right)^{\dagger} = \left(a^{\nu}_{0}\gamma^{0} - \sum_{k=1}^{3} a^{\nu}_{k}\gamma^{k}\right) = [\hat{S}^{-1}(\boldsymbol{\Lambda})\boldsymbol{\gamma}^{\nu}\hat{S}(\boldsymbol{\Lambda})]^{\dagger} \tag{84}$$

since $\boldsymbol{\gamma}^{0}$ is Hermitian and $\boldsymbol{\gamma}^{k}$ is anti-Hermitian.

Multiplying eq. (84) from the left and right by $\boldsymbol{\gamma}^{0}$, we obtain

$$\boldsymbol{\gamma}^{0}\left(a^{\nu}_{0}\gamma^{0} - \sum_{k=1}^{3} a^{\nu}_{k}\gamma^{k}\right)\boldsymbol{\gamma}^{0} = \boldsymbol{\gamma}^{0}(a^{\nu}_{\mu}\boldsymbol{\gamma}^{\mu})^{\dagger}\boldsymbol{\gamma}^{0} = a^{\nu}_{\mu}\boldsymbol{\gamma}^{0}(\boldsymbol{\gamma}^{\mu})^{\dagger}\boldsymbol{\gamma}^{0} = a^{\nu}_{\mu}(\boldsymbol{\gamma}^{0}\boldsymbol{\gamma}^{\mu}\boldsymbol{\gamma}^{0})^{\dagger}$$

$$= a^{\nu}_{\mu}(\boldsymbol{\gamma}^{\mu\dagger})^{\dagger} = a^{\nu}_{\mu}\boldsymbol{\gamma}^{\mu} = \boldsymbol{\gamma}^{0}[\hat{S}^{-1}(\boldsymbol{\Lambda})\boldsymbol{\gamma}^{\nu}\hat{S}(\boldsymbol{\Lambda})]^{\dagger}\boldsymbol{\gamma}^{0} = \boldsymbol{\gamma}^{0}\hat{S}^{\dagger}(\boldsymbol{\Lambda})\boldsymbol{\gamma}^{\nu\dagger}[\hat{S}^{-1}(\boldsymbol{\Lambda})]^{\dagger}\boldsymbol{\gamma}^{0}$$

$$= \boldsymbol{\gamma}^{0}\hat{S}^{\dagger}(\boldsymbol{\Lambda})\boldsymbol{\gamma}^{0}\boldsymbol{\gamma}^{\nu}\boldsymbol{\gamma}^{0}[\hat{S}^{\dagger}(\boldsymbol{\Lambda})]^{-1}\boldsymbol{\gamma}^{0} = [\boldsymbol{\gamma}^{0}\hat{S}^{\dagger}(\boldsymbol{\Lambda})\boldsymbol{\gamma}^{0}]\boldsymbol{\gamma}^{\nu}[\boldsymbol{\gamma}^{0}\hat{S}^{\dagger}(\boldsymbol{\Lambda})\boldsymbol{\gamma}^{0}]^{-1} \tag{85}$$

Combining eqs. (83) and (85), we get

$$a^{\nu}_{\mu}\boldsymbol{\gamma}^{\mu} = \hat{S}^{-1}(\boldsymbol{\Lambda})\boldsymbol{\gamma}^{\nu}(\boldsymbol{\Lambda}) = [\boldsymbol{\gamma}^{0}\hat{S}^{\dagger}(\boldsymbol{\Lambda})\boldsymbol{\gamma}^{0}]\boldsymbol{\gamma}^{\nu}[\boldsymbol{\gamma}^{0}\hat{S}^{\dagger}(\boldsymbol{\Lambda})\boldsymbol{\gamma}^{0}]^{-1}$$

$$\Rightarrow \qquad [\hat{S}\ (\boldsymbol{\Lambda})\boldsymbol{\gamma}^{0}\ \hat{S}^{\dagger}(\boldsymbol{\Lambda})\boldsymbol{\gamma}^{0}]\boldsymbol{\gamma}^{\nu} = \boldsymbol{\gamma}^{\nu}[\hat{S}(\boldsymbol{\Lambda})\boldsymbol{\gamma}^{0}\hat{S}^{\dagger}(\boldsymbol{\Lambda})\boldsymbol{\gamma}^{0}] \tag{86}$$

Therefore, $\hat{S}(\boldsymbol{\Lambda})\boldsymbol{\gamma}^{0}\hat{S}^{\dagger}(\boldsymbol{\Lambda})\boldsymbol{\gamma}^{0}$ commutes with all the $\boldsymbol{\gamma}^{\mu}$ and, hence, is a constant multiple of identity, i.e.

$$\hat{S}(\boldsymbol{\Lambda})\boldsymbol{\gamma}^{0}\hat{S}^{\dagger}(\boldsymbol{\Lambda})\boldsymbol{\gamma}^{0} = k\mathbf{I} \Rightarrow \hat{S}^{\dagger}(\boldsymbol{\Lambda})\boldsymbol{\gamma}^{0} = k[\hat{S}(\boldsymbol{\Lambda})\boldsymbol{\gamma}^{0}]^{-1}$$

$$= k(\boldsymbol{\gamma}^{0})^{-1}\hat{S}^{-1}(\boldsymbol{\Lambda}) = k\boldsymbol{\gamma}^{0}\hat{S}^{-1}(\boldsymbol{\Lambda}) \tag{87}$$

Since $\boldsymbol{\gamma}^{0}$ is Hermitian and $[\hat{S}(\boldsymbol{\Lambda})\boldsymbol{\gamma}^{0}\hat{S}^{\dagger}(\boldsymbol{\Lambda})]^{\dagger} = [\hat{S}^{\dagger}(\boldsymbol{\Lambda})]^{\dagger}\boldsymbol{\gamma}^{0\dagger}\hat{S}^{\dagger}(\boldsymbol{\Lambda}) = \hat{S}(\boldsymbol{\Lambda})\boldsymbol{\gamma}^{0}\hat{S}^{\dagger}(\boldsymbol{\Lambda})$ so that $[\hat{S}(\boldsymbol{\Lambda})\boldsymbol{\gamma}^{0}\hat{S}^{\dagger}(\boldsymbol{\Lambda})]$ is also Hermitian. Therefore, k must be real, i.e., $k = k^{*}$. If the transformation matrix $\hat{S}$ is to be normalized to $\det \hat{S} = 1$, then $\det k\mathbf{I} = k^{4} = 1$, since $\det \boldsymbol{\gamma}^{0} = 1$, $\det \hat{S}^{\dagger} = \det \hat{S} = 1$. Since $k = k^{*}$, this constraint implies $k = \pm 1$. We shall now, examine under what circumstances $k = +1$ and $k = -1$. For this consider

$$\hat{S}^{\dagger}\hat{S} = \hat{S}^{\dagger}\boldsymbol{\gamma}^{0}\boldsymbol{\gamma}^{0}\hat{S} = k\boldsymbol{\gamma}^{0}\hat{S}^{-1}\boldsymbol{\gamma}^{0}\hat{S} = k\boldsymbol{\gamma}^{0}a^{0}_{\nu}\boldsymbol{\gamma}^{\nu} = k\boldsymbol{\gamma}^{0}\left(a^{\nu}_{0}\gamma^{0} + \sum_{k=1}^{3} a^{0}_{k}\gamma^{k}\right)$$

$$= k\left(a^{0}_{0}\mathbf{I} + \sum_{k=1}^{3} a^{0}_{k}\gamma^{0}\gamma^{k}\right) = k\left(a^{0}_{0}\mathbf{I} + \sum_{k=1}^{3} a^{0}_{k}\alpha^{k}\right) \tag{88}$$

Now $\hat{S}^\dagger\hat{S}$ must necessarily be positive definite, being the product of an operator and its Hermitian conjugate. Its eigenvalues will also be real and positive definite (since $\hat{S}^\dagger\hat{S}$ is Hermitian). Therefore, $Tr\ \hat{S}^\dagger\hat{S} > 0$. Also $Tr\ \boldsymbol{\alpha}^k = 0$, $Tr\ \mathbf{I} = 4$. Therefore, from eq. (88), we have $Tr\ \hat{S}^\dagger\hat{S} = 4ka_0^0 > 0$. Hence, if

$$a_0^0 \geq 1,\ k = +1,\ a_0^0 \leq -1,\ k = -1 \tag{89}$$

Now, we have

$$\begin{aligned}\overline{\psi}'(x') &= \psi'(x')\gamma'^0 = \boldsymbol{\psi}'^\dagger(x')\boldsymbol{\gamma}^0 = [\hat{S}\,\psi(x)]^\dagger\ \boldsymbol{\gamma}^0 \\ &= \psi^\dagger(x)\hat{S}^\dagger\boldsymbol{\gamma}^0 = k\boldsymbol{\psi}^\dagger(x)\boldsymbol{\gamma}^0\hat{S}^{-1} = k\overline{\psi}(x)\hat{S}^{-1}\end{aligned} \tag{90}$$

Therefore, we infer the following properties for $\overline{\psi}(x)$:

(i) In the case of those transformations that do not involve time reversal, i.e., $a^0{}_0 \geq 1$, we have

$$\overline{\psi}'(x') = \overline{\psi}(x)\hat{S}^{-1} \tag{91}$$

(ii) In the case of Lorentz transformations that involve time reversal, i.e., $a_0^0 \leq -1$, $\overline{\psi}(x)$ transforms as

$$\overline{\psi}'(x') = -\overline{\psi}(x)\hat{S}^{-1} \tag{92}$$

We now explore the transformation properties of various bilinear combinations of the Dirac spinors $\psi(x)$, $\overline{\psi}(x)$ and the 16 $\boldsymbol{\Gamma}$ matrices $\boldsymbol{\Gamma}_i$, $i = 1, 2, 3,\ldots,16$. In the sequel, the positive sign $(k = +1)$ holds when the transformation does not involve time reversal, i.e., $a_0^0 \geq 1$ and the negative sign $(k = -1)$ holds in case of a transformation involving time reversal $a_0^0 \leq -1$.

(a) $\overline{\psi}(x)\boldsymbol{\psi}(x)$

We have

$$\overline{\psi}'(x')\boldsymbol{\psi}'(x') = k\overline{\psi}(x)\hat{S}^{-1}\hat{S}\,\boldsymbol{\psi}(x) = k\overline{\psi}(x)\boldsymbol{\psi}(x) = \pm\overline{\psi}(x)\boldsymbol{\psi}(x) \tag{93}$$

Hence, $\overline{\psi}(x)\boldsymbol{\psi}(x)$ is a scalar, covariant with Lorentz transformations involving no time reversal and a pseudoscalar under time inversion.

(b) $\overline{\psi}(x)\boldsymbol{\gamma}^\mu\,\boldsymbol{\psi}(x)$

We have

$$\begin{aligned}\overline{\psi}'(x')\boldsymbol{\gamma}^\mu\boldsymbol{\psi}'(x') &= k\overline{\psi}(x)\hat{S}^{-1}\boldsymbol{\gamma}^\mu\hat{S}\,\boldsymbol{\psi}(x) = ka^\mu_\nu\,\overline{\psi}(x)\boldsymbol{\gamma}^\nu\boldsymbol{\psi}(x) \\ &= \pm\,a^\mu_\nu\,\overline{\psi}(x)\boldsymbol{\gamma}^\nu\boldsymbol{\psi}(x)\end{aligned} \tag{94}$$

Hence, $\overline{\psi}(x)\boldsymbol{\gamma}^\mu\boldsymbol{\psi}(x)$ is a vector, covariant with Lorentz transformations involving no time reversal and a pseudovector under time inversion.

(c) $\bar{\psi}(x)\boldsymbol{\sigma}^{\mu\nu}\psi(x)$ where $\boldsymbol{\sigma}^{\mu\nu} = \dfrac{1}{2i}(\boldsymbol{\gamma}^{\mu}\boldsymbol{\gamma}^{\nu} - \boldsymbol{\gamma}^{\nu}\boldsymbol{\gamma}^{\mu})$

We have

$$\bar{\psi}'(x')\boldsymbol{\sigma}^{\mu\nu}\boldsymbol{\psi}'(x') = k\bar{\psi}(x)\hat{S}^{-1}\boldsymbol{\sigma}^{\mu\nu}\hat{S}\psi(x)$$

$$= k\bar{\psi}(x)\hat{S}^{-1}\left[\frac{1}{2i}(\gamma^{\mu}\gamma^{\nu} - \gamma^{\nu}\gamma^{\mu})\right]\hat{S}\psi(x)$$

$$= k\bar{\psi}(x)\left\{\frac{1}{2i}[(\hat{S}^{-1}\gamma^{\mu}\hat{S})(\hat{S}^{-1}\gamma^{\nu}\hat{S}) - (\hat{S}^{-1}\gamma^{\nu}\hat{S})(\hat{S}^{-1}\gamma^{\mu}\hat{S})]\right\}\psi(x)$$

$$= k\bar{\psi}(x)\left\{\frac{1}{2i}[(a^{\mu}_{\kappa}\boldsymbol{\gamma}^{\kappa})(a^{\mu}_{\lambda}\boldsymbol{\gamma}^{\lambda}) - (a^{\nu}_{\lambda}\boldsymbol{\gamma}^{\lambda})(a^{\mu}_{\kappa}\boldsymbol{\gamma}^{\kappa})]\right\}\psi(x)$$

$$= ka^{\mu}_{\kappa}a^{\nu}_{\lambda}\bar{\psi}(x)\left[\frac{1}{2i}(\gamma^{\kappa}\gamma^{\lambda} - \gamma^{\lambda}\gamma^{\kappa})\right]\psi(x) = ka^{\mu}_{\kappa}a^{\nu}_{\lambda}\bar{\psi}(x)\sigma^{\kappa\lambda}\psi(x) \qquad (95)$$

Hence, the bilinear $\bar{\psi}(x)\sigma^{\mu\nu}\psi(x)$ transforms as an antisymmetric second rank tensor under Lorentz transformations involving no time inversion.

(d) $\bar{\psi}(x)\boldsymbol{\gamma}^{5}\boldsymbol{\psi}(x)$, where $\boldsymbol{\gamma}^{5} = i\boldsymbol{\gamma}^{0}\boldsymbol{\gamma}^{1}\boldsymbol{\gamma}^{2}\boldsymbol{\gamma}^{3}$

We have

$$\bar{\psi}'(x')\boldsymbol{\gamma}^{5}\boldsymbol{\psi}'(x') = ik\bar{\psi}(x)\hat{S}^{-1}\boldsymbol{\gamma}^{0}\boldsymbol{\gamma}^{1}\boldsymbol{\gamma}^{2}\boldsymbol{\gamma}^{3}\hat{S}\boldsymbol{\psi}(x)$$

$$= ik\bar{\psi}(x)(\hat{S}^{-1}\boldsymbol{\gamma}^{0}\hat{S})(\hat{S}^{-1}\boldsymbol{\gamma}^{1}\hat{S})(\hat{S}^{-1}\boldsymbol{\gamma}^{2}\hat{S})(\hat{S}^{-1}\boldsymbol{\gamma}^{3}\hat{S})\psi(x)$$

$$= ika^{0}_{\kappa}\, a^{1}_{\lambda}a^{2}_{\mu}a^{3}_{\nu}\bar{\psi}(x)\boldsymbol{\gamma}^{\kappa}\boldsymbol{\gamma}^{\lambda}\boldsymbol{\gamma}^{\mu}\boldsymbol{\gamma}^{\nu}\boldsymbol{\psi}(x)$$

$$= i\frac{k}{4!}\boldsymbol{\varepsilon}_{\rho\sigma\zeta\tau}a^{\rho}_{\kappa}\, a^{\sigma}_{\lambda}a^{\zeta}_{\mu}a^{\tau}_{\nu}\bar{\psi}(x)\boldsymbol{\gamma}^{\kappa}\boldsymbol{\gamma}^{\lambda}\boldsymbol{\gamma}^{\mu}\boldsymbol{\gamma}^{\nu}\boldsymbol{\psi}(x) = k|\boldsymbol{\Lambda}|\bar{\psi}(x)\boldsymbol{\gamma}^{5}\boldsymbol{\psi}(x) \qquad (96)$$

Hence, $\bar{\psi}(x)\boldsymbol{\gamma}^{5}\boldsymbol{\psi}(x)$ transforms like a scalar under proper Lorentz transformations but as a pseudoscalar under time inversion.

(e) $\bar{\psi}(x)\boldsymbol{\gamma}^{\lambda}\boldsymbol{\gamma}^{\mu}\boldsymbol{\gamma}^{\nu}\boldsymbol{\psi}(x)$, $\lambda < \mu < \nu$

Defining $\boldsymbol{\gamma}^{5} = i\boldsymbol{\gamma}^{0}\boldsymbol{\gamma}^{1}\boldsymbol{\gamma}^{2}\boldsymbol{\gamma}^{3}$, we have

(i) $\boldsymbol{\gamma}^{0}\boldsymbol{\gamma}^{1}\boldsymbol{\gamma}^{2} = -\boldsymbol{\gamma}^{0}\boldsymbol{\gamma}^{1}\boldsymbol{\gamma}^{2}(\boldsymbol{\gamma}^{3})^{2} = i\boldsymbol{\gamma}^{5}\boldsymbol{\gamma}^{3}$

(ii) $\boldsymbol{\gamma}^{0}\boldsymbol{\gamma}^{1}\boldsymbol{\gamma}^{3} = -\boldsymbol{\gamma}^{0}\boldsymbol{\gamma}^{1}\boldsymbol{\gamma}^{3}(\boldsymbol{\gamma}^{2})^{2} = \boldsymbol{\gamma}^{0}\boldsymbol{\gamma}^{1}\boldsymbol{\gamma}^{2}\boldsymbol{\gamma}^{3}\boldsymbol{\gamma}^{2} = -i\boldsymbol{\gamma}^{5}\boldsymbol{\gamma}^{2}$

(iii) $\boldsymbol{\gamma}^{0}\boldsymbol{\gamma}^{2}\boldsymbol{\gamma}^{3} = -\boldsymbol{\gamma}^{0}\boldsymbol{\gamma}^{2}\boldsymbol{\gamma}^{3}(\boldsymbol{\gamma}^{1})^{2} = -\boldsymbol{\gamma}^{0}\boldsymbol{\gamma}^{1}\boldsymbol{\gamma}^{2}\boldsymbol{\gamma}^{3}\boldsymbol{\gamma}^{1} = i\boldsymbol{\gamma}^{5}\boldsymbol{\gamma}^{1}$

(iv) $\gamma^1\gamma^2\gamma^3 = \gamma^1\gamma^2\gamma^3(\gamma^0)^2 = -\gamma^0\gamma^1\gamma^2\gamma^3\gamma^0 = i\gamma^5\gamma^0$

Now,

$$\bar{\psi}'(x')\gamma^5\gamma^\mu\psi'(x') = k\bar{\psi}(x)\hat{S}^{-1}\gamma^5\gamma^\mu\hat{S}\psi(x) = k\bar{\psi}(x)(\hat{S}^{-1}\gamma^5\hat{S})(\hat{S}^{-1}\gamma^\mu\hat{S})\psi(x)$$
$$= k|\Lambda|a^\mu_\nu\bar{\psi}(x)\gamma^5\gamma^\nu\psi(x) = \pm\, a^\mu_\nu|\Lambda|\bar{\psi}(x)\gamma^5\gamma^\nu\psi(x) \quad (97)$$

so that $\bar{\psi}(x)\gamma^5\gamma^\mu\psi(x)$, $\mu = 0, 1, 2, 3$ behaves as a tensor of the third rank, antisymmetric in all the three indices. This shows that the four products $\bar{\psi}(x)\,\gamma^5\gamma^\mu\psi(x)$, $\mu = 0, 1, 2, 3$ transform like the components of a 4-vector, except that, on multiplying by a 4-vector, like a coordinate vector, we get a pseudoscalar and not a scalar. Hence, $\bar{\psi}(x)\gamma^5\gamma^\mu\psi(x)$, $\mu = 0, 1, 2, 3$ is sometimes called a pseudovector.

5.10 SCALAR PRODUCT IN DIRAC FRAMEWORK

We define the invariant scalar product of two Dirac spinors $\psi(x)$ and $\phi(x)$ at time $t \equiv x^0$ as

$$\langle\phi|\psi\rangle_{x^0} = \int_{x^0} d\mathbf{x}\bar{\phi}(x)\gamma^0\psi(x) = \int_{x^0} d\mathbf{x}[\phi^\dagger(x)\gamma^0]\gamma^0\psi(x)$$
$$= \int_{x^0} d\mathbf{x}\phi^\dagger(x)\psi(x) \quad (98)$$

This product is Lorentz invariant which is shown as follows:

$$\langle\phi'|\psi'\rangle_{x^0} = \int_{x^0} d\mathbf{x}'\bar{\phi}'(x')\gamma^0\psi'(x') = \int_{x^0} dx'^1dx'^2dx'^3\bar{\phi}'(x')\gamma^0\psi'(x')$$
$$= ka^1_{\ \kappa}a^2_{\ \lambda}a^3_{\ \mu}\int_{x^0} dx^\kappa dx^\lambda dx^\mu\bar{\phi}(x)\hat{S}^{-1}\gamma^0\hat{S}\psi(x)$$
$$= ka^1_{\ \kappa}a^2_{\lambda}a^3_{\ \mu}a^0_{\ \nu}\int_{x^0} dx^\kappa dx^\lambda dx^\mu\bar{\phi}(x)\gamma^\nu\psi(x)$$
$$= \frac{k}{4!}\varepsilon_{\rho\sigma\zeta\tau}a^\rho_{\ \kappa}a^\sigma_{\ \lambda}a^\zeta_{\ \mu}a^\tau_{\ \nu}\int_{x^0} dx^\kappa dx^\lambda dx^\mu\bar{\phi}(x)\gamma^\nu\psi(x)$$
$$= k\,|\Lambda|\int_{x^0} dx^1dx^2dx^3\bar{\phi}(x)\gamma^0\psi(x)$$
$$= k|\Lambda|\int_{x^0} d\mathbf{x}\bar{\phi}(x)\gamma^0\psi(x) \quad (99)$$

The scalar product is also time independent, e.g.

$$\partial_0\langle\phi|\psi\rangle = \int d\mathbf{x}\partial_0[\bar{\phi}(x)\gamma^0\psi(x)]$$

$$= \int d\mathbf{x}\{[\partial_0\bar{\phi}(x)]\gamma^0\psi(x) + \bar{\phi}(x)\gamma^0[\partial_0\psi(x)]\} \tag{100}$$

Since the Dirac spinors $\psi(x)$ and $\phi(x)$ obey the Dirac equation, we have $(i\gamma^\mu\partial_\mu - m)\psi(x) = 0$ for $\psi(x)$, and for the adjoint spinor $\bar{\phi}(x)$

$$0 = (i\gamma^\mu\partial_\mu - m)\phi(x) = [(i\gamma^\mu\partial_\mu - m)\phi(x)]^\dagger = -i[\partial_\mu\phi(x)]^\dagger(\gamma^\mu) - m\phi^\dagger(x)$$

$$= -i[\partial_\mu\phi(x)]^\dagger(\gamma^\mu)^\dagger\gamma^0 - m\phi^\dagger(x)\gamma^0 = -i[\partial_\mu\phi(x)]^\dagger\gamma^0\gamma^\mu - m\phi^\dagger(x)\gamma^0$$

$$= -i\partial_\mu\bar{\phi}(x)\gamma^\mu - m\bar{\phi}(x) = i\partial_\mu\bar{\phi}(x)\gamma^\mu + m\bar{\phi}(x) \tag{101}$$

From the above expressions, we have

$$[\partial_0\bar{\phi}(x)]\gamma^0\psi(x) + \bar{\phi}(x)\gamma^0[\partial_0\psi(x)] = [im\bar{\phi}(x) - \partial_k\bar{\phi}(x)\gamma^k]\psi(x) + \bar{\phi}(x)$$

$$[-im\psi(x) - \gamma^k\partial_k\psi(x)] = -\partial_k[\bar{\phi}(x)\gamma^k\psi(x)] \tag{102}$$

Hence, eq. (100) reduces to

$$\partial_0\langle\phi|\psi\rangle = -\int d\mathbf{x}\partial_k[\bar{\phi}(x)\gamma^0\psi(x)]. \tag{103}$$

which vanishes because of being the volume integral of a divergence.

The above scalar product may also be defined with reference to a spacelike surface σ as

$$\langle\phi|\psi\rangle_\sigma = \int_\sigma d\sigma_\mu(x)\bar{\phi}(x)\gamma^\mu\psi(x) \tag{104}$$

We can show that this scalar product is independent of σ if the spinors $\psi(x)$ and $\phi(x)$ obey the Dirac equation for

$$\frac{\delta}{\delta(x)}\langle\phi|\psi\rangle_\sigma = \frac{\delta}{\delta(x)}\int_\sigma d\sigma_\mu(x)\bar{\phi}(x)\gamma^\mu\psi(x) = \partial_\mu[\bar{\phi}(x)\gamma^\mu\psi(x)] \tag{104a}$$

$$= \partial_\mu[\bar{\phi}(x)\gamma^\mu]\psi(x) + \bar{\phi}(x)\gamma^\mu\partial_\mu\psi(x)$$

$$= im\bar{\phi}(x)\psi(x) + \bar{\phi}(x)[-im\psi(x)] = 0 \tag{105}$$

Hence, the scalar product defined by eq. (104) is independent of the spacelike surface σ. Therefore, we may choose to evaluate this product on a surface $\sigma = t \equiv x^0 =$ constant.

Let us now consider an arbitrary Lorentz transformation $x' = \Lambda x + a$ under which the spinor $\psi(x)$ transforms as $\psi'(x') = \hat{S}\psi(x) = \hat{S}\psi[\Lambda^{-1}(x' - a)]$. Similarly, for the adjoint spinor $\overline{\psi}'(x') = k\overline{\psi}(x)\hat{S}^{-1}$. Hence, we have

$$\langle\phi'|\psi'\rangle_\sigma = \langle\hat{S}\phi|\hat{S}\psi\rangle_\sigma = \langle\hat{S}^\dagger\hat{S}\phi|\psi\rangle_\sigma = \int_\sigma d\sigma_\mu(x)\overline{\phi}'(x')\gamma^\mu\psi'(x')$$

$$= k\int_\sigma d\sigma_\mu(x)\overline{\phi}(x)\hat{S}^{-1}\gamma^\mu\hat{S}\psi(x) = k\int_\sigma d\sigma_\mu(x)a^\mu_\nu\overline{\phi}(x)\gamma^\nu\psi(x)$$

$$= k\int_\sigma d\sigma_\nu(x)\overline{\phi}(x)\gamma^\nu\psi(x) = k\langle\phi|\psi\rangle_\sigma \tag{106}$$

showing the unitarity of the spinor transformation operator with respect to this scalar product, provided that the transformation does not involve a time inversion. In the case of a time inversion transformation, the spinor transformation operator is antiunitary.

5.11 INVARIANCE ALGEBRA OF THE DIRAC EQUATION (FUSHCHICH, 1995)

In this section, we establish that the maximal invariance algebra of the Dirac equation is the Poincare algebra. We define the quantities $\gamma_4 = \gamma_0\gamma_1\gamma_2\gamma_3$ and $S_{\mu\nu} = \frac{i}{4}[\gamma_\mu, \gamma_\nu]$. We, then, have the expressions:

(a) $\gamma_\mu\gamma_\nu = g_{\mu\nu} - 2iS_{\mu\nu}$ (107)

(b) $S_{\mu\nu}\gamma_4 = \frac{1}{2}\varepsilon_{\mu\nu\rho\sigma}S^{\rho\sigma}$ (108)

(c) $S_{\mu\nu}\gamma_\lambda = \frac{i}{2}(g_{\nu\lambda}\gamma_\mu - g_{\mu\lambda}\gamma_\nu + \varepsilon_{\mu\nu\lambda\sigma}\gamma_4\gamma^\sigma)$ (109)

for $S_{\mu\nu}\gamma_\lambda = \frac{i}{4}(\gamma_\mu\gamma_\nu\gamma_\lambda - \gamma_\nu\gamma_\mu\gamma_\lambda) = \frac{i}{4}[\gamma_\mu(2g_{\nu\lambda} - \gamma_\lambda\gamma_\nu) - \gamma_\nu(2g_{\mu\lambda} - \gamma_\lambda\gamma_\mu)]$

$$= \frac{i}{4}[2g_{\nu\lambda}\gamma_\mu - 2g_{\mu\lambda}\gamma_\nu - \gamma_\mu\gamma_\lambda\gamma_\nu + \gamma_\nu\gamma_\lambda\gamma_\mu] = \frac{i}{2}[2g_{\nu\lambda}\gamma_\mu - 2g_{\mu\lambda}\gamma_\nu + \varepsilon_{\mu\nu\lambda\sigma}\gamma_4\gamma^\sigma]$$

(d) $[S_{\mu\nu}, S_{\lambda\sigma}] = i(g_{\mu\sigma}S_{\nu\lambda} + g_{\nu\lambda}S_{\mu\sigma} - g_{\mu\lambda}S_{\nu\sigma} - g_{\nu\sigma}S_{\mu\lambda})$ (110)

This can be proved easily by direct computation. We have the following possible combinations of indices μ, v, λ, σ for which $S_{\mu v}$, $S_{\lambda\sigma}$ are both non-vanishing, viz.
(i) $\mu \neq v \neq \lambda \neq \sigma$, (ii) $\mu \neq v = \lambda \neq \sigma$, (iii) $v \neq \mu = \lambda \neq \sigma$, (iv) $v \neq \mu = \sigma \neq \lambda$, (v) $\mu \neq v = \sigma \neq \lambda$. We shall consider each of these cases separately.

$$\text{(i) } [S_{\mu v}, S_{\lambda\sigma}] = (S_{\mu v}S_{\lambda\sigma} - S_{\lambda\sigma}S_{\mu v}) = -\frac{1}{16}\begin{bmatrix} (\gamma_\mu\gamma_v - \gamma_v\gamma_\mu)(\gamma_\lambda\gamma_\sigma - \gamma_\sigma\gamma_\lambda) \\ -(\gamma_\lambda\gamma_\sigma - \gamma_\sigma\gamma_\lambda)(\gamma_\mu\gamma_v - \gamma_v\gamma_\mu) \end{bmatrix}$$

$$= -\frac{1}{4}(\gamma_\mu\gamma_v\gamma_\lambda\gamma_\sigma - \gamma_\lambda\gamma_\sigma\gamma_\mu\gamma_v) = -\frac{1}{4}(\gamma_\mu\gamma_v\gamma_\lambda\gamma_\sigma - \gamma_\sigma\gamma_\mu\gamma_\lambda\gamma_v)$$

$$= -\frac{1}{4}(\gamma_\mu\gamma_v\gamma_\lambda\gamma_\sigma - \gamma_\mu\gamma_\lambda\gamma_\sigma\gamma_v) = -\frac{1}{4}(\gamma_\mu\gamma_v\gamma_\lambda\gamma_\sigma - \gamma_\mu\gamma_v\gamma_\lambda\gamma_\sigma) = 0$$

$$\text{(ii) } [S_{\mu v}, S_{v\sigma}] = (S_{\mu v}S_{v\sigma} - S_{v\sigma}S_{\mu v}) = -\frac{1}{4}(\gamma_\mu\gamma_v\gamma_v\gamma_\sigma - \gamma_v\gamma_\sigma\gamma_\mu\gamma_v) = -\frac{1}{4}(\gamma_\mu\gamma_v\gamma_v\gamma_\sigma - \gamma_\mu\gamma_v\gamma_\sigma\gamma_v)$$

$$= -\frac{1}{2}\gamma_\mu\gamma_v\gamma_v\gamma_\sigma = -\frac{1}{2}g_{vv}\gamma_\mu\gamma_\sigma = ig_{vv}S_{\mu\sigma}$$

Similarly,

(iii) $[S_{\mu v}, S_{\mu\sigma}] = -ig_{\mu\mu}S_{v\sigma}$, (iv) $[S_{\mu v}, S_{\lambda\mu}] = ig_{\mu\mu}S_{v\lambda}$, (v) $[S_{\mu v}, S_{\lambda v}] = -ig_{vv}S_{\mu\lambda}$.

Combining these results, we get eq. (110).

(e) $[\gamma^\mu, S_{\lambda\sigma}] = i(g^\mu_\lambda\gamma_\sigma - g^\mu_\sigma\gamma_\lambda)$ (111)

Since $\quad [\gamma^\mu, S_{\lambda\sigma}] = \frac{i}{4}(\gamma^\mu\gamma_\lambda\gamma_\sigma - \gamma^\mu\gamma_\sigma\gamma_\lambda - \gamma_\lambda\gamma_\sigma\gamma^\mu + \gamma_\sigma\gamma_\lambda\gamma^\mu)$

$$= \frac{i}{4}[\gamma^\mu\gamma_\lambda\gamma_\sigma - \gamma^\mu(2g_{\lambda\sigma} - \gamma_\lambda\gamma_\sigma) - \gamma_\lambda\gamma_\sigma\gamma^\mu + (2g_{\lambda\sigma} - \gamma_\lambda\gamma_\sigma)\gamma^\mu]$$

$$= \frac{i}{2}(\gamma^\mu\gamma_\lambda\gamma_\sigma - \gamma_\lambda\gamma_\sigma\gamma^\mu) = \frac{i}{2}[\gamma^\mu\gamma_\lambda\gamma_\sigma - \gamma_\lambda(2g^\mu_\sigma - \gamma^\mu\gamma_\sigma)]$$

$$= \frac{i}{2}[\gamma^\mu\gamma_\lambda\gamma_\sigma - 2g^\mu_\sigma\gamma_\lambda + (2g^\mu_\lambda - \gamma^\mu\gamma_\lambda)\gamma_\sigma] = i(g^\mu_\lambda\gamma_\sigma - g^\mu_\sigma\gamma_\lambda)$$

Using the above commutators, we obtain

$$[\gamma^\mu P_\mu, P_v] = 0 \tag{112}$$

and

$$[\gamma^\mu P_\mu, J_{\lambda\sigma}] = [\gamma^\mu P_\mu, x_\lambda P_\sigma - x_\sigma P_\lambda + S_{\lambda\sigma}]$$
$$= [\gamma^\mu P_\mu, x_\lambda P_\sigma] - [\gamma^\mu P_\mu, x_\sigma P_\lambda] + [\gamma^\mu P_\mu, S_{\lambda\sigma}] \tag{113}$$

Now

$$[\gamma^\mu P_\mu, x_\lambda P_\sigma] - \gamma^\mu [P_\mu, x_\lambda] P_\sigma = i g_{\mu\lambda} \gamma^\mu P_\sigma \tag{114}$$

$$[\gamma^\mu P_\mu, x_\sigma P_\lambda] = \gamma^\mu [P_\mu, x_\sigma] P_\lambda = i g_{\mu\sigma} \gamma^\mu P_\lambda \tag{115}$$

$$[\gamma^\mu P_\mu, S_{\lambda\sigma}] = [\gamma^\mu, S_{\lambda\sigma}] P_\mu = i(g^\mu_\lambda \gamma_\sigma - g^\mu_\sigma \gamma_\lambda) P_\mu \tag{116}$$

Substituting from eqs. (114)-(116) in eq. (113), we obtain

$$[\gamma^\mu P_\mu, J_{\lambda\sigma}] = 0 \tag{117}$$

It follows from eqs. (112) & (117) that the Dirac operator

$$L \equiv \gamma^\mu P_\mu - m \tag{118}$$

commutes with all the generators of the Poincare group so that the Dirac equation is invariant under the Poincare algebra.

We now prove that the Poincare algebra constitutes the maximally extensive invariant algebra of the Dirac operator.

In terms of the inequivalent products of the Dirac matrices, $\gamma_0, \gamma_1, \gamma_2, \gamma_3, \gamma_4$, we can write a basis for the 16-dimensional vector space as

$$\mathbf{I},\ \gamma_0,\ i\gamma_1,\ i\gamma_2,\ i\gamma_3,\ i\gamma_4,\ \gamma_4\gamma_0,\ i\gamma_4\gamma_1,\ i\gamma_4\gamma_2,\ i\gamma_4\gamma_3,\ \gamma_0\gamma_1,\ \gamma_0\gamma_2,\ \gamma_0\gamma_3,\ i\gamma_1\gamma_2,\ i\gamma_2\gamma_3,\ i\gamma_3\gamma_1 \tag{119}$$

As in the case of the Klein Gordon equation, we introduce a vector space of complex valued functions $\psi(x)$ that are solutions of the Dirac equation, V^4 V^4 is defined on an open and connected set D of the real four-dimensional space. The Dirac operator $L \equiv \gamma^\mu P_\mu - m$ is defined on D and satisfies $L\psi \in V^4$ for $\psi \in V^4$. Let G^4 denote the space of 4×4 matrices whose matrix elements belong to V.

A linear first order differential operator $Q = A^\mu P_\mu + B$, $A^\mu \in V$, $B \in G^4$ is defined as a Symmetry Operator (SO) of the Dirac equation if $[Q, L] = \alpha_Q L$, $\alpha_Q \in G^4$. We write Q, α_Q as linear combinations of the basis elements (119) as

$$Q = \mathbf{I}(K^\mu P_\mu + a^0) + i\gamma_4 a^1 + \gamma_\mu b^\mu + S_{\mu\nu} f^{[\mu\nu]} + \gamma_4\gamma_\mu c^\mu \tag{120}$$

$$\alpha_Q = \mathbf{I} d^0 + i\gamma_4 d^1 + \gamma_\mu h^\mu + \gamma_4\gamma_\mu k^\mu + S_{\mu\nu} n^{[\mu\nu]} \tag{121}$$

We, then have

$$\begin{aligned}
&\gamma^\lambda P_\lambda [\mathbf{I}(K^\mu P_\mu + a^0) + i\gamma_4 a^1 + \gamma_\mu b^\mu + S_{\mu\nu} f^{[\mu\nu]} + \gamma_4\gamma_\mu c^\mu]\psi \\
&= i\gamma^\lambda \partial_\lambda [\mathbf{I}(iK^\mu \partial_\mu + a^0) + i\gamma_4 a^1 + \gamma_\mu b^\mu + S_{\mu\nu} f^{[\mu\nu]} + \gamma_4\gamma_\mu c^\mu]\psi \\
&= -\gamma^\lambda K^\mu_\lambda \partial_\mu \psi - \gamma^\lambda K^\mu \partial^2_{\lambda\mu}\psi + i\gamma^\lambda a^0_\lambda \psi + i\gamma^\lambda a^0 \partial_\lambda \psi - \gamma^\lambda \gamma_4 a^1_\lambda \psi - \gamma^\lambda \gamma_4 a^1 \partial_\lambda \psi + i\gamma^\lambda \gamma_\mu b^\mu_\lambda \psi \\
&\quad + i\gamma^\lambda \gamma_\mu b^\mu \partial_\lambda \psi + i\gamma^\lambda S_{\mu\nu} f^{[\mu\nu]}\psi + i\gamma^\lambda S_{\mu\nu} f^{[\mu\nu]} \partial_\lambda \psi + i\gamma^\lambda \gamma_4 \gamma_\mu c^\mu_\lambda \psi + i\gamma^\lambda \gamma_4 \gamma_\mu c^\mu \partial_\lambda \psi
\end{aligned} \tag{122}$$

and

$$\begin{aligned}
&[\mathbf{I}(K^\mu P_\mu + a^0) + i\gamma_4 a^1 + \gamma_\mu b^\mu + S_{\mu\nu} f^{[\mu\nu]} + \gamma_4\gamma_\mu c^\mu]\gamma^\lambda P_\lambda \psi \\
&= i[\mathbf{I}(iK^\mu \partial_\mu + a^0) + i\gamma_4 a^1 + \gamma_\mu b^\mu + S_{\mu\nu} f^{[\mu\nu]} + \gamma_4\gamma_\mu c^\mu]\gamma^\lambda \partial_\lambda \psi \\
&= -\gamma^\lambda K^\mu \partial^2_{\lambda\mu}\psi + i\gamma^\lambda a^0 \partial_\lambda \psi - \gamma_4 \gamma^\lambda a^1 \partial_\lambda \psi + i\gamma_\mu \gamma^\lambda b^\mu \partial_\lambda \psi \\
&\quad + iS_{\mu\nu}\gamma^\lambda f^{[\mu\nu]} \partial_\lambda \psi + i\gamma_4 \gamma_\mu \gamma^\lambda c^\mu \partial_\lambda \psi
\end{aligned} \tag{123}$$

so that

$$
\begin{aligned}
[\gamma^\lambda P_\lambda, Q]\psi &= -\gamma^\lambda K^\mu_\lambda \partial_\mu \psi + i\gamma^\lambda a^0_\lambda \psi - \gamma^\lambda \gamma_4 a^1_\lambda \psi - (\gamma^\lambda \gamma_4 - \gamma_4 \gamma^\lambda) a^1 \partial_\lambda \psi \\
&+ i\gamma^\lambda \gamma_\mu b^\mu_\lambda \psi + i(\gamma^\lambda \gamma_\mu - \gamma_\mu \gamma^\lambda) b^\mu \partial_\mu \psi + i\gamma^\lambda S_{\mu\nu} f^{[\mu\nu]}_\lambda \psi \\
&+ i(\gamma^\lambda S_{\mu\nu} - S_{\mu\nu}\gamma^\lambda) f^{[\mu\nu]} \partial_\lambda \psi + i\gamma^\lambda \gamma_4 \gamma_\mu c^\mu_\lambda \psi + i(\gamma^\lambda \gamma_4 \gamma_\mu - \gamma_4 \gamma_\mu \gamma^\lambda) c^\mu \partial_\lambda \psi \\
&= -\gamma^\lambda K^\mu_\lambda \partial_\mu \psi + i\gamma^\lambda a^0_\lambda \psi - \gamma^\lambda \gamma_4 a^1_\lambda \psi + 2\gamma_4 \gamma^\lambda a^1 \partial_\lambda \psi + i(g^\lambda_\mu - 2iS^\lambda_\mu) b^\mu_\lambda \psi \\
&+ 4S^\lambda_\mu b^\mu \partial_\lambda \psi - \frac{1}{2}(g^\lambda_\mu \gamma_\nu - g^\lambda_\nu \gamma_\mu + \varepsilon^\lambda_{\mu\nu\sigma}\gamma_4 \gamma^\sigma) f^{[\mu\nu]}_\lambda \psi - (g^\lambda_\mu \gamma_\nu - g^\lambda_\nu \gamma_\mu) f^{[\mu\nu]} \partial_\lambda \psi \\
&- i\gamma_4 (g^\lambda_\mu - 2iS^\lambda_\mu) c^\mu_\lambda \psi - 2i\gamma_4 g^\lambda_\mu c^\mu \partial_\lambda \psi \\
&= i\gamma^\lambda K^\mu_\lambda P_\mu \psi + i\gamma^\lambda a^0_\lambda \psi - \gamma_4 \gamma^\lambda a^1_\lambda \psi - 2i\gamma_4 \gamma^\lambda a^1 P_\lambda \psi + ib^\lambda_\lambda \psi + 2S^\lambda_\mu b^\mu_\lambda \psi \\
&- 4iS^\lambda_\mu b^\mu P_\lambda \psi - \frac{1}{2}(\gamma_\nu f^{[\lambda\nu]}_\lambda - \gamma_\mu f^{[\mu\lambda]}_\lambda + \varepsilon^\lambda_{\mu\nu\sigma}\gamma_4 \gamma^\lambda f^{[\mu\nu]}_\lambda)\, \psi \\
&+ i\,(\gamma_\nu f^{[\lambda\nu]} - \gamma_\mu f^{[\mu\lambda]})\, P_\lambda \psi - \gamma_4 (ic^\lambda_\lambda + 2c^\lambda P_\lambda)\psi - 2\gamma_4 S^\lambda_\mu c^\mu_\lambda \psi \\
&= i\gamma^\lambda K^\mu_\lambda P_\mu \psi + i\gamma^\lambda a^0_\lambda \psi - 2i\gamma_4 \gamma^\lambda a^1 P_\lambda \psi + \gamma_4 \gamma^\lambda a^1_\lambda \psi + ib^\lambda_\lambda \psi \\
&- 4iS^\lambda_\mu b^\mu P_\lambda \psi + 2S^\lambda_\mu b^\mu_\lambda \psi + \gamma_\mu f^{[\mu\lambda]}_\lambda \psi - \frac{1}{2}\varepsilon^\lambda_{\mu\nu\sigma}\gamma_4 \gamma^\sigma f^{[\mu\nu]}_\lambda \psi \\
&- 2i\gamma_\mu f^{[\mu\lambda]} P_\lambda \psi - \gamma_4 (ic^\lambda_\lambda + 2c^\lambda P_\lambda)\psi - \varepsilon^\lambda_{\mu\nu\sigma} S^{\nu\sigma} c^\mu_\lambda \psi
\end{aligned}
\tag{124}
$$

Hence,

$$
\begin{aligned}
[Q, L]\psi &= -i\gamma^\lambda K^\mu_\lambda P_\mu \psi - i\gamma^\lambda a^0_\lambda \psi + 2i\gamma_4 \gamma^\lambda a^1 P_\lambda \psi - \gamma_4 \gamma^\lambda a^1_\lambda \psi - ib^\lambda_\lambda \psi + 4iS^\lambda_\mu b^\mu p_\lambda \psi \\
&- 2\,S^\lambda_\mu b^\mu_\lambda \psi - \gamma_\mu f^{[\mu\lambda]}_\lambda \psi + \frac{1}{2}\varepsilon^\lambda_{\mu\nu\sigma}\gamma_4 \gamma^\sigma f^{[\mu\nu]}_\lambda \psi + 2i\gamma_\mu f^{[\mu\lambda]} P_\lambda \psi \\
&+ \gamma_4 (ic^\lambda_\lambda + 2c^\lambda P_\lambda)\psi + \varepsilon^\lambda_{\mu\nu\sigma} S^{\nu\sigma} c^\mu_\lambda \psi
\end{aligned}
\tag{125}
$$

Also

$$
\alpha_Q L\psi = \begin{bmatrix} \mathbf{I}(d^0\gamma^\lambda P_\lambda - d^0 m) + id^1\gamma_4\gamma^\lambda P_\lambda - i\gamma_4 m d^1 + \gamma_\mu \gamma^\lambda h^\mu P_\lambda - m\gamma_\mu h^\mu \\ +\gamma_4\gamma_\mu\gamma^\lambda k^\mu P_\lambda - m\gamma_4\gamma_\mu k^\mu + S_{\mu\nu}\gamma^\lambda n^{[\mu\nu]} P_\lambda - mS_{\mu\nu} n^{[\mu\nu]} \end{bmatrix}\psi
$$

$$
\alpha_Q L\psi = \begin{bmatrix} \mathbf{I}(d^0\gamma^\lambda P_\lambda - d^0 m + h^\lambda P_\lambda) - m\gamma_\mu h^\mu \\ -m\gamma_4\gamma_\mu k^\mu + \gamma_4 k^\lambda P_\lambda - 2i\gamma_4 S^\lambda_\mu k^\mu P_\lambda + \frac{i}{2}(g^\lambda_\nu \gamma_\mu - g^\lambda_\nu \gamma_\nu + \varepsilon^\lambda_{\mu\nu\sigma}\gamma_4\gamma^\sigma) \end{bmatrix}
$$

$$
\begin{matrix} - 2iS^\lambda_\mu h^\mu P_\lambda - i\gamma_4 m d^1 + i\gamma_4\gamma^\lambda d^1 P_\lambda \\ n^{[\mu\nu]} P_\lambda - mS_{\mu\nu} n^{[\mu\nu]} \end{matrix} \Bigg]\psi
$$

$$= \begin{bmatrix} \mathrm{I}(d^0 g^\lambda P_\lambda - d^0 m + h^\lambda P_\lambda) - m g_\mu h^\mu - 2iS_\mu^\lambda h^\mu P_\lambda - i g_4 m d^1 + i g_4 g^\lambda d^1 P_\lambda \\ - m g_4 g_\mu k^\mu + g_4 k^\lambda P_\lambda - i e_{\mu\nu\sigma}^\lambda S^{\nu\sigma} k^\mu P_\lambda + i g_\mu n^{[\mu\lambda]} P_\lambda + \frac{i}{2}\varepsilon_{\mu\nu\sigma}^\lambda g_4 g^\sigma n^{[\mu\nu]} P_\lambda - m S_{\mu\nu} n^{[\mu\nu]} \end{bmatrix} \psi \quad (126)$$

Substituting eqs. (125) and (126) in the expression $[Q, L] = \alpha_Q L$ and then equating the coefficients of the linearly independent matrices and differential operators, we obtain the following system of equations:

$$md^0 = 0 \quad (127)$$

$$K_\nu^\mu - 2g_{\lambda\sigma} f^{[\sigma\mu]} = i g_\lambda^\mu d^0 \quad (128)$$

$$a_\lambda^0 + i g_{\lambda\sigma} f_\nu^{[\sigma\mu]} = 0 \quad (129)$$

$$a_\lambda^1 - \frac{1}{2}\varepsilon_{\mu\nu\lambda}^\sigma f_\sigma^{[\mu\nu]} = 0 \quad (130)$$

$$d^1 = 2a^1 \quad (131)$$

$$md^1 = 0 \quad (132)$$

$$h^\mu = b^\mu = k^\mu = n^{[\mu\nu]} = c^\mu = 0 \quad (133)$$

Making use of the symmetry of the metric tensor and the antisymmetry of $f^{[\sigma\mu]}$, we can write eq. (128) as

$$K_\mu^\nu + K_\nu^\mu - g_\nu^\mu K_\mu^\mu = 0 \quad (134)$$

together with

$$f^{\sigma\mu} = \frac{1}{2} g^{\sigma\lambda} K_\lambda^\mu,\ d^0 = -iK_\nu^\nu \quad (135)$$

Equation (134) has the solution

$$K^\mu = 2x^\mu x_\nu F^\nu - F^\mu x_\nu x^\nu + C^{[\mu\sigma]} x_\sigma + Dx^\mu + E^\mu \quad (136)$$

where F^μ, $C^{[\mu\sigma]}$, D, E^μ are arbitrary constants. Substituting eq. (136) in eq. (135) and making use of eqs. (127), (131) and (132), we get

$$K^\mu = C^{[\mu\nu]} x_\nu + \varphi^\mu,\ f^{[\mu\nu]} = \frac{1}{2} C^{[\mu\nu]},\ D = 0 \quad (137)$$

Using the solution (137) together with eqs. (127), (132) and (133) in eq. (120), (121) we obtain

$$Q = \mathbf{I}(a_0 + C^{[\mu\nu]} x_\mu P_\nu + \varphi^\mu P_\mu) + \frac{1}{2} C^{[\mu\nu]} S_{\mu\nu} \quad (138)$$

$$\alpha_Q = 0 \quad (139)$$

where $C^{[\mu\nu]}$, φ^μ, a^0 are arbitrary complex numbers. It is easily seen that the symmetry operator Q is a linear combination of the generators of the Poincare group together with the identity operator. It follows that the Poincare group generators form a basis for the maximally extensive Invariance Algebra of the Dirac equation in the class M_1. In fact, the set of Poincare generators constitutes the maximally invariance algebra of the Dirac equation in the class of operators that are first order differential operators with non-matrix complex valued coefficients and that form a Lie algebra. This set (of the Poincare group generators) exhausts all the symmetry operators that can be obtained by the classical Lie approach. It, however, does not exhaust all the symmetry operators of the Dirac equation – there exist symmetry operators of the Dirac equation (i) that are first order differential operators with matrix coefficients but that nevertheless form a Lie algebra and (ii) that are symmetry operators but do not form the basis of a Lie algebra.

We now investigate the existence of symmetry operators that are first order differential operators with matrix coefficients and form a Lie algebra. For the purpose, we transform the Dirac operator, $L \equiv (\gamma_\mu P^\mu - m)$, into a diagonal form by introducing the transformations

$$L \to L' = WLV^{-1},\ \psi \to \psi' = V\psi \tag{140}$$

where
$$V = \exp\left[-\frac{1}{2m}(1 - i\gamma_4)\gamma^\mu P_\mu\right] \equiv 1 - \frac{1}{2m}(1 - i\gamma_4)\gamma^\mu P_\mu \tag{141}$$

$$W = \exp\left[\frac{1}{2m}(1 + i\gamma_4)\gamma^\mu P_\mu\right] \equiv 1 + \frac{1}{2m}(1 + i\gamma_4)\gamma^\mu P_\mu \tag{142}$$

so that

$$L' = WLV^{-1} = \left[1 + \frac{1}{2m}(1 + i\gamma_4)\gamma^\mu P_\mu\right](\gamma_\mu P^\mu - m)\left[1 + \frac{1}{2m}(1 - i\gamma_4)\gamma^\mu P_\mu\right]$$

$$= \left[\gamma^\mu P_\mu - m + \frac{1}{2m}(1 + i\gamma_4)(\gamma^\mu\gamma_\mu)P^\mu P_\mu - \frac{1}{2}(1 + i\gamma_4)\gamma^\mu P_\mu\right]$$

$$\left[1 + \frac{1}{2m}(1 - i\gamma_4)\gamma^\mu P_\mu\right]$$

$$= \left[\frac{1}{2}(1 - i\gamma_4)\gamma^\mu P_\mu - m + \frac{4}{2m}(1 + i\gamma_4)P^\mu P_\mu\right]\left[1 + \frac{1}{2m}(1 - i\gamma_4)\gamma^\mu P_\mu\right]$$

$$= \frac{1}{2}(1 - i\gamma_4)\gamma^\mu P_\mu - m + \frac{4}{2m}(1 + i\gamma_4)P^\mu P_\mu + \left[\frac{1}{2}(1 - i\gamma_4)\gamma^\mu P_\mu\right]$$

$$\left[\frac{1}{2m}(1 - i\gamma_4)\gamma^\mu P_\mu\right]$$

$$-\frac{1}{2}(1 - i\gamma_4)\gamma^\mu P_\mu + \left[\frac{4}{2m}(1 + i\gamma_4)P^\mu P_\mu\right]\left[\frac{1}{2m}(1 - i\gamma_4)\gamma^\mu P_\mu\right]$$

$$= \frac{2}{m}(1 + i\gamma_4)P^\mu P_\mu - m \tag{143}$$

where we have used

$$\left[\frac{1}{2}(1 - i\gamma_4)\gamma^\mu P_\mu\right]\left[\frac{1}{2m}(1 - i\gamma_4)\gamma^\mu P_\mu\right]\left[\frac{1}{2}\gamma^\mu P_\mu(1 + i\gamma_4)\right]$$

$$\left[\frac{1}{2m}(1-i\gamma_4)\gamma^\mu P_\mu\right]=0$$

$$\left[\frac{4}{2m}(1+i\gamma_4)P^\mu P_\mu\right]\left[\frac{1}{2m}(1-i\gamma_4)\gamma^\mu P_\mu\right]=\left[\frac{4}{2m}P^\mu P_\mu(1+i\gamma_4)\right]$$

$$\left[\frac{1}{2m}(1-i\gamma_4)\gamma^\mu P_\mu\right]=0$$

Equation (143) is the diagonal form of the Dirac equation in the chiral representation in which $\boldsymbol{\gamma}_4$ is diagonal. Furthermore, it is obvious that the Dirac equation, (143), is invariant under all arbitrary matrix transformations that commute with $\boldsymbol{\gamma}_4$. It immediately follows that the corresponding symmetry operators will be linear combination of the basis elements

$$Q'_{\mu\nu}=\frac{1}{4}[\boldsymbol{\gamma}_\mu,\boldsymbol{\gamma}_\nu]=-iS_{\mu\nu},\ Q'_0=\mathbf{I},\ Q'_1=\boldsymbol{\gamma}_4 \tag{144}$$

These basis elements constitute a basis of an eight-dimensional Lie algebra and are linearly independent of the field of real numbers.

It is simple to establish the following commutator, using eq. (110):

$$[Q'_{\mu\nu},Q'_{\lambda\sigma}]=i(g_{\mu\lambda}Q'_{\nu\sigma}+g_{\nu\sigma}Q'_{\mu\lambda}-g_{\mu\sigma}Q'_{\nu\lambda}-g_{\nu\lambda}Q'_{\mu\sigma}) \tag{145}$$

We also have

$$[Q'_1,Q'_0]=[Q'_1,Q'_{\mu\nu}]=[Q'_0,Q'_{\mu\nu}]=0 \tag{146}$$

The explicit form of the basis elements (144) of the symmetry operators in the original Dirac representation is obtained by the transformation $Q'\to Q=V^{-1}Q'V$ and we get:

$$\begin{aligned}Q_1=V^{-1}\boldsymbol{\gamma}_4V&=\left[1+\frac{1}{2m}(1-i\gamma_4)\gamma^\mu P_\mu\right]\boldsymbol{\gamma}_4\left[1-\frac{1}{2m}(1-i\gamma_4)\gamma^\mu P_\mu\right]\\&=\boldsymbol{\gamma}_4\left[1-\frac{1}{2m}(1-i\gamma_4)\gamma^\mu P_\mu\right]^2=\boldsymbol{\gamma}_4\left[1-\frac{1}{m}(1-i\gamma_4)\gamma^\mu P_\mu\right]\\&=\boldsymbol{\gamma}_4-\frac{1}{m}\boldsymbol{\gamma}_4\boldsymbol{\gamma}^\mu P_\mu-\frac{i}{m}\boldsymbol{\gamma}^\mu P_\mu\end{aligned} \tag{147}$$

since $(1-i\boldsymbol{\gamma}_4)\boldsymbol{\gamma}^\mu P_\mu(1-i\boldsymbol{\gamma}_4)\boldsymbol{\gamma}^\mu P_\mu=\boldsymbol{\gamma}^\mu P_\mu(1+i\boldsymbol{\gamma}_4)(1-i\boldsymbol{\gamma}_4)\boldsymbol{\gamma}^\mu P_\mu=0$.

We also have

$$V^{-1}\boldsymbol{\gamma}_\lambda\boldsymbol{\gamma}_\sigma V=\left[1+\frac{1}{2m}(1-i\gamma_4)\gamma^\mu P_\mu\right]\boldsymbol{\gamma}_\lambda\boldsymbol{\gamma}_\sigma\left[1-\frac{1}{2m}(1-i\gamma_4)\gamma^\mu P_\mu\right]$$

$$= \left[\gamma_\lambda\gamma_\sigma + \frac{1}{2m}(1 - i\gamma_4)(2g_\lambda^\mu\gamma_\sigma - 2g_\sigma^\mu\gamma_\lambda + \gamma_\lambda\gamma_\sigma\gamma^\mu)P_\mu\right]$$

$$\left[1 - \frac{1}{2m}(1 - i\gamma_4)\gamma^\mu P_\mu\right]$$

$$= \left[\gamma_\lambda\gamma_\sigma + \frac{1}{2m}(2g_\lambda^\mu\gamma_\sigma - 2g_\sigma^\mu\gamma_\lambda + \gamma_\lambda\gamma_\sigma\gamma^\mu)(1 + i\gamma_4)P_\mu\right]$$

$$\left[1 - \frac{1}{2m}\gamma^\mu(1 + i\gamma_4)P_\mu\right]$$

$$= \gamma_\lambda\gamma_\sigma + \frac{1}{2m}(2g_\lambda^\mu\gamma_\sigma - 2g_\sigma^\mu\gamma_\lambda + \gamma_\lambda\gamma_\sigma\gamma^\mu)(1 + i\gamma_4)P_\mu$$

$$- \frac{1}{2m}\gamma_\lambda\gamma_\sigma\gamma^\mu(1 + i\gamma_4)P_\mu$$

$$- \frac{1}{4m^2}(2g_\lambda^\mu\gamma_\sigma - 2g_\sigma^\mu\gamma_\lambda + \gamma_\lambda\gamma_\sigma\gamma^\mu)(1 + i\gamma_4)\gamma^\mu(1 + i\gamma_4)P_\mu P_\mu$$

$$= \gamma_\lambda\gamma_\sigma + \frac{1}{2m}(2g_\lambda^\mu\gamma_\sigma - 2g_\sigma^\mu\gamma_\lambda)(1 + i\gamma_4)P_\mu$$

$$= \gamma_\lambda\gamma_\sigma + \frac{1}{m}(\gamma_\sigma P_\lambda - \gamma_\lambda P_\sigma)(1 + i\gamma_4) = \gamma_\lambda\gamma_\sigma + \frac{1}{m}(1 - i\gamma_4)(\gamma_\sigma P_\lambda - \gamma_\lambda P_\sigma) \tag{148}$$

where we have used $(1 + i\gamma_4)\gamma^\mu(1 + i\gamma_4) = \gamma^\mu(1 - i\gamma_4)(1 + i\gamma_4) = 0$. Similarly,

$$V^{-1}\gamma_\sigma\gamma_\lambda V = \gamma_\sigma\gamma_\lambda + \frac{1}{m}(1 - i\gamma_4)(\gamma_\lambda P_\sigma - \gamma_\sigma P_\lambda) \text{ so that}$$

$$Q_{\lambda\sigma} = V^{-1}\frac{1}{4}[\gamma_\lambda, \gamma_\sigma]V = \frac{1}{4}[\gamma_\lambda, \gamma_\sigma] + \frac{1}{2m}(1 - i\gamma_4)(\gamma_\sigma P_\lambda - \gamma_\lambda P_\sigma) \tag{149}$$

We also have the following commutators that indicate the fact that the operators given by eqs. (147) and (149) in addition to the unit matrix are symmetry operators for the Dirac equation:

(i) $[Q_{\mu\nu}, L] = -\frac{1}{m}(\gamma_\mu P_\nu - \gamma_\nu P_\mu)L$ (150)

(ii) $[Q_1, L] = -\frac{2}{m}\gamma_4\gamma^\mu P_\mu L$ (151)

(iii) $[Q_0, L] = 0$ (152)

To prove (i) we have

$$
\begin{aligned}
[[\gamma_\mu, \gamma_\nu], \gamma^\lambda P_\lambda] &= [(\gamma_\mu\gamma_\nu - \gamma_\nu\gamma_\mu), \gamma^\lambda P_\lambda] \\
&= (\gamma_\mu\gamma_\nu - \gamma_\nu\gamma_\mu)\gamma^\lambda P_\lambda - \gamma^\lambda P_\lambda(\gamma_\mu\gamma_\nu - \gamma_\nu\gamma_\mu) \\
&= \gamma_\mu\gamma_\nu\gamma^\lambda P_\lambda - (2g_{\mu\nu} - \gamma_\mu\gamma_\nu)\gamma^\lambda P_\lambda - (2g_\mu^\lambda - \gamma_\mu\gamma^\lambda)\gamma_\nu P_\lambda + (2g_\nu^\lambda - \gamma_\nu\gamma^\lambda)\gamma_\mu P_\lambda \\
&= 2\gamma_\mu\gamma_\nu\gamma^\lambda P_\lambda - 2g_{\mu\nu}\gamma^\lambda P_\lambda - 2g_\mu^\lambda\gamma_\nu P_\lambda + \gamma_\mu\gamma^\lambda\gamma_\nu P_\lambda + 2g_\nu^\lambda\gamma_\mu P_\lambda - \gamma_\nu\gamma^\lambda\gamma_\mu P_\lambda \\
&= \gamma_\mu\gamma_\nu\gamma^\lambda P_\lambda - 2g_{\mu\nu}\gamma^\lambda P_\lambda - 2g_\mu^\lambda\gamma_\nu P_\lambda + 2g_\nu^\lambda\gamma_\mu P_\lambda + 2g_\nu^\lambda\gamma_\mu P_\lambda - \gamma_\nu(2g_\mu^\lambda - \gamma_\mu\gamma^\lambda)P_\lambda \\
&= \gamma_\mu\gamma_\nu\gamma^\lambda P_\lambda - 2g_{\mu\nu}\gamma^\lambda P_\lambda - 4g_\mu^\lambda\gamma_\nu P_\lambda + 4g_\nu^\lambda\gamma_\mu P_\lambda + (2g_{\mu\nu} - \gamma_\mu\gamma_\nu)\gamma^\lambda P_\lambda \\
&= 4g_\nu^\lambda\gamma_\mu P_\lambda - 4g_\mu^\lambda\gamma_\nu P_\lambda = 4\gamma_\mu P_\nu - 4\gamma_\nu P_\mu
\end{aligned} \tag{153}
$$

$$
\begin{aligned}
&[(1 - i\gamma_4)(\gamma_\nu P_\mu - \gamma_\mu P_\nu), \gamma^\lambda P_\lambda] \\
&= (1 - i\gamma_4)(\gamma_\nu P_\mu - \gamma_\mu P_\nu), \gamma^\lambda P_\lambda - \gamma^\lambda P_\lambda(1 - i\gamma_4)(\gamma_\nu P_\mu - \gamma_\mu P_\nu) \\
&= (1 - i\gamma_4)(\gamma_\nu P_\mu - \gamma_\mu P_\nu)\gamma^\lambda P_\lambda - (1 + i\gamma_4)\gamma^\lambda P_\lambda(\gamma_\nu P_\mu - \gamma_\mu P_\nu) \\
&= (1 - i\gamma_4)(\gamma_\nu P_\mu - \gamma_\mu P_\nu)\gamma^\lambda P_\lambda - (1 + i\gamma_4) \\
&\quad [(2g_\nu^\lambda - \gamma_\nu\gamma^\lambda)P_\lambda P_\mu - (2g_\mu^\lambda - \gamma_\mu\gamma^\lambda)P_\lambda P_\nu] \\
&= (1 - i\gamma_4)(\gamma_\nu P_\mu - \gamma_\mu P_\nu)\gamma^\lambda P_\lambda - (1 + i\gamma_4)(\gamma_\mu P_\nu - \gamma_\nu P_\mu)\gamma^\lambda P_\lambda \\
&= -2(\gamma_\mu P_\nu - \gamma_\nu P_\mu)\gamma^\lambda P_\lambda
\end{aligned} \tag{154}
$$

From eqs. (153) and (154) we obtain

$$
\begin{aligned}
[Q_{\mu\nu}, L] &= \left[\frac{1}{4}[\gamma_\mu, \gamma_\nu] + \frac{1}{2m}(1 - i\gamma_4)(\gamma_\nu P_\mu - \gamma_\mu P_\nu), \gamma^\lambda P_\lambda\right] \\
&= \gamma_\mu P_\nu - \gamma_\nu P_\mu - \frac{1}{m}(\gamma_\mu P_\nu - \gamma_\nu P_\mu)\gamma^\lambda P_\lambda = -\frac{1}{m}(\gamma_\mu P_\nu - \gamma_\nu P_\mu)L
\end{aligned} \tag{155}
$$

For (ii), we have

$$
\begin{aligned}
[Q_1, L] &= \left[\gamma_4 - \frac{1}{m}\gamma_4\gamma^\mu P_\mu - \frac{i}{m}\gamma^\mu P_\mu, \gamma^\lambda P_\lambda\right] \\
&= [\gamma_4, \gamma^\lambda P_\lambda] - \frac{1}{m}[\gamma_4\gamma^\mu P_\mu, \gamma^\lambda P_\lambda] = 2\gamma_4\gamma^\lambda P_\lambda - \frac{2}{m}\gamma_4(\gamma^\mu P_\mu)(\gamma^\lambda P_\lambda) \\
&= -\frac{2}{m}\gamma_4\gamma^\mu P_\mu(\gamma^\lambda P_\lambda - m) = -\frac{2}{m}\gamma_4\gamma^\mu P_\mu L
\end{aligned} \tag{156}
$$

where we have used the fact that λ, μ are dummy indices so that $\gamma^\mu P_\mu = \gamma^\lambda P_\lambda$.

From eq. (145), we obtain

$$
\begin{aligned}
V[Q_{\mu\nu}, Q_{\lambda\sigma}]V^{-1} &= [VQ_{\mu\nu}V^{-1}, VQ_{\lambda\sigma}V^{-1}] \\
&= i(g_{\mu\lambda}VQ_{\nu\sigma}V^{-1} + g_{\nu\sigma}VQ_{\mu\lambda}V^{-1} - g_{\mu\sigma}VQ_{\nu\lambda}V^{-1} - g_{\nu\lambda}VQ_{\mu\sigma}V^{-1})
\end{aligned}
$$

$$= iV(g_{\mu\lambda}Q_{\nu\sigma} + g_{\nu\sigma}Q_{\mu\lambda} - g_{\mu\sigma}Q_{\nu\lambda} - g_{\nu\lambda}Q_{\mu\sigma})V^{-1}$$

whence

$$[Q_{\mu\nu}, Q_{\lambda\sigma}] = i(g_{\mu\lambda}Q_{\nu\sigma} + g_{\nu\sigma}Q_{\mu\lambda} - g_{\mu\sigma}Q_{\nu\lambda} - g_{\nu\lambda}Q_{\mu\sigma}) \tag{157}$$

and similarly, from eq. (146)

$$[Q_1, Q_0] = [Q_1, Q_{\mu\nu}] = [Q_0, Q_{\mu\nu}] = 0 \tag{158}$$

e.g.,

$$[Q_1, Q_{\lambda\sigma}] = \left[\gamma_4 - \frac{1}{m}\gamma_4\gamma^\mu P_\mu - \frac{i}{m}\gamma^\mu P_\mu, \frac{1}{4}[\gamma_\lambda, \gamma_\sigma] + \frac{1}{2m}(1 - i\gamma_4)(\gamma_\sigma P_\lambda - \gamma_\lambda P_\sigma)\right]$$

$$= \left[\gamma_4 - \frac{i}{m}(1 - i\gamma_4)\gamma^\mu P_\mu, \frac{1}{2}\gamma_\lambda\gamma_\sigma + \frac{1}{2m}(1 - i\gamma_4)(\gamma_\sigma P_\lambda - \gamma_\lambda P_\sigma)\right]$$

Now, $[\gamma_4, \gamma_\lambda\gamma_\sigma] = 0$, $[\gamma_4, (1 - i\gamma_4)(\gamma_\sigma P_\lambda - \gamma_\lambda P_\sigma)]$

$$= \gamma_4(1 - i\gamma_4)(\gamma_\sigma P_\lambda - \gamma_\lambda P_\sigma) - (1 - i\gamma_4)(\gamma_\sigma P_\lambda - \gamma_\lambda P_\sigma)\gamma_4$$

$$= \gamma_4(1 - i\gamma_4)(\gamma_\sigma P_\lambda - \gamma_\lambda P_\sigma) + \gamma_4(1 - i\gamma_4)$$

$$(\gamma_\sigma P_\lambda - \gamma_\lambda P_\sigma) = 2\gamma_4(1 - i\gamma_4)(\gamma_\sigma P_\lambda - \gamma_\lambda P_\sigma),$$

$$[(1 - i\gamma_4)\gamma^\mu P_\mu, \gamma_\lambda\gamma_\sigma] = (1 - i\gamma_4)\gamma^\mu P_\mu\gamma_\lambda\gamma_\sigma - \gamma_\lambda\gamma_\sigma(1 - i\gamma_4)\gamma^\mu P_\mu$$

$$= (1 - i\gamma_4)[\gamma^\mu P_\mu\gamma_\lambda\gamma_\sigma - \gamma_\lambda\gamma_\sigma\gamma^\mu P_\mu] = 2(1 - i\gamma_4)(\gamma_\sigma P_\lambda - \gamma_\lambda P_\sigma),$$

$$[(1 - i\gamma_4)\gamma^\mu P_\mu, (1 - i\gamma_4)(\gamma_\sigma P_\lambda - \gamma_\lambda P_\sigma)]$$

$$= (1 - i\gamma_4)\gamma^\mu P_\mu(1 - i\gamma_4)(\gamma_\sigma P_\lambda - \gamma_\lambda P_\sigma) - (1 - i\gamma_4)(\gamma_\sigma P_\lambda - \gamma_\lambda P_\sigma)(1 - i\gamma_4)\gamma^\mu P_\mu$$

$$= (1 - i\gamma_4)(1 + i\gamma_4)[\gamma^\mu P_\mu(\gamma_\sigma P_\lambda - \gamma_\lambda P_\sigma) - (\gamma_\sigma P_\lambda - \gamma_\lambda P_\sigma)\gamma^\mu P_\mu] = 0$$

so that

$$[Q_1, Q_{\lambda\sigma}] = \frac{1}{m}(\gamma_4 - i)(1 - i\gamma_4)(\gamma_\sigma P_\lambda - \gamma_\lambda P_\sigma) = \frac{1}{m}\gamma_4(1 + i\gamma_4)$$

$(1 - i\gamma_4)(\gamma_\sigma P_\lambda - \gamma_\lambda P_\sigma) = 0.$

It follows from the above that the Dirac equation is also invariant under the eight-dimensional algebra (with matrix coefficients) having the basis set $\{Q_{\mu\nu}, Q_0, Q_1\}$. Incidentally, this algebra is also a Lie algebra A_8 defined over the field of real numbers with the commutators given by eqs. (157) and (158).

Furthermore, this algebra is isomorphic to the Lie algebra of the group GL(2,**C**) for $m \neq 0$. The isomorphism is defined by the identification

$$\lambda_{11} = Q_0 + Q_{12},\ \lambda_{22} = Q_0 - Q_{12},\ \lambda_{12} = Q_{23} - Q_{02}$$

$$\lambda_{21} = Q_{23} + Q_{02},\ \tilde{\lambda}_{11} = Q_1 + Q_{03},\ \tilde{\lambda}_{22} = Q_1 - Q_{03} \tag{159}$$

$$\tilde{\lambda}_{12} = -Q_{01} + Q_{31},\ \tilde{\lambda}_{21} = Q_{31} + Q_{01}$$

where λ_{ab}, $\tilde{\lambda}_{ab}$ are the basis elements of the Lie algebra of the group $GL(2,)$. These basis elements satisfy the commutators

$$[\lambda_{ab}, \lambda_{cd}] = -[\tilde{\lambda}_{ab}, \tilde{\lambda}_{cd}] = \delta_{ac}\lambda_{bd} + \delta_{bc}\lambda_{ad} - \delta_{ad}\lambda_{bc} - \delta_{bd}\lambda_{ac} \tag{160}$$

$$[\tilde{\lambda}_{ab}, \tilde{\lambda}_{cd}] = \delta_{ac}\tilde{\lambda}_{bd} + \delta_{bc}\tilde{\lambda}_{ad} - \delta_{ad}\tilde{\lambda}_{bc} - \delta_{bd}\tilde{\lambda}_{ac} \tag{161}$$

Consistency between eqs. (157), (158) and eqs. (160), (161) is easily established by direct calculations.

Hence, in addition to being invariant under the Poincare group algebra, the Dirac equation is also invariant under the matrix algebra A_8. However, since the basis elements of this algebra do not belong to the class M_1, they cannot be treated as Lie derivatives. Nevertheless, these basis elements do, indeed, form a Lie algebra. Additionally, they satisfy

$$Q_{0j}^2 = -Q_{ij}^2 = Q_0^2 = -Q_1^2 = 1 \tag{162}$$

It is instructive to compute the commutators of the group algebra A_8 with the ten generators of the Poincare group. We have, straightaway,

$$[P_\lambda, Q_{\mu\nu}] = [P_\lambda, Q_0] = [P_\lambda, Q_1] = 0 \tag{163}$$

To compute $[J_{\mu\nu}, Q_{\lambda\sigma}]$, we make use of the following:

(i) From eq. (149) and the fact that $J_{\mu\nu} = \frac{i}{4}[\gamma_\mu, \gamma_\nu] + (x_\mu P_\nu - x_\nu P_\mu)$, we obtain

$$J_{\mu\nu} = iQ_{\mu\nu} + \frac{i}{2m}(1 - i\gamma_4)(\gamma_\mu P_\nu - \gamma_\nu P_\mu) + (x_\mu P_\nu - x_\nu P_\mu) \tag{164}$$

(ii) From eq. (157)

$$[iQ_{\mu\nu}, Q_{\lambda\sigma}] = -(g_{\mu\lambda}Q_{\nu\sigma} + g_{\nu\sigma}Q_{\mu\lambda} - g_{\mu\sigma}Q_{\nu\lambda} - g_{\nu\lambda}Q_{\mu\sigma}) \tag{165}$$

(iii) $[(1 - i\gamma_4)(\gamma_\mu P_\nu - \gamma_\nu P_\mu), \gamma_\lambda\gamma_\sigma]$

$$= (1 - i\gamma_4)(\gamma_\mu P_\nu - \gamma_\nu P_\mu)\gamma_\lambda\gamma_\sigma - \gamma_\lambda\gamma_\sigma(1 - i\gamma_4)(\gamma_\mu P_\nu - \gamma_\nu P_\mu)$$

$$= (1 - i\gamma_4)[(\gamma_\mu P_\nu - \gamma_\nu P_\mu)\gamma_\lambda\gamma_\sigma - \gamma_\lambda\gamma_\sigma(\gamma_\mu P_\nu - \gamma_\nu P_\mu)$$

$$= (1 - i\gamma_4)[(\gamma_\mu P_\nu - \gamma_\nu P_\mu)\gamma_\lambda\gamma_\sigma - \gamma_\lambda(2g_{\sigma\mu} - \gamma_\mu\gamma_\sigma)P_\nu + \gamma_\lambda(2g_{\sigma\nu} - \gamma_\nu\gamma_\sigma)P_\mu]$$

$$= (1 - i\gamma_4)\begin{bmatrix}(\gamma_\mu P_\nu - \gamma_\nu P_\mu)\gamma_\lambda\gamma_\sigma - 2g_{\sigma\mu}\gamma_\lambda P_\nu + (2g_{\lambda\mu} - \gamma_\mu\gamma_\lambda)\gamma_\sigma P_\nu \\ +2g_{\sigma\nu}\gamma_\lambda P_\mu - (2g_{\lambda\nu} - \gamma_\nu\gamma_\lambda)\gamma_\sigma P_\mu\end{bmatrix}$$

$$= (1 - i\gamma_4)[-2g_{\sigma\mu}\gamma_\lambda P_\nu + 2g_{\lambda\mu}\gamma_\sigma P_\nu + 2g_{\sigma\nu}\gamma_\lambda P_\mu - 2g_{\lambda\nu}\gamma_\sigma P_\mu]$$

so that

$$\left[\frac{i}{2m}(1 - i\gamma_4)(\gamma_\mu P_\nu - \gamma_\nu P_\mu), \frac{1}{4}[\gamma_\mu, \gamma_\sigma]\right]$$

$$= \frac{i}{2m}(1 - i\gamma_4)[g_{\lambda\mu}\gamma_\sigma P_\nu - g_{\sigma\mu}\gamma_\lambda P_\nu + g_{\sigma\nu}\gamma_\lambda P_\mu - \gamma_{\lambda\nu}\gamma_\sigma P_\mu] \tag{166}$$

(iv) $[(1 - i\gamma_4)(\gamma_\mu P_\nu - \gamma_\nu P_\mu), (1 - i\gamma_4)(\gamma_\lambda P_\sigma - \gamma_\sigma P_\mu)]$

$$= (1 - i\gamma_4)(\gamma_\mu P_\nu - \gamma_\nu P_\mu)(1 - i\gamma_4)(\gamma_\lambda P_\sigma - \gamma_\sigma P_\lambda)$$
$$- (1 - i\gamma_4)(\gamma_\lambda P_\sigma - \gamma_\sigma P_\lambda)(1 - i\gamma_4)(\gamma_\mu P_\nu - \gamma_\nu P_\mu)$$
$$(1 - i\gamma_4)(1 + i\gamma_4)(\gamma_\mu P_\nu - \gamma_\nu P_\mu)(\gamma_\lambda P_\sigma - \gamma_\sigma P_\lambda)$$
$$- (1 - i\gamma_4)(1 + i\gamma_4)(\gamma_\lambda P_\sigma - \gamma_\sigma P_\lambda)(\gamma_\mu P_\nu - \gamma_\nu P_\mu) = 0$$

(v) $[(x_\mu P_\nu - x_\nu P_\mu), [\gamma_\lambda, \gamma_\sigma]] = 0$ (167)

since both x_μ, P_μ commute with the Dirac matrices.

(vi) $\left[(x_\mu P_\nu - x_\nu P_\mu), \frac{1}{2m}(1 - i\gamma_4)(\gamma_\lambda P_\sigma - \gamma_\sigma P_\lambda)\right]$

$$= \frac{1}{2m}(1 - i\gamma_4)[(x_\mu P_\nu - x_\nu P_\mu), (\gamma_\lambda P_\sigma - \gamma_\sigma P_\lambda)]$$

$$= \frac{1}{2m}(1 - i\gamma_4)\{[(x_\mu P_\nu, \gamma_\lambda P_\sigma) - [x_\mu P_\nu, \gamma_\sigma P_\lambda] - [x_\nu P_\mu, \gamma_\lambda P_\sigma] + [x_\nu P_\mu, \gamma_\sigma P_\lambda]\}$$

$$= \frac{i}{2m}(1 - i\gamma_4)(-g_{\mu\sigma}\gamma_\lambda P_\nu + g_{\mu\lambda}\gamma_\sigma P_\nu + g_{\nu\sigma}\gamma_\lambda P_\mu - g_{\nu\lambda}\gamma_\sigma P_\mu) \tag{168}$$

Making use of (i)-(vi) above, we finally obtain

$$[J_{\mu\nu}, Q_{\lambda\sigma}] = -(g_{\mu\lambda}Q_{\nu\sigma} + g_{\nu\sigma}Q_{\mu\lambda} - g_{\mu\sigma}Q_{\nu\lambda} - g_{\nu\lambda}Q_{\mu\sigma}) \tag{169}$$

The commutators (163) and (169) establish that the Dirac equation is not only invariant under the Poincare group and the matrix group A_8 but it is also invariant under the combined 18-dimensional Lie algebra of the two groups. In other words, the Dirac equation is invariant under the 18 parameter group that includes the Lorentz transformations and the transformations in the group A_8.

Cardinal properties of the transformations of A_8 are that (i) the transformed functions also depend on the derivatives of the wavefunction with respect to the independent variables and (ii) unlike the Lorentz transformations, these transformations do not transform the independent variables, i.e., they do not affect the space-time continuum. Finite transformations corresponding to these additional generators can be obtained, as usual, by exponentiation, e.g.,

$$\psi \to \psi' = (\cos\theta_{ij} + \gamma_i\gamma_j \sin\theta_{ij})\psi - \frac{i}{m}\sin\theta_{ij}(1 - i\gamma_4)\left(\gamma_j \frac{\partial\psi}{\partial x_i} - \gamma_i \frac{\partial\psi}{\partial x_j}\right) \tag{170}$$

$$\psi \to \psi'' = (\cosh\theta_{0j} + \boldsymbol{\gamma}_0\boldsymbol{\gamma}_j \sinh\theta_{0j})\psi - \frac{i}{m}\sinh\theta_{0j}(1 - i\boldsymbol{\gamma}_4)\left(\gamma_j \frac{\partial\psi}{\partial x_0} - \gamma_0 \frac{\partial\psi}{\partial x_j}\right) \quad (171)$$

$$\psi \to \psi''' = (\cos\theta_1 + \boldsymbol{\gamma}_4 \sin\theta_1)\psi + \frac{i}{m}\sin\theta_1 (1 - i\boldsymbol{\gamma}_4)\boldsymbol{\gamma}_\mu \frac{\partial\psi}{\partial x_\mu} \quad (172)$$

5.12 COMPLETE SET OF FIRST ORDER SYMMETRIES OF THE DIRAC EQUATION (FUSHCHICH, 1995)

Finally, we obtain the complete set of first order symmetry operators of the Dirac equation. This amounts to obtaining the general solution of the operator equation

$$[L, Q] = \alpha_Q L \quad (173)$$

where Q and α_Q are both first order operators that can be expressed as

$$Q = A^\mu P_\mu + B \quad (174)$$

$$\alpha_Q = C^\mu P_\mu + D \quad (175)$$

with A_μ, B, C_μ, D being 4×4 matrices that are functions of the space-time coordinates x^μ.

Now, in the solution space of the Dirac equation, $(\boldsymbol{\gamma}^\mu P_\mu - m)\psi = 0$, we can express P_0 in terms of the spatial components P_i with matrix coefficients as

$$P_0\psi = H\psi \equiv \boldsymbol{\gamma}_0 m\psi - \boldsymbol{\gamma}_0\boldsymbol{\gamma}^i P_i\psi \quad (176)$$

We, therefore, need consider only those symmetry operators that are of the form

$$Q = \mathbf{E}.\mathbf{P} + G \quad (177)$$

where $\mathbf{E}$, G are 4×4 matrices depending on the coordinates. Further, we can write

$$[L, Q] = \alpha_Q L = [P_0 - H, Q] = \alpha_Q(P_0 - H) = 0 \quad (178)$$

where the last step follows from the fact that the commutator $[P_0 - H, Q]$ should not depend on P_0 if we choose Q as $Q = \mathbf{E}.\mathbf{P} + G$.

We now expand the operators $\mathbf{E}$, G that are 4×4 matrices in terms of the Dirac basis, (119) as

$$\mathbf{E} = \mathbf{I}\mathbf{d}^0 + i\boldsymbol{\gamma}_4 d^1 + \boldsymbol{\gamma}_\nu \mathbf{n}^\nu + S_{\mu\nu}\mathbf{m}^{\mu\nu} + \boldsymbol{\gamma}_4\boldsymbol{\gamma}_\nu \mathbf{b}^\nu \quad (179)$$

$$G = \mathbf{I}a^0 + i\boldsymbol{\gamma}_4 a^1 + \boldsymbol{\gamma}_\nu c^\nu + S_{\mu\nu} f^{\mu\nu} + \boldsymbol{\gamma}_4\boldsymbol{\gamma}_\nu h^\nu \quad (180)$$

where the coefficients of the basis vectors are functions of space-time coordinates.

Substituting from eqs. (177), (179) and (180) in eq. (178) and equating the coefficients of the basis vectors and the differential operators, we obtain the following set of equations:

$$\mathbf{n}^0 = \mathbf{b}^0 = 0,\ n_j^i = i\varepsilon_{ijk} d_k^2,\ b_j^i = i\varepsilon_{ijk} d_k^3,\ m_k^{ij} = \varepsilon_{ijk}\mathbf{A}^1,\ m_j^{0i} = i\delta_{ij}\mathbf{A}^0,\ i, j, k = 1, 2, 3 \quad (181)$$

$$\frac{\partial d_i^\mu}{\partial x_j} = -\frac{\partial d_j^\mu}{\partial x_i},\ \frac{\partial d_i^\mu}{\partial x_i} = \frac{\partial d_j^\mu}{\partial x_j},\ i \neq j,\ m\ \mathbf{div}\ \mathbf{d}^0 = 0,\ m\ \mathbf{div}\ \mathbf{d}^1 = 2ima^1 \tag{182}$$

$$\mathbf{d}^2 = -\frac{1}{2}\mathbf{rot}\ \mathbf{d}^3,\ \mathbf{d}^3 = \frac{1}{2}\mathbf{rot}\ \mathbf{d}^2,\ \mathbf{d}^i = -\mathbf{grad}\ A^i,\ \mathbf{div}\ \mathbf{d}^i = -3\mathbf{A}^i,\ i = 0, 1 \tag{183}$$

$$c^i = -\frac{1}{2}(\mathbf{rot}\ \mathbf{d}^2)_i,\ c^0 = -\frac{1}{3}\mathbf{div}\ \mathbf{d}^3 + m\mathbf{A}^0,\ h^0 = \frac{1}{3}\mathbf{div}\ \mathbf{d}^2 \tag{184}$$

$$h^i = -\frac{1}{2}(\mathbf{rot}\ \mathbf{d}^3)_i - im\mathbf{d}_i^1,\ \dot{a}^0 = -\frac{1}{2}i\mathbf{div}\ \mathbf{d}^0,\ \mathbf{grad}\ a^0 = -\frac{3}{2}i\ddot{\mathbf{d}}0 \tag{185}$$

$$a^1 = -\frac{1}{2}i\mathbf{div}\ \dot{\mathbf{d}}^1 + \frac{1}{3}m\mathbf{div}\ \mathbf{d}^2,\ \mathbf{grad}\ a^1 = -m\dot{\mathbf{d}}^2 - \frac{3}{2}i\ddot{\mathbf{d}}^1 \tag{186}$$

$$f^{0i} = \frac{1}{2}\dot{\mathbf{d}}_i^0 - \frac{1}{4}i(\mathbf{rot}\ \mathbf{d}^1)_i, f^{ij} = \varepsilon_{ijk}\left[\frac{1}{2}i\mathbf{d}_k^1 + \frac{1}{4}(\mathbf{rot}\ \mathbf{d}^0)_k + m\mathbf{d}_k^2\right] \tag{187}$$

In the above, a dot represents derivative with respect to x_0, there is no summation over repeated indices and the symbol $\mathbf{d}^\mu$ denotes a vector with components $(d_1^\mu, d_2^\mu, d_3^\mu)$. The above set of equations admits the solution (for $m \neq 0$):

$$\mathbf{d}^0 = \mathbf{x} \times \boldsymbol{\eta} + \boldsymbol{\rho}x_0 + \mathbf{v},\ \mathbf{d}^1 = \boldsymbol{\xi} + \lambda\mathbf{x},\ \mathbf{d}^2 = \mathbf{x} \times \boldsymbol{\varepsilon} + \boldsymbol{\zeta},\ \mathbf{d}^3 = \boldsymbol{\varepsilon}x_0 + \mu\mathbf{x} + \boldsymbol{\sigma},\ c^i = \varepsilon^i,$$

$$h^0 = 0,\ h^i = -im(\xi_i + \lambda x_i), f^{0i} = \frac{1}{2}\rho_i, f^{ij} = \frac{1}{2}\varepsilon_{ijk}(2m\zeta_k - \eta_k) + m(x_i\varepsilon_j - x_j\varepsilon_i),$$

$$c^0 = -\mu - m(\boldsymbol{\rho}.\mathbf{x} + \boldsymbol{\kappa}),\ A^0 = -\boldsymbol{\rho}.\mathbf{x} - \boldsymbol{\kappa},\ A^1 = -\lambda x^0 + \omega,\ a^0 = \Omega,\ a^1 = -\frac{3}{2}i\lambda \tag{188}$$

This solution set contains 26 arbitrary parameters (the Greek letters) and hence, there exist 26 linearly independent symmetry operators of the Dirac equation. Substituting from eqs. (179), (180), (188) in eq. (177), we obtain the most general expression for the symmetry operators of the Dirac equation. It is seen that the generalized symmetry operators can be expressed as a linear combination of the ten Poincare group generators, the trivial identity operator and fifteen other operators given by:

$$\eta_\mu = \frac{1}{4}i\gamma_4(P_\mu - m\gamma_\mu),\quad \omega_{\mu\nu} = mS_{\mu\nu} + \frac{1}{2}i(\gamma_\mu P_\nu - \gamma_\nu P_\mu),$$

$$A_\mu = \omega_{\mu\nu}x^\nu + x^\nu\omega_{\mu\nu} - i\gamma^\mu,\quad B = i\gamma_4(D - m\gamma_\mu x^\mu) \tag{189}$$

where $\quad D = x^\mu P_\mu + \frac{3}{2}i,\ S_{\mu\nu} = \frac{1}{4}i[\gamma_\mu, \gamma_\nu]$ with $\mu, \nu = 0, 1, 2, 3$.

An equivalent expression for the above set of fifteen operators is ($\gamma_5 = i\gamma_4$):

$$W_\mu = \frac{1}{2}\gamma_5(P_\mu - m\gamma_\mu), \quad W_{\mu\nu} = \frac{1}{2}\gamma_5(\gamma_\mu P_\nu - \gamma_\nu P_\mu),$$

$$A_\mu = \frac{1}{2}(\gamma_\mu + \varepsilon_{\mu\nu\rho\sigma}J^{\nu\rho}\gamma_5\gamma^\sigma), \quad B = \gamma_5\left(x^\mu P_\mu + \frac{3}{2}i - \gamma_\mu x^\mu m\right) \tag{190}$$

with $$J_{\mu\nu} = x_\mu P_\nu - x_\nu P_\mu + \frac{i}{4}[\gamma_\mu, \gamma_\nu].$$

The above set of symmetry operators does not form a Lie algebra for, we have,

$$[\omega_{\mu\nu}, \omega_{\lambda\sigma}] = \left[mS_{\mu\nu} + \frac{1}{2}i(\gamma_\mu P_\nu - \gamma_\nu P_\mu), mS_{\lambda\sigma} + \frac{1}{2}i(\gamma_\lambda P_\sigma - \gamma_\sigma P_\lambda)\right]$$

$$= m^2[S_{\mu\nu}, S_{\lambda\sigma}] + \frac{1}{2}im[S_{\mu\nu}, (\gamma_\lambda P_\sigma - \gamma_\sigma P_\lambda)] + \frac{1}{2}im[(\gamma_\mu P_\nu - \gamma_\nu P_\mu), S_{\lambda\sigma}]$$

$$-\frac{1}{4}[\gamma_\mu P_\nu - \gamma_\nu P_\mu), (\gamma_\lambda P_\sigma - \gamma_\sigma P_\lambda)]$$

The first commutator on the RHS is expressible as products of purely Dirac matrices, i.e., of the form $g_{\alpha\beta}\gamma_\eta\gamma_\kappa$, the second and third ones as products of Dirac matrices and the translation generators, e.g., $g_{\alpha\beta}\gamma_\eta P_\kappa$. However, an evaluation of the fourth commutator yields terms involving products of translation generators, e.g., $\gamma_\alpha\gamma_\beta P_\eta P_\kappa$, $g_{\alpha\beta}P_\eta P_\kappa$. Hence, the overall commutator $[\omega_{\mu\nu}, \omega_{\lambda\sigma}]$ shall necessarily contain terms that are of the form $\gamma_\alpha\gamma_\beta P_\eta P_\kappa$, $g_{\alpha\beta}P_\eta P_\kappa$. Thus, the first set of symmetry operators is not closed under the Lie bracket. Similarly,

$$[W_{\mu\nu}, W_{\lambda\rho}] = \frac{1}{4}[\gamma_5(\gamma_\mu P_\nu - \gamma_\nu P_\mu).\gamma_5(\gamma_\lambda P_\sigma - \gamma_\sigma P_\lambda) - \gamma_5(\gamma_\lambda P_\sigma - \gamma_\sigma P_\lambda).\gamma_5(\gamma_\mu P_\nu - \gamma_\nu P_\mu)]$$

$$= -\frac{1}{4}[\gamma_\mu P_\nu - \gamma_\nu P_\mu)(\gamma_\lambda P_\sigma - \gamma_\sigma P_\lambda) - (\gamma_\lambda P_\sigma - \gamma_\sigma P_\lambda)(\gamma_\mu P_\nu - \gamma_\nu P_\mu)]$$

$$= -\frac{1}{4}\begin{pmatrix}\gamma_\mu\gamma_\lambda P_\nu P_\sigma - \gamma_\mu\gamma_\sigma P_\nu P_\lambda - \gamma_\nu\gamma_\lambda P_\mu P_\sigma + \gamma_\nu\gamma_\sigma P_\mu P_\lambda - \\ \gamma_\lambda\gamma_\mu P_\sigma P_\nu - \gamma_\lambda\gamma_\nu P_\sigma P_\mu + \gamma_\sigma\gamma_\mu P_\lambda P_\nu - \gamma_\sigma\gamma_\nu P_\lambda P_\mu\end{pmatrix}$$

$$= -\frac{1}{2}\begin{pmatrix}\gamma_\mu\gamma_\lambda P_\nu P_\sigma - \gamma_\mu\gamma_\sigma P_\nu P_\lambda - \gamma_\nu\gamma_\lambda P_\mu P_\sigma + \gamma_\nu\gamma_\sigma P_\mu P_\lambda \\ -g_{\mu\lambda}P_\nu P_\sigma + g_{\mu\sigma}P_\nu P_\lambda + g_{\nu\lambda}P_\mu P_\sigma - g_{\nu\sigma}P_\mu P_\lambda\end{pmatrix}$$

which shows that $[W_{\mu\nu}, W_{\lambda\rho}]$ is also not closed under the Lie bracket operation. Thus, the complete set of symmetry operators of the Dirac equation does not constitute a Lie algebra.

To investigate the structure further, let us evaluate the anticommutators

$$\{W_\mu, W_\nu\} = \frac{1}{4}[\gamma_5(P_\mu - m\gamma_\mu].\gamma_5(P_\nu - m\gamma_\nu) + \gamma_5(P_\nu - m\gamma_\nu).\gamma_5(P_\mu - m\gamma_\mu)]$$

$$= \frac{1}{4}[(P_\mu + m\boldsymbol{\gamma}_\mu)(P_\nu - m\boldsymbol{\gamma}_\nu) + (P_\nu + m\boldsymbol{\gamma}_\nu)(P_\mu - m\boldsymbol{\gamma}_\mu)]$$

$$= \frac{1}{2}[P_\mu P_\nu - m^2 g_{\mu\nu}] \tag{191}$$

$$\{W_{\mu\nu}, W_{\lambda\sigma}\} = \frac{1}{4}[\boldsymbol{\gamma}_5(\boldsymbol{\gamma}_\mu P_\nu - \boldsymbol{\gamma}_\nu P_\mu)\cdot\boldsymbol{\gamma}_5(\boldsymbol{\gamma}_\lambda P_\sigma - \boldsymbol{\gamma}_\sigma P_\lambda) + \boldsymbol{\gamma}_5(\boldsymbol{\gamma}_\lambda P_\sigma - \boldsymbol{\gamma}_\sigma P_\lambda)\cdot\boldsymbol{\gamma}_5(\boldsymbol{\gamma}_\mu P_\nu - \boldsymbol{\gamma}_\nu P_\mu)]$$

$$= -\frac{1}{4}[(\boldsymbol{\gamma}_\mu P_\nu - \boldsymbol{\gamma}_\nu P_\mu)(\boldsymbol{\gamma}_\lambda P_\sigma - \boldsymbol{\gamma}_\sigma P_\lambda) + (\boldsymbol{\gamma}_\lambda P_\sigma - \boldsymbol{\gamma}_\sigma P_\lambda)(\boldsymbol{\gamma}_\mu P_\nu - \boldsymbol{\gamma}_\nu P_\mu)]$$

$$= -\frac{1}{4}\begin{pmatrix} \gamma_\mu\gamma_\lambda P_\nu P_\sigma - \gamma_\mu\gamma_\sigma P_\nu P_\lambda - \gamma_\nu\gamma_\lambda P_\mu P_\sigma + \gamma_\nu\gamma_\sigma P_\mu P_\lambda + \\ \gamma_\lambda\gamma_\mu P_\sigma P_\nu - \gamma_\lambda\gamma_\nu P_\sigma P_\mu + \gamma_\sigma\gamma_\mu P_\lambda P_\nu + \gamma_\sigma\gamma_\nu P_\lambda P_\mu \end{pmatrix}$$

$$= -\frac{1}{2}(g_{\mu\lambda}P_\nu P_\sigma - g_{\mu\sigma}P_\nu P_\lambda - g_{\nu\lambda}P_\mu P_\sigma + g_{\nu\sigma}P_\mu P_\lambda) \tag{192}$$

The above anticommutators are extremely instructive insofar as they point to the possibility of the existence of a superstructure within this algebra by incorporating therein the ten second order operators $P_\mu P_\nu$. We also have the following:

$$\{W_\mu, W_{\lambda\sigma}\} = \frac{1}{4}[\boldsymbol{\gamma}_5(P_\mu - m\boldsymbol{\gamma}_\mu)\cdot\boldsymbol{\gamma}_5(\boldsymbol{\gamma}_\lambda P_\sigma - \boldsymbol{\gamma}_\sigma P_\lambda) + \boldsymbol{\gamma}_5(\boldsymbol{\gamma}_\lambda P_\sigma - \boldsymbol{\gamma}_\sigma P_\lambda)\cdot\boldsymbol{\gamma}_5(P_\mu - m\boldsymbol{\gamma}_\mu)]$$

$$= \frac{1}{4}[(P_\mu + m\boldsymbol{\gamma}_\mu)(\boldsymbol{\gamma}_\lambda P_\sigma - \boldsymbol{\gamma}_\sigma P_\lambda) - (\boldsymbol{\gamma}_\lambda P_\sigma - \boldsymbol{\gamma}_\sigma P_\lambda)(P_\mu - m\boldsymbol{\gamma}_\mu)]$$

$$= \frac{m}{4}(\boldsymbol{\gamma}_\mu\boldsymbol{\gamma}_\lambda P_\sigma - \boldsymbol{\gamma}_\mu\boldsymbol{\gamma}_\sigma P_\lambda + \boldsymbol{\gamma}_\lambda\boldsymbol{\gamma}_\mu P_\sigma - \boldsymbol{\gamma}_\sigma\boldsymbol{\gamma}_\mu P_\lambda) = \frac{1}{2}m(g_{\mu\lambda}P_\sigma - g_{\mu\sigma}P_\lambda) \tag{193}$$

$$[P_\mu, W_\nu] = \left[P_\mu, \frac{1}{2}\gamma_5(P_\nu - m\gamma_\nu)\right] = 0,\ [P_\mu, W_{\rho\lambda}] = \left[P_\mu, \frac{1}{2}\gamma_5(\gamma_\rho P_\lambda - \gamma_\lambda P_\rho)\right] = 0,$$

$$[P_\mu P_\nu, P_\rho] = 0,\ [P_\mu P_\nu, W_\lambda] = \left[p_\mu\ p_\nu, \frac{1}{2}\gamma_5(p_\lambda - m\gamma_\lambda)\right] = 0,$$

$$[P_\mu P_\nu, J_{\rho\lambda}] = P_\mu[P_\nu, J_{\rho\lambda}] + [P_\mu, J_{\rho\lambda}]P_\nu = i(g_{\nu\rho}P_\mu P_\lambda + g_{\mu\rho}P_\nu P_\lambda - g_{\nu\lambda}P_\mu P_\rho - g_{\mu\lambda}P_\nu P_\rho),$$

$$[P_\mu P_\nu, W_{\rho\sigma}] = \left[P_\mu P_\nu, \frac{1}{2}\gamma_5(\gamma_\rho P_\sigma - \gamma_\sigma P_\rho)\right] = 0$$

$$[J_{\mu\nu}, W_\rho] = \left[J_{\mu\nu}, \frac{1}{2}\gamma_5(P_\rho - m\gamma_\rho)\right] = \frac{1}{2}\boldsymbol{\gamma}_5\{[J_{\mu\nu}, P_\rho] - m[J_{\mu\nu}, \boldsymbol{\gamma}_\rho]\}$$

$$= \frac{1}{2}\boldsymbol{\gamma}_5\left\{i(g_{\rho\nu}P_\mu - g_{\rho\mu}P_\nu) - \frac{im}{4}[[\gamma_\mu, \gamma_\nu], \gamma_\rho]\right\}$$

$$= \frac{1}{2}\gamma_5\left\{i(g_{\rho\nu}P_\mu - g_{\rho\mu}P_\nu) - \frac{im}{2}[\gamma_\mu\gamma_\nu - g_{\mu\nu}, \gamma_\rho]\right\}$$

$$= \frac{1}{2}\gamma_5\left\{i(g_{\rho\nu}P_\mu - g_{\rho\mu}P_\nu) - \frac{im}{2}[\gamma_\mu\gamma_\nu, \gamma_\rho]\right\}$$

$$= \frac{1}{2}\gamma_5\left\{i(g_{\rho\nu}P_\mu - g_{\rho\mu}P_\nu) - \frac{im}{2}[\gamma_\mu\gamma_\nu, \gamma_\rho - (2g_{\rho\mu} - \gamma_\mu\gamma_\rho)\gamma_\nu]\right\}$$

$$= \frac{1}{2}\gamma_5[i(g_{\rho\nu}P_\mu - g_{\rho\mu}P_\nu) + im(g_{\rho\mu}\gamma_\nu - g_{\rho\nu}\gamma_\mu)] = i(g_{\nu\rho}W_\mu - g_{\mu\rho}W_\nu)$$

$$[J_{\mu\nu}, W_{\rho\sigma}] = \left[J_{\mu\nu}, \frac{1}{2}\gamma_5(\gamma_\rho P_\sigma - \gamma_\sigma P_\rho)\right]$$

$$= \frac{1}{2}\gamma_5\left[\left(x_\mu P_\nu - x_\nu P_\mu + \frac{i}{2}\gamma_\mu\gamma_\nu\right), (\gamma_\rho P_\sigma - \gamma_\sigma P_\rho)\right]$$

Now $[x_\mu P_\nu, \gamma_\rho P_\sigma] = \gamma_\rho[x_\mu, P_\sigma]P_\nu = -ig_{\mu\sigma}\gamma_\rho P_\nu$, $[x_\mu P_\nu, \gamma_\sigma P_\rho] = -ig_{\mu\rho}\gamma_\sigma P_\nu$,

$[x_\nu P_\mu, \gamma_\rho P_\sigma] = -ig_{\nu\sigma}\gamma_\rho P_\mu$, $[x_\nu P_\mu, \gamma_\sigma P_\rho] = -ig_{\nu\rho}\gamma_\sigma P_\mu$,

$[\gamma_\mu\gamma_\nu, \gamma_\rho P_\sigma] = [\gamma_\mu\gamma_\nu\gamma_\rho - (2g_{\mu\rho} - \gamma_\mu\gamma_\rho)\gamma_\nu]P_\sigma = (2g_{\nu\rho}\gamma_\mu - 2g_{\mu\rho}\gamma_\nu)P_\sigma$

$[\gamma_\mu\gamma_\nu, \gamma_\sigma P_\rho] = (2g_{\nu\sigma}\gamma_\mu - 2g_{\mu\sigma}\gamma_\nu)P_\rho$ so that

$$[J_{\mu\nu}, W_{\rho\sigma}] = i(g_{\mu\sigma}W_{\nu\rho} + g_{\nu\sigma}W_{\rho\mu} + g_{\mu\rho}W_{\sigma\nu} + g_{\nu\rho}W_{\mu\sigma}) \tag{194}$$

The above (anti)commutators vindicate the existence of a Lie superalgebra among the set of first order symmetry operators of the Dirac equation as extended by the inclusion of the ten second order operators $P_\mu P_\nu$. This Lie superalgebra is spanned by a basis comprising of the twenty even operators $\{P_\mu, J_{\mu\nu}, P_\mu P_\nu\}$ and the ten odd ones $\{W_\mu, W_{\mu\nu}\}$ in addition to the trivial identity operator.

These odd operators show an intimate connection with relativistic dynamics of massive particles and play an important role in the study of supersymmetry among elementary particles.

CHAPTER 6

Solutions of the Dirac Equation

6.1 FREE PARTICLE SOLUTIONS

The free particle Dirac equation can be written as

$$i\partial_0\psi(\mathbf{x}, x^0) = (\boldsymbol{\alpha}\hat{\mathbf{p}} + \beta m)\,\psi(\mathbf{x}, x^0) \tag{1}$$

The time dependencies can be separated by writing

$$\psi(\mathbf{x}, x^0) = \psi(\mathbf{x})\,\exp(-\,ip_0x^0) \tag{2}$$

with p_0 being the eigenvalue of $i\partial_0$, i.e.,

$$i\partial_0\,\psi(\mathbf{x}, x^0) = i\partial_0\,\psi(\mathbf{x})\,\exp(-\,ip_0x^0)$$

$$= p_0\,\psi(\mathbf{x})\,\exp(-ip_0x^0) = p_0\,\psi(\mathbf{x}, x^0)$$

p_0 is unrestricted in sign and $\psi(\mathbf{x})$ is a solution of

$$p_0\,\psi(\mathbf{x}) = (\boldsymbol{\alpha}\hat{\mathbf{p}} + \beta m)\,\psi(\mathbf{x}) \tag{3}$$

As mentioned earlier, $\boldsymbol{\alpha}^k$, $\boldsymbol{\beta}$ are 4×4 matrices so that $\psi(\mathbf{x})$ must necessarily be a 4-component spinor, i.e., a 4-component column vector $\psi(\mathbf{x}) = (\psi_1, \psi_2, \psi_3, \psi_4)^T$ which may be split into 2-component spinors as

$$\phi = (\phi_1\ \phi_2)^T = (\psi_1\ \psi_2)^T,\ \chi = (\chi_1\ \chi_2)^T = (\psi_3\ \psi_4)^T$$

Substituting these expressions in eq. (3) and using the representation of the $\boldsymbol{\alpha}^k$, $\boldsymbol{\beta}$ matrices

$$\boldsymbol{\alpha} = \begin{pmatrix} \mathbf{0} & \sigma \\ \sigma & \mathbf{0} \end{pmatrix},\ \boldsymbol{\beta} = \begin{pmatrix} \mathbf{I} & 0 \\ 0 & -\mathbf{I} \end{pmatrix} \tag{4}$$

we obtain

$$p_0\begin{pmatrix} \phi \\ \chi \end{pmatrix} = \begin{pmatrix} \mathbf{0} & \sigma \\ \sigma & \mathbf{0} \end{pmatrix}\hat{\mathbf{p}}\begin{pmatrix} \phi \\ \chi \end{pmatrix} + m\begin{pmatrix} \mathbf{I} & 0 \\ 0 & -\mathbf{I} \end{pmatrix}\begin{pmatrix} \phi \\ \chi \end{pmatrix} \tag{5}$$

yielding the following pair of equations:

$$p_0\phi = \boldsymbol{\sigma}\mathbf{p}\chi + m\phi,\ p_0\chi = \boldsymbol{\sigma}\mathbf{p}\phi - m\chi \tag{6}$$

where we have substituted the momentum operator $\hat{\mathbf{p}}$ by its eigenvalue $\mathbf{p}$ using

$$\hat{\mathbf{p}}\psi(x) = \hat{\mathbf{p}}\begin{pmatrix}\phi\\ \chi\end{pmatrix} = \mathbf{p}\ \psi(x) = \mathbf{p}\begin{pmatrix}\phi\\ \chi\end{pmatrix}$$

The second of these equations gives

$$\chi = \frac{\boldsymbol{\sigma}\mathbf{p}}{p_0 + m}\ \phi \tag{7}$$

which on substituting in the first equation gives

$$\left[p_0 - m - \frac{(\boldsymbol{\sigma}\mathbf{p})^2}{p_0 + m}\right]\phi = 0 \quad \Rightarrow p_0^2 = \mathbf{p}^2 + m^2 \tag{8}$$

which is nothing but the relativistic momentum-energy relationship. In other words, if the Dirac equation is to embody the correct relativistic momentum-energy relationship, then eq. (8) is identically satisfied. This, in turn, implies that ϕ, χ are linearly independent.

This linear independence can also be established by assuming that

$$a\phi + b\chi = 0 \tag{9}$$

and then showing that eq. (9) implies that $a = b = 0$. We have, from eqs. (7) and (9)

$$0 = \begin{pmatrix}0 & 0\\ 0 & 0\end{pmatrix} = \left(a\mathbf{I} + \frac{b}{p_0 + m}\boldsymbol{\sigma}\mathbf{p}\right)$$

$$\phi = a\begin{pmatrix}1 & 0\\ 0 & 1\end{pmatrix} + \frac{b}{p_0 + m}\begin{pmatrix}p_3 & p_1 - ip_2\\ p_1 + ip_2 & -p_3\end{pmatrix}$$

$$= \begin{pmatrix} a + \dfrac{bp_3}{p_0 + m} & \dfrac{b(p_1 - ip_2)}{p_0 + m} \\ \dfrac{b(p_1 + ip_2)}{p_0 + m} & a - \dfrac{bp_3}{p_0 + m}\end{pmatrix} \tag{10}$$

which can only hold if $a = b = 0$. Therefore, there are two linearly independent solutions for each momentum state $\mathbf{p}$ for positive p_0 (positive energy solutions) and similarly, there exist two linearly independent solutions for each momentum state $\mathbf{p}$ for negative p_0 (negative energy solutions). For states with definite momentum $\mathbf{p}$, we have

$$\hat{\mathbf{p}}\phi(\mathbf{x}) = -\,i\nabla\phi(\mathbf{x}) = \mathbf{p}\phi(\mathbf{x}) \tag{11}$$

whence

$$\phi(\mathbf{x}) = \phi_0 \exp(i\mathbf{px}) \tag{12}$$

and similarly

$$\chi(\mathbf{x}) = \chi_0 \exp(i\mathbf{px}) \tag{13}$$

so that

$$\begin{pmatrix} \phi \\ \chi \end{pmatrix} = \begin{pmatrix} \phi_0 \\ \chi_0 \end{pmatrix} \exp(i\mathbf{px}) \tag{14}$$

and

$$p_0 \phi_0 = \boldsymbol{\sigma}\mathbf{p}\chi_0 + m\phi_0,\ p_0 \chi_0 = \boldsymbol{\sigma}\mathbf{p}\phi_0 - m\chi_0 \tag{15}$$

the second of which gives

$$\chi_0 = \frac{\boldsymbol{\sigma}\mathbf{p}}{p_0 + m}\phi_0 \tag{16}$$

Writing $\phi_0 = (\phi_{01}\ \phi_{02})^T$, $\chi_0 = (\chi_{01}\ \chi_{02})^T$ we end up with the following explicit solution of the Dirac equation:

$$\psi_{\mathbf{p},\varepsilon}(\mathbf{x}, x^0) = N \begin{pmatrix} \phi_0 \\ \dfrac{\boldsymbol{\sigma}\mathbf{p}}{\varepsilon E_p + m}\phi_0 \end{pmatrix} \exp[i(\mathbf{px} - \varepsilon E_p x^0)] \tag{17}$$

where $\varepsilon = \pm 1$, $p_0 = \varepsilon E_p = \varepsilon\sqrt{\mathbf{p}^2 + m^2} = \pm\sqrt{\mathbf{p}^2 + m^2}$ and N is a normalization constant to be determined from the normalization condition

$$\int d\mathbf{x}\psi^{\dagger}_{\mathbf{p},\varepsilon}(\mathbf{x}, x^0)\ \psi_{\mathbf{p}',\varepsilon'}(\mathbf{x}, x^0) = \delta_{\varepsilon\varepsilon'}\ \delta(\mathbf{p} - \mathbf{p}') \tag{18}$$

On using eqs. (17) and (18), we have

$$N^2 \left[\phi_0^{\dagger}\phi_0 + \phi_0^{\dagger}\frac{(\boldsymbol{\sigma}\mathbf{p})(\boldsymbol{\sigma}\mathbf{p})}{(\varepsilon E_p + m)^2}\phi_0\right]$$

$$= N^2\left[1 + \frac{(\boldsymbol{\sigma}\mathbf{p})(\boldsymbol{\sigma}\mathbf{p})}{(\varepsilon E_p + m)^2}\right]\phi_0^{\dagger}\phi_0 = 1 \tag{19}$$

If we prescribe the normalization condition $\phi_0^{\dagger}\phi_0 = 1$ for the 2-component spinor ϕ_0, then from eq. (19), we have

$$N = \left[1 + \frac{(\boldsymbol{\sigma}\mathbf{p})(\boldsymbol{\sigma}\mathbf{p})}{(\varepsilon E_p + m)^2}\right]^{-\frac{1}{2}} = \left[1 + \frac{\mathbf{p}^2}{(\varepsilon E_p + m)^2}\right]^{-\frac{1}{2}} = \left[\frac{(\varepsilon E_p + m)^2 + \mathbf{p}^2}{(\varepsilon E_p + m)^2}\right]^{-\frac{1}{2}}$$

$$= \left[\frac{E_p^2 + 2\varepsilon m E_p + m^2 + \mathbf{p}^2}{(\varepsilon E_p + m)^2}\right]^{-\frac{1}{2}} = \left[\frac{2E_p^2 + 2\varepsilon m E_p}{(\varepsilon E_p + m)^2}\right]^{-\frac{1}{2}} = \left[\frac{2\varepsilon^2 E_p^2 + 2\varepsilon m E_p}{(\varepsilon E_p + m)^2}\right]^{-\frac{1}{2}}$$

$$= \left[\frac{2\varepsilon E_p}{(\varepsilon E_p + m)}\right]^{-\frac{1}{2}} = \left[\frac{(\varepsilon E_p + m)}{2\varepsilon E_p}\right]^{\frac{1}{2}} \tag{20}$$

where, in the first step, we have used

$$(\boldsymbol{\sigma}\mathbf{A})\cdot(\boldsymbol{\sigma}\mathbf{B}) = \mathbf{AB}\cdot\mathbf{I} + i\sigma\,(\mathbf{A}\times\mathbf{B}) \tag{21}$$

so that

$$(\boldsymbol{\sigma}\mathbf{p})^2 = \mathbf{p}^2\,\mathbf{I} \tag{22}$$

The fact that there exist two different kinds of solutions, viz. positive energy solutions ($\varepsilon = +1$) and negative energy solutions ($\varepsilon = -1$) can be arrived at in another way as follows:

Consider the helicity operator

$$\hat{s}(\mathbf{p}) = \frac{\hat{\Sigma}.\mathbf{p}}{|\mathbf{p}|} \tag{23}$$

where $\Sigma = \begin{pmatrix} \hat{\sigma} & \mathbf{0} \\ \mathbf{0} & \hat{\sigma} \end{pmatrix}$.

The eigenvalues of the operator $\hat{s}(\mathbf{p})$ are called the "helicity" of the particle states. We have

$$[\alpha\hat{\mathbf{p}},\ \hat{\Sigma}\mathbf{p}] = (\alpha\hat{\mathbf{p}})\,(\hat{\Sigma}\mathbf{p}) - (\hat{\Sigma}\mathbf{p})\,(\alpha\hat{\mathbf{p}})$$

$$= \begin{pmatrix} \mathbf{0} & \sigma\hat{\mathbf{p}} \\ \sigma\hat{\mathbf{p}} & \mathbf{0} \end{pmatrix}\begin{pmatrix} \hat{\sigma}\mathbf{p} & \mathbf{0} \\ \mathbf{0} & \hat{\sigma}\mathbf{p} \end{pmatrix} - \begin{pmatrix} \hat{\sigma}\mathbf{p} & \mathbf{0} \\ \mathbf{0} & \hat{\sigma}\mathbf{p} \end{pmatrix}\begin{pmatrix} \mathbf{0} & \sigma\hat{\mathbf{p}} \\ \sigma\hat{\mathbf{p}} & \mathbf{0} \end{pmatrix}$$

$$= \begin{pmatrix} \mathbf{0} & \sigma\hat{\mathbf{p}}.\hat{\sigma}\mathbf{p} \\ \sigma\hat{\mathbf{p}}.\hat{\sigma}\mathbf{p} & \mathbf{0} \end{pmatrix} - \begin{pmatrix} \mathbf{0} & \hat{\sigma}\mathbf{p}.\sigma\hat{\mathbf{p}} \\ \hat{\sigma}\mathbf{p}.\sigma\hat{\mathbf{p}} & \mathbf{0} \end{pmatrix} = \begin{pmatrix} \mathbf{0} & \mathbf{0} \\ \mathbf{0} & \mathbf{0} \end{pmatrix} \tag{24}$$

where the final step follows from the fact that we are considering here well defined momentum states so that the momentum operator can be replaced by its eigenstate, i.e., $\hat{\mathbf{p}}\psi(\mathbf{p}) = \mathbf{p}\psi(\mathbf{p})$.

Similarly,

$$[\beta\hat{\mathbf{p}},\ \hat{\Sigma}\mathbf{p}] = (\beta\hat{\mathbf{p}})\,(\hat{\Sigma}\mathbf{p}) - (\hat{\Sigma}\mathbf{p})\,(\beta\hat{\mathbf{p}})$$

$$= \begin{pmatrix} \hat{\mathbf{p}} & \mathbf{0} \\ \mathbf{0} & -\hat{\mathbf{p}} \end{pmatrix}\begin{pmatrix} \hat{\sigma}\mathbf{p} & \mathbf{0} \\ \mathbf{0} & \hat{\sigma}\mathbf{p} \end{pmatrix} - \begin{pmatrix} \hat{\sigma}\mathbf{p} & \mathbf{0} \\ \mathbf{0} & \hat{\sigma}\mathbf{p} \end{pmatrix}\begin{pmatrix} \hat{\mathbf{p}} & \mathbf{0} \\ \mathbf{0} & -\hat{\mathbf{p}} \end{pmatrix}$$

$$= \begin{pmatrix} \hat{\mathbf{p}}.\hat{\sigma}\mathbf{p} & \mathbf{0} \\ \mathbf{0} & -\hat{\mathbf{p}}.\hat{\sigma}\mathbf{p} \end{pmatrix} - \begin{pmatrix} \hat{\sigma}\mathbf{p}.\mathbf{p} & \mathbf{0} \\ \mathbf{0} & \hat{\sigma}\mathbf{p}.\mathbf{p} \end{pmatrix} = \begin{pmatrix} \mathbf{0} & \mathbf{0} \\ \mathbf{0} & \mathbf{0} \end{pmatrix} \tag{25}$$

Also

$$[\hat{\mathbf{p}}, \hat{s}(\mathbf{p})] = [\hat{\mathbf{p}}, \hat{\Sigma}\mathbf{p}]$$

$$= \left[\hat{\mathbf{p}}\begin{pmatrix} \mathbf{I} & \mathbf{0} \\ \mathbf{0} & \mathbf{I} \end{pmatrix}, \begin{pmatrix} \hat{\sigma} & \mathbf{0} \\ \mathbf{0} & \hat{\sigma} \end{pmatrix}\mathbf{p}\right] = \left[\begin{pmatrix} \hat{\mathbf{p}}\mathbf{I} & \mathbf{0} \\ \mathbf{0} & \hat{\mathbf{p}}\mathbf{I} \end{pmatrix}, \begin{pmatrix} \hat{\sigma}\mathbf{p} & \mathbf{0} \\ \mathbf{0} & \hat{\sigma}\mathbf{p} \end{pmatrix}\right]$$

$$= \begin{pmatrix} \hat{\mathbf{p}}\hat{\sigma}\mathbf{p} - \hat{\sigma}\mathbf{p}\hat{\mathbf{p}} & \mathbf{0} \\ \mathbf{0} & \hat{\mathbf{p}}\hat{\sigma}\mathbf{p} - \hat{\sigma}\mathbf{p}\hat{\mathbf{p}} \end{pmatrix} = \begin{pmatrix} \hat{\sigma}\hat{\mathbf{p}}\mathbf{p} - \hat{\sigma}\mathbf{p}\hat{\mathbf{p}} & \mathbf{0} \\ \mathbf{0} & \hat{\sigma}\hat{\mathbf{p}}\mathbf{p} - \hat{\sigma}\mathbf{p}\hat{\mathbf{p}} \end{pmatrix}$$

$$= \begin{pmatrix} \hat{\sigma}\mathbf{p}\hat{\mathbf{p}} - \hat{\sigma}\mathbf{p}\hat{\mathbf{p}} & \mathbf{0} \\ \mathbf{0} & \hat{\sigma}\mathbf{p}\hat{\mathbf{p}} - \hat{\sigma}\mathbf{p}\hat{\mathbf{p}} \end{pmatrix} = \begin{pmatrix} \mathbf{0} & \mathbf{0} \\ \mathbf{0} & \mathbf{0} \end{pmatrix} \tag{26}$$

Hence, the Dirac Hamiltonian $\hat{H} = \alpha\hat{\mathbf{p}} + \beta m$ commutes with the helicity operator

$$\hat{s}(\mathbf{p}) = \frac{\hat{\Sigma}.\mathbf{p}}{|\mathbf{p}|}, \text{ i.e., } [\hat{H}, \hat{s}(\mathbf{p})] = 0 = [\hat{H}, \hat{\Sigma}] \tag{27}$$

Therefore, $\hat{s}(\mathbf{p})$, $\hat{H}$ can be diagonalized simultaneously and we can have solutions of the Dirac equation that are simultaneous eigenfunctions of $\hat{s}(\mathbf{p})$, $\hat{H}$.

Furthermore,

$$\hat{s}^2(\mathbf{p}) = \frac{1}{|\mathbf{p}|^2}\begin{pmatrix} \hat{\sigma}\mathbf{p} & 0 \\ 0 & \hat{\sigma}\mathbf{p} \end{pmatrix}\begin{pmatrix} \hat{\sigma}\mathbf{p} & 0 \\ 0 & \hat{\sigma}\mathbf{p} \end{pmatrix}$$

$$= \frac{1}{|\mathbf{p}|^2}\begin{pmatrix} (\hat{\sigma}\mathbf{p})^2 & 0 \\ 0 & (\hat{\sigma}\mathbf{p})^2 \end{pmatrix} = \frac{1}{|\mathbf{p}|^2}\begin{pmatrix} \mathbf{p}^2 & \mathbf{0} \\ \mathbf{0} & \mathbf{p}^2 \end{pmatrix} = \mathbf{I} \tag{28}$$

Hence, the eigenvalues of $\hat{s}(\mathbf{p})$ are ± 1. The corresponding eigenfunctions are calculated by solving the equation

$$\pm\begin{pmatrix} \varphi \\ \xi \end{pmatrix} = \hat{s}(\mathbf{p})\begin{pmatrix} \varphi \\ \xi \end{pmatrix} = \frac{1}{|\mathbf{p}|}\begin{pmatrix} \hat{\sigma}\mathbf{p} & 0 \\ 0 & \hat{\sigma}\mathbf{p} \end{pmatrix}\begin{pmatrix} \varphi \\ \xi \end{pmatrix} = \frac{1}{|\mathbf{p}|}\begin{pmatrix} \hat{\sigma}\mathbf{p}\varphi \\ \hat{\sigma}\mathbf{p}\xi \end{pmatrix} \tag{29}$$

For a particle moving along the x^3 direction $p_1 = p_2 = 0$, $p_3 = \mathbf{p}$. Using $\sigma^3 \equiv \sigma^3 = \begin{pmatrix} 1 & 0 \\ 0 & -1 \end{pmatrix}$, $\varphi = (\varphi_1\ \varphi_2)^T$, $\xi = (\xi_1\ \xi_2)^T$, eq. (29) gives, for helicity + 1, $\begin{pmatrix} \varphi_1 \\ \varphi_2 \end{pmatrix} = \frac{p_3}{|p_3|}\begin{pmatrix} \varphi_1 \\ -\varphi_2 \end{pmatrix}$ leading to the two linearly independent solutions $\begin{pmatrix}1\\0\end{pmatrix}, \begin{pmatrix}0\\1\end{pmatrix}$ corresponding to the $p_3 > 0$ and $p_3 < 0$ momentum states, respectively. These solutions also meet the normalization requirement $\varphi^\dagger \varphi = 1$. Proceeding similarly, we can show that for helicity −1, the linearly independent solutions $\begin{pmatrix}0\\1\end{pmatrix}, \begin{pmatrix}1\\0\end{pmatrix}$ correspond, respectively, to the $p_3 > 0$ and $p_3 < 0$ momentum states. Exactly same results hold for the two-component spinor $\xi = (\xi_1\ \xi_2)^T$. Hence, we conclude with the following eigenfunctions of the helicity operator $\hat{s}(\mathbf{p})$:

Momentum ↓ Helicity →	*+1*	*−1*
$p_3 > 0$	$(1\ 0\ 1\ 0)^T$	$(0\ 1\ 0\ 1)^T$
$p_3 < 0$	$(0\ 1\ 0\ 1)^T$	$(1\ 0\ 1\ 0)^T$

In the sequel, unless explicitly stated, we shall assume throughout that the orientation of the axes is such that the motion of the particle is along the positive direction of the x^3 axes so that $p_3 > 0$.

However, vector $(\varphi\ \xi)^T$ must also satisfy eq. (17) being solution of the free particle Dirac equation. Therefore, we can write the solution of the Dirac equation for a free particle, with helicity of +1, moving in the positive x^3 direction with well-defined momentum state p_3 as

$$\psi_{p_3,\ \varepsilon,\ +1} = N \begin{pmatrix} \begin{pmatrix}1\\0\end{pmatrix} \\ \dfrac{\hat{\sigma}^3\ p_3}{\varepsilon E_p + m}\begin{pmatrix}1\\0\end{pmatrix} \end{pmatrix} \exp[i(p_3 x^3 - \varepsilon E_p x^0)] \tag{30}$$

where $\varepsilon = +1$ for positive energy solutions and $\varepsilon = -1$ for negative energy solutions. Similarly, for helicity −1, we have the solution

$$\psi_{p_3,\ \varepsilon,\ -1} = N \begin{pmatrix} \begin{pmatrix}0\\1\end{pmatrix} \\ \dfrac{\hat{\sigma}^3\ p_3}{\varepsilon E_p + m}\begin{pmatrix}0\\1\end{pmatrix} \end{pmatrix} \exp[i(p_3 x^3 - \varepsilon E_p x^0)] \tag{31}$$

The two positive energy solutions corresponding to helicity states +1 and −1 are mutually orthogonal, i.e.,

$$\langle \psi_{p_3,\ \varepsilon,\ r} | \psi_{p_3,\ \varepsilon, s} \rangle = N^2 \delta_{rs},\ r, s = \pm 1 \tag{32}$$

So far we have obtained the simultaneous eigenfunctions of the Hamiltonian and the helicity operator for the specific case a free Dirac particle moving in the positive direction of the x^3 axis. We now obtain the eigenfunctions of the helicity operator of a free Dirac particle moving in a direction given by the unit vector $\mathbf{n} \equiv (n_1, n_2, n_3)$, where $\mathbf{n} \equiv (n_1, n_2, n_3) = \dfrac{\mathbf{p}}{|\mathbf{p}|}$ represents the direction of the 3-momentum $\mathbf{p}$.

As in the previous case, we assume that the eigenfunctions of the spin (helicity) operator $\hat{s}(\mathbf{p}) = \dfrac{\hat{\Sigma}.\mathbf{p}}{|\mathbf{p}|}$ are given by $(\varphi\ \xi)^T$, where $\varphi \equiv (\varphi_1\ \varphi_2)^T$, $\xi \equiv (\xi_1\ \xi_2)^T$. Then, the eigenvalue equation is given by eq. (29) which implies

$$\pm\begin{pmatrix}\varphi\\ \xi\end{pmatrix} = \hat{s}(\mathbf{p})\begin{pmatrix}\varphi\\ \xi\end{pmatrix} = \hat{\Sigma}.\mathbf{n}\begin{pmatrix}\varphi\\ \xi\end{pmatrix} \tag{33}$$

where $\mathbf{n} = \dfrac{\mathbf{p}}{|\mathbf{p}|}$ is the unit vector in the direction of $\mathbf{p}$. Since $\hat{\Sigma} = \begin{pmatrix}\hat{\sigma} & 0\\ 0 & \hat{\sigma}\end{pmatrix}$ we have, from eq. (33)

$$\hat{\sigma}.\mathbf{n}\varphi = \pm\,\varphi,\ \hat{\sigma}.\mathbf{n}\xi = \pm\,\xi \tag{34}$$

or equivalently

$$\left(\sum_{k=1}^{3}\hat{\sigma}^k\, n_k\right)\varphi = \pm\,\varphi,\ \left(\sum_{k=1}^{3}\hat{\sigma}^k\, n_k\right)\xi = \pm\,\xi \tag{35}$$

The first of these equations, on substituting the values of $\hat{\sigma}^k$, gives

$$\begin{pmatrix} n_3 & n_1 - in_2\\ n_1 + in_2 & -n_3\end{pmatrix}\begin{pmatrix}\varphi_1\\ \varphi_2\end{pmatrix} = \pm\begin{pmatrix}\varphi_1\\ \varphi_2\end{pmatrix} \tag{36}$$

the second of which gives

$$\varphi_2 = \frac{n_1 + in_2}{n_3 \pm 1}\varphi_1 = \frac{(n_1 + in_2).(n_1 - in_2)}{(n_3 \pm 1)(n_1 - in_2)}\varphi_1 = \frac{n_1^2 + n_2^2}{(n_3 \pm 1)(n_1 - in_2)}\varphi_1$$

$$= \frac{1 - n_3^2}{(n_3 \pm 1)(n_1 - in_2)}\varphi_1 = \frac{1 \mp n_3}{\pm(n_1 - in_2)}\varphi_1 \tag{37}$$

Eliminating φ_2 in eq. (36), we get $\left(\dfrac{n_1^2 + n_2^2 + n_3^2 - 1}{n_3 \pm 1}\right)\varphi_1 = 0$ which is identically satisfied for all values of φ_1 because $\mathbf{n}^2 = 1$. We again have two linearly independent solutions (for helicity +1) and we can choose them conveniently as $A\begin{pmatrix} n_3 + 1\\ n_1 + in_2\end{pmatrix}$, $A\begin{pmatrix} n_1 - in_2\\ 1 - n_3\end{pmatrix}$ for the motion of the particle along

n and $-$**n**, respectively. We shall, however, confine ourselves to the former case only in line with the assumed definition of **n** being a unit vector along the direction of motion of the free particle. Similarly, for a particle of helicity -1 moving along **n**, we can choose the 2-component vector $\varphi = A\begin{pmatrix} -n_1 + in_2 \\ n_3 + 1 \end{pmatrix}$, where A is a normalization constant to be determined from $\varphi^\dagger\varphi = 1$. We have

$$\varphi^\dagger\varphi = A^2\begin{pmatrix} n_3 + 1 \\ n_1 + in_2 \end{pmatrix}^\dagger \begin{pmatrix} n_3 + 1 \\ n_1 + in_2 \end{pmatrix} = A^2((n_3 + 1)^* \; (n_1 + in_2)^*)\begin{pmatrix} n_3 + 1 \\ n_1 + in_2 \end{pmatrix}$$

$$= A^2((n_3 + 1)(n_1 - in_2))\begin{pmatrix} n_3 + 1 \\ n_1 + in_2 \end{pmatrix}$$

$$= A^2[(n_3 + 1)^2 + (n_1 - in_2)(n_1 + in_2)] = A^2(2 + 2n_3) \tag{38}$$

thereby giving

$$A = \frac{1}{\sqrt{2(1 + n_3)}} \tag{39}$$

Exactly same results hold for the 2-component spinor ξ so that we have the following solution for the eigenfunctions of the "helicity operator":

Helicity	*Eigenfunction*
+1	$A(n_3 + 1, n_1 + in_2, n_3 + 1, n_1 + in_2)^T$
-1	$A(-n_1 + in_2, n_3 + 1, -n_1 + in_2, n_3 + 1)^T$

with $A = \dfrac{1}{\sqrt{2(1 + n_3)}}$.

To obtain simultaneous eigenfunctions of $\hat{H}$, $\hat{s}(\mathbf{p})$, the above must also satisfy the Dirac equation which leads to eqs. (16) and (17). Therefore, the normalized simultaneous eigenfunctions of $\hat{H}$, $\hat{s}(\mathbf{p})$ are given by

$$\psi_{\mathbf{p},\,\varepsilon,\,+1} = \frac{1}{\sqrt{2(1 + n_3)}}\sqrt{\frac{\varepsilon E_p + m}{2\varepsilon E_p}}\begin{pmatrix} \begin{pmatrix} n_3 + 1 \\ n_1 + in_2 \end{pmatrix} \\ \dfrac{\hat{\sigma}\mathbf{p}}{\varepsilon E_p + m}\begin{pmatrix} n_3 + 1 \\ n_1 + in_2 \end{pmatrix} \end{pmatrix} \exp[i(\mathbf{px} - \varepsilon E_p x^0)] \tag{40}$$

$$\psi_{\mathbf{p},\varepsilon,-1} = \frac{1}{\sqrt{2(1+n_3)}}\sqrt{\frac{\varepsilon E_p + m}{2\varepsilon E_p}}\begin{pmatrix} \begin{pmatrix} -n_1 + in_2 \\ n_3 + 1 \end{pmatrix} \\ \dfrac{\hat{\sigma}\mathbf{p}}{\varepsilon E_p + m}\begin{pmatrix} -n_1 + in_2 \\ n_3 + 1 \end{pmatrix} \end{pmatrix} \exp[i(\mathbf{px} - \varepsilon E_p x^0)] \quad (41)$$

Now

$$\hat{\sigma}\mathbf{p}\begin{pmatrix} n_3 + 1 \\ n_1 + in_2 \end{pmatrix} = \hat{\sigma}\mathbf{n}|\mathbf{p}| = \begin{pmatrix} n_3 + 1 \\ n_1 + in_2 \end{pmatrix} = |\mathbf{p}|\,\hat{\sigma}\mathbf{n}\begin{pmatrix} n_3 + 1 \\ n_1 + in_2 \end{pmatrix} = |\mathbf{p}|\left(\sum_{k=1}^{3}\hat{\sigma}^k n_k\right)\begin{pmatrix} n_3 + 1 \\ n_1 + in_2 \end{pmatrix}$$

$$= |\mathbf{p}|\begin{pmatrix} n_3 & n_1 - in_2 \\ n_1 + in_2 & -n_3 \end{pmatrix}\begin{pmatrix} n_3 + 1 \\ n_1 + in_2 \end{pmatrix}$$

$$= |\mathbf{p}|\begin{pmatrix} n_3(n_3+1) + (n_1 - in_2)(n_1 + in_2) \\ (n_1 + in_2)(n_3 + 1) - n_3(n_1 + in_2) \end{pmatrix} = |\mathbf{p}|\begin{pmatrix} n_3 + 1 \\ n_1 + in_2 \end{pmatrix} \quad (42)$$

and similarly

$$\hat{\sigma}\mathbf{p}\begin{pmatrix} -n_1 + in_2 \\ n_3 + 1 \end{pmatrix} = \hat{\sigma}\mathbf{n}|\mathbf{p}|\begin{pmatrix} -n_1 + in_2 \\ n_3 + 1 \end{pmatrix} = |\mathbf{p}|\hat{\sigma}\mathbf{n}\begin{pmatrix} -n_1 + in_2 \\ n_3 + 1 \end{pmatrix} = |\mathbf{p}|\left(\sum_{k=1}^{3}\hat{\sigma}^k n_k\right)\begin{pmatrix} -n_1 + in_2 \\ n_3 + 1 \end{pmatrix}$$

$$= |\mathbf{p}|\begin{pmatrix} n_3 & n_1 - in_2 \\ n_1 + in_2 & -n_3 \end{pmatrix}\begin{pmatrix} -n_1 + in_2 \\ n_3 + 1 \end{pmatrix}$$

$$= |\mathbf{p}|\begin{pmatrix} n_3(-n_1 + in_2) + (n_1 - in_2)(n_3 + 1) \\ (n_1 + in_2)(-n_1 + in_2) - n_3(n_3 + 1) \end{pmatrix} = -|\mathbf{p}|\begin{pmatrix} -n_1 + in_2 \\ n_3 + 1 \end{pmatrix} \quad (43)$$

Hence, eqs. (40) and (41) can be written as

$$\psi_{\mathbf{p},\varepsilon,+1} = \frac{1}{\sqrt{2(1+n_3)}}\sqrt{\frac{\varepsilon E_p + m}{2\varepsilon E_p}}\begin{pmatrix} \begin{pmatrix} n_3 + 1 \\ n_1 + in_2 \end{pmatrix} \\ \dfrac{|\mathbf{p}|}{\varepsilon E_p + m}\begin{pmatrix} n_3 + 1 \\ n_1 + in_2 \end{pmatrix} \end{pmatrix} \exp[i(\mathbf{px} - \varepsilon E_p x^0)] \quad (44)$$

$$\psi_{\mathbf{p},\varepsilon,-1} = \frac{1}{\sqrt{2(1+n_3)}}\sqrt{\frac{\varepsilon E_p + m}{2\varepsilon E_p}}\begin{pmatrix} \begin{pmatrix} -n_1 + in_2 \\ n_3 + 1 \end{pmatrix} \\ \dfrac{-|\mathbf{p}|}{\varepsilon E_p + m}\begin{pmatrix} -n_1 + in_2 \\ n_3 + 1 \end{pmatrix} \end{pmatrix} \exp[i(\mathbf{px} - \varepsilon E_p x^0)] \quad (45)$$

For a free particle at rest the above equations get modified to

$$\begin{aligned}\psi_{0,+1,+1} &= (1\ 0\ 0\ 0)^T \exp(-iE_p x^0)\\ \psi_{0,+1,-1} &= (0\ 1\ 0\ 0)^T \exp(-iE_p x^0)\end{aligned} \tag{46}$$

for positive rest energies $E = \varepsilon E_p = E_p\sqrt{\mathbf{p}^2 + m^2} = m$ and

$$\begin{aligned}\psi_{0,-1,+1} &= (0\ 0\ 1\ 0)^T \exp(iE_p x^0)\\ \psi_{0,+1,-1} &= (0\ 0\ 0\ 1)^T \exp(iE_p x^0)\end{aligned} \tag{47}$$

for negative energies.

6.2 ORTHOGONALITY OF DIRAC SPINORS

We shall, henceforth, use the notation $u^{(r)}(\mathbf{p})$, $r = 1, 2, 3, 4$ to specify these spinors. To study various orthogonality relations among the Dirac spinors, we write the spinors in abbreviated form as follows:

Abbreviation	*Energy sign*	*Helicity*	*Expression*
$u^{(1)}(\mathbf{p})$	+	+1	$\frac{1}{\sqrt{2(1+n_3)}}\sqrt{\frac{E_p+m}{2E_p}}\begin{pmatrix}\begin{pmatrix}n_3+1\\ n_1+in_2\end{pmatrix}\\ \frac{\lvert\mathbf{p}\rvert}{E_p+m}\begin{pmatrix}n_3+1\\ n_1+in_2\end{pmatrix}\end{pmatrix}$
$u^{(2)}(\mathbf{p})$	+	−1	$\frac{1}{\sqrt{2(1+n_3)}}\sqrt{\frac{E_p+m}{2E_p}}\begin{pmatrix}\begin{pmatrix}-n_1+in_2\\ n_3+1\end{pmatrix}\\ \frac{-\lvert\mathbf{p}\rvert}{E_p+m}\begin{pmatrix}-n_1+in_2\\ n_3+1\end{pmatrix}\end{pmatrix}$
$u^{(3)}(\mathbf{p})$	−	+1	$\frac{1}{\sqrt{2(1+n_3)}}\sqrt{\frac{E_p-m}{2E_p}}\begin{pmatrix}\begin{pmatrix}n_3+1\\ n_1+in_2\end{pmatrix}\\ \frac{-\lvert\mathbf{p}\rvert}{E_p-m}\begin{pmatrix}n_3+1\\ n_1+in_2\end{pmatrix}\end{pmatrix}$
$u^{(4)}(\mathbf{p})$	−	−1	$\frac{1}{\sqrt{2(1+n_3)}}\sqrt{\frac{E_p-m}{2E_p}}\begin{pmatrix}\begin{pmatrix}-n_1+in_2\\ n_3+1\end{pmatrix}\\ \frac{\lvert\mathbf{p}\rvert}{E_p-m}\begin{pmatrix}-n_1+in_2\\ n_3+1\end{pmatrix}\end{pmatrix}$

(48)

The spinors can be collectively written as

$$u_{\varepsilon,s}(\mathbf{p}) = \frac{1}{\sqrt{2(1+n_3)}}\sqrt{\frac{\varepsilon E_p + m}{2\varepsilon E_p}}\begin{pmatrix} k\begin{pmatrix} n_3 + s \\ n_1 + in_2 \end{pmatrix} \\ \frac{s|\mathbf{p}|}{\varepsilon E_p + m}k\begin{pmatrix} n_3 + s \\ n_1 + in_2 \end{pmatrix} \end{pmatrix} \tag{49}$$

where

$\varepsilon = +1$ for positive energy

$\varepsilon = -1$ for positive energy

$k = +1$ for $s = +1$

$$k = \frac{n_3 + 1}{n_1 + in_2} \text{ for } s = -1 \tag{50}$$

Then, we have the following orthogonality relations:

$$u^{\dagger}_{\varepsilon,s}(\mathbf{p})u_{\varepsilon,s}(\mathbf{p}) = \frac{1}{2(1+n_3)}\frac{\varepsilon E_p + m}{2\varepsilon E_p}\begin{pmatrix} k\begin{pmatrix} n_3 + s \\ n_1 + in_2 \end{pmatrix} \\ \frac{s|\mathbf{p}|}{\varepsilon E_p + m}k\begin{pmatrix} n_3 + s \\ n_1 + in_2 \end{pmatrix} \end{pmatrix}^{\dagger}\begin{pmatrix} k\begin{pmatrix} n_3 + s \\ n_1 + in_2 \end{pmatrix} \\ \frac{s|\mathbf{p}|}{\varepsilon E_p + m}k\begin{pmatrix} n_3 + s \\ n_1 + in_2 \end{pmatrix} \end{pmatrix}$$

$$= \frac{1}{2(1+n_3)}\frac{\varepsilon E_p + m}{2\varepsilon E_p}k^*\left((n_3+s),(n_1 - in_2),\frac{s|\mathbf{p}|}{\varepsilon E_p + m}(n_3+s),\frac{s|\mathbf{p}|}{\varepsilon E_p + m}(n_1 - in_2)\right)$$

$$\begin{pmatrix} k\begin{pmatrix} n_3 + s \\ n_1 + in_2 \end{pmatrix} \\ \frac{s|\mathbf{p}|}{\varepsilon E_p + m}k\begin{pmatrix} n_3 + s \\ n_1 + in_2 \end{pmatrix} \end{pmatrix}$$

$$= \frac{1}{2(1+n_3)}\frac{\varepsilon E_p + m}{2\varepsilon E_p}k^*k\left((n_3+s)^2 + (n_1^2 + n_2^2) + \left(\frac{s|p|}{\varepsilon E_p + m}\right)^2\left[(n_3+s)^2 + (n_1^2 + n_2^2)\right]\right)$$

$$= \frac{1}{(1+n_3)}\frac{\varepsilon E_p + m}{2\varepsilon E_p}k^*k\,(1 + sn_3)\left[1 + \left(-\frac{s|\mathbf{p}|}{\varepsilon E_p + m}\right)^2\right]$$

$$= \frac{1}{(1+n_3)} \cdot \frac{1}{2\varepsilon E_p} \; k^* k \,(1 + s n_3) \left(\frac{E_p^2 + 2\varepsilon E_p m + m^2 + |\mathbf{p}|^2}{\varepsilon E_p + m} \right)$$

$$= \frac{1}{(1+n_3)} \cdot \frac{1}{2\varepsilon E_p} \; k^* k \,(1 + s n_3) \left(\frac{2E_p^2 + 2\varepsilon E_p m}{\varepsilon E_p + m} \right)$$

$$= \frac{1}{(1+n_3)} \; k^* k \,(1 + s n_3) = 1 \tag{51}$$

where the last step follows from $k = +1$ for $s = +1$ and $k = \dfrac{n_3 + 1}{n_1 + i n_2}$ for $s = -1$ so that

$$k^* k = \frac{[(n_3 + 1)(n_1 + i n_2)][(n_3 + 1)(n_1 - i n_2)]}{(n_1^2 + n_2^2)^2}$$

$$= \frac{(n_3 + 1)^2}{(n_1^2 + n_2^2)} = \frac{(n_3 + 1)^2}{(1 - n_3^2)} = \frac{(n_3 + 1)}{(1 - n_3)} \tag{52}$$

$$u^\dagger_{\varepsilon', s'}(\mathbf{p}) u_{\varepsilon, s}(\mathbf{p}) = \frac{1}{2(1+n_3)} \left(\frac{\varepsilon' E_p + m}{2\varepsilon' E_p} \right)^{\frac{1}{2}} \left(\frac{\varepsilon E_p + m}{2\varepsilon E_p} \right)^{\frac{1}{2}}$$

$$\begin{pmatrix} k' \begin{pmatrix} n_3 + s' \\ n_1 + i n_2 \end{pmatrix} \\ \dfrac{s'|\mathbf{p}|}{\varepsilon' E_p + m} k' \begin{pmatrix} n_3 + s' \\ n_1 + i n_2 \end{pmatrix} \end{pmatrix}^\dagger \begin{pmatrix} k \begin{pmatrix} n_3 + s \\ n_1 + i n_2 \end{pmatrix} \\ \dfrac{s|\mathbf{p}|}{\varepsilon E_p + m} k \begin{pmatrix} n_3 + s \\ n_1 + i n_2 \end{pmatrix} \end{pmatrix}$$

$$= \frac{1}{2(1+n_3)} \left(\frac{\varepsilon' E_p + m}{2\varepsilon' E_p} \right)^{\frac{1}{2}} \left(\frac{\varepsilon E_p + m}{2\varepsilon E_p} \right)^{\frac{1}{2}} k^* k$$

$$\left((n_3 + s'), (n_1 - i n_2), \frac{s'|\mathbf{p}|}{\varepsilon' E_p + m} (n_3 + s'), \frac{s'|\mathbf{p}|}{\varepsilon' E_p + m} (n_1 - i n_2) \right)$$

$$\begin{pmatrix} \begin{pmatrix} n_3 + s \\ n_1 + i n_2 \end{pmatrix} \\ \dfrac{s|\mathbf{p}|}{\varepsilon E_p + m} \begin{pmatrix} n_3 + s \\ n_1 + i n_2 \end{pmatrix} \end{pmatrix}$$

$$= \frac{1}{2(1+n_3)} \left(\frac{\varepsilon' E_p + m}{2\varepsilon' E_p} \right)^{\frac{1}{2}} \left(\frac{\varepsilon E_p + m}{2\varepsilon E_p} \right)^{\frac{1}{2}} k'^* k$$

$$\left\{ (n_3 + s')(n_3 + s) + (n_1^2 + n_2^2) + \left(\frac{s' |\mathbf{p}|}{\varepsilon' E_p + m} \right) \right.$$

$$\left. \left(\frac{s|\mathbf{p}|}{\varepsilon E_p + m} \right) [(n_3 + s')(n_3 + s) + (n_1^2 + n_2^2)] \right\}$$

$$= \frac{1}{2(1+n_3)} \left(\frac{\varepsilon' E_p + m}{2\varepsilon' E_p} \right)^{\frac{1}{2}} \left(\frac{\varepsilon E_p + m}{2\varepsilon E_p} \right)^{\frac{1}{2}} k'^* k$$

$$[1 + (s + s')n_3 + ss'] \left[1 + \left(\frac{s'|\mathbf{p}|}{\varepsilon' E_p + m} \right) \left(\frac{s|\mathbf{p}|}{\varepsilon E_p + m} \right) \right]$$

$$= \frac{1}{2(1+n_3)} \left(\frac{\varepsilon' E_p + m}{2\varepsilon' E_p} \right)^{\frac{1}{2}} \left(\frac{\varepsilon E_p + m}{2\varepsilon E_p} \right)^{\frac{1}{2}} k'^* k$$

$$[1 + (s + s')n_3 + ss'] \left(\frac{\varepsilon' \varepsilon E_p^2 + (\varepsilon' + \varepsilon) E_p m + m^2 + s' s |\mathbf{p}|^2}{(\varepsilon' E_p + m)(\varepsilon E_p + m)} \right)$$

$$= \frac{1}{2(1+n_3)} \frac{1}{2E_p} k'^* k \, [1 + (s + s')n_3 + ss']$$

$$\left[\frac{(\varepsilon' \varepsilon + 1) E_p^2 + (\varepsilon' + \varepsilon) E_p m + (s' s - 1) p^2}{\sqrt{(E_p + \varepsilon' m)(E_p + \varepsilon m)}} \right] = \delta_{s' s} \delta_{\varepsilon' \varepsilon} \tag{53}$$

where the last step follows from the fact that $\varepsilon = \pm 1$ so that if $\varepsilon' \neq \varepsilon$ then $\varepsilon' + \varepsilon = 0$, $\varepsilon' \varepsilon = -1$ and similarly $s' + s = 0$, $s' s = -1$ if $s' \neq s$.

$$u^\dagger_{\varepsilon', s'}(-\mathbf{p}) \, u_{\varepsilon, s}(\mathbf{p}) = \frac{1}{2(1+n_3)} \left(\frac{\varepsilon' E_p + m}{2\varepsilon' E_p} \right)^{\frac{1}{2}} \left(\frac{\varepsilon E_p + m}{2\varepsilon E_p} \right)^{\frac{1}{2}}$$

$$\begin{pmatrix} k'\begin{pmatrix} n_3+s' \\ n_1+in_2 \end{pmatrix} \\ \dfrac{-s'|\mathbf{p}|}{\varepsilon' E_p+m}k'\begin{pmatrix} n_3+s' \\ n_1+in_2 \end{pmatrix} \end{pmatrix}^{\dagger} \begin{pmatrix} k\begin{pmatrix} n_3+s \\ n_1+in_2 \end{pmatrix} \\ \dfrac{s|\mathbf{p}|}{\varepsilon E_p+m}k\begin{pmatrix} n_3+s \\ n_1+in_2 \end{pmatrix} \end{pmatrix}$$

$$= \frac{1}{2(1+n_3)}\left(\frac{\varepsilon' E_p+m}{2\varepsilon' E_p}\right)^{\frac{1}{2}}\left(\frac{\varepsilon E_p+m}{2\varepsilon E_p}\right)^{\frac{1}{2}} k'^{*}k$$

$$\left((n_3+s'),(n_1-in_2),\frac{-s'|\mathbf{p}|}{\varepsilon' E_p+m}(n_3+s'),\frac{-s'|p|}{\varepsilon' E_p+m}(n_1-in_2)\right)$$

$$\begin{pmatrix} \begin{pmatrix} n_3+s \\ n_1+in_2 \end{pmatrix} \\ \dfrac{s|\mathbf{p}|}{\varepsilon E_p+m}\begin{pmatrix} n_3+s \\ n_1+in_2 \end{pmatrix} \end{pmatrix}$$

$$= \frac{1}{2(1+n_3)}\left(\frac{\varepsilon' E_p+m}{2\varepsilon' E_p}\right)^{\frac{1}{2}}\left(\frac{\varepsilon E_p+m}{2\varepsilon E_p}\right)^{\frac{1}{2}} k'^{*}k$$

$$\left\{(n_3+s')(n_3+s)(n_1^2+n_2^2)-\left(\frac{s'|\mathbf{p}|}{\varepsilon' E_p+m}\right)\left(\frac{s|\mathbf{p}|}{\varepsilon E_p+m}\right)[(n_3+s')(n_3+s)+(n_1^2+n_2^2)]\right\}$$

$$= \frac{1}{2(1+n_3)}\left(\frac{\varepsilon' E_p+m}{2\varepsilon' E_p}\right)^{\frac{1}{2}}\left(\frac{\varepsilon E_p+m}{2\varepsilon E_p}\right)^{\frac{1}{2}} k'^{*}k$$

$$[1+(s+s')\,n_3+ss']\left[1-\left(\frac{s'|\mathbf{p}|}{\varepsilon' E_p+m}\right)\left(\frac{s|\mathbf{p}|}{\varepsilon E_p+m}\right)\right]$$

$$= \frac{1}{2(1+n_3)}\left(\frac{\varepsilon' E_p+m}{2\varepsilon' E_p}\right)^{\frac{1}{2}}\left(\frac{\varepsilon E_p+m}{2\varepsilon E_p}\right)^{\frac{1}{2}} k'^{*}k$$

$$[1+(s+s')n_3+ss']\left[\frac{\varepsilon'\varepsilon E_p^2+(\varepsilon'+\varepsilon)E_p\,m+m^2-s's|\mathbf{p}|^2}{(\varepsilon' E_p+m)(\varepsilon E_p+m)}\right]$$

$$= \delta_{s's}\left[\left(\delta_{\varepsilon',\varepsilon}\frac{m}{\varepsilon E_p}\right) \pm \left(\delta_{\varepsilon',-\varepsilon}\frac{\mathbf{p}}{E_p}\right)\right] \tag{54}$$

We have, so far, evaluated the orthogonality relations between $u_{\varepsilon,\, s}(\mathbf{p})$ and $u^{\dagger}_{\varepsilon,\, s}(\mathbf{p})$. We now examine the orthogonality properties of $u_{\varepsilon,\, s}(\mathbf{p})$ and $\bar{u}_{\varepsilon,\, s}(\mathbf{p})$. We have

$$\bar{u}_{\varepsilon,\, s}(\mathbf{p}) = u^{\dagger}_{\varepsilon,\, s}(\mathbf{p})\gamma^0 = \frac{1}{\sqrt{2(1+n_3)}}\sqrt{\frac{\varepsilon E_p + m}{2\varepsilon E_p}}$$

$$\begin{pmatrix} k\begin{pmatrix} n_3 + s \\ n_1 + in_2 \end{pmatrix} \\ \dfrac{s|\mathbf{p}|}{\varepsilon E_p + m}k\begin{pmatrix} n_3 + s \\ n_1 + in_2 \end{pmatrix} \end{pmatrix}^{\dagger} \begin{pmatrix} 1 & 0 & 0 & 0 \\ 0 & 1 & 0 & 0 \\ 0 & 0 & -1 & 0 \\ 0 & 0 & 0 & -1 \end{pmatrix}$$

$$= \frac{1}{\sqrt{2(1+n_3)}}\sqrt{\frac{\varepsilon E_p + m}{2\varepsilon E_p}} \begin{pmatrix} k^*\begin{pmatrix} n_3 + s \\ n_1 - in_2 \end{pmatrix} \\ \dfrac{-s|\mathbf{p}|}{\varepsilon E_p + m}k^*\begin{pmatrix} n_3 + s \\ n_1 - in_2 \end{pmatrix} \end{pmatrix}^{T} \tag{55}$$

so that

$$\bar{u}_{\varepsilon,\, s}(\mathbf{p})u_{\varepsilon,\, s}(\mathbf{p}) = \frac{1}{2(1+n_3)}\frac{\varepsilon E_p + m}{2\varepsilon E_p}$$

$$\begin{pmatrix} k^*\begin{pmatrix} n_3 + s \\ n_1 - in_2 \end{pmatrix} \\ \dfrac{-s|\mathbf{p}|}{\varepsilon E_p + m}k^*\begin{pmatrix} n_3 + s \\ n_1 - in_2 \end{pmatrix} \end{pmatrix}^{T} \begin{pmatrix} k\begin{pmatrix} n_3 + s \\ n_1 + in_2 \end{pmatrix} \\ \dfrac{s|\mathbf{p}|}{\varepsilon E_p + m}k\begin{pmatrix} n_3 + s \\ n_1 + in_2 \end{pmatrix} \end{pmatrix}$$

$$= \frac{1}{2(1+n_3)}\frac{\varepsilon E_p + m}{2\varepsilon E_p}k^*k\left((n_3 + s), (n_1 - in_2), \frac{-s|\mathbf{p}|}{\varepsilon E_p + m}\right.$$

$$\left.(n_3 + s), \frac{-s|\mathbf{p}|}{\varepsilon E_p + m}(n_1 - in_2)\right) \begin{pmatrix} \begin{pmatrix} n_3 + s \\ n_1 + in_2 \end{pmatrix} \\ \dfrac{s|\mathbf{p}|}{\varepsilon E_p + m}\begin{pmatrix} n_3 + s \\ n_1 + in_2 \end{pmatrix} \end{pmatrix}$$

$$= \frac{1}{2(1+n_3)} \frac{\varepsilon E_p + m}{2\varepsilon E_p} k^* k \Big\{(n_3+s)^2 + (n_1^2+n_2^2)$$

$$- \left(\frac{s|\mathbf{p}|}{\varepsilon E_p + m}\right)^2 [(n_3+s)^2 + (n_1^2+n_2^2)]\Big\}$$

$$= \frac{1}{(1+n_3)} \frac{\varepsilon E_p + m}{2\varepsilon E_p} k^* k\,(1+sn_3) \left[1 - \left(\frac{s|\mathbf{p}|}{\varepsilon E_p + m}\right)^2\right]$$

$$= \frac{1}{(1+n_3)} \cdot \frac{1}{2\varepsilon E_p} k^* k\,(1+sn_3) \left(\frac{E_p^2 + 2\varepsilon E_p m + m^2 - s^2|\mathbf{p}|^2}{\varepsilon E_p + m}\right)$$

$$= \frac{1}{(1+n_3)} \cdot \frac{1}{2\varepsilon E_p} k^* k\,(1+sn_3) \left(\frac{2\varepsilon E_p m + 2m^2}{\varepsilon E_p + m}\right)$$

$$= \frac{1}{(1+n_3)} \frac{m}{\varepsilon E_p} k^* k\,(1+sn_3) = \frac{m}{\varepsilon E_p} \tag{56}$$

$$\bar{u}_{\varepsilon', s'}(\mathbf{p})\, u_{\varepsilon, s}(\mathbf{p}) = \frac{1}{2(1+n_3)} \left(\frac{\varepsilon' E_p + m}{2\varepsilon' E_p}\right)^{\frac{1}{2}} \left(\frac{\varepsilon E_p + m}{2\varepsilon E_p}\right)^{\frac{1}{2}}$$

$$\begin{pmatrix} k' \begin{pmatrix} n_3 + s' \\ n_1 - in_2 \end{pmatrix} \\ \dfrac{-s'|\mathbf{p}|}{\varepsilon' E_p + m} k' \begin{pmatrix} n_3 + s' \\ n_1 + in_2 \end{pmatrix} \end{pmatrix}^\dagger \begin{pmatrix} k \begin{pmatrix} n_3 + s \\ n_1 + in_2 \end{pmatrix} \\ \dfrac{s|\mathbf{p}|}{\varepsilon E_p + m} k \begin{pmatrix} n_3 + s \\ n_1 + in_2 \end{pmatrix} \end{pmatrix}$$

$$= \frac{1}{2(1+n_3)} \left(\frac{\varepsilon' E_p + m}{2\varepsilon' E_p}\right)^{\frac{1}{2}} \left(\frac{\varepsilon E_p + m}{2\varepsilon E_p}\right)^{\frac{1}{2}} k'^* k$$

$$\left((n_3 + s'), (n_1 - in_2), \frac{-s'|\mathbf{p}|}{\varepsilon' E_p + m}(n_3 + s'), \frac{-s'|\mathbf{p}|}{\varepsilon' E_p + m}(n_1 - in_2)\right)$$

$$\begin{pmatrix} \begin{pmatrix} n_3 + s \\ n_1 + in_2 \end{pmatrix} \\ \dfrac{s|\mathbf{p}|}{\varepsilon E_p + m} k \begin{pmatrix} n_3 + s \\ n_1 + in_2 \end{pmatrix} \end{pmatrix}$$

$$= \frac{1}{2(1+n_3)} \left(\frac{\varepsilon' E_p + m}{2\varepsilon' E_p}\right)^{\frac{1}{2}} \left(\frac{\varepsilon E_p + m}{2\varepsilon E_p}\right)^{\frac{1}{2}} k'^* k$$

$$\begin{Bmatrix} (n_3 + s) + (n_1^2 + n_2^2) - \left(\dfrac{s' |\mathbf{p}|}{\varepsilon' E_p + m}\right)\left(\dfrac{s |\mathbf{p}|}{\varepsilon' E_p + m}\right) \\ [(n_3 + s')(n_3 + s) + (n_1^2 + n_2^2)] \end{Bmatrix}$$

$$= \frac{1}{2(1+n_3)} \left(\frac{\varepsilon' E_p + m}{2\varepsilon' E_p}\right)^{\frac{1}{2}} \left(\frac{\varepsilon E_p + m}{2\varepsilon E_p}\right)^{\frac{1}{2}} k'^* k$$

$$[1 + (s + s')n_3 + ss']\left[1 - \left(\frac{s'|\mathbf{p}|}{\varepsilon' E_p + m}\right)\left(\frac{s|\mathbf{p}|}{\varepsilon E_p + m}\right)\right]$$

$$= \frac{1}{2(1+n_3)} \left(\frac{\varepsilon' E_p + m}{2\varepsilon' E_p}\right)^{\frac{1}{2}} \left(\frac{\varepsilon E_p + m}{2\varepsilon E_p}\right)^{\frac{1}{2}} k'^* k\, [1 + (s + s')n_3 + ss']$$

$$\left[\frac{\varepsilon'\varepsilon E_p^2 + (\varepsilon' + \varepsilon) E_p m + m^2 - s's|\mathbf{p}|^2}{(\varepsilon' E_p + m)(\varepsilon E_p + m)}\right]$$

$$= \delta_{s's}\left[\left(\delta_{\varepsilon'\varepsilon}\frac{m}{\varepsilon E_p}\right) \pm \left(\delta_{\varepsilon',-\varepsilon}\frac{\mathbf{p}}{E_p}\right)\right] \tag{57}$$

$$\bar{u}_{\varepsilon',s'}(-\mathbf{p})u_{\varepsilon,s}(\mathbf{p}) = \frac{1}{2(1+n_3)} \left(\frac{\varepsilon' E_p + m}{2\varepsilon' E_p}\right)^{\frac{1}{2}} \left(\frac{\varepsilon E_p + m}{2\varepsilon E_p}\right)^{\frac{1}{2}}$$

$$\begin{pmatrix} k'\begin{pmatrix} n_3 + s' \\ n_1 + in_2 \end{pmatrix} \\ \dfrac{s'|\mathbf{p}|}{\varepsilon' E_p + m} k' \begin{pmatrix} n_3 + s' \\ n_1 + in_2 \end{pmatrix} \end{pmatrix}^{\dagger} \begin{pmatrix} k\begin{pmatrix} n_3 + s \\ n_1 + in_2 \end{pmatrix} \\ \dfrac{s|\mathbf{p}|}{\varepsilon E_p + m} k \begin{pmatrix} n_3 + s \\ n_1 + in_2 \end{pmatrix} \end{pmatrix}$$

$$= \frac{1}{2(1+n_3)} \left(\frac{\varepsilon' E_p + m}{2\varepsilon' E_p}\right)^{\frac{1}{2}} \left(\frac{\varepsilon E_p + m}{2\varepsilon E_p}\right)^{\frac{1}{2}} k'^* k$$

$$\left((n_3+s'),(n_1-in_2),\frac{s'|\mathbf{p}|}{\varepsilon' E_p+m}(n_3+s'),\frac{s'|\mathbf{p}|}{\varepsilon' E_p+m}(n_1-in_2)\right)$$

$$\begin{pmatrix} \begin{pmatrix} n_3+s \\ n_1+in_2 \end{pmatrix} \\ \frac{s|\mathbf{p}|}{\varepsilon E_p+m}\begin{pmatrix} n_3+s \\ n_1+in_2 \end{pmatrix} \end{pmatrix}$$

$$=\frac{1}{2(1+n_3)}\left(\frac{\varepsilon' E_p+m}{2\varepsilon' E_p}\right)^{\frac{1}{2}}\left(\frac{\varepsilon E_p+m}{2\varepsilon E_p}\right)^{\frac{1}{2}} k'^{*}k\,\{(n_3+s')$$

$$(n_3+s)+(n_1^2+n_2^2)+\left(\frac{s'|\mathbf{p}|}{\varepsilon' E_p+m}\right)\left(\frac{s|\mathbf{p}|}{\varepsilon E_p+m}\right)$$

$$[(n_3+s')(n_3+s)+(n_1^2+n_2^2)]\}$$

$$=\frac{1}{2(1+n_3)}\left(\frac{\varepsilon' E_p+m}{2\varepsilon' E_p}\right)^{\frac{1}{2}}\left(\frac{\varepsilon E_p+m}{2\varepsilon E_p}\right)^{\frac{1}{2}} k'^{*}k\,[1+(s+s'))(n_3+ss')]$$

$$\left[1+\left(\frac{s'|\mathbf{p}|}{\varepsilon' E_p+m}\right)\left(\frac{s|\mathbf{p}|}{\varepsilon E_p+m}\right)\right]$$

$$=\frac{1}{2(1+n_3)}\left(\frac{\varepsilon' E_p+m}{2\varepsilon' E_p}\right)^{\frac{1}{2}}\left(\frac{\varepsilon E_p+m}{2\varepsilon E_p}\right)^{\frac{1}{2}} k'^{*}k\,[1+(s+s')\,n_3+ss']$$

$$\left[\frac{\varepsilon'\varepsilon E_p^2+(\varepsilon'+\varepsilon)E_p m+m^2+s's|\mathbf{p}|^2}{(\varepsilon' E_p+m)(\varepsilon E_p+m)}\right]$$

$$=\frac{1}{2(1+n_3)}\,\frac{1}{2E_p}\,k'^{*}k\,[1+(s+s')n_3+ss']$$

$$\left[\frac{(\varepsilon'\varepsilon+1)E_p^2+(\varepsilon'+\varepsilon)E_p m+(s's-1)|p|^2}{\sqrt{(E_p+\varepsilon' m)(E_p+\varepsilon m)}}\right]=\delta_{s's}\,\delta_{\varepsilon'\varepsilon} \tag{58}$$

The normalization of the adjoint spinor $\bar{u}^{(r)}(\mathbf{p})$ can also be obtained directly from the Dirac equation as follows:

The spinor $u^{(r)}(\mathbf{p})$ is a solution of the equation:

$$0 = (\gamma^{\mu} p_{\mu} - m)\, u^{(r)}(\mathbf{p}) = \left(\gamma^0 p_0 + \sum_{k=1}^{3} \gamma^k p_k - m\right) u^{(r)}(\mathbf{p}) \tag{59}$$

Taking the Hermitian conjugate of eq. (59)

$$0 = u^{(r)\dagger}(\mathbf{p})\, (\gamma^{\mu\dagger} p_{\mu} - m) = u^{(r)\dagger}(\mathbf{p}) \left(\gamma^0 p_0 + \sum_{k=1}^{3} \gamma^{k^\dagger} p_k - m\right)$$

$$= u^{(r)\dagger}(\mathbf{p}) \left(\gamma^0 p_0 - \sum_{k=1}^{3} \gamma^k p_k - m\right) \tag{60}$$

Multiplying eq. (59) on the left by $u^{(r)\dagger}(\mathbf{p})$ and eq. (60) on the right by $u^{(r)}(\mathbf{p})$, and adding the resulting equations, we get

$$0 = \bar{u}^{(r)}(\mathbf{p})\, p_0 u^{(r)}(\mathbf{p}) - m u^{(r)\dagger}(\mathbf{p}) u^{(r)}(\mathbf{p}) = p_0\, \bar{u}^{(r)}(\mathbf{p})\, u^{(r)}(\mathbf{p}) - m$$

or

$$\bar{u}^{(r)}(\mathbf{p}) u^{(r)}(\mathbf{p}) = \frac{m}{p_0} = \frac{m}{\varepsilon E_p} = \varepsilon \frac{m}{E_p} \tag{61}$$

It is, sometimes preferred to normalize

$$u^{(r)\dagger}(\mathbf{p})\, u^{(r)}(\mathbf{p}) = \varepsilon \frac{p_0}{m} = \frac{E_p}{m} \tag{62}$$

Abbreviation	Energy sign	Helicity	Expression
$u^{(1)}(\mathbf{p})$	+	+ 1	$\frac{1}{\sqrt{2(1+n_3)}} \sqrt{\frac{E_p + m}{2m}} \begin{pmatrix} \begin{pmatrix} n_3 + 1 \\ n_1 + i n_2 \end{pmatrix} \\ \frac{\lvert\mathbf{p}\rvert}{E_p + m} \begin{pmatrix} n_3 + 1 \\ n_1 + i n_2 \end{pmatrix} \end{pmatrix}$
$u^{(2)}(\mathbf{p})$	+	− 1	$\frac{1}{\sqrt{2(1+n_3)}} \sqrt{\frac{E_p + m}{2m}} \begin{pmatrix} \begin{pmatrix} -n_1 + i n_2 \\ n_3 + 1 \end{pmatrix} \\ \frac{-\lvert\mathbf{p}\rvert}{E_p + m} \begin{pmatrix} -n_1 + i n_2 \\ n_3 + 1 \end{pmatrix} \end{pmatrix}$

Contd...

$u^{(3)}(\mathbf{p})$	–	$+1$	$\frac{1}{\sqrt{2(1+n_3)}}\sqrt{\frac{E_p-m}{2m}}\begin{pmatrix}\begin{pmatrix}n_3+1\\ n_1+in_2\end{pmatrix}\\ \frac{-\lvert\mathbf{p}\rvert}{E_p-m}\begin{pmatrix}n_3+1\\ n_1+in_2\end{pmatrix}\end{pmatrix}$
$u^{(4)}(\mathbf{p})$	–	-1	$\frac{1}{\sqrt{2(1+n_3)}}\sqrt{\frac{E_p-m}{2m}}\begin{pmatrix}\begin{pmatrix}-n_1+in_2\\ n_3+1\end{pmatrix}\\ \frac{\lvert\mathbf{p}\rvert}{E_p-m}\begin{pmatrix}-n_1+in_2\\ n_3+1\end{pmatrix}\end{pmatrix}$ (63)

which has the advantage that both sides transform as the fourth component of a 4-vector under a Lorentz transformation. Hence, this normalization is invariant under Lorentz transformation. In this case $\bar{u}^{(r)}(\mathbf{p})\, u^{(r)}(\mathbf{p})$ gets normalized to $\bar{u}^{(r)}(\mathbf{p})u^{(r)}(\mathbf{p}) = \varepsilon(p_0)$, where $\varepsilon(p_0)$ is the sign of the energy.

With the normalization (62), the modified expressions for the various spinors are given in the above table.

We have shown in eq. (58) that spinors with momentum $-\mathbf{p}$ and negative energy are orthogonal to their positive energy counterparts with momentum $\mathbf{p}$. This can also be shown more elegantly directly from the Dirac equation as follows:

The conjugate spinor $u^{(r)\dagger}(\mathbf{p})$ satisfies the Dirac eq. (60). Multiplying eq. (60) from the right by $\boldsymbol{\gamma}^0$, we get

$$0 = u^{(r)\dagger}(\mathbf{p})\left(\gamma^0 p_0 - \sum_{k=1}^{3}\gamma^k p_k - m\right)\gamma^0$$

$$= u^{(r)\dagger}(p)\left(\gamma^0 p_0 \gamma^0 - \sum_{k=1}^{3}\gamma^k \gamma^0 p_k - m\gamma^0\right)$$

$$= u^{(r)\dagger}(\mathbf{p})\left[(\gamma^0)^2 p_0 + \gamma^0\sum_{k=1}^{3}\gamma^k p_k - m\gamma^0\right]$$

$$= u^{(r)\dagger}(\mathbf{p})\boldsymbol{\gamma}^0\left[\gamma^0 p_0 + \sum_{k=1}^{3}\gamma^k p_k - m\right] = \bar{u}^{(r)}(\mathbf{p})\,(\boldsymbol{\gamma}^\mu p_\mu - m) \tag{64}$$

For momentum $-\mathbf{p}$ and negative energy, the spinor $u^{(r)}(\mathbf{p})$ will satisfy

$$(\boldsymbol{\gamma}^\mu p_\mu + m)\, u^{(r)}(-\mathbf{p}) = 0 \tag{65}$$

and the adjoint spinor with momentum $-\mathbf{p}$ and negative energy will satisfy

$$\bar{u}^{(r)}(-\mathbf{p})\,(\gamma^\mu p_\mu + m) = 0 \tag{66}$$

Multiplying eq. (59) on the left by $\bar{u}^{(r)}(-\mathbf{p})$, we get

$$\bar{u}^{(r)}(-\mathbf{p})\gamma^{\mu} p_{\mu} u^{(r)}(\mathbf{p}) = m\bar{u}^{(r)}(-\mathbf{p})u^{(r)}(\mathbf{p}) \tag{67}$$

Similarly, multiplying eq. (66) on the right by $u^{(r)}(\mathbf{p})$, we get

$$\bar{u}^{(r)}(-\mathbf{p})\gamma^{\mu} p_{\mu} u^{(r)}(\mathbf{p}) = -m\bar{u}^{(r)}(-\mathbf{p})\, u^{(r)}(\mathbf{p}) \tag{68}$$

From eqs. (67) and (68), we have

$$\bar{u}^{(r)}(-\mathbf{p})\, u^{(r)}(\mathbf{p}) = 0 \tag{69}$$

We, now proceed to compute the matrix elements of various Γ_i; $i = 1, 2, 3, \ldots, 16$ in a given quantum mechanical state represented by the spinor $\bar{u}^{(r)}(\mathbf{p})$. For example, we have, on multiplying eq. (59) by $\bar{u}^{(r)}(\mathbf{p})\gamma^{\mu}$ from the left

$$\bar{u}^{(r)}(\mathbf{p})\gamma^{\mu}(\gamma^{\nu} p_{\nu} - m)\, u^{(r)}(\mathbf{p}) = 0 \tag{70}$$

Similarly, multiplying eq. (64) on the right by $\gamma^{\mu} u^{(r)}(\mathbf{p})$, we get

$$\bar{u}^{(r)}(\mathbf{p})\, (\gamma^{\nu} p_{\nu} - m)\gamma^{\mu} u^{(r)}(\mathbf{p}) = 0 \tag{71}$$

Adding eqs. (70) and (71), we obtain

$$\begin{aligned}
&\bar{u}^{(r)}(\mathbf{p})\, (\gamma^{\mu}\gamma^{\nu} + \gamma^{\nu}\gamma^{\mu})p_{\nu} u^{(r)}(\mathbf{p}) - 2m\, \bar{u}^{(r)}(\mathbf{p})\gamma^{\mu} u^{(r)}(\mathbf{p}) = 0 \\
&= 2\bar{u}^{(r)}(\mathbf{p})g^{\mu\nu} p_{\nu} u^{(r)}(\mathbf{p}) - 2m\, \bar{u}^{(r)}(\mathbf{p})\, \gamma^{\mu} u^{(r)}(\mathbf{p}) \\
&= 2\bar{u}^{(r)}(\mathbf{p})\, p^{\mu} u^{(r)}(\mathbf{p}) - 2m\bar{u}^{(r)}(\mathbf{p})\, \gamma^{\mu} u^{(r)}(\mathbf{p}) \\
&= 2p^{\mu}\, \bar{u}^{(r)}(\mathbf{p})u^{(r)}(\mathbf{p}) - 2m\bar{u}^{(r)}(\mathbf{p})\gamma^{\mu} u^{(r)}(\mathbf{p})
\end{aligned}$$

so that

$$\bar{u}^{(r)}(\mathbf{p})\gamma^{\mu} u^{(r)}(\mathbf{p}) = \frac{1}{m}\, p^{\mu}\, \bar{u}^{(r)}(\mathbf{p})u^{(r)}(\mathbf{p}) = \frac{\varepsilon(p_0)}{m}\, p^{\mu} \tag{72}$$

in case the normalization (62) is adopted for the adjoint spinor.

Proceeding similarly, we have

$$\begin{aligned}
&\bar{u}^{(r)}(\mathbf{p})\, (\gamma^{5}\gamma^{\nu} + \gamma^{\nu}\gamma^{5})p_{\nu} u^{(r)}(\mathbf{p}) - 2m\bar{u}^{(r)}(\mathbf{p})\gamma^{5} u^{(r)}(\mathbf{p}) = 0 \\
&= -2m\bar{u}^{(r)}(\mathbf{p})\gamma^{5} u^{(r)}(\mathbf{p}) \quad \Rightarrow \bar{u}^{(r)}(\mathbf{p})\gamma^{5} u^{(r)}(\mathbf{p}) = 0
\end{aligned} \tag{73}$$

because $\{\gamma^{\mu}, \gamma^{5}\} = 0$.

If the initial and final states are different, then we have

$$\begin{aligned}
&\bar{u}^{(r)}(\mathbf{p}_2)\, (\gamma^{5}\gamma^{\nu} p_{1\nu} + \gamma^{\nu}\gamma^{5} p_{2\nu})\, u^{(r)}(\mathbf{p}_1) - 2m\bar{u}^{(r)}(\mathbf{p}_2)\gamma^{5} u^{(r)}(\mathbf{p}_1) = 0 \\
&= \bar{u}^{(r)}(\mathbf{p}_2)\, (\gamma^{5}\gamma^{\nu} p_{1\nu} - \gamma^{5}\gamma^{\nu} p_{2\nu})u^{(r)}(\mathbf{p}_1) - 2m\bar{u}^{(r)}(\mathbf{p}_2)\gamma^{5} u^{(r)}(\mathbf{p}_1) \\
&= \bar{u}^{(r)}(\mathbf{p}_2)\, (p_{1\nu} - p_{2\nu})\gamma^{5}\gamma^{\nu} u^{(r)}(\mathbf{p}_1) - 2m\bar{u}^{(r)}(\mathbf{p}_2)\gamma^{5} u^{(r)}(\mathbf{p}_1)
\end{aligned}$$

$$\Rightarrow \quad \bar{u}^{(r)}(\mathbf{p}_2)\gamma^{5} u^{(r)}(\mathbf{p}_1) = \frac{1}{2m}\, \bar{u}^{(r)}(\mathbf{p}_2)\, (p_{1\nu} - p_{2\nu})\gamma^{5}\gamma^{\nu} u^{(r)}(\mathbf{p}_1) \tag{74}$$

6.3 TRANSFORMATION OF SPINORS UNDER LORENTZ TRANSFORMATIONS

The Dirac equation for the free particle can be written as

$$(\hat{\mathbf{p}} - \varepsilon_r m)\psi(x) = 0 \tag{75}$$

which has the solution

$$\psi^{(r)}(x) = \omega^{(r)}(\mathbf{p})\exp(-i\varepsilon_r p_\mu x^\mu),\ r = 1, 2, 3, 4 \tag{76}$$

where

$$\varepsilon_r = \begin{cases} +1 \text{ for } r = 1, 2 \\ -1 \text{ for } r = 3, 4 \end{cases} \tag{77}$$

and
$$\omega^{(1)}(\mathbf{p}) = u^{(1)}(\mathbf{p}),\ \omega^{(2)}(\mathbf{p}) = u^{(2)}(\mathbf{p}),\ \omega^{(3)}(\mathbf{p}) = u^{(3)}(-\mathbf{p}) \text{ and } \omega^{(4)}(\mathbf{p}) = u^{(4)}(-\mathbf{p}) \tag{78}$$

In the rest frame of the Dirac particle, eq. (76) becomes

$$\psi^{(r)}(\mathbf{0}, x^0) = \omega^{(r)}(\mathbf{0})\exp(-i\varepsilon_r p_0 x^0),\ r = 1, 2, 3, 4 \tag{79}$$

with the spinors being represented by

Abbreviation	*Energy sign*	*Helicity*	*Expression*
$\omega^{(1)}(\mathbf{0})$	+	+1	$(1\,0\,0\,0)^T$
$\omega^{(2)}(\mathbf{0})$	+	–1	$(0\,1\,0\,0)^T$
$\omega^{(3)}(\mathbf{0})$	–	+1	$(0\,0\,1\,0)^T$
$\omega^{(4)}(\mathbf{0})$	–	–1	$(0\,0\,0\,1)^T$ (80)

$\omega^{(r)}(\mathbf{0})$ are eigenfunctions of the spin operator (in the orientation where the x^3 axis is chosen along the direction of the motion of the particle)

$$s^3(\mathbf{p}) = \frac{\Sigma^3\mathbf{p}}{|\mathbf{p}|} = \begin{pmatrix} \sigma^3 & \mathbf{0} \\ \mathbf{0} & \sigma^3 \end{pmatrix}\frac{\mathbf{p}}{|\mathbf{p}|} = \frac{1}{|\mathbf{p}|}\begin{pmatrix} \sigma^3 & \mathbf{0} \\ \mathbf{0} & \sigma^3 \end{pmatrix}\mathbf{p} \tag{81}$$

because, for example,

$$s^3(\mathbf{p})\,\omega^{(1)}(\mathbf{0}) = \frac{1}{|\mathbf{p}|}\begin{pmatrix} \sigma^3 & \mathbf{0} \\ \mathbf{0} & \sigma^3 \end{pmatrix}\mathbf{p}\omega^{(1)}(\mathbf{0}) = \frac{\mathbf{p}}{|\mathbf{p}|} \times$$
$$diag\,(1, -1, 1, -1)\,(1\ 0\ 0\ 0)^T = \pm\,(1\ 0\ 0\ 0)^T \tag{82}$$

etc.

In order to construct the spinors corresponding to the wavefunction of a Dirac particle moving with a velocity **v** with respect to its rest frame, we compute the spinors in a frame which is moving with a velocity – **v** with respect to the rest frame. Considering in the first instance, motion parallel to the direction of the x^1 axis, the spinor transformation operator, as obtained in Chapter 5, is given by :

$$\hat{S} = \exp\left[-\frac{1}{2}i\omega\hat{\sigma}_{01}\right] = \exp\left\{-\frac{1}{2}i\omega\left[\frac{i}{2}(\gamma_0\gamma_1 - \gamma_1\gamma_0)\right]\right\} = \exp\left(\frac{1}{2}\omega\gamma_0\gamma_1\right)$$

$$= \exp\left(-\frac{1}{2}\omega\gamma^0\gamma^1\right) = \exp\left(-\frac{1}{2}\omega\alpha^1\right)$$

$$= \mathbf{I}\left[1 + \left(-\frac{1}{2}\omega\right)(\alpha^1) + \frac{1}{2!}\left(-\frac{1}{2}\omega\right)^2 (\alpha^1)^2 + \ldots\right]$$

$$= \mathbf{I}\left[1 + \frac{1}{2!}\left(\frac{1}{2}\omega\right)^2 + \ldots\right] - (\boldsymbol{\alpha})^1\left[\left(\frac{1}{2}\omega\right) + \frac{1}{3!}\left(\frac{1}{2}\omega\right)^3 + \ldots\right]$$

$$= \mathbf{I}\cosh\left(\frac{\omega}{2}\right) - (\boldsymbol{\alpha}^1)\sinh\left(\frac{\omega}{2}\right) \qquad (83)$$

where $\left(\sigma_{\mu\nu} = \frac{i}{2}[\gamma_\mu, \gamma_\nu]\right)$.

Since, the moving frame is moving with a velocity $-v_1$ with respect to the rest frame we have

$$\omega = \tanh^{-1}\left(-\frac{v_1}{c}\right) = \tanh^{-1}\left\{-\left[\frac{mv_1 c\left(1 - \frac{v_1^2}{c^2}\right)^{-\frac{1}{2}}}{mc^2\left(1 - \frac{v_1^2}{c^2}\right)^{-\frac{1}{2}}}\right]\right\} = \tanh^{-1}\left(-\frac{p_1 c}{E}\right) \qquad (84)$$

so that

$$\tanh\left(\frac{\omega}{2}\right) = \frac{\left(\frac{1+\tanh\omega}{1-\tanh\omega}\right)^{\frac{1}{2}} - 1}{\left(\frac{1+\tanh\omega}{1-\tanh\omega}\right)^{\frac{1}{2}} + 1} = \frac{\left(1 - \frac{p_1 c}{E}\right)^{\frac{1}{2}}\left(1 + \frac{p_1 c}{E}\right)^{-\frac{1}{2}} - 1}{\left(1 - \frac{p_1 c}{E}\right)^{\frac{1}{2}}\left(1 + \frac{p_1 c}{E}\right)^{-\frac{1}{2}} + 1} = \frac{\left(\frac{E - p_1c}{E + p_1c}\right)^{\frac{1}{2}} - 1}{\left(\frac{E - p_1c}{E + p_1c}\right)^{\frac{1}{2}} + 1}$$

$$= \frac{\left(\frac{E - p_1c}{mc^2}\right) - 1}{\left(\frac{E - p_1c}{mc^2}\right) + 1} = \frac{E - p_1c - mc^2}{E - p_1c + mc^2} = \frac{E - p_1c - mc^2}{E - p_1c + mc^2}\,\frac{E + p_1c + mc^2}{E + p_1c + mc^2}$$

$$= \frac{E^2 - (p_1c + mc^2)^2}{(E + mc^2)^2 - (p_1c)^2} = -\frac{p_1c}{(E + mc^2)} \tag{85}$$

Also $\cosh\left(\frac{\omega}{2}\right) = \frac{1}{\left[1 - \tan^2\frac{\omega}{2}\right]^{\frac{1}{2}}} = \frac{1}{\left[1 - \left(\frac{E - p_1c - mc^2}{E - p_1c + mc^2}\right)^2\right]^{\frac{1}{2}}}$

$$= \frac{1}{\left[\frac{4mc^2(E - p_1c)}{E^2 + p_1^2c^2 + m^2c^4 - 2Ep_1c + 2mc^2(E - p_1c)}\right]^{\frac{1}{2}}}$$

$$= \frac{1}{\left[\frac{4mc^2(E - p_1c)}{2E^2 - 2Ep_1c + 2mc^2(E - p_1c)}\right]^{\frac{1}{2}}} = \sqrt{\frac{E + mc^2}{2mc^2}} \tag{86}$$

which, in natural units is

$$\cosh\left(\frac{\omega}{2}\right) = \sqrt{\frac{E + m}{2m}} \tag{87}$$

Further,

$$\sinh\left(\frac{\omega}{2}\right) = \tanh\left(\frac{\omega}{2}\right)\cosh\left(\frac{\omega}{2}\right) = -\frac{p_1c}{(E + mc^2)}\left[\frac{E + mc^2}{2mc^2}\right]^{\frac{1}{2}}$$

$$= -\frac{p_1c}{\sqrt{2mc^2(E + mc^2)}} = -\sqrt{\frac{E - mc^2}{2mc^2}} \tag{88}$$

and in natural units

$$\sinh\left(\frac{\omega}{2}\right) = -\sqrt{\frac{E - m}{2m}} \tag{89}$$

Substituting these values in eq. (83), we get

$$\hat{S}(p_3) = \frac{1}{\sqrt{2m}}\begin{pmatrix} (E+m)^{1/2} & 0 & 0 & (E-m)^{1/2} \\ 0 & (E+m)^{1/2} & (E-m)^{1/2} & 0 \\ 0 & (E-m)^{1/2} & (E+m)^{1/2} & 0 \\ (E-m)^{1/2} & 0 & 0 & (E+m)^{1/2} \end{pmatrix} \tag{90}$$

Hence,

$$\omega^r(\mathbf{p}) = \hat{S}\,\omega^r(\mathbf{0}) = \frac{1}{\sqrt{2m}}\begin{pmatrix} (E+m)^{1/2} & 0 & 0 & (E-m)^{1/2} \\ 0 & (E+m)^{1/2} & (E-m)^{1/2} & 0 \\ 0 & (E-m)^{1/2} & (E+m)^{1/2} & 0 \\ (E-m)^{1/2} & 0 & 0 & (E+m)^{1/2} \end{pmatrix}\omega^r(\mathbf{0}) \tag{91}$$

6.4 SPINOR TRANSFORMATIONS IN ARBITRARY DIRECTION

In the preceding part, we have considered the transformation of a spinor from the rest frame to a frame moving with a velocity $-\mathbf{v}_1$ in a direction parallel to the x^1 axis or equivalently, we have constructed a spinor corresponding to a wave particle moving with a velocity of $+\mathbf{v}_1$ in a direction parallel to the x^1 axis from a spinor in the rest frame. We shall now consider motion of the wave particle travelling with a velocity $\mathbf{v} \equiv (v_1\ v_2\ v_3)$ in an arbitrary direction represented by the unit vector $\mathbf{n} \equiv (n_1\ n_2\ n_3)$, i.e., $\mathbf{v} = |\mathbf{v}|\ \mathbf{n} = |\mathbf{v}|\ (\cos\alpha\ \cos\beta\ \cos\gamma)$. The generators of unit Lorentz transformations along the direction of the unit vector $\mathbf{n} \equiv (n_1\ n_2\ n_3)$ are given by

$$(I_n)^{\nu}_{\mu} = \begin{pmatrix} 0 & -\cos\alpha & -\cos\beta & -\cos\gamma \\ -\cos\alpha & 0 & 0 & 0 \\ -\cos\beta & 0 & 0 & 0 \\ -\cos\gamma & 0 & 0 & 0 \end{pmatrix} \tag{92}$$

and the infinitesimal Lorentz transformations along $\mathbf{n} \equiv (n_1\ n_2\ n_3)$ are

$$x'^{\nu} = a^{\nu}_{\mu}\, x^{\mu} = x^{\nu} + \Delta\omega^{\nu}_{\mu}\, x^{\mu} = x^{\nu} + \Delta\,\omega(I_{\mathbf{n}})^{\nu}_{\mu}\, x^{\mu} \tag{93}$$

The spinor transformation operator $\hat{S}$, as obtained in the previous chapter corresponding to the above Lorentz transformation is $\left(\sigma_{\mu\nu} = \frac{i}{2}[\gamma_{\mu}, \gamma_{\nu}]\right)$

$$\hat{S}(\mathrm{p}) = \exp\left[-\frac{i}{4}\omega\hat{\sigma}_{\mu\nu}(I_n)^{\mu\nu}\right] = \exp\left\{-\frac{i\omega}{4}[(\hat{\sigma}_{01}I^{01}_{\mathbf{n}} + \hat{\sigma}_{02}I^{02}_{\mathbf{n}} + \hat{\sigma}_{03}I^{03}_{\mathbf{n}} + (\hat{\sigma}_{10}I^{10}_{\mathbf{n}} + \hat{\sigma}_{20}I^{20}_{\mathbf{n}} + \hat{\sigma}_{30}I^{30}_{\mathbf{n}})]\right\}$$

because all the other elements of $I_{\mathbf{n}}$ vanish. Now, since $\sigma_{\mu\nu} = -\sigma_{\nu\mu}$ and $I^{10}_{\mathbf{n}} = g^{00}\,(I_{\mathbf{n}})^1_0 = (I_{\mathbf{n}})^1_0$ $(I_{\mathbf{n}})^0_1 = -\,g^{11}\,(I_{\mathbf{n}})^0_1 = -\,I^{01}_{\mathbf{n}}$, etc., we have

$$\hat{S}(\mathrm{p}) = \exp\left[-\frac{i\omega}{2}(\hat{\sigma}_{01}I^{01}_{\mathbf{n}} + \hat{\sigma}_{02}I^{02}_{\mathbf{n}} + \hat{\sigma}_{03}I^{03}_{\mathbf{n}})\right] \tag{94}$$

$$= \exp\left[-\frac{i\omega}{2}(\hat{\sigma}_{01}\cos\alpha + \hat{\sigma}_{02}\cos\beta + \hat{\sigma}_{03}\cos\gamma)\right] \tag{95}$$

where we have used $I_{\mathbf{n}}^{01} = g^{11}(I_{\mathbf{n}})_1^0 = -(I_{\mathbf{n}})_1^0 = \cos\alpha$, etc. This gives

$$\hat{S}(\text{p}) = \exp\left(-\frac{\omega}{2}\alpha.\mathbf{n}\right) = \exp\left(-\frac{\omega}{2}\frac{\alpha.\mathbf{v}}{|\mathbf{v}|}\right) \tag{96}$$

since $$\sigma_{0i} = \frac{i}{2}[\gamma_0, \gamma_i] = i\gamma_0\,\gamma_i = i\gamma^0\,\gamma^i\,\gamma_{ii} = -\,i\gamma^0\,\gamma^i = -\,i\alpha_i.$$

Now,

$$(\boldsymbol{\alpha}.\mathbf{v})^2 = (\boldsymbol{\alpha}^i\,v_i)\,(\boldsymbol{\alpha}^j\,v_j) = \boldsymbol{\alpha}^i\,\boldsymbol{\alpha}^j\,v_i\,v_j = \frac{1}{2}\,(\boldsymbol{\alpha}^i\,\boldsymbol{\alpha}^j\,v_i\,v_j + \boldsymbol{\alpha}^j\,\boldsymbol{\alpha}^i\,v_j\,v_i)$$

$$= \frac{1}{2}\,(\boldsymbol{\alpha}^i\,\boldsymbol{\alpha}^j + \boldsymbol{\alpha}^j\,\boldsymbol{\alpha}^i)\,v_i\,v_j = \delta^{ij}\,v_i\,v_j = \mathbf{v}^2 \tag{97}$$

Therefore, $\dfrac{(\alpha.\mathbf{v})^2}{|\mathbf{v}|^2} = 1$ and on expanding the exponential of eq. (96) as a power series in $\left(-\dfrac{\omega}{2}\dfrac{\alpha.\mathbf{v}}{|\mathbf{v}|}\right)$ and collecting even and odd powers in $\dfrac{\omega}{2}$, we obtain

$$\hat{S}(\text{p}) = \exp\left(-\frac{\omega}{2}\frac{\alpha.\mathbf{v}}{|\mathbf{v}|}\right) = \mathbf{I}\left[1 + \left(-\frac{1}{2}\omega\right)\left(\frac{\alpha.\mathbf{v}}{|\mathbf{v}|}\right) + \frac{1}{2!}\left(-\frac{1}{2}\omega\right)^2\left(\frac{\alpha.\mathbf{v}}{|\mathbf{v}|}\right)^2 + \ldots\right]$$

$$= \mathbf{I}\left[1 + \frac{1}{2!}\left(\frac{1}{2}\omega\right)^2 + \ldots\right] - \left(\frac{\alpha.\mathbf{v}}{|\mathbf{v}|}\right)\left[\left(\frac{1}{2}\omega\right) + \frac{1}{3!}\left(\frac{1}{2}\omega\right)^3 + \ldots\right]$$

$$= \mathbf{1}\cosh\left(\frac{\omega}{2}\right) - \left(\frac{\alpha.\mathbf{v}}{|\mathbf{v}|}\right)\sinh\left(\frac{\omega}{2}\right)$$

$$= 1\cosh\left(\frac{\omega}{2}\right) - \left(\frac{\text{a}.\mathbf{p}}{|\mathbf{p}|}\right)\sinh\left(\frac{\omega}{2}\right) \tag{98}$$

Using

$$\boldsymbol{\alpha}.\mathbf{v} = \sum_{k=1}^{3}\alpha^k\,v_k = \sum_{k=1}^{3}\begin{pmatrix}0 & \sigma^k\,v_k\\ \sigma^k\,v_k & 0\end{pmatrix},\ \frac{v_k}{|\mathbf{v}|} = \frac{p_k}{|\mathbf{p}|},\ \cosh\left(\frac{\omega}{2}\right)$$

$$= \sqrt{\frac{E+m}{2m}},\ \sinh\left(\frac{\omega}{2}\right) = -\sqrt{\frac{E-m}{2m}} = -\frac{|\text{p}|}{\sqrt{2m(E+m)}}$$

we get

$$\hat{S}(\text{p}) = \frac{1}{\sqrt{2m}}\begin{pmatrix} (E+m)^{1/2} & 0 & \frac{p_3}{(E+m)^{1/2}} & \frac{(p_1-ip_2)}{(E+m)^{1/2}} \\ 0 & (E+m)^{1/2} & \frac{(p_1+ip_2)}{(E+m)^{1/2}} & \frac{-p_3}{(E+m)^{1/2}} \\ \frac{p_3}{(E+m)^{1/2}} & \frac{(p_1-ip_2)}{(E+m)^{1/2}} & (E+m)^{1/2} & 0 \\ \frac{(p_1+ip_2)}{(E+m)^{1/2}} & \frac{-p_3}{(E+m)^{1/2}} & 0 & (E+m)^{1/2} \end{pmatrix} \tag{99}$$

6.5 FURTHER PROPERTIES OF DIRAC SPINORS

On the basis of the foregoing discussion, we prove some more properties of the Dirac spinors. We first prove that

$$\omega^{(r)\dagger}(\varepsilon_r\,\mathbf{p})\,\omega^{(r')}\,(\varepsilon_{r'}\,\mathbf{p}) = \delta_{rr'}\,\frac{E}{m} \tag{100}$$

To prove this, we note that the Dirac spinors satisfy $(\mathbf{p} - \varepsilon_{r'}\,m)\,\omega^{(r')}\,(\mathbf{p}) = 0$ and

$$\bar{\omega}^{(r)}\,(\mathbf{p})(\gamma^{\mu}p_{\mu} - \varepsilon_r\,m) = 0 \tag{101}$$

Now

$$\omega^{(r)}\,(\mathbf{p}) = \hat{S}\,(\mathbf{p})\,\omega^{(r)}\,(\mathbf{0}),\ \omega^{(r)}\,(-\mathbf{p}) = \hat{S}\,(-\mathbf{p})\,\omega^{(r)}\,(\mathbf{0})$$

so that

$$\omega^{(r)}\,(\varepsilon_r\,\mathbf{p}) = \hat{S}\,(\varepsilon_r\,\mathbf{p})\,\omega^{(r)}\,(\mathbf{0}) \tag{102}$$

Hence,

$$\omega^{(r)\dagger}\,(\varepsilon_r\,\mathbf{p}) = \omega^{(r)\dagger}\,(\mathbf{0})\,\hat{S}^{\dagger}(\varepsilon_r\,\mathbf{p}) \tag{103}$$

so that

$$\omega^{(r)\dagger}\,(\varepsilon_r\,\mathbf{p})\,\omega^{(r')}\,(\varepsilon_{r'}\,\mathbf{p}) = \omega^{(r)\dagger}\,(\mathbf{0})\hat{S}^{\dagger}\,(\varepsilon_r\,\mathbf{p})\,\hat{S}\,(\varepsilon_{r'}\,\mathbf{p})\,\omega^{(r')}\,(\mathbf{0})$$

Now

$$\hat{S}\,(\varepsilon_r\,\mathbf{p}) = \exp\left(-\varepsilon_r\,\frac{\omega}{2}\,\frac{\boldsymbol{\alpha}.\mathbf{v}}{|\mathbf{v}|}\right)$$

is Hermitian since the $\boldsymbol{\alpha}$ matrices are Hermitian. Therefore

$$\hat{S}^{\dagger}(\varepsilon_r\,\mathbf{p})\,\hat{S}\,(\varepsilon_{r'}\,\mathbf{p}) = \hat{S}\,(\varepsilon_r\,\mathbf{p})\,\hat{S}\,(\varepsilon_{r'}\,\mathbf{p}) = \exp\left[-\frac{1}{2}(\varepsilon_r + \varepsilon_{r'})\omega\frac{\boldsymbol{\alpha}.\mathbf{v}}{|\mathbf{v}|}\right]$$

$$= \mathbf{I} \cosh \frac{\Omega}{2} - \frac{\alpha . \mathbf{v}}{|\mathbf{v}|} \sinh \frac{\Omega}{2} \tag{104}$$

where $\Omega = (\varepsilon_r + \varepsilon_{r'})\omega.$

Thus,

$$\omega^{(r)\dagger} (\varepsilon_r \mathbf{p})\omega^{(r')} (\varepsilon_{r'} \mathbf{p}) = \omega^{(r)\dagger} (\mathbf{0}) \left(\mathbf{I} \cosh \frac{\Omega}{2} - \frac{\alpha . \mathbf{v}}{|\mathbf{v}|} \sinh \frac{\Omega}{2} \right) \omega^{(r')} (\mathbf{0})$$

$$= \cosh \frac{\Omega}{2} \, \omega^{(r)\dagger} (\mathbf{0}) \, \omega^{(r')} (\mathbf{0}) - \sinh \frac{\Omega}{2} \, \omega^{(r)\dagger} (\mathbf{0}) \, \frac{\alpha . \mathbf{v}}{|\mathbf{v}|} \, \omega^{(r')} (\mathbf{0}) \tag{105}$$

Now

$$\cosh \frac{\Omega}{2} \, \omega^{(r)\dagger} (\mathbf{0})\omega^{(r')} (\mathbf{0}) = \delta_{rr'} \cosh \frac{\Omega}{2} \tag{106}$$

and

$$\sinh \frac{\Omega}{2} \, \omega^{(r)\dagger} (\mathbf{0}) \, \frac{\alpha . \mathbf{v}}{|\mathbf{v}|} \, \omega^{(r')} (\mathbf{0}) = \sinh \frac{\Omega}{2} \, (\omega_1^{(r)*} (\mathbf{0}), \omega_2^{(r)*} (\mathbf{0}), \omega_3^{(r)*} (\mathbf{0}), \omega_4^{(r)*} (\mathbf{0}))$$

$$\begin{pmatrix} 0 & 0 & p_3|\mathbf{p}|^{-1} & (p_1 - ip_2)|\mathbf{p}|^{-1} \\ 0 & 0 & (p_1 + ip_2)|\mathbf{p}|^{-1} & -p_3|\mathbf{p}|^{-1} \\ p_3|\mathbf{p}|^{-1} & (p_1 - ip_2)|\mathbf{p}|^{-1} & 0 & 0 \\ (p_1 + ip_2)|\mathbf{p}|^{-1} & -p_3|\mathbf{p}|^{-1} & 0 & 0 \end{pmatrix} \begin{pmatrix} \omega_1^{(r')} (\mathbf{0}) \\ \omega_2^{(r')} (\mathbf{0}) \\ \omega_3^{(r')} (\mathbf{0}) \\ \omega_4^{(r')} (\mathbf{0}) \end{pmatrix}$$

$$= \frac{1}{|\mathbf{p}|} \sinh \frac{\Omega}{2} \left[\begin{array}{l} \omega_3^{(r)*} (\mathbf{0}) \, p_3 \omega_1^{(r')} (\mathbf{0}) + \omega_4^{(r)*} (\mathbf{0})(p_1 + ip_2) \omega_1^{(r')} (\mathbf{0}) \\ + \omega_1^{(r')} (\mathbf{0}) \, p_3 \, \omega_3^{(r')} (\mathbf{0}) + \omega_2^{(r)*} (\mathbf{0})(p_1 + ip_2) \omega_3^{(r')} (\mathbf{0}) \\ + \omega_3^{(r)*} (\mathbf{0})(p_1 - ip_2) \omega_2^{(r')} (\mathbf{0}) - \omega_4^{(r)*} (\mathbf{0}) \, p_3 \omega_2^{(r')} (\mathbf{0}) \\ + \omega_1^{(r)*} (\mathbf{0})(p_1 - ip_2) \omega_4^{(r')} (\mathbf{0}) - \omega_2^{(r)*} (\mathbf{0}) \, p_3 \omega_4^{(r')} (\mathbf{0}) \end{array} \right] = 0 \tag{107}$$

where the last step follows from the following:

(a) The expression within square brackets is non-vanishing if and only if either $r = 1, 2; r' = 3, 4$ or vice versa, i.e., if the two spinors have opposite energy signs.

(b) If the two spinors are of opposite energy signs, then $\varepsilon_r + \varepsilon_{r'} = 0$ so that $\Omega = 0$ and, therefore, $\sinh \Omega = 0$.

Furthermore, eq. (106) will be non-vanishing if and only if $r = r'$ in which case $\Omega = 2\varepsilon_r \omega$ so that $\cosh \frac{\Omega}{2} = \cosh (\varepsilon_r \omega) = \cosh (\omega) = 2 \cosh^2 \frac{\omega}{2} - 1 = \frac{E}{m}$ and we finally have

$$\omega^{(r)\dagger} (\varepsilon_r \mathbf{p}) \, \omega^{(r')} (\varepsilon_{r'} \mathbf{p}) = \delta_{rr'} \frac{E}{m} \tag{108}$$

which agrees with the normalization (62).

We now prove the orthogonality relation for the adjoint spinor

$$\bar{\omega}^{(r)}(\mathbf{p})\omega^{(r')}(\mathbf{p}) = \delta_{rr'}\varepsilon_r \tag{109}$$

To prove this, we note that the Dirac spinors satisfy $\omega^{(r)}(\mathbf{p}) = \hat{S}(\mathbf{p})\,\omega^{(r)}(\mathbf{0})$, so that

$$\bar{\omega}^{(r)}(\varepsilon_r\mathbf{p}) = \omega^{(r)\dagger}(\varepsilon_r\mathbf{p})\boldsymbol{\gamma}^0 = \omega^{(r)\dagger}(\mathbf{0})\,\hat{S}^{\dagger}(\varepsilon_r\mathbf{p})\boldsymbol{\gamma}^0 = \omega^{(r)\dagger}(\mathbf{0})\,\hat{S}(\varepsilon_r\mathbf{p})\boldsymbol{\gamma}^0$$

$$= \omega^{(r)\dagger}(\mathbf{0})\exp\left(-\varepsilon_r\frac{\omega}{2}\frac{\boldsymbol{\alpha}.\mathbf{v}}{|\mathbf{v}|}\right)\boldsymbol{\gamma}^0 = \omega^{(r)\dagger}(\mathbf{0})\left(\mathbf{I}\cosh\frac{\varepsilon_r\omega}{2} - \frac{\boldsymbol{\alpha}.\mathbf{v}}{|\mathbf{v}|}\sinh\frac{\varepsilon_r\omega}{2}\right)\boldsymbol{\gamma}^0$$

$$= \omega^{(r)\dagger}(\mathbf{0})\boldsymbol{\gamma}^0\left(\mathbf{I}\cosh\frac{\varepsilon_r\omega}{2} + \frac{\boldsymbol{\alpha}.\mathbf{v}}{|\mathbf{v}|}\sinh\frac{\varepsilon_r\omega}{2}\right)$$

$$= \omega^{(r)\dagger}(\mathbf{0})\boldsymbol{\gamma}^0\,\hat{S}(-\varepsilon_r\mathbf{p}) = \bar{\omega}^{(r)\dagger}(\mathbf{0})\hat{S}(-\varepsilon_r\mathbf{p}) \tag{110}$$

so that

$$\bar{\omega}^{(r)}(\varepsilon_r\mathbf{p})\omega^{(r')}(\varepsilon_r\mathbf{p}) = \omega^{(r)\dagger}(\mathbf{0})\boldsymbol{\gamma}^0\,\hat{S}(-\varepsilon_r\mathbf{p})\,\hat{S}(\varepsilon_{r'}\mathbf{p})\omega^{(r')}(\mathbf{0}) \tag{111}$$

Using eq. (110), this reduces to

$$\bar{\omega}^{(r)}(\varepsilon_r\mathbf{p})\omega^{(r')}(\varepsilon_{r'}\mathbf{p}) = \omega^{(r)\dagger}(\mathbf{0})\boldsymbol{\gamma}^0\,\hat{S}(-\varepsilon_r\mathbf{p})\,\hat{S}(\varepsilon_{r'}\mathbf{p})\omega^{(r')}(\mathbf{0})$$

$$= \omega^{(r)\dagger}(\mathbf{0})\boldsymbol{\gamma}^0\left(\mathbf{I}\cosh\frac{\varepsilon_r\omega}{2} + \frac{\boldsymbol{\alpha}.\mathbf{v}}{|\mathbf{v}|}\sinh\frac{\varepsilon_r\omega}{2}\right)$$

$$\left(\mathbf{I}\cosh\frac{\varepsilon_{r'}\omega}{2} - \frac{\boldsymbol{\alpha}.\mathbf{v}}{|\mathbf{v}|}\sinh\frac{\varepsilon_{r'}\omega}{2}\right)\omega^{r'}(\mathbf{0})$$

$$= \omega^{(r)\dagger}(\mathbf{0})\boldsymbol{\gamma}^0\left[\mathbf{I}\cosh\frac{1}{2}(\varepsilon_r - \varepsilon_{r'})\omega + \frac{\boldsymbol{\alpha}.\mathbf{v}}{|\mathbf{v}|}\right.$$

$$\left.\sinh\frac{1}{2}(\varepsilon_r - \varepsilon_{r'})\omega\right]\omega^{(r')}(\mathbf{0}) \tag{112}$$

Now,

$$\cosh\left[\frac{1}{2}(\varepsilon_r - \varepsilon_{r'})\omega\right]\omega^{(r)\dagger}(\mathbf{0})\boldsymbol{\gamma}^0\,\omega^{(r')}(\mathbf{0})$$

$$= \cosh\left[\frac{1}{2}(\varepsilon_r - \varepsilon_{r'})\omega\right] \begin{pmatrix} \omega_1^{(r)*}(\mathbf{0}) \\ \omega_2^{(r)*}(\mathbf{0}) \\ -\omega_3^{(r)*}(\mathbf{0}) \\ -\omega_4^{(r)*}(\mathbf{0}) \end{pmatrix}^T \begin{pmatrix} \omega_1^{(r')}(\mathbf{0}) \\ \omega_2^{(r')}(\mathbf{0}) \\ \omega_3^{(r')}(\mathbf{0}) \\ \omega_4^{(r')}(\mathbf{0}) \end{pmatrix}$$

$$= \cosh\left[\frac{1}{2}(\varepsilon_r - \varepsilon_{r'})\omega\right]\left[\sum_{\alpha=1}^{4} \varepsilon_r\, \omega_\alpha^{(r)*}(\mathbf{0})\omega_a^{(r')}(\mathbf{0})\right]$$

$$= \varepsilon_r\, \delta_{rr'} \cosh\left[\frac{1}{2}(\varepsilon_r - \varepsilon_{r'})\omega\right] = \varepsilon_r\, \delta_{rr} \tag{113}$$

and

$$\sinh\left[\frac{1}{2}(\varepsilon_r - \varepsilon_{r'})\omega\right] \omega^{(r)\dagger}(\mathbf{0})\boldsymbol{\gamma}^0 \frac{\boldsymbol{\alpha}.\mathbf{v}}{|\mathbf{v}|}\, \omega^{(r')}(\mathbf{0})$$

$$= \sinh\left[\frac{1}{2}(\varepsilon_r - \varepsilon_{r'})\omega\right] (\omega_1^{(r)*}(\mathbf{0}),\ \omega_2^{(r)*}(\mathbf{0}),\ -\omega_3^{(r)*}(\mathbf{0}),\ -\omega_4^{(r)*}(\mathbf{0}))$$

$$\begin{pmatrix} 0 & 0 & p_3|\mathbf{p}|^{-1} & (p_1 - ip_2)|\mathbf{p}|^{-1} \\ 0 & 0 & (p_1 + ip_2)|\mathbf{p}|^{-1} & -p_3|\mathbf{p}|^{-1} \\ p_3|\mathbf{p}|^{-1} & (p_1 - ip_2)|\mathbf{p}|^{-1} & 0 & 0 \\ (p_1 + ip_2)|\mathbf{p}|^{-1} & -p_3|\mathbf{p}|^{-1} & 0 & 0 \end{pmatrix} \begin{pmatrix} \omega_1^{(r')}(\mathbf{0}) \\ \omega_2^{(r')}(\mathbf{0}) \\ \omega_3^{(r')}(\mathbf{0}) \\ \omega_4^{(r')}(\mathbf{0}) \end{pmatrix}$$

$$= \frac{1}{|\mathbf{p}|} \sinh\left[\frac{1}{2}(\varepsilon_r - \varepsilon_{r'})\omega\right]\left[\begin{matrix} -\omega_3^{(r)*}(\mathbf{0})\, p_3\omega_1^{(r')}(\mathbf{0}) - \omega_4^{(r)*}(\mathbf{0})(p_1 + ip_2)\omega_1^{(r')}(\mathbf{0}) \\ +\omega_1^{(r)*}(\mathbf{0})\, p_3\, \omega_3^{(r')}(\mathbf{0}) + \omega_2^{(r)*}(\mathbf{0})(p_1 + ip_2)\omega_3^{(r')}(\mathbf{0}) \end{matrix}\right.$$

$$\left.\begin{matrix} -\omega_3^{(r)*}(\mathbf{0})(p_1 - ip_2)\omega_2^{(r')}(\mathbf{0}) + \omega_4^{(r)*}(\mathbf{0})\, p_3\omega_2^{(r')}(\mathbf{0}) \\ +\omega_1^{(r)*}(\mathbf{0})(p_1 - ip_2)\omega_4^{(r')}(\mathbf{0}) - \omega_2^{(r)*}(\mathbf{0})\, p_3\omega_4^{(r')}(\mathbf{0}) \end{matrix}\right] \tag{114}$$

Now

(a) The expression within square brackets is non-vanishing if and only if either $r = 1, 2$; $r' = 3, 4$ or vice versa, i.e., if the two spinors have opposite energy signs.

(b) If the two spinors are of opposite energy signs, then the expression within square brackets reduces to $\varepsilon_r\, \omega_r^{(r)*}(\mathbf{0})\, (\boldsymbol{\alpha}.\mathbf{p})_{rr'}\, \omega_{r'}^{(r')}(\mathbf{0}) = \varepsilon_r\, (\boldsymbol{\alpha}.\mathbf{p})_{rr'}$.

(c) If the two spinors are of opposite energy sign, i.e., $\varepsilon_{r'} = -\varepsilon_r$, then

$$\sinh\left[\frac{1}{2}(\varepsilon_r - \varepsilon_{r'})\omega\right] = \sinh(\varepsilon_r\, \omega) = \varepsilon_r \sinh\omega = \varepsilon_r\, \frac{\mathbf{p}}{m}$$

Hence,

$$\sinh\left[\frac{1}{2}(\varepsilon_r - \varepsilon_{r'})\omega\right] \omega^{(r)\dagger}(\mathbf{0})\boldsymbol{\gamma}^0 \frac{\boldsymbol{\alpha}.\mathbf{v}}{|\mathbf{v}|}\omega^{(r')}(\mathbf{0}) = \begin{cases} 0 \text{ for } \varepsilon_r = \varepsilon_{r'} \\ \dfrac{\mathbf{p}}{m|\mathbf{p}|}(\boldsymbol{\alpha}.\mathbf{p})_{rr'} \text{ for } \varepsilon_r \neq \varepsilon_{r'} \end{cases} \tag{115}$$

Combining eqs. (113) and (115), we get the following normalization for the adjoint spinor:

$$\bar{\omega}^{(r)}(\varepsilon_r\mathbf{p})\omega^{(r')}(\varepsilon_{r'}\mathbf{p}) = \begin{cases} \varepsilon_r \text{ for } r = r' \\ 0 \text{ for } r \neq r', \varepsilon_r = \varepsilon_{r'} \\ \dfrac{\mathbf{p}}{m|\mathbf{p}|}(\boldsymbol{\alpha}.\mathbf{p})_{rr'} \ \textit{for}\ \varepsilon_r \neq \varepsilon_{r'} \end{cases} \tag{116}$$

6.6 CLOSURE RELATIONS AMONG DIRAC SPINORS

We, now, prove the closure property of Dirac spinors viz.

$$\sum_{r=1}^{4} \varepsilon_r \omega_\alpha^{(r)}(\mathbf{p})\,\bar{\omega}_\beta^{(r)}(\mathbf{p}) = \delta_{\alpha\beta} \tag{117}$$

We have

$$\bar{\omega}^{(r)}(\mathbf{0}) = \omega^{(r)\dagger}(\mathbf{0})\boldsymbol{\gamma}^0 \tag{118}$$

so that

$$\bar{\omega}^{(1)}(\mathbf{0}) = \omega^{(1)\dagger}(\mathbf{0})\boldsymbol{\gamma}^0 = (1\ 0\ 0\ 0)\begin{pmatrix} 1 & 0 & 0 & 0 \\ 0 & 1 & 0 & 0 \\ 0 & 0 & -1 & 0 \\ 0 & 0 & 0 & -1 \end{pmatrix} = (1\ 0\ 0\ 0) = \omega^{(1)}(\mathbf{0}) \tag{119}$$

$$\bar{\omega}^{(2)}(\mathbf{0}) = (0\ 1\ 0\ 0) = \omega^{(2)}(\mathbf{0}) \tag{120}$$

$$\bar{\omega}^{(3)}(\mathbf{0}) = (0\ 0\ {-1}\ 0) = -\,\omega^{(3)}(\mathbf{0}) \tag{121}$$

$$\bar{\omega}^{(4)}(\mathbf{0}) = (0\ 0\ 0\ {-1}) = -\,\omega^{(4)}(\mathbf{0}) \tag{122}$$

that is,

$$\bar{\omega}^{(r)}(\mathbf{0}) = \varepsilon_r\,\omega^{(r)}(\mathbf{0}) \tag{123}$$

It is trivially seen that

$$\sum_{r=1}^{4} \varepsilon_r \omega_\alpha^{(r)}(\mathbf{0})\,\bar{\omega}_\beta^{(r)}(\mathbf{0}) = \delta_{\alpha\beta} \tag{124}$$

Now

$$\omega^{(r)\dagger}(\mathbf{p}) = [\hat{S}\,\omega^{(r)}(\mathbf{0})]^\dagger = \omega^{(r)\dagger}(\mathbf{0})\hat{S}^\dagger$$

$$= \omega^{(r)\dagger}(\mathbf{0})\boldsymbol{\gamma}^0 S^{-1}\boldsymbol{\gamma}^0 = \bar{\omega}^{(r)}(\mathbf{0})\hat{S}^{-1}\boldsymbol{\gamma}^0 \tag{125}$$

so that

$$\bar{\omega}^{(r)}(\mathbf{p}) = \omega^{(r)\dagger}(\mathbf{p})\boldsymbol{\gamma}^0 = \bar{\omega}^{(r)}(\mathbf{0})\,\hat{S}^{-1} \tag{126}$$

Therefore,

$$\sum_{r=1}^{4}\varepsilon_r\,\omega_\alpha^{(r)}(\mathbf{p})\,\bar{\omega}_\beta^{(r)}(\mathbf{p}) = \sum_{r=1}^{4}\varepsilon_r\left[\sum_{\kappa=1}^{4}\hat{S}_{\alpha\kappa}\,\omega_\kappa^{(r)}(\mathbf{0})\right]\left[\sum_{\lambda=1}^{4}\bar{\omega}_\lambda^{(r)}(\mathbf{0})\hat{S}_{\lambda\beta}^{-1}\right]$$

$$= \sum_{r,\kappa,\lambda=1}^{4}(\hat{S}_{\alpha\kappa})[\varepsilon_r\,\omega_\kappa^{(r)}(\mathbf{0})\,\bar{\omega}_\lambda^{(r)}(\mathbf{0})]\,(\hat{S}_{\lambda\,\beta}^{-1})$$

$$= \sum_{\kappa,\lambda=1}^{4}(\hat{S}_{\alpha\kappa}\,\delta_{\kappa\lambda}\,\hat{S}_{\lambda\beta}^{-1}) = \sum_{\kappa=1}^{4}(\hat{S}_{\alpha\kappa}\,\hat{S}_{\kappa\beta}^{-1}) = \delta_{\alpha\beta} \tag{127}$$

Another closure relation of importance is

$$\sum_{r=1}^{4}\omega_\alpha^{(r)\dagger}(\varepsilon_r\,\mathbf{p})\,\omega_\beta^{(r)}(\varepsilon_r\,\mathbf{p}) = \frac{E}{m}\,\delta_{\alpha\beta} \tag{128}$$

To prove this relation, we note that

$$\omega^{(r)}(\varepsilon_r\,\mathbf{p}) = \hat{S}(\varepsilon_r\,\mathbf{p})\omega^{(r)}(\mathbf{0})$$

and

$$\omega^{(r)\dagger}(\varepsilon_r\,\mathbf{p}) = \omega^{(r)\dagger}(\mathbf{0})\hat{S}^\dagger(\varepsilon_r\,\mathbf{p}) = \omega^{(r)\dagger}(\mathbf{0})\boldsymbol{\gamma}^0\,\hat{S}^{-1}(\varepsilon_r\,\mathbf{p})\boldsymbol{\gamma}^0$$

so that

$$\omega^{(r)\dagger}(\varepsilon_r\,\mathbf{p})\,\omega^{(r)}(\varepsilon_r\,\mathbf{p}) = \omega^{(r)\dagger}(\mathbf{0})\boldsymbol{\gamma}^0\,\hat{S}^{-1}(\varepsilon_r\,\mathbf{p})\boldsymbol{\gamma}^0\,\hat{S}(\varepsilon_r\,\mathbf{p})\omega^{(r)}(\mathbf{0})$$

$$= \omega^{(r)\dagger}(\mathbf{0})\boldsymbol{\gamma}^0\left(\mathbf{I}\cosh\frac{\varepsilon_r\,\omega}{2} - \frac{\alpha.\mathbf{p}}{|\mathbf{p}|}\sinh\frac{\varepsilon_r\,\omega}{2}\right)^{-1}$$

$$\boldsymbol{\gamma}^0\left(\mathbf{I}\cosh\frac{\varepsilon_r\,\omega}{2} - \frac{\alpha.\mathbf{p}}{|\mathbf{p}|}\sinh\frac{\varepsilon_r\,\omega}{2}\right)\omega^{(r)}(\mathbf{0})$$

$$= \omega^{(r)\dagger}(\mathbf{0})\,\boldsymbol{\gamma}^0\left(\mathbf{I}\cosh\frac{\varepsilon_r\,\omega}{2} + \frac{\mathrm{a}.\mathbf{p}}{|\,\mathbf{p}\,|}\sinh\frac{\varepsilon_r\,\omega}{2}\right)$$

$$\boldsymbol{\gamma}^0 \left(\mathbf{I}\cosh\frac{\varepsilon_r\,\omega}{2} - \frac{\boldsymbol{\alpha}.\mathbf{p}}{|\mathbf{p}|}\sinh\frac{\varepsilon_r\,\omega}{2} \right) \omega^{(r)}\,(\mathbf{0})$$

$$= \omega^{(r)\dagger}\,(\mathbf{0}) \left(\mathbf{I}\cosh\frac{\varepsilon_r\,\omega}{2} - \frac{\boldsymbol{\alpha}.\mathbf{p}}{|\mathbf{p}|}\sinh\frac{\varepsilon_r\,\omega}{2} \right)^2 \omega^{(r)}\,(\mathbf{0})$$

$$= \omega^{(r)\dagger}\,(\mathbf{0}) \left(\mathbf{I}\cosh\varepsilon_r\,\omega - \frac{\boldsymbol{\alpha}.\mathbf{p}}{|\mathbf{p}|}\sinh\varepsilon_r\,\omega \right) \omega^{(r)}\,(\mathbf{0})$$

$$= \omega^{(r)\dagger}\,(\mathbf{0}) \left(\mathbf{I}\cosh\omega - \frac{\varepsilon_r\,\boldsymbol{\alpha}.\mathbf{p}}{|\mathbf{p}|}\sinh\omega \right) \omega^{(r)}\,(\mathbf{0})$$

$$= \frac{\omega^{(r)\dagger}\,(\mathbf{0})}{m} \left(E\mathbf{I} + \frac{\varepsilon_r\,\boldsymbol{\alpha}.\mathbf{p}}{|\mathbf{p}|}\mathbf{p} \right) \omega^{(r)}\,(\mathbf{0})$$

$$= \frac{1}{m}\begin{pmatrix} \omega_1^{(r)*}(\mathbf{0}) \\ \omega_2^{(r)*}(\mathbf{0}) \\ \omega_3^{(r)*}(\mathbf{0}) \\ \omega_4^{(r)*}(\mathbf{0}) \end{pmatrix}^T \begin{pmatrix} E & 0 & \varepsilon_r\,p_3\frac{\mathrm{p}}{|\mathrm{p}|} & \varepsilon_r\,(p_1 - ip_2)\frac{\mathrm{p}}{|\mathrm{p}|} \\ 0 & E & \varepsilon_r\,(p_1 + ip_2)\frac{\mathrm{p}}{|\mathrm{p}|} & -\varepsilon_r\,p_3\frac{\mathbf{p}}{|\mathbf{p}|} \\ \varepsilon_r\,p_3\frac{\mathrm{p}}{|\mathrm{p}|} & \varepsilon_r\,(p_1 - ip_2)\frac{\mathrm{p}}{|\mathrm{p}|} & E & 0 \\ \varepsilon_r\,(p_1 + ip_2)\frac{\mathrm{p}}{|\mathrm{p}|} & -\varepsilon_r\,p_3\frac{\mathbf{p}}{|\mathbf{p}|} & 0 & E \end{pmatrix} \begin{pmatrix} \omega_1^{(r)}(\mathbf{0}) \\ \omega_2^{(r)}(\mathbf{0}) \\ \omega_3^{(r)}(\mathbf{0}) \\ \omega_4^{(r)}(\mathbf{0}) \end{pmatrix}$$

$$= \left\{ \frac{1}{m} \begin{bmatrix} \omega_1^{(r)*}(\mathbf{0})\,E\omega_1^{(r)}(\mathbf{0}) + \varepsilon_r\,\omega_3^{(r)*}(\mathbf{0})\,p_3\frac{\mathbf{p}}{|\mathbf{p}|}\omega_1^{(r)}(\mathbf{0}) + \varepsilon_r\,\omega_4^{(r)*}(\mathbf{0})(p_1 + ip_2)\frac{\mathbf{p}}{|\mathbf{p}|}\omega_1^{(r)}(\mathbf{0}) + \\ \omega_2^{(r)*}(\mathbf{0})\,E\omega_2^{(r)}(\mathbf{0}) + \varepsilon_r\,\omega_3^{(r)*}(\mathbf{0})(p_1 - ip_2)\frac{\mathbf{p}}{|\mathbf{p}|}\omega_2^{(r)}(\mathbf{0}) + \varepsilon_r\,\omega_4^{(r)*}(\mathbf{0})\,p_3\frac{\mathbf{p}}{|\mathbf{p}|}\omega_2^{(r)}(\mathbf{0}) + \\ \varepsilon_r\omega_1^{(r)*}(\mathbf{0})\,p_3\frac{\mathbf{p}}{|\mathbf{p}|}\omega_3^{(r)}(\mathbf{0}) + \varepsilon_r\,\omega_2^{(r)*}(\mathbf{0})(p_1 + ip_2)\frac{\mathbf{p}}{|\mathbf{p}|}\omega_3^{(r)}(\mathbf{0}) + \omega_3^{(r)*}(\mathbf{0})\,E\omega_3^{(r)}(\mathbf{0}) + \\ \varepsilon_r\omega_1^{(r)*}(\mathbf{0})(p_1 - ip_2)\frac{\mathbf{p}}{|\mathbf{p}|}\omega_4^{(r)}(\mathbf{0}) - \varepsilon_r\,\omega_2^{(r)*}(\mathbf{0})\,p_3\frac{\mathbf{p}}{|\mathbf{p}|}\omega_4^{(r)}(\mathbf{0}) + \omega_4^{(r)*}(\mathbf{0})\,E\omega_4^{(r)}(\mathbf{0}) \end{bmatrix} \right\} \tag{129}$$

Now the expressions of the form $\varepsilon_r\,\omega_a^{(r)*}(\mathbf{0})\,p_3\,\frac{\mathbf{p}}{|\mathbf{p}|}\,\omega_b^{(r)}(\mathbf{0})$, $\varepsilon_r\,\omega_a^{(r)*}(\mathbf{0})\,(p_1 \pm ip_2)\,\frac{\mathbf{p}}{|\mathbf{p}|}\,\omega_b^{(r)}(\mathbf{0})$ with $a \neq b$ will all vanish. Therefore, only the diagonal elements of the form $\omega_r^{(r)*}(\mathbf{0})\,E\omega_r^{(r)}(\mathbf{0})$ would remain when a summation is done over the index $r = 1, 2, 3, 4$. Hence, we have

$$\sum_{r=1}^{4} \omega_\alpha^{(r)\dagger}(\varepsilon_r\,\mathbf{p})\,\omega_\beta^{(r)}(\varepsilon_r\,\mathbf{p}) = \delta_{\alpha\beta}\frac{E}{m}\sum_{r=1}^{4}\omega_\alpha^{(r)*}(\mathbf{0})\,\omega_\beta^{(r)}(\mathbf{0}) = \delta_{\alpha\beta}\frac{E}{m} \tag{130}$$

because

$$\sum_{r=1}^{4} \omega_\alpha^{(r)\dagger}(\mathbf{0})\,\omega_\alpha^{(r)}(\mathbf{0}) = 1 \tag{131}$$

6.7 NORMALIZATION OF DIRAC WAVES

Consider two plane waves with well defined momentum states $\mathbf{p}$, $\mathbf{p}'$ represented by the wavefunctions

$$\psi_{\mathbf{p}}^{(r)}(x) = \omega^{(r)}(\mathbf{p})\exp(-i\varepsilon_r p_\mu x^\mu) \tag{132}$$

and

$$\psi_{\mathbf{p}'}^{(r')}(x) = \omega^{(r')}(\mathbf{p}')\exp(-i\varepsilon_{r'} p'_\mu x^\mu) \tag{133}$$

Then

$$\begin{aligned}\langle\psi_{\mathbf{p}'}^{(r')}(x)|\psi_{\mathbf{p}}^{(r)}(x)\rangle &= \int d\mathbf{x}\,\psi_{\mathbf{p}'}^{(r')\dagger}(x)\psi_{\mathbf{p}}^{(r)}(x)\\ &= \int d\mathbf{x}\,\omega^{(r')\dagger}(\mathbf{p}')\omega^{(r)}(\mathbf{p})\exp[-i(\varepsilon_r p_\mu x^\mu - \varepsilon_{r'} p'_\mu x^\mu)]\end{aligned} \tag{134}$$

The integration over the spatial coordinates leads to $\delta^3(\mathbf{p}-\mathbf{p}')$ so that eq. (134) reduces to

$$\begin{aligned}\langle\psi_{\mathbf{p}'}^{(r')}(x)|\psi_{\mathbf{p}}^{(r)}(x)\rangle &= \omega^{(r')\dagger}(\mathbf{p}')\omega^{(r)}(\mathbf{p})\exp\\ &\quad[-i(\varepsilon_r p_0 x^0 - \varepsilon_{r'} p'_0 x^0)].\,(2\pi)^3\,\delta^3(\varepsilon_r\,\mathbf{p}-\varepsilon_{r'}\,\mathbf{p}')\end{aligned} \tag{135}$$

The scalar product will vanish unless either $\varepsilon_r = \varepsilon_{r'}$ and $\mathbf{p}' = +\,\mathbf{p}$ or $\varepsilon_r = -\,\varepsilon_{r'}$ and $\mathbf{p}' = -\,\mathbf{p}$. Now, the normalization of spinors (eq. (108)) gives $\omega^{(r')\dagger}(\varepsilon_{r'}\,\mathbf{p})\,\omega^{(r)}(\varepsilon_r\,\mathbf{p}) = \delta_{rr'}\dfrac{E}{m}$. Therefore, in case $\varepsilon_r = -\,\varepsilon_{r'}$ and $\mathbf{p}' = -\,\mathbf{p}$, we have $\omega^{(r')\dagger}(\mathbf{p}')\,\omega^{(r)}(\mathbf{p}) = \omega^{(r')\dagger}(-\mathbf{p})\omega^{(r)}(\mathbf{p}) = 0$. Hence, the scalar product will be non-zero only if $r = r'$ which implies $\varepsilon_r = \varepsilon_{r'}$ and $\mathbf{p}' = +\,\mathbf{p}$. In this situation, $p'_0 = (\mathbf{p}'^2 + m^2)^{\frac{1}{2}} = (\mathbf{p}^2 + m^2)^{\frac{1}{2}} = p_0$ as well. Hence, the scalar product will be

$$\langle\psi_{\mathbf{p}'}^{(r')}(x)|\psi_{\mathbf{p}}^{(r)}(x)\rangle = \frac{E}{m}(2\pi)^3\,\delta^3(\mathbf{p}-\mathbf{p}')\,\delta_{rr'} \tag{136}$$

6.8 PROJECTION OPERATORS: ENERGY PROJECTION OPERATORS

A projection operator is an operator that, when acting on an arbitrary free particle solution of the Dirac equation, projects out the spinor that corresponds to a given sign of energy and/or a given polarization.

In the Dirac theory, there will be four fundamental projection operators that will project out positive/negative energy with spin up/down out of a plane wave with arbitrary but well-defined momentum state **p**.

Thus, these four projection operators $\hat{P}_r(\mathbf{p})$ must satisfy

$$\hat{P}_r(\mathbf{p})\,\omega^{(r')}(\mathbf{p}) = \delta_{rr'}\,\omega^{(r')}(\mathbf{p}) \tag{137}$$

and

$$\hat{P}_r(\mathbf{p})\,\hat{P}_{r'}(\mathbf{p}) = \delta_{rr'}\,\hat{P}_r(\mathbf{p}) \tag{138}$$

where eq. (138) follows from

$$\begin{aligned}\hat{P}_r(\mathbf{p})\,\hat{P}_{r'}(\mathbf{p})\,\omega^{(r'')}(\mathbf{p}) &= \delta_{r'r''}\,\hat{P}_r(\mathbf{p})\omega^{(r'')}(\mathbf{p}) = \delta_{r'r''}\,\delta_{rr''}\,\omega^{(r'')}(\mathbf{p})\\ &= \delta_{rr'}\,\delta_{rr''}\,\omega^{(r'')}(\mathbf{p}) = \delta_{rr'}\,p_r(\mathbf{p})\omega^{(r'')}(\mathbf{p})\end{aligned}$$

To find the explicit expression for the two energy projection operators we note that

$$\begin{aligned}(\not{p} - m)\,\omega^{(r)}(\mathbf{p}) &= 0, \quad r = 1, 2\\ (\not{p} + m)\,\omega^{(r)}(\mathbf{p}) &= 0, \quad r = 3, 4\end{aligned} \tag{139}$$

or equivalently

$$(\not{p} - \varepsilon_r\, m)\,\omega^{(r)}(\mathbf{p}) = 0 = (\varepsilon_r \not{p} - m)\,\omega^{(r)}(\mathbf{p}), \quad r = 1, 2, 3, 4 \tag{140}$$

where $\varepsilon_r = +1$ for $r = 1, 2$ and $\varepsilon_r = -1$ for $r = 3, 4$.

Let us define

$$\hat{\Lambda}_{\text{sgn}\,\varepsilon_r}(\mathbf{p}) = \frac{\varepsilon_r \not{p} + m}{2m} \tag{141}$$

where, on the right-hand side, we have used the momentum p (which is the eigenvalue of the momentum operator $\hat{p}$ and is a real number) in the definition of the projection operator (141) and not the momentum operator $\hat{p}$ itself. Then, we have

$$\hat{\Lambda}_{\text{sgn}\,\varepsilon_r}(\mathbf{p})\omega^{(r)}(\mathbf{p}) = \frac{\varepsilon_r \not{p} + m}{2m}\,\omega^{(r)}(\mathbf{p}) = \frac{2m}{2m}\,\omega^{(r)}(\mathbf{p}) = \omega^{(r)}(\mathbf{p}) \tag{142}$$

where, in the penultimate step, we have used eq. (140). Also

$$\hat{\Lambda}_{\text{sgn}\,\varepsilon_r}(\mathbf{p})\omega^{(r')}(\mathbf{p}) = \frac{\varepsilon_r \not{p} + m}{2m}\,\omega^{(r')}(\mathbf{p}) = \frac{\varepsilon_r \not{p} + \varepsilon_{r'} \not{p}}{2m}\,\omega^{(r')}(\mathbf{p})$$

$$= \frac{(\varepsilon_r + \varepsilon_{r'})\, \not{p}}{2m}\, \omega^{(r')}(\mathbf{p}) = \delta_{\varepsilon_r \varepsilon_{r'}}\, \omega^{(r')}(\mathbf{p}) \tag{143}$$

Furthermore,

$$\hat{\Lambda}_{\text{sgn}\,\varepsilon_r}(\mathbf{p})\, \hat{\Lambda}_{\text{sgn}\,\varepsilon_{r'}}(\mathbf{p}) = \frac{(\varepsilon_r \not{p} + m)(\varepsilon_{r'} \not{p} + m)}{4m^2}$$

$$= \frac{\varepsilon_r\, \varepsilon_{r'}\, \not{p}\not{p} + (\varepsilon_r + \varepsilon_{r'})\, \not{p} m + m^2}{4m^2} = \frac{(\varepsilon_r\, \varepsilon_{r'} + 1)m^2 + (\varepsilon_r + \varepsilon_{r'})\, \not{p} m}{4m^2}$$

$$= \delta_{\varepsilon_r \varepsilon_{r'}}\, \frac{m + \varepsilon_r \not{p}}{2m} = \delta_{\varepsilon_r \varepsilon_{r'}}\, \hat{\Lambda}_{\text{sgn}\,\varepsilon_r}(\mathbf{p}) \tag{144}$$

where we have used

$$\not{p}\not{p} = \boldsymbol{\gamma}^{\mu}\, p_{\mu}\, \boldsymbol{\gamma}^{\nu}\, p_{\nu} = \boldsymbol{\gamma}^{\mu}\, \boldsymbol{\gamma}^{\nu}\, p_{\mu} p_{\nu} = \boldsymbol{\gamma}^{\nu}\, \boldsymbol{\gamma}^{\mu}\, p_{\nu} p_{\mu} = \boldsymbol{\gamma}^{\nu}\, \boldsymbol{\gamma}^{\mu}\, p_{\mu} p_{\nu}$$

$$= \frac{1}{2}(\boldsymbol{\gamma}^{\mu}\, \boldsymbol{\gamma}^{\nu} + \boldsymbol{\gamma}^{\nu}\, \boldsymbol{\gamma}^{\mu})\, p_{\mu} p_{\nu} = g^{\mu\nu}\, p_{\mu} p_{\nu} = p^{\nu}\, p_{\nu} = p_0^2 - \mathbf{p}^2 = m^2 \tag{145}$$

Now

$$\gamma^0\, \hat{\Lambda}_{\text{sgn}\,\varepsilon_r}(\mathbf{p})\, \gamma^0 = \frac{\gamma^0 (\varepsilon_r \not{p} + m)\gamma^0}{2m} = \frac{\gamma^0 (\varepsilon_r\, \gamma^{\mu}\, p_{\mu})\gamma^0 + m}{2m}$$

$$= \frac{\varepsilon_r\, \gamma^0\, \gamma^{\mu}\, p_{\mu}\, \gamma^0 + m}{2m} = \frac{\varepsilon_r\, (2g^{0\mu} - \gamma^{\mu}\, \gamma^0)\, p_{\mu}\, \gamma^0 + m}{2m}$$

$$= \frac{\varepsilon_r\, (2g^{0\mu} - \gamma^{\mu}\, \gamma^0)\gamma^0\, p_{\mu} + m}{2m}$$

$$= \frac{\varepsilon_r\, (2g^{0\mu}\, \gamma^0 - \gamma^{\mu})\, p_{\mu} + m}{2m} = \frac{\varepsilon_r\, (2\gamma^0\, p^0 - \gamma^{\mu}\, p_{\mu}) + m}{2m}$$

$$= \frac{\varepsilon_r\, (2\gamma^0\, p_0 - \gamma^{\mu}\, p_{\mu}) + m}{2m}$$

$$= \frac{\varepsilon_r\, (\gamma^0\, p_0 - \gamma^k\, p_k) + m}{2m} = \frac{\varepsilon_r\, (\gamma^{0\dagger}\, p_0 + \gamma^{k\dagger}\, p_k) + m}{2m}$$

$$= \frac{\varepsilon_r\, (\gamma^{0\dagger}\, p_0^{\dagger} + \gamma^{k\dagger}\, p_k^{\dagger}) + m}{2m}$$

$$= \frac{\varepsilon_r \gamma^{\mu\dagger} p_m^\dagger + m}{2m} = \frac{\varepsilon_r p_\mu^\dagger \gamma^{\mu\dagger} + m}{2m}$$

$$= \frac{\varepsilon_r (\gamma^\mu p_\mu)^\dagger + m}{2m} = [\hat{\Lambda}_{\text{sgn}\,\varepsilon_r}(\mathbf{p})]^\dagger \tag{146}$$

6.9 SPIN PROJECTION OPERATORS

We first obtain the explicit representation of the spin projection operator with respect to the x^3 axis in the rest frame of the Dirac particle. The unit 4-vector in the direction of the x^3 axis in the rest frame of the particle is given by $(n_3^\mu)_{RF} = (0, 0, 0, 1) \equiv (0, \mathbf{n}_3)$. In analogy with the non-relativistic expression for the spin projection operator with respect to the polarization about the x^3 axis which is given by, $\hat{\sigma}_s = \frac{1 + \varepsilon_s \sigma_3}{2}$, we write, in the relativistic framework

$$(\hat{\Sigma}(\varepsilon_s n_3))_{RF} = \frac{\mathbf{I} + \varepsilon_s \Sigma_3}{2} = \frac{1}{2}\left[\mathbf{I} + \varepsilon_s \begin{pmatrix} \sigma_3 & \mathbf{0} \\ \mathbf{0} & \sigma_3 \end{pmatrix}\right]$$

$$= \frac{1}{2}\left[\mathbf{I} + i\varepsilon_s \begin{pmatrix} -\sigma_1\sigma_2 & \mathbf{0} \\ \mathbf{0} & -\sigma_1\sigma_2 \end{pmatrix}\right]$$

$$= \frac{1}{2}\left[\mathbf{I} + i\varepsilon_s \begin{pmatrix} \mathbf{0} & \sigma_1 \\ -\sigma_1 & \mathbf{0} \end{pmatrix}\begin{pmatrix} \mathbf{0} & \sigma_2 \\ -\sigma_2 & \mathbf{0} \end{pmatrix}\right]$$

$$= \frac{1}{2}(\mathbf{I} + i\varepsilon_s \gamma^1 \gamma^2) = \frac{1}{2}[\mathbf{I} - i\varepsilon_s (\gamma^0)^2 \gamma^1 \gamma^2 (\gamma^3)^2]$$

$$= \frac{1}{2}[\mathbf{I} - \varepsilon_s i(\gamma^0)^2 \gamma^1 \gamma^2 (\gamma^3)^2] = \frac{1}{2}[\mathbf{1} - \varepsilon_s i\gamma^0 \gamma^1 \gamma^2 (\gamma^3)^2 \gamma^0]$$

$$= \frac{1}{2}(\mathbf{I} - \varepsilon_s \gamma^5 \gamma^3 \gamma^0) = \frac{1}{2}(\mathbf{I} + \varepsilon_s \gamma^5 \gamma_3 \gamma_0) = \frac{1}{2}[\mathbf{I} + \varepsilon_s \gamma^5 (n_3^3)_{RF} \gamma_3 \gamma_0]$$

$$= \frac{1}{2}[\mathbf{I} + \varepsilon_s \gamma^5 (n_3^\mu)_{RF} \gamma_\mu \gamma_0] = \frac{1}{2}[\mathbf{I} + \varepsilon_s \gamma^5 (\not{n}_3)_{RF} \gamma_0] \tag{147}$$

where we have used the fact $n_3^\mu = (n_3^0, n_3^1, n_3^2, n_3^3) = (0, 0, 0, 1)$ so that $(n_3^\mu)\gamma_\mu = (n_3^0)\gamma_0 + (n_3^1)\gamma_1 + (n_3^2)\gamma_2 + (n_3^3)\gamma_3 = (n_3^3)\gamma_3$. The effect of this operator on the components of an arbitrary rest spinor $\omega^{(r)}(\mathbf{0})$ is given by

$$(\hat{\Sigma}(\varepsilon_s n_3))_{RF}\, \omega^{(r)}(\mathbf{0}) = \frac{\mathbf{I} + \varepsilon_s \Sigma_3}{2}\, \omega^{(r)}(\mathbf{0}) = \frac{1}{2}\left[\mathbf{I} + \varepsilon_s \begin{pmatrix} \sigma_3 & \mathbf{0} \\ \mathbf{0} & \sigma_3 \end{pmatrix}\right] \omega^{(r)}(\mathbf{0})$$

$$= \frac{1}{2}\begin{pmatrix} 1+\varepsilon_s & 0 & 0 & 0 \\ 0 & 1-\varepsilon_s & 0 & 0 \\ 0 & 0 & 1+\varepsilon_s & 0 \\ 0 & 0 & 0 & 1-\varepsilon_s \end{pmatrix}\begin{pmatrix} \omega_1^{(r)}(\mathbf{0}) \\ \omega_2^{(r)}(\mathbf{0}) \\ \omega_3^{(r)}(\mathbf{0}) \\ \omega_4^{(r)}(\mathbf{0}) \end{pmatrix} = \frac{1}{2}\begin{pmatrix} (1+\varepsilon_s)\omega_1^{(r)}(\mathbf{0}) \\ (1-\varepsilon_s)\omega_2^{(r)}(\mathbf{0}) \\ (1+\varepsilon_s)\omega_3^{(r)}(\mathbf{0}) \\ (1-\varepsilon_s)\omega_4^{(r)}(\mathbf{0}) \end{pmatrix} \tag{148}$$

which leads to the following expressions for the impact of the spin projection operator in the rest frame on the rest spinors

$$(\hat{\Sigma}(\varepsilon_s\, n_3))_{RF}\begin{pmatrix} \omega^{(1)}(\mathbf{0}) \\ \omega^{(2)}(\mathbf{0}) \\ \omega^{(3)}(\mathbf{0}) \\ \omega^{(4)}(\mathbf{0}) \end{pmatrix} = \frac{1}{2}\begin{pmatrix} (1+\varepsilon_s)\omega^{(1)}(\mathbf{0}) \\ (1-\varepsilon_s)\omega^{(2)}(\mathbf{0}) \\ (1+\varepsilon_s)\omega^{(3)}(\mathbf{0}) \\ (1-\varepsilon_s)\omega^{(4)}(\mathbf{0}) \end{pmatrix} \tag{149}$$

We now proceed to obtain covariant generalization of the spin projection operator. The projection operator defined by eq. (147) is not relativistically covariant due to the presence of $\boldsymbol{\gamma}_0$ term in the numerator although the other two terms, viz. $\boldsymbol{\gamma}^5$ and $(\not{n}_3)_{RF}$ are relativistically covariant. In this context, we note that the 4-component vector $(n_3^{\mu})_{RF} = (0, 0, 0, 1) = (0, \mathbf{n}_3)$ with $\mathbf{n}_3 = (0, 0, 1)$ can be transformed into an arbitrary inertial frame by an appropriate Lorentz transformation, i.e.,

$$n_3^{\mu} = a_{\nu}^{\mu}\,(n_3^{\nu})_{RF} \tag{150}$$

Furthermore, since $\boldsymbol{\gamma}^0 = \boldsymbol{\gamma}_0 = \begin{pmatrix} \mathbf{I} & \mathbf{0} \\ \mathbf{0} & -\mathbf{I} \end{pmatrix}$ the effect of $\boldsymbol{\gamma}^0$ on the rest spinors $\omega^{(r)}(\mathbf{0})$ is given by

$$\boldsymbol{\gamma}^0\,\omega^{(r)}(\mathbf{0}) = \boldsymbol{\gamma}_0\,\omega^{(r)}(\mathbf{0}) = \varepsilon_r\,\omega^{(r)}(\mathbf{0}) \tag{151}$$

We, therefore, consider the spin projection operator about the x^3 axis in an arbitrary inertial frame as

$$\hat{\Sigma}(\varepsilon_s\, n_3) = \frac{1}{2}\,[\mathbf{I} + \varepsilon_s\,\gamma^5(\not{n}_3)] = \frac{\mathbf{I} + \varepsilon_s\Sigma_3\gamma_0}{2}$$

$$= \frac{1}{2}\left[\mathbf{I} + \varepsilon_s\begin{pmatrix} \sigma_3 & \mathbf{0} \\ \mathbf{0} & \sigma_3 \end{pmatrix}\begin{pmatrix} \mathbf{I} & \mathbf{0} \\ \mathbf{0} & -\mathbf{I} \end{pmatrix}\right] = \frac{1}{2}\left[\mathbf{I} + \varepsilon_s\begin{pmatrix} \sigma_3 & \mathbf{0} \\ \mathbf{0} & -\sigma_3 \end{pmatrix}\right]$$

$$= \frac{1}{2}\left[\mathbf{I} + \varepsilon_s\begin{pmatrix} 1 & 0 & 0 & 0 \\ 0 & -1 & 0 & 0 \\ 0 & 0 & -1 & 0 \\ 0 & 0 & 0 & 1 \end{pmatrix}\right] = \frac{1}{2}\begin{pmatrix} 1+\varepsilon_s & 0 & 0 & 0 \\ 0 & 1-\varepsilon_s & 0 & 0 \\ 0 & 0 & 1-\varepsilon_s & 0 \\ 0 & 0 & 0 & 1+\varepsilon_s \end{pmatrix} \tag{152}$$

The action of this operator on the components of an arbitrary rest spinor $\omega^{(r)}(\mathbf{0})$ is seen by

$$\hat{\Sigma}(\varepsilon_s\, n_3)\omega^{(r)}(\mathbf{0}) = \frac{1}{2}\begin{pmatrix} 1+\varepsilon_s & 0 & 0 & 0 \\ 0 & 1-\varepsilon_s & 0 & 0 \\ 0 & 0 & 1-\varepsilon_s & 0 \\ 0 & 0 & 0 & 1+\varepsilon_s \end{pmatrix}$$

$$\left(\begin{pmatrix} \omega_1^{(r)}(\mathbf{0}) \\ \omega_2^{(r)}(\mathbf{0}) \\ \omega_3^{(r)}(\mathbf{0}) \\ \omega_4^{(r)}(\mathbf{0}) \end{pmatrix}\right) = \frac{1}{2}\begin{pmatrix} (1+\varepsilon_s)\omega_1^{(r)}(\mathbf{0}) \\ (1-\varepsilon_s)\omega_2^{(r)}(\mathbf{0}) \\ (1-\varepsilon_s)\omega_3^{(r)}(\mathbf{0}) \\ (1+\varepsilon_s)\omega_4^{(r)}(\mathbf{0}) \end{pmatrix} \tag{153}$$

which leads us to the following explicit relations:

$$\hat{\Sigma}(\varepsilon_s\, n_3)\begin{pmatrix} \omega^{(1)}(\mathbf{0}) \\ \omega^{(2)}(\mathbf{0}) \\ \omega^{(3)}(\mathbf{0}) \\ \omega^{(4)}(\mathbf{0}) \end{pmatrix} = \frac{1}{2}\begin{pmatrix} (1-\varepsilon_s)\omega^{(1)}(\mathbf{0}) \\ (1-\varepsilon_s)\omega^{(2)}(\mathbf{0}) \\ (1-\varepsilon_s)\omega^{(3)}(\mathbf{0}) \\ (1+\varepsilon_s)\omega^{(4)}(\mathbf{0}) \end{pmatrix} \tag{154}$$

If we define

$$\begin{aligned} &\omega^{(1)}(\mathbf{p}) = u(\mathbf{p}, \mathbf{n}_3),\ \omega^{(2)}(\mathbf{p}) = u(\mathbf{p}, -\mathbf{n}_3) \\ &\omega^{(3)}(\mathbf{p}) = u(-\mathbf{p}, -\mathbf{n}_3) = v(\mathbf{p}, -\mathbf{n}_3), \\ &\omega^{(4)}(\mathbf{p}) = u(-\mathbf{p}, \mathbf{n}_3) = v(\mathbf{p}, \mathbf{n}_3) \end{aligned} \tag{155}$$

then

$$\hat{\Sigma}(\varepsilon_s\, n_3)\begin{pmatrix} u(\mathbf{0}, \mathbf{n}_3) \\ u(\mathbf{0}, -\mathbf{n}_3) \\ v(\mathbf{0}, -\mathbf{n}_3) \\ v(\mathbf{0}, \mathbf{n}_3) \end{pmatrix} = \frac{1}{2}\begin{pmatrix} (1+\varepsilon_s)\, u(\mathbf{0}, \mathbf{n}_3) \\ (1-\varepsilon_s)\, u(\mathbf{0}, -\mathbf{n}_3) \\ (1-\varepsilon_s)\, v(\mathbf{0}, -\mathbf{n}_3) \\ (1+\varepsilon_s)\, v(\mathbf{0}, \mathbf{n}_3) \end{pmatrix} \tag{156}$$

Furthermore, using eq. (152), we get

$$\hat{\Sigma}(\varepsilon_s\, n_3)\begin{pmatrix} \omega_1^{(r)}(\mathbf{p}) \\ \omega_2^{(r)}(\mathbf{p}) \\ \omega_3^{(r)}(\mathbf{p}) \\ \omega_4^{(r)}(\mathbf{p}) \end{pmatrix} = \frac{1}{2}\,(I + \varepsilon_s\, \Sigma_3\, \gamma_0).\begin{pmatrix} \omega_1^{(r)}(\mathbf{p}) \\ \omega_2^{(r)}(\mathbf{p}) \\ \omega_3^{(r)}(\mathbf{p}) \\ \omega_4^{(r)}(\mathbf{p}) \end{pmatrix}$$

$$= \frac{1}{2}\begin{pmatrix} 1+\varepsilon_s & 0 & 0 & 0 \\ 0 & 1-\varepsilon_s & 0 & 0 \\ 0 & 0 & 1-\varepsilon_s & 0 \\ 0 & 0 & 0 & 1+\varepsilon_s \end{pmatrix} \cdot \begin{pmatrix} \omega_1^{(r)}(\mathbf{p}) \\ \omega_2^{(r)}(\mathbf{p}) \\ \omega_3^{(r)}(\mathbf{p}) \\ \omega_4^{(r)}(\mathbf{p}) \end{pmatrix}$$

$$= \frac{1}{2}\begin{pmatrix} (1+\varepsilon_s)\omega_1^{(r)}(\mathbf{p}) \\ (1-\varepsilon_s)\omega_2^{(r)}(\mathbf{p}) \\ (1-\varepsilon_s)\omega_3^{(r)}(\mathbf{p}) \\ (1+\varepsilon_s)\omega_4^{(r)}(\mathbf{p}) \end{pmatrix} \tag{157}$$

Actually, the very fact that the spin projection operator is in a relativistically covariant form implies that the results of eq. (156) will hold in any moving frame as well. Hence, we will have

$$\hat{\Sigma}(\varepsilon_t\, n_3)\begin{pmatrix} u(\mathbf{p},\mathbf{n}_3) \\ u(\mathbf{p},-\mathbf{n}_3) \\ v(\mathbf{p},-\mathbf{n}_3) \\ v(\mathbf{p},\mathbf{n}_3) \end{pmatrix} = \frac{1}{2}\begin{pmatrix} (1+\varepsilon_s)\,u(\mathbf{p},\mathbf{n}_3) \\ (1-\varepsilon_s)\,u(\mathbf{p},-\mathbf{n}_3) \\ (1-\varepsilon_s)\,v(\mathbf{p},-\mathbf{n}_3) \\ (1+\varepsilon_s)\,v(\mathbf{p},\mathbf{n}_3) \end{pmatrix} \tag{158}$$

It is important to note that the spin projection operator $\hat{\Sigma}(\varepsilon_s n_3)$ does not commute with either the spinor transformation operator $\hat{S} = \mathbf{I}\cosh\left(\frac{\omega}{2}\right) - \left(\frac{\alpha.\mathbf{v}}{|\mathbf{v}|}\right)\sinh\left(\frac{\omega}{2}\right)$ or, indeed, the Dirac Hamiltonian so that we cannot have simultaneous eigenfunctions for the spin projection operator and either $\hat{S}$ or the Hamiltonian. It is instructive to observe the consequences of the non-commutativity of the spin projection operator with the spinor transformation operator. We have

$$\hat{\Sigma}(\varepsilon_s\, n_3)\begin{pmatrix} \omega_1^{(r)}(\mathbf{p}) \\ \omega_2^{(r)}(\mathbf{p}) \\ \omega_3^{(r)}(\mathbf{p}) \\ \omega_4^{(r)}(\mathbf{p}) \end{pmatrix} = \frac{1}{2}(\mathbf{I} + \varepsilon_s\,\Sigma_3\,\gamma_0).\left(\mathbf{I}\cosh\frac{\omega}{2} - \frac{\alpha.\mathbf{p}}{|\mathbf{p}|}\sinh\frac{\omega}{2}\right)\begin{pmatrix} \omega_1^{(r)}(\mathbf{0}) \\ \omega_2^{(r)}(\mathbf{0}) \\ \omega_3^{(r)}(\mathbf{0}) \\ \omega_4^{(r)}(\mathbf{0}) \end{pmatrix}$$

$$= \frac{1}{2}\begin{pmatrix} 1+\varepsilon_s & 0 & 0 & 0 \\ 0 & 1-\varepsilon_s & 0 & 0 \\ 0 & 0 & 1-\varepsilon_s & 0 \\ 0 & 0 & 0 & 1+\varepsilon_s \end{pmatrix}.$$

$$\begin{pmatrix} \kappa & 0 & -\frac{p_3}{|\mathbf{p}|}\lambda & -\frac{p_1 - ip_2}{|\mathbf{p}|}\lambda \\ 0 & \kappa & -\frac{p_1 + ip_2}{|\mathbf{p}|}\lambda & \frac{p_3}{|\mathbf{p}|}\lambda \\ -\frac{p_3}{|\mathbf{p}|}\lambda & -\frac{p_1 - ip_2}{|\mathbf{p}|}\lambda & \kappa & 0 \\ -\frac{p_1 + ip_2}{|\mathbf{p}|}\lambda & \frac{p_3}{|\mathbf{p}|}\lambda & 0 & \kappa \end{pmatrix} \begin{pmatrix} \omega_1^{(r)}(\mathbf{0}) \\ \omega_2^{(r)}(\mathbf{0}) \\ \omega_3^{(r)}(\mathbf{0}) \\ \omega_4^{(r)}(\mathbf{0}) \end{pmatrix}$$

$$= \frac{1}{2} \begin{pmatrix} (1+\varepsilon_s)\kappa & 0 & -(1+\varepsilon_s)\frac{p_3}{|\mathbf{p}|}\lambda & -(1+\varepsilon_s)\frac{p_1 - ip_2}{|\mathbf{p}|}\lambda \\ 0 & (1-\varepsilon_s)\kappa & -(1-\varepsilon_s)\frac{p_1 + ip_2}{|\mathbf{p}|}\lambda & (1-\varepsilon_s)\frac{p_3}{|\mathbf{p}|}\lambda \\ -(1-\varepsilon_s)\frac{p_3}{|\mathbf{p}|}\lambda & -(1-\varepsilon_s)\frac{p_1 - ip_2}{|\mathbf{p}|}\lambda & (1-\varepsilon_s)\kappa & 0 \\ -(1+\varepsilon_s)\frac{p_1 + ip_2}{|\mathbf{p}|}\lambda & (1+\varepsilon_s)\frac{p_3}{|\mathbf{p}|}\lambda & 0 & (1+\varepsilon_s)\kappa \end{pmatrix}$$

$$\begin{pmatrix} \omega_1^{(r)}(\mathbf{0}) \\ \omega_2^{(r)}(\mathbf{0}) \\ \omega_3^{(r)}(\mathbf{0}) \\ \omega_4^{(r)}(\mathbf{0}) \end{pmatrix}$$

$$= \frac{1}{2} \left\{ \begin{matrix} (1+\varepsilon_s)\kappa\omega_1^{(r)}(\mathbf{0}) - (1+\varepsilon_s)\frac{p_3}{|\mathbf{p}|}\lambda\omega_3^{(r)}(\mathbf{0}) - (1+\varepsilon_s)\frac{p_1 - ip_2}{|\mathbf{p}|}\lambda\omega_4^{(r)}(\mathbf{0}) \\ (1-\varepsilon_s)\kappa\omega_2^{(r)}(\mathbf{0}) - (1-\varepsilon_s)\frac{p_1 + ip_2}{|\mathbf{p}|}\lambda\omega_3^{(r)}(\mathbf{0}) + (1-\varepsilon_s)\frac{p_3}{|\mathbf{p}|}\lambda\omega_4^{(r)}(\mathbf{0}) \\ -(1-\varepsilon_s)\frac{p_3}{|\mathbf{p}|}\lambda\omega_1^{(r)}(\mathbf{0}) - (1-\varepsilon_s)\frac{p_1 - ip_2}{|\mathbf{p}|}\lambda\omega_2^{(r)}(\mathbf{0}) + (1-\varepsilon_s)\kappa\omega_3^{(r)}(\mathbf{0}) \\ -(1+\varepsilon_s)\frac{p_1 + ip_2}{|\mathbf{p}|}\lambda\omega_1^{(r)}(\mathbf{0}) + (1+\varepsilon_s)\frac{p_3}{|\mathbf{p}|}\lambda\omega_2^{(r)}(\mathbf{0}) + (1+\varepsilon_s)\kappa\omega_4^{(r)}(\mathbf{0}) \end{matrix} \right\} \tag{159}$$

This may be contrasted with

$$\hat{S}\,\hat{\Sigma}(\varepsilon_s\, n_3) \begin{pmatrix} \omega_1^{(r)}(\mathbf{0}) \\ \omega_2^{(r)}(\mathbf{0}) \\ \omega_3^{(r)}(\mathbf{0}) \\ \omega_4^{(r)}(\mathbf{0}) \end{pmatrix} = \frac{1}{2}\left(\mathbf{I}\cosh\frac{\omega}{2} - \frac{\boldsymbol{\alpha}.\mathbf{p}}{|\mathbf{p}|}\sinh\frac{\omega}{2}\right)(\mathbf{I} + \varepsilon_s\,\Sigma_3\,\boldsymbol{\gamma}_0) \begin{pmatrix} \omega_1^{(r)}(\mathbf{0}) \\ \omega_2^{(r)}(\mathbf{0}) \\ \omega_3^{(r)}(\mathbf{0}) \\ \omega_4^{(r)}(\mathbf{0}) \end{pmatrix}$$

$$= \frac{1}{2}\begin{pmatrix} \kappa & 0 & -\frac{p_3}{|\mathbf{p}|}\lambda & -\frac{p_1 - ip_2}{|\mathbf{p}|}\lambda \\ 0 & \kappa & -\frac{p_1 + ip_2}{|\mathbf{p}|}\lambda & \frac{p_3}{|\mathbf{p}|}\lambda \\ -\frac{p_3}{|\mathbf{p}|}\lambda & -\frac{p_1 - ip_2}{|\mathbf{p}|}\lambda & \kappa & 0 \\ -\frac{p_1 + ip_2}{|\mathbf{p}|}\lambda & \frac{p_3}{|\mathbf{p}|}\lambda & 0 & \kappa \end{pmatrix}$$

$$\begin{pmatrix} 1+\varepsilon_s & 0 & 0 & 0 \\ 0 & 1-\varepsilon_s & 0 & 0 \\ 0 & 0 & 1-\varepsilon_s & 0 \\ 0 & 0 & 0 & 1+\varepsilon_s \end{pmatrix}\begin{pmatrix} \omega_1^{(r)}(\mathbf{0}) \\ \omega_2^{(r)}(\mathbf{0}) \\ \omega_3^{(r)}(\mathbf{0}) \\ \omega_4^{(r)}(\mathbf{0}) \end{pmatrix}$$

$$= \frac{1}{2}\left\{\begin{matrix} (1+\varepsilon_s)\kappa & 0 & -(1-\varepsilon_s)\frac{p_3}{|\mathbf{p}|}\lambda & -(1+\varepsilon_s)\frac{p_1 - ip_2}{|\mathbf{p}|}\lambda \\ 0 & (1-\varepsilon_s)\kappa & -(1-\varepsilon_s)\frac{p_1 + ip_2}{|\mathbf{p}|}\lambda & (1+\varepsilon_s)\frac{p_3}{|\mathbf{p}|}\lambda \\ -(1+\varepsilon_s)\frac{p_3}{|\mathbf{p}|}\lambda & -(1-\varepsilon_s)\frac{p_1 - ip_2}{|\mathbf{p}|}\lambda & (1-\varepsilon_s)\kappa & 0 \\ -(1-\varepsilon_s)\frac{p_1 + ip_2}{|\mathbf{p}|}\lambda & (1-\varepsilon_s)\frac{p_3}{|\mathbf{p}|}\lambda & 0 & (1+\varepsilon_s)\kappa \end{matrix}\right\}$$

$$\begin{pmatrix} \omega_1^{(r)}(\mathbf{0}) \\ \omega_2^{(r)}(\mathbf{0}) \\ \omega_3^{(r)}(\mathbf{0}) \\ \omega_4^{(r)}(\mathbf{0}) \end{pmatrix}$$

$$= \frac{1}{2}\begin{pmatrix} (1+\varepsilon_s)\kappa\omega_1^{(r)}(\mathbf{0}) - (1-\varepsilon_s)\frac{p_3}{|\mathbf{p}|}\lambda\omega_3^{(r)}(\mathbf{0}) - (1+\varepsilon_s)\frac{p_1 - ip_2}{|\mathbf{p}|}\lambda\omega_4^{(r)}(\mathbf{0}) \\ (1-\varepsilon_s)\kappa\omega_2^{(r)}(\mathbf{0}) - (1-\varepsilon_s)\frac{p_1 + ip_2}{|\mathbf{p}|}\lambda\omega_3^{(r)}(\mathbf{0}) + (1+\varepsilon_s)\frac{p_3}{|\mathbf{p}|}\lambda\omega_4^{(r)}(\mathbf{0}) \\ -(1+\varepsilon_s)\frac{p_3}{|\mathbf{p}|}\lambda\omega_1^{(r)}(\mathbf{0}) - (1-\varepsilon_s)\frac{p_1 - ip_2}{|\mathbf{p}|}\lambda\omega_2^{(r)}(\mathbf{0}) + (1-\varepsilon_s)\kappa\omega_3^{(r)}(\mathbf{0}) \\ -(1+\varepsilon_s)\frac{p_1 + ip_2}{|\mathbf{p}|}\lambda\omega_1^{(r)}(\mathbf{0}) + (1-\varepsilon_s)\frac{p_3}{|\mathbf{p}|}\lambda\omega_2^{(r)}(\mathbf{0}) + (1+\varepsilon_s)\kappa\omega_4^{(r)}(\mathbf{0}) \end{pmatrix} \tag{160}$$

where we have abbreviated $\cosh\frac{\omega}{2}$ κ, $\sinh\frac{\omega}{2} = \lambda$. A comparison of eqs. (159) and (160) explicitly shows the non-commutativity of the spin projection operator and the spinor transformation operator. The results of eqs. (159) and (160), however, do agree in the special case when $p_3 = 0$, i.e. the particle has no motion in the direction of the orientation axis. A perusal of eq. (160) further shows that if the spin projection operator is applied first on the zero spinors and then transformed to the desired momentum state, there is no mixing of the spinors with differing spin orientations. This is seen from the fact that, whereas $\omega_1^{(r)}(\mathbf{0})$ and $\omega_4^{(r)}(\mathbf{0})$ carry the coefficient $(1 + \varepsilon_s)$, the spinors $\omega_2^{(r)}(\mathbf{0})$ and $\omega_3^{(r)}(\mathbf{0})$ carry the coefficient $(1 - \varepsilon_s)$ which is commensurate with their orientations being opposite as seen in eq. (154). This is to be expected since the prior application of the spin projection operator would segregate the spinors with different spins, i.e., $\omega_1^{(r)}(\mathbf{0})$ and $\omega_4^{(r)}(\mathbf{0})$ would be segregated from $\omega_2^{(r)}(\mathbf{0})$ and $\omega_3^{(r)}(\mathbf{0})$. However, when the spin projection operator is applied after the spinors are transformed to the desired momentum state, then the segregation of spins would take effect on the transformed spinors $\omega^{(r)}(\mathbf{p})$ and not $\omega^{(r)}(\mathbf{0})$ as in the previous case. The output would, therefore, be different in both cases.

Equation (159) also gives

$$\hat{\Sigma}(\varepsilon_s\, n_3)\omega^1(\mathbf{p}) = \frac{1}{2}\,\varepsilon \begin{pmatrix} (1+\varepsilon_s)\kappa \\ 0 \\ -(1-\varepsilon_s)\frac{p_3}{|\mathbf{p}|}\lambda \\ -(1+\varepsilon_s)\frac{p_1+ip_2}{|\mathbf{p}|}\lambda \end{pmatrix},$$

$$\hat{\Sigma}(\varepsilon_s\, n_3)\omega^2(\mathbf{p}) = \frac{1}{2} \begin{pmatrix} 0 \\ (1-\varepsilon_s)\kappa \\ -(1-\varepsilon_s)\frac{p_1-ip_2}{|\mathbf{p}|}\lambda \\ (1+\varepsilon_s)\frac{p_3}{|\mathbf{p}|}\lambda \end{pmatrix}$$

$$\hat{\Sigma}(\varepsilon_s\, n_3)\omega^3(\mathbf{p}) = \frac{1}{2} \begin{pmatrix} -(1+\varepsilon_s)\frac{p_3}{|\mathbf{p}|}\lambda \\ -(1-\varepsilon_s)\frac{p_1+ip_2}{|\mathbf{p}|}\lambda \\ (1-\varepsilon_s)\kappa \\ 0 \end{pmatrix},$$

$$\hat{\Sigma}(\varepsilon_s\, n_3)\omega^4(\mathbf{p}) = \frac{1}{2} \begin{pmatrix} -(1+\varepsilon_s)\frac{p_1-ip_2}{|\mathbf{p}|}\lambda \\ (1-\varepsilon_s)\frac{p_3}{|\mathbf{p}|}\lambda \\ 0 \\ (1+\varepsilon_s)\kappa \end{pmatrix} \tag{161}$$

We have so far explicitly obtained an expression for the spin projection operator about the $\mathbf{n}_3$ axis. We now generalize the expression for this operator for spin about an arbitrary axis along the direction of the unit vector $n = (0, \mathbf{n}) = (0, n_1, n_2, n_3)$.

We know from elementary quantum mechanics that the rotation operator in three dimensions is given by

$$\hat{S}_R(\phi) = \exp\left(-\frac{i}{2}\phi\boldsymbol{\sigma}.\mathbf{n}\right) \tag{162}$$

where $\boldsymbol{\sigma} = (\boldsymbol{\sigma}_1, \boldsymbol{\sigma}_2, \boldsymbol{\sigma}_3)$ is the set of three Pauli matrices and ϕ the angle of rotation around the axis represented by the unit vector $\mathbf{n} = (n_1, n_2, n_3)$ which is, sometimes referred to as the polarization of the particle. This is viewing the rotation in the active form wherein the state vector is rotated about a well, defined fixed axis of rotation. The same physical phenomenon can be viewed in the passive form whence the state vector remains unchanged and the axis of rotation, and, therefore, the basis is deemed to be rotating rather than the state vector. It is usually more convenient to study spin related physical phenomena by considering rotations in the passive form. The corresponding rotation operator will obviously become

$$\hat{S}_R(\phi) = \exp\left(\frac{i}{2}\phi\boldsymbol{\sigma}.\mathbf{n}\right) \tag{163}$$

The above expression for the rotation operator in three dimensions admits an obvious generalization in the relativistic framework as

$$\hat{S}_R(\phi) = \exp\left(\frac{i}{2}\phi\hat{\Sigma}_R.\mathbf{n}\right) \tag{164}$$

where $\hat{\Sigma}_R = \begin{pmatrix} \sigma & \mathbf{0} \\ \mathbf{0} & \sigma \end{pmatrix}$ and $\mathbf{n}$ is the unit vector along the rotating axis.

When the operator $\hat{S}_R$ acts on the rest spinors $\omega^{(r)}(\mathbf{0})$, we will obtain states that are polarized in the direction of the unit vector $\mathbf{n}$. Actually, the eigenstates of the operator $\boldsymbol{\Sigma}_R.\mathbf{n}$ and, therefore, of the rotation operator $\hat{S}_R$ would be the states that are polarized in the direction of the unit vector $\mathbf{n}$, i.e., $(\hat{\Sigma}_R.\mathbf{n})\omega^{(r)}(\mathbf{0}) = \pm\,\omega^{(r)}(\mathbf{0})$. In fact

$$(\hat{\Sigma}_R.\mathbf{n})\omega^{(r)}(\mathbf{0}) = (\hat{\Sigma}_1.\mathbf{n}_1 + \hat{\Sigma}_2.\,\mathbf{n}_2 + \hat{\Sigma}_3.\,\mathbf{n}_3)\omega^{(r)}(\mathbf{0})$$

$$= \begin{pmatrix} \boldsymbol{\sigma}_1.\mathbf{n}_1 + \boldsymbol{\sigma}_2.\mathbf{n}_2 + \boldsymbol{\sigma}_3.\mathbf{n}_3 & 0 \\ 0 & \boldsymbol{\sigma}_1.\mathbf{n}_1 + \boldsymbol{\sigma}_2.\mathbf{n}_2 + \boldsymbol{\sigma}_3.\mathbf{n}_3 \end{pmatrix}\omega^{(r)}(\mathbf{0})$$

$$= \begin{pmatrix} n_3 & n_1 - in_2 & 0 & 0 \\ n_1 + in_2 & -n_3 & 0 & 0 \\ 0 & 0 & n_3 & n_1 - in_2 \\ 0 & 0 & n_1 + in_2 & -n_3 \end{pmatrix}\begin{pmatrix} \omega_1^{(r)}(\mathbf{0}) \\ \omega_2^{(r)}(\mathbf{0}) \\ \omega_3^{(r)}(\mathbf{0}) \\ \omega_4^{(r)}(\mathbf{0}) \end{pmatrix}$$

$$= \begin{pmatrix} n_3\,\omega_1^{(r)}(\mathbf{0}) + (n_1 - in_2)\omega_2^{(r)}(\mathbf{0}) \\ (n_1 + in_2)\omega_1^{(r)}(\mathbf{0}) - n_3\,\omega_2^{(r)}(\mathbf{0}) \\ n_3\,\omega_3^{(r)}(\mathbf{0}) + (n_1 - in_2)\omega_4^{(r)}(\mathbf{0}) \\ (n_1 + in_2)\omega_3^{(r)}(\mathbf{0}) - n_3\,\omega_4^{(r)}(\mathbf{0}) \end{pmatrix} \tag{165}$$

which gives

$$(\hat{\Sigma}_R.\mathbf{n})\omega^{(1)}(\mathbf{0}) = \begin{pmatrix} n_3 \\ n_1 + in_2 \\ 0 \\ 0 \end{pmatrix} \tag{166}$$

so that we have $(\hat{\Sigma}_R.\mathbf{n}_3)\omega^{(1)}(\mathbf{0}) = (1\ 0\ 0\ 0)^T = \omega^{(1)}(\mathbf{0})$ showing that $\omega^{(1)}(\mathbf{0})$ is polarized in the direction of $\mathbf{n}_3$ axis. Similarly, we can show that $\omega^{(3)}(\mathbf{0})$ is also polarized in the direction of $\mathbf{n}_3$ axis, while $\omega^{(2)}(\mathbf{0})$ and $\omega^{(4)}(\mathbf{0})$ are polarized in the direction of $-\mathbf{n}_3$.

The operator $\hat{\Sigma}_R.\mathbf{n}$ has the eigenvalues $\pm 1, \pm i$ with the eigenvectors corresponding to the two real eigenvalues being $(n_3 + 1, n_1 + in_2, n_3 + 1, n_1 + in_1)^T$ and $(n_3 - 1, n_1 + in_2, n_3 - 1, n_1 + in_1)^T$, respectively.

Now, in the relativistic theory, we need to transform the polarization 3-vector to a 4-vector. In the rest frame, this can be easily accomplished by writing

$$(n^\mu)_{RF} = (n^0, n^1, n^2, n^3) = (0, \mathbf{n}) = (0, \mathbf{n}_1, \mathbf{n}_2, \mathbf{n}_3) \tag{167}$$

The corresponding spin 4-vector in an arbitrary inertial frame can be obtained by applying the appropriate Lorentz transformation on $(n^\mu)_{RF}$, i.e.

$$n^\mu = a^\mu_\nu (n^\nu)_{RF} = a^\mu_k (n^k)_{RF} \tag{168}$$

Similarly, for the 4-momentum, we have

$$p^\mu = a^\mu_\nu (p^\nu)_{RF} = a^\mu_0 m \tag{169}$$

because $p^0 = (\mathbf{p}^2 + m^2)^{1/2} = m$ since $\mathbf{p} = p^k = 0$ in the rest frame.

Also

$$\sum_{\mu=0}^{3} (n^\mu)^2_{RF} = \mathbf{n}^2 = 1$$

and

$$\begin{aligned} n_\mu n^\mu &= a^\sigma_\mu (n_\sigma)_{RF}\, a^\mu_\rho (n^\rho)_{RF} = a^\sigma_\mu a^\mu_\rho (n_\sigma)_{RF} (n^\rho)_{RF} \\ &= \delta^\sigma_\rho (n_\sigma)_{RF} (n^\rho)_{RF} = (n_\sigma)_{RF} (n^\sigma)_{RF} = -\mathbf{n}^2 = -1 \end{aligned} \tag{170}$$

since $\quad n_\mu = g_{\mu\nu} n^\nu = (n^0, -n^1, -n^2, -n^3)$.

Further

$$p^\mu n_\mu = a^\mu_\rho (p^\rho)_{RF}\, a^\sigma_\mu (n_\sigma)_{RF}$$
$$= a^\mu_\rho\, a^\sigma_\mu (p^\rho)_{RF}\, (n_\sigma)_{RF} = \delta^\sigma_\rho\, (p^\rho)_{RF}\, (n_\sigma)_{RF}$$
$$= (p^\sigma)_{RF}\, (n_\sigma)_{RF} = (m\ 0\ 0\ 0)\ (0 - n^1 - n^2 - n^3)^T = 0 \tag{171}$$

since $\qquad (n_\mu)_{RF} = g_{\mu\nu}(n^\nu)_{RF} = (0, -n^1, -n^2, -n^3).$

The spin projection operator generalized for an arbitrary spin vector n^μ with $n^\mu\, p_\mu = n_\mu\, p^\mu = 0$ will take the form

$$\hat{\Sigma}(n) = \frac{1 + \gamma^5 \not{n}}{2} \tag{172}$$

We now show that the spin projection operator defined by eq. (172) has various properties of a projection operator. We have

$$\hat{\Sigma}(\varepsilon_1\, n)\ \hat{\Sigma}(\varepsilon_2\, n) = \frac{1 + \varepsilon_1\, \gamma^5 \not{n}}{2}\ \frac{1 + \varepsilon_2\, \gamma^5 \not{n}}{2}$$

$$= \frac{1 + (\varepsilon_1 + \varepsilon_2)\gamma_5 \not{n} \ + \varepsilon_1\, \varepsilon_2\, (\gamma^5 \not{n}\,)^2}{4}$$

$$= \frac{1 + (\varepsilon_1 + \varepsilon_2)\gamma^5 \not{n} \ + \varepsilon_1 \varepsilon_2}{4}$$

$$= \delta_{\varepsilon_1\varepsilon_2}\, \frac{1 + \varepsilon\gamma^5 \not{n}}{2} = \delta_{\varepsilon_1\varepsilon_2}\, \hat{\Sigma}(\varepsilon n) \tag{173}$$

because

$$(\gamma^5.\not{n})^2 = (\gamma^5\, \gamma_\mu\, n^\mu)\ (\gamma^5\, \gamma_\nu\, n^\nu)$$
$$= \gamma^5\, \gamma_\mu\, \gamma^5\, \gamma_\nu\, (n^\mu\, n^\nu) = -\, (\gamma^5)^2\, \gamma_\mu\, \gamma_\nu\, (n^\mu\, n^\nu)$$
$$= -\, \gamma_\mu\, \gamma_\nu\, (n^\mu\, n^\nu) = -\left(\frac{\gamma_\mu\, \gamma_\nu + \gamma_\nu\, \gamma_\mu}{2}\right)(n^\mu\, n^\nu)$$
$$= -\, g_{\mu\nu}\, (n^\mu\, n^\nu) = -\, n_\nu\, n^\nu = +\, 1 \tag{174}$$

Also

$$\hat{\Sigma}(\varepsilon_1\, n) + \hat{\Sigma}(\varepsilon_2\, n) = \frac{1 + \varepsilon_1\, \gamma^5 \not{n}}{2} + \frac{1 + \varepsilon_2\, \gamma^5 \not{n}}{2}$$

$$= \frac{2 + (\varepsilon_1 + \varepsilon_2)\gamma^5 \not{n}}{2} = 2\delta_{\varepsilon_1\varepsilon_2}\, \hat{\Sigma}(\varepsilon n) + \delta_{\varepsilon_1, -\varepsilon_2} \tag{175}$$

The energy and spin projection operators commute, e.g.,

$$[\hat{\Sigma}(\varepsilon_s\, n), \Lambda_{\text{sgn}\,\varepsilon_r}(p)] = \frac{1+\varepsilon_s\gamma^5 \not{n}}{2}\,\frac{\varepsilon_r \not{\mathbf{p}} + m}{2} - \frac{\varepsilon_r \not{\mathbf{p}} + m}{2}\,\frac{1+\varepsilon_s\gamma^5 \not{n}}{2}$$

$$= \frac{\varepsilon_r \not{\mathbf{p}} + \varepsilon_r \varepsilon_s \gamma^5 \not{n}\not{p} + m + \varepsilon_s m\gamma^5 \not{n}}{2}$$

$$- \frac{\varepsilon_r \not{\mathbf{p}} + m + \varepsilon_r \varepsilon_s \not{p}\gamma^5 \not{n} + \varepsilon_s m\gamma^5 \not{n}}{2}$$

$$\varepsilon_r\,\varepsilon_s \frac{\gamma^5 \not{n}\not{p} - \not{p}\gamma^5 \not{n}}{2} = \varepsilon_r\,\varepsilon_s\,\boldsymbol{\gamma}^5 \frac{\not{n}\not{\mathbf{p}} + \not{\mathbf{p}}\not{n}}{2}$$

$$= \varepsilon_r\,\varepsilon_s\,\boldsymbol{\gamma}^5 \frac{(\gamma_\mu n^\mu)(\gamma_\nu p^\nu) + (\gamma_\rho p^\rho)(\gamma_\sigma n^\sigma)}{2}$$

$$= \varepsilon_r\,\varepsilon_s\,\boldsymbol{\gamma}^5 \frac{\gamma_\mu \gamma_\nu n^\mu p^\nu + \gamma_\rho \gamma_\sigma p^\rho n^\sigma}{2}$$

$$= \varepsilon_r\,\varepsilon_s\,\boldsymbol{\gamma}^5 \frac{\gamma_\mu \gamma_\nu n^\mu p^\nu + (2g_{\rho\sigma} - \gamma_\sigma \gamma_\rho) n^\sigma p^\rho}{2}$$

$$= \varepsilon_r\,\varepsilon_s\,\boldsymbol{\gamma}^5 \frac{\gamma_\mu \gamma_\nu n^\mu p^\nu - \gamma_\sigma \gamma_\rho n^\sigma p^\rho + 2g_{\rho\sigma} n^\sigma p^\rho}{2}$$

$$= \varepsilon_r\,\varepsilon_s\,\boldsymbol{\gamma}^5 (g_{\rho\sigma} n^\sigma p^\rho) = \varepsilon_r\,\varepsilon_s\,\boldsymbol{\gamma}^5 (n_\rho p^\rho) = 0 \tag{176}$$

The effect of the covariant spin projection operators on the four Dirac spinors can be summarized as

$$\hat{\Sigma}(n) \begin{pmatrix} u(p,n) = \omega^{(1)}(\mathbf{p}) \\ u(p,-n) = \omega^{(2)}(\mathbf{p}) \\ v(p,-n) = u(-p,-n) = \omega^{(3)}(\mathbf{p}) \\ v(p,n) = u(-p,n) = \omega^{(4)}(\mathbf{p}) \end{pmatrix} = \begin{pmatrix} u(p,n) \\ 0 \\ 0 \\ v(p,n) = u(-p,n) \end{pmatrix} \tag{177}$$

or simply

$$\hat{\Sigma}(\varepsilon_s\, n)\, u(\varepsilon_p\, p, \varepsilon_s'\, n) = \delta_{\varepsilon_s \varepsilon_s'}\, u(\varepsilon_p\, p, \varepsilon_s'\, n) \tag{178}$$

It is very important to note that the covariant spin operator results in spin projections of negative energy states that are contragradient to the spin projection of the positive energy states. This is explicitly seen in eq. (154) and also in eqs. (157) and (178) wherein $\hat{\Sigma}(n)\,\omega^{(3)}(\mathbf{p}) = 0$,

$\hat{\Sigma}(n)\omega^{(4)}(\mathbf{p}) = \omega^{(4)}(\mathbf{p})$ for negative energies as against $\hat{\Sigma}(n)\omega^{(1)}(\mathbf{p}) = \omega^{(1)}(\mathbf{p})$ and $\hat{\Sigma}(n)\omega^{(2)}(\mathbf{p}) = 0$ for the two positive energy spinors.

A justification of this paradoxical result can be found in the "hole theory" of Dirac and the consequential existence of antiparticles. While the "normal" particles (positive 4-momentum, i.e., positive energy and positive **p**) spinors $u(p, \varepsilon n)$ satisfy the Dirac equation $(\not{p} - m)u(p, \varepsilon n) = 0$ and also $\mathbf{\Sigma.n}u(p, n) = u(p, n)$, the "antiparticles" (negative 4-momentum, i.e., negative energy and negative **p**) $v(p, \varepsilon n)$ satisfy the Dirac equation $(\not{p} + m)\, v(p, \varepsilon n) = 0$ and also $\mathbf{\Sigma.n}v(p, n) = -v(p, n)$. This is because the antiparticles are defined with opposite spin, e.g., we interpret an electron with negative energy, momentum $-\mathbf{p}$ and spin $\downarrow$ as a positron with positive energy, momentum $+\mathbf{p}$ and spin $\uparrow$.

An explicit representation of the covariant spin operator $\hat{\Sigma}(n)$ in the standard representation of the $\boldsymbol{\gamma}$ matrices, viz. $\boldsymbol{\gamma}^k = \begin{pmatrix} \mathbf{0} & \sigma^k \\ -\sigma^k & \mathbf{0} \end{pmatrix}$, $\boldsymbol{\gamma}^5 = \begin{pmatrix} \mathbf{0} & \mathbf{I} \\ \mathbf{I} & \mathbf{0} \end{pmatrix}$ is given by

$$\hat{\Sigma}(n) = \frac{1+\gamma^5 \not{n}}{2} = \frac{1}{2}\left\{\begin{pmatrix} \mathbf{I} & \mathbf{0} \\ \mathbf{0} & \mathbf{I} \end{pmatrix} + \begin{pmatrix} \mathbf{0} & \mathbf{I} \\ \mathbf{I} & \mathbf{0} \end{pmatrix}\left[\begin{pmatrix} \mathbf{I} & \mathbf{0} \\ \mathbf{0} & -\mathbf{I} \end{pmatrix} n_0 + \begin{pmatrix} \mathbf{0} & \sigma^k \\ -\sigma^k & \mathbf{0} \end{pmatrix} n_k\right]\right\}$$

$$= \frac{1}{2}\begin{pmatrix} \mathbf{I} - \sigma^k n_k & -\mathbf{I}n_0 \\ \mathbf{I}n_0 & \mathbf{I} + \sigma^k n_k \end{pmatrix} = \frac{1}{2}\begin{pmatrix} \mathbf{I} + \sum_{k=1}^{3}\sigma^k n^k & -\mathbf{I}n^0 \\ \mathbf{I}n^0 & \mathbf{I} - \sum_{k=1}^{3}\sigma^k n^k \end{pmatrix}$$

$$= \frac{1}{2}\begin{pmatrix} 1+n^3 & n^1 - in^2 & -n^0 & 0 \\ n^1 + in^2 & 1-n^3 & 0 & -n^0 \\ n^0 & 0 & 1-n^3 & -n^1 + in^2 \\ 0 & n^0 & -n^1 - in^2 & 1+n^3 \end{pmatrix} \tag{179}$$

It is pertinent to mention here that the real eigenvalues of the covariant spin projection operator $\hat{\Sigma}(n)$ are $+1, 0$ as they should be because the eigenvalues of the spin operator component $\boldsymbol{\gamma}^5\not{n}$ should be -1 since, spin should not alter the magnitude of the state vector. The corresponding eigenvectors are found to be $(n^3 + 1\ n^1 + in^2\ n^0\ 0)^T$ and $(n^3 - 1\ n^1 + in^2\ n^0\ 0)^T$, respectively.

6.10 SIMULTANEOUS ENERGY AND SPIN PROJECTION OPERATORS

We can define simultaneous projection operators for energy and momentum by their action on the various Dirac spinors as

$$Q(\varepsilon_r p, \varepsilon_s n) = \Lambda_{\text{sgn}\,\varepsilon_r}(p)\,\hat{\Sigma}(\varepsilon_s n)\,u(\varepsilon_{r'} p, \varepsilon_{s'} n) = \delta_{\varepsilon_r \varepsilon_{r'}}\delta_{\varepsilon_s \varepsilon_{s'}}\,u(\varepsilon_r p, \varepsilon_s n) \tag{180}$$

CHAPTER 7

Further Properties of the Dirac Equation

7.1 FOLDY-WOUTHUYSEN REPRESENTATION FOR FREE PARTICLES

The Dirac theory envisages that a free particle with positive energy has two independent states associated with each value of the well-defined momentum corresponding to the two possible directions of the particle spin. This implies that each set of physical states of a free particle should be represented by exactly two vectors in Hilbert space. Hence, the Dirac wave function for a free particle should be a 2-component vector. The standard Dirac theory that has been discussed so far, however, envisages a 4-component wave function. There must, therefore, exist a redundancy in the Dirac theory expounded so far and a representation thereof must exist in which the free particle wave function with well-defined momentum is a 2-component vector.

The existence of such a representation was confirmed theoretically and the transformation operators discovered by Foldy and Wouthuysen in 1950. They found a transformation in which the $\boldsymbol{\alpha}^k$'s. which are so-called "odd" operators (i.e., operators that couple the upper pair of components with the lower pair of components) can be eliminated from the Hamiltonian. In this representation called the Foldy-Wouthuysen representation (the Φ representation) the Hamiltonian consists only of "even" operators (i.e., the operators that do not couple the upper pair of components with the lower pair of components). Therefore, in this representation, the eigenfunctions of the energy eigenvalue equation can be expressed as 2-component spinors as mandated by the spin degrees of freedom.

A clue to the possible existence of such a representation is also provided by the fact that in the non-relativistic limit of the Dirac equation, the lower (upper) pair of components becomes insignificant in the limit $\frac{v}{c} \to 0$ for positive (negative) energy particles. There should, therefore, exist a representation in which the same structure holds even for non-vanishing $\frac{v}{c}$.

We now derive the explicit form of the Foldy-Wouthuysen transformation operators. Let us assume that $\hat{U}$ constitutes the transformation operator that removes all the "odd" operators from the free particle Dirac Hamiltonian and consequently decouples the upper pair of components from the lower pair of components in the 4-component Dirac wave function. Let

$$\hat{U} = \exp(i\hat{S}) \tag{1}$$

Since we want $\hat{U}$ to be unitary so that the underlying physics does not change due to the envisaged transformation, $\hat{S}$ must necessarily be Hermitian because unitarity of $\hat{U}$ implies that

$$\begin{aligned} \exp(-i\hat{S}) &= \hat{U}^{-1} = \hat{U}^{\dagger} \\ &= [\exp(i\hat{S})]^{\dagger} = \exp(i\hat{S})^{\dagger} \\ &= \exp(i\hat{S}^{\dagger}) \Rightarrow \hat{S} = \hat{S}^{\dagger} \end{aligned} \tag{2}$$

Further, let

$$\phi = \hat{U}\psi \tag{3}$$

where ϕ is the Foldy-Wouthuysen wave function and ψ the usual 4-component Dirac wave function. Since, the underlying physics is to be unaffected by the transformation, ϕ must also satisfy the Dirac equation in the Foldy-Wouthuysen representation, i.e.,

$$i\partial_0 \phi = \hat{H}_{\Phi}\phi \tag{4}$$

$\hat{H}_{\Phi}$ being the free particle Dirac Hamiltonian in the $\mathbf{\Phi}$ representation.

Using eqs. (1)-(4), we have

$$\begin{aligned} \hat{H}_{\Phi}e^{i\hat{S}}\psi &= i\partial_0 e^{i\hat{S}}\psi = e^{i\hat{S}}(i\partial_0 - \partial_0\hat{S})\psi \\ &= e^{i\hat{S}}(\hat{H} - \partial_0\hat{S})\psi \\ &= e^{i\hat{S}}(\hat{H} - \partial_0\hat{S})e^{-i\hat{S}}e^{i\hat{S}}\psi \end{aligned}$$

$$\Rightarrow \qquad \hat{H}_{\Phi} = e^{i\hat{S}}\hat{H}e^{-i\hat{S}} - \partial_0\hat{S} \tag{5}$$

For time independent transformations

$$\hat{H}_{\Phi} = e^{i\hat{S}}\hat{H}e^{-i\hat{S}} \tag{6}$$

We make the assumption that the momentum and Hamiltonian operators act on plane waves so that the momentum operator $\hat{\mathbf{p}}$ can be replaced by the corresponding eigenvalue $\mathbf{p}$, i.e., $\hat{\mathbf{p}} = \mathbf{p}$ and so

$$\hat{H} = \boldsymbol{\alpha}.\mathbf{p} + \boldsymbol{\beta} m \tag{7}$$

We postulate the ansatz

$$\hat{S} = -\frac{i}{2m}\boldsymbol{\beta\alpha}\mathbf{p}\omega\left(\frac{|\mathbf{p}|}{m}\right) \tag{8}$$

where $\omega\left(\frac{|\mathbf{p}|}{m}\right)$ is real and is to be so determined that $\hat{S}$ satisfies our requirements, i.e., eliminates the "odd" operators from the Hamiltonian.

We also have

$$\hat{S}^{+} = \frac{i}{2m}\left[\beta\alpha\mathbf{p}\omega\left(\frac{|\mathbf{p}|}{m}\right)\right]^{\dagger} = \frac{i}{2m}\boldsymbol{\alpha}^{\dagger}\boldsymbol{\beta}^{\dagger}\mathbf{p}\omega\left(\frac{|\mathbf{p}|}{m}\right)$$

$$= \frac{i}{2m}\boldsymbol{\alpha\beta}\mathbf{p}\omega\left(\frac{|\mathbf{p}|}{m}\right) = -\frac{i}{2m}\boldsymbol{\beta\alpha}\mathbf{p}\omega\left(\frac{|\mathbf{p}|}{m}\right) \tag{9}$$

so that $\hat{S}$ is Hermitian.

Now

$$\hat{H}_{\Phi} = e^{i\hat{S}}\ \hat{H}e^{-i\hat{S}} = e^{i\hat{S}}(\boldsymbol{\alpha}.\mathbf{p} + \boldsymbol{\beta}m)e^{-i\hat{S}}$$

$$= e^{i\hat{S}}\boldsymbol{\beta}(\boldsymbol{\beta\alpha}.\mathbf{p} + m)e^{-i\hat{S}} = e^{i\hat{S}}\beta e^{-i\hat{S}}\ \beta\ (\boldsymbol{\alpha}.\mathbf{p} + \beta m) \tag{10}$$

where the last step follows from

$$\hat{S}\boldsymbol{\beta\alpha}\mathbf{p} = -\frac{i}{2m}\ \beta\alpha\mathbf{p}\omega\left(\frac{|\mathbf{p}|}{m}\right).\ (\beta\alpha\mathbf{p}) = \beta\alpha\mathbf{p}\left[-\frac{i}{2m}\ \omega\left(\frac{|\mathbf{p}|}{m}\right)(\beta\alpha\mathbf{p})\right]$$

$$= \boldsymbol{\beta\alpha}\mathbf{p}\left[-\frac{i}{2m}(\boldsymbol{\beta\alpha}\mathbf{p})\omega\left(\frac{|\mathbf{p}|}{m}\right)\right] = \boldsymbol{\beta\alpha}\mathbf{p}.\hat{S} \tag{11}$$

Now consider the expression

$$\boldsymbol{\beta}(\boldsymbol{\beta\alpha}\mathbf{p})^{n} = \boldsymbol{\beta}(\boldsymbol{\beta\alpha}\mathbf{p})(\boldsymbol{\beta\alpha}\mathbf{p})^{n-1}$$

$$= -(\boldsymbol{\beta\alpha\beta}\mathbf{p})(\boldsymbol{\beta\alpha}\mathbf{p})^{n-1}$$

$$= -(\boldsymbol{\beta\alpha}\mathbf{p})\boldsymbol{\beta}(\boldsymbol{\beta\alpha}\mathbf{p})^{n-1}$$

$$= -(\boldsymbol{\beta\alpha}\mathbf{p})\boldsymbol{\beta}(\boldsymbol{\beta\alpha}\mathbf{p})(\boldsymbol{\beta\alpha}\mathbf{p})^{n-2}$$

$$= (-1)^{2}(\boldsymbol{\beta\alpha}\mathbf{p})^{2}\boldsymbol{\beta}(\boldsymbol{\beta\alpha}\mathbf{p})^{n-2} \tag{12}$$

Iterating eq. (12) *n* times, we get

$$\boldsymbol{\beta}(\boldsymbol{\beta\alpha}\mathbf{p})^{n} = (-1)^{n}(\boldsymbol{\beta\alpha}\mathbf{p})^{n}\ \boldsymbol{\beta} \tag{13}$$

Hence,

$$\beta e^{-i\hat{S}} = \beta\sum_{n=0}^{\infty}\frac{1}{n!}\left\{(-i)\left[-\frac{i}{2m}\beta\alpha\mathbf{p}\omega\left(\frac{|\mathbf{p}|}{m}\right)\right]\right\}^{n}$$

$$= \beta \sum_{n=0}^{\infty} \frac{1}{n!} \left[\left(-\frac{1}{2m} \right) \beta\alpha\mathbf{p}\omega\left(\frac{|\mathbf{p}|}{m} \right) \right]^n$$

$$= \sum_{n=0}^{\infty} \frac{1}{n!} \left(-\frac{1}{2m} \right)^n \beta(\beta\alpha\mathbf{p})^n \omega^n\left(\frac{|\mathbf{p}|}{m} \right)$$

$$= \sum_{n=0}^{\infty} \frac{1}{n!} \left(-\frac{1}{2m} \right)^n (-1)^n (\beta\alpha\mathbf{p})^n \beta\omega^n\left(\frac{|\mathbf{p}|}{m} \right)$$

$$= \sum_{n=0}^{\infty} \frac{1}{n!} \left(\frac{1}{2m} \right)^n (\beta\alpha\mathbf{p})^n \omega^n\left(\frac{|\mathbf{p}|}{m} \right) \beta$$

$$= \sum_{n=0}^{\infty} \frac{1}{n!} \left\{ (i) \left[-\frac{i}{2m} \beta\alpha\mathbf{p}\omega\left(\frac{|\mathbf{p}|}{m} \right) \right] \right\}^n \beta = e^{i\hat{S}}\beta \tag{14}$$

From eqs. (10) and (14), we have

$$\hat{H}_\Phi = e^{i\hat{S}}\beta e^{-i\hat{S}}\beta(\alpha.\mathbf{p} + \beta m) = e^{2i\hat{S}}\beta^2(\alpha.\mathbf{p} + \beta m)$$

$$= e^{2i\hat{S}}(\alpha.\mathbf{p} + \beta m) \tag{15}$$

Now

$$(\beta\alpha\mathbf{p})^2 = (\beta\alpha\mathbf{p})(\beta\alpha\mathbf{p}) = -(\alpha\beta\mathbf{p})(\beta\alpha\mathbf{p}) = -(\mathbf{p}\alpha\beta)(\beta\alpha\mathbf{p}) = -\mathbf{p}^2 \tag{16}$$

Therefore,

$$e^{2i\hat{S}} = \sum_{n=0}^{\infty} \frac{1}{n!} \left\{ (2i) \left[-\frac{i}{2m} \beta\alpha\mathbf{p}\omega\left(\frac{|\mathbf{p}|}{m} \right) \right] \right\}^n = \sum_{n=0}^{\infty} \frac{1}{n!} \left[\frac{1}{m} \beta\alpha\mathbf{p}\omega\left(\frac{|\mathbf{p}|}{m} \right) \right]^n$$

$$= \sum_{l=0}^{\infty} \frac{1}{(2l)!} \left[\frac{1}{m} \beta\alpha\mathbf{p}\omega\left(\frac{|\mathbf{p}|}{m} \right) \right]^{2l} + \sum_{l=0}^{\infty} \frac{1}{(2l+1)!} \left[\frac{1}{m} \beta\alpha\mathbf{p}\omega\left(\frac{|\mathbf{p}|}{m} \right) \right]^{2l+1}$$

$$= \sum_{l=0}^{\infty} \frac{1}{(2l)!} \left[-\frac{\mathbf{p}^2}{m^2} \omega^2\left(\frac{|\mathbf{p}|}{m} \right) \right]^l + (\beta\alpha\mathbf{p}) \sum_{l=0}^{\infty} \frac{(\beta\alpha\mathbf{p})^{2l}}{(2l+1)!} \left[\frac{1}{m} \omega\left(\frac{|\mathbf{p}|}{m} \right) \right]^{2l+1}$$

$$= \sum_{l=0}^{\infty} \frac{1}{(2l)!} \left[-\frac{\mathbf{p}^2}{m^2} \omega^2\left(\frac{|\mathbf{p}|}{m} \right) \right]^l + \frac{(\beta\alpha\mathbf{p})}{\mathbf{p}} \sum_{l=0}^{\infty} \frac{(-\mathbf{p}^2)^l \mathbf{p}}{(2l+1)!} \left[\frac{1}{m} \omega\left(\frac{|\mathbf{p}|}{m} \right) \right]^{2l+1}$$

$$= \sum_{l=0}^{\infty} \frac{(-1)^l}{(2l)!} \left[\frac{\mathbf{p}}{m} \omega\left(\frac{|\mathbf{p}|}{m} \right) \right]^{2l} + \frac{(\beta\alpha\mathbf{p})}{\mathbf{p}} \sum_{l=0}^{\infty} \frac{(-1)^l}{(2l+1)!} \left[\frac{\mathbf{p}}{m} \omega\left(\frac{|\mathbf{p}|}{m} \right) \right]^{2l+1}$$

$$= \cos\left[\frac{\mathbf{p}}{m}\omega\left(\frac{|\mathbf{p}|}{m}\right)\right] + \frac{(\beta\alpha\mathbf{p})}{\mathbf{p}}\sin\left[\frac{\mathbf{p}}{m}\omega\left(\frac{|\mathbf{p}|}{m}\right)\right] \tag{17}$$

Hence,

$$\hat{H}_{\Phi} = \left\{\cos\left[\frac{\mathbf{p}}{m}\omega\left(\frac{|\mathbf{p}|}{m}\right)\right] + \frac{(\beta\alpha\mathbf{p})}{\mathbf{p}}\sin\left[\frac{\mathbf{p}}{m}\omega\left(\frac{|\mathbf{p}|}{m}\right)\right]\right\}(\alpha\mathbf{p} + \beta m)$$

$$= \beta\left\{m\cos\left[\frac{\mathbf{p}}{m}\omega\left(\frac{|\mathbf{p}|}{m}\right)\right] + \mathbf{p}\sin\left[\frac{\mathbf{p}}{m}\omega\left(\frac{|\mathbf{p}|}{m}\right)\right]\right\}$$

$$+ \frac{(\alpha\mathbf{p})}{\mathbf{p}}\left\{\mathbf{p}\cos\left[\frac{\mathbf{p}}{m}\omega\left(\frac{|\mathbf{p}|}{m}\right)\right] + m\sin\left[\frac{\mathbf{p}}{m}\omega\left(\frac{|\mathbf{p}|}{m}\right)\right]\right\} \tag{18}$$

From eq. (18) it is obvious that the "odd" operators are contained in the second term through the coefficient $\boldsymbol{\alpha}$. Therefore, if we choose $\omega\left(\frac{|\mathbf{p}|}{m}\right)$ such that

$$\mathbf{p}\cos\left[\frac{\mathbf{p}}{m}\omega\left(\frac{|\mathbf{p}|}{m}\right)\right] - m\sin\left[\frac{\mathbf{p}}{m}\omega\left(\frac{|\mathbf{p}|}{m}\right)\right] = 0$$

or equivalently

$$\omega = \frac{m}{\mathbf{p}}\tan^{-1}\left(\frac{\mathbf{p}}{m}\right) \tag{19}$$

then the free particle Dirac Hamiltonian becomes liberated of the "odd" operators, as desired. Equation (19) implies

$$\cos\left(\frac{\mathbf{p}}{m}\omega\right) = \frac{m}{\sqrt{\mathbf{p}^2 + m^2}} = \frac{m}{\sqrt{\mathbf{p}^2 + m^2}} = \frac{m}{E_p}$$

and

$$\sin\left(\frac{\mathbf{p}}{m}\omega\right) = \tan\left(\frac{\mathbf{p}}{m}\omega\right)\cos\left(\frac{\mathbf{p}}{m}\omega\right) = \frac{\mathbf{p}}{E_p} \tag{20}$$

Using eq. (20) in (18), we get

$$\hat{H}_{\Phi} = \beta\left(\frac{m^2}{E_p} + \frac{\mathbf{p}^2}{E_p}\right) = \beta E_p \tag{21}$$

We also have from eq. (17)

$$\hat{U} = e^{i\hat{S}} = \cos\left[\frac{\mathbf{p}}{2m}\omega\left(\frac{|\mathbf{p}|}{m}\right)\right] + \frac{(\beta\alpha\mathbf{p})}{\mathbf{p}}\sin\left[\frac{\mathbf{p}}{2m}\omega\left(\frac{|\mathbf{p}|}{m}\right)\right]$$

$$= \left\{\frac{1+\cos\left[\frac{\mathbf{p}}{m}\,\omega\left(\frac{|\mathbf{p}|}{m}\right)\right]}{2}\right\}^{\frac{1}{2}} + \frac{(\beta\alpha\mathbf{p})}{\mathbf{p}}\left\{\frac{1-\cos\left[\frac{\mathbf{p}}{m}\,\omega\left(\frac{|\mathbf{p}|}{m}\right)\right]}{2}\right\}^{\frac{1}{2}}$$

$$= \left(\frac{1+\frac{m}{E_p}}{2}\right)^{\frac{1}{2}} + \frac{(\beta\alpha\mathbf{p})}{\mathbf{p}}\left(\frac{1-\frac{m}{E_p}}{2}\right)^{\frac{1}{2}} = \left(\frac{E_p+m}{2E_p}\right)^{\frac{1}{2}} + \frac{(\beta\alpha\mathbf{p})}{\mathbf{p}}\left(\frac{E_p-m}{2E_p}\right)^{\frac{1}{2}}$$

$$= \frac{(E_p+m)+\frac{(\beta\alpha\mathbf{p})}{(\mathbf{p})}(E_p^2-m^2)^{\frac{1}{2}}}{[2E_p(E_p+m)]^{\frac{1}{2}}} = \frac{(E_p+m)+(\beta\alpha\mathbf{p})}{[2E_p+E_p+m]^{\frac{1}{2}}} \tag{22}$$

It can be explicitly verified that $\hat{U}^\dagger\hat{U} = \mathbf{I}$ for

$$\hat{U}^\dagger\hat{U} = \left\{\cos\left[\frac{\mathbf{p}}{2m}\,\omega\left(\frac{|\mathbf{p}|}{m}\right)\right] + \frac{(\beta\alpha\mathbf{p})}{\mathbf{p}}\sin\left[\frac{|\mathbf{p}|}{2m}\,\omega\left(\frac{|\mathbf{p}|}{m}\right)\right]\right\}^\dagger$$

$$\left\{\cos\left[\frac{\mathbf{p}}{2m}\,\omega\left(\frac{|\mathbf{p}|}{m}\right)\right] + \frac{(\beta\alpha\mathbf{p})}{\mathbf{p}}\sin\left[\frac{\mathbf{p}}{2m}\,\omega\left(\frac{|\mathbf{p}|}{m}\right)\right]\right\}$$

$$= \left\{\cos\left[\frac{\mathbf{p}}{2m}\,\omega\left(\frac{|\mathbf{p}|}{m}\right)\right] + \frac{(\beta\alpha\mathbf{p})^\dagger}{\mathbf{p}}\sin\left[\frac{\mathbf{p}}{2m}\,\omega\left(\frac{|\mathbf{p}|}{m}\right)\right]\right\}$$

$$\left\{\cos\left[\frac{\mathbf{p}}{2m}\,\omega\left(\frac{|\mathbf{p}|}{m}\right)\right] + \frac{(\beta\alpha\mathbf{p})}{\mathbf{p}}\sin\left[\frac{\mathbf{p}}{2m}\,\omega\left(\frac{|\mathbf{p}|}{m}\right)\right]\right\}$$

$$= \left\{\cos\left[\frac{\mathbf{p}}{2m}\,\omega\left(\frac{|\mathbf{p}|}{m}\right)\right] + \frac{(\alpha\beta\mathbf{p})}{\mathbf{p}}\sin\left[\frac{\mathbf{p}}{2m}\,\omega\left(\frac{|\mathbf{p}|}{m}\right)\right]\right\}$$

$$\left\{\cos\left[\frac{\mathbf{p}}{2m}\,\omega\left(\frac{|\mathbf{p}|}{m}\right)\right] + \frac{(\beta\alpha\mathbf{p})}{\mathbf{p}}\sin\left[\frac{\mathbf{p}}{2m}\,\omega\left(\frac{|\mathbf{p}|}{m}\right)\right]\right\}$$

$$= \cos^2\left[\frac{\mathbf{p}}{2m}\,\omega\left(\frac{|\mathbf{p}|}{m}\right)\right] + \frac{(\alpha\beta\mathbf{p})}{\mathbf{p}}\frac{(\beta\alpha\mathbf{p})}{\mathbf{p}}\sin^2\left[\frac{\mathbf{p}}{2m}\,\omega\left(\frac{|\mathbf{p}|}{m}\right)\right]$$

$$+\left[\frac{(\alpha\beta\mathbf{p})}{\mathbf{p}}+\frac{(\beta\alpha\mathbf{p})}{\mathbf{p}}\right]\cos\left[\frac{\mathbf{p}}{2m}\,\omega\left(\frac{|\mathbf{p}|}{m}\right)\right]\sin\left[\frac{\mathbf{p}}{2m}\,\omega\left(\frac{|\mathbf{p}|}{m}\right)\right]$$

$$=\cos^2\left[\frac{\mathbf{p}}{2m}\,\omega\left(\frac{|\mathbf{p}|}{m}\right)\right]+\sin^2\left[\frac{\mathbf{p}}{2m}\,\omega\left(\frac{|\mathbf{p}|}{m}\right)\right]$$

$$+\left[\frac{(\alpha\beta+\beta\alpha)\mathbf{p}}{\mathbf{p}}\right]\cos\left[\frac{\mathbf{p}}{2m}\,\omega\left(\frac{|\mathbf{p}|}{m}\right)\right]\sin\left[\frac{\mathbf{p}}{2m}\,\omega\left(\frac{|\mathbf{p}|}{m}\right)\right]=1 \tag{23}$$

The Hamiltonian in the Foldy-Wouthuysen representation can also be explicitly calculated as

$$\hat{H}_\Phi=\hat{U}\hat{H}\hat{U}^\dagger=\frac{(E_p+m)+(\beta\alpha\mathbf{p})}{[2E_p(E_p+m)]^{\frac{1}{2}}}(\alpha\mathbf{p}+\beta m)\left\{\frac{(E_p+m)+(\beta\alpha\mathbf{p})}{[2E_p(E_p+m)]^{\frac{1}{2}}}\right\}^\dagger$$

$$=\frac{(E_p+m)+(\beta\alpha\mathbf{p})}{[2E_p(E_p+m)]^{\frac{1}{2}}}(\alpha\mathbf{p}+\beta m)\,\frac{(E_p+m)+(\alpha\beta\mathbf{p})}{[2E_p(E_p+m)]^{\frac{1}{2}}}$$

$$=\frac{1}{[2E_p(E_p+m)]}\left\{\begin{matrix}(E_p+m)^2\,(\alpha\mathbf{p}+\beta m)+(\beta\alpha\mathbf{p})(\alpha\mathbf{p}+\beta m)(\alpha\beta\mathbf{p})\\ +(E_p+m)[(\beta\alpha\mathbf{p})(\alpha\mathbf{p}+\beta m)+(\alpha\mathbf{p}+\beta m)(\alpha\beta\mathbf{p})]\end{matrix}\right\}$$

$$=\frac{(E_p+m)^2\,(\alpha\mathbf{p}+\beta m)+2(E_p+m)(\beta\mathbf{p}^2-\alpha m\mathbf{p})-(\alpha\mathbf{p}+\beta m)\mathbf{p}^2}{[2E_p(E_p+m)]}$$

$$=\frac{[(E_p+m)^2-\mathbf{p}^2](\alpha\mathbf{p}+\beta m)+2(E_p+m)(\beta\mathbf{p}^2-\alpha m\mathbf{p})}{[2E_p(E_p+m)]}$$

$$=\frac{2m(E_p+m)(\alpha\mathbf{p}+\beta m)+2(E_p+m)(\beta\mathbf{p}^2-\alpha m\mathbf{p})}{[2E_p(E_p+m)]}$$

$$=\frac{m(\alpha\mathbf{p}+\beta m)+(\beta\mathbf{p}^2-\alpha m\mathbf{p})}{E_p}=\frac{\beta(\mathbf{p}^2+m^2)}{E_p}=\beta E_p \tag{24}$$

We now compute the "sign" operator $\tilde{\hat{\Lambda}}_\Phi=\dfrac{\hat{H}}{\sqrt{(\hat{H})^2}}$ in the Φ representation.

We have

$$\tilde{\hat{\Lambda}}_\Phi=\hat{U}\frac{\hat{H}}{\sqrt{(\hat{H})^2}}\hat{U}^\dagger=\hat{U}\,\frac{(\alpha\mathbf{p}+\beta m)}{E_p}\,\hat{U}^\dagger=\frac{\beta E_p}{E_p}=\boldsymbol{\beta} \tag{25}$$

or simply as

$$\tilde{\hat{\Lambda}}_\Phi = \frac{\hat{H}\Phi}{\sqrt{(\hat{H}_\Phi)^2}} = \frac{\beta E_p}{E_p} = \beta$$

Similarly,

$$\alpha_\Phi = \hat{U}\alpha\hat{U}^\dagger = \frac{(E_p+m)+(\beta\alpha\mathbf{p})}{[2E_p(E_p+m)]^{\frac{1}{2}}}\,\boldsymbol{\alpha}\left[\frac{(E_p+m)+(\boldsymbol{\alpha\beta}\mathbf{p})}{[2E_p(E_p+m)]^{\frac{1}{2}}}\right]^\dagger$$

$$= \frac{(E_p+m)+(\beta\alpha\mathbf{p})}{[2E_p(E_p+m)]^{\frac{1}{2}}}\,\boldsymbol{\alpha}\,\frac{(E_p+m)+(\alpha\beta\mathbf{p})}{[2E_p(E_p+m)]^{\frac{1}{2}}}$$

$$= \frac{(E_p+m)^2\,\alpha + (E_p+m)[(\beta\alpha\mathbf{p})\alpha + \alpha\,(\alpha\beta\mathbf{p})] + (\beta\alpha\mathbf{p})\,\alpha\,(\alpha\beta\mathbf{p})}{[2E_p(E_p+m)]}$$

$$= \frac{(E_p+m)^2\,\alpha + 2(E_p+m)\beta\mathbf{p} - \alpha\mathbf{p}^2}{[2E_p(E_p+m)]}$$

$$= \frac{[(E_p+m)^2 - \mathbf{p}^2]\,\alpha + 2(E_p+m)\beta\mathbf{p}}{[2E_p(E_p+m)]}$$

$$= \frac{2m(E_p+m)\,\alpha + 2(E_p+m)\beta\mathbf{p}}{[2E_p(E_p+m)]} = \frac{m\alpha+\beta\mathbf{p}}{E_p} \tag{26}$$

and

$$\boldsymbol{\beta}_\Phi = \hat{U}\boldsymbol{\beta}\hat{U}^\dagger = \frac{(E_p+m)+(\beta\alpha\mathbf{p})}{[2E_p(E_p+m)]^{\frac{1}{2}}}\,\boldsymbol{\beta}\left\{\frac{(E_p+m)+(\beta\alpha\mathbf{p})}{[2E_p(E_p+m)]^{\frac{1}{2}}}\right\}^\dagger$$

$$= \frac{(E_p+m)+(\beta\alpha\mathbf{p})}{[2E_p(E_p+m)]^{\frac{1}{2}}}\,\beta\,\frac{(E_p+m)+(\alpha\beta\mathbf{p})}{[2E_p(E_p+m)]^{\frac{1}{2}}}$$

$$= \frac{(E_p+m)^2\,\boldsymbol{\beta} + (E_p+m)[(\boldsymbol{\beta\alpha}\mathbf{p})\,\boldsymbol{\beta} + \boldsymbol{\beta}\,(\boldsymbol{\alpha\beta}\mathbf{p})] + (\boldsymbol{\beta\alpha}\mathbf{p})\boldsymbol{\beta}(\boldsymbol{\alpha\beta}\mathbf{p})}{[2E_p(E_p+m)]}$$

$$= \frac{(E_p+m)^2\,\beta - 2\,(E_p+m)\,\alpha\mathbf{p} - \beta\mathbf{p}^2}{[2E_p(E_p+m)]}$$

$$= \frac{[(E_p + m)^2 - \mathbf{p}^2]\beta - 2(E_p + m)\alpha\mathbf{p}}{[2E_p(E_p + m)]}$$

$$= \frac{2m[(E_p + m)]\beta - 2(E_p + m)\alpha\mathbf{p}}{[2E_p(E_p + m)]} = \frac{m\beta - \alpha\mathbf{p}}{E_p} \tag{27}$$

Further, in the momentum representation, noting that

$$\mathbf{x} \equiv i\nabla_{\mathbf{p}} \equiv i\left(\frac{\partial}{\partial p_1}, \frac{\partial}{\partial p_2}, \frac{\partial}{\partial p_3}\right)$$

we have

$$\hat{\mathbf{x}}_\Phi = \hat{U}\,\hat{\mathbf{x}}\,\hat{U}^\dagger = i\hat{U}\,\nabla\mathbf{p}\hat{U}^\dagger = i\,\frac{(E_p + m) + (\beta\alpha\mathbf{p})}{[2E_p(E_p + m)]^{\frac{1}{2}}}\nabla_{\mathbf{p}}\left\{\frac{(E_p + m) + (\beta\alpha\mathbf{p})}{[2E_p(E_p + m)]^{\frac{1}{2}}}\right\}^\dagger$$

$$= i\,\frac{(E_p + m) + (\beta\alpha\mathbf{p})}{[2E_p(E_p + m)]^{\frac{1}{2}}}\nabla_{\mathbf{p}}\frac{(E_p + m) + (\alpha\beta\mathbf{p})}{[2E_p(E_p + m)]^{\frac{1}{2}}}$$

$$= i\,\frac{(E_p + m) + (\beta\alpha\mathbf{p})}{[2E_p(E_p + m)]^{\frac{1}{2}}}\nabla_{\mathbf{p}}\left[\frac{(E_p + m)}{[2E_p(E_p + m)]^{\frac{1}{2}}} + \frac{(\alpha\beta\mathbf{p})}{[2E_p(E_p + m)]^{\frac{1}{2}}}\right]$$

$$= \frac{(E_p + m) + (\beta\alpha\mathbf{p})}{[2E_p(E_p + m)]^{\frac{1}{2}}}\left\{\frac{(E_p + m)}{[2E_p(E_p + m)]^{\frac{1}{2}}}\nabla_{\mathbf{p}}\right.$$

$$\left. + \frac{[2E_p(E_p + m)]^{\frac{1}{2}}\dfrac{\mathrm{p}}{E_p} - (E_p + m)\dfrac{(4E_p + 2m)}{2[2E_p(E_p + m)]^{\frac{1}{2}}}\dfrac{\mathrm{p}}{E_p}}{[2E_p(E_p + m)]}\right\}$$

$$+ \frac{(E_p + m) + (\beta\alpha\mathbf{p})}{[2E_p(E_p + m)]^{\frac{1}{2}}}\left\{\frac{(\alpha\beta\mathbf{p})}{[2E_p(E_p + m)]^{\frac{1}{2}}}\nabla_{\mathbf{p}}\right.$$

$$+\frac{[2E_p(E_p+m)]^{\frac{1}{2}}\,\mathrm{ab}-\mathrm{abp}\,\dfrac{(4E_p+2m)}{2[2E_p+m]^{\frac{1}{2}}}\dfrac{\mathrm{p}}{E_p}}{[2E_p(E_p+m)]}\Bigg\}$$

$$=i\,\frac{(E_p+m)+(\boldsymbol{\beta\alpha}\mathbf{p})}{[2E_p(E_p+m)]^{\frac{1}{2}}}\left\{\begin{array}{c}\left[\dfrac{E_p+m}{[2E_p(E_p+m)]^{\frac{1}{2}}}+\dfrac{(\boldsymbol{\alpha\beta}\mathbf{p})}{[2E_p(E_p+m)]^{\frac{1}{2}}}\right]\nabla_{\mathbf{p}}\\ \dfrac{-(E_p+m)m\dfrac{\mathrm{p}}{E_p}+[2E_p(E_p+m)]\boldsymbol{\alpha\beta}-\boldsymbol{\alpha\beta}(2E_p+m)\dfrac{\mathrm{p}^2}{E_p}}{[2E_p(E_p+m)]^{\frac{3}{2}}}\end{array}\right\}$$

$$=i\,\frac{(E_p+m)+(\boldsymbol{\beta\alpha}\mathbf{p})}{[2E_p(E_p+m)]}(E_p+m+\boldsymbol{\alpha\beta}\mathbf{p})\,\nabla_{\mathbf{p}}$$

$$+i\,\frac{(E_p+m)+(\boldsymbol{\beta\alpha}\mathbf{p})}{[2E_p(E_p+m)]^{\frac{1}{2}}}\left\{\frac{-(E_p+m)m\dfrac{\mathbf{p}}{E_p}+\dfrac{(E_p+m)^2m}{E_p}\alpha\beta}{[2E_p(E_p+m)]^{\frac{3}{2}}}\right\}$$

$$=i\,\frac{(E_p+m)^2+\mathbf{p}^2}{[2E_p(E_p+m)]}\nabla_{\mathbf{p}}+i\,\frac{\dfrac{(E_p+m)^3m}{E_p}\alpha\beta+(E_p+m)m\dfrac{\mathbf{p}}{E_p}(\alpha\beta\mathbf{p})}{[2E_p(E_p+m)]^2}$$

$$=i\nabla_{\mathbf{p}}+i\,\frac{m}{2E_p^2}\boldsymbol{\alpha\beta}=\hat{\mathbf{x}}+i\,\frac{m}{2E_p^2}\,\boldsymbol{\alpha\beta} \tag{28}$$

It is to be noted that since the momentum operator $\hat{\mathbf{p}}$ commutes with the transformation operator, we have

$$\hat{\mathbf{p}}_{\Phi}=\hat{U}\hat{\mathbf{p}}\hat{U}^{\dagger}=\hat{\mathbf{p}}\hat{U}\hat{U}^{\dagger}=\hat{\mathbf{p}} \tag{29}$$

7.2 DIRAC WAVES IN THE ϕ REPRESENTATION

We have already proved that $\hat{H}_\Phi = \boldsymbol{\beta} E_p$ so that

$$\hat{H}_\Phi \psi = \beta E_p \psi \tag{30}$$

Furthermore, we can write the Dirac equation in the Φ representation as

$$\hat{H}_\Phi \phi = (\hat{U}\hat{H}\hat{U}^\dagger)\hat{U}\psi = \hat{U}\hat{H}\psi = \hat{U}\varepsilon E_p \psi = \varepsilon E_p \hat{U}\psi = \varepsilon E_p \phi \tag{31}$$

where ϕ is the 4-component Dirac spinor in the Φ representation wherein the upper pair of components is decoupled from the lower pair. This is explicitly seen as follows:

$$\phi = \hat{U}\psi = \frac{(E_p + m) + (\boldsymbol{\beta\alpha p})}{[2E_p(E_p + m)]^{\frac{1}{2}}}\begin{pmatrix} u^r(\mathbf{p}) \\ v^r(\mathbf{p}) \end{pmatrix}$$

$$= \frac{1}{[2E_p(E_p + m)]^{\frac{1}{2}}}\begin{pmatrix} (E_p + m) & \boldsymbol{\sigma p} \\ -\boldsymbol{\sigma p} & (E_p + m) \end{pmatrix}$$

$$\left(\mathbf{I}\cosh\left(\frac{\omega}{2}\right) - \left(\frac{\boldsymbol{\alpha}.\mathbf{p}}{|\mathbf{p}|}\right)\sinh\left(\frac{\omega}{2}\right)\right)\begin{pmatrix} u^r(\mathbf{0}) \\ v^r(\mathbf{0}) \end{pmatrix}$$

$$= \frac{1}{[2E_p(E_p + m)]^{\frac{1}{2}}}\begin{pmatrix} (E_p + m) & \boldsymbol{\sigma p} \\ -\boldsymbol{\sigma p} & (E_p + m) \end{pmatrix}$$

$$\begin{pmatrix} \sqrt{\dfrac{E_p + m}{2m}} & \dfrac{1}{|\mathbf{p}|}\sqrt{\dfrac{E_p - m}{2m}}\boldsymbol{\sigma p} \\ \dfrac{1}{|\mathbf{p}|}\sqrt{\dfrac{E_p - m}{2m}}\boldsymbol{\sigma p} & \sqrt{\dfrac{E_p + m}{2m}} \end{pmatrix}\begin{pmatrix} u^r(\mathbf{0}) \\ v^r(\mathbf{0}) \end{pmatrix}$$

$$= \frac{1}{[2E_p(E_p + m)]^{\frac{1}{2}}}\begin{pmatrix} (E_p + m) & \boldsymbol{\sigma p} \\ -\boldsymbol{\sigma p} & (E_p + m) \end{pmatrix}$$

$$\begin{pmatrix} \sqrt{\dfrac{E_p + m}{2m}}u^r(\mathbf{0}) + \dfrac{1}{|\mathbf{p}|}\sqrt{\dfrac{E_p - m}{2m}}\boldsymbol{\sigma p}v^r(\mathbf{0}) \\ \sqrt{\dfrac{E_p + m}{2m}}v^r(\mathbf{0}) + \dfrac{1}{|\mathbf{p}|}\sqrt{\dfrac{E_p - m}{2m}}\boldsymbol{\sigma p}u^r(\mathbf{0}) \end{pmatrix}$$

$$= \frac{1}{[2E_p(E_p+m)]^{\frac{1}{2}}} \begin{pmatrix} (E_p+m)\sqrt{\frac{E_p+m}{2m}}u^r(\mathbf{0}) + (E_p+m)\frac{1}{|\mathbf{p}|}\sqrt{\frac{E_p-m}{2m}} \\ -\boldsymbol{\sigma}\mathbf{p}\sqrt{\frac{E_p+m}{2m}}v^r(\mathbf{0}) - \boldsymbol{\sigma}\mathbf{p}\frac{1}{|\mathbf{p}|}\sqrt{\frac{E_p-m}{2m}} \end{pmatrix}$$

$$\begin{pmatrix} \boldsymbol{\sigma}\mathbf{p}v^r(\mathbf{0}) + \boldsymbol{\sigma}\mathbf{p}\sqrt{\frac{E_p+m}{2m}}v^r(\mathbf{0}) + \boldsymbol{\sigma}\mathbf{p}\frac{1}{|\mathbf{p}|}\sqrt{\frac{E_p-m}{2m}}\boldsymbol{\sigma}\mathbf{p}u^r(\mathbf{0}) \\ \boldsymbol{\sigma}\mathbf{p}v^r(\mathbf{0}) + (E_p+m)\sqrt{\frac{E_p+m}{2m}}v^r(\mathbf{0}) + (E_p+m)\frac{1}{|\mathbf{p}|}\sqrt{\frac{E_p-m}{2m}}\boldsymbol{\sigma}\mathbf{p}u^r(\mathbf{0}) \end{pmatrix}$$

$$= \frac{1}{2[E_pm]^{\frac{1}{2}}(E_p+m)} \begin{pmatrix} [(E_p+m)^2 + \varepsilon_\mathbf{p}\mathbf{p}^2]u^r(\mathbf{0}) + (E_p+m)(\varepsilon_\mathbf{p}+1)\boldsymbol{\sigma}\mathbf{p}v^r(\mathbf{0}) \\ [(E_p+m)^2 - \varepsilon_\mathbf{p}\mathbf{p}^2]v^r(\mathbf{0}) + (\varepsilon_p-1)(E_p+m)\boldsymbol{\sigma}\mathbf{p}u^r(\mathbf{0}) \end{pmatrix}$$

$$= \frac{1}{2[E_pm]^{\frac{1}{2}}(E_p+m)} \left\{ \begin{pmatrix} [(E_p+m)^2 + \varepsilon_\mathbf{p}\mathbf{p}^2] \\ (\varepsilon_\mathbf{p}-1)(E_p+m)\boldsymbol{\sigma}\mathbf{p} \end{pmatrix} u^r(\mathbf{0}) \right.$$

$$\left. + \begin{pmatrix} (E_p+m)(1+\varepsilon_\mathbf{p})\boldsymbol{\sigma}\mathbf{p} \\ [(E_p+m)^2 - \varepsilon_\mathbf{p}\mathbf{p}^2] \end{pmatrix} v^r(\mathbf{0}) \right\} \tag{32}$$

where we have introduced $\sqrt{(E_p^2 - m^2)} = \varepsilon_\mathbf{p}|\mathbf{p}|$.

Now, in case of positive energy solutions $v^r(\mathbf{0}) = (0\ 0)^T$ and negative energy solutions $u^r(\mathbf{0}) = (0\ 0)^T$. Therefore, we find from the above expression for the wave function ϕ in the Φ representation that there is no coupling of the two 2-component spinors, viz. the upper and lower pair of components $u^r(\mathbf{p})$ and $v^r(\mathbf{p})$, respectively.

Furthermore, in the case of "normal" particles, i.e., particles with non-negative 4-momentum, i.e., non-negative energy ($E = E_p$) and non-negative 3-momentum ($\varepsilon_\mathbf{p} = +1$), the above expression for the Dirac spinors in the Φ representation will take the simple form

$$\phi^{(+)} = \frac{1}{2[E_pm]^{\frac{1}{2}}(E_p+m)} \left\{ \begin{pmatrix} [(E_p+m)^2 + \mathbf{p}^2] \\ \mathbf{0} \end{pmatrix} u^r(\mathbf{0}) \right\} = \left(\frac{E_p}{m}\right)^{\frac{1}{2}} \begin{pmatrix} u^r(\mathbf{0}) \\ \mathbf{0} \end{pmatrix} \tag{33}$$

Similarly, in the case of "antiparticles", i.e., particles with negative 4-momentum, i.e., negative energy ($E = -E_p$) and negative 3-momentum ($\varepsilon_\mathbf{p} = -1$), the above expression for the Dirac spinors in the Φ representation will take the form

$$\phi^{(-)} = \frac{1}{2[E_pm]^{\frac{1}{2}}(E_p+m)} \left\{ \begin{pmatrix} \mathbf{0} \\ [(E_p+m)^2 + \mathbf{p}^2] \end{pmatrix} v^r(\mathbf{0}) \right\} = \left(\frac{E_p}{m}\right)^{\frac{1}{2}} \begin{pmatrix} \mathbf{0} \\ v^r(\mathbf{0}) \end{pmatrix} \tag{34}$$

In other words, $\phi^{(+)}$ has $(0\ 0)^T$ as its lower pair of components and $\phi^{(-)}$ has $(0\ 0)^T$ as its upper pair.

We can, therefore, write in this representation

$$\phi = \phi^{(+)} + \phi^{(-)} \tag{35}$$

corresponding to the positive and negative energy solutions of the Dirac equation.

Then, we have

$$\hat{H}_\Phi \phi^{(\pm)} = \pm E_p \phi^{(\pm)} = \varepsilon E_p \phi^{(\pm)} \tag{36}$$

which gives

$$\hat{H}_\Phi \phi = E_p \phi^{(+)} - E_p \phi^{(-)} \tag{37}$$

We also have from eq. (30), since the said equation holds for all Dirac wave functions, $\hat{H}_\Phi \phi = \boldsymbol{\beta} E_p \phi$. Combining this with eq. (31), we conclude that $\boldsymbol{\beta}\phi = \varepsilon\phi$. This can also be inferred from the explicit form of ϕ.

Now

$$\phi^{(\text{sgn}\,\varepsilon)} = \hat{U}\psi^{(\text{sgn}\,\varepsilon)} = \frac{(E_p + m) + (\boldsymbol{\beta\alpha}\mathbf{p})}{[2E_p(E_p + m)]^{\frac{1}{2}}}\psi^{(\text{sgn}\,\varepsilon)}$$

$$= \frac{(E_p + m)\psi^{(\text{sgn}\,\varepsilon)} + \boldsymbol{\beta}[(\boldsymbol{\alpha}\mathbf{p} + \boldsymbol{\beta}m) - \boldsymbol{\beta}m]\psi^{(\text{sgn}\,\varepsilon)}}{[2E_p(E_p + m)]^{\frac{1}{2}}}$$

$$= \frac{(E_p + m)\psi^{(\text{sgn}\,\varepsilon)} + \boldsymbol{\beta}(\varepsilon E_p - \boldsymbol{\beta}m)\psi^{(\text{sgn}\,\varepsilon)}}{[2E_p(E_p + m)]^{\frac{1}{2}}}$$

$$= \frac{E_p}{[2E_p(E_p + m)]^{\frac{1}{2}}}(\mathbf{I} + \varepsilon\boldsymbol{\beta})\psi^{(\text{sgn}\,\varepsilon)} \tag{38}$$

where we have used the fact that ψ satisfies the Dirac equation so that $(\boldsymbol{\alpha}\mathbf{p} + \boldsymbol{\beta}m)\,\psi^{(\text{sgn}\,\varepsilon)} = i\partial_0\psi^{(\text{sgn}\,\varepsilon)} = \varepsilon E_p \psi^{(\text{sgn}\,\varepsilon)}$.

The above expression gives another form of the Dirac wave function ϕ in the Φ representation.

Now

$$\frac{1}{2}(\mathbf{I} + \varepsilon'\boldsymbol{\beta})\phi^{(\text{sng}\,\varepsilon)} = \frac{E_p}{2[2E_p(E_p + m)]^{\frac{1}{2}}}(\mathbf{I} + \varepsilon'\boldsymbol{\beta})(\mathbf{I} + \varepsilon\boldsymbol{\beta})\psi^{(\text{sgn}\,\varepsilon)}$$

$$= \frac{E_p}{2[2E_p(E_p + m)]^{\frac{1}{2}}}[\mathbf{I} + (\varepsilon + \varepsilon')\boldsymbol{\beta} + \varepsilon\varepsilon']\psi^{(\text{sgn}\,\varepsilon)}$$

$$= \delta_{\varepsilon\varepsilon'} \frac{E_p}{[2E_p(E_p + m)]^{\frac{1}{2}}} [\mathbf{I} + \varepsilon\boldsymbol{\beta}]\psi^{(\text{sgn } \varepsilon)} = \delta_{\varepsilon\varepsilon'}\phi^{(\text{sgn } \varepsilon)} \tag{39}$$

This expression shows that the operators

$$\Pi^{(+)} = \frac{1}{2}(\mathbf{I} + \beta),\ \Pi^{(-)} = \frac{1}{2}(\mathbf{I} - \boldsymbol{\beta}) \tag{40}$$

act as projection operators for positive and negative energy states, respectively. These operators project out the positive and negative wave functions from an arbitrary Dirac wave function in the $\boldsymbol{\Phi}$ representation, e.g.,

$$\frac{1}{2}(\mathbf{I} + \beta)\,\phi = \frac{1}{2}(\mathbf{I} + \beta)(\phi^{(+)} + \phi^{(-)}) = \phi^{(+)},\ \frac{1}{2}(\mathbf{I} - \beta)\phi$$
$$= \frac{1}{2}(\mathbf{I} - \beta)(\phi^{(+)} + \phi^{(-)}) = \phi^{(-)}$$

7.3 ISSUE OF SPIN OF DIRAC PARTICLES

As we know, the free particle Hamiltonian for a Dirac particle is given by

$$\hat{H} = \boldsymbol{\alpha}\hat{\mathbf{p}} + \boldsymbol{\beta}m = -i\boldsymbol{\alpha}\nabla + \boldsymbol{\beta}m \tag{41}$$

Further, the 1st component of the orbital angular momentum is

$$\hat{L}_1 = i(x^2\partial_3 - x^3\partial_2) \tag{42}$$

Then we have

$$[\hat{H},\ \hat{L}_1] = -(\boldsymbol{\alpha}.\nabla + i\boldsymbol{\beta}.m)(x^2\partial_3 - x^3\partial_2) + (x^2\partial_3 - x^3\partial_2)(\boldsymbol{\alpha}.\nabla + i\boldsymbol{\beta}m)$$
$$= (x^2\partial_3 - x^3\partial_2)(\boldsymbol{\alpha}^1\partial_1 + \boldsymbol{\alpha}^2\partial_2 + \boldsymbol{\alpha}^3\partial_3)$$
$$-(\boldsymbol{\alpha}^1\partial_1 + \boldsymbol{\alpha}^2\partial_2 + \boldsymbol{\alpha}^3\partial_3)\,(x^2\partial_3 - x^3\partial_2)$$
$$= (\boldsymbol{\alpha}^3\partial_2 - \boldsymbol{\alpha}^2\partial_3) \neq 0 \tag{43}$$

Similarly,

$$[\hat{H},\ \hat{L}_2] = (\boldsymbol{\alpha}^1\partial_3 - \boldsymbol{\alpha}^3\partial_1) \neq 0 \tag{44}$$

$$[\hat{H},\ \hat{L}_3] = (\boldsymbol{\alpha}^2\partial_1 - \boldsymbol{\alpha}^1\partial_2) \neq 0 \tag{45}$$

$$[\hat{H},\ \hat{L}] = (\boldsymbol{\alpha}^3\partial_2 + \boldsymbol{\alpha}^1\partial_3 + \boldsymbol{\alpha}^2\partial_1) - (\boldsymbol{\alpha}^2\partial_3 + \boldsymbol{\alpha}^3\partial_1 + \boldsymbol{\alpha}^1\partial_2) \neq 0 \tag{46}$$

This shows that the orbital angular momentum does not commute with the Hamiltonian of the system and, therefore, the orbital angular momentum is not a constant of the motion.

However, the requirement of rotational invariance of the system mandates that the total angular momentum be a constant of motion of the system and, therefore, commute with the system Hamiltonian.

There must, therefore, be an additional rotational motion of the Dirac particle (called the particle "spin") leading to a third angular momentum such that

$$\hat{J} = \hat{L} + \hat{S} \tag{47}$$

where $\hat{J}$ is the total angular momentum operator of the system, $\hat{L}$ its orbital angular momentum operator and $\hat{S}$ the spin angular momentum operator. $\hat{S}$ does not depend on the state of the system and possesses the following properties:

(i) $\hat{S}$ must necessarily satisfy the usual commutation rules of an angular momentum operator and also commute with $\hat{L}$ in order that the total angular momentum $\hat{J}$ satisfies the usual commutation rules of angular momentum, e.g.,

$$[\hat{J}_k, \hat{J}_l] = i\varepsilon_{klm}\hat{J}_m \tag{48}$$

$$[\hat{L}_k + \hat{S}_k, \hat{L}_l + \hat{S}_l] = [\hat{L}_k, \hat{L}_l] + [\hat{L}_k, \hat{S}_l]$$

$$+ [\hat{S}_k, \hat{L}_l] + [\hat{S}_k, \hat{S}_l] = i\varepsilon_{klm}(\hat{L}_m + \hat{S}_m) \tag{49}$$

The equality of the left-hand side with the last term on the right-hand side mandates that

$$[\hat{L}_k, \hat{L}_l] = i\varepsilon_{klm}\hat{L}_m \tag{50}$$

$$[\hat{S}_k, \hat{S}_l] = i\varepsilon_{klm}\hat{S}_m \tag{51}$$

$$[\hat{L}_k, \hat{S}_l] = [\hat{S}_k, \hat{L}_l] = 0 \tag{52}$$

(ii) If we define the spin angular momentum by

$$\hat{S} = \frac{1}{2}\Sigma = \frac{1}{2}\begin{pmatrix} \boldsymbol{\sigma} & \mathbf{0} \\ \mathbf{0} & \boldsymbol{\sigma} \end{pmatrix} \tag{53}$$

so that

$$[\hat{S}_k, \hat{S}_l] = \frac{1}{4}\left[\begin{pmatrix} \sigma_k & \mathbf{0} \\ \mathbf{0} & \sigma_k \end{pmatrix}\begin{pmatrix} \sigma_l & \mathbf{0} \\ \mathbf{0} & \sigma_l \end{pmatrix} - \begin{pmatrix} \sigma_l & \mathbf{0} \\ \mathbf{0} & \sigma_l \end{pmatrix}\begin{pmatrix} \sigma_k & \mathbf{0} \\ \mathbf{0} & \sigma_k \end{pmatrix}\right]$$

$$= \frac{1}{2}\left[\begin{pmatrix} \sigma_k & \mathbf{0} \\ \mathbf{0} & \sigma_k \end{pmatrix}\begin{pmatrix} \sigma_l & \mathbf{0} \\ \mathbf{0} & \sigma_l \end{pmatrix}\right] = \frac{1}{2}\left[\begin{pmatrix} \sigma_k\sigma_l & \mathbf{0} \\ \mathbf{0} & \sigma_k\sigma_l \end{pmatrix}\right]$$

$$= i\varepsilon_{klm}\frac{1}{2}\begin{pmatrix} \sigma_m & \mathbf{0} \\ \mathbf{0} & \sigma_m \end{pmatrix} = i\varepsilon_{klm}\hat{S}_m \tag{54}$$

so that eq. (51) is satisfied. The requirements of eq. (52) are trivially satisfied by the above defined spin angular momentum operator.

(iii) We also have

$$[\hat{H},\ \hat{S}_1] = \left[-i\alpha\nabla + \beta m, \frac{1}{2}\Sigma_1\right] = -\frac{i}{2}\,[\boldsymbol{\alpha}\nabla,\ \Sigma_1] + \frac{1}{2}\,[\beta m,\ \Sigma_1]$$

$$= -\frac{i}{2}\left[\begin{pmatrix} \mathbf{0} & \sum_{j=1}^{3}\sigma_j\partial_j \\ \sum_{j=1}^{3}\sigma_j\partial_j & \mathbf{0}\end{pmatrix}, \begin{pmatrix}\sigma_1 & \mathbf{0} \\ \mathbf{0} & \sigma_1\end{pmatrix}\right] + \frac{m}{2}\left[\begin{pmatrix}\mathbf{I} & \mathbf{0} \\ \mathbf{0} & -\mathbf{I}\end{pmatrix}, \begin{pmatrix}\sigma_1 & \mathbf{0} \\ \mathbf{0} & \sigma_1\end{pmatrix}\right]$$

$$= -\frac{i}{2}\left[\begin{pmatrix} \mathbf{0} & \partial_1 - i\sigma_3\partial_2 + i\sigma_2\partial_3 \\ \partial_1 - i\sigma_3\partial_2 + i\sigma_2\partial_3 & \mathbf{0}\end{pmatrix} - \begin{pmatrix} \mathbf{0} & \partial_1 + i\sigma_3\partial_2 - i\sigma_2\partial_3 \\ \partial_1 + i\sigma_3\partial_2 - i\sigma_2\partial_3 & \mathbf{0}\end{pmatrix}\right]$$

$$= \begin{pmatrix} \mathbf{0} & -\boldsymbol{\sigma}_3\partial_2 + \boldsymbol{\sigma}_2\partial_3 \\ -\boldsymbol{\sigma}_3\partial_2 + \boldsymbol{\sigma}_2\partial_3 & \mathbf{0}\end{pmatrix}$$

$$= -\begin{pmatrix}\mathbf{0} & \sigma_3 \\ \sigma_3 & \mathbf{0}\end{pmatrix}\partial_2 + \begin{pmatrix}\mathbf{0} & \sigma_2 \\ \sigma_2 & \mathbf{0}\end{pmatrix}\partial_3 = -\boldsymbol{\alpha}^3\partial_2 + \boldsymbol{\alpha}^2\partial_3 \tag{55}$$

Adding eqs. (43) and (55), we find that $[\hat{H},\ \hat{J}_1] = 0$ as required by rotational symmetry. Similarly, $[\hat{H},\ \hat{J}_2] = [\hat{H},\ \hat{J}_3] = 0$ and so $[\hat{H},\ \hat{J}] = 0$.

(iv) It is pertinent to mention that the eigenvalues of Σ_k for all values of $k = 1, 2, 3$ are $\pm\frac{1}{2}$ as can be explicitly verified. Therefore, the Dirac equation describes particles of spin $\pm\frac{1}{2}$. Hence, the mathematical formulation of the Dirac equation automatically incorporates the degrees of freedom corresponding to the particle spin.

7.4 DIRAC PARTICLE IN EXTERNAL FIELDS

Consider a Dirac particle with spin $\pm\frac{1}{2}$ having a charge e and interacting with an electromagnetic field specified by the 4-vector potential $A_\mu(x) \equiv \{A_0, -\mathbf{A}\}$ or $A^\mu(x) \equiv \{A^0, \mathbf{A}\}$. The 4-potential satisfies

$$\partial_\mu A^\mu(x) = 0 \tag{56}$$

in the Lorentz gauge.

We can then obtain the Dirac equation for the particle by making the gauge invariant replacement

$$\hat{p}_\mu\,(\hat{p}_0, -\hat{\mathbf{p}}) \equiv i\partial_\mu\,(i\partial_0, i\nabla) \rightarrow \hat{p}_\mu - eA_\mu\,(\hat{p}_0 - eA_0, -\hat{\mathbf{p}} + e\mathbf{A})$$
$$\equiv i\partial_\mu - eA_\mu\,(i\partial_0 - eA_0, i\nabla + e\mathbf{A}) \tag{57}$$

or equivalently $\hat{p}_0 \rightarrow i\partial_0 - eA_0$, $\hat{\mathbf{p}} \rightarrow -i\nabla - e\mathbf{A}$ in the free particle Dirac equation $(i\gamma^\mu\partial_\mu - m)$ $\psi(x) = 0$ to get

$$\gamma^\mu[i\partial_\mu - eA_\mu(x)]\;\psi(x) = m\psi(x) \tag{58}$$

as the Dirac equation for a particle in an electromagnetic field. Equation (58) can also be written as

$$(i\partial_0 - eA_0)\psi(x) = \{[\alpha(\hat{\mathbf{p}} - e\mathbf{A})] + \beta m\}\psi(x) \equiv \hat{H}\,\psi(x) \tag{59}$$

To investigate the non-relativistic limit of the Dirac equation for a particle in an electromagnetic field, we write the 4-component Dirac wave function $\psi(x)$ as a pair of 2-component wave functions, e.g.,

$$\psi = (\phi_{nr}\;\chi_{nr})^T \tag{60}$$

Then, we have

$$i\partial_0\begin{pmatrix}\phi_{nr}\\ \chi_{nr}\end{pmatrix} = [\boldsymbol{\alpha}(\hat{\mathbf{p}} - e\mathbf{A}) + eA_0 + \boldsymbol{\beta}m]\begin{pmatrix}\phi_{nr}\\ \chi_{nr}\end{pmatrix} \tag{61}$$

Noting that, $\boldsymbol{\alpha} = \begin{pmatrix}\mathbf{0} & \boldsymbol{\sigma}\\ \boldsymbol{\sigma} & \mathbf{0}\end{pmatrix}$, $\boldsymbol{\beta} = \begin{pmatrix}\mathbf{I} & 0\\ 0 & -\mathbf{I}\end{pmatrix}$ and writting $\hat{\mathbf{P}} = \hat{\mathbf{p}} - e\mathbf{A}$, we have

$$i\partial_0\begin{pmatrix}\phi_{nr}\\ \chi_{nr}\end{pmatrix} = \begin{pmatrix}\boldsymbol{\sigma}\hat{\mathbf{P}}\chi_{nr}\\ \boldsymbol{\sigma}\hat{\mathbf{P}}\phi_{nr}\end{pmatrix} + eA_0\begin{pmatrix}\phi_{nr}\\ \chi_{nr}\end{pmatrix} + m\begin{pmatrix}\phi_{nr}\\ -\chi_{nr}\end{pmatrix} \tag{62}$$

If, as we did when investigating the non-relativistic limit of the Klein Gordon equation, we separate the contribution to the wave function due to the rest energy and write

$$\begin{pmatrix}\phi_{nr}\\ \chi_{nr}\end{pmatrix} = \psi = \psi_0\,\exp(-imx^0) = \begin{pmatrix}\phi_{0nr}\\ \chi_{0nr}\end{pmatrix}\exp(-imx^0) \tag{63}$$

then

$$i\partial_0\psi = i\partial_0(\psi_0 e^{-imx^0}) = i(\partial_0\psi_0)e^{-imx^0} + m\psi_0 e^{-imx^0} \tag{64}$$

so that eq. (62) becomes

$$i\partial_0\begin{pmatrix}\phi_{0nr}\\ \chi_{0nr}\end{pmatrix} = \begin{pmatrix}\boldsymbol{\sigma}\hat{\mathbf{P}}\chi_{0nr}\\ \boldsymbol{\sigma}\hat{\mathbf{P}}\phi_{0nr}\end{pmatrix} + eA_0\begin{pmatrix}\phi_{0nr}\\ \chi_{0nr}\end{pmatrix} - 2m\begin{pmatrix}0\\ \chi_{0nr}\end{pmatrix} \tag{65}$$

leading to the following pair of coupled differential equations:

$$i\partial_0\phi_{0nr} = \boldsymbol{\sigma}\hat{\mathbf{P}}\,\chi_{0nr} + eA_0\phi_{0nr} \tag{66}$$

and

$$i\partial_0\chi_{0nr} = \sigma\hat{\mathbf{P}}\,\phi_{0nr} + eA_0\chi_{0nr} - 2m\chi_{0nr} \tag{67}$$

In the non-relativistic limit, which we assume to be represented by the following conditions:

$$|i\partial_0\chi| << |m\chi|, |eA_0\chi| << |m\chi| \tag{68}$$

i.e., the kinetic energy as well as the potential energy are small compared to the rest energy, we obtain from eq. (67)

$$\chi_{0nr} = \frac{\sigma\hat{\mathbf{P}}}{2m}\phi_{0nr} \Rightarrow \chi_{0nr} << \phi_{0nr} \tag{69}$$

i.e., χ represents the smaller component of ψ. Substituting the value of χ from eq. (69) in (66), we get

$$i\partial_0\phi_{0nr} = \sigma\hat{\mathbf{P}}.\frac{\sigma\hat{\mathbf{P}}}{2m}\phi_{0nr} + eA_0\phi_{0nr} \tag{70}$$

Now, for any function $\varphi(x)$, we have

$$\begin{aligned}
\sigma\hat{\mathbf{P}}.\sigma\hat{\mathbf{P}}\,\varphi &= \hat{\mathbf{P}}.\hat{\mathbf{P}}\,\mathbf{I}\varphi + i\sigma(\hat{\mathbf{P}}\times\hat{\mathbf{P}})\varphi \\
&= (\hat{\mathbf{p}} - e\mathbf{A})^2\,\varphi + i\boldsymbol{\sigma}[(\hat{\mathbf{p}} - e\mathbf{A})\times(\hat{\mathbf{p}} - e\mathbf{A})]\varphi \\
&= (\hat{\mathbf{p}} - e\mathbf{A})^2\,\varphi + i\boldsymbol{\sigma}[(-i\nabla - e\mathbf{A})\times(-i\nabla\varphi - e\mathbf{A}\varphi)] \\
&= (\hat{\mathbf{p}} - e\mathbf{A})^2\,\varphi + i\boldsymbol{\sigma}[-(\nabla\times\nabla\varphi)\times ie(\mathbf{A}\times\nabla\varphi) \\
&\quad + ie(\nabla\times\mathbf{A}\varphi) + e^2(\mathbf{A}\times\mathbf{A}\varphi)] \\
&= (\hat{\mathbf{p}} - e\mathbf{A})^2\,\varphi - e\boldsymbol{\sigma}\{(\mathbf{A}\times\nabla\varphi) + [(\nabla\times\mathbf{A})\varphi - (\mathbf{A}\times\nabla\varphi)]\} \\
&= (\hat{\mathbf{p}} - e\mathbf{A})^2\,\varphi - e\boldsymbol{\sigma}(\nabla\times\mathbf{A})\varphi \\
&= (\hat{\mathbf{p}} - e\mathbf{A})^2\,\varphi - e\boldsymbol{\sigma}.\mathbf{B}\varphi
\end{aligned} \tag{71}$$

Hence, from eqs. (70) and (71)

$$i\partial_0\phi_{0nr} = \left[\frac{1}{2m}(\hat{\mathbf{p}} - e\mathbf{A})^2 - \frac{1}{2m}e\boldsymbol{\sigma}.\mathbf{B} + eA_0\right]\phi_{0nr} \tag{72}$$

which is exactly the Pauli equation for the non-relativistic dynamics of a spin $\frac{1}{2}$ particle in an electromagnetic field.

The two components of ϕ_{0nr} are attributed to the spin degrees of freedom. Hence, it is established that the Dirac equation has the correct non-relativistic limit in the form of the Pauli equation.

Furthermore, since the spin degrees of freedom are independent of the velocity, it follows that the Dirac equation constitutes the relativistic extension of the Pauli equation for the dynamics of spin $\frac{1}{2}$ particles.

Since the Dirac equation constitutes a linearization of the relativistic momentum-energy relationship (that is described by the Klein Gordon equation in the quadratic form), it follows that the emergence of spin may be attributed to the linearization of the Klein Gordon equation.

Now consider a homogeneous magnetic field described by

$$\mathbf{B} = \nabla \times \mathbf{A},\ \mathbf{A} = \frac{1}{2}\ \mathbf{B} \times \mathbf{x} \tag{73}$$

We then have

$$\begin{aligned}(\hat{\mathbf{p}} - e\mathbf{A})^2 &= \left(\hat{\mathbf{p}} - \frac{e}{2}\mathbf{B}\times\mathbf{x}\right)^2 \approx \hat{\mathbf{p}}^2 - e\hat{\mathbf{p}}\,(\mathbf{B}\times\mathbf{x}) \\ &= \hat{\mathbf{p}}^2 - e(\mathbf{B}\times\mathbf{x})\hat{\mathbf{p}} = \hat{\mathbf{p}}^2 - e\mathbf{B}(\mathbf{x}\times\hat{\mathbf{p}}) \\ &= \hat{\mathbf{p}}^2 - e\mathbf{B}.\hat{\mathbf{L}}\end{aligned} \tag{74}$$

where $\hat{\mathbf{L}} = \mathbf{x}\times\hat{\mathbf{p}}$ is the orbital angular momentum operator. Hence, eq. (72) reduces to

$$i\partial_0\phi_{0nr} = \left[\frac{1}{2m}(\hat{\mathbf{p}}^2 - e\mathbf{B}.\hat{\mathbf{L}}) - \frac{1}{2m}e\boldsymbol{\sigma}.\mathbf{B} + eA_0\right]\phi_{0nr}$$

$$i\partial_0\phi_{0nr} = \left[\frac{1}{2m}\hat{\mathbf{p}}^2 - \frac{1}{2m}e\mathbf{B}.(\hat{\mathbf{L}} + 2\hat{S}) + eA_0\right]\phi_{0nr} \tag{75}$$

where $\hat{S} = \dfrac{1}{2}\boldsymbol{\sigma}$ is the spin operator. This expression formally establishes that the gyromagnetic ratio $g = 2$.

7.5 NON-RELATIVISTIC LIMIT OF OPERATORS

We can also evaluate the non-relativistic limit of various operators arising in the Dirac framework.

Given an operator $\hat{O}$, its non-relativistic limit $\hat{O}_{nr}$ will have the matrix elements given by $\langle\phi^a_{nr}|\hat{O}|\phi^b_{nr}\rangle$ for positive energy and $\langle\chi^a_{nr}|\hat{O}|\chi\phi^b_{nr}\rangle$ for negative energy, respectively.

(i) **Sign Projection Operators**

We have

$$\hat{\Lambda} = \frac{\hat{H}}{\sqrt{\hat{H}^2}} = \frac{\hat{H}}{E_p} = \frac{\alpha\hat{\mathbf{p}} + \beta m}{E_p} \approx \frac{m}{E_p}\boldsymbol{\beta} \approx \boldsymbol{\beta} \tag{76}$$

because in the non-relativistic limit, $|\hat{\mathbf{p}}| << m$ and $E_p \approx m$.

(ii) **Energy Projection Operators**

We can define the energy projection operators as follows:

$$\hat{\Lambda}_{\text{sgn}\,\varepsilon} = \frac{1}{2}\,(1 + \text{e}\hat{\Lambda}) \tag{77}$$

because, for an arbitrary Dirac wave function $\psi^{\varepsilon'}(x)$

$$\hat{\Lambda}_{\text{sgn}\,\varepsilon}\,\psi^{\varepsilon'}(x) = \frac{1}{2}\,(1 + \varepsilon\hat{\Lambda})\psi^{\varepsilon'}(x) = \frac{1}{2}\left(1 + \varepsilon\frac{\hat{H}}{E_p}\right)\psi^{\varepsilon'}(x)$$

$$= \frac{1}{2}\left(1 + \varepsilon\frac{\varepsilon' E_p}{E_p}\right)\psi^{\varepsilon'}(x) = \frac{1}{2}\,(1 + \varepsilon\varepsilon')\psi^{\varepsilon'}(x)$$

$$= \delta_{\varepsilon\varepsilon'}\psi^{\varepsilon'}(x) \tag{78}$$

as desired.

Let us assume that $\psi_+ = (\phi_+\ \chi_+)^T$ is a positive energy Dirac wave function. Then, we have

$$\begin{pmatrix}\phi_+\\ \chi_+\end{pmatrix} = \psi_+(x) = \hat{\Lambda}_+\psi_+(x) = \frac{1}{2}(1 + \hat{\Lambda})\begin{pmatrix}\phi_+\\ \chi_+\end{pmatrix}$$

$$= \frac{1}{2}\,(1 + \beta)\begin{pmatrix}\phi_+\\ \chi_+\end{pmatrix} = \begin{pmatrix}\mathbf{I} & \mathbf{0}\\ \mathbf{0} & \mathbf{0}\end{pmatrix}\begin{pmatrix}\phi_+\\ \chi_+\end{pmatrix} = \begin{pmatrix}\phi_+\\ \mathbf{0}\end{pmatrix} \tag{79}$$

showing explicitly that in the non-relativistic limit, the lower pair of components of a positive energy Dirac wave function vanish. Similarly, the upper pair of components vanish in the case of a negative energy wave function in the non-relativistic limit.

(iii) **Non-Relativistic Limits of the Matrices α, β**

The matrix element of $\boldsymbol{\beta} = \begin{pmatrix}\mathbf{I} & \mathbf{0}\\ \mathbf{0} & -\mathbf{I}\end{pmatrix}$ in the non-relativistic limit will be given by

$$(\boldsymbol{\beta}_{nr})^+_{ab} = \langle\phi^a_{nr}|\beta|\phi^b_{nr}\rangle = (\phi^{a*}_{nr}\ \mathbf{0})\begin{pmatrix}\mathbf{1} & \mathbf{0}\\ \mathbf{0} & -\mathbf{1}\end{pmatrix}\begin{pmatrix}\phi^b_{nr}\\ \mathbf{0}\end{pmatrix}$$

$$= \phi^{a*}_{nr}\,\phi^b_{nr} = \langle\phi^a_{nr}|\phi^b_{nr}\rangle = \delta_{ab} \tag{80}$$

so that

$$\boldsymbol{\beta}^+_{nr} = \mathbf{I} \tag{81}$$

Similarly,

$$(\boldsymbol{\beta}_{nr})^-_{ab} = \langle\chi^a_{nr}|\boldsymbol{\beta}|\chi^b_{nr}\rangle = (\mathbf{0}\ \chi^{a*}_{nr})\begin{pmatrix}\mathbf{1} & \mathbf{0}\\ \mathbf{0} & -\mathbf{1}\end{pmatrix}\begin{pmatrix}\mathbf{0}\\ \chi^b_{nr}\end{pmatrix}$$

$$= -\,\chi^{a*}_{nr}\chi^b_{nr} = -\,\langle\chi^a_{nr}|\chi^b_{nr}\rangle = -\,\delta_{ab} \tag{82}$$

so that

$$\boldsymbol{\beta}^-_{nr} = -\,\mathbf{I} \tag{83}$$

Furthermore, the $\boldsymbol{\alpha}$ matrices vanish in the non-relativistic limit, e.g.,

$$(\boldsymbol{\alpha}_{nr})^+_{ab} = \langle\phi^a_{nr}|\boldsymbol{\alpha}|\phi^b_{nr}\rangle = (\phi^{a*}_{nr}\ \mathbf{0})\begin{pmatrix}\mathbf{0} & \sigma\\ \sigma & \mathbf{0}\end{pmatrix}\begin{pmatrix}\phi^b_{nr}\\ \mathbf{0}\end{pmatrix} = \mathbf{0} \tag{84}$$

and similarly $(\boldsymbol{\alpha}_{nr})^-_{ab} = 0$.

The spin operator reduces to the Pauli matrices in the non-relativistic limit, e.g.,

$$(\hat{S}_{nr})^+_{ab} = \langle \phi^a_{nr} | \hat{S}_{nr} | \phi^a_{nr} \rangle = \frac{1}{2} \,(\phi^{a*}_{nr} \; \mathbf{0}) \begin{pmatrix} \sigma & \mathbf{0} \\ \mathbf{0} & \sigma \end{pmatrix} \begin{pmatrix} \phi^b_{nr} \\ \mathbf{0} \end{pmatrix}$$

$$= \frac{1}{2} \,\phi^{a*}_{nr} \,\sigma \phi^b_{nr} = \frac{1}{2} \,\langle \phi^a_{nr} | \sigma | \phi^b_{nr} \rangle = \frac{1}{2} \,\delta_{ab} \boldsymbol{\sigma} \tag{85}$$

with $\quad (\hat{S}_{nr})^-_{ab} = \frac{1}{2} \,\delta_{ab} \boldsymbol{\sigma}$ as well.

7.6 SPIN-ORBIT COUPLING

The Dirac equation for a spin $\frac{1}{2}$ particle in the presence of an electromagnetic field represented by the 4-vector A_μ is given by eq. (62) which, on operating on a wave function with well-defined energy state E can be written as

$$\begin{pmatrix} \sigma \hat{\mathbf{P}} \chi \\ \sigma \hat{\mathbf{P}} \phi \end{pmatrix} + m \begin{pmatrix} \phi \\ -\chi \end{pmatrix} = (E - eA_0) \begin{pmatrix} \phi \\ \chi \end{pmatrix} \tag{86}$$

Now we consider an electrostatic field such that $\mathbf{A} = 0$. We then have $\hat{\mathbf{P}} = \hat{\mathbf{p}}$ and so, from eq. (86)

$$\begin{pmatrix} \sigma \hat{\mathbf{p}} \chi \\ \sigma \hat{\mathbf{p}} \phi \end{pmatrix} = \begin{pmatrix} (E - eA_0 - m)\phi \\ (E - eA_0 + m)\chi \end{pmatrix} \tag{87}$$

or

$$\chi = \frac{\sigma \hat{\mathbf{p}}}{(E - eA_0 + m)} \,\phi = \left(1 + \frac{E - eA_0 - m}{2m}\right)^{-1} \frac{\sigma \hat{\mathbf{p}}}{2m} \,\phi \approx \left(1 - \frac{E - eA_0 - m}{2m}\right) \frac{\sigma \hat{\mathbf{p}}}{2m} \phi \tag{88}$$

Substituting this value of χ in the first of eq. (87), we have

$$(E - eA_0 - m)\,\phi = \boldsymbol{\sigma} \hat{\mathbf{p}} \left(1 - \frac{E - eA_0 - m}{2m}\right) \frac{\sigma \hat{\mathbf{p}}}{2m} \phi$$

$$= \left(1 - \frac{E - m}{2m}\right) \frac{\hat{\mathbf{p}}^2}{2m} \,\phi + \frac{e}{4m^2} \,\boldsymbol{\sigma} \hat{\mathbf{p}} \, A_0 \boldsymbol{\sigma} \hat{\mathbf{p}} \,\phi \tag{89}$$

where we have used $(\boldsymbol{\sigma}.\mathbf{A})\,(\boldsymbol{\sigma}.\mathbf{B}) = \mathbf{A}.\mathbf{B} + i\sigma(\mathbf{A} \times \mathbf{B})$.

Now

$$\boldsymbol{\sigma}\hat{\mathbf{p}} A_0 \boldsymbol{\sigma}\hat{\mathbf{p}}\,\phi = -(\boldsymbol{\sigma}\nabla)\, A_0(\boldsymbol{\sigma}\nabla\phi)$$

$$= -\left(\sum_{k=1}^{3}\sigma^k \partial_k\right) A_0 \left(\sum_{l=1}^{3}\sigma^l \partial_l \phi\right) = -\left(\sum_{k=1}^{3}\sigma^k \partial_k\right)\left(\sum_{l=1}^{3}\sigma^l A_0 \partial_l \phi\right)$$

$$= -\sum_{k=1}^{3}\sigma^k \sum_{l=1}^{3}\sigma^l\,[(\partial_k A_0)(\partial_l \phi) + A_0 \partial_k \partial_l \phi]$$

$$= -(\nabla A_0)(\nabla\phi) - A_0 \nabla^2\phi - i\boldsymbol{\sigma}(\nabla A_0 \times \nabla\phi)$$

$$= -i(\nabla A_0)(\hat{\mathbf{p}}\,\phi) + A_0 \hat{\mathbf{p}}^2\phi + \boldsymbol{\sigma}(\nabla A_0 \times \hat{\mathbf{p}}\,\phi) \tag{90}$$

Using eq. (90), we can write eq. (89) as

$$(E-m)\phi = \left(1 - \frac{E - m - eA_0}{2m}\right)\frac{\hat{\mathbf{p}}^2}{2m}\,\phi + \frac{e}{4m^2}$$

$$[-i(\nabla A_0)(\hat{\mathbf{p}}\,\phi) + \boldsymbol{\sigma}(\nabla A_0 \times \hat{\mathbf{p}}\,\phi)] + eA_0\phi \tag{91}$$

The normalization is obtained as

$$1 = N^2 \int \psi^\dagger \psi d\mathbf{x} = N^2 \int \begin{pmatrix}\phi\\ \chi\end{pmatrix}^\dagger \begin{pmatrix}\phi\\ \chi\end{pmatrix} d\mathbf{x} = N^2 \int (\phi^* \; \chi^*)\begin{pmatrix}\phi\\ \chi\end{pmatrix} d\mathbf{x}$$

$$= N^2 \int (\phi^*\phi + \chi^*\chi)\, d\mathbf{x} \approx N^2 \int \phi^* \left(1 + \frac{1}{4m^2}\hat{\mathbf{p}}^2\right)\phi d\mathbf{x}$$

$$= N^2\left(1 + \frac{1}{4m^2}\hat{\mathbf{p}}^2\right) \tag{92}$$

where the last step follows from the first order approximation

$$\chi \approx \left(1 - \frac{E - eA_0 - m}{2m}\right)\frac{\boldsymbol{\sigma}\hat{\mathbf{p}}}{2m}\,\phi \approx \frac{\boldsymbol{\sigma}\hat{\mathbf{p}}}{2m}\phi \tag{93}$$

so that

$$\chi^*\chi \approx \frac{1}{4m^2}(\boldsymbol{\sigma}\hat{\mathbf{p}}\,\phi)^*(\boldsymbol{\sigma}\hat{\mathbf{p}}\,\phi) = \frac{1}{4m^2}\;\hat{\mathbf{p}}^2\phi^*\phi \tag{94}$$

Equation (92) yields the normalization

$$N = \left(1 + \frac{1}{4m^2}\hat{\mathbf{p}}^2\right)^{-\frac{1}{2}} = \left(1 - \frac{1}{8m^2}\hat{\mathbf{p}}^2\right).$$

Hence, the normalized eigenvector is given by

$$\psi_n = N\psi = \left(1 - \frac{1}{4m^2}\hat{\mathbf{p}}^2\right)^{\frac{1}{2}}\begin{pmatrix}\phi\\ \chi\end{pmatrix} \approx \left(1 - \frac{1}{8m^2}\hat{\mathbf{p}}^2\right)\begin{pmatrix}\phi\\ \chi\end{pmatrix} = N\begin{pmatrix}\phi\\ \chi\end{pmatrix} \tag{95}$$

If we write

$$\hat{H} = \left(1 - \frac{E - m - eA_0}{2m}\right) \frac{\hat{\mathbf{p}}^2}{2m} + \frac{e}{4m^2}$$

$$[-i(\nabla A_0)\hat{\mathbf{p}} + \boldsymbol{\sigma}(\nabla A_0 + \hat{\mathbf{p}})] + eA_0 \tag{96}$$

then we have from eq. (91)

$$\hat{H}\phi = (E - m)\phi \tag{97}$$

Further

$$\hat{H}\chi = \hat{H}\frac{\boldsymbol{\sigma}\hat{\mathbf{p}}}{2m}\phi = \frac{\boldsymbol{\sigma}\hat{\mathbf{p}}}{2m}\hat{H}\phi = \frac{\boldsymbol{\sigma}\hat{\mathbf{p}}}{2m}(E - m)$$

$$\phi = (E - m)\frac{\boldsymbol{\sigma}\hat{\mathbf{p}}}{2m}\phi = (E - m)\chi$$

which follows from the fact that $[\hat{H}, \boldsymbol{\sigma}\mathbf{p}] = 0$. Hence, we have

$$\hat{H}\psi = \hat{H}\begin{pmatrix}\phi \\ \chi\end{pmatrix} = (E - m)\begin{pmatrix}\phi \\ \chi\end{pmatrix} = (E - m)\psi$$

Thus,

$$(E - m)\psi_n = (E - m)N\psi = N(E - m)\psi = N\hat{H}\psi = N\hat{H}N^{-1}N\psi = \hat{H}'N\psi = \hat{H}\psi_n \tag{98}$$

where

$$\hat{H}' = N\hat{H}N^{-1} = \left(1 - \frac{1}{8m^2}\hat{\mathbf{p}}^2\right)\left\{\begin{matrix}\left(1 - \frac{E - m - eA_0}{2m}\right)\frac{\hat{\mathbf{p}}^2}{2m} + \frac{e}{4m^2} \\ [-i(\nabla A_0)\hat{\mathbf{p}} + \sigma(\nabla A_0 \times \hat{\mathbf{p}})] + eA_0\end{matrix}\right\}\left(1 - \frac{1}{8m^2}\hat{\mathbf{p}}^2\right)^{-1}$$

$$\approx \left(1 - \frac{1}{8m^2}\hat{\mathbf{p}}^2\right)\left\{\left(1 - \frac{E - m - eA_0}{2m}\right)\frac{\hat{\mathbf{p}}^2}{2m} + \frac{e}{4m^2}[-i(\nabla A_0)\hat{\mathbf{p}} + \sigma(\nabla A_0 \times \hat{\mathbf{p}})] + eA_0\right\}$$

$$\left(1 + \frac{1}{8m^2}\hat{\mathbf{p}}^2\right) \tag{99}$$

Retaining terms of the order of $\hat{\mathbf{p}}^2$, we have

$$\hat{H} \approx \left(1 - \frac{E - m - eA_0}{2m}\right)\frac{\hat{\mathbf{p}}^2}{2m} + \frac{e}{4m^2}[-i(\nabla A_0)\hat{\mathbf{p}} + \boldsymbol{\sigma}(\nabla A_0 + \hat{\mathbf{p}})]$$

$$+ eA_0 - \left(\frac{1}{8m^2}\hat{\mathbf{p}}^2\right)eA_0 + eA_0\left(\frac{1}{8m^2}\hat{\mathbf{p}}^2\right) \tag{100}$$

Now, for an arbitrary wave function $\varphi(x)$, we have

$$(\hat{\mathbf{p}}^2 A_0 - A_0\hat{\mathbf{p}}^2)\varphi = -(\nabla^2 A_0 - A_0\nabla^2)\varphi = -\nabla(\nabla A_0.\varphi + A_0\nabla\varphi) + A_0\nabla^2\varphi$$

$$= -\nabla^2 A_0.\varphi - 2\nabla A_0\nabla\varphi = -\nabla^2 A_0.\varphi - 2i\nabla A_0\hat{\mathbf{p}}\varphi \tag{101}$$

From eqs. (100) and (101), we obtain

$$\hat{H}' \approx \left(1 - \frac{E - m - eA_0}{2m}\right)\frac{\hat{\mathbf{p}}^2}{2m} + \frac{e}{4m^2}\,[-i(\nabla A_0)\hat{\mathbf{p}} + \boldsymbol{\sigma}(\nabla A_0 \times \hat{\mathbf{p}})] + eA_0$$

$$+ \frac{e}{8m^2}(\nabla^2 A_0 + 2i\nabla A_0\hat{\mathbf{p}})$$

$$= \frac{\hat{\mathbf{p}}^2}{2m} - \left(\frac{E - m - eA_0}{2m}\right)\frac{\hat{\mathbf{p}}^2}{2m} + eA_0 + \frac{e}{8m^2}\,\nabla^2 A_0 + \frac{e}{4m^2}\,\boldsymbol{\sigma}(\nabla A_0 \times \hat{\mathbf{p}}) \tag{102}$$

The first and third terms correspond to the non-relativistic Hamiltonian. The term $\hat{C}_1 = -\left(\frac{E - m - eA_0}{2m}\right)\frac{\hat{\mathbf{p}}^2}{2m}$ represents the correction to the kinetic energy operator that is due to the variation of particle mass with velocity. The term $\hat{C}_2 = \frac{e}{4m^2}\,\boldsymbol{\sigma}(\nabla A_0 \times \hat{\mathbf{p}})$ is the spin-orbit interaction operator. The remaining term $\hat{C}_3 = \frac{e}{8m^2}\,\nabla A_0$ is the correction arising due to the mixing of the positive and negative energy states and is referred to as Zitterbewegung.

7.7 ZITTERBEWEGUNG

The Hamiltonian of a Dirac particle in an electromagnetic field $A_\mu \equiv \{A_0, -\mathbf{A}\}$ is given by

$$\hat{H} = \boldsymbol{\alpha}\,(\hat{\mathbf{p}} - eA) + \boldsymbol{\beta}m + eA_0 \tag{103}$$

The commutator of this Hamiltonian with the position operator is given by

$$[\hat{H}, \hat{\mathbf{x}}] = [\boldsymbol{\alpha}\hat{\mathbf{p}}, \hat{\mathbf{x}}] - e[\boldsymbol{\alpha}\mathbf{A}, \hat{\mathbf{x}}] + m[\boldsymbol{\beta}, \hat{\mathbf{x}}] + e[A_0, \hat{\mathbf{x}}] \tag{104}$$

Now

$$[A_0, \hat{\mathbf{x}}]\,\varphi = (A_0\hat{\mathbf{x}} - \hat{\mathbf{x}}A_0)\,\varphi = A_0(\hat{\mathbf{x}}\,\varphi) - \hat{\mathbf{x}}A_0\varphi = A_0(\mathbf{x}\,\varphi) - \hat{\mathbf{x}}A_0\varphi$$

$$= \mathbf{x}A_0\varphi - \hat{\mathbf{x}}A_0\varphi = \hat{\mathbf{x}}A_0\varphi - \hat{\mathbf{x}}A_0\varphi = 0 \tag{105}$$

Similarly

$$[\beta, \hat{\mathbf{x}}]\,\varphi = [\boldsymbol{\alpha}\mathbf{A}, \hat{\mathbf{x}}]\,\varphi = 0 \tag{106}$$

Further

$$[\boldsymbol{\alpha}\hat{\mathbf{p}}, \hat{\mathbf{x}}]\,\varphi = -i[\boldsymbol{\alpha}\nabla, \hat{\mathbf{x}}]\,\varphi = i\hat{\mathbf{x}}.\boldsymbol{\alpha}\nabla\varphi - i\alpha\nabla(\hat{\mathbf{x}}\,\varphi) = i\hat{\mathbf{x}}.\boldsymbol{\alpha}\nabla\varphi - i\boldsymbol{\alpha}\nabla(\mathbf{x}\varphi)$$

$$= i\mathbf{x}\boldsymbol{\alpha}\nabla\varphi - i\boldsymbol{\alpha}(\nabla\mathbf{x})\varphi - i\boldsymbol{\alpha}\mathbf{x}(\nabla\varphi)$$

$$= i\mathbf{x}\boldsymbol{\alpha}\nabla\varphi - i\boldsymbol{\alpha}\varphi - i\mathbf{x}\boldsymbol{\alpha}(\nabla\varphi) = -i\boldsymbol{\alpha}\varphi \tag{107}$$

Therefore,

$$[\hat{H},\ \hat{\mathbf{x}}] = -i\boldsymbol{\alpha} \tag{108}$$

But, in the Heisenberg formulation of the non-relativistic quantum mechanics, the velocity of a quantum particle is given by

$$\hat{\mathbf{v}} = \frac{1}{i}[\hat{\mathbf{x}},\ \hat{H}] = i\,[\hat{H},\ \hat{\mathbf{x}}] = \boldsymbol{\alpha} \tag{109}$$

so that the velocity operator of a Dirac particle is given by $\hat{\mathbf{v}} = \boldsymbol{\alpha}$. Now each of the $\boldsymbol{\alpha}^k$, $k = 1, 2, 3$ has real eigenvalues ±1. Hence, we conclude that the absolute value of the velocity of a Dirac particle shall invariably equal the velocity of light (we are working here in natural units wherein $\hbar = c = 1$). Besides, since the various components $\boldsymbol{\alpha}^k$, $k = 1, 2, 3$ do not commute among themselves, the components of velocity are not simultaneously measurable. Let us explore this further to resolve this paradox.

We have

$$\begin{aligned}
\frac{d\hat{\alpha}}{dx^0} &= i\,[\hat{H},\ \hat{\alpha}] = i(\hat{H}\hat{\alpha} - \hat{\alpha}\hat{H}) \\
&= i(\hat{H}\hat{\alpha} + \hat{\alpha}\hat{H}) - 2i\hat{\alpha}\hat{H} \\
&= i[(\boldsymbol{\alpha}\hat{\mathbf{p}})\hat{\alpha} + \hat{\alpha}(\boldsymbol{\alpha}\hat{\mathbf{p}})] - 2i\hat{\alpha}\hat{H} \\
&= i[(\boldsymbol{\alpha})\hat{\alpha}\hat{\mathbf{p}} + \hat{\alpha}(\boldsymbol{\alpha}\hat{\mathbf{p}})] - 2i\hat{\alpha}\hat{H} \\
&= 2i\boldsymbol{\alpha}^2\hat{\mathbf{p}} - 2i\hat{\alpha}\hat{H} = 2i\hat{\mathbf{p}} - 2i\hat{\alpha}\hat{H}
\end{aligned} \tag{110}$$

Integrating this expression and noting that $\hat{H}$, $\hat{\mathbf{p}}$ are constant in time, we have

$$2ix^0 = 2i\int dx^0 = \int \frac{d\hat{\alpha}}{(\hat{\mathbf{p}} - \hat{\alpha}\hat{H})} = -\frac{1}{\hat{H}}\ln(\hat{\mathbf{p}} - \hat{\alpha}\hat{H}) + C \tag{111}$$

where $C = \frac{1}{\hat{H}}\ln[\hat{\mathbf{p}} - \hat{\alpha}(0)\hat{H}]$ so that

$$\hat{\alpha}(x^0) = \frac{1}{\hat{H}}[(\hat{\alpha}(0)\hat{H} - \hat{\mathbf{p}})\exp(-2i\hat{H}x^0) + p] \tag{112}$$

Substituting this value of α in eq. (109) and performing the integration with respect to the time coordinate x^0, we get

$$\hat{\mathbf{x}}(x^0) = \hat{\mathbf{x}}(0) + \frac{1}{\hat{H}}\hat{\mathbf{p}}x^0 + \left(\hat{\alpha}(0) - \frac{1}{\hat{H}}\hat{\mathbf{p}}\right)\frac{i}{2\hat{H}}\exp(-2i\hat{H}x^0) \tag{113}$$

Comparing this with the classical equation of motion, viz.

$$\hat{\mathbf{x}}(x^0) = \hat{\mathbf{x}}(0) + \frac{1}{E}\mathbf{p}x^0 \tag{114}$$

we find that the trajectory of the free Dirac particle consists of a superposition of:

(i) Uniform classical motion represented by the first two terms on the right-hand side of eq. (113).

(ii) A rapidly oscillating motion represented by the third term of eq. (113) called the "Zitterbewegung".

The amplitude of this oscillating motion is of the order of $\dfrac{1}{2E} \approx \dfrac{1}{2m}$ since $|\boldsymbol{\alpha}| = 1 >> \dfrac{\mathbf{p}}{E_p}$ while the frequency is of the order of $2E_p = 2\sqrt{\mathbf{p}^2 - m^2} \approx 2m$.

As stated above, this oscillating motion is referred to as the "Zitterbewegung". It vanishes if the wave packets consist exclusively of either positive energy or negative energy wave functions. To prove this, we define the positive and negative energy projection operators as

$$\hat{O}_{\text{sgn}\,\varepsilon} = \frac{1}{2}\left(1 + \varepsilon\frac{\hat{H}}{E_p}\right) \tag{115}$$

We then have

$$\begin{aligned}\hat{O}_{\text{sgn}\varepsilon}\,\psi_{\varepsilon'} &= \frac{1}{2}\left(1+\varepsilon\frac{\hat{H}}{E_p}\right)\psi_{\varepsilon'} = \frac{1}{2}\left(1+\varepsilon\frac{\varepsilon' E_p}{E_p}\right)\psi_{\varepsilon'}\\ &= \frac{1}{2}(1+\varepsilon\varepsilon')\psi_{\varepsilon'} = \delta_{\varepsilon\varepsilon'}\psi_{\varepsilon'}\end{aligned} \tag{116}$$

Also

$$\begin{aligned}(\hat{O}_{\text{sgn}\,\varepsilon})\,(\hat{O}_{\text{sgn}\,\varepsilon'})\,\psi_{\varepsilon''} &= \frac{1}{4}\left(1+\varepsilon\frac{\hat{H}}{E_p}\right)\left(1+\varepsilon'\frac{\hat{H}}{E_p}\right)\psi_{\varepsilon''}\\ &= \frac{1}{4}\left(1+\varepsilon\frac{\hat{H}}{E_p}\right)\left(\psi_{\varepsilon''}+\varepsilon'\frac{\hat{H}\psi_{\varepsilon''}}{E_p}\right) = \frac{1}{4}\left(1+\varepsilon\frac{\hat{H}}{E_p}\right)(1+\varepsilon'\varepsilon'')\,\psi_{\varepsilon''}\\ &= \frac{1}{4}(1+\varepsilon'\varepsilon'')\left(1+\varepsilon\frac{\hat{H}}{E_p}\right)\psi_{\varepsilon''}\\ &= \frac{1}{4}(1+\varepsilon'\varepsilon'')\,(1+\varepsilon'\varepsilon'')\psi_{\varepsilon''}\\ &= \frac{1}{4}(1+\varepsilon'\varepsilon''+\varepsilon\varepsilon''+\varepsilon\varepsilon'\varepsilon''^2)\psi_{\varepsilon''} = \frac{1}{4}[1+(\varepsilon+\varepsilon')\varepsilon''+\varepsilon\varepsilon']\psi_{\varepsilon''}\\ &= \frac{1}{2}\delta_{\varepsilon\varepsilon'}(1+\varepsilon\varepsilon'')\psi_{\varepsilon''} = \delta_{\varepsilon\varepsilon'}\hat{O}_{\text{sgn}\,\varepsilon}\psi_{\varepsilon''}\end{aligned} \tag{117}$$

We also obtain the following commutators:

$$[\hat{O}_{\text{sgn}\,\varepsilon}, \hat{\alpha}] = \frac{1}{2}\left[1+\varepsilon\frac{\hat{H}}{E_p}, \hat{\alpha}\right] = \frac{1}{2E_p}[\varepsilon\hat{H}, \hat{\alpha}] = \frac{\varepsilon}{2E_p}(\hat{H}\hat{\alpha} - \hat{\alpha}\hat{H})$$

$$= \frac{\varepsilon}{2E_p}[(\boldsymbol{\alpha}\hat{\mathbf{p}} + \boldsymbol{\beta}m)\hat{\alpha} - \hat{\alpha}(\boldsymbol{\alpha}\hat{\mathbf{p}} + \boldsymbol{\beta}m)]$$

$$= \frac{\varepsilon m}{2E_p}(\boldsymbol{\beta}\hat{\alpha} - \hat{\alpha}\boldsymbol{\beta}) = -\frac{\varepsilon m}{E_p}\hat{\alpha}\boldsymbol{\beta}$$

$$= -\frac{\varepsilon}{E_p}\hat{\alpha}(\hat{H} - \boldsymbol{\alpha}\hat{\mathbf{p}}) = \frac{\varepsilon}{E_p}\hat{\mathbf{p}} - \frac{\varepsilon}{E_p}\hat{\alpha}\hat{H} \tag{118}$$

$$[\hat{O}_{\text{sgn}\,\varepsilon}, \hat{H}] = \frac{1}{2}\left[1+\varepsilon\frac{\hat{H}}{E_p}, \hat{H}\right] = \frac{1}{2E_p}[\varepsilon\hat{H}, \hat{H}] = 0 \tag{119}$$

$$\hat{H}\hat{O}_{\text{sgn}\varepsilon}\psi_{\varepsilon'} = \frac{1}{2}\hat{H}\left(1+\varepsilon\frac{\hat{H}}{E_p}\right)\psi_{\varepsilon'} = \frac{1}{2}\hat{H}\left(\psi_{\varepsilon'} + \varepsilon\frac{\hat{H}\psi_{\varepsilon'}}{E_p}\right) = \frac{1}{2}\hat{H}(1+\varepsilon\varepsilon')\psi_{\varepsilon'}$$

$$= \frac{1}{2}(1+\varepsilon\varepsilon')\hat{H}\psi_{\varepsilon'} = \frac{1}{2}(1+\varepsilon\varepsilon')\varepsilon' E_p\psi_{\varepsilon'}$$

$$= \frac{1}{2}(\varepsilon'+\varepsilon)E_p\psi_{\varepsilon'} = \varepsilon\delta_{\varepsilon\varepsilon'}E_p\psi_{\varepsilon'} = \varepsilon E_p\hat{O}_{\text{sgn}\varepsilon}\psi_{\varepsilon'} \tag{120}$$

From eqs. (119) and (120), we have

$$\hat{H}\hat{O}_{\text{sgn}\varepsilon} = \hat{O}_{\text{sgn}\,\varepsilon}\hat{H} = \varepsilon E_p\hat{O}_{\text{sgn}\varepsilon} \tag{121}$$

The above properties enable us to evaluate the expression

$$\hat{O}_{\text{sgn}\varepsilon}\left(\hat{\alpha}(0) - \frac{1}{\hat{H}}\hat{\mathbf{p}}\right)\frac{i}{2\hat{H}}\exp(-2i\hat{H}x^0)\,\hat{O}_{\text{sgn}\,\varepsilon'}$$

$$= \hat{O}_{\text{sgn}\,\varepsilon}\left(\hat{\alpha}(0) - \frac{1}{\hat{H}}\hat{\mathbf{p}}\right)\frac{i}{2\hat{H}}\hat{O}_{\text{sgn}\,\varepsilon'}\exp(-2i\hat{H}x^0)$$

$$= \left(\hat{O}_{\text{sgn}\varepsilon}\hat{\alpha}(0) - \hat{O}_{\text{sgn}\varepsilon}\frac{1}{\hat{H}}\hat{\mathbf{p}}\right)\frac{i}{2\hat{H}}\hat{O}_{\text{sgn}\,\varepsilon'}\exp(-2i\hat{H}x^0)$$

$$= \left(\hat{\alpha}(0)\hat{O}_{\text{sgn}\varepsilon} + \frac{\varepsilon}{E_p}\hat{\mathbf{p}} - \frac{\varepsilon}{E_p}\hat{\alpha}(0)\hat{H} - \hat{O}_{\text{sgn}\,\varepsilon}\frac{1}{\hat{H}}\hat{\mathbf{p}}\right)\frac{i}{2\hat{H}}$$

$$\hat{O}_{\text{sgn}\,\varepsilon'} \exp(-2i\hat{H}x^0)$$

$$= \left[\hat{\alpha}(0)(\hat{O}_{\text{sgn}\,\varepsilon})(\hat{O}_{\text{sgn}\,\varepsilon'}) + \frac{\varepsilon}{E_p}\hat{\mathbf{p}}\hat{O}_{\text{sgn}\,\varepsilon'} - \frac{\varepsilon}{E_p}\hat{\alpha}(0)\hat{H}\hat{O}_{\text{sgn}\,\varepsilon'}\right.$$

$$\left. - \hat{O}_{\text{sgn}\,\varepsilon}\frac{1}{\hat{H}}\hat{\mathbf{p}}\hat{O}_{\text{sgn}\,\varepsilon'}\right]\frac{i}{2\hat{H}}\exp(-2i\hat{H}x^0)$$

$$= \left[\delta_{\varepsilon\varepsilon'}\hat{\alpha}(0)\hat{O}_{\text{sgn}\,\varepsilon'} + \frac{\varepsilon}{E_p}\hat{\mathbf{p}}\hat{O}_{\text{sgn}\,\varepsilon'} - \frac{\varepsilon\varepsilon' E_p}{E_p}\hat{\alpha}(0)\hat{O}_{\text{sgn}\,\varepsilon'}\right.$$

$$\left. - \delta_{\varepsilon\varepsilon'}\frac{1}{\varepsilon E_p}\hat{\mathbf{p}}\hat{O}_{\text{sgn}\,\varepsilon'}\right]\frac{i}{2\hat{H}}\exp(-2i\hat{H}x^0)$$

$$= \left[(\delta_{\varepsilon\varepsilon'} - \varepsilon\varepsilon')\hat{\alpha}(0)\hat{O}_{\text{sgn}\,\varepsilon'} - (\delta_{\varepsilon\varepsilon'} - 1)\frac{\varepsilon}{E_p}\hat{\mathbf{p}}\hat{O}_{\text{sgn}\,\varepsilon'}\right]\frac{i}{2\hat{H}}\exp(-2i\hat{H}x^0)$$

$$= \left[(1 - \delta_{\varepsilon\varepsilon'})\hat{\alpha}(0)\hat{O}_{\text{sgn}\,\varepsilon'} + (1 - \delta_{\varepsilon\varepsilon'})\frac{\varepsilon}{E_p}\hat{\mathbf{p}}\hat{O}_{\text{sgn}\,\varepsilon'}\right]\frac{i}{2\hat{H}}\exp(-2i\hat{H}x^0)$$

$$= (1 - \delta_{\varepsilon\varepsilon'})\left[\hat{\alpha}(0)\hat{O}_{\text{sgn}\,\varepsilon'} + \frac{\varepsilon}{E_p}\hat{\mathbf{p}}\hat{O}_{\text{sgn}\,\varepsilon'}\right]\frac{i}{2\hat{H}}\exp(-2i\hat{H}x^0) \tag{122}$$

Let us now examine the physical implications of this result. We apply the operator $\hat{O}$ on a mixed wave function, i.e., a wave function with both positive and negative energy components, the output is a pure positive energy or pure negative energy wave function depending on the sign of $\hat{O}$.

Next, the operation with the operator $\left(\hat{\alpha}(0) - \frac{1}{\hat{H}}\hat{\mathbf{p}}\right)\frac{i}{2\hat{H}}\exp(-2i\hat{H}x^0)$ on the sterilized wave function would create the "Zitterbewegung" if possible, in the sterilized wave function.

After undergoing Zitterbewegung, we again apply the energy projection operator of the same sign as the initial projection operator on the output of $\hat{O}$ followed by $\left(\hat{\alpha}(0) - \frac{1}{\hat{H}}\hat{\mathbf{p}}\right)$ $\frac{i}{2\hat{H}}\exp(-2i\hat{H}x^0)$. It is seen from the above that application of $\hat{O}$ a second time results in a complete annihilation of the wave function. This means that the wave function that was the output from the Zitterbewegung operator did not contain components of the same energy sign as the input to the Zitterbewegung operator, i.e., the Zitterbewegung resulted in a transformation of energy states from those of one sign to the other.

To reduce the abstractness in the above analysis, we take up a concrete scenario. Consider an arbitrary wave function that may consist of both positive and negative energy states. Let us apply the operator $\hat{O}_+$ to this arbitrary wave function. Then the output will be a pure wave function of positive energy states alone. We input this positive energy wave function to the Zitterbewegung operator. The output from the Zitterbewegung operator is then again operated on by $\hat{O}_+$. We find that $\hat{O}_+$ completely annihilates the wave function. Therefore, we conclude that the output from the Zitterbewegung operator did not contain any positive energy states components. Hence, the Zitterbewegung operator must have transformed the positive energy input wave function into negative energy wave functions which were annihilated by $\hat{O}_+$. The phenomenon of Zitterbewegung is purely of quantum mechanical origin and has no classical analog.

7.8 EQUATION OF MOTION FOR THE KINETIC MOMENTUM

Let us now explore the equation of motion for the kinetic momentum. We have $\hat{\mathbf{P}} = \hat{\mathbf{p}} - e\mathbf{A}$ and the Hamiltonian

$$\hat{H} = \boldsymbol{\alpha}(\hat{\mathbf{p}} - e\mathbf{A}) + \boldsymbol{\beta}m + eA_0 = \boldsymbol{\alpha}\hat{\mathbf{P}} + \boldsymbol{\beta}m + eA_0 \tag{123}$$

so that

$$[\hat{H},\ \hat{\mathbf{P}}] = [\hat{H},\ \hat{\mathbf{p}}] - e\,[\hat{H},\ \mathbf{A}] \tag{124}$$

Now

$$\begin{aligned}[\hat{H},\ \hat{\mathbf{P}}] &= [\boldsymbol{\alpha}\hat{\mathbf{p}},\ \hat{\mathbf{p}}] - e[\boldsymbol{\alpha}\mathbf{A},\ \hat{\mathbf{p}}] + [\boldsymbol{\beta},\ \hat{\mathbf{p}}]\,m + e[A_0,\ \hat{\mathbf{p}}] \\ &= \boldsymbol{\alpha}[\hat{\mathbf{p}},\ \hat{\mathbf{p}}] + [\boldsymbol{\alpha},\ \hat{\mathbf{p}}]\hat{\mathbf{p}} - e\boldsymbol{\alpha}\,[\mathbf{A},\ \hat{\mathbf{p}}] - e\,[\boldsymbol{\alpha},\ \hat{\mathbf{p}}]\mathbf{A} + e[A_0,\ \hat{\mathbf{p}}]\end{aligned} \tag{125}$$

and

$$[\hat{\mathbf{p}},\ \hat{\mathbf{p}}] = [\boldsymbol{\alpha},\ \hat{\mathbf{p}}] = 0 \tag{126}$$

Also

$$\boldsymbol{\alpha}[\mathbf{A},\ \hat{\mathbf{p}}]\,\varphi = -\,i\boldsymbol{\alpha}[\mathbf{A},\ \nabla]\,\varphi = -i\boldsymbol{\alpha}[\mathbf{A}\nabla\varphi - \nabla(\mathbf{A}\varphi)] = i\boldsymbol{\alpha}(\nabla\mathbf{A})\,\varphi = -\,\boldsymbol{\alpha}(\hat{\mathbf{p}}\,\mathbf{A})\,\varphi \tag{127}$$

$$[A_0,\ \hat{\mathbf{p}}]\,\varphi = -\,i[A_0,\ \nabla]\,\varphi = -\,i[A_0\nabla\varphi - \nabla(A_0\varphi)] = i(\nabla A_0)\,\varphi = -\,(\hat{\mathbf{p}}\,A_0)\,\varphi \tag{128}$$

Using the above results, we have

$$[\hat{H},\ \hat{\mathbf{P}}] = -\,ie\boldsymbol{\alpha}(\nabla\mathbf{A}) + ie(\nabla A_0) \tag{129}$$

Again

$$\begin{aligned}[\hat{H},\ \mathbf{A}] &= [\boldsymbol{\alpha}\hat{\mathbf{p}},\ \mathbf{A}] - e[\boldsymbol{\alpha}\mathbf{A},\ \mathbf{A}] + [\boldsymbol{\beta},\ \mathbf{A}]m + e[\mathbf{A}_0,\ \mathbf{A}] = [\boldsymbol{\alpha}\hat{\mathbf{p}},\ \mathbf{A}] \\ &= \boldsymbol{\alpha}[\hat{\mathbf{p}},\ \mathbf{A}] + [\boldsymbol{\alpha},\ \mathbf{A}]\hat{\mathbf{p}} = \boldsymbol{\alpha}[\hat{\mathbf{p}},\ \mathbf{A}] = -\,i\boldsymbol{\alpha}[\nabla,\ \mathbf{A}] = -\,i\boldsymbol{\alpha}(\nabla\mathbf{A})\end{aligned} \tag{130}$$

Now

$$\frac{d\hat{\mathbf{P}}}{dx^0} = \frac{\partial \hat{\mathbf{P}}}{\partial x^0} + i[\hat{H}, \hat{\mathbf{p}}] = -e\frac{\partial \mathbf{A}}{\partial x^0} + i[\hat{H}, \hat{\mathbf{p}}] - ie[\hat{H}, \mathbf{A}]$$

$$= -e\frac{\partial \mathbf{A}}{\partial x^0} + e\boldsymbol{\alpha}(\nabla \mathbf{A}) - e(\nabla A_0) - e\boldsymbol{\alpha}(\nabla \mathbf{A})$$

$$= -e\frac{\partial \mathbf{A}}{\partial x^0} - e(\nabla A_0) \tag{131}$$

which constitutes the quantum mechanical equation for kinetic momentum operator for a Dirac particle in an electromagnetic field.

CHAPTER 8

Dirac Equation: Charge Conjugation and Time Reversal

8.1 DIRAC'S HOLE THEORY

Initially, when Paul Dirac propounded his theory of relativistic quantum mechanics, the existence of the negative-energy solutions to the Dirac equation were believed to be a serious impediment to the success of the theory. For example, the obvious issue that warranted explanation in a theory that postulated the presence of negative energy solutions for free particles was:

"What stops a positive-energy electron from radiating a photon and falling into a negative-energy (lower-energy) state?"

If this phenomenon was allowed to occur then it would result in a "radiation catastrophe" since nature prefers the lowest energy state.

A plausible explanation to counter this aspect was initiated by Dirac who provided a first solution to this problem by reinterpreting the vacuum state. To understand his philosophy, we consider the energy level diagram (Fig. 8.1) for the free Dirac particle.

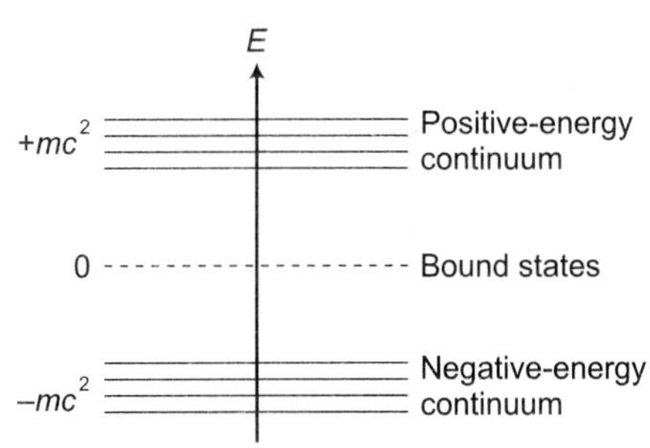

Fig. 8.1 Spectrum of energy states of the free particle Dirac equation.

The vacuum state is normally defined as the lowest energy state or ground state of the system. However, Dirac reinterpreted this definition of "vacuum state" to mean the state that has all the negative-energy levels filled and all positive-energy levels empty, as shown in Fig. 8.2. In other words, and to reiterate, the vacuum state was construed by Dirac as the state that has all the negative-energy states occupied. This ensures that the positive-energy electrons cannot fall into the lower-energy unoccupied states since there are no vacancies available in the negative energy states. The negative-energy levels are filled according to the Pauli's exclusion principle. The stability of the system directly follows from the fact that no more electrons can be accommodated in the negative-energy sea.

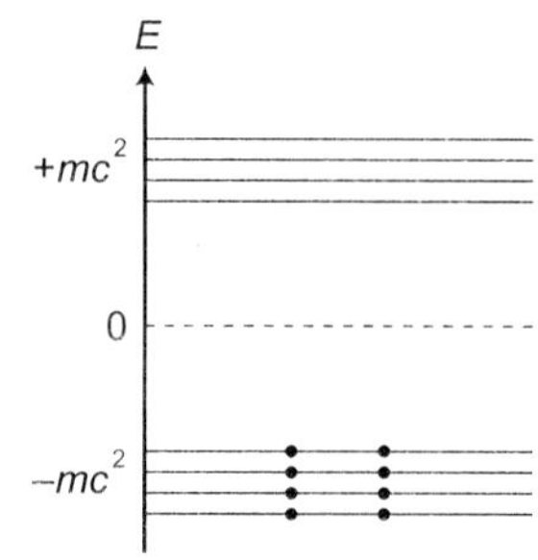

Fig. 8.2 Negative-energy states are completely occupied (Dirac Sea). '0' is the vacuum state.

The above proposition of Dirac is well known as the "hole theory". The cardinal result of the "hole theory" was the reinterpretation of the so-called "vacuum state". The vacuum should, as per the usual interpretation, be unobservable by having zero energy and no charge. However, as per the Dirac proposition the "vacuum" obviously, does not have these properties. The states occupied with Dirac particles of negative energy together will result in an infinitely large negative energy and infinitely large negative charge.

It is here that Dirac's ingenuity becomes evident. To tackle the issue of the infinite negative energy and infinite negative charge, he introduced the concept of "renormalization of vacuum". He mandated that the infinite negative energy and infinite negative charge be renormalized to zero. In other words, the zero point of energy and charge are chosen in such a way that the Dirac sea (consisting of negatively energized and negatively charged particles) has the equivalence of no mass and no charge.

Therefore, it follows as a corollary to the above renormalization of vacuum that the net energy and charge of a given state must be defined with respect to this "renormalized vacuum". Since a positive-energy electron has charge $-e$, a "hole" state (defined as the vacancy of an electron in the negative sea) will behave as if it has charge $e_h = [Q_{\text{vacuum}} - (-e)] - Q_{\text{vacuum}} = +e$. Similarly, if the momentum of the electron (absence of which from the Dirac sea results in the creation of the hole) is $\mathbf{p}$, then the "hole" may be interpreted as a particle with momentum $-\mathbf{p}$. It can be shown, in complete analogy, that the energy and spin of the "hole" will also be opposite in sign to those of the corresponding electron.

Thus, a "hole" state behaves as a positive-energy, positive-charge state of momentum $-\mathbf{p}$ and spin $-\frac{1}{2}\Sigma$.

An important consequence of the renormalization of vacuum is that the vacuum can be modified by external stimulus, e.g., by the influence of external fields. If the applied fields have energy less than $2mc^2$, they will result in a redistribution of the charge of the occupied negative-energy states, thereby causing a vacuum polarization with respect to the original renormalized state (without external fields).

On the other hand, if strong fields, i.e., field with energy exceeding $2mc^2$ are applied, it may cause the jumping of a negative-energy electron from the Dirac sea into a positive-energy state thereby altering the vacuum state. This situation is physically equivalent to a negative-energy electron absorbing radiation and jumping into a positive-energy state. The fallout of this would be the creation of a "hole" in the negative energy Dirac sea manifesting the vacancy created by the jumping electron. The energy of the electron shall also change from $-E$ to $+E$. The hole that personifies the absence of the jumped electron from the Dirac sea is physically equivalent, relative to the vacuum, as the presence of a particle of charge $+e$ and energy $+E$ and is referred to as the "positron". This is the "hole-theory" interpretation of pair production by intense external fields and is diagrammatically shown in Fig. 8.3.

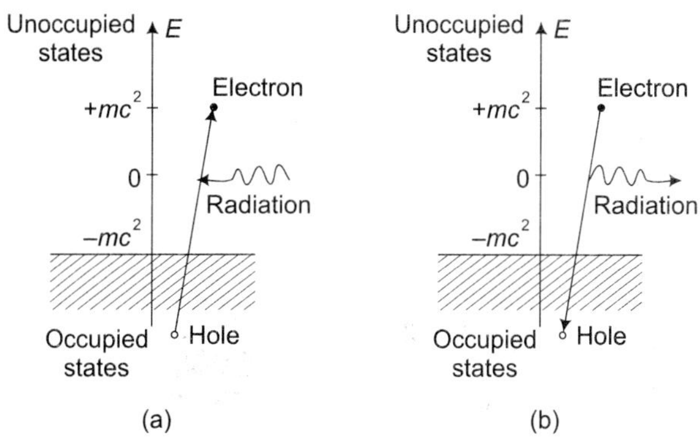

Fig. 8.3 Electron transitions in the hole theory: (a) pair production and (b) pair annihilation.

Conversely, we may also interpret a "hole" in the negative-energy sea, or a positron, as a "trap" for a positive-energy electron. The positive-energy electron may lose energy by emitting radiation and fall into the lower negative-energy hole. This would provide the explanation for the observed physical phenomenon of electron-positron pair annihilation with the emission of radiation.

It is pertinent to mention here that the "hole theory" explicitly acknowledges the necessity of "many particle" framework for describing particles of both signs of charge and also justifying the existence of negative energy solutions to the Dirac equation.

A critical implication of this "many particle" interpretation is that the wave function loses the simple probability based interpretation of the one-particle theory.

8.2 CHARGE CONJUGATION

The Dirac equation for a particle of charge e and mass m with negative energy is given by

$$(\not{p} - m)\,\psi = 0 \tag{1}$$

where ψ is a negative energy wave function and

$$\not{p} = \gamma^{\mu}\,\hat{P}_{\mu} = \gamma^{\mu}\,(\hat{p}_{\mu} - eA_{\mu}) = (\hat{\not{p}} - e\not{A}) = \gamma^{\mu}\,(i\partial_{\mu} - eA_{\mu}) = (i\not{\partial} - e\not{A}) \tag{2}$$

so that eq. (1) becomes

$$(i\not{\partial} - e\not{A} - m)\psi = 0 \tag{3}$$

Now, as per the contemporary theory of elementary particles, every known "fundamental particle" has an "antiparticle" that has the same mass as the original particle, but has energy, momentum, charge and spin with opposite sign to that of the original particle. Thus, if ψ_c is the wave function of the "antiparticle" then ψ_c will also be a solution of the Dirac equation

$$(i\not{\partial} - e\not{A} - m)\psi_c = 0 \tag{4}$$

ψ_c would be a solution of eq. (4) with positive energy.

We now attempt to find the operator that links ψ_c with ψ, i.e., that associates the wave functions of the "particle" and its "antiparticle". For this, let us take the complex conjugate of eq. (3). We have

$$\begin{aligned}0 &= [(i\not{\partial} - e\not{A} - m)\psi]^{*} = [(i\gamma^{\mu}\,\partial_{\mu} - e\gamma^{\mu}\,A_{\mu} - m)\psi]^{*}\\ &= (-i\gamma^{\mu *}\,\partial_{\mu} - e\gamma^{\mu *}\,A_{\mu} - m)\,\psi^{*}\\ &= (i\gamma^{\mu *}\,\partial_{\mu} + e\gamma^{\mu *}\,A_{\mu} + m)\psi^{*}\end{aligned} \tag{5}$$

Let us now assume that there exists a non-singular operator $\hat{U}$ that satisfies

$$\hat{U}\gamma^{\mu *}\hat{U}^{-1} = -\gamma^{\mu} \tag{6}$$

Then, from eqs. (5) and (6), we have

$$\begin{aligned}0 &= [\hat{U}\gamma^{\mu *}\hat{U}^{-1}\,(i\partial_{\mu} + eA_{\mu}) + m]\,\hat{U}\psi^{*} = [-\gamma^{\mu}\,(i\partial_{\mu} + eA_{\mu}) + m]\hat{U}\psi^{*}\\ &= [\gamma^{\mu}\,(i\partial_{\mu} + eA_{\mu}) - m]\hat{U}\psi^{*} = (i\not{\partial} + e\not{A} - m)\hat{U}\psi^{*} = 0\end{aligned} \tag{7}$$

Comparing eqs. (4) and (7), we get

$$\psi_c = \hat{U}\psi^{*} = \hat{U}\,\hat{K}\psi \tag{8}$$

To determine $\hat{U}$ explicitly, we adopt the following representation of the γ matrices:

$$\gamma^{k} = \begin{pmatrix} 0 & \sigma_k \\ -\sigma_k & 0 \end{pmatrix},\ \gamma^{0} = \begin{pmatrix} I & 0 \\ 0 & -I \end{pmatrix} \tag{9}$$

We have for $k = 1, 3$

$$\gamma^{0}\,(\gamma^{k})^{T}\,\gamma^{0} = \begin{pmatrix} I & 0 \\ 0 & -I \end{pmatrix}\begin{pmatrix} 0 & -(\sigma_k)^{T} \\ (\sigma_k)^{T} & 0 \end{pmatrix}\begin{pmatrix} I & 0 \\ 0 & -I \end{pmatrix}$$

$$= \begin{pmatrix} \mathbf{0} & (\sigma_k)^T \\ -(\sigma_k)^T & \mathbf{0} \end{pmatrix} = \begin{pmatrix} \mathbf{0} & \sigma_k \\ -\sigma_k & \mathbf{0} \end{pmatrix} = \boldsymbol{\gamma}^k = (\boldsymbol{\gamma}^k)^*$$

$$\boldsymbol{\gamma}^0 (\boldsymbol{\gamma}^2)^T \boldsymbol{\gamma}^0 = \begin{pmatrix} \mathbf{I} & \mathbf{0} \\ \mathbf{0} & -\mathbf{I} \end{pmatrix} \begin{pmatrix} \mathbf{0} & -(\sigma_2)^T \\ (\sigma_2)^T & \mathbf{0} \end{pmatrix} \begin{pmatrix} \mathbf{I} & \mathbf{0} \\ \mathbf{0} & -\mathbf{I} \end{pmatrix}$$

$$= \begin{pmatrix} \mathbf{0} & (\sigma_2)^T \\ -(\sigma_2)^T & \mathbf{0} \end{pmatrix} = \begin{pmatrix} \mathbf{0} & -\sigma_2 \\ \sigma_2 & \mathbf{0} \end{pmatrix} = -\boldsymbol{\gamma}^2 = (\boldsymbol{\gamma}^2)^*$$

and

$$\boldsymbol{\gamma}^0 (\boldsymbol{\gamma}^0)^T \boldsymbol{\gamma}^0 = \begin{pmatrix} \mathbf{I} & \mathbf{0} \\ \mathbf{0} & -\mathbf{I} \end{pmatrix} \begin{pmatrix} \mathbf{I} & \mathbf{0} \\ \mathbf{0} & -\mathbf{I} \end{pmatrix} \begin{pmatrix} \mathbf{I} & \mathbf{0} \\ \mathbf{0} & -\mathbf{I} \end{pmatrix} = \begin{pmatrix} \mathbf{I} & \mathbf{0} \\ \mathbf{0} & -\mathbf{I} \end{pmatrix} = \boldsymbol{\gamma}^0 = (\boldsymbol{\gamma}^0)^*$$

so that

$$\boldsymbol{\gamma}^0 (\boldsymbol{\gamma}^\mu)^T \boldsymbol{\gamma}^0 = (\boldsymbol{\gamma}^\mu)^* \text{ or } \boldsymbol{\gamma}^0 (\boldsymbol{\gamma}^\mu)^* \boldsymbol{\gamma}^0 = (\boldsymbol{\gamma}^\mu)^T \tag{10}$$

Now, we can write

$$-\boldsymbol{\gamma}^\mu = \hat{U}\boldsymbol{\gamma}^{\mu *}\hat{U}^{-1} = \hat{U}\boldsymbol{\gamma}^0\boldsymbol{\gamma}^0\boldsymbol{\gamma}^{\mu *}\boldsymbol{\gamma}^0\boldsymbol{\gamma}^0\hat{U}^{-1} = \hat{U}\boldsymbol{\gamma}^0\boldsymbol{\gamma}^{\mu T}\boldsymbol{\gamma}^0\hat{U}^{-1} = \hat{U}\boldsymbol{\gamma}^0\boldsymbol{\gamma}^{\mu T}(\hat{U}\boldsymbol{\gamma}^0)^{-1} \tag{11}$$

which gives

$$[(\hat{U}\boldsymbol{\gamma}^0)^{-1}]^T \boldsymbol{\gamma}^\mu (\hat{U}\boldsymbol{\gamma}^0)^T = -\boldsymbol{\gamma}^{\mu T} \tag{12}$$

But

$$\boldsymbol{\gamma}^{1T} = -\boldsymbol{\gamma}^1,\ \boldsymbol{\gamma}^{2T} = \boldsymbol{\gamma}^2,\ \boldsymbol{\gamma}^{3T} = -\boldsymbol{\gamma}^3,\ \boldsymbol{\gamma}^{0T} = \boldsymbol{\gamma}^0 \tag{13}$$

Therefore,

$$[(\hat{U}\boldsymbol{\gamma}^0)^{-1}]^T \boldsymbol{\gamma}^{(1,3)} (\hat{U}\boldsymbol{\gamma}^0)^T = \boldsymbol{\gamma}^{(1,3)} \text{ or } \boldsymbol{\gamma}^{(1,3)} (\hat{U}\boldsymbol{\gamma}^0)^T = (\hat{U}\boldsymbol{\gamma}^0)^T \boldsymbol{\gamma}^{(1,3)} \tag{14}$$

and

$$[(\hat{U}\boldsymbol{\gamma}^0)^{-1}]^T \boldsymbol{\gamma}^{(0,2)} (\hat{U}\boldsymbol{\gamma}^0)^T = -\boldsymbol{\gamma}^{(0,2)} \text{ or } \boldsymbol{\gamma}^{(0,2)} (\hat{U}\boldsymbol{\gamma}^0)^T = -(\hat{U}\boldsymbol{\gamma}^0)^T \boldsymbol{\gamma}^{(0,2)} \tag{15}$$

i.e., $(\hat{U}\boldsymbol{\gamma}^0)^T$ must commute with $\boldsymbol{\gamma}^1$, $\boldsymbol{\gamma}^3$ and anticommute with $\boldsymbol{\gamma}^0$, $\boldsymbol{\gamma}^2$.

Now

$$\hat{U} = i\boldsymbol{\gamma}^2 \tag{16}$$

satisfies the above conditions because

$$\begin{aligned}\boldsymbol{\gamma}^{(1,3)} (i\boldsymbol{\gamma}^2\boldsymbol{\gamma}^0)^T &= \boldsymbol{\gamma}^{(1,3)} (i\boldsymbol{\gamma}^{0T}\boldsymbol{\gamma}^{2T}) = \boldsymbol{\gamma}^{(1,3)} (i\boldsymbol{\gamma}^0\boldsymbol{\gamma}^2) = -(i\boldsymbol{\gamma}^0\boldsymbol{\gamma}^{(1,3)}\boldsymbol{\gamma}^2) \\ &= (i\boldsymbol{\gamma}^0\boldsymbol{\gamma}^2)\boldsymbol{\gamma}^{(1,3)} = (i\boldsymbol{\gamma}^{0T}\boldsymbol{\gamma}^{2T})\boldsymbol{\gamma}^{(1,3)} = (i\boldsymbol{\gamma}^2\boldsymbol{\gamma}^0)^T \boldsymbol{\gamma}^{(1,3)}\end{aligned} \tag{17}$$

$$\begin{aligned}\boldsymbol{\gamma}^2 (i\boldsymbol{\gamma}^2\boldsymbol{\gamma}^0)^T &= \boldsymbol{\gamma}^2 (i\boldsymbol{\gamma}^{0T}\boldsymbol{\gamma}^{2T}) = \boldsymbol{\gamma}^2 (i\boldsymbol{\gamma}^0\boldsymbol{\gamma}^2) = -(i\boldsymbol{\gamma}^0\boldsymbol{\gamma}^2)\boldsymbol{\gamma}^2 \\ &= -(i\boldsymbol{\gamma}^{0T}\boldsymbol{\gamma}^{2T})\boldsymbol{\gamma}^2 = -(i\boldsymbol{\gamma}^2\boldsymbol{\gamma}^0)^T\boldsymbol{\gamma}^2\end{aligned} \tag{18}$$

$$\begin{aligned}\boldsymbol{\gamma}^0 (i\boldsymbol{\gamma}^2\boldsymbol{\gamma}^0)^T &= \boldsymbol{\gamma}^0(i\boldsymbol{\gamma}^{0T}\boldsymbol{\gamma}^{2T}) = \boldsymbol{\gamma}^0(i\boldsymbol{\gamma}^0\boldsymbol{\gamma}^2) = -(i\boldsymbol{\gamma}^0\boldsymbol{\gamma}^2)\boldsymbol{\gamma}^0 \\ &= -(i\boldsymbol{\gamma}^{0T}\boldsymbol{\gamma}^{2T})\boldsymbol{\gamma}^0 = -(i\boldsymbol{\gamma}^2\boldsymbol{\gamma}^0)^T \boldsymbol{\gamma}^0\end{aligned} \tag{19}$$

as desired.Furthermore, if we write

$$\hat{C} = \hat{U}\gamma^0 = i\gamma^2\gamma^0 = \hat{\mathbb{C}}\hat{K}\gamma^0 \tag{20}$$

then

$$\hat{C} = i\gamma^2\gamma^0 = -\hat{C}^{-1} = -\hat{C}^\dagger = -\hat{C}^T \tag{21}$$

since

$$-\hat{C}^{-1} = -(i\gamma^2\gamma^0)^{-1} = +i(\gamma^0)^{-1}(\gamma^2)^{-1} = -i\gamma^0\gamma^2 = +i\gamma^2\gamma^0 = \hat{C} \tag{22}$$

$$-\hat{C}^\dagger = -(i\gamma^2\gamma^0)^\dagger = +i(\gamma^0)^\dagger(\gamma^2)^\dagger = -i\gamma^0\gamma^2 = +i\gamma^2\gamma^0 = \hat{C} \tag{23}$$

$$-\hat{C}^T = -(i\gamma^2\gamma^0)^T = -i(\gamma^0)^T(\gamma^2)^T = -i\gamma^0\gamma^2 = +i\gamma^2\gamma^0 = \hat{C} \tag{24}$$

8.3 PROPERTIES OF THE CHARGE CONJUGATION OPERATOR $\hat{C}$

(i) The charge conjugation operator is unitary ($\hat{C}^\dagger = C^{-1}$) as seen from eqs. (22) and (23), skew-symmetric ($\hat{C} = -C^T$) from eq. (24), and skew-Hermitian ($\hat{C} = -C^\dagger$) from eq. (23).

(ii) The charge conjugation operator is non-singular since $\hat{C}^{-1} = -\hat{C} \neq 0$.

(iii) The charge conjugation operator $\hat{C}$ can be obtained in other representations of the Dirac matrices through a unitary transformation. This is because any two representations of the γ matrices are connected through a unitary transformation, i.e., if γ, γ' are any two representations of the Dirac matrices, then we have

$$(\gamma')^\mu = \hat{S}^{-1}\gamma^\mu\hat{S}\gamma^\mu = \hat{S}(\gamma')^\mu\hat{S}^{-1} \tag{25}$$

so that

$$\hat{C}' = i(\gamma')^2(\gamma')^0 = i\hat{S}^{-1}\gamma^2\hat{S}\hat{S}^{-1}\gamma^0\hat{S} = \hat{S}^{-1}(i\gamma^2\gamma^0)\hat{S} = \hat{S}^{-1}\hat{C}\hat{S} \tag{26}$$

(iv) The charge conjugated wave function ψ_c is given by

$$\begin{aligned}\psi_c(x) &= \hat{U}\psi^*(x) = \hat{C}\gamma^0\psi^*(x) = \hat{C}\gamma^0(\psi^{*T})^T \\ &= \hat{C}\gamma^{0T}(\psi^{*T})^T = \hat{C}(\psi^{*T}\gamma^0)^T = \hat{C}(\psi^\dagger\gamma^0)^T = \hat{C}\overline{\psi}^T\end{aligned} \tag{27}$$

Furthermore

$$\psi_c = \hat{C}\gamma^0\psi^* = (i\gamma^2\gamma^0)\gamma^0\psi^* = i\gamma^2\psi^* = i\gamma^2\hat{K}\psi = (i\gamma^2\gamma^0)\gamma^0\hat{K}\psi = \hat{C}\gamma^0\hat{K}\psi \tag{28}$$

where $\hat{K}$ is the complex conjugation operator.

(v) We have

$$\begin{aligned}(\psi_c)_c &= (\hat{U}\psi^*(x))_c = (i\gamma^2\psi^*(x))_c = i\gamma^2(i\gamma^2\psi^*(x))^* \\ &= \gamma^2(\gamma^2)^*\psi(x) = -(\gamma^2)^2\psi(x) = \psi(x)\end{aligned} \tag{29}$$

Hence, we find that the correspondence between ψ and ψ_c is reciprocal.

(vi) **Expectation values of various operators in charge conjugated states**

We have

$$\langle\hat{Q}\rangle_c = \langle\psi_c|\hat{Q}|\psi_c\rangle = \int d\mathbf{x}\psi_c^\dagger\hat{Q}\psi_c = \int d\mathbf{x}\,(i\boldsymbol{\gamma}^2\psi^*)^\dagger\;\hat{Q}\;(i\boldsymbol{\gamma}^2\psi^*)$$

$$= \int d\mathbf{x}\,(\psi^*)^\dagger\;\boldsymbol{\gamma}^{2\dagger}\hat{Q}\;(\boldsymbol{\gamma}^2\psi^*) = -\int d\mathbf{x}\,(\psi^*)^\dagger\;\boldsymbol{\gamma}^2\hat{Q}\boldsymbol{\gamma}^2\psi^*$$

$$= -\left(\int d\mathbf{x}\psi^\dagger\,(\boldsymbol{\gamma}^2\hat{Q}\boldsymbol{\gamma}^2)^*\;\psi\right)^*$$

$$= -\langle\psi|\boldsymbol{\gamma}^{2*}\;\hat{Q}^*\boldsymbol{\gamma}^{2*}|\psi\rangle^* = -\langle\psi|\boldsymbol{\gamma}^2\hat{Q}^*\boldsymbol{\gamma}^2|\psi\rangle^* \tag{30}$$

because $\boldsymbol{\gamma}^{2*} = -\boldsymbol{\gamma}^2$. We give below some examples on the calculation of expectation values in charge conjugated states. We have

$$\langle\beta\rangle_c = -\langle\psi|\boldsymbol{\gamma}^2\beta^*\boldsymbol{\gamma}^2|\psi\rangle^* = -\langle\psi|\boldsymbol{\gamma}^2\beta\boldsymbol{\gamma}^2|\psi\rangle^*$$

$$= -\langle\psi|\boldsymbol{\gamma}^2\boldsymbol{\gamma}^0\boldsymbol{\gamma}^2|\psi\rangle^* = \langle\psi|\boldsymbol{\gamma}^2\boldsymbol{\gamma}^2\boldsymbol{\gamma}^0|\psi\rangle^*$$

$$= -\langle\psi|\boldsymbol{\gamma}^0|\psi\rangle^* = -\langle\psi|\boldsymbol{\beta}|\psi\rangle^* = -\langle\psi|\boldsymbol{\beta}|\psi\rangle = -\langle\boldsymbol{\beta}\rangle \tag{31}$$

$$\langle\boldsymbol{\alpha}^{1,3}\rangle_c = -\langle\psi|\boldsymbol{\gamma}^2\,(\boldsymbol{\alpha}^{1,3})^*\;\boldsymbol{\gamma}^2|\psi\rangle^* = -\langle\psi|\boldsymbol{\gamma}^2\,(\boldsymbol{\beta}\boldsymbol{\gamma}^{1,3})^*\;\boldsymbol{\gamma}^2|\psi\rangle^*$$

$$= -\langle\psi|\boldsymbol{\gamma}^2\,(\boldsymbol{\gamma}^0\boldsymbol{\gamma}^{1,3})^*\;\boldsymbol{\gamma}^2|\psi\rangle^*$$

$$= -\langle\psi|\boldsymbol{\gamma}^2\,(\boldsymbol{\gamma}^0)^*\;(\boldsymbol{\gamma}^{1,3})^*\;\boldsymbol{\gamma}^2|\psi\rangle^* = -\langle\psi|\boldsymbol{\gamma}^2\boldsymbol{\gamma}^0\boldsymbol{\gamma}^{1,3}\boldsymbol{\gamma}^2|\psi\rangle^* = \langle\psi|\boldsymbol{\gamma}^0\boldsymbol{\gamma}^{1,3}|\psi\rangle^*$$

$$= \langle\psi|\boldsymbol{\alpha}^{1,3}|\psi\rangle^* = \langle\psi|\boldsymbol{\alpha}^{1,3}|\psi\rangle = \langle\boldsymbol{\alpha}^{1,3}\rangle \tag{32}$$

$$\langle\alpha^2\rangle_c = -\langle\psi|\boldsymbol{\gamma}^2\,(\boldsymbol{\alpha}^2)^*\boldsymbol{\gamma}^2|\psi\rangle^* = -\langle\psi|\boldsymbol{\gamma}^2\,(\boldsymbol{\beta}\boldsymbol{\gamma}^2)^*\;\boldsymbol{\gamma}^2|\psi\rangle^*$$

$$= -\langle\psi|\boldsymbol{\gamma}^2\,(\boldsymbol{\gamma}^0\boldsymbol{\gamma}^2)^*\boldsymbol{\gamma}^2|\psi\rangle^*$$

$$= -\langle\psi|\boldsymbol{\gamma}^2\,(\boldsymbol{\gamma}^0)^*\;(\boldsymbol{\gamma}^2)^*\boldsymbol{\gamma}^2|\psi\rangle^* = \langle\psi|\boldsymbol{\gamma}^2\boldsymbol{\gamma}^0\boldsymbol{\gamma}^2\boldsymbol{\gamma}^2|\psi\rangle^* = -\langle\psi|\boldsymbol{\gamma}^2\boldsymbol{\gamma}^0|\psi\rangle^*$$

$$= \langle\psi|\boldsymbol{\gamma}^0\boldsymbol{\gamma}^2|\psi\rangle^* = \langle\psi|\boldsymbol{\alpha}^2|\psi\rangle^* = \langle\psi|\boldsymbol{\alpha}^2|\psi\rangle = \langle\boldsymbol{\alpha}^2\rangle \tag{33}$$

Therefore,

$$\langle\boldsymbol{\alpha}\rangle_c = \langle\boldsymbol{\alpha}\rangle \tag{34}$$

$$\langle\hat{\mathbf{x}}\rangle_c = -\langle\psi|\boldsymbol{\gamma}^2\hat{\mathbf{x}}^*\boldsymbol{\gamma}^2|\psi\rangle^* = \langle\psi|\hat{\mathbf{x}}^*|\psi\rangle^* = \langle\psi|\hat{\mathbf{x}}|\psi\rangle^*$$

$$= \mathbf{x}^*\,\langle\psi|\psi\rangle^* = \mathbf{x}\,\langle\psi|\psi\rangle = \langle\psi|\mathbf{x}|\psi\rangle = \langle\psi|\hat{\mathbf{x}}|\psi\rangle = \langle\hat{\mathbf{x}}\rangle \tag{35}$$

$$\langle\hat{\mathbf{p}}\rangle_c = -\langle\psi|\boldsymbol{\gamma}^2\hat{\mathbf{p}}^*\boldsymbol{\gamma}^2|\psi\rangle^* = \langle\psi|\hat{\mathbf{p}}^*|\psi\rangle^* = -\langle\psi|\hat{\mathbf{p}}|\psi\rangle^* = -\langle\psi|\mathbf{p}|\psi\rangle^*$$

$$= -\mathbf{p}^*\,\langle\psi|\psi\rangle^* = -\mathbf{p}\,\langle\psi|\psi\rangle = -\langle\psi|\mathbf{p}|\psi\rangle = -\langle\psi|\hat{\mathbf{p}}|\psi\rangle = -\langle\hat{\mathbf{p}}\rangle \tag{36}$$

$$\psi_c^\dagger\psi_c = (\hat{U}\psi^*)^\dagger\;\hat{U}\psi^* = \psi^{*\dagger}U^\dagger U\psi^* = \psi^T\,(i\boldsymbol{\gamma}^2)^\dagger\;(i\boldsymbol{\gamma}^2)\psi^*$$

$$= \psi^T\,(-i\boldsymbol{\gamma}^{2\dagger})\;(i\boldsymbol{\gamma}^2)\psi^*$$

$$= \psi^T\boldsymbol{\gamma}^{2\dagger}\boldsymbol{\gamma}^2\psi^* = -\psi^T\,(\boldsymbol{\gamma}^2)^2\;\psi^* = \psi^T\psi^* = (\psi^\dagger\psi)^* = \psi^\dagger\psi \tag{37}$$

since $\psi^\dagger\psi$ is real

$$\psi_c^\dagger\hat{\alpha}\,\psi_c = (\hat{U}\psi^*)^\dagger\;\hat{\alpha}\;(\hat{U}\psi^*) = (i\boldsymbol{\gamma}^2\psi^*)^\dagger\;\hat{\alpha}\;(i\boldsymbol{\gamma}^2\psi^*) = \psi^{*\dagger}\,(i\boldsymbol{\gamma}^2)^\dagger\;\hat{\alpha}\;(i\boldsymbol{\gamma}^2\psi^*) \tag{38}$$

Now

$$\psi^{*\dagger}(i\boldsymbol{\gamma}^2)^\dagger\ \hat{\alpha}^{1,3}(i\boldsymbol{\gamma}^2\psi^*) = -\psi^{*\dagger}\boldsymbol{\gamma}^2\hat{\alpha}^{1,3}\boldsymbol{\gamma}^2\psi^* = -\psi^{*\dagger}\hat{\alpha}^{1,3}(\boldsymbol{\gamma}^2)^2\,\psi^*$$

$$= \psi^{*\dagger}\hat{\alpha}^{1,3}\psi^* = (\psi^\dagger\hat{\alpha}^{1,3}\psi)^* = (\psi^\dagger\hat{\alpha}^{1,3}\psi) \tag{39}$$

because $(\psi^\dagger\hat{\alpha}^{1,3}\psi)$ is real.

Further

$$\psi^{*\dagger}(i\boldsymbol{\gamma}^2)^\dagger\hat{\alpha}^2\,(i\boldsymbol{\gamma}^2\psi^*) = -\psi^{*\dagger}\boldsymbol{\gamma}^2\hat{\alpha}^2\boldsymbol{\gamma}^2\psi^* = +\psi^{*\dagger}\hat{\alpha}^2(\boldsymbol{\gamma}^2)^2\psi^* = -\psi^{*\dagger}\hat{\alpha}^2\psi^*$$

$$= -(\psi^\dagger\hat{\alpha}^{2*}\psi)^* = (\psi^\dagger\hat{\alpha}^2\psi)^* = (\psi^\dagger\hat{\alpha}^2\psi) \tag{40}$$

because $\hat{\alpha}^2$ is imaginary.

From eqs. (39) and (40), we have

$$\psi_c^\dagger\hat{\alpha}\psi_c = \psi^\dagger\hat{\alpha}\psi \tag{41}$$

$$\langle\hat{\Sigma}_{1,3}\rangle_c = -\langle\psi|\boldsymbol{\gamma}^2\hat{\Sigma}_{1,3}{}^*\boldsymbol{\gamma}^2|\psi\rangle^* = -\left\langle\psi\left|\begin{pmatrix}0 & \sigma_2\\ -\sigma_2 & 0\end{pmatrix}\begin{pmatrix}\sigma_{1,3} & 0\\ 0 & \sigma_{1,3}\end{pmatrix}^*\begin{pmatrix}0 & \sigma_2\\ -\sigma_2 & 0\end{pmatrix}\right|\psi\right\rangle^*$$

$$= -\left\langle\psi\left|\begin{pmatrix}0 & \sigma_2\\ -\sigma_2 & 0\end{pmatrix}\begin{pmatrix}\sigma_{1,3} & 0\\ 0 & \sigma_{1,3}\end{pmatrix}\begin{pmatrix}0 & \sigma_2\\ -\sigma_2 & 0\end{pmatrix}\right|\psi\right\rangle^*$$

$$= -\left\langle\psi\left|\begin{pmatrix}\sigma_{1,3} & 0\\ 0 & \sigma_{1,3}\end{pmatrix}\right|\psi\right\rangle^*$$

$$= -\left\langle\psi\left|\begin{pmatrix}\sigma_{1,3} & 0\\ 0 & \sigma_{1,3}\end{pmatrix}\right|\psi\right\rangle = -\langle\psi|\hat{\Sigma}_{1,3}|\psi\rangle = -\langle\hat{\Sigma}_{1,3}\rangle \tag{42}$$

$$\langle\hat{\Sigma}_2\rangle_c = -\langle\psi|\boldsymbol{\gamma}^2\hat{\Sigma}_2{}^*\boldsymbol{\gamma}^2|\psi\rangle^* = -\left\langle\psi\left|\begin{pmatrix}0 & \sigma_2\\ -\sigma_2 & 0\end{pmatrix}\begin{pmatrix}\sigma_2 & 0\\ 0 & \sigma_2\end{pmatrix}^*\begin{pmatrix}0 & \sigma_2\\ -\sigma_2 & 0\end{pmatrix}\right|\psi\right\rangle^*$$

$$= -\left\langle\psi\left|\begin{pmatrix}0 & \sigma_2\\ -\sigma_2 & 0\end{pmatrix}\begin{pmatrix}-\sigma_2 & 0\\ 0 & -\sigma_2\end{pmatrix}\begin{pmatrix}0 & \sigma_2\\ -\sigma_2 & 0\end{pmatrix}\right|\psi\right\rangle^* = -\left\langle\psi\left|\begin{pmatrix}\sigma_2 & 0\\ 0 & \sigma_2\end{pmatrix}\right|\psi\right\rangle^*$$

$$= -\left\langle\psi\left|\begin{pmatrix}\sigma_2 & 0\\ 0 & \sigma_2\end{pmatrix}\right|\psi\right\rangle = -\langle\psi|\hat{\Sigma}_2|\psi\rangle = -\langle\hat{\Sigma}_2\rangle \tag{43}$$

where in the last step we have used for $\psi = (\phi\quad\chi)^T$ so that

$$\left\langle \psi \left| \begin{pmatrix} \sigma_2 & 0 \\ 0 & \sigma_2 \end{pmatrix} \right| \psi \right\rangle^* = \left[(\phi^\dagger \ \chi^\dagger \begin{pmatrix} \sigma_2 & 0 \\ 0 & \sigma_2 \end{pmatrix} \begin{pmatrix} \phi \\ \chi \end{pmatrix} \right]^* = (\phi^\dagger \boldsymbol{\sigma}_2 \phi + \chi^\dagger \boldsymbol{\sigma}_2 \chi)^*$$

$$= \phi^T \sigma_2^* \phi^* + \chi^T \sigma_2^* \chi^* = -\phi^T \sigma_2 \phi^* - \chi^T \sigma_2 \chi^* = (-\phi^T \sigma_2 \phi^* - \chi^T \sigma_2 \chi^*)^T$$

(because $-\phi^T \sigma_2 \phi^* - \chi^T \sigma_2 \chi^*$ is a number, i.e., a 1×1 matrix)

$$= -\phi^{*T} \sigma_2^T \phi - \chi^{*T} \sigma_2^T \chi = \phi^{*T} \sigma_2 \phi + \chi^{*T} \sigma_2 \chi = \phi^\dagger \sigma_2 \phi + \chi^\dagger \sigma_2 \chi$$

$$= \left[(\phi^\dagger \ \chi^\dagger \begin{pmatrix} \sigma_2 & 0 \\ 0 & \sigma_2 \end{pmatrix} \begin{pmatrix} \phi \\ \chi \end{pmatrix} \right] = \left\langle \psi \left| \begin{pmatrix} \sigma_2 & 0 \\ 0 & \sigma_2 \end{pmatrix} \right| \psi \right\rangle \tag{44}$$

Hence,

$$\langle \hat{\Sigma} \rangle_c = -\langle \hat{\Sigma} \rangle \tag{45}$$

$$\langle \hat{\mathbf{L}} \rangle_c = -\langle \psi | \boldsymbol{\gamma}^2 \hat{\mathbf{L}}^* \boldsymbol{\gamma}^2 | \psi \rangle^* = -\langle \psi | \boldsymbol{\gamma}^2 (\hat{\mathbf{x}} \times \hat{\mathbf{p}})^* \boldsymbol{\gamma}^2 | \psi \rangle^* = \langle \psi | (\hat{\mathbf{x}} \times \hat{\mathbf{p}})^* \psi | \rangle^*$$

$$\langle \psi | -(\hat{\mathbf{x}} \times \hat{\mathbf{p}}) | \psi \rangle^* = \langle \psi | -\hat{\mathbf{L}} | \psi \rangle^* = \langle \psi | -\hat{\mathbf{L}} | \psi \rangle = -\langle \hat{\mathbf{L}} \rangle \tag{46}$$

$$\langle \hat{\mathbf{j}} \rangle_c = \left\langle \hat{\mathbf{L}} + \frac{1}{2} \hat{\Sigma} \right\rangle_c = -\left\langle \hat{\mathbf{L}} + \frac{1}{2} \hat{\Sigma} \right\rangle = -\langle \hat{\mathbf{J}} \rangle \tag{47}$$

$$\langle \hat{\alpha} . \hat{\mathbf{p}} \rangle_c = -\langle \psi | \boldsymbol{\gamma}^2 (\hat{\alpha} . \hat{\mathbf{p}})^* \boldsymbol{\gamma}^2 | \psi \rangle^* = -\langle \psi | \boldsymbol{\gamma}^2 (\hat{\alpha})^* . (\hat{\mathbf{p}})^* \boldsymbol{\gamma}^2 | \psi \rangle^*$$

$$= -\langle \psi | -\boldsymbol{\gamma}^2 (\hat{\alpha})^* \boldsymbol{\gamma}^2 . \boldsymbol{\gamma}^2 (\hat{\mathbf{p}})^* \boldsymbol{\gamma}^2 | \psi \rangle^*$$

$$= [-\langle \psi | \boldsymbol{\gamma}^2 (\hat{\alpha})^* \boldsymbol{\gamma}^2 | \psi \rangle^*][-\langle \psi | \boldsymbol{\gamma}^2 (\hat{\mathbf{p}})^* \boldsymbol{\gamma}^2 | \psi \rangle^*]$$

$$= \langle \hat{\alpha} \rangle_c \langle \hat{\mathbf{p}} \rangle_c = -\langle \hat{\alpha} \rangle \langle \hat{\mathbf{p}} \rangle = -\langle \hat{\alpha} . \hat{\mathbf{p}} \rangle \tag{48}$$

If we define the Hamiltonian

$$\hat{H}(e) = \hat{\alpha} (\hat{\mathbf{p}} - e\mathbf{A}) + eA_0 + \hat{\beta} m \tag{49}$$

then

$$\hat{H}(-e) = \hat{\alpha} (\hat{\mathbf{p}} + e\mathbf{A}) - eA_0 + \hat{\beta} m \tag{50}$$

$$\langle \hat{H}(-e) \rangle_c = \langle \hat{\alpha} \hat{\mathbf{p}} \rangle_c + e \langle \hat{\alpha} \rangle_c \mathbf{A} - eA_0 + \langle \hat{\beta} \rangle_c m$$

$$= -\langle \hat{\alpha} \hat{\mathbf{p}} \rangle + e \langle \hat{\alpha} \rangle \mathbf{A} - eA_0 - \langle \hat{\beta} \rangle m$$

$$= -\langle \hat{\alpha} \hat{\mathbf{p}} - e\hat{\alpha} \mathbf{A} + eA_0 + \hat{\beta} m \rangle = -\hat{H}(e) \tag{51}$$

8.4 IMPACT OF THE CHARGE CONJUGATION OPERATOR ON THE DIRAC WAVE FUNCTION

Let $\psi(x)$ be an arbitrary wave function of a Dirac particle. Then, $\psi(x)$ is a solution of

$$(\not{P} - m)\,\psi(x) = 0 \tag{52}$$

Using energy and spin projection operators, we can construct from an arbitrary Dirac wave function $\psi(x)$, solutions of a specific energy sign and spin direction. The Dirac wave function of a particle with energy εE_p, momentum **p** and spin polarization along the direction of vector **s** is obtained as

$$\psi_{\text{sgn}\,\boldsymbol{\varepsilon},\boldsymbol{p},\boldsymbol{s}} = \left(\frac{m+\varepsilon\not{p}}{2m}\right)\left(\frac{1+\gamma_5\not{s}}{2}\right)\psi(x) \tag{53}$$

where $\varepsilon = +1$ corresponds to positive energy states and $\varepsilon = -1$ corresponds to negative energy states, respectively. The spin of the plane wave is oriented in the direction of **s**.

Now

$$(\psi_{\text{sgn},\,\varepsilon,\mathbf{p},\mathbf{s}})_c = \hat{U}\,\psi^*_{\text{sgn}\varepsilon,\mathbf{p},\mathbf{s}} = \hat{C}\boldsymbol{\gamma}^0\psi^*_{\varepsilon,\,\mathbf{p},\,\mathbf{s}} = \hat{C}\boldsymbol{\gamma}^0\left(\frac{m+\varepsilon\,\not{p}}{2m}\right)^*\left(\frac{1+\gamma_5\not{s}}{2}\right)^*\psi^*(x)$$

$$\hat{C}\boldsymbol{\gamma}^0\left(\frac{m+\varepsilon\,\not{p}}{2m}\right)^*\left(\frac{1+\gamma_5\not{s}}{2m}\right)^*(\hat{C}\boldsymbol{\gamma}^0)^{-1}\,(\hat{C}\boldsymbol{\gamma}^0)\psi^*(x) \tag{54}$$

Further

$$\hat{U}\boldsymbol{\gamma}^{(0,1,3)*}\hat{U}^{-1} = (i\boldsymbol{\gamma}^2)\boldsymbol{\gamma}^{(0,1,3)*}\,(i\boldsymbol{\gamma}^2)^{-1} = -\,\boldsymbol{\gamma}^2\boldsymbol{\gamma}^{(0,1,3)*}\,\boldsymbol{\gamma}^2 = -\,\boldsymbol{\gamma}^2\boldsymbol{\gamma}^{(0,1,3)}\boldsymbol{\gamma}^2 = -\,\boldsymbol{\gamma}^{(0,1,3)} \tag{55}$$

and

$$\hat{U}\boldsymbol{\gamma}^{2*}\hat{U}^{-1} = (i\boldsymbol{\gamma}^2)\boldsymbol{\gamma}^{2*}\,(i\boldsymbol{\gamma}^2)^{-1} = -\,\boldsymbol{\gamma}^2\boldsymbol{\gamma}^{2*}\boldsymbol{\gamma}^2 = \boldsymbol{\gamma}^2\boldsymbol{\gamma}^2\boldsymbol{\gamma}^2 = -\,\boldsymbol{\gamma}^2 \tag{56}$$

Therefore,

$$\hat{U}\boldsymbol{\gamma}^{\mu*}\hat{U}^{-1} = -\,\boldsymbol{\gamma}^{\mu} \tag{57}$$

which gives

$$(\hat{C}\boldsymbol{\gamma}^0)\boldsymbol{\gamma}^{\mu*}(\hat{C}\boldsymbol{\gamma}^0)^{-1} = -\,\boldsymbol{\gamma}^{\mu} \tag{58}$$

since

$$\hat{C} = \hat{U}\boldsymbol{\gamma}^0 = i\boldsymbol{\gamma}^2\boldsymbol{\gamma}^0 \Rightarrow \hat{C}\boldsymbol{\gamma}^0 = i\boldsymbol{\gamma}^2 = \hat{U} \tag{59}$$

Also $p^*_\mu = p_\mu$ since the projection operator involves the momentum eigenvalues (which are real numbers) and not the momentum operator $\hat{p}_\mu$. Hence,

$$\hat{C}\boldsymbol{\gamma}^0\not{p}^*\,(\hat{C}\boldsymbol{\gamma}^0)^{-1} = -\,\not{p} \tag{60}$$

Similarly,

$$\hat{C}\boldsymbol{\gamma}^0\not{s}^*\,(\hat{C}\boldsymbol{\gamma}^0)^{-1} = -\,\not{s} \tag{61}$$

Now

$$\hat{C}\gamma^0\gamma^*_5(\hat{C}\gamma^0)^{-1} = \hat{C}\gamma^0(i\gamma^0\gamma^1\gamma^2\gamma^3)^*(\hat{C}\gamma^0)^{-1} = -i\hat{C}\gamma^0(\gamma^{0*}\gamma^{1*}\gamma^{2*}\gamma^{3*})(\hat{C}\gamma^0)^{-1}$$
$$= -i[(\hat{C}\gamma^0)\gamma^{0*}(\hat{C}\gamma^0)^{-1}(\hat{C}\gamma^0)\gamma^{1*}(\hat{C}\gamma^0)^{-1}$$
$$(\hat{C}\gamma^0)\gamma^{2*}(\hat{C}\gamma^0)^{-1}(\hat{C}\gamma^0)\gamma^{3*}(\hat{C}\gamma^0)^{-1}]$$
$$= -i\gamma^0\gamma^1\gamma^2\gamma^3 = -\gamma_5 \tag{62}$$

so that

$$(\psi_{\text{sgn}\,\varepsilon,\mathbf{p},\mathbf{s}})_c = \left(\frac{m-\varepsilon\not{p}}{2m}\right)\left(\frac{1+\gamma_5\not{s}}{2}\right)\hat{C}\gamma^0\psi^*(x) = \left(\frac{m-\varepsilon\not{p}}{2m}\right)\left(\frac{1+\gamma_5\not{s}}{2}\right)\psi_c \tag{63}$$

Hence, we find that the charge conjugated solution has opposite energy and momentum compared to the original wave function but has the same spin polarization.

The fact that the spin polarization does not change under charge conjugation is because the spin projection of particles with negative energy is itself defined with inverted sign—this is an intrinsic property of the spin projection operator.

8.5 TIME REVERSAL OF A FREE PARTICLE WAVE FUNCTION

Before we examine the impact of PCT inversion on a Dirac wave function, let us review the effect of time reversal on an arbitrary Dirac wave function through the time inversion operator

$$\hat{T} = -i\hat{\alpha}_1\hat{\alpha}_3\hat{K} = i\gamma^1\gamma^3\hat{K} = \hat{T}_0\hat{K} \tag{64}$$

Consider $\psi(\mathbf{x}, x^0)$ as a free particle solution of the Dirac equation. Then, as shown above, we can construct therefrom solutions pertaining to a particular energy sign and spin polarization by applying the relevant projection operators. We have

$$\psi_{\text{sgn},\varepsilon,\mathbf{p},\mathbf{s}}(\mathbf{x}, x^0) = \left(\frac{m+\varepsilon\not{p}}{2m}\right)\left(\frac{1+\gamma_5\not{s}}{2}\right)\psi(\mathbf{x}, x^0) \tag{65}$$

Operating on this wave function with the time reversal operator, we get

$$\hat{T}\psi_{\text{sgn},\varepsilon,\mathbf{p},\mathbf{s}}(\mathbf{x}, x^0) = \hat{T}\left[\left(\frac{m+\varepsilon\not{p}}{2m}\right)\left(\frac{1+\gamma_5\not{s}}{2}\right)\right]\psi(\mathbf{x}, x^0)$$
$$= \hat{T}\left[\left(\frac{m+\varepsilon\not{p}}{2m}\right)\hat{T}^{-1}\hat{T}\left(\frac{1+\gamma_5\not{s}}{2}\right)\right]\hat{T}^{-1}\hat{T}\psi(\mathbf{x}, x^0) \tag{66}$$

Now

$$\psi'(x') = \psi'(\mathbf{x}', x'^0) = \psi'(\mathbf{x}, -x^0) = \hat{T}\psi(\mathbf{x}, x^0) = \hat{T}_0\hat{K}\psi(\mathbf{x}, x^0)$$
$$= \hat{T}_0\psi^*(\mathbf{x}, x^0) \tag{67}$$

$$\hat{T}\gamma^0\hat{T}^{-1} = (i\gamma^1\gamma^3\hat{K})\gamma^0(i\gamma^1\gamma^3\hat{K})^{-1} = \gamma^1\gamma^3(\hat{K}\gamma^0\hat{K}^{-1})\gamma^3\gamma^1 = \gamma^1\gamma^3\gamma^{0*}\gamma^3\gamma^1$$

$$= \boldsymbol{\gamma}^1\boldsymbol{\gamma}^3\boldsymbol{\gamma}^0\boldsymbol{\gamma}^3\boldsymbol{\gamma}^1 = \boldsymbol{\gamma}^1\boldsymbol{\gamma}^3\boldsymbol{\gamma}^3\boldsymbol{\gamma}^1\boldsymbol{\gamma}^0 = \boldsymbol{\gamma}^0 \tag{68}$$

$$\hat{T}\boldsymbol{\gamma}^1\hat{T}^{-1} = (i\boldsymbol{\gamma}^1\boldsymbol{\gamma}^3\hat{K})\boldsymbol{\gamma}^1(i\boldsymbol{\gamma}^1\boldsymbol{\gamma}^3\hat{K})^{-1} = \boldsymbol{\gamma}^1\boldsymbol{\gamma}^3(\hat{K}\boldsymbol{\gamma}^1\hat{K}^{-1})\boldsymbol{\gamma}^3\boldsymbol{\gamma}^1 = \boldsymbol{\gamma}^1\boldsymbol{\gamma}^3\boldsymbol{\gamma}^{1*}\boldsymbol{\gamma}^3\boldsymbol{\gamma}^1$$

$$= \boldsymbol{\gamma}^1\boldsymbol{\gamma}^3\boldsymbol{\gamma}^1\boldsymbol{\gamma}^3\boldsymbol{\gamma}^1 = -\boldsymbol{\gamma}^1 \tag{69}$$

$$\hat{T}\boldsymbol{\gamma}^2\hat{T}^{-1} = (i\boldsymbol{\gamma}^1\boldsymbol{\gamma}^3\hat{K})\boldsymbol{\gamma}^2(i\boldsymbol{\gamma}^1\boldsymbol{\gamma}^3\hat{K})^{-1} = \boldsymbol{\gamma}^1\boldsymbol{\gamma}^3(\hat{K}\boldsymbol{\gamma}^2\hat{K}^{-1})\boldsymbol{\gamma}^3\boldsymbol{\gamma}^1 = \boldsymbol{\gamma}^1\boldsymbol{\gamma}^3\boldsymbol{\gamma}^{2*}\boldsymbol{\gamma}^3\boldsymbol{\gamma}^1$$

$$= -\boldsymbol{\gamma}^1\boldsymbol{\gamma}^3\boldsymbol{\gamma}^2\boldsymbol{\gamma}^3\boldsymbol{\gamma}^1 = -\boldsymbol{\gamma}^2$$

$$\hat{T}\boldsymbol{\gamma}^3\hat{T}^{-1} = (i\boldsymbol{\gamma}^1\boldsymbol{\gamma}^3\hat{K})\boldsymbol{\gamma}^3(i\boldsymbol{\gamma}^1\boldsymbol{\gamma}^3\hat{K})^{-1} = \boldsymbol{\gamma}^1\boldsymbol{\gamma}^3(\hat{K}\boldsymbol{\gamma}^3\hat{K}^{-1})\boldsymbol{\gamma}^3\boldsymbol{\gamma}^1 = \boldsymbol{\gamma}^1\boldsymbol{\gamma}^3\boldsymbol{\gamma}^{3*}\boldsymbol{\gamma}^3\boldsymbol{\gamma}^1$$

$$= \boldsymbol{\gamma}^1\boldsymbol{\gamma}^3\boldsymbol{\gamma}^3\boldsymbol{\gamma}^3\boldsymbol{\gamma}^1 = -\boldsymbol{\gamma}^1\boldsymbol{\gamma}^3\boldsymbol{\gamma}^1 = -\boldsymbol{\gamma}^3 \tag{70}$$

Also $p^*_\mu = p_\mu$, since the projection operator involves the momentum eigenvalues and not the momentum operator $\hat{p}_\mu$. Hence, $\not{p} = \boldsymbol{\gamma}^0 p_0 - \boldsymbol{\gamma}.\mathbf{p}$ or $p_\mu \equiv \{p_0, p_1, p_2, p_3\} \equiv \{p_0, -\mathbf{p}\}$ gets transformed under time inversion as

$$\hat{T}\not{p}\,\hat{T}^{-1} = \not{p}' = \boldsymbol{\gamma}^0 p_0 + \boldsymbol{\gamma}.\mathbf{p} \tag{71}$$

or

$$(p')_\mu \equiv \{p_0, -p_1, -p_2, -p_3\} \equiv \{p_0, \mathbf{p}\} \tag{72}$$

i.e, the sign of the 3-momentum changes while that of the energy remains the same.

Similarly,

$$\hat{T}\not{s}\hat{T}^{-1} = \not{s}' \tag{73}$$

where

$$(s')_\mu = \{s_0, \mathbf{s}\} \tag{74}$$

and

$$\hat{T}\boldsymbol{\gamma}_5 T^{-1} = \hat{T}(i\boldsymbol{\gamma}^0\boldsymbol{\gamma}^1\boldsymbol{\gamma}^2\boldsymbol{\gamma}^3)\hat{T}^{-1} = (\hat{T}i\hat{T}^{-1})(\hat{T}\boldsymbol{\gamma}^0\hat{T}^{-1})(\hat{T}\boldsymbol{\gamma}^1\hat{T}^{-1})$$

$$(\hat{T}\boldsymbol{\gamma}^2\hat{T}^{-1})(\hat{T}\boldsymbol{\gamma}^3\hat{T}^1)$$

$$= (\hat{T}i\hat{T}^{-1})(\hat{T}\boldsymbol{\gamma}^0\hat{T}^{-1})(\hat{T}\boldsymbol{\gamma}^1\hat{T}^{-1})(\hat{T}\boldsymbol{\gamma}^2\hat{T}^{-1})(\hat{T}\boldsymbol{\gamma}^3\hat{T}^{-1})$$

$$= (-i)(\boldsymbol{\gamma}^0)(-\boldsymbol{\gamma}^1)(-\boldsymbol{\gamma}^2)(-\boldsymbol{\gamma}^3) = \boldsymbol{\gamma}_5 \tag{75}$$

Hence,

$$\hat{T}\psi_{\text{sgn}\varepsilon,\mathbf{p},\mathbf{s}} = \hat{T}\left[\left(\frac{m+\varepsilon\not{p}}{2m}\right)\hat{T}^{-1}\hat{T}\left(\frac{1+\gamma_5\not{s}}{2}\right)\right]\hat{T}^{-1}\hat{T}\psi(\mathbf{x}, x^0)$$

$$= \left[\hat{T}\left(\frac{m+\varepsilon\not{p}}{2m}\right)\hat{T}^{-1}\hat{T}\left(\frac{1+\gamma_5\not{s}}{2}\right)\hat{T}^{-1}\right]\hat{T}\psi(\mathbf{x}, x^0)$$

$$= \left(\frac{m+\varepsilon\not{p}'}{2m}\right)\left(\frac{1+\gamma_5\not{s}'}{2}\right)\psi'(\mathbf{x}, -x^0) \tag{76}$$

Thus, the impact of time inversion on a free particle wave function is to

(i) reverse the direction of the spatial momentum;

(ii) reverse the direction of the spin polarization; and

(iii) leave the energy unchanged.

8.6 $\hat{P}\,\hat{\mathbb{C}}\,\hat{T}$ INVERSION

We now examine the impact of a sequential operation of time inversion followed by charge conjugation and then by parity inversion on a Dirac wave function. We define the charge conjugation operator by

$$\hat{\mathbb{C}} \equiv \hat{U}\,\hat{K} \equiv \hat{C}\gamma_0\,\hat{K} \tag{77}$$

We have

$$\begin{aligned}
\psi_{P\,\mathbb{C}\,T}(x') &= \hat{P}\,\hat{\mathbb{C}}\,\hat{T}\psi\,(\mathbf{x}, x^0) = \hat{P}(\hat{C}\gamma^0\hat{K})\,[\hat{T}\psi(\mathbf{x}, x^0)] = \hat{P}(\hat{C}\gamma^0)\,[\hat{T}\psi\,(\mathbf{x}, x^0)]^* \\
&= \hat{P}\,(\hat{C}\gamma^0)\,[i\gamma^1\gamma^3\hat{K}\psi\,(\mathbf{x}, x^0)]^* = \hat{P}\,(\hat{C}\gamma^0)\,[i\gamma^1\gamma^3\psi^*\,(\mathbf{x}, x^0)]^* \\
&= -\,i\hat{P}\,\hat{C}\gamma^0\gamma^{1*}\gamma^{3*}\psi\,(\mathbf{x}, x^0) = -\,i\hat{P}\,\hat{C}\gamma^0\gamma^1\gamma^3\psi\,(\mathbf{x}, x^0) \\
&= -\,i\hat{P}\,(i\gamma^2\gamma^0)\gamma^0\gamma^1\gamma^3\psi\,(\mathbf{x}, x^0) \\
&= \hat{P}\gamma^2\gamma^1\gamma^3\psi\,(\mathbf{x}, x^0) = (e^{i\phi}\gamma^0)\gamma^2\gamma^1\gamma^3\psi\,(\mathbf{x}, x^0) \\
&= -\,e^{i\phi}\gamma^0\gamma^1\gamma^2\gamma^3\psi\,(\mathbf{x}, x^0) \\
&= ie^{i\phi}\,(i\gamma^0\gamma^1\gamma^2\gamma^3)\,\psi(\mathbf{x}, x^0) = ie^{i\phi}\gamma_5\psi\,(\mathbf{x}, x^0) \\
&= ie^{i\phi}\gamma_5\psi\,(-\mathbf{x}', -x'^0) = ie^{i\phi}\gamma_5\psi\,(-x')
\end{aligned} \tag{78}$$

where the last step follows from

$$\mathbf{x}' = \hat{P}\mathbf{x} = -\,\mathbf{x},\; x'^0 = \hat{T}x^0 = -\,x^0 \tag{79}$$

8.7 IMPACT OF $\hat{P}\,\hat{\mathbb{C}}\,\hat{T}$ INVERSION ON FREE PARTICLE DIRAC WAVE FUNCTION

To investigate the impact of $\hat{P}\,\hat{\mathbb{C}}\,\hat{T}$ inversion on a free particle Dirac wave function, consider an arbitrary Dirac wave function $\psi\,(x)$. We generate a plane wave from $\psi\,(x)$ of desired energy sign and spin polarization by operating thereon by the corresponding projection operators. We have

$$\psi_{\text{sgn}\,\varepsilon,\mathbf{p},\mathbf{s}}(\mathbf{x},x^0) = \left[\left(\frac{m+\varepsilon\not{p}}{2m}\right)\left(\frac{1+\gamma_5\not{s}}{2}\right)\right]\psi(\mathbf{x},x^0) \tag{80}$$

This plane wave has energy sign indicated by ε, momentum **p** and spin polarization **s**. We operate on $\psi_{\text{sgn}\,\varepsilon,\mathbf{p},\mathbf{s}}(\mathbf{x},x^0)$ by the $\hat{P}\,\hat{\mathbb{C}}\,\hat{T}$ operator to obtain

$$\psi_{\text{sgn}\,\varepsilon,\mathbf{p},\mathbf{s}(P\,\mathbb{C}\,T)}(\mathbf{x}',x'^0) = [\psi_{\text{sgn}\,\varepsilon,\mathbf{p},\mathbf{s}}(\mathbf{x},x^0)]_{P\,\mathbb{C}\,T}$$

$$= \hat{P}\,\hat{\mathbb{C}}\,\hat{T}\left[\left(\frac{m+\varepsilon\not{p}}{2m}\right)\left(\frac{1+\gamma_5\not{s}}{2}\right)\right]\psi(\mathbf{x},x^0)$$

$$= ie^{i\phi}\boldsymbol{\gamma}_5\left[\left(\frac{m+\varepsilon\not{p}}{2m}\right)\left(\frac{1+\gamma_5\not{s}}{2}\right)\right]\psi(\mathbf{x},x^0)$$

$$= ie^{i\phi}\boldsymbol{\gamma}_5\left[\left(\frac{m+\varepsilon\not{p}}{2m}\right)\left(\frac{1+\gamma_5\not{s}}{2}\right)\right]\psi(-\mathbf{x}',-x'^0) \tag{81}$$

Now, using the fact that $\boldsymbol{\gamma}_5$ anticommutes with $\boldsymbol{\gamma}^\mu$, μ = 0, 1, 2, 3, i.e., $\{\boldsymbol{\gamma}_5, \boldsymbol{\gamma}^\mu\} = 0$, we have

$$\psi_{\text{sgn}\,\varepsilon,\mathbf{p},\mathbf{s}(P\,\mathbb{C}\,T)}(\mathbf{x}',x'^0) = [\psi_{\text{sgn}\,\varepsilon,\mathbf{p},\mathbf{s}}(\mathbf{x},x^0)]_{P\,\mathbb{C}\,T}$$

$$= \left[\left(\frac{m-\varepsilon\not{p}}{2m}\right)\left(\frac{1-\gamma_5\not{s}}{2}\right)\right] ie^{i\phi}\boldsymbol{\gamma}_5\psi(\mathbf{x},x^0)$$

$$= \left[\left(\frac{m-\varepsilon\not{p}}{2m}\right)\left(\frac{1-\gamma_5\not{s}}{2}\right)\right]\hat{P}\,\hat{\mathbb{C}}\,\hat{T}\psi(\mathbf{x},x^0)$$

$$= \left[\left(\frac{m-\varepsilon\not{p}}{2m}\right)\left(\frac{1-\gamma_5\not{s}}{2}\right)\right]\psi\,\hat{P}\,\hat{\mathbb{C}}\,\hat{T}(\mathbf{x},x^0)$$

$$= \left[\left(\frac{m-\varepsilon\not{p}}{2m}\right)\left(\frac{1-\gamma_5\not{s}}{2}\right)\right]\psi_{P\,\mathbb{C}\,T}(-\mathbf{x}',-x'^0) \tag{82}$$

Thus, from eqs. (81) and (82), it is explicitly seen that when a plane wave of energy sign indicated by ε, momentum **p** and spin polarization **s** is operated on by the $\hat{P}\,\hat{\mathbb{C}}\,\hat{T}$ operator, the resulting plane wave has energy sign indicated by −ε, momentum −**p** and spin polarization −**s**.

8.8 IMPACT OF $\hat{P}\,\hat{\mathbb{C}}\,\hat{T}$ INVERSION ON DIRAC WAVE FUNCTION IN ELECTROMAGNETIC FIELD

The Dirac equation for a particle in an electromagnetic field can be written as

$$[\hat{\alpha}\,(-i\nabla - e\mathbf{A}) + \hat{\beta}\,m + eA_0]\,\psi(x) = \varepsilon E_p\psi(x) \tag{83}$$

Operating on eq. (83) by the combined $\hat{P}\hat{\mathbb{C}}\hat{T}$ operator, we have

$$\hat{P}\hat{\mathbb{C}}\hat{T}\varepsilon E_p\psi(x) = \hat{P}\hat{\mathbb{C}}\hat{T}\,[\hat{\alpha}\,(-i\nabla - e\mathbf{A}) + \hat{\beta}m + eA_0]\,\psi(x)$$

$$= \left\{\begin{matrix} \hat{P}\hat{\mathbb{C}}\hat{T}\hat{\alpha}(\hat{P}\hat{\mathbb{C}}\hat{T})^{-1}[\hat{P}\hat{\mathbb{C}}\hat{T}(-i\nabla)(\hat{P}\hat{\mathbb{C}}\hat{T})^{-1} + \hat{P}\hat{\mathbb{C}}\hat{T}(-e\mathrm{A})(\hat{P}\hat{\mathbb{C}}\hat{T})^{-1}] \\ +\hat{P}\hat{\mathbb{C}}\hat{T}(\hat{\beta}m)(\hat{P}\hat{\mathbb{C}}\hat{T})^{-1} + \hat{P}\hat{\mathbb{C}}\hat{T}(eA_0)(\hat{P}\hat{\mathbb{C}}\hat{T})^{-1} \end{matrix}\right\}\hat{P}\hat{\mathbb{C}}\hat{T}\psi(x) \tag{84}$$

We now evaluate each term separately. We have

$$\hat{P}\hat{\mathbb{C}}\hat{T}\hat{\alpha}\,(\hat{P}\hat{\mathbb{C}}\hat{T})^{-1} = e^{i\phi}\gamma_5\hat{\alpha}\,(e^{i\phi}\gamma_5)^{-1} = \gamma_5\hat{\alpha}\gamma_5 = \gamma_5\hat{\beta}\,\hat{\gamma}\gamma_5 = \gamma_5\gamma^0\hat{\gamma}\gamma_5 = -\gamma_5\hat{\gamma}^0\gamma_5\hat{\gamma}$$

$$= \gamma_5\gamma_5\hat{\gamma}^0\hat{\gamma} = \hat{\gamma}^0\hat{\gamma} = \hat{\alpha} \tag{85}$$

$$\hat{P}\hat{\mathbb{C}}\hat{T}\hat{\beta}\,(\hat{P}\hat{\mathbb{C}}\hat{T})^{-1} = e^{i\phi}\gamma_5\hat{\beta}\,(e^{i\phi}\gamma_5)^{-1} = \gamma_5\hat{\beta}\gamma_5$$

$$= \gamma_5\hat{\gamma}^0\gamma_5 = -\gamma_5\gamma_5\hat{\gamma}^0 = -\hat{\gamma}^0 = -\hat{\beta} \tag{86}$$

$$\hat{P}\hat{\mathbb{C}}\hat{T}(-i\nabla)(\hat{P}\hat{\mathbb{C}}\hat{T})^{-1} = ie^{i\phi}\gamma_5\,(-i\nabla)_{-\mathbf{x}}\,(ie^{i\phi}\gamma_5)^{-1}$$

$$= \gamma_5\,(-i\nabla)_{-\mathbf{x}}\gamma_5 = (-i\nabla)_{-\mathbf{x}} = (i\nabla) \tag{87}$$

$$\hat{P}\hat{\mathbb{C}}\hat{T}\,(e\mathbf{A})\,(\hat{P}\hat{\mathbb{C}}\hat{T})^{-1} = ie^{i\phi}\gamma_5\,(e\mathbf{A}_0)_{-\mathrm{x},-x^0}\,(ie^{i\phi}\gamma_5)^{-1}$$

$$= (e\mathbf{A}_0)_{-\mathrm{x},-x^0} = e\mathbf{A} = e\mathbf{A}'(x') \tag{88}$$

The last step follows because the vector potential $\mathbf{A}\,(\mathbf{x}, x^0)$ is related to the electric currents $\mathbf{j}(\mathbf{x}, x^0)$: $(\mathbf{j}\,(\mathbf{x}, x^0) = \partial_0^2\mathbf{A} - \nabla_0^2\mathbf{A}$ in the Lorentz gauge.

Now, a change in sign of x^0 would cause a change in sign of the vector current $\mathbf{j}(\mathbf{x}, x^0)$. This can only happen if the sign of the vector potential $\mathbf{A}$ changes. We, therefore, conclude that time inversion of the vector potential would reverse the sign of $\mathbf{A}$, i.e., $\mathbf{A}'\,(\mathbf{x}', x'^0) = \hat{T}\mathbf{A}\,(\mathbf{x}, x^0)\hat{T}^{-1} = -\mathbf{A}\,(\mathbf{x}, x^0) = -\mathbf{A}\,(\mathbf{x}', -x'^0)$.

Furthermore, the parity operator changes the sign of the space coordinates which will cause a change in the vector current and, therefore, of the vector potential, i.e., $\mathbf{A}'\,(\mathbf{x}', x'^0) = \hat{P}\mathbf{A}\,(\mathbf{x}, x^0)\,\hat{P}^{-1} = -\mathbf{A}\,(\mathbf{x}, x^0) = -\mathbf{A}\,(-\mathbf{x}', x'^0)$.

Hence, under a $\hat{P}\hat{\mathbb{C}}\hat{T}$ inversion, we have

$$\hat{P}\hat{\mathbb{C}}\hat{T}\mathbf{A}\,(\mathbf{x}, x^0)\,(\hat{P}\hat{\mathbb{C}}\hat{T})^{-1} = \hat{P}\hat{\mathbb{C}}\hat{T}\mathbf{A}\,(\mathbf{x}, x^0)\hat{T}^{-1}\hat{\mathbb{C}}^{-1}\hat{P}^{-1}$$

$$= \hat{P}\hat{\mathbb{C}}\,[\hat{T}\mathbf{A}\,(\mathbf{x}, x^0)\hat{T}^{-1}]\hat{\mathbb{C}}^{-1}\hat{P}^{-1}$$

$$\hat{P}\hat{\mathbb{C}}\,[-\mathbf{A}\,(\mathbf{x}, x^0)]\hat{\mathbb{C}}^{-1}\hat{P}^{-1} = \hat{P}\hat{\mathbb{C}}\hat{\mathbb{C}}^{-1}\,[-\mathbf{A}\,(\mathbf{x}, x^0)]\,\hat{P}^{-1} = \hat{P}\,[-\mathbf{A}\,(\mathbf{x}, x^0)]\,\hat{P}^{-1}$$

$$= \mathbf{A}\,(\mathbf{x}, x^0) = \mathbf{A}'\,(-\mathbf{x}', -x'^0)$$

Similarly,

$$\hat{P}\hat{\mathbb{C}}\hat{T}\,(eA_0)\,(\hat{P}\hat{\mathbb{C}}\hat{T})^{-1} = ie^{i\phi}\gamma_5\,(eA_0)_{-\mathbf{x},-x^0}\,(ie^{i\phi}\gamma_5)^{-1}$$

$$= (eA_0)_{-\mathbf{x},-x^0} = eA_0 = eA_0'(x') \tag{89}$$

(The last step, in this case, follows because the scalar potential $A_0\,(\mathbf{x}, x^0)$ is related to the electric charge density $\rho\,(\mathbf{x}, x^0)$, $\rho\,(\mathbf{x}, x^0) = \partial_0^2 A_0 - \nabla^2 A_0$ in the Lorentz gauge.)

Now, a change in sign of x^0 would NOT cause a change in sign of the scalar charge density $\rho\,(\mathbf{x}, x^0)$. This can happen if the sign of the scalar potential A_0 does not change under time inversion. We, therefore, conclude that time inversion of the scalar potential would NOT reverse the sign of A_0, i.e.,

$$A_0'\,(\mathbf{x}', x'^0) = \hat{T}A_0\,(\mathbf{x}, x^0)\hat{T}^{-1} = A_0\,(\mathbf{x}, x^0) = A_0\,(\mathbf{x}', -x'^0)$$

Furthermore, the parity operator changes the sign of the space coordinates ONLY which will NOT cause a change in the scalar charge density and therefore of the scalar potential, i.e., $A_0'\,(\mathbf{x}', x'^0) = \hat{P}A_0\,(\mathbf{x}, x^0)\,\hat{P}^{-1} = A_0\,(\mathbf{x}, x^0) = A_0\,(-\mathbf{x}', -x'^0)$.

Hence, under a $\hat{P}\hat{\mathbb{C}}\hat{T}$ inversion, we have

$$\hat{P}\hat{\mathbb{C}}\hat{T}A_0\,(\mathbf{x}, x^0)\,(\hat{P}\hat{\mathbb{C}}\hat{T})^{-1} = \hat{P}\hat{\mathbb{C}}\hat{T}A_0\,(\mathbf{x}, x^0)\hat{T}^{-1}\hat{\mathbb{C}}^{-1}\hat{P}^{-1}$$

$$= \hat{P}\hat{\mathbb{C}}\,[\hat{T}A_0\,(\mathbf{x}, x^0)\hat{T}^{-1}]\,\hat{\mathbb{C}}^{-1}\hat{P}^{-1}$$

$$\hat{P}\hat{\mathbb{C}}\,[A_0\,(\mathbf{x}, x^0)]\,\hat{\mathbb{C}}^{-1}\hat{P}^{-1} = \hat{P}\hat{\mathbb{C}}\hat{\mathbb{C}}^{-1}\,[A_0\,(\mathbf{x}, x^0)]\,\hat{P}^{-1}$$

$$= \hat{P}\,[A_0\,(\mathbf{x}, x^0)]\,\hat{P}^{-1} = A_0\,(\mathbf{x}, x^0) = A_0'(-\mathbf{x}', -x'^0)$$

Substituting all these values in eq. (84), we get

$$\varepsilon E_p \hat{P}\hat{\mathbb{C}}\hat{T}\psi\,(x) = [\hat{\alpha}\,(i\nabla - e\boldsymbol{A}) - \hat{\beta}m + eA_0]\,\hat{P}\hat{\mathbb{C}}\hat{T}\psi\,(x) \tag{90}$$

or equivalently

$$\varepsilon E_p \psi_{P\mathbb{C}T}(x') = [\hat{\alpha}\,(-i\nabla' - e\mathbf{A}') - \hat{\beta}m + eA'_0]\,\psi_{P\mathbb{C}T}(x') \tag{91}$$

or

$$-\varepsilon E_p \psi_{P\mathbb{C}T}(x') = [\hat{\alpha}\,(i\nabla' + e\mathbf{A}') + \hat{\beta}m - eA'_0]\,\psi_{P\mathbb{C}T}(x') \tag{92}$$

with $x' = -x$.

Comparison of eqs. (83) and (92) explicitly shows that if the input plane wave is of energy εE_p, charge e, momentum $\mathbf{p}$ moving forward in space time, then the $\hat{P}\hat{\mathbb{C}}\hat{T}$ inverted wave corresponds to a particle with the same mass having charge $-e$, energy $-\varepsilon E_p$ travelling backwards in space-time $x' = -x$ with reversed momentum $-\mathbf{p}$.

CHAPTER 9

Relativistic Dynamics of Massless Particles

9.1 DIRAC EQUATION FOR MASSLESS PARTICLES AND THE WEYL EQUATION

Experimental studies on β-decay processes have confirmed the existence of (near) zero mass spin $\frac{1}{2}$ particles called the "neutrinos". Theoretical predictions of the existence of such particles was made way back in 1930 by Pauli to account for momentum and energy conservation in β-decay processes. This chapter constructs the fundamental equations that describe the quantum mechanical dynamics of zero mass spin $\frac{1}{2}$ particles.

The Dirac equation for a particle of zero rest mass reduces to

$$i\partial_0\phi(x) = -i\boldsymbol{\alpha}.\nabla\phi(x) = \boldsymbol{\alpha}.\hat{\mathbf{p}}\phi(x) \tag{1}$$

This equation does not contain the $\boldsymbol{\beta}$-matrix, since its coefficient in the Dirac equation is the rest mass of the particle, which is assumed zero in the present case. Furthermore, the components of $\boldsymbol{\alpha}$ satisfy certain anticommutation relations because eq. (1) must be compatible with the corresponding Klein Gordon equation for massless particles, viz.,

$$(\partial_0^2 - \nabla^2)\,\phi(x) = 0 \tag{2}$$

Squaring the operators of eq. (1), we have

$$\begin{aligned} 0 &= \partial_0^2\phi(x) - (\boldsymbol{\alpha}.\nabla)^2\,\phi(x) \\ &= \partial_0^2\phi(x) - \sum_{i,j=1}^{3}(\boldsymbol{\alpha}^i\partial_i)(\boldsymbol{\alpha}^j\partial_j)\,\phi(x) \\ &= \partial_0^2\phi(x) - \sum_{i,j=1}^{3}(\alpha^i\alpha^j)\,(\partial_i\partial_j)\,\phi(x) \end{aligned} \tag{3}$$

Equation (3) will correspond to eq. (2) if and only if

$$\{\boldsymbol{\alpha}^i, \boldsymbol{\alpha}^j\} = 2\delta^{ij}\mathbf{I} \tag{4}$$

which coincides with the corresponding relations for massive particles. We also require that $(\boldsymbol{\alpha}^i)^\dagger = \boldsymbol{\alpha}^i$ in order that the Dirac Hamiltonian $\hat{H} = -i\boldsymbol{\alpha}\cdot\nabla = \boldsymbol{\alpha}.\hat{\mathbf{p}}$ be Hermitian. It can be easily seen, as in the case of the massive Dirac equation, that α matrices must be even dimensional traceless matrices for

(i) $\boldsymbol{\alpha}^i\boldsymbol{\alpha}^j = (-\mathbf{I})\,\boldsymbol{\alpha}^j\boldsymbol{\alpha}^i$ (5)

so that on taking determinants of both sides

$$(\det \boldsymbol{\alpha}^i)\,(\det \boldsymbol{\alpha}^j) = (-1)^n\,(\det \boldsymbol{\alpha}^j)\,(\det \boldsymbol{\alpha}^i) = (-1)^n\,(\det \boldsymbol{\alpha}^i)\,(\det \boldsymbol{\alpha}^j) \tag{6}$$

we establish that n is even because $(\det \boldsymbol{\alpha}^i)\,(\det \boldsymbol{\alpha}^j) \neq 0$.

(ii) To establish that these matrices are traceless, we note that all the $\boldsymbol{\alpha}^i$ are Hermitian matrices and hence, can be diagonalized (but not simultaneously, i.e., not in the same basis). We consider a basis in which $\boldsymbol{\alpha}^1$ is diagonal, i.e., $\boldsymbol{\alpha}^1 = \mathrm{diag}(a_1, a_2,\ldots, a_n)$. Since $(\boldsymbol{\alpha}^1)^2 = \mathbf{I}$, $a_i^2 = 1\ \forall\, i = 1, 2, 3,\ldots, n$ or $a_i = \pm 1\ \forall i = 1, 2, 3,\ldots, n$. Now, eq. (5) implies that

$$(\boldsymbol{\alpha}^j)^{-1}\,\boldsymbol{\alpha}^1\boldsymbol{\alpha}^j = (-\mathbf{I})\boldsymbol{\alpha}^1 \tag{7}$$

Taking trace of both sides, we obtain

$$Tr\,[(\boldsymbol{\alpha}^j)^{-1}\,\boldsymbol{\alpha}^1\boldsymbol{\alpha}^j] = Tr\,[\boldsymbol{\alpha}^j\,(\boldsymbol{\alpha}^j)^{-1}\,\boldsymbol{\alpha}^1] = Tr\,(\boldsymbol{\alpha}^1) = Tr\,[(-\mathbf{I})\,\boldsymbol{\alpha}^1] = -Tr\,(\boldsymbol{\alpha}^1) \tag{8}$$

showing that $Tr\,(\boldsymbol{\alpha}^1)$ vanishes. Similarly, we can establish that $Tr\,(\boldsymbol{\alpha}^2) = Tr\,(\boldsymbol{\alpha}^3) = 0$.

In summary, the $\boldsymbol{\alpha}^i$ must possess the following properties:

(a) $\{\boldsymbol{\alpha}^i, \boldsymbol{\alpha}^j\} = 2\delta^{ij}\mathbf{I}$

(b) $(\boldsymbol{\alpha}^i)^\dagger = \boldsymbol{\alpha}^i$

(c) $Tr\,(\boldsymbol{\alpha}^i) = 0$

(d) All $\boldsymbol{\alpha}^i$ must be even dimensional matrices

(e) There exist bases in which at least one of the $\boldsymbol{\alpha}^i$'s can be diagonalized. In this basis, the diagonal elements of the diagonal $\boldsymbol{\alpha}^i$ will be only ± 1 with the number of $+1, -1$ being equal.

The set of three Pauli matrices $\boldsymbol{\sigma}^i$, $i = 1, 2, 3$ that is well known to physicists in connection with the representations of angular momentum operators and particle spin operators in non-relativistic quantum mechanics meet all the aforesaid requirements. The three $\boldsymbol{\sigma}^i$, $i = 1, 2, 3$ are given by

$$\boldsymbol{\sigma}^1 = \begin{pmatrix} 0 & 1 \\ 1 & 0 \end{pmatrix},\ \boldsymbol{\sigma}^2 = \begin{pmatrix} 0 & -i \\ i & 0 \end{pmatrix},\ \boldsymbol{\sigma}^3 = \begin{pmatrix} 1 & 0 \\ 0 & -1 \end{pmatrix} \tag{9}$$

Additionally, these matrices also satisfy

$$\boldsymbol{\sigma}^i\boldsymbol{\sigma}^j = i\varepsilon^{ijk}\boldsymbol{\sigma}^k,\ i, j, k = 1, 2, 3 \tag{10}$$

It needs to be emphasized here that any set of three matrices obtained from $\boldsymbol{\sigma}^i$, $i = 1, 2, 3$

through a unitary transformation $\hat{U}$, e.g.,

$$(\boldsymbol{\sigma}')^i = \hat{U}\boldsymbol{\sigma}^i\hat{U}^{-1} \tag{11}$$

would also satisfy our requirements and hence could be used as representations for the $\boldsymbol{\alpha}$ operators. This is easily proved as follows:

$$\text{(a) } \{(\boldsymbol{\sigma}')^i, (\boldsymbol{\sigma}')^j\} = \{\hat{U}\boldsymbol{\sigma}^i\hat{U}^{-1}, \hat{U}\boldsymbol{\sigma}^j\hat{U}^{-1}\} = \hat{U}2\delta^{ij}\mathbf{I}\hat{U}^{-1} = 2\delta^{ij}\mathbf{I} = \{\boldsymbol{\sigma}^i, \boldsymbol{\sigma}^j\} \tag{12}$$

$$\begin{aligned}\text{(b) } (\boldsymbol{\sigma}')^{i\dagger} &= (\hat{U}\boldsymbol{\sigma}^i\hat{U}^{-1})^\dagger = (\hat{U}^{-1})^\dagger\,(\boldsymbol{\sigma})^{i\dagger}\,(\hat{U})^\dagger = (\hat{U}^\dagger)^{-1}\,(\boldsymbol{\sigma})^{i\dagger}\hat{U}^\dagger \\ &= (\hat{U}^{-1})^{-1}\,(\boldsymbol{\sigma})^i\hat{U}^{-1} = \hat{U}\,(\boldsymbol{\sigma})^i\hat{U}^{-1} = (\boldsymbol{\sigma}')^i\end{aligned} \tag{13}$$

$$\text{(c) } Tr\,(\boldsymbol{\sigma}')^i = Tr\,[\hat{U}\boldsymbol{\sigma}^i\hat{U}^{-1}] = Tr\,[\hat{U}^{-1}\hat{U}\boldsymbol{\sigma}^i] = Tr\,(\boldsymbol{\sigma}^i) = 0 \tag{14}$$

Furthermore, the set of the three Pauli matrices together with the identity matrix $\mathbf{I}_{2\times 2}$ forms a linearly independent set and hence, can be used as a basis for representing any 2×2 matrix.

We, therefore, arrive at a very important result. The Dirac equation for massless particles gets reduced to a 2-dimensional wave equation and the Dirac wave function, in this case, is a 2-component vector. We can write the Dirac equation for massless particles as

$$i\partial_0\Phi\,(x) = \boldsymbol{\sigma}\hat{\mathbf{p}}\Phi\,(x) = -i\boldsymbol{\sigma}.\nabla\Phi\,(x) \tag{15}$$

where $\Phi\,(x) = (\phi_1\ \phi_2)^T$ and $\boldsymbol{\sigma} = \{\boldsymbol{\sigma}^1, \boldsymbol{\sigma}^2, \boldsymbol{\sigma}^3\}$ is the set of Pauli matrices. Equation (15) is known as the Weyl's equation for massless particles.

9.2 LORENTZ INVARIANCE OF THE WEYL EQUATION

We now establish the form invariance of the Weyl equation under Lorentz transformations. Introducing $\boldsymbol{\sigma}^0 = \mathbf{I}$, we can write the Weyl equation in a symmetric form as

$$\boldsymbol{\sigma}^\mu\partial_\mu\Phi\,(x) = 0 \tag{16}$$

Equation (16) will be invariant under Lorentz transformation if $\Phi\,(x)$ transforms according to

$$\Phi'\,(x') = \hat{S}\,(\boldsymbol{\Lambda})\Phi(x) = \hat{S}\,(\boldsymbol{\Lambda})\Phi(\boldsymbol{\Lambda}^{-1}x') \tag{17}$$

under a Lorentz transformation

$$(x')^\mu = \boldsymbol{\Lambda}x^\mu = a^\mu_\nu x^\nu \tag{18}$$

Since $\Phi(x)$ is a 2-component spinor, $\hat{S}\,(\boldsymbol{\Lambda})$ will be a 2×2 matrix that will operate on the spinor $\Phi\,(x) = (\phi_1\ \phi_2)^T$. Also, $\hat{S}\,(\boldsymbol{\Lambda})$ will be a function of the parameters of the Lorentz transformation a^μ_ν and shall also depend on the relative velocities and spatial orientations of the two frames.

Now

$$\partial_\mu = \frac{\partial}{\partial x^\mu} = \frac{\partial x'^\nu}{\partial x^\mu}\frac{\partial}{\partial x'^\nu} = a^\nu_\mu\frac{\partial}{\partial x'^\nu} = a^\nu_\mu\partial'_\nu,\ \text{since } (x')^\nu = \sum_{\mu=0}^{3} a^\nu_\mu x^\mu \equiv a^\nu_\mu x^\mu \tag{19}$$

Hence, we can write the Weyl equation (16) as

$$0 = \boldsymbol{\sigma}^{\mu}\, a_{\mu}^{\nu}\, \partial_{\nu}' [\hat{S}^{-1}(\boldsymbol{\Lambda})\, \Phi'(x')]$$

$$= a_{\mu}^{\nu}\boldsymbol{\sigma}^{\mu}\partial_{\nu}' [\hat{S}^{-1}(\boldsymbol{\Lambda})\, \Phi'(x')] = \hat{S}(\boldsymbol{\Lambda}) a_{\mu}^{\nu}\boldsymbol{\sigma}^{\mu}\partial_{\nu}' [\hat{S}^{-1}(\boldsymbol{\Lambda})\, \Phi'(x')] \tag{20}$$

Now, form invariance of the Weyl equation decrees that eq. (20) should have the same form as eq. (16), i.e.,

$$\boldsymbol{\sigma}^{\lambda}\partial_{\lambda}'\Phi'(x') = 0 \tag{21}$$

On comparing eqs. (20) and (21), we obtain

$$a_{\mu}^{\nu}\hat{S}(\boldsymbol{\Lambda})\sigma^{\mu}\hat{S}^{-1}(\boldsymbol{\Lambda}) = \boldsymbol{\sigma}^{\nu} \text{ or } a_{\mu}^{\nu}\boldsymbol{\sigma}^{\mu} = \hat{S}^{-1}(\boldsymbol{\Lambda})\boldsymbol{\sigma}^{\nu}\hat{S}(\boldsymbol{\Lambda}) \tag{22}$$

Noting that $\boldsymbol{\sigma}^{0} = \mathbf{I}$, eq. (22) implies that

$$a_{0}^{0} = 1,\ a_{k}^{0} = 0,\ k = 1, 2, 3 \tag{23}$$

Furthermore,

$$\hat{S}^{-1}(\boldsymbol{\Lambda})\, \Phi'(\boldsymbol{\Lambda}x) = \hat{S}^{-1}(\boldsymbol{\Lambda})\, \boldsymbol{\Phi}'(x') = \Phi(x) = \hat{S}(\boldsymbol{\Lambda}^{-1})\, \Phi'(x')$$

$$= \hat{S}(\boldsymbol{\Lambda}^{-1})\, \Phi'(\boldsymbol{\Lambda}x) \Rightarrow \hat{S}^{-1}(\boldsymbol{\Lambda}) = \hat{S}(\boldsymbol{\Lambda}^{-1}) \tag{24}$$

Since a_{ν}^{μ} are all real and $\boldsymbol{\sigma}^{\mu\dagger} = \boldsymbol{\sigma}^{\mu}$, we also have

$$\hat{S}^{-1}(\boldsymbol{\Lambda})\boldsymbol{\sigma}^{\nu}\hat{S}(\boldsymbol{\Lambda}) = a_{\mu}^{\nu}\boldsymbol{\sigma}^{\mu} = (a_{\mu}^{\nu}\boldsymbol{\sigma}^{\mu})^{\dagger} = [\hat{S}^{-1}(\boldsymbol{\Lambda})\boldsymbol{\sigma}^{\nu}\hat{S}(\boldsymbol{\Lambda})]^{\dagger}$$

$$= \hat{S}^{\dagger}(\boldsymbol{\Lambda})\, (\boldsymbol{\sigma}^{\nu})^{\dagger}\, [\hat{S}^{-1}(\boldsymbol{\Lambda})]^{\dagger} = \hat{S}^{\dagger}(\boldsymbol{\Lambda})\boldsymbol{\sigma}^{\nu}\, [\hat{S}^{\dagger}(\boldsymbol{\Lambda})]^{-1} \tag{25}$$

so that

$$\hat{S}^{-1}(\boldsymbol{\Lambda})\boldsymbol{\sigma}^{\nu}\hat{S}(\boldsymbol{\Lambda}) = \hat{S}^{\dagger}(\boldsymbol{\Lambda})\boldsymbol{\sigma}^{\nu}\, [\hat{S}^{\dagger}(\boldsymbol{\Lambda})]^{-1}$$

or

$$\boldsymbol{\sigma}^{\nu} = \hat{S}(\boldsymbol{\Lambda})\hat{S}^{\dagger}(\boldsymbol{\Lambda})\boldsymbol{\sigma}^{\nu}\, [\hat{S}^{\dagger}(\boldsymbol{\Lambda})]^{-1}\, \hat{S}^{-1}(\boldsymbol{\Lambda})$$

$$= \hat{S}(\boldsymbol{\Lambda})\hat{S}^{\dagger}(\boldsymbol{\Lambda})\boldsymbol{\sigma}^{\nu}\, [\hat{S}(\boldsymbol{\Lambda})\hat{S}^{\dagger}(\boldsymbol{\Lambda})]^{-1} \tag{26}$$

Hence, the matrix $\hat{S}(\boldsymbol{\Lambda})\hat{S}^{\dagger}(\boldsymbol{\Lambda})$ commutes with all the Pauli matrices $\boldsymbol{\sigma}^{\mu}$ and is, therefore, a constant multiple of the identity, i.e., $\hat{S}(\boldsymbol{\Lambda})\hat{S}^{\dagger}(\boldsymbol{\Lambda}) = b\mathbf{I}$, whence

$$\hat{S}^{\dagger}(\boldsymbol{\Lambda}) = b\hat{S}^{-1}(\boldsymbol{\Lambda}) \tag{27}$$

Further

$$b^{*}\mathbf{I} = (b\mathbf{I})^{\dagger} = [\hat{S}(\boldsymbol{\Lambda})\hat{S}^{\dagger}(\boldsymbol{\Lambda})]^{\dagger} = \hat{S}(\boldsymbol{\Lambda})\hat{S}^{\dagger}(\boldsymbol{\Lambda})$$

$$= b\mathbf{I} \text{ so that } b = b^{*} \tag{28}$$

and hence, b is real.

If we now desire the normalization $\det \hat{S} = 1$, then

$$b^{2} = \det(b\mathbf{I}) = \det(\hat{S}\,\hat{S}^{\dagger}) = \det(\hat{S})\det(\hat{S}^{\dagger})$$

$$= (\det \hat{S})^{2} = 1 \text{ so that } b = \pm 1 \tag{29}$$

We also have, from eq. (27)

$$\hat{S}^{\dagger}\hat{S} = b\mathbf{I} \tag{30}$$

Now $\hat{S}^{\dagger}\hat{S}$ is positive definite, being the product of a matrix and its Hermitian conjugate. Its eigenvalues are also real and positive definite (because $\hat{S}^{\dagger}\hat{S}$ is Hermitian and positive definite). Thus,

$$Tr\hat{S}^{\dagger}\hat{S} > 0 \tag{31}$$

Also $Tr\,\mathbf{I} = 2$. We, therefore, conclude that $b = +1$.

We consider the case of the infinitesimal proper orthochronous Lorentz transformation given by $\mathbf{\Lambda} = \mathbf{I} + \varepsilon\lambda$ under which the coordinates transform as $(x')^{\mu} = x^{\mu} + \varepsilon\lambda^{\mu}_{\nu}x^{\nu}$ with

$$\lambda^{\mu\nu} = -\lambda^{\nu\mu} \tag{32}$$

Hence, the diagonal elements vanish. Furthermore, since the massless particles travel with the velocity of light, we do not have Lorentz boosts in such cases so that $\lambda^{0\mu} = -\lambda^{\mu 0} = 0$.

We can write

$$\hat{S}(\mathbf{\Lambda}) = \hat{S}(\mathbf{I} + \varepsilon\lambda) = \mathbf{I} + \varepsilon\hat{T} \tag{33}$$

to first order so that $\hat{S}^{-1}(\mathbf{\Lambda}) = \mathbf{I} - \varepsilon\hat{T}$.

Hence, in the infinitesimal case

$$\begin{aligned}\hat{S}^{-1}(\mathbf{\Lambda})\boldsymbol{\sigma}^{\nu}\hat{S}(\mathbf{\Lambda}) &= (\mathbf{I} - \varepsilon\hat{T})\boldsymbol{\sigma}^{\nu}(\mathbf{I} + \varepsilon\hat{T})\,\boldsymbol{\sigma}^{\nu} + \varepsilon(\sigma^{\nu}\hat{T} - \hat{T}\boldsymbol{\sigma}^{\nu})\\ &= a^{\nu}_{\mu}\boldsymbol{\sigma}^{\mu} = (\delta^{\nu}_{\mu} + \varepsilon\lambda^{\nu}_{\mu})\boldsymbol{\sigma}^{\mu} = \boldsymbol{\sigma}^{\nu} + \varepsilon\lambda^{\nu}_{\mu}\boldsymbol{\sigma}^{\mu}\end{aligned}$$

Hence, $\hat{T}$ must satisfy

$$(\boldsymbol{\sigma}^{\nu}\hat{T} - \hat{T}\sigma^{\nu}) = \lambda^{\nu}_{\mu}\sigma^{\mu} \tag{35}$$

Furthermore, since we demand

$$1 = \det\hat{S} = \det(\mathbf{I} + \varepsilon\hat{T}) \simeq 1 + \varepsilon Tr\,\hat{T} \Rightarrow Tr\,\hat{T} = 0 \tag{36}$$

Hence, by imposing this additional requirement, we can determine $\hat{T}$ uniquely.

Let us consider the traceless matrix

$$\hat{T} = \frac{i}{2}(\lambda^{1}_{2}\boldsymbol{\sigma}^{3} + \lambda^{2}_{3}\sigma^{1} + \lambda^{3}_{1}\boldsymbol{\sigma}^{2}) \tag{37}$$

as a candidate for $\hat{T}$. It can be shown that this matrix meets our requirements through explicit calculations, e.g.,

$$\begin{aligned}\boldsymbol{\sigma}^{1}\hat{T} - \hat{T}\boldsymbol{\sigma}^{1} &= \frac{i\boldsymbol{\sigma}^{1}}{2}(\lambda^{1}_{2}\boldsymbol{\sigma}^{3} + \lambda^{2}_{3}\sigma^{1} + \lambda^{3}_{1}\boldsymbol{\sigma}^{2}) - \frac{i}{2}(\lambda^{1}_{2}\boldsymbol{\sigma}^{3} + \lambda^{2}_{3}\boldsymbol{\sigma}^{1} + \lambda^{3}_{1}\sigma^{2})\boldsymbol{\sigma}^{1}\\ &= \frac{i}{2}[(\lambda^{1}_{2}\boldsymbol{\sigma}^{1}\boldsymbol{\sigma}^{3} + \lambda^{2}_{3}\boldsymbol{\sigma}^{1}\boldsymbol{\sigma}^{1} + \lambda^{3}_{1}\boldsymbol{\sigma}^{1}\boldsymbol{\sigma}^{2}) - (\lambda^{1}_{2}\boldsymbol{\sigma}^{3}\boldsymbol{\sigma}^{1} + \lambda^{2}_{3}\boldsymbol{\sigma}^{1}\boldsymbol{\sigma}^{1} + \lambda^{3}_{1}\boldsymbol{\sigma}^{2}\boldsymbol{\sigma}^{1})]\end{aligned}$$

$$= \frac{i}{2}[(-i\lambda_2^1\boldsymbol{\sigma}^2 + i\lambda_1^3\boldsymbol{\sigma}^3) - (i\lambda_2^1\boldsymbol{\sigma}^2 - i\lambda_1^3\boldsymbol{\sigma}^3)]$$

$$= \frac{i}{2}(-2i\lambda_2^1\boldsymbol{\sigma}^2 + 2i\lambda_1^3\boldsymbol{\sigma}^3)$$

$$= \frac{i}{2}(-2i\lambda_2^1\boldsymbol{\sigma}^2 - 2i\lambda_3^1\boldsymbol{\sigma}^3) = \lambda_2^1\boldsymbol{\sigma}^2 + \lambda_3^1\boldsymbol{\sigma}^3$$

$$= \lambda_0^1\boldsymbol{\sigma}^0 + \lambda_1^1\boldsymbol{\sigma}^1 + \lambda_2^1\boldsymbol{\sigma}^2 + \lambda_3^1\boldsymbol{\sigma}^3 = \lambda_v^1\boldsymbol{\sigma}^v \tag{38}$$

where the last step follows because

$$\lambda_0^1 = \lambda_1^1 = 0 \tag{39}$$

$$\boldsymbol{\sigma}^2\hat{T} - \hat{T}\boldsymbol{\sigma}^2 = \frac{i\sigma^2}{2}(\lambda_2^1\boldsymbol{\sigma}^3 + \lambda_3^2\boldsymbol{\sigma}^1 + \lambda_1^3\boldsymbol{\sigma}^2) - \frac{i}{2}(\lambda_2^1\boldsymbol{\sigma}^3 + \lambda_3^2\boldsymbol{\sigma}^1 + \lambda_1^3\boldsymbol{\sigma}^2)\boldsymbol{\sigma}^2$$

$$= \frac{i}{2}[(\lambda_2^1\boldsymbol{\sigma}^2\boldsymbol{\sigma}^3 + \lambda_3^2\boldsymbol{\sigma}^2\boldsymbol{\sigma}^1 + \lambda_1^3\boldsymbol{\sigma}^2\boldsymbol{\sigma}^2)$$
$$- (\lambda_2^1\sigma^3\boldsymbol{\sigma}^2 + \lambda_3^2\boldsymbol{\sigma}^1\boldsymbol{\sigma}^2 + \lambda_1^3\boldsymbol{\sigma}^2\boldsymbol{\sigma}^2)]$$

$$= \frac{i}{2}[(i\lambda_2^1\boldsymbol{\sigma}^1 - i\lambda_3^2\boldsymbol{\sigma}^3) - (-i\lambda_2^1\boldsymbol{\sigma}^1 + i\lambda_3^2\boldsymbol{\sigma}^3)] = \frac{i}{2}(2i\lambda_2^1\boldsymbol{\sigma}^1 - 2i\lambda_3^2\boldsymbol{\sigma}^3)$$

$$= \frac{i}{2}(-2i\lambda_1^2\boldsymbol{\sigma}^1 - 2i\lambda_3^2\boldsymbol{\sigma}^3) = \lambda_1^2\boldsymbol{\sigma}^1 + \lambda_3^2\boldsymbol{\sigma}^3$$

$$= \lambda_0^2\boldsymbol{\sigma}^0 + \lambda_1^2\boldsymbol{\sigma}^1 + \lambda_2^2\boldsymbol{\sigma}^2 + \lambda_3^2\boldsymbol{\sigma}^3 = \lambda_v^2\boldsymbol{\sigma}^v \tag{40}$$

where the last step follows because $\lambda_0^2 = \lambda_2^2 = 0$.

$$\boldsymbol{\sigma}^3\hat{T} - \hat{T}\boldsymbol{\sigma}^3 = \frac{i\sigma^3}{2}(\lambda_2^1\boldsymbol{\sigma}^3 + \lambda_3^2\boldsymbol{\sigma}^1 + \lambda_1^3\boldsymbol{\sigma}^2) - \frac{i}{2}(\lambda_2^1\boldsymbol{\sigma}^3 + \lambda_3^2 + \lambda_1^3\boldsymbol{\sigma}^2)\boldsymbol{\sigma}^3$$

$$= \frac{i}{2}[\lambda_2^1\boldsymbol{\sigma}^3\boldsymbol{\sigma}^3 + \lambda_3^2\boldsymbol{\sigma}^3\boldsymbol{\sigma}^1 + \lambda_1^3\boldsymbol{\sigma}^3\boldsymbol{\sigma}^2)$$
$$- (\lambda_2^1\boldsymbol{\sigma}^3\boldsymbol{\sigma}^3 - \lambda_3^2\boldsymbol{\sigma}^1\boldsymbol{\sigma}^3 + \lambda_1^3\boldsymbol{\sigma}^2\boldsymbol{\sigma}^3)]$$

$$= \frac{i}{2}[(i\lambda_3^2\boldsymbol{\sigma}^2 - i\lambda_1^3\boldsymbol{\sigma}^1) - (-i\lambda_3^2\boldsymbol{\sigma}^2 + i\lambda_1^3\boldsymbol{\sigma}^1)] = \frac{i}{2}(2i\lambda_3^2\boldsymbol{\sigma}^2 - 2i\lambda_1^3\boldsymbol{\sigma}^1)$$

$$= \frac{i}{2}(-2i\lambda_2^3\boldsymbol{\sigma}^2 - 2i\lambda_1^3\boldsymbol{\sigma}^1) = \lambda_1^3\boldsymbol{\sigma}^1 + \lambda_2^3\boldsymbol{\sigma}^2$$

$$= \lambda_0^3\boldsymbol{\sigma}^0 + \lambda_1^3\boldsymbol{\sigma}^1 + \lambda_2^3\boldsymbol{\sigma}^2 + \lambda_3^3\boldsymbol{\sigma}^3 = \lambda_v^3\boldsymbol{\sigma}^v \tag{41}$$

where the last step follows because $\lambda_0^3 = \lambda_3^3 = 0$.

$$\boldsymbol{\sigma}^0\hat{T} - \hat{T}\boldsymbol{\sigma}^0 = \hat{T} - \hat{T} = 0 = \lambda_0^0\boldsymbol{\sigma}^0 + \lambda_1^0\boldsymbol{\sigma}^1 + \lambda_2^0\boldsymbol{\sigma}^2 + \lambda_3^0\boldsymbol{\sigma}^3 = \lambda_v^0\boldsymbol{\sigma}^v \tag{42}$$

where the last step follows because $\lambda^0_\nu = 0$.

In the case of infinitesimal rotations about the x^1 axis, by an infinitesimal angle ε, we have $\lambda^2_3 = -1, \lambda^3_2 = +1$ with all the other λ^μ_ν vanishing so that $\hat{T} = -\frac{i}{2}\boldsymbol{\sigma}^1$. The corresponding operator for rotations about a finite angle θ about the x^1 axis is then

$$\hat{S}(R_1) = e^{\theta \hat{T}(R_1)} = e^{-\frac{i}{2}\boldsymbol{\sigma}_1 \theta} = \cos\frac{\theta}{2} - i\boldsymbol{\sigma}_1 \sin\frac{\theta}{2} \tag{43}$$

Similar expressions can be obtained for the rotation operators about the other axes.

9.3 SPACE INVERSION OF THE WEYL EQUATION

Under a space inversion $x^k \to (x')^k = -x^k, x^0 \to (x')^0 = x^0$. If $\hat{S}(i_s)$ represents the space inversion operator, then for form invariance of the Dirac equation, we must have

$$\hat{S}^{-1}(i_s)\boldsymbol{\sigma}^k \hat{S}(i_s) = -\boldsymbol{\sigma}^k, \; k = 1, 2, 3 \tag{44}$$

Equation (44) gives

$$\boldsymbol{\sigma}^k \hat{S}(i_s) = -\hat{S}(i_s)\boldsymbol{\sigma}^k, \; k = 1, 2, 3 \tag{45}$$

i.e., in order to provide for space inversion, the space inversion operator $\hat{S}(i_s)$ must anticommute with each of the Pauli matrices. However, it is well known that $\boldsymbol{\sigma}^1$, $\boldsymbol{\sigma}^2$, $\boldsymbol{\sigma}^3$ are the only 2×2 matrices that anticommute with each other.

Therefore, it is not possible to define a space inversion operator for massless particles. Hence, space parity is violated for massless Dirac particles as has been experimentally vindicated for neutrinos and anti-neutrinos.

Why it is Sufficient to Have Two Component Spinors for Describing Massless Spin $\frac{1}{2}$ Particles

We define a 2-component spinor as an object that transforms according to the $D^{\left(\frac{1}{2}, 0\right)}$ irreducible representation of the Lorentz group. Similarly, a conjugate spinor transforms according to the $D^{\left(0, \frac{1}{2}\right)}$ representation of this group.

Now, the infinitesimal generator of rotations about the x^i axis $\mathbf{M}_i = \frac{1}{2}\boldsymbol{\sigma}_i$, is same for the spinor as well as its conjugate. In other words, the process of pure rotations does not differentiate between $D^{\left(\frac{1}{2}, 0\right)}$ irreducible representation and $D^{\left(0, \frac{1}{2}\right)}$ representation of Lorentz group. However, for infinitesimal Lorentz boosts along the x^i axis the generators are $\mathbf{N}_i = \frac{1}{2} i\boldsymbol{\sigma}_i$ for objects transforming according to the $D^{\left(\frac{1}{2},0\right)}$ representation and $\mathbf{N}_i = -\frac{1}{2} i\boldsymbol{\sigma}_i$ for the $D^{\left(0,\frac{1}{2}\right)}$ representation. There is, therefore, a reversal of sign between the transformation, by way of a Lorentz boost, of the spinor and its conjugate.

Now a spatial reflection (space inversion) carries objects from $D^{\left(\frac{1}{2},0\right)}$ representation to the $D^{\left(0,\frac{1}{2}\right)}$ representation and vice versa, i.e., a space inversion is characterized by $\mathbf{M} \to \mathbf{M}$, $\mathbf{N} \to -\mathbf{N}$. Under a space inversion, therefore, a spinor transforms to its conjugate and vice versa.

Now, if a theory is to be invariant under space inversion, underlying representation space cannot be either only $D^{\left(\frac{1}{2},0\right)}$ representation or only $D^{\left(0,\frac{1}{2}\right)}$ representation because, in either case, a space inverted object would not have any representable state, e.g., the $D^{\left(\frac{1}{2},0\right)}$ spinor would get carried to $D^{\left(0,\frac{1}{2}\right)}$ representation by space inversion which, if the underlying representation is $D^{\left(\frac{1}{2},0\right)}$, would not exist in any representable state within the theory thereby violating space parity. Hence, in order to accommodate space parity it becomes necessary to consider the direct sum of $D^{\left(\frac{1}{2},0\right)}$ and $D^{\left(0,\frac{1}{2}\right)}$ as the underlying representation space because then the space inverted objects that would transform from $D^{\left(\frac{1}{2},0\right)}$ to $D^{\left(0,\frac{1}{2}\right)}$ representation would still find a representation within the theory and space parity would be conserved. In other words, if space parity is to be conserved, the representations $D^{\left(\frac{1}{2},0\right)}$ and $D^{\left(0,\frac{1}{2}\right)}$ no longer remain irreducible and it is only their direct sum that becomes irreducible.

Now, both the $D^{\left(\frac{1}{2},0\right)}$ and $D^{\left(0,\frac{1}{2}\right)}$ representations correspond to 2-component objects (spinors). Therefore, the direct sum $D^{\left(\frac{1}{2},0\right)} \oplus D^{\left(0,\frac{1}{2}\right)}$ representation would naturally correspond to a 4-component spinor, as has already been established in the case of massive Dirac equation.

It has been proved above and also experimentally established that in the case of massless Dirac particles (neutrinos) space parity is violated and we cannot explicitly define a space inversion operator. Hence, in this case the issue highlighted above necessitating the use of the direct sum of $D^{\left(\frac{1}{2},0\right)}$ and $D^{\left(0,\frac{1}{2}\right)}$ representations becomes redundant and we can use either of the $D^{\left(\frac{1}{2},0\right)}$ or

$D^{\left(0,\frac{1}{2}\right)}$ representations each of which corresponds to a 2-component spinor. Consequently, we need only a 2-component spinor for the massless Dirac theory in contrast to a 4-component spinor for the massive theory.

Thus, the spinor $\Phi(x)$ transforms as the $D^{\left(\frac{1}{2},0\right)}$ representation of the homogeneous Lorentz group while its conjugate spinor $\Psi(x)$ transforms as the $D^{\left(0,\frac{1}{2}\right)}$ representation and is a solution of the Weyl equation

$$i\partial_0\Psi(x) = i\boldsymbol{\sigma}.\nabla\Psi(x) \tag{46}$$

9.4 FREE PARTICLE SOLUTIONS OF THE WEYL EQUATION

The generalized Weyl equation for massless Dirac spinors and conjugate spinors can be written as $i\partial_0\Psi(x) = -i\varepsilon\boldsymbol{\sigma}.\nabla\Psi(x)$, where $\varepsilon = +1$ for Weyl spinors and $\varepsilon = -1$ for conjugate spinors. This equation admits the plane wave solution

$$\Phi(x) = \frac{1}{\sqrt{2(2\pi)^3 p_0}} u(\mathbf{p})e^{-ipx} = Nu(\mathbf{p})e^{-ipx} \tag{47}$$

with

$$p = \{p_0 . -\mathbf{p}\}, x = \{x^0, \mathbf{x}\}, px = p_0x^0 - \mathbf{px} \tag{48}$$

The above plane wave solutions are normalized such that the norms remain invariant under Lorentz transformations.

The $u(\mathbf{p})$ is a 2-component spinor/conjugate spinor that satisfies $p_0u(\mathbf{p}) = \varepsilon\boldsymbol{\sigma}.\mathbf{p}u(\mathbf{p})$ as is shown below.

$i\partial_0\Phi(x) = iN(-ip_0)u(\mathbf{p})e^{-ipx} = \varepsilon\boldsymbol{\sigma}\hat{\mathbf{p}}\Phi(x) = -\varepsilon\boldsymbol{\sigma}i\nabla\Phi(x) = -\varepsilon\boldsymbol{\sigma}Ni(i\mathbf{p})u(\mathbf{p})e^{-ipx}$ or $p_0u(\mathbf{p}) = \varepsilon\boldsymbol{\sigma}\mathbf{p}u(\mathbf{p})$ or equivalently $\varepsilon p_0u(\mathbf{p}) = \boldsymbol{\sigma}\mathbf{p}u(\mathbf{p})$ (49)

Writing $u(\mathbf{p}) = (u_1\ u_2)^T$, we obtain

$$\varepsilon p_0\begin{pmatrix} u_1 \\ u_2 \end{pmatrix} = \sum_{k=1}^{3}\sigma^k p_k\begin{pmatrix} u_1 \\ u_2 \end{pmatrix} = \begin{pmatrix} p_3 & p_1 - ip_2 \\ p_1 + ip_2 & -p_3 \end{pmatrix}\begin{pmatrix} u_1 \\ u_2 \end{pmatrix}$$

$$= \begin{pmatrix} p_3u_1 + (p_1 - ip_2)u_2 \\ (p_1 + ip_2)u_1 - p_3u_2 \end{pmatrix} \tag{50}$$

Solving this set of equations, we have

$$u_2 = \frac{\varepsilon p_0 - p_3}{p_1 - ip_2}u_1 \tag{51}$$

and

$$(-p_0^2 + p_1^2 + p_2^2 + p_3^2)u_1 = 0 \tag{52}$$

Equation (52) is trivially true, being the momentum-energy relation for massless particles. Hence, we can take $u_1 = 1$ so that $u_2 = \dfrac{\varepsilon p_0 - p_3}{p_1 - ip_2}$.

On multiplying both sides of eq. (50) by the helicity operator $\hat{s}(\mathbf{p}) = \frac{1}{|\mathbf{p}|}\boldsymbol{\sigma}.\hat{\mathbf{p}}$, we obtain

$$\frac{1}{|\mathbf{p}|}\boldsymbol{\sigma}.\hat{\mathbf{p}}\varepsilon p_0\begin{pmatrix} u_1 \\ u_2 \end{pmatrix} = \varepsilon p_0\frac{1}{|\mathbf{p}|}\boldsymbol{\sigma}.\mathbf{p}\begin{pmatrix} u_1 \\ u_2 \end{pmatrix}\ \varepsilon p_0\frac{1}{|\mathbf{p}|}\boldsymbol{\sigma}.\mathbf{p}\begin{pmatrix} u_1 \\ u_2 \end{pmatrix}$$

$$= \varepsilon p_0\frac{1}{|\mathbf{p}|}\begin{pmatrix} p_3 & p_1 - ip_2 \\ p_1 + ip_2 & -p_3 \end{pmatrix}\begin{pmatrix} u_1 \\ u_2 \end{pmatrix}$$

$$= \frac{\varepsilon p_0}{|\mathbf{p}|}\begin{pmatrix} p_3 u_1 + (p_1 - ip_2)u_2 \\ (p_1 + ip_2)u_1 - p_3 u_2 \end{pmatrix} = \frac{\boldsymbol{\sigma}\hat{\mathbf{p}}\boldsymbol{\sigma}\mathbf{p}}{|\mathbf{p}|}u(\mathbf{p})$$

$$= \frac{\boldsymbol{\sigma}\hat{\mathbf{p}}\boldsymbol{\sigma}\mathbf{p}}{|\mathbf{p}|}u(\mathbf{p}) = \frac{(\boldsymbol{\sigma}\mathbf{p})^2}{|\mathbf{p}|}u(\mathbf{p}) = \frac{\mathbf{p}^2}{|\mathbf{p}|}\begin{pmatrix} u_1 \\ u_2 \end{pmatrix} \tag{53}$$

From eqs. (51) and (53), we get

$$\frac{\varepsilon p_0}{|\mathbf{p}|}\begin{pmatrix} p_3 u_1 + (p_1 - ip_2)\dfrac{\varepsilon p_0 - p_3}{p_1 - ip_2}u_1 \\ (p_1 + ip_2)u_1 - p_3\dfrac{\varepsilon p_0 - p_3}{p_1 - ip_2}u_1 \end{pmatrix}$$

$$= \frac{p_0^2}{|\mathbf{p}|}\begin{pmatrix} u_1 \\ \dfrac{\varepsilon p_0 - p_3}{p_1 - ip_2}u_1 \end{pmatrix} = \frac{\mathbf{p}^2}{|\mathbf{p}|}\begin{pmatrix} u_1 \\ \dfrac{\varepsilon p_0 - p_3}{p_1 - ip_2}u_1 \end{pmatrix} \tag{54}$$

showing that non-vanishing solution for $u(\mathbf{p})$ exist only when $p_0 = \pm\,\mathbf{p}$ which, incidentally, is the energy-momentum relation for relativistic dynamics of massless particles. This also confirms that massless particles travel with the speed of light.

Multiplying the wave function $u(\mathbf{p})$ by the helicity operator $\hat{s}(\mathbf{p}) = \frac{1}{|\mathbf{p}|}\boldsymbol{\sigma}.\hat{\mathbf{p}}$ to investigate the spin states, we obtain

$$\hat{s}(\mathbf{p})u(\mathbf{p}) = \frac{\boldsymbol{\sigma}.\hat{\mathbf{p}}}{|\mathbf{p}|}\begin{pmatrix} u_1 \\ u_2 \end{pmatrix} = \frac{\boldsymbol{\sigma}.\hat{\mathbf{p}}}{|\mathbf{p}|}\begin{pmatrix} u_1 \\ u_2 \end{pmatrix} = \frac{1}{|\mathbf{p}|}\begin{pmatrix} p_3 & p_1 - ip_2 \\ p_1 + ip_2 & -p_3 \end{pmatrix}\begin{pmatrix} u_1 \\ u_2 \end{pmatrix}$$

$$= \frac{1}{|\mathbf{p}|}\begin{pmatrix} p_3 u_1 + (p_1 - ip_2)u_2 \\ (p_1 + ip_2)u_1 - p_3 u_2 \end{pmatrix}$$

which, on using eq. (51), reduces to

$$\frac{1}{|\mathbf{p}|}\begin{pmatrix} p_3u_1 + (p_1 - ip_2)\dfrac{\varepsilon p_0 - p_3}{p_1 - ip_2}u_1 \\ (p_1 + ip_2)u_1 - p_3\dfrac{\varepsilon p_0 - p_3}{p_1 - ip_2}u_1 \end{pmatrix}$$

$$= \frac{\varepsilon p_0}{|\mathbf{p}|}\begin{pmatrix} u_1 \\ \dfrac{\varepsilon p_0 - p_3}{p_1 - ip_2}u_1 \end{pmatrix} = \pm\varepsilon\frac{\mathbf{p}}{|\mathbf{p}|}u\,(\mathbf{p}) \tag{55}$$

Now, if we take the direction of motion as positive so that $\mathbf{p} = |\mathbf{p}|$, then $\hat{s}\,(\mathbf{p})u(\mathbf{p}) = \pm\,\varepsilon u(\mathbf{p})$ with (in the case of $D^{\left(\frac{1}{2},0\right)}$ spinors) the positive or negative sign being determined by whether $p_0 = +\,\mathbf{p} > 0$ (i.e., positive energy solutions) or $p_0 = -\,\mathbf{p} < 0$ (i.e., negative energy solutions), respectively.

In the case of conjugate spinors ($D^{\left(0,\frac{1}{2}\right)}$ spinors) the case is converse because $\varepsilon = -\,1$ so that the positive or negative sign is determined by whether $p_0 = -\,\mathbf{p} < 0$ (i.e., negative energy solutions) or $p_0 = +\,\mathbf{p} > 0$ (i.e., positive energy solutions), respectively. In other words, we conclude that

(a) the fact that eq. (15) contains the spin operator (the Pauli matrices $\boldsymbol{\sigma}$) mandates that for a given sign of the energy there corresponds a specific orientation of the spin with respect to the linear momentum.

(b) for positive energy solutions of the Weyl equation for the spinor, if we take the direction of motion as positive ($\mathbf{p} > 0$), then the helicity operator has a positive eigenvalue and hence, the spin is directed parallel to the momentum which implies a right-handed screw model; and

(c) for positive energy solutions of the Weyl equation for the **conjugate** spinor, if we take the direction of motion as positive ($\mathbf{p} > 0$) then the helicity operator has a negative eigenvalue and hence, the spin is directed anti-parallel to the momentum which implies a left-handed screw model (the neutrino state).

Left-hand screw

For $D^{\left(\frac{1}{2},0\right)}$ spinor $p_0 = -\,|\mathbf{p}|$

For $D^{\left(0,\frac{1}{2}\right)}$ conjugate spinor $p_0 = |\mathbf{p}|$

(Neutrino state)

Right-hand screw

For $D^{\left(\frac{1}{2},0\right)}$ spinor $p_0 = |\mathbf{p}|$

For $D^{\left(0,\frac{1}{2}\right)}$ conjugate spinor $p_0 = -|\mathbf{p}|$

(Anti-neutrino state)

9.5 THE NEUTRINO CURRENT

The equation for the neutrino wave function, i.e., the conjugate spinor eq. (46), can also be written as

$$i\partial_0 \Psi(x) - i\boldsymbol{\sigma}.\nabla\Psi(x) = 0 \tag{56}$$

Introducing $\boldsymbol{\sigma}^0 = \boldsymbol{\sigma}_0 = \mathbf{I}$, $\boldsymbol{\sigma}^\mu = \{\mathbf{I}, \boldsymbol{\sigma}\}$ and noting that $\partial_0 = \partial^0$, $\partial_k = -\partial^k = +\nabla$, we can write eq. (56) as

$$\boldsymbol{\sigma}_\mu \partial^\mu \Psi(x) = 0 \tag{57}$$

The Hermitian conjugate of eq. (57) is

$$\partial^\mu \Psi^\dagger(x)(\boldsymbol{\sigma}_\mu)^\dagger = \partial^\mu \Psi^\dagger(x)\boldsymbol{\sigma}_\mu = 0 \tag{58}$$

Multiplying eq. (57) by $\Psi^\dagger$ from the left and eq. (58) by Ψ from the right, and adding, we obtain

$$\Psi^\dagger(x)\sigma_\mu\partial^\mu\Psi(x) + \partial^\mu\Psi^\dagger(x)\sigma_\mu\Psi(x) = 0 = \partial^\mu\,[\Psi^\dagger(x)\sigma_\mu\Psi(x)] \tag{59}$$

Equation (59) is in the nature of a continuity equation of the form $\partial^\mu j_\mu = 0$ so that the 4-current j_μ may be identified as

$$g_{\mu\nu} j^\nu = j_\mu = \Psi^\dagger(x)\sigma_\mu \Psi(x) \tag{60}$$

leading to the identification of charge and current densities as

$$j_0 = j^0 = \rho = \Psi^\dagger(x)\Psi(x) \tag{61}$$

$$j_k = -j^k = \Psi^\dagger(x)\,\sigma_k\Psi(x) \tag{62}$$

The normalization of Ψ may be obtained as $\int \rho d\mathbf{x} = 1$.

9.6 DIRAC EQUATION FOR NEUTRINOS IN THE STANDARD REPRESENTATION

The Dirac equation for massless particles is given by

$$i\partial_0 \Psi(x) = -i\alpha\cdot\nabla\Psi(x) = \hat{H}\Psi(x) \tag{63}$$

The plane wave solutions of eq. (63) may be obtained by writing

$$\Psi(x) = u(\mathbf{p})e^{-ipx} = u(\mathbf{p})e^{-i(p_0 x^0 - \mathbf{px})} \tag{64}$$

where $u(\mathbf{p})$ satisfies

$$Eu(\mathbf{p}) = \boldsymbol{\alpha}\hat{\mathbf{p}}\,u(\mathbf{p}) = \boldsymbol{\alpha}\mathbf{p}u(\mathbf{p}) \tag{65}$$

with

$$E = p_0 = \varepsilon E_p = \pm|\mathbf{p}|.$$

Writing $\boldsymbol{\gamma}^5 = \boldsymbol{\gamma}_5 = i\boldsymbol{\gamma}^0\boldsymbol{\gamma}^1\boldsymbol{\gamma}^2\boldsymbol{\gamma}^3 = \begin{pmatrix} \mathbf{0} & \mathbf{I} \\ \mathbf{I} & \mathbf{0} \end{pmatrix}$ in the standard representation of the Dirac matrices, we have

$$\boldsymbol{\gamma}^5\Sigma = \begin{pmatrix} \mathbf{0} & \mathbf{I} \\ \mathbf{I} & \mathbf{0} \end{pmatrix}\begin{pmatrix} \sigma & \mathbf{0} \\ \mathbf{0} & \sigma \end{pmatrix} = \begin{pmatrix} 0 & \sigma \\ \sigma & \mathbf{0} \end{pmatrix} = \boldsymbol{\alpha}$$

$$= \begin{pmatrix} \boldsymbol{\sigma} & \mathbf{0} \\ \mathbf{0} & \boldsymbol{\sigma} \end{pmatrix} \begin{pmatrix} \mathbf{0} & \mathbf{I} \\ \mathbf{I} & \mathbf{0} \end{pmatrix} = \Sigma\boldsymbol{\gamma}^5 \tag{66}$$

Hence,

$$\hat{H} = \boldsymbol{\alpha}\hat{p} = \boldsymbol{\gamma}^5\Sigma\hat{p} = \Sigma\hat{p}\boldsymbol{\gamma}^5 \tag{67}$$

so that the Hamiltonian $\hat{H}$ commutes with both the spin operator $\hat{\Sigma}$ as well as $\boldsymbol{\gamma}^5$. Hence, we can have simultaneous eigenfunctions for the Hamiltonian $\hat{H}$, the spin operator $\hat{\Sigma}$ as well as $\boldsymbol{\gamma}^5$. Now, writing $u(\mathbf{p}) = (u_a\ u_b)^T$, where u_a, u_b are both 2-component spinors, i.e., $u_a = (u_{a1}\ u_{a2})^T$, $u_b = (u_{b1}\ u_{b2})^T$, we have, from eq. (65)

$$E\begin{pmatrix} u_a \\ u_b \end{pmatrix} = \begin{pmatrix} 0 & \boldsymbol{\sigma} \\ \boldsymbol{\sigma} & \mathbf{0} \end{pmatrix} \mathbf{p} \begin{pmatrix} u_a \\ u_b \end{pmatrix} = \begin{pmatrix} \boldsymbol{\sigma}\mathbf{p}u_b \\ \boldsymbol{\sigma}\mathbf{p}u_a \end{pmatrix} \tag{68}$$

which gives

$$Eu_a = \boldsymbol{\sigma}\mathbf{p}u_b = \boldsymbol{\sigma}\mathbf{p}\frac{\boldsymbol{\sigma}\mathbf{p}}{E}u_a = \frac{\mathbf{p}^2}{E}u_a \text{ or } (E^2 - \mathbf{p}^2)u_a = 0 \tag{69}$$

This is the usual momentum-energy relationship for massless particles. Here again, we have positive and negative energy solutions according as $p_0 = E = +\,\mathbf{p}$ and, $p_0 = E = -\,\mathbf{p}$, respectively. The eigenvalues of the Hamiltonian operator $\hat{H} = \boldsymbol{\alpha}\hat{\mathbf{p}} = \begin{pmatrix} 0 & \boldsymbol{\sigma} \\ \boldsymbol{\sigma} & \mathbf{0} \end{pmatrix}\mathbf{p}$ are easily found to be $\pm\,\mathbf{p}$ with the corresponding eigenfunctions being

$$\begin{pmatrix} u_a \\ u_b \end{pmatrix} = \frac{\boldsymbol{\sigma}\mathbf{p}}{|\mathbf{p}|}\begin{pmatrix} u_b \\ u_a \end{pmatrix} \text{ since } \hat{H}\begin{pmatrix} u_a \\ u_b \end{pmatrix} = \begin{pmatrix} \mathbf{0} & \boldsymbol{\sigma}\mathbf{p} \\ \boldsymbol{\sigma}\mathbf{p} & \mathbf{0} \end{pmatrix}\frac{\boldsymbol{\sigma}\mathbf{p}}{|\mathbf{p}|}\begin{pmatrix} u_b \\ u_a \end{pmatrix}$$

$$= \pm\,\mathbf{p}\begin{pmatrix} u_a \\ u_b \end{pmatrix} = \pm\,\mathbf{p}\frac{\boldsymbol{\sigma}\mathbf{p}}{|\mathbf{p}|}\begin{pmatrix} u_b \\ u_a \end{pmatrix} \tag{70}$$

Let us now examine the spin states of the massless Dirac particles. Since

$$\hat{s}^2(\mathbf{p}) = \frac{1}{|\mathbf{p}|^2}\begin{pmatrix} \hat{\sigma}\mathbf{p} & \mathbf{0} \\ \mathbf{0} & \hat{\sigma}\mathbf{p} \end{pmatrix}\begin{pmatrix} \hat{\sigma}\mathbf{p} & \mathbf{0} \\ \mathbf{0} & \hat{\sigma}\mathbf{p} \end{pmatrix} = \frac{1}{|\mathbf{p}|^2}\begin{pmatrix} (\hat{\sigma}\mathbf{p})^2 & \mathbf{0} \\ \mathbf{0} & (\hat{\sigma}\mathbf{p})^2 \end{pmatrix}$$

$$= \frac{1}{|\mathbf{p}|^2}\begin{pmatrix} \mathbf{p}^2 & \mathbf{0} \\ \mathbf{0} & \mathbf{p}^2 \end{pmatrix} = \begin{pmatrix} \mathbf{I} & \mathbf{0} \\ \mathbf{0} & \mathbf{I} \end{pmatrix} \tag{71}$$

the eigenvalues of $\hat{s}(\mathbf{p})$ are $\pm$ 1.

The corresponding eigenfunctions are calculated by solving the equation

$$\pm\begin{pmatrix} u_a \\ u_b \end{pmatrix} = \hat{s}(\mathbf{p})\begin{pmatrix} u_a \\ u_b \end{pmatrix} = \frac{1}{|\mathbf{p}|}\begin{pmatrix} \hat{\sigma}\mathbf{p} & \mathbf{0} \\ \mathbf{0} & \hat{\sigma}\mathbf{p} \end{pmatrix}$$

$$\begin{pmatrix} u_a \\ u_b \end{pmatrix} = \frac{1}{|\mathbf{p}|}\begin{pmatrix} \hat{\sigma}\mathbf{p}u_a \\ \hat{\sigma}\mathbf{p}u_b \end{pmatrix} \tag{72}$$

Using eq. (69), which must also be satisfied by $(u_a\ u_b)^T$ since $(u_a\ u_b)^T$ is also a solution of the Dirac equation (65), we have

$$\pm\begin{pmatrix} u_a \\ u_b \end{pmatrix} = \hat{s}(\mathbf{p})\begin{pmatrix} u_a \\ u_b \end{pmatrix} = \frac{1}{|\mathbf{p}|}\begin{pmatrix} \hat{\sigma}\mathbf{p} & \mathbf{0} \\ \mathbf{0} & \hat{\sigma}\mathbf{p} \end{pmatrix}\begin{pmatrix} u_a \\ u_b \end{pmatrix}$$

$$= \frac{1}{|\mathbf{p}|}\begin{pmatrix} \hat{\sigma}\mathbf{p}.\dfrac{\hat{\sigma}\mathbf{p}}{E}u_b \\ \hat{\sigma}\mathbf{p}\dfrac{\hat{\sigma}\mathbf{p}}{E}u_a \end{pmatrix} = \frac{|\mathbf{p}|}{E}\begin{pmatrix} u_b \\ u_a \end{pmatrix} \tag{73}$$

Now, for massless particles, the momentum-energy relation implies $E = \pm\ \mathbf{p}$. Hence, from eq. (73) we conclude that $\pm\begin{pmatrix} u_a \\ u_b \end{pmatrix} = \begin{pmatrix} u_b \\ u_a \end{pmatrix}$ for positive energy solutions and $\pm\begin{pmatrix} u_a \\ u_b \end{pmatrix} = -\begin{pmatrix} u_b \\ u_a \end{pmatrix}$ for negative energy solutions. In other words, we have the following classification:

Energy ↓ Helicity →	+ 1	− 1
Positive	$\begin{pmatrix} u_a \\ u_b \end{pmatrix} = \begin{pmatrix} u_b \\ u_a \end{pmatrix} = \begin{pmatrix} u_a \\ u_a \end{pmatrix} = \begin{pmatrix} u_b \\ u_b \end{pmatrix}$	$\begin{pmatrix} u_a \\ u_b \end{pmatrix} = -\begin{pmatrix} u_b \\ u_a \end{pmatrix} = \begin{pmatrix} u_a \\ -u_a \end{pmatrix} = \begin{pmatrix} -u_b \\ u_b \end{pmatrix}$
Negative	$\begin{pmatrix} u_a \\ u_b \end{pmatrix} = -\begin{pmatrix} u_b \\ u_a \end{pmatrix} = \begin{pmatrix} u_a \\ -u_a \end{pmatrix} = \begin{pmatrix} -u_b \\ u_b \end{pmatrix}$	$\begin{pmatrix} u_a \\ u_b \end{pmatrix} = \begin{pmatrix} u_b \\ u_a \end{pmatrix} = \begin{pmatrix} u_a \\ u_a \end{pmatrix} = \begin{pmatrix} u_b \\ u_b \end{pmatrix}$

It is pertinent to mention here that the above eigenfunctions can be explicitly verified to be eigenfunctions of γ^5 as well.

To study the components of u_a, u_b we note that the first of eq. (72) gives

$$\frac{1}{|\mathbf{p}|}\begin{pmatrix} p_3 & p_1 - ip_2 \\ p_1 + ip_2 & -p_3 \end{pmatrix}\begin{pmatrix} u_{a1} \\ u_{a2} \end{pmatrix} = \pm\begin{pmatrix} u_{a1} \\ u_{a2} \end{pmatrix} \tag{74}$$

the second of which gives

$$u_{a2} = \frac{p_1 + ip_2}{p_3 \pm |\mathbf{p}|}u_{a1} = \frac{(p_1 + ip_2).(p_1 - ip_2)}{(p_3 \pm |\mathbf{p}|).(p_1 - ip_2)}u_{a1} = \frac{p_1^2 + p_2^2}{(p_3 \pm |\mathbf{p}|).(p_1 - ip_2)}u_{a1}$$

$$= \frac{|\mathbf{p}|^2 - p_3^2}{(p_3 \pm |\mathbf{p}|)(p_1 - ip_2)} u_{a1} = \frac{|\mathbf{p}| \mp p_3}{(\pm p_1 \mp ip_2)} u_{a1} \tag{75}$$

Eliminating u_{a2} in eq. (74), we get

$$\frac{1}{|\mathbf{p}|}\left(\frac{p_1^2 + p_2^2 + p_3^2 - |\mathbf{p}|^2}{p_3 \pm |\mathbf{p}|}\right) u_{a1} = 0$$

which is identically satisfied for all the values of u_{a1}. We can choose the solution corresponding to helicity + 1 conveniently as $A\begin{pmatrix} p_3 + |\mathbf{p}| \\ p_1 + ip_2 \end{pmatrix}$ and for a particle of helicity – 1 moving along **p**, we can choose the 2-component vector as $A\begin{pmatrix} -p_1 + ip_2 \\ p_3 + |\mathbf{p}| \end{pmatrix}$, where A is a normalization constant to be determined from $u_a^\dagger u_a = 1$. We have

$$\begin{aligned}
u_a^\dagger u_a &= A^2 \begin{pmatrix} p_3 + |\mathbf{p}| \\ p_1 + ip_2 \end{pmatrix}^\dagger \begin{pmatrix} p_3 + |\mathbf{p}| \\ p_1 + ip_2 \end{pmatrix} \\
&= A^2 \left((p_3 + |\mathbf{p}|)^* (p_1 + ip_2)^*\right) \begin{pmatrix} p_3 + |\mathbf{p}| \\ p_1 + ip_2 \end{pmatrix} \\
&= A^2 \left((p_3 + |\mathbf{p})|(p_1 - ip_2)\right) \begin{pmatrix} p_3 + |\mathbf{p}| \\ p_1 + ip_2 \end{pmatrix} \\
&= A^2 \left[(p_3 + |\mathbf{p}|)^2 + (p_1 - ip_2)(p_1 + ip_2)\right] \\
&= A^2 |\mathbf{p}|(2|\mathbf{p}| + 2p_3)
\end{aligned} \tag{76}$$

thereby giving

$$A = \frac{1}{[2|\mathbf{p}|(|\mathbf{p}| + p_3)]^{\frac{1}{2}}} \tag{77}$$

Exactly same results hold for the 2-component spinor u_b so that we have the following solution for the eigenfunctions of the "helicity operator":

Helicity ↓	
+ 1	$A(p_3 + |\mathbf{p}|, p_1 + ip_2, p_3 + |\mathbf{p}|, p_1 + ip_2)^T$
– 1	$A(-p_1 + ip_2, p_3 + |\mathbf{p}|, -p_1 + ip_2, p_3 + |\mathbf{p}|)^T$

To obtain simultaneous eigenfunctions of $\hat{H}, \hat{s}\,(\mathbf{p})$, the above must also satisfy the Dirac equation which leads to eqs. (69) and (73). Therefore, the simultaneous eigenfunctions of $\hat{H}, \hat{s}\,(\mathbf{p})$ are given by the following:

Energy → Helicity ↓	+ 1	− 1
Positive	$A\begin{pmatrix} p_3+\|\mathbf{p}\| \\ p_1+ip_2 \\ p_3+\|\mathbf{p}\| \\ p_1+ip_2 \end{pmatrix}$	$A\begin{pmatrix} -p_1+ip_2 \\ p_3+\|\mathbf{p}\| \\ p_1-ip_2 \\ -p_3-\|\mathbf{p}\| \end{pmatrix}$
Negative	$A\begin{pmatrix} p_3+\|\mathbf{p}\| \\ p_1+ip_2 \\ -p_3-\|\mathbf{p}\| \\ -p_1-ip_2 \end{pmatrix}$	$A\begin{pmatrix} -p_1+ip_2 \\ p_3+\|\mathbf{p}\| \\ -p_1+ip_2 \\ p_3+\|\mathbf{p}\| \end{pmatrix}$

In the orientation, where we choose the direction of **p** along the x^3 axis, we have $p_1 = p_2 = 0$, $p_3 = \mathbf{p}$ so that the above eigenfunctions get modified to

Energy → Helicity ↓	+ 1	− 1
Positive	$\begin{pmatrix} 1 \\ 0 \\ 1 \\ 0 \end{pmatrix}$	$\begin{pmatrix} 0 \\ 1 \\ 0 \\ -1 \end{pmatrix}$
Negative	$\begin{pmatrix} 1 \\ 0 \\ -1 \\ 0 \end{pmatrix}$	$\begin{pmatrix} 0 \\ 1 \\ 0 \\ 1 \end{pmatrix}$

We can also obtain the dynamics of the massless Dirac particles, more succinctly in the following representation of the Dirac matrices:

$$\boldsymbol{\alpha}_i = \begin{pmatrix} \sigma_i & \mathbf{0} \\ \mathbf{0} & -\sigma_i \end{pmatrix},\ \beta = \begin{pmatrix} \mathbf{0} & -\mathbf{I} \\ -\mathbf{I} & \mathbf{0} \end{pmatrix} \tag{78}$$

In this representation, the Dirac equation for massless particles will be

$$i\partial_0 \Psi = -i\boldsymbol{\alpha}\nabla\Psi \tag{79}$$

Writing $\Psi = (\psi^+, \psi^-)^T$, we have

$$i\partial_0 \begin{pmatrix} \psi^+ \\ \psi^- \end{pmatrix} = -i\begin{pmatrix} \sigma & \mathbf{0} \\ \mathbf{0} & -\sigma \end{pmatrix} \nabla \begin{pmatrix} \psi^+ \\ \psi^- \end{pmatrix} \tag{80}$$

or

$$i\partial_0\psi^+ = -\,i\sigma\nabla\psi^+ \tag{81}$$

$$i\partial_0\psi^- = +\,i\sigma\nabla\psi^- \tag{82}$$

ψ^+ and ψ^- describe right-handed and left-handed particles, respectively.

We may also define the following projection operators in this representation

$$\hat{P}_\pm = \hat{P}_{\text{sgn}\,\varepsilon} = \frac{1}{2}\,(\mathbf{I} + \varepsilon\gamma_5) = \frac{1}{2}\,(\mathbf{I} \pm \gamma_5) \tag{83}$$

with
$$\gamma_5 = \gamma^5 i\gamma^0\gamma^1\gamma^2\gamma^3 = \begin{pmatrix} \mathbf{I} & \mathbf{0} \\ \mathbf{0} & -\mathbf{I} \end{pmatrix}.$$

We have

$$\begin{aligned}
\hat{P}_{\text{sgn}\,\varepsilon}\,\hat{P}_{\text{sgn}\,\varepsilon'} &= \frac{1}{4}\,(\mathbf{I} + \varepsilon\gamma_5)(\mathbf{I} + \varepsilon\gamma_5) \\
&= \frac{1}{4}\,[\mathbf{I} + (\varepsilon + \varepsilon')\gamma_5 + \varepsilon\varepsilon'(\gamma_5)^2] \\
&= \delta_{\varepsilon\varepsilon'}\,\frac{1}{2}(\mathbf{I} + \varepsilon\gamma_5) = \delta_{\varepsilon\varepsilon'}\,\hat{P}_{\text{sgn}\,\varepsilon}
\end{aligned} \tag{84}$$

We also have

$$\begin{aligned}
\Psi^{\text{sgn}\,\varepsilon} &= (\delta_{\varepsilon,+1}\psi^+ \;\; \delta_{\varepsilon,-1}\psi^-)^T \\
&= \hat{P}_{\text{sgn}\,\varepsilon}\Psi = \hat{P}_{\text{sgn}\,\varepsilon}(\psi^+ \;\; \psi^-)^T
\end{aligned}$$

and
$$\hat{P}_{\text{sgn}\,\varepsilon}\,\Psi^{\text{sgn}\,\varepsilon'} = \delta_{\varepsilon,\varepsilon'}\psi^{\text{sgn}\,\varepsilon'} \tag{85}$$

We also have

$$\begin{aligned}
\boldsymbol{\alpha}\hat{P}_{\text{sgn}\,\varepsilon} &= \begin{pmatrix} \sigma & \mathbf{0} \\ \mathbf{0} & -\sigma \end{pmatrix}\left\{\frac{1}{2}\left[\mathbf{I} + \varepsilon\begin{pmatrix} \mathbf{I} & \mathbf{0} \\ \mathbf{0} & -\mathbf{I} \end{pmatrix}\right]\right\} \\
&= \frac{1}{2}\begin{pmatrix} \sigma & \mathbf{0} \\ \mathbf{0} & -\sigma \end{pmatrix}\begin{pmatrix} \mathbf{I}+\varepsilon\mathbf{I} & \mathbf{0} \\ \mathbf{0} & \mathbf{I}-\varepsilon\mathbf{I} \end{pmatrix} \\
&= \frac{1}{2}\begin{pmatrix} \sigma(\mathbf{I}+\varepsilon\mathbf{I}) & \mathbf{0} \\ \mathbf{0} & -\sigma(\mathbf{I}-\varepsilon\mathbf{I}) \end{pmatrix} \\
&= \frac{1}{2}\begin{pmatrix} (\mathbf{I}+\varepsilon\mathbf{I})\sigma & \mathbf{0} \\ \mathbf{0} & -(\mathbf{I}-\varepsilon\mathbf{I})\sigma \end{pmatrix} = \hat{P}_{\text{sgn}\,\varepsilon}\boldsymbol{\alpha}
\end{aligned} \tag{86}$$

Therefore,

$$i\partial_0\,\Psi^+ = i\partial_0(\psi^+ \;\; \mathbf{0})^T = \hat{P}_+(i\partial_0\Psi)$$

$$= i\partial_0 \hat{P}_+ \Psi = \hat{P}_+(-i\boldsymbol{\alpha}\nabla\Psi)$$

$$= -i(\hat{P}_+ \boldsymbol{\alpha}\nabla\Psi) = -i\boldsymbol{\alpha}\nabla(\hat{P}_+\Psi)$$

$$= -i\boldsymbol{\alpha}\nabla(\psi^+ \ \mathbf{0})^T = -i\boldsymbol{\alpha}\nabla\Psi^+$$

and similarly,

$$i\partial_0\Psi^- = -i(\hat{P}_-\alpha\nabla\Psi) = -\ i\alpha\nabla(\hat{P}_-\Psi) = -i\alpha\nabla\Psi^- \tag{87}$$

which gives

$$i\partial_0\Psi^{\text{sgn}\,\varepsilon} = i\partial_0(\delta_{\varepsilon\varepsilon_1}\psi^{\text{sgn}\,\varepsilon_1}\,\delta_{\varepsilon\varepsilon_2}\psi^{\text{sgn}\,\varepsilon_2})^T$$

$$= -i\begin{pmatrix} \sigma & 0 \\ 0 & -\sigma \end{pmatrix}\nabla(\delta_{\varepsilon\varepsilon_1}\psi^{\text{sgn}\,\varepsilon_1}\delta_{\varepsilon\varepsilon_2}\psi^{\text{sgn}\,\varepsilon_2})^T$$

$$= -i\sigma\nabla(\delta_{\varepsilon\varepsilon_1}\psi^{\text{sgn}\,\varepsilon_1} - \delta_{\varepsilon\varepsilon_2}\psi^{\text{sgn}\,\varepsilon_2})^T \tag{88}$$

We also have

$$\gamma^5\Psi^{\text{sgn}\,\varepsilon} = \gamma^5(\delta_{\varepsilon\varepsilon_1}\psi^{\text{sgn}\,\varepsilon_1}\,\delta_{\varepsilon\varepsilon_2}\psi^{\text{sgn}\,\varepsilon_2})^T$$

$$= \begin{pmatrix} \mathbf{I} & \mathbf{0} \\ \mathbf{0} & -\mathbf{I} \end{pmatrix}(\delta_{\varepsilon\varepsilon_1}\psi^{\text{sgn}\,\varepsilon_1}\,\delta_{\varepsilon\varepsilon_2}\psi^{\text{sgn}\,\varepsilon_2})^T$$

$$= (\delta_{\varepsilon\varepsilon_1}\psi^{\text{sgn}\,\varepsilon_1}, -\,\delta_{\varepsilon\varepsilon_2}\psi^{\text{sgn}\,\varepsilon_2})^T \tag{89}$$

Furthermore, from eq. (86), we have

$$\boldsymbol{\alpha}\varepsilon(2\hat{P} - \mathbf{I}) = \boldsymbol{\alpha\gamma}_5 = \varepsilon(2\hat{P} - \mathbf{I})\ \boldsymbol{\alpha} = \boldsymbol{\gamma}_5\boldsymbol{\alpha} \tag{90}$$

Hence,

$$\boldsymbol{\gamma}_5 i\partial_0\boldsymbol{\Psi} = i\partial_0(\boldsymbol{\gamma}_5\Psi) = -\ \boldsymbol{\gamma}_5 i\nabla\boldsymbol{\alpha\Psi}$$

$$= -\ i\nabla\boldsymbol{\alpha}(\boldsymbol{\gamma}_5\Psi) \tag{91}$$

so that the Dirac equation for massless particles may be construed as a 4-component Dirac equation on which the following constraint is imposed:

$$\Psi \equiv \boldsymbol{\gamma}_5\Psi \tag{92}$$

Consider the following:

$$\overline{\Psi}^{\text{sgn}\varepsilon}\boldsymbol{p\!\!/}\Psi^{\text{sgn}\varepsilon} = \Psi^{\text{sgn}\varepsilon\dagger}\boldsymbol{\gamma}_0\boldsymbol{p\!\!/}\Psi^{\text{sgn}\varepsilon} = \left[\frac{\mathbf{I}+\varepsilon\gamma_5}{2}\boldsymbol{\Psi}\right]^\dagger \gamma_0\boldsymbol{p\!\!/}\left[\frac{\mathbf{I}+\varepsilon\gamma_5}{2}\boldsymbol{\Psi}\right]$$

$$= \boldsymbol{\Psi}^\dagger\left(\frac{\mathbf{I}+\varepsilon\gamma_5}{2}\right)\gamma_0\boldsymbol{p\!\!/}\left(\frac{\mathbf{I}+\varepsilon\gamma_5}{2}\right)\boldsymbol{\Psi}$$

$$= \boldsymbol{\Psi}^\dagger\boldsymbol{\gamma}_0\left(\frac{\mathbf{I}-\varepsilon\gamma_5}{2}\right)\boldsymbol{p\!\!/}\left(\frac{\mathbf{I}+\varepsilon\gamma_5}{2}\right)\boldsymbol{\Psi}$$

$$= \boldsymbol{\Psi}^\dagger\boldsymbol{\gamma}_0\boldsymbol{p\!\!/}\left(\frac{\mathbf{I}+\varepsilon\gamma_5}{2}\right)\left(\frac{\mathbf{I}+\varepsilon\gamma_5}{2}\right)\boldsymbol{\Psi}$$

$$= \Psi^{\dagger}\gamma_0 \not{\mathbf{p}}\left(\frac{\mathbf{I}+\varepsilon\gamma_5}{2}\right)\Psi = \Psi^{\dagger}\gamma_0 \not{\mathbf{p}}\,\Psi^{\mathrm{sgn}\varepsilon} = \overline{\Psi}\not{\mathbf{p}}^{\mathrm{sgn}\varepsilon} \tag{93}$$

This shows that the massless Dirac equation does not involve any mixing of the right-handed and left-handed states.

We also note that

$$\overline{\Psi}^{\mathrm{sgn}\varepsilon}\gamma_5\Psi^{\mathrm{sgn}\varepsilon} = \overline{\Psi}^{\mathrm{sgn}\varepsilon}\gamma_5\left(\frac{\mathbf{I}+\varepsilon\gamma_5}{2}\right)\Psi = \overline{\Psi}^{\mathrm{sgn}\varepsilon}\left(\frac{\gamma_5+\varepsilon}{2}\right)\Psi$$

$$= \varepsilon\overline{\Psi}^{\mathrm{sgn}\varepsilon}\left(\frac{\varepsilon\gamma_5+\mathbf{I}}{2}\right)\Psi = \varepsilon\overline{\Psi}^{\mathrm{sgn}\varepsilon}\Psi^{\mathrm{sgn}\varepsilon} \tag{94}$$

The left-hand side of eq. (94) is a pseudoscalar since $\overline{\Psi}^{\mathrm{sgn}\,\varepsilon}\gamma_5\Psi^{\mathrm{sgn}\,\varepsilon}$ transforms like a Lorentz scalar under proper Lorentz transformations but reverses its sign under improper Lorentz transformations while the right-hand side is a scalar density because it has the same value in every Lorentz frame. We see this explicitly by observing the impact of the parity operator, e.g.,

$$\begin{aligned}
(\overline{\Psi}'^{\mathrm{sgn}\varepsilon}\gamma_5\Psi'^{\mathrm{sgn}\varepsilon}) &= (\Psi'^{\mathrm{sgn}\varepsilon\dagger}\gamma^0\gamma_5\Psi'^{\mathrm{sgn}\varepsilon}) = (\hat{P}\,\Psi'^{\mathrm{sgn}\varepsilon})^{\dagger}\,\gamma^0\gamma_5\hat{P}\,\Psi^{\mathrm{sgn}\varepsilon}\\
&= (\Psi^{\mathrm{sgn}\varepsilon})^{\dagger}\,\hat{P}^{\dagger}\gamma^0\gamma_5\hat{P}\,\Psi^{\mathrm{sgn}\varepsilon}\\
&= (\Psi^{\mathrm{sgn}\varepsilon})^{\dagger}\,\hat{P}^{-1}\gamma^0\gamma_5\hat{P}\,\Psi^{\mathrm{sgn}\varepsilon}\\
&= (\Psi^{\mathrm{sgn}\varepsilon})^{\dagger}\,\hat{P}^{-1}\gamma^0(i\gamma^0\gamma^1\gamma^2\gamma^3)\hat{P}\,\Psi^{\mathrm{sgn}\varepsilon}\\
&= i(\Psi^{\mathrm{sgn}\varepsilon})^{\dagger}\,\hat{P}^{-1}\gamma^0\hat{P}\hat{P}^{-1}\gamma^0\hat{P}\hat{P}^{-1}\gamma^1\hat{P}\hat{P}^{-1}\gamma^2\hat{P}\hat{P}^{-1}\gamma^3\hat{P}\,\Psi^{\mathrm{sgn}\varepsilon}\\
&= i(\Psi^{\mathrm{sgn}\varepsilon})^{\dagger}\gamma^0.\gamma^0(-\gamma^1)(-\gamma^2)(-\gamma^3)\Psi^{\mathrm{sgn}\varepsilon}\\
&= -(\Psi^{\mathrm{sgn}\varepsilon})^{\dagger}\gamma^0.\gamma_5\Psi^{\mathrm{sgn}\varepsilon} = -\,(\overline{\Psi}^{\mathrm{sgn}\varepsilon}\gamma_5\Psi^{\mathrm{sgn}\varepsilon})
\end{aligned} \tag{95}$$

However,

$$\begin{aligned}
(\overline{\Psi}'^{\mathrm{sgn}\varepsilon}\Psi'^{\mathrm{sgn}\varepsilon}) &= (\Psi'^{\mathrm{sgn}\varepsilon\dagger}\gamma^0\Psi'^{\mathrm{sgn}\varepsilon})\\
&= (\hat{P}\Psi^{\mathrm{sgn}\varepsilon})^{\dagger}\,\boldsymbol{\gamma}^0\,\hat{P}\Psi^{\mathrm{sgn}\varepsilon} = (\Psi^{\mathrm{sgn}\varepsilon})^{\dagger}\,\hat{P}^{\dagger}\boldsymbol{\gamma}^0\,\hat{P}\Psi^{\mathrm{sgn}\varepsilon}\\
&= (\Psi^{\mathrm{sgn}\varepsilon})^{\dagger}\,\hat{P}^{-1}\boldsymbol{\gamma}^0\,\hat{P}\Psi^{\mathrm{sgn}\varepsilon}\\
&= (\Psi^{\mathrm{sgn}\varepsilon})^{\dagger}\,\boldsymbol{\gamma}^0\,\Psi^{\mathrm{sgn}\varepsilon} = \overline{\Psi}^{\mathrm{sgn}\varepsilon}\Psi^{\mathrm{sgn}\varepsilon}
\end{aligned} \tag{96}$$

This shows that the helicity states are not parity eigenstates for the massless Dirac equation. This is also explicitly established by

$$\begin{aligned}
\hat{P}\Psi^{\mathrm{sgn}\varepsilon} &= e^{i\phi}\gamma^0\Psi^{\mathrm{sgn}\varepsilon} = e^{i\phi}\beta\Psi^{\mathrm{sgn}\varepsilon}\\
&= e^{i\phi}\begin{pmatrix}\mathbf{0} & -\mathbf{I}\\ -\mathbf{I} & \mathbf{0}\end{pmatrix}\begin{pmatrix}\delta_{\varepsilon\varepsilon_1}\psi^{\varepsilon_1}\\ \delta_{\varepsilon\varepsilon_2}\psi^{\varepsilon_2}\end{pmatrix} = -e^{i\phi}\begin{pmatrix}\delta_{\varepsilon\varepsilon_2}\psi^{\varepsilon_2}\\ \delta_{\varepsilon\varepsilon_1}\psi^{\varepsilon_1}\end{pmatrix}
\end{aligned} \tag{97}$$

This is justified on physical grounds as follows:

A space inversion is characterized by the following set of transformations:

$$\mathbf{x} \to -\mathbf{x},\ x^0 \to x^0,\ \mathbf{p} \to -\mathbf{p},\ \boldsymbol{\sigma} \to \boldsymbol{\sigma} \tag{98}$$

Under space inversion, therefore, a state with energy p_0, momentum $\mathbf{p}$ and helicity $\dfrac{\sigma\mathbf{p}}{|\mathbf{p}|}$ will get transformed to a state with energy p_0, momentum $-\mathbf{p}$ and, therefore, helicity $-\dfrac{\sigma\mathbf{p}}{|\mathbf{p}|}$. However, such a state does not exist in the 2-component description of massless particles.

9.7 $\hat{C}\hat{P}$ AS A SYMMETRY FOR MASSLESS PARTICLES

We consider the Dirac matrices in the standard representation

$$\boldsymbol{\gamma}^k = \begin{pmatrix} \mathbf{0} & \sigma^k \\ -\sigma^k & \mathbf{0} \end{pmatrix},\ \boldsymbol{\gamma}^0 = \begin{pmatrix} \mathbf{I} & \mathbf{0} \\ \mathbf{0} & -\mathbf{I} \end{pmatrix} \tag{99}$$

The charge conjugated wave function is defined by

$$\boldsymbol{\Psi}_c(x) = \hat{C}\boldsymbol{\gamma}^0\boldsymbol{\Psi}^*(x) \tag{100}$$

where the charge conjugation operator is $\hat{C} = i\boldsymbol{\gamma}^2\boldsymbol{\gamma}^0$ in the standard representation.

The action of the parity operator is

$$\boldsymbol{\Psi}_p(x') = e^{i\phi}\boldsymbol{\gamma}^0\boldsymbol{\Psi}(-\mathbf{x}, x^0) \tag{101}$$

Hence, we have

$$\Psi_{cp}(x') = \hat{C}\hat{P}\Psi(x) = \hat{C}[e^{i\phi}\boldsymbol{\gamma}^0\boldsymbol{\Psi}(-\mathbf{x}, x^0)] = ie^{i\phi}\gamma^2\gamma^0\boldsymbol{\Psi}^*(-\mathbf{x}, x^0) \tag{102}$$

We now examine whether $\boldsymbol{\Psi}_{cp}(x)$ satisfies the Dirac equation for massless particles. We have

$$\partial_\mu\boldsymbol{\Psi}_{cp}(x') = ie^{i\phi}\boldsymbol{\gamma}^\mu\boldsymbol{\gamma}^2\boldsymbol{\gamma}^0\partial_\mu\boldsymbol{\Psi}^*(-\mathbf{x}, x^0) \tag{103}$$

Now

$$\boldsymbol{\gamma}^0\boldsymbol{\gamma}^2\boldsymbol{\gamma}^0 = -\boldsymbol{\gamma}^2\boldsymbol{\gamma}^0\boldsymbol{\gamma}^0,\ \boldsymbol{\gamma}^1\boldsymbol{\gamma}^2\boldsymbol{\gamma}^0 = \boldsymbol{\gamma}^2\boldsymbol{\gamma}^0\boldsymbol{\gamma}^1,\ \boldsymbol{\gamma}^2\boldsymbol{\gamma}^2\boldsymbol{\gamma}^0 = -\boldsymbol{\gamma}^2\boldsymbol{\gamma}^0\boldsymbol{\gamma}^2,\ \boldsymbol{\gamma}^3\boldsymbol{\gamma}^2\boldsymbol{\gamma}^0 = \boldsymbol{\gamma}^2\boldsymbol{\gamma}^0\boldsymbol{\gamma}^3$$

Therefore, eq. (103) reduces to

$$\partial\boldsymbol{\Psi}_{cp}(x') = -ie^{i\phi}\boldsymbol{\gamma}^2\boldsymbol{\gamma}^0(\boldsymbol{\gamma}^0\partial_0 - \boldsymbol{\gamma}^1\partial_1 + \boldsymbol{\gamma}^2\partial_2 - \boldsymbol{\gamma}^3\partial_3)\ \Psi^*(-\mathbf{x}, x^0) \tag{104}$$

Now, since the matrix $\boldsymbol{\gamma}^2$ is purely imaginary in the standard representation, we have $\boldsymbol{\gamma}^{2*} = -\boldsymbol{\gamma}^2$. However, the other three matrices being real, we have $\boldsymbol{\gamma}^{0*} = \boldsymbol{\gamma}^0$, $\boldsymbol{\gamma}^{1*} = \boldsymbol{\gamma}^1$, $\boldsymbol{\gamma}^{3*} = \boldsymbol{\gamma}^3$. Hence, eq. (104) gets modified to

$$\partial\Psi_{cp}(x') = -ie^{i\phi}\boldsymbol{\gamma}^2\boldsymbol{\gamma}^0(\boldsymbol{\gamma}^{0*}\partial_0 - \boldsymbol{\gamma}^{1*}\partial_1 - \boldsymbol{\gamma}^{2*}\partial_2 - \boldsymbol{\gamma}^{3*}\partial_3)\ \boldsymbol{\Psi}^*(-\mathbf{x}, x^0)$$

Writing $\mathbf{x}' \equiv -\mathbf{x},\ x'^0 \equiv x^0$, we obtain

$$\begin{aligned}\partial\Psi_{cp}(x') &= -ie^{i\phi}\boldsymbol{\gamma}^2\boldsymbol{\gamma}^0(\boldsymbol{\gamma}^{0*}\partial_{0'} + \boldsymbol{\gamma}^{1*}\partial_{1'} + \boldsymbol{\gamma}^{2*}\partial_{2'} + \boldsymbol{\gamma}^{3*}\partial_{3'})\ \Psi^*(x') \\ &= -ie^{i\phi}\boldsymbol{\gamma}^2\boldsymbol{\gamma}^0\boldsymbol{\gamma}^{\mu*}\partial_{\mu'}\Psi^*(x') = -ie^{i\phi}\boldsymbol{\gamma}^2\boldsymbol{\gamma}^0[\boldsymbol{\gamma}^\mu\partial_{\mu'}\Psi(x')]^* = 0\end{aligned} \tag{105}$$

showing that the $\hat{C}\hat{P}$ transformed wave function $\Psi_{cp}(x)$ satisfies the Dirac equation and, therefore, the combined $\hat{C}\hat{P}$ operation can be considered a symmetry transformation with regard to the Dirac equation for massless particles.

9.8 POLARIZATION OF MASSLESS PARTICLES

1. We define a Lorentz transformation as a linear, homogeneous transformation in Minkowki spacetime that leaves the distance between two spacetime points x and x' represented by the coordinates $x = \{x^\mu, \mu = 0, 1, 2, 3\}$, $x' = \{x'^\mu, \mu = 0, 1, 2, 3\}$ measured with reference to the metric

$$ds^2 = g_{\mu\nu} dx^\mu dx^\nu \tag{106}$$

invariant.

Such a Lorentz transformation $\mathbf{\Lambda} = \{a^\mu_\nu, \mu, \nu = 0, 1, 2, 3\}$ when operated on a spacetime point $x = \{x^\mu, \mu = 0, 1, 2, 3\}$ gives

$$(x')^\nu = a^\mu_\nu x^\nu \tag{107}$$

where the coefficients a^μ_ν, μ, $\nu = 0, 1, 2, 3$ depend only on the relative velocities and the spatial orientation of the two frames and also satisfy

$$a^\mu_\nu g_{\mu\rho} a^\rho_\sigma = g_{\nu\sigma} \tag{108}$$

Now, Lorentz transformations can be segregated into two distinct classes, viz.

(a) The "pure" rotations that leave the differential line element in 3-dimensional space $ds^2 = g_{ij} dx^i dx^j$, $i, j = 1, 2, 3$ invariant under the transformation.

(b) The Lorentz "boosts" that, when acting in the direction of the coordinate axis x^k (k fixed), leave the differential line element in 2-dimensional space-time $ds^2 = g_{\mu\nu} dx^\mu dx^\nu$, $\mu, \nu = 0, k$ invariant under the transformation.

In fact, any arbitrary Lorentz transformation can be conveniently split into a composition of a "pure" rotation followed by a "pure" boost, i.e., $\mathbf{\Lambda} = \mathbf{LR}$.

It is convenient to parameterize Lorentz "boosts" in terms of the relative velocity of the two frames of reference ($0 \le v < c$). In terms of the 2×2 Hermitian Pauli matrices, "pure" boosts can be represented by

$$\mathbf{L}(\omega, \mathbf{f}) = \mathbf{I}\cosh\frac{1}{2}\omega - \mathbf{f}.\boldsymbol{\sigma}\sinh\frac{1}{2}\omega \tag{109}$$

where $\tanh \omega = \dfrac{v}{c}$ and $\mathbf{f}$ is a unit vector along the direction of the boost. A 4-vector, in the representation space spanned by the 2×2 identity matrix and the three Pauli matrices is conveniently represented as $x = x^0\mathbf{I} + \sum_{i=1}^{3} x^i \sigma_i$.

The action of this Lorentz boost on a spacetime point $x = \{x^\mu, \mu = 0, 1, 2, 3\}$ is explicitly given by

$$(x')^0 = \gamma x^0 - \frac{\gamma}{c}(\mathbf{v.x}) \tag{110}$$

$$\mathbf{x}' = \mathbf{x} - \frac{\gamma}{c}\,\mathbf{v}x^0 + \frac{\gamma^2}{(1+\gamma)c^2}(\mathbf{v.x})\mathbf{v} \tag{111}$$

where
$$\gamma = \left(1 - \frac{v^2}{c^2}\right)^{\frac{1}{2}} \tag{112}$$

In contrast to boosts, we characterize "pure" rotations in terms of a unit vector **e** along the axis of rotation and the angle of rotation θ (taken positive when the rotation is counter clockwise, looking along the direction of **e**). In terms of the Pauli matrices, we can represent "pure" rotations as

$$\mathbf{R}(\theta,\ \mathbf{e}) = \mathbf{I}\cos\frac{1}{2}\theta - i(\boldsymbol{\sigma}.\mathbf{e})\sin\frac{1}{2}\theta \tag{113}$$

Hence, an arbitrary Lorentz transformation may be represented as

$$\begin{aligned}
\mathbf{L}(\omega,\ \mathbf{f})\ \mathbf{R}(\theta,\ \mathbf{e}) &= \left[\mathbf{I}\cosh\frac{1}{2}\omega - (\boldsymbol{\sigma}.\mathbf{f})\sinh\frac{1}{2}\omega\right]\left[\mathbf{I}\cos\frac{1}{2}\theta - i(\boldsymbol{\sigma}.\mathbf{e})\sin\frac{1}{2}\theta\right] \\
&= \mathbf{I}\cosh\frac{1}{2}\omega\cos\frac{1}{2}\theta - (\boldsymbol{\sigma}.\mathbf{f})\sinh\frac{1}{2}\omega\cos\frac{1}{2}\theta \\
&\quad - i(\boldsymbol{\sigma}.\mathbf{e})\cosh\frac{1}{2}\omega\sin\frac{1}{2}\theta + i(\boldsymbol{\sigma}.\mathbf{f})(\boldsymbol{\sigma}.\mathbf{e})\sinh\frac{1}{2}\omega\sin\frac{1}{2}\theta \\
&= \mathbf{I}\cosh\frac{1}{2}\omega\cos\frac{1}{2}\theta - (\boldsymbol{\sigma}.\mathbf{f})\sinh\frac{1}{2}\omega\cos\frac{1}{2}\theta \\
&\quad - i(\boldsymbol{\sigma}.\mathbf{e})\cosh\frac{1}{2}\omega\sin\frac{1}{2}\theta \\
&\quad + i[(\mathbf{f.e}) + i\boldsymbol{\sigma}.(\mathbf{f}\times\mathbf{e})]\sinh\frac{1}{2}\omega\sin\frac{1}{2}\theta
\end{aligned} \tag{114}$$

Equation (114) gives

$$\begin{aligned}
\mathbf{L}(\omega,\ \mathbf{e})\ \mathbf{R}(\theta,\ \mathbf{e}) &= \mathbf{I}\cosh\frac{1}{2}\omega\cos\frac{1}{2}\theta - (\boldsymbol{\sigma}.\mathbf{e})\sinh\frac{1}{2}\omega\cos\frac{1}{2}\theta \\
&\quad - i(\boldsymbol{\sigma}.\mathbf{e})\cosh\frac{1}{2}\omega\sin\frac{1}{2}\theta + \mathbf{I}\,i\sinh\frac{1}{2}\omega\sin\frac{1}{2}\theta \\
&= \mathbf{I}\cos\frac{1}{2}i\omega\cos\frac{1}{2}\theta + \mathbf{I}\sin\frac{1}{2}\omega\sin\frac{1}{2}\theta \\
&\quad + i(\boldsymbol{\sigma}.\mathbf{e})\sin\frac{1}{2}i\omega\cos\frac{1}{2}\theta - i(\boldsymbol{\sigma}.\mathbf{e})\cos\frac{1}{2}i\omega\sin\frac{1}{2}\theta
\end{aligned}$$

$$= \mathbf{I}\cos\frac{1}{2}(\theta - i\omega) - i(\boldsymbol{\sigma}.\mathbf{e})\sin\frac{1}{2}(\theta - i\omega)$$

$$= \mathbf{R}(\theta, \mathbf{e})\mathbf{L}(\omega, \mathbf{e}) \tag{115}$$

that is

$$[\mathbf{R}(\theta, \mathbf{e})\mathbf{L}(\omega, \mathbf{e})] = 0 \tag{116}$$

The fact that $\mathbf{R}(\theta, \mathbf{e})$ and $\mathbf{L}(\omega, \mathbf{e})$ both commute has important physical ramifications. It implies that the sequence of performing the "pure" boost and the "pure" rotation is immaterial provided that the boost and the rotation are directed along the same axis. In other words, one will obtain the same "final state" of a system irrespective of whether one:

(a1) first, imparts a velocity v to a given "initial state" along a given axis and then

(a2) rotates the resulting state through an angle θ about the same axis, or

(b1) first rotates the given initial state through an angle θ about a given axis and then

(b2) imparts a velocity v along the same axis, i.e. along the axis of rotation.

(2) The impact on a state vector of a Lorentz transformation can be viewed in two ways, viz. active and passive. These views are mutually indistinguishable in line with the "principle of relativity".

In the "active" view, the transformation matrix operates on the coordinates of the spacetime point representing the "initial (given) state" to produce the "final state" represented by the coordinates of the transformed spacetime point. The important point here is that the coordinates of both, the initial spacetime point and the final spacetime point are measured with respect to the same frame of reference (which is assumed fixed and unaffected by the transformation).

In the passive view, the state remains unchanged. The effect of the transformation is embodied into the system through a transformation of the frame of reference. This is done by transforming the frame of reference with the "opposite" transformation of the spacetime point as seen in the "active" view.

As a specific example, a state moving with a velocity v with reference to a frame of reference ***I*** can be obtained either by

(a) operating on the "rest state" with respect to the frame ***I*** by a Lorentz transformation imparting a velocity v to this rest state, or

(b) by keeping the "rest state" of the particle unchanged and imparting a velocity $-v$ to the frame of reference ***I***, i.e. observing the particle from a frame of reference ***I′*** that is moving with a velocity $-v$ with respect to the frame ***I***.

(3) In view of (2), all the "states" of a system can be obtained by looking at a "standard state" from an appropriate coordinate system. In fact, instead of defining a "state" with respect to the particle, we can equally well and indistinguishably define the "state" with respect to the frame of reference from which the particle is being observed.

This "standard state" is usually defined by choosing an arbitrary but fixed Lorentz frame and then prescribing "definite properties" to that "standard state" with respect to the fixed Lorentz frame. Conventionally, this "standard state' is defined with respect to the Lorentz frame in which the particle is "at rest", i.e., with respect to the "rest frame" and in which the spin is directed along the x^3 axis.

The corollaries that follow to the above are:

(a) Two states of the same particle defined with reference to two identical Lorentz frames are identical.

(b) Two states defined by Lorentz frames that differ by an infinitesimal Lorentz transformation are infinitesimally different from each other.

(c) The conventional "standard state" can be defined only for massive particles since for massless particles there is no rest frame.

(d) Since states are defined with respect to Lorentz frames, comparison of "states" is independent of the intrinsic properties of the particle like mass, spin, etc. In this framework, comparison of "states" boils down to comparison of the defining Lorentz frames.

(e) Since a "standard state" is defined in the "rest frame" with spin directed along the x^3 axis, the corresponding Lorentz transformation is

$$\begin{aligned}\mathbf{R}(\boldsymbol{\theta}, \mathbf{e}^3)\mathbf{L}(0) &= \mathbf{I}\cos\frac{1}{2}\theta - i(\sigma_3\mathbf{e}^3)\sin\frac{1}{2}\theta \\ &= \begin{pmatrix} \cos\frac{1}{2}\theta - i\sin\frac{1}{2}\theta & 0 \\ 0 & \cos\frac{1}{2}\theta + i\sin\frac{1}{2}\theta \end{pmatrix} \\ &= \begin{pmatrix} e^{-i\frac{\theta}{2}} & 0 \\ 0 & e^{i\frac{\theta}{2}} \end{pmatrix}\end{aligned} \tag{117}$$

(4) In the conventional representation of the dynamics of relativistic particles in 4 × 4 dimensions, the transformation to a Lorentz frame in which the particle is moving with a velocity v along the direction of the x^3 axis and has the spin orientation in that direction is represented by the matrix

$$\mathbf{L}(\omega, \mathbf{f}^3) = \begin{pmatrix} \cosh\omega & 0 & 0 & \sinh\omega \\ 0 & 1 & 0 & 0 \\ 0 & 0 & 1 & 0 \\ \sinh\omega & 0 & 0 & \cosh\omega \end{pmatrix} \tag{118}$$

because the transformed frame moves with a velocity $-v$ along the direction of the x^3 axis with respect to the original Lorentz frame with reference to which the "standard state" is defined. Here, $\tanh\omega = \dfrac{v}{c}$.

Similarly, a rotation about the x^3 axis of the particle state by an angle θ corresponds to a rotation of the frame of reference by an angle $-\theta$ about the same axis and is, therefore, represented by the 4×4 matrix

$$\mathbf{R}(-\theta, \mathbf{e}^3) = \begin{pmatrix} 1 & 0 & 0 & 0 \\ 0 & \cos\theta & -\sin\theta & 0 \\ 0 & \sin\theta & \cos\theta & 0 \\ 0 & 0 & 0 & 1 \end{pmatrix} \tag{119}$$

Suppose, now, that we want to obtain a "state" of a particle in which

(i) it is moving with a velocity v in the direction of a unit vector $\mathbf{n}$ in the x^2-x^3 plane and making an angle ϕ with the x^3 axis, and

(ii) making a rotation about this unit vector $\mathbf{n}$ as the rotation axis, i.e., the spin polarization is also along this axis.

We can obtain this state by performing the following two operations on the "standard state":

(i) Impart a velocity $-v$ in the direction of the unit vector $\mathbf{e}^3$, i.e., the x^3 axis by performing the transformation

$$\mathbf{L}(\omega, \mathbf{e}^3) = \begin{pmatrix} \cosh\omega & 0 & 0 & \sinh\omega \\ 0 & 1 & 0 & 0 \\ 0 & 0 & 1 & 0 \\ \sinh\omega & 0 & 0 & \cosh\omega \end{pmatrix}$$

(ii) Rotate the coordinate system $\boldsymbol{I}$ through an angle $+\phi$ in the $x^2 - x^3$ plane along the x^1 axis to align the x^3 axis of $\boldsymbol{I}$ with the direction of unit vector $\mathbf{n}$ (this is equivalent to rotating the "standard state" by an angle $-\phi$), i.e., perform the operation $\mathbf{R}(\phi, \mathbf{e}^1)$ that is represented by the matrix.

$$\mathbf{R}(\phi, \mathbf{e}^1) = \begin{pmatrix} 1 & 0 & 0 & 0 \\ 0 & 1 & 0 & 0 \\ 0 & 0 & \cos\phi & \sin\phi \\ 0 & 0 & -\sin\phi & \cos\phi \end{pmatrix}$$

The composite transformation is given by

$$\mathbf{T}(\phi, v, \mathbf{n}) = \mathbf{R}(\phi, \mathbf{e}^1)\mathbf{L}(v, \mathbf{e}^3)$$

$$= \begin{pmatrix} 1 & 0 & 0 & 0 \\ 0 & 1 & 0 & 0 \\ 0 & 0 & \cos\phi & \sin\phi \\ 0 & 0 & -\sin\phi & \cos\phi \end{pmatrix} \cdot \begin{pmatrix} \cosh\omega & 0 & 0 & \sinh\omega \\ 0 & 1 & 0 & 0 \\ 0 & 0 & 1 & 0 \\ \sinh\omega & 0 & 0 & \cosh\omega \end{pmatrix}$$

$$= \begin{pmatrix} \cosh\omega & 0 & 0 & \sinh\omega \\ 0 & 1 & 0 & 0 \\ \sin\phi\sinh\omega & 0 & \cos\phi & \sin\phi\cosh\omega \\ \cos\phi\sinh\omega & 0 & -\sin\phi & \cos\phi\cosh\omega \end{pmatrix}$$

$$= \begin{pmatrix} \cosh\omega & 0 & \sin\phi\sinh\omega & \cos\phi\sinh\omega \\ 0 & 1 & 0 & 0 \\ \sin\phi\sinh\omega & 0 & \cos^2\phi + \sin^2\phi\cos\omega & \sin\phi\cos\phi\,(\cosh\omega - 1) \\ \cos\phi\sinh\omega & 0 & \sin\phi\cos\phi\,(\cosh\omega - 1) & \sin^2\phi + \cos^2\phi\cosh\omega \end{pmatrix}$$

$$\begin{pmatrix} 1 & 0 & 0 & 0 \\ 0 & 1 & 0 & 0 \\ 0 & 0 & \cos\phi & \sin\phi \\ 0 & 0 & -\sin\phi & \cos\phi \end{pmatrix}$$

This shows that we can obtain the same state by performing the operations aforesaid in the reverse order, i.e., by performing the following operations in sequence:

(i) Rotate the coordinate system ***I*** through an angle ϕ in the x^2–x^3 plane along the x^1 axis to align the x^3 axis of ***I*** with the direction of unit vector $\mathbf{n}$, i.e., perform the operation $\mathbf{R}(\phi, \mathbf{e}^1)$ that is represented by the matrix

$$\mathbf{R}(\phi, \mathbf{e}^1) = \begin{pmatrix} 1 & 0 & 0 & 0 \\ 0 & 1 & 0 & 0 \\ 0 & 0 & \cos\phi & \sin\phi \\ 0 & 0 & -\sin\phi & \cos\phi \end{pmatrix}$$

(ii) Impart a velocity $-v$ in the direction of the unit vector $\mathbf{n}$ represented by the matrix transformation

$$\mathbf{L}(\omega,\mathbf{n}) = \begin{pmatrix} \cosh\omega & 0 & \sin\phi\sinh\omega & \cos\phi\sinh\omega \\ 0 & 1 & 0 & 0 \\ \sin\phi\sinh\omega & 0 & \cos^2\phi + \sin^2\phi\cos\omega & \sin\phi\cos\phi\,(\cosh\omega - 1) \\ \cos\phi\sinh\omega & 0 & \sin\phi\cos\phi\,(\cosh\omega - 1) & \sin^2\phi + \cos^2\phi\cosh\omega \end{pmatrix}$$

Hence, we have

$$\mathbf{T}(\phi, v, \mathbf{n}) = \mathbf{L}(v, \mathbf{n})\,\mathbf{R}(\phi, \mathbf{e}^1) = \mathbf{R}(\phi, \mathbf{e}^1)\,\mathbf{L}(v, \mathbf{e}^3) \tag{120}$$

Now, for further rotations/boosts, we have

$$\mathbf{R}(\varphi, \mathbf{e}^1)\,\mathbf{T}(\phi, v, \mathbf{n}) = \mathbf{R}(\varphi, \mathbf{e}^1)\,\mathbf{R}(\phi, \mathbf{e}^1)\,\mathbf{L}(v, \mathbf{e}^3)$$

$$= \mathbf{R}(\varphi + \phi, \mathbf{e}^1)\, \mathbf{L}(v, \mathbf{e}^3)$$
$$= \mathbf{L}(v, \mathbf{n}')\, \mathbf{R}(\phi + \varphi, \mathbf{e}^1) \tag{121}$$

and

$$\mathbf{L}(v', \mathbf{n})\, \mathbf{T}(\phi, v, \mathbf{n}) = \mathbf{L}(v', \mathbf{n})\, \mathbf{L}(v, \mathbf{n})\, \mathbf{R}(\phi, \mathbf{e}^1)$$
$$= \mathbf{L}(v + v', \mathbf{n})\, \mathbf{R}(\phi, \mathbf{e}^1)$$
$$= \mathbf{R}(\phi, \mathbf{e}^1)\, \mathbf{L}(v + v', \mathbf{e}^3) \tag{122}$$

From the above results, we conclude that if the spin of a particle is parallel (antiparallel) to its velocity in a particular frame of reference, then it remains parallel (antiparallel) with respect to further rotations or further boosts in the same direction.

Let us now examine a different scenario. So far we have considered the case of the motion of the frame along the same (opposite) direction to the motion of the particle. We now, imagine a particle moving in the x^3 direction in frame I with a velocity u. The spin polarization of the particle as observed from I is also along this direction.

We identify an observer in a frame I' moving along the x^2 axis of frame I with a relative velocity $-v$ with respect to I. Then the state of the particle, as seen from the observer in I' is generated by the composite Lorentz transformation operating on the rest frame. The velocity u of the particle in frame I is accounted for by imparting a velocity $-u$ to the rest frame in the x^3 direction. Thus, the transformation $\mathbf{L}(u, \mathbf{e}^3)$ accommodates the motion of the particle in x^3 direction in the frame I with a velocity u. The transformation $\mathbf{L}(v, \mathbf{e}^2)$ provides for the motion of the frame I' with respect to I in the direction of the x^2 axis of frame I with a relative velocity $-v$ with respect to I. The composite transformation, therefore, is

$$\mathbf{T}' = \mathbf{L}(v, \mathbf{e}^2)\mathbf{L}(u, \mathbf{e}^3)$$

$$= \begin{pmatrix} \cosh\omega' & 0 & \sinh\omega' & 0 \\ 0 & 1 & 0 & 0 \\ \sinh\omega' & 0 & \cosh\omega' & 0 \\ 0 & 0 & 0 & 1 \end{pmatrix} \begin{pmatrix} \cosh\omega & 0 & 0 & \sinh\omega \\ 0 & 1 & 0 & 0 \\ 0 & 0 & 1 & 0 \\ \sinh\omega & 0 & 0 & \cosh\omega \end{pmatrix}$$

$$\cosh\omega = \begin{pmatrix} \cosh\omega'\cosh\omega & 0 & \sinh\omega' & \cosh\omega'\sinh\omega \\ 0 & 1 & 0 & 0 \\ \sinh\omega'\cosh\omega & 0 & \cosh\omega' & \sinh\omega'\sinh\omega \\ \sinh\omega & 0 & 0 & \cosh\omega \end{pmatrix} \tag{123}$$

The above composite transformation leads to the following transformation equations for the differential length elements dx^μ, $\mu = 0, 1, 2, 3$ of the rest frame:

$$dx'^0 = \cosh\omega \cosh\omega'\, dx^0 + \sinh\omega' dx^2 + \sinh\omega \cosh\omega'\, dx^3 \tag{124}$$
$$dx'^1 = dx^1 \tag{125}$$
$$dx'^2 = \cosh\omega \sinh\omega'\, dx^0 + \cosh\omega'\, dx^2 + \sinh\omega \sinh\omega' dx^3 \tag{126}$$
$$dx'^3 = \sinh\omega dx^0 + \cosh\omega dx^3 \tag{127}$$

From the above transformation equations, we can deduce the velocity components as observed from the primed frame as

$$u_1' = \frac{dx'^1}{dx'^0} = \frac{dx^1}{\cosh\omega\cosh\omega' dx^0 + \sinh\omega' dx^2 + \sinh\omega\cosh\omega' dx^3}$$

$$= \frac{u_1}{\cosh\omega\cosh\omega' + \sinh\omega' u_2 + \sinh\omega\cosh\omega' u_3} = 0 \qquad (128)$$

where the last step follows from the fact that, in the rest frame $u_1 = u_2 = u_3 = 0$.

Similarly,

$$u_2' = \frac{dx'^2}{dx'^0} = \frac{\cosh\omega\sinh\omega' dx^0 + \cosh\omega' dx^2 + \sinh\omega\sinh\omega' dx^3}{\cosh\omega\cosh\omega' dx^0 + \sinh\omega' dx^2 + \sinh\omega\cosh\omega' dx^3}$$

$$= \frac{\cosh\omega\sinh\omega' + \cosh\omega' u_2 + \sinh\omega\sinh\omega' u_3}{\cosh\omega\cosh\omega' + \sinh\omega' u_2 + \sinh\omega\cosh\omega' u_3} \qquad (129)$$

and

$$u_3' = \frac{dx'^3}{dx'^0} = \frac{\sinh\omega dx^0 + \cosh\omega dx^3}{\cosh\omega\cosh\omega' dx^0 + \sinh\omega' dx^2 + \sinh\omega\cosh\omega' dx^3}$$

$$= \frac{\sinh\omega + \cosh\omega u_3}{\cosh\omega\cosh\omega' + \sinh\omega' u_2 + \sinh\omega\cosh\omega' u_3} = \frac{\tanh\omega}{\cosh\omega'} \qquad (130)$$

In other words, as viewed from the primed frame, the particle will appear to be travelling with a velocity $\mathbf{v}' = \left\{0, \tanh\omega', \frac{\tanh\omega}{\cosh\omega'}\right\}$, i.e., moving in the $x^2 - x^3$ plane making an angle ε given by

$$\tan\varepsilon = \frac{\tanh\omega'\cosh\omega'}{\tanh\omega} = \frac{v}{u}(1 - v^2)^{-\frac{1}{2}}.$$

It is, however, to be noted that the spin polarization of the particle continues to be along the x'^3 axis of I' (which, incidentally, continues to be parallel to the x^3 axis of I as the relative motion between I' and I consists only of linear motion along the x^2 axis of I).

In view of the above, it is clear that, as viewed from the primed frame, the velocity of the particle and its spin polarization no longer coincided with each other. In fact, the spin polarization makes an angle ε given by $\tan\varepsilon = \frac{\tanh\omega'\cosh\omega'}{\tanh\omega} = \frac{v}{u}(1 - v^2)^{-\frac{1}{2}}$ with the velocity of the particle. As a corollary, it would be necessary to rotate the polarization axis by this angle, i.e., ε to bring the spin polarization axis and the velocity vector in tandem with each other.

Now, the angle between the velocity vector and spin polarization in I' is given by $\tan\varepsilon = \frac{\tanh\omega'\cosh\omega'}{\tanh\omega} = \frac{v}{u}(1 - v^2)^{-\frac{1}{2}}$. Obviously, for massless particles that travel with the velocity of light tan ε and hence, $\varepsilon \to 0$. Therefore, in the case of massless particles that travel with the velocity of light, the spin polarization vector and the velocity vector remain parallel to each other even when viewed from different frames of reference that are in relative motion with each other (at non-relativistic relative velocities).

We, therefore, arrive at a very important conclusion that "the spin polarization is parallel (antiparallel) to the velocity vector of massless particles travelling with the velocity of light is INVARIANT with respect to any Lorentz frames of reference provided that the relative motion between the frames of reference is confined to non-relativistic speeds".

Hence, in the case of massless particles, there can be only two spin polarization states, viz., parallel to and antiparallel to the velocity vector.

The two states with spin parallel and antiparallel to the velocity vector are mirror images of each other and hence, their simultaneous existence is mandated by invariance of the theory under space inversion. However, it is now well established, both experimentally and theoretically, that the theory of massless spin $\frac{1}{2}$ particles violates space parity and hence, is not invariant under space inversion. In view of this, there exists only one state of spin polarization of the massless spin $\frac{1}{2}$ particles with the spin axis orientation along the velocity vector.

Therefore, we conclude as follows:

(a) (i) The angular momentum of a massless particle is parallel to its velocity.

(ii) The parallelism between the angular momentum and velocity is relativistically invariant, i.e., for massless particles, the angular momentum and velocity vector will remain parallel irrespective of the relative motion of Lorentz frames from which the particle is observed.

(b) (i) In the case of massive particles, there can be Lorentz frames in which the angular momentum of the particles may appear to be parallel to the velocity vector.

(ii) However, this property is not relativistically invariant, i.e., it is not universal with respect to observers in different Lorentz frames. While there may exist a Lorentz frame in which the spin axis and the velocity vector may appear parallel, there would be other Lorentz frames in which the two vectors would not be parallel, e.g., in the rest frame, the particle would have zero velocity but may still possess spin. In such a case, obviously, there is no question of the spin vector being parallel to the velocity since the velocity vector would be the null vector.

CHAPTER 10

Wave Packets of Dirac Waves

Wave packets are constituted by superposition of plane waves to yield localized wave functions in spacetime. As per the superposition principle, the wave functions representing the wave packets are also solutions of the free particle Dirac equation (the Dirac equation is a linear differential equation so that linear combinations of solutions of the Dirac equation will themselves become solutions thereof). We can, therefore, write the expression for the wave function of a wave packet of plane waves with positive energy as

$$\psi^{(+)}(x) = \frac{1}{(2\pi)^{3/2}} \sqrt{\frac{m}{E}} \int d\mathbf{p} \sum_{\pm s} b(p, s)\, u(p, s)\, e^{-ip_\mu x^\mu} \tag{1}$$

10.1 NORMALIZATION OF POSITIVE ENERGY WAVE PACKETS

The normalizing condition implies that

$$\begin{aligned}\int \psi^{(+)\dagger}(x)\psi^{(+)}(x)\, d\mathbf{x} = 1 &= \left(\frac{m}{E}\right)^{1/2} \left(\frac{m}{E'}\right)^{1/2} \int d\mathbf{p} \int d\mathbf{p}' \sum_{\pm s}\sum_{\pm s'} \\ &\times b^\dagger(p, s)\, b(p', s')\, u^\dagger(p, s)\, u(p', s')\, e^{-i(p_0 - p_0')x^0} \\ &\times \left[\frac{1}{(2\pi)^{3/2}} \cdot \frac{1}{(2\pi)^{3/2}} \int d\mathbf{x} e^{i(\mathbf{p}-\mathbf{p}').\mathbf{x}}\right] \\ &= \left(\frac{m^2}{EE'}\right)^{1/2} \int d\mathbf{p} \int d\mathbf{p}' \sum_{\pm s}\sum_{\pm s'} b^\dagger(p, s)\, b(p', s')\, u^\dagger \\ &(p, s)\, u(p', s')\, e^{-i(p_0 - p_0')x^0}\, \delta(\mathbf{p} - \mathbf{p}')\end{aligned}$$

$$= \frac{m}{E} \int d\mathbf{p} \sum_{\pm s} \sum_{\pm s'} b^{\dagger} \ (p, s) \ b \ (p, s') \ u^{\dagger} \ (p, s) \ u \ (p, s')$$

$$= \frac{m}{E} \int d\mathbf{p} \sum_{\pm s} \sum_{\pm s'} b^{\dagger} \ (p, s) \ b \ (p, s') \left(\frac{E}{m} \delta_{ss'} \right)$$

$$= \int d\mathbf{p} \sum_{\pm s} b^{\dagger} \ (p, s) \ b \ (p, s) = \int d\mathbf{p} \sum_{\pm s} |b \ (p, s)|^2 = 1 \tag{2}$$

where we have used the following

$$p_0 = E = \sqrt{\mathbf{p}^2 + m^2} = \sqrt{\mathbf{p}'^2 + m^2} = E' = p_0' \tag{3}$$

and so

$$\exp \ [i \ (p_0 - p_0) \ x^0] = \exp \ [i \ (E - E') \ x^0) = 1 \tag{4}$$

The amplitudes $b \ (p, s)$ are the linear coefficients that determine the admixture of the plane waves $u \ (p, s) \exp(- ipx)$ in the wave packet. We have also used, in the above derivative, the property of the Dirac spinors, viz.

$$u^{\dagger} \ (p, s) \ u \ (p, s') = \frac{E}{m} \delta_{ss'} \tag{5}$$

10.2 THE GORDON DECOMPOSITION

Let ψ_1 and ψ_2 represent wave functions of a free Dirac particle. Then they satisfy the free particle Dirac equations

$$(\hat{\not{\mathbf{p}}} - m) \ \psi_1 = 0 \tag{6}$$

and

$$(\hat{\not{\mathbf{p}}} - m) \ \psi_2 \tag{7}$$

The adjoint spinors $\bar{\psi}_1$, $\bar{\psi}_2$ then satisfy

$$\bar{\psi}_1 \ (- \overleftarrow{\hat{\not{\mathbf{p}}}} - m) = 0 \tag{8}$$

and

$$\bar{\psi}_2 \ (- \overleftarrow{\hat{\not{\mathbf{p}}}} - m) = 0 \tag{9}$$

since

$$0 = 0^{\dagger} = [(\hat{\not{\mathbf{p}}} - m) \ \psi_1]^{\dagger} = [\gamma^0 (\hat{\not{\mathbf{p}}} - m) \ \psi_1]^{\dagger} = \psi_1^{\dagger} \ (\hat{\not{\mathbf{p}}}^{\dagger} - m) \ \boldsymbol{\gamma}^{0\dagger}$$

$$= \psi_1^{\dagger} \boldsymbol{\gamma}^0 \ (\boldsymbol{\gamma}^0 \hat{\not{\mathbf{p}}}^{\dagger} - \boldsymbol{\gamma}^0 m) \ \boldsymbol{\gamma}^0 = \psi_1^{\dagger} \boldsymbol{\gamma}^0 \ (\boldsymbol{\gamma}^0 \hat{p}_{\mu}^{\dagger} \boldsymbol{\gamma}^{\mu\dagger} - \boldsymbol{\gamma}^0 m) \ \boldsymbol{\gamma}^0$$

$$= \bar{\psi}_1 (\hat{p}_\mu^\dagger \boldsymbol{\gamma}^0 \boldsymbol{\gamma}^{\mu\dagger} - \boldsymbol{\gamma}^0 m)\, \boldsymbol{\gamma}^0 = \bar{\psi}_1 (\hat{p}_\mu^\dagger \boldsymbol{\gamma}^\mu \boldsymbol{\gamma}^0 - m\boldsymbol{\gamma}^0)\, \boldsymbol{\gamma}^0$$

$$= \bar{\psi}_1 (\hat{p}_\mu^\dagger \boldsymbol{\gamma}^\mu - m) = \bar{\psi}_1 (- \overleftarrow{\hat{\mathbf{p}}} - m)$$

where $\qquad \bar{\psi}\overleftarrow{\hat{\mathbf{p}}} \equiv (\hat{p}_\mu \bar{\psi})\, \boldsymbol{\gamma}^\mu.$

Multiplying eq. (6) on the left by $\bar{\psi}_2 \boldsymbol{\gamma}^\mu$ and eq. (9) by $\boldsymbol{\gamma}^\mu \psi_1$ on the right and adding, we get

$$\bar{\psi}_2 \boldsymbol{\gamma}^\mu (\hat{\mathbf{p}} - m)\, \psi_1 + \bar{\psi}_2 (- \overleftarrow{\hat{\mathbf{p}}} - m)\, \boldsymbol{\gamma}^\mu \psi_1 = 0$$

$$= (\bar{\psi}_2 \boldsymbol{\gamma}^\mu \hat{\mathbf{p}}\, \bar{\psi}_1 - \bar{\psi}_2 \overleftarrow{\hat{\mathbf{p}}}\, \boldsymbol{\gamma}^\mu \bar{\psi}_1) - 2m\, \bar{\psi}_2 \boldsymbol{\gamma}^\mu \psi_1 \qquad (10)$$

whence

$$\bar{\psi}_2 \boldsymbol{\gamma}^\mu \psi_1 = \frac{1}{2m} (\bar{\psi}_2 \boldsymbol{\gamma}^\mu \hat{\mathbf{p}}\, \psi_1 - \bar{\psi}_2 \overleftarrow{\hat{\mathbf{p}}}\, \boldsymbol{\gamma}^\mu \psi_1)$$

$$= \frac{1}{2m} (\bar{\psi}_2 \boldsymbol{\gamma}^\mu \boldsymbol{\gamma}^\nu \hat{p}_\nu \psi_1 - \hat{p}_\nu \bar{\psi}_2 \boldsymbol{\gamma}^\nu \boldsymbol{\gamma}^\mu \psi_1) \qquad (11)$$

Now, substituting

$$\boldsymbol{\gamma}^\mu \boldsymbol{\gamma}^\nu = \frac{1}{2} (\boldsymbol{\gamma}^\mu \boldsymbol{\gamma}^\nu + \boldsymbol{\gamma}^\nu \boldsymbol{\gamma}^\mu) + \frac{1}{2} (\boldsymbol{\gamma}^\mu \boldsymbol{\gamma}^\nu - \boldsymbol{\gamma}^\nu \boldsymbol{\gamma}^\mu) = g^{\mu\nu} - i\hat{\sigma}^{\mu\nu} \qquad (12)$$

(because $\frac{i}{2} (\boldsymbol{\gamma}^\mu \boldsymbol{\gamma}^\nu - \boldsymbol{\gamma}^\nu \boldsymbol{\gamma}^\mu) = \hat{\sigma}^{\mu\nu}$) and similarly $\boldsymbol{\gamma}^\nu \boldsymbol{\gamma}^\mu = g^{\nu\mu} - i\hat{\sigma}^{\nu\mu}$ in eq. (11), we get

$$\bar{\psi}_2 \boldsymbol{\gamma}^\mu \psi_1 = \frac{1}{2m} [\bar{\psi}_2 (g^{\mu\nu} - i\hat{\sigma}^{\mu\nu})\, \hat{p}_\nu \psi_1 - \hat{p}_\nu \bar{\psi}_2 (g^{\nu\mu} - i\hat{\sigma}^{\nu\mu})\, \psi_1)]$$

$$= \frac{1}{2m} [\bar{\psi}_2\, \hat{p}^\mu \psi_1 - (\hat{p}^\mu \bar{\psi}_2)\, \psi_1 - i\bar{\psi}_2\, \hat{\sigma}^{\mu\nu}\, \hat{p}_\nu \psi_1 + i\, (\hat{p}_\nu \bar{\psi}_2)\, \hat{\sigma}^{\nu\mu}\, \psi_1]$$

$$= \frac{1}{2m} [\bar{\psi}_2\, \hat{p}^\mu \psi_1 - (\hat{p}^\mu \bar{\psi}_2)\, \psi_1 - i\bar{\psi}_2\, \hat{\sigma}^{\mu\nu}\, \hat{p}_\nu \psi_1 - i\, (\hat{p}_\nu \bar{\psi}_2)\, \hat{\sigma}^{\mu\nu} \psi_1]$$

$$= \frac{1}{2m} [\bar{\psi}_2\, \hat{p}^\mu \psi_1 - (\hat{p}^\mu \bar{\psi}_2)\, \psi_1] - \frac{i}{2m}\, \hat{p}_\nu (\bar{\psi}_2\, \hat{\sigma}^{\mu\nu} \psi_1) \qquad (13)$$

This constitutes the "Gordon Decomposition" of the Dirac current density $\bar{\psi}_2\, \boldsymbol{\gamma}^\mu\, \psi_1$. The implication of this decomposition is that the Dirac current density $\bar{\psi}_2\, \boldsymbol{\gamma}^\mu\, \psi_1$ can be split into two components, viz.

(a) the convection current density $\dfrac{1}{2m}\, [\bar{\psi}_2\, \hat{p}^\mu \psi_1 - (\hat{p}^\mu \bar{\psi}_2)\, \psi_1]$ and (14)

(b) the spin current density $-\dfrac{i}{2m}\,\hat{p}_\nu\,(\bar{\psi}_2\,\hat{\sigma}^{\mu\nu}\,\psi_1)$ (15)

10.3 CURRENT OF A POSITIVE ENERGY DIRAC WAVE PACKET

The current of Dirac wave packet is given by the expectation value of the velocity operator $\hat{\alpha}$. For a positive energy wave packet, the expression for the current, therefore, becomes

$$J^{(+)} = \int \psi^{(+)\dagger}\;\hat{\alpha}\,\psi^{(+)}\,d\mathbf{x} \tag{16}$$

We compute the component-wise expectation value of this current. We have

$$J^{k(+)} = \int \psi^{(+)\dagger}\;\hat{\alpha}_k\psi^{(+)}\,d\mathbf{x} = \int \psi^{(+)\dagger}\;\gamma^0\gamma^k\psi^{(+)}\,d\mathbf{x} = \int \overline{\psi}^{(+)}\;\gamma^k\psi^{(+)}\,d\mathbf{x}$$

$$= \int d\mathbf{x} \iint \frac{d\mathbf{p}\,d\mathbf{p}'}{(2\pi)^{3/2}\,(2\pi)^{3/2}}\sqrt{\frac{m}{E}}\sqrt{\frac{m}{E'}}$$

$$\times \sum_{\pm s,\pm s'} b^*(p',s')b(p,s)e^{i(p'^\mu - p^\mu)x_\mu}\bar{u}(p',s')\gamma^k u(p,s) \tag{17}$$

Using the Gordon decomposition, and noting that $\hat{p}^\mu u\,(p,s) = p^\mu u\,(p,s)$, $\hat{p}^\mu u\,(p',s') = p'^\mu\,u\,(p',s')$, where p^μ, p'^μ are the respective eigenvalues of the momentum operator, we can write

$$\bar{u}\,(p',s')\,\boldsymbol{\gamma}^k u\,(p,s) = \frac{1}{2m}\,\bar{u}\,(p',s')\,[p^k + p'^k]\,u\,(p,s)$$

$$-\frac{i}{2m}\,\bar{u}\,(p',s')\,[p_\nu - p'_\nu]\,\hat{\sigma}^{k\nu}u\,(p,s)$$

$$= \frac{1}{2m}\,\bar{u}\,(p',s')\,[(p^k + p'^k] - i\,(p_\nu - p'_\nu)\,\hat{\sigma}^{k\nu}]\,u\,(p,s) \tag{18}$$

Furthermore, we also have

$$\frac{1}{(2\pi)^3}\int e^{-i(\mathbf{p}'-\mathbf{p})\mathbf{x}}\,d\mathbf{x} = \delta\,(\mathbf{p}'-\mathbf{p}) \tag{19}$$

$$\int d\mathbf{p}'\;e^{i(p'_0 - p_0)x^0}\,\delta\,(\mathbf{p}'-\mathbf{p}) = 1 \tag{20}$$

and

$$\iint d\mathbf{p}\,d\mathbf{p}'\sqrt{\frac{m}{E}}\sqrt{\frac{m}{E'}}\;\sum_{\pm s,\pm s'} b^*(p',s')\,b\,(p,s)$$

$$e^{i(p'^0 - p^0)\,x_0}\,\bar{u}\,(p',s')\,\boldsymbol{\gamma}^k u\,(p,s)\,\delta\,(\mathbf{p}'-\mathbf{p})$$

$$= \frac{m}{E}\int d\mathbf{p}\;\sum_{\pm s,\pm s'} b^*\;(p,s')\,b\,(p,s)\,\bar{u}\,(p,s')\,\boldsymbol{\gamma}^k u\,(p,s) \tag{21}$$

Using eqs. (18), (19), (20) and (21), eq. (17) redces to

$$\boldsymbol{J}^{k(+)} = \frac{m}{E}\int d\mathbf{p} \sum_{\pm s,\pm s'} b^*(p, s')\, b(p, s)\, \bar{u}(p, s')\, \boldsymbol{\gamma}^k u(p, s)$$

$$= \frac{m}{E}\int d\mathbf{p} \sum_{\pm s,\pm s'} b^*(p, s')\, b(p, s)\left[\frac{1}{2m}\bar{u}(p, s')(2p^k)\, u(p, s)\right]$$

$$= \frac{m}{E}\int d\mathbf{p} \sum_{\pm s,\pm s'} b^*(p, s')\, b(p, s)\left[\frac{p^k}{m}\delta_{ss'}\right]$$

$$= \frac{1}{E}\int d\mathbf{p}\, p^k \sum_{\pm s} |b(p, s)|^2 \tag{22}$$

In view of the normalization (2), the final term on the right-hand side of eq. (22) can be written as the expectation $\left\langle \frac{p^k}{E} \right\rangle_+$ in the wave packet consisting of positive energy states.

Thus, the mean current of an arbitrary wave packet of plane waves of positive energy is equal to the expectation value of classical group velocity $\frac{\mathbf{p}}{E}$. This is the equivalence of Ehrenfest's theorem in relativistic quantum mechanics. It is also seen that the mean current is independent of the spin current density.

10.4 THE GENERALIZED DIRAC WAVE PACKET

Wave packets consisting of both positive and negative energy plane waves may be represented by

$$\psi(x) = \frac{1}{(2\pi)^{3/2}}\sqrt{\frac{m}{E}}\int d\mathbf{p}$$

$$\sum_{\pm s}\left[b(p, s)\, u(p, s)\, e^{-ip_\mu x^\mu} + d^*(p, s)\, v(p, s)\, e^{+ip_\mu x^\mu}\right] \tag{23}$$

The coefficients $b(p, s)$ represent the probability amplitudes for positive energy plane waves and $d^*(p, s)$ are the probability amplitudes for the negative energy waves. In the sequel, we shall use $u(-p, s) = u((\mathbf{p}^2 + m^2)^{1/2}, -\mathbf{p}, s)$ etc.

10.5 NORMALIZATION OF A GENERALIZED WAVE PACKET

The zero component of the current represents the probability current so that

$$J^0 = \int d\mathbf{x}\bar{\psi}(x)\gamma^0\psi(x) = \int d\mathbf{x}\psi^\dagger(x)\gamma^0.\gamma^0\psi(x)$$

$$= \int d\mathbf{x} \psi^{\dagger}(x)\, \psi(x) \tag{24}$$

Using eq. (23) in (24), we get

$$J^0 = \iiint \frac{d\mathbf{x}\, d\mathbf{p}\, d\mathbf{p}'}{(2\pi)^3} \frac{m}{\sqrt{EE'}}$$

$$\times \sum_{\pm s, \pm s'} \begin{bmatrix} b^*(p', s') b(p, s) u^{\dagger}(p', s') u(p, s) e^{-i(p-p')x} \\ + b^*(p', s') d^*(p, s) u^{\dagger}(p', s') \mathrm{v}(p, s) e^{i(p+p')x} \\ + d(p', s') b(p, s) \mathrm{v}^{\dagger}(p', s') u(p, s) e^{-i(p+p')x} \\ + d(p', s') d^*(p, s) \mathrm{v}^{\dagger}(p', s') \mathrm{v}(p, s) e^{i(p-p')x} \end{bmatrix} \tag{25}$$

When we integrate the above expression with respect to **x**, we get $\frac{1}{(2\pi)^3} \int e^{-i(\mathbf{p}'\pm\mathbf{p})\mathbf{x}}\, d\mathbf{x} = \delta$ $(\mathbf{p}' \pm \mathbf{p})$. On substituting this Dirac δ function in eq. (25), and thereafter integrating with respect to $\mathbf{p}'$, making use of the property of the Dirac δ function, viz. $\int f(x)\ \delta\,(x - a)\, dx = f(a)$, the result would be the substitution of $\mathbf{p}' = +\,\mathbf{p}$ and $\mathbf{p}' = -\,\mathbf{p}$ depending on whether $\delta(\mathbf{p}' - \mathbf{p})$ or $\delta(\mathbf{p}' + \mathbf{p})$ is associated with that term respectively. However, we further have $p_0 = E = \sqrt{\mathbf{p}^2 + m^2} = \sqrt{\mathbf{p}'^2 + m^2}$ $= E' = p_0'$, i.e., $\mathbf{p}' = \pm\,\mathbf{p} \Rightarrow p_0' = p_0$. Hence, on performing the **x** and $\mathbf{p}'$ integrations in eq. (25) we end up with

$$J^0 = \frac{m}{E} \int d\mathbf{p} \sum_{\pm s, \pm s'} \begin{bmatrix} b^*(p, s') b(p, s) u^{\dagger}(p, s') u(p, s) \\ + b^*(-p, s') d^*(p, s) u^{\dagger}(-p, s') \mathrm{v}(p, s) e^{2ip_0x^0} \\ + d(-p, s') b(p, s) \mathrm{v}^{\dagger}(-p, s') u(p, s) e^{-2ip_0x^0} \\ + d(p, s') d^*(p, s) \mathrm{v}^{\dagger}(p, s') \mathrm{v}(p, s) \end{bmatrix} \tag{26}$$

where we have used

$$\iiint \frac{d\mathbf{x}\, d\mathbf{p}\, d\mathbf{p}'}{(2\pi)^3} \frac{m}{\sqrt{EE'}}\ b^*(p', s')\, b(p, s)\, u^{\dagger}(p', s')\, u(p, s)\, e^{-i(p-p')x}$$

$$= \iiint \frac{d\mathbf{x}\, d\mathbf{p}\, d\mathbf{p}'}{(2\pi)^3} \frac{m}{\sqrt{EE'}}\ b^*(p', s')\, b(p, s)\, u^{\dagger}(p', s')\, u(p, s)\, e^{-i(p_0 - p_0')x^0 + i(\mathbf{p} - \mathbf{p}')\mathbf{x}}$$

$$= \iint d\mathbf{p}\, d\mathbf{p}' \frac{m}{\sqrt{EE'}}\ b^*(p', s')\, b(p, s)\, u^{\dagger}(p', s')\, u(p, s)\, e^{-i(p_0 - p_0')x^0}\, \boldsymbol{\delta}\,(\mathbf{p} - \mathbf{p}')$$

$$= \frac{m}{E} \int d\mathbf{p}\, b^* (p, s')\, b\, (p, s)\, u^\dagger\, (p, s')\, u\, (p, s) \tag{27}$$

$$\iiint \frac{d\mathbf{x}\, d\mathbf{p}\, d\mathbf{p}'}{(2\pi)^3} \frac{m}{\sqrt{EE'}}\, b^* (p', s')\, d^* (p, s)\, u^\dagger\, (p', s')\, \nu\, (p, s)\, e^{i(p+p')x}$$

$$= \iiint \frac{d\mathbf{x}\, d\mathbf{p}\, d\mathbf{p}'}{(2\pi)^3} \frac{m}{\sqrt{EE'}}\, b^* (p', s')\, d^* (p, s)\, u^\dagger\, (p', s')$$

$$\nu\, (p, s)\, e^{i(p_0 + p_0')x^0 - i(\mathbf{p} + \mathbf{p}')\mathbf{x}}$$

$$= \iint d\mathbf{p}\, d\mathbf{p}' \frac{m}{\sqrt{EE'}}\, b^* (p', s')\, d^* (p, s)\, u^\dagger\, (p', s')\, \nu\, (p, s)\, e^{i(p_0 + p_0')x^0}\, \boldsymbol{\delta}(\mathbf{p} + \mathbf{p}')$$

$$= \frac{m}{E} \int d\mathbf{p}\, b^* (-p, s')\, d^* (p, s)\, u^\dagger\, (-p, s')\, \nu\, (p, s)\, e^{2ip_0 x^0} \tag{28}$$

and similarly for the remaining two terms.

We have proved in earlier chapters, the following relations for the Dirac spinors

$$u^\dagger\, (p, s')\, u\, (p, s) = \nu^\dagger\, (p, s')\, \nu\, (p, s) = \frac{E}{m}\, \delta_{ss'} \tag{29}$$

$$u^\dagger\, (-p, s')\, \nu\, (p, s) = \nu^\dagger\, (-p, s')\, u\, (p, s) = 0 \tag{30}$$

and, therefore

$$J^0 = \int d\mathbf{p} \sum_{\pm s} \left\{ |b(p, s)|^2 + |d\,(p, s)^2 \right\} = 1 \tag{31}$$

10.6 CURRENT DENSITY OF GENERALIZED WAVE PACKET

We now proceed to calculate the space components of the current J^k of the generalized Dirac wave packet. We have

$$J^k = \int \psi^\dagger\, (x)\, \hat{\alpha}_k\, \psi(x)\, d\mathbf{x} = \int \overline{\psi}(x)\, \gamma^k\, \psi(x)\, d\mathbf{x} \tag{32}$$

which, using the Gordon decomposition, becomes

$$J^k = \frac{1}{2m} \int d\mathbf{x} \left[\overline{\psi} \hat{p}^k \psi - \left(\hat{p}^k \overline{\psi} \right) \psi - i \hat{p}_\nu \left(\overline{\psi} \hat{\sigma}^{k\nu} \psi \right) \right] \tag{33}$$

Substituting from eq. (23), we get

$$J^k = \frac{1}{(2\pi)^3} \frac{1}{2m} \frac{m}{\sqrt{EE'}} \iiint \frac{d\mathbf{x}\, d\mathbf{p}\, d\mathbf{p}'}{(2\pi)^3}$$

$$\times\sum_{\pm s,s'}\begin{bmatrix} b^*(p',s')b(p,s)\bar{u}(p',s')u(p,s)\,p^k\,e^{-i(p-p')x} \\ -b^*(p',s')d^*(p,s)\bar{u}(p',s')v(p,s)\,p^k\,e^{i(p+p')x} \\ +d(p',s')b(p,s)\bar{v}(p',s')u(p,s)\,p^k\,e^{-i(p+p')x} \\ -d(p',s')d^*(p,s)\bar{v}(p',s')v(p,s)\,p^k\,e^{i(p-p')x} \\ +b^*(p',s')b(p,s)\bar{u}(p',s')u(p,s)\,p'^k\,e^{-i(p-p')x} \\ -b^*(p',s')d^*(p,s)\bar{u}(p',s')v(p,s)\,p'^k\,e^{i(p+p')x} \\ +d(p',s')b(p,s)\bar{v}(p',s')u(p,s)\,p'^k\,e^{-i(p+p')x} \\ -d(p',s')d^*(p,s)\bar{v}(p',s')v(p,s)\,p'^k\,e^{i(p-p')x} \\ -ib^*(p',s')b(p,s)\bar{u}(p',s')\hat{\sigma}^{k\nu}u(p,s)(p_\nu-p'_\nu)e^{-i(p-p')x} \\ +ib^*(p',s')d^*(p,s)\bar{u}(p',s')\hat{\sigma}^{k\nu}v(p,s)(p_\nu+p'_\nu)e^{i(p+p')x} \\ -id(p',s')b(p,s)\bar{v}(p',s')\hat{\sigma}^{k\nu}u(p,s)(p_\nu+p'_\nu)e^{-i(p+p')x} \\ +id(p',s')d^*(p,s)\hat{v}(p',s')\hat{\sigma}^{k\nu}v(p,s)(p_\nu-p'_\nu)e^{i(p-p')x} \end{bmatrix} \tag{34}$$

Proceeding in the same manner as in the previous case, we obtain, on integrating with respect to **x**, the Dirac delta function δ (**p′** ± **p**). Substituting this in eq. (34) and thereafter integrating with respect to **p′**, we obtain

$$J^k = \frac{1}{2m}\int d\mathbf{p}\,\frac{m}{E}\sum_{\pm s,\pm s'}\left\{\begin{array}{l} 2p^k\begin{bmatrix} b^*(p,s')b(p,s)\bar{u}(p,s')u(p,s) \\ -d(p,s')d^*(p,s)\bar{v}(p,s')v(p,s) \end{bmatrix} \\ +2p^k\begin{bmatrix} d(-p,s')b(p,s)\bar{v}(-p,s')\,u(p,s)e^{-2ip_0x^0} \\ -b^*(-p,s')d^*(p,s)\bar{u}(-p,s')v(p,s)e^{-2ip_0x^0} \end{bmatrix} \\ +2iE\begin{bmatrix} b^*(-p,s')d^*(p,s)\bar{u}(-p,s')\hat{\sigma}^{k0}v(p,s)e^{2ip_0x^0} \\ -d(-p,s')b^*(p,s)\bar{v}(-p,s')\hat{\sigma}^{k0}u(p,s)e^{-2ip_0x^0} \end{bmatrix} \end{array}\right\} \tag{35}$$

The Dirac spinors satisfy the following orthogonality relations:

$$\bar{u}(p,s')\,u\,(p,s) = \delta_{ss'} = -\,\bar{v}(p,s')\,v(p,s) \tag{36}$$

and

$$\bar{u}\,(-p,s')\;\;v(p,s) = 0 = \bar{v}(-p,s')\,u\,(p,s) \tag{37}$$

Using these orthogonality relations in eq. (35), we obtain

$$J^k = \int d\mathrm{p} \left\{ \begin{array}{l} \sum\limits_{\pm s} \left[|b(p,s')|^2 + |d(p,s)|^2 \right] \dfrac{p^k}{E} \\ +i \sum\limits_{\pm s, \pm s'} \begin{bmatrix} b^*(-p,s')d^*(p,s)\bar{u}(-p,s')\hat{\sigma}^{k0} v(p,s) e^{2ip_0 x^0} \\ -d(-p,s')b(p,s)\bar{v}(-p,s')\hat{\sigma}^{k0} u(p,s) e^{-2ip_0 x^0} \end{bmatrix} \end{array} \right\} \tag{38}$$

The first term on the right-hand side is the analogue of eq. (22), whereas the remaining two terms are a fallout of the interference of the positive and negative energy solutions – they constitute the Zitterbewegung. These interferences between positive and negative energy solutions result in time-dependent oscillations as a result of the factors $e^{\pm 2ip_0 x^0}$.

CHAPTER 11

Angular Momentum and the Dirac Equation for the Hydrogen Atom

11.1 ORBITAL, SPIN AND TOTAL ANGULAR MOMENTUM

In this chapter, we obtain the exact solution of the central force Dirac equation for a spin $\frac{1}{2}$ particle (electron) in a spherically symmetric Coulomb potential

$$V(\mathbf{r}) = -\frac{Ze^2}{\mathbf{r}} \tag{1}$$

The Dirac Hamiltonian, in this case, gets modified to

$$\hat{H} = \boldsymbol{\alpha}\hat{\mathbf{p}} + \boldsymbol{\beta} m + V(\mathbf{r}) \tag{2}$$

with $V(\mathbf{r})$ being given by eq. (1).

We have already shown that the free particle Dirac Hamiltonian

$$\hat{H}_0 = \boldsymbol{\alpha}\hat{\mathbf{p}} + \boldsymbol{\beta} m \tag{3}$$

commutes with the total angular momentum operator $\hat{\mathbf{J}} = \hat{\mathbf{L}} + \hat{\mathbf{S}}$, where $\hat{\mathbf{L}} = \mathbf{r} \times \hat{\mathbf{p}}$ is the orbital angular momentum operator and $\hat{\mathbf{S}} = \frac{1}{2}\Sigma = \frac{1}{2}\begin{pmatrix} \sigma & 0 \\ 0 & \sigma \end{pmatrix}$ is the spin angular momentum that is intrinsic to the particle, i.e.,

$$[\hat{H}_0, \hat{J}_1] = [\boldsymbol{\alpha}\hat{\mathbf{p}} + \boldsymbol{\beta} m, \hat{L}_1 + \hat{S}_1] = 0 \tag{4}$$

for

$$[\hat{H}_0, \hat{L}_1] = (-i\boldsymbol{\alpha}\nabla + \boldsymbol{\beta} m)\,[(-i)\,(x^2\partial_3 - x^3\partial_2)]$$

$$
\begin{aligned}
&- [(- i)\,(x^2 \partial_3 - x^3 \partial_2)]\,(- i\boldsymbol{\alpha}\boldsymbol{\nabla} + \boldsymbol{\beta} m) \\
&= - (\boldsymbol{\alpha}^1 \partial_1 + \boldsymbol{\alpha}^2 \partial_2 + \boldsymbol{\alpha}^3 \partial_3)\,(x^2 \partial_3 - x^3 \partial_2) \\
&\quad + (x^2 \partial_3 - x^3 \partial_2)\,(\boldsymbol{\alpha}^1 \partial_1 \ \ \boldsymbol{\alpha}^2 \partial_2 + \boldsymbol{\alpha}^3 \partial_3) \\
&= (\boldsymbol{\alpha}^3 \partial_2 - \boldsymbol{\alpha}^2 \partial_3)
\end{aligned} \tag{5}
$$

and

$$
[\hat{H}_0, \hat{S}_1] = \left[-i\alpha\nabla + \beta m, \frac{1}{2}\Sigma_1 \right] = - \frac{i}{2}\,[\boldsymbol{\alpha}\boldsymbol{\nabla}, \boldsymbol{\Sigma}_1] + \frac{1}{2}\,[\boldsymbol{\beta} m, \boldsymbol{\Sigma}_1]
$$

$$
= - \frac{i}{2} \left[\begin{pmatrix} \mathbf{0} & \sum_{j=1}^{3} \hat{\sigma}_j \partial_j \\ \sum_{j=1}^{3} \sigma_j \partial_j & \mathbf{0} \end{pmatrix}, \begin{pmatrix} \sigma_1 & \mathbf{0} \\ \mathbf{0} & \sigma_1 \end{pmatrix} \right] + \frac{m}{2} \left[\begin{pmatrix} \mathbf{I} & \mathbf{0} \\ \mathbf{0} & -\mathbf{I} \end{pmatrix}, \begin{pmatrix} \sigma_1 & \mathbf{0} \\ \mathbf{0} & \sigma_1 \end{pmatrix} \right]
$$

$$
= - \frac{i}{2} \left[\begin{pmatrix} \mathbf{0} & \partial_1 - i\sigma_3\partial_2 + i\sigma_2\partial_3 \\ \partial_1 - i\sigma_3\partial_2 + i\sigma_2\partial_3 & \mathbf{0} \end{pmatrix} \right.
$$

$$
\left. - \begin{pmatrix} \mathbf{0} & \partial_1 + i\sigma_3\partial_2 - i\sigma_2\partial_3 \\ \partial_1 - i\sigma_3\partial_2 - i\sigma_2\partial_3 & \mathbf{0} \end{pmatrix} \right]
$$

$$
= \begin{pmatrix} \mathbf{0} & -\sigma_3\partial_2 + \sigma_2\partial_3 \\ -\sigma_3\partial_2 + \sigma_2\partial_3 & \mathbf{0} \end{pmatrix}
$$

$$
= - \begin{pmatrix} \mathbf{0} & \sigma_3 \\ \sigma_3 & \mathbf{0} \end{pmatrix} \partial_2 + \begin{pmatrix} \mathbf{0} & \sigma_2 \\ \sigma_2 & \mathbf{0} \end{pmatrix} \partial_3 = - \boldsymbol{\alpha}^3 \partial_2 + \boldsymbol{\alpha}^2 \partial_3 \tag{6}
$$

Adding eqs. (5) and (6), we get eq. (4) and hence, conclude that the x^1 component of the total angular momentum operator commutes with the free particle Hamiltonian. Similarly, it can be shown that $[\hat{H}_0, \hat{J}_2] = [\hat{H}_0, \hat{J}_3] = 0$ so that the total angular momentum commutes with the free particle Hamiltonian. Furthermore,

$$
\begin{aligned}
[V(\mathbf{r}), \hat{\mathbf{L}}]\,\phi(x) &= [V(\mathbf{r}), \mathbf{r} \times \hat{\mathbf{p}}]\,\phi(x) = - i\,[V(\mathbf{r}), \mathbf{r} \times \boldsymbol{\nabla}]\,\phi(x) \\
&= i\,\{\mathbf{r} \times \boldsymbol{\nabla}\,[V(\mathbf{r})\,\phi(x)] - V(\mathbf{r})\,\mathbf{r} \times \boldsymbol{\nabla}\phi(x)\} \\
&= i\,\{\mathbf{r} \times [\boldsymbol{\nabla} V(\mathbf{r})]\,\phi(x) + \mathbf{r} \times [V(\mathbf{r})]\,\boldsymbol{\nabla}\phi(x) - V(\mathbf{r})\,\mathbf{r} \times \boldsymbol{\nabla}\phi(x)\} \\
&= i\,[V(\mathbf{r})\,\mathbf{r} \times \boldsymbol{\nabla}\phi(x) - V(\mathbf{r})\,\mathbf{r} \times \boldsymbol{\nabla}\phi(x)] = 0
\end{aligned} \tag{7}
$$

and

$$
[V(\mathbf{r}), \hat{\mathbf{S}}] = 0 \tag{8}
$$

because $[V(\mathbf{r}), \boldsymbol{\sigma}] = 0$ (since the potential is spherically symmetric) so that the Dirac Hamiltonian of eq. (2) also commutes with the total angular momentum operator.

Now

$$[\hat{H}, \hat{J}_3] = [\hat{H}_0 + V(\mathbf{r}), \hat{J}_3] = [\hat{H}_0, \hat{J}_3] + [V(\mathbf{r}), \hat{J}_3]$$
$$= \left[V(\mathbf{r}), \hat{J}_3\right] = \left[V(\mathbf{r}), \hat{L}_3 + \hat{S}_3\right]$$
$$= \left[V(\mathbf{r}), \hat{L}_3\right] + \left[V(\mathbf{r}), \hat{S}_3\right] = 0 \tag{9}$$

where the last step follows in analogy with the derivations in eqs. (7) and (8).

Also

$$\left[\hat{H}, \hat{\mathbf{J}}^2\right] = \left[\hat{H}, \hat{\mathbf{J}}\right] \hat{\mathbf{J}} + \hat{\mathbf{J}} \left[\hat{H}, \hat{\mathbf{J}}\right] = 0 \tag{10}$$

Hence, we can compute simultaneous eigenfunctions for the three operators $\hat{H}$, $\hat{\mathbf{J}}^2$, $\hat{J}_3$ that are also constants of motion.

11.2 EIGENVALUE PROBLEM OF $\hat{\mathbf{J}}^2$, $\hat{\mathbf{J}}_3$

The eigenvalue problem of finding eigenvalues and simultaneous eigenfunctions of $\hat{\mathbf{J}}^2$, $\hat{J}_3$ is well documented and can be accessed in any book on non-relativistic theory of angular momentum. However, we give below the essentials of the solution to facilitate continuity.

Let us identify the common eigenfunctions of $\hat{\mathbf{J}}^2$ and $\hat{J}_3$ in terms of their respective eigenvalues α, β so that

$$\hat{\mathbf{J}}^2 |\alpha, \beta\rangle = \alpha |\alpha, \beta\rangle \tag{11}$$

and

$$\hat{J}_3 |\alpha, \beta\rangle = \beta |\alpha, \beta\rangle \tag{12}$$

We define the operators $\hat{J}_\pm = \hat{J}_1 \pm i\hat{J}_2$ so that

$$[\hat{J}_3, \hat{J}_\pm] = [\hat{J}_3, \hat{J}_1 \pm i\hat{J}_2] = [\hat{J}_3, \hat{J}_1]$$
$$\pm i [\hat{J}_3, \hat{J}_2] = i\hat{J}_2 \mp i^2\hat{J}_1 = \pm (\hat{J}_1 \pm i\hat{J}_2) = \pm \hat{J}_\pm \tag{13}$$

and

$$\left[\hat{\mathbf{J}}^2, \hat{J}_\pm\right] = \left[\hat{\mathbf{J}}^2, \hat{J}_1 \pm i\hat{J}_2\right] = \left[\hat{\mathbf{J}}^2, \hat{J}_1\right] \pm i \left[\hat{\mathbf{J}}^2, \hat{J}_2\right]$$
$$= \hat{\mathbf{J}} \left[\hat{\mathbf{J}}, \hat{J}_1\right] + \left[\hat{\mathbf{J}}, \hat{J}_1\right] \hat{\mathbf{J}} \pm i\hat{\mathbf{J}} \left[\hat{\mathbf{J}}, \hat{J}_2\right] \pm i \left[\hat{\mathbf{J}}, \hat{J}_2\right] \hat{\mathbf{J}} = 0 \tag{14}$$

We also have

$$\hat{J}_3 \left(\hat{J}_\pm |\alpha, \beta\rangle\right) = \left(\hat{J}_\pm \hat{J}_3 \pm \hat{J}_\pm\right) |\alpha, \beta\rangle = \hat{J}_\pm (\beta \pm 1) |\alpha, \beta\rangle = (\beta \pm 1) \hat{J}_\pm |\alpha, \beta\rangle \tag{15}$$

so that $\hat{J}_\pm$ raises/lowers the eigenvalue of $\hat{J}_3$ by one unit. However, the eigenvalue of $\hat{\mathbf{J}}^2$ is left unchanged, e.g.,

$$\hat{\mathbf{J}}^2 \left(\hat{J}_\pm |\alpha, \beta\rangle\right) = \hat{J}_\pm \hat{\mathbf{J}}^2 |\alpha, \beta\rangle = \hat{J}_\pm \alpha |\alpha, \beta\rangle = \alpha \hat{J}_\pm |\alpha, \beta\rangle \tag{16}$$

From eqs. (15) and (16) we infer that $\hat{J}_\pm |\alpha, \beta\rangle$ is an eigenstate of $\hat{J}_3$ with eigenvalue $\beta \pm 1$ and of $\hat{\mathbf{J}}^2$ with eigenvalue α, i.e.,

$$\hat{J}_\pm |\alpha, \beta\rangle = C_\pm (\alpha, \beta) |\alpha, \beta \pm 1\rangle \tag{17}$$

From the properties of $\hat{J}_\pm$ of eq. (15) it is also seen that given an eigenstate $|\alpha, \beta\rangle$ of the operators $\hat{J}_3$, $\mathbf{J}^2$ there would also exist states $|\alpha, \beta + 1\rangle$, $|\alpha, \beta + 2\rangle$, $|\alpha, \beta + 3\rangle$, ... and also $|\alpha, \beta - 1\rangle$, $|\alpha, \beta - 2\rangle$, $|\alpha, \beta - 3\rangle$, and so on, that can be generated from the eigenstate $|\alpha, \beta\rangle$ by successive application of the raising operator $\hat{J}_+$ and lowering operator $\hat{J}_-$ respectively.

This situation of unboundedness of the eigenvalues of $\hat{J}_3$ (given a value of the total angular momentum) is obviously, classically unacceptable for, we must have $j_3 \leq \left(j^2\right)^{\frac{1}{2}}$. Analogically, in quantum mechanics, we have

$$\begin{aligned} \langle \alpha, \beta | \alpha - \beta^2 | \alpha, \beta\rangle &= \langle \alpha, \beta | \alpha^* - \beta^2 | \alpha, \beta\rangle \\ &= \langle \alpha, \beta | \hat{\mathbf{J}}^2 - \hat{J}_3^2 | \alpha, \beta\rangle \\ &= \langle \alpha, \beta | \hat{\mathbf{J}}_1^2 + \hat{J}_2^2 | \alpha, \beta\rangle \end{aligned} \tag{18}$$

The last expression must necessarily be positive definite. Hence, we must have

$$\alpha - \beta^2 \geq 0 \tag{19}$$

This implies that β must be bounded. Therefore, there must exist an eigenstate, say $|\alpha, \beta_{max}\rangle$, that cannot be raised by the raising operator $\hat{J}_+$, i.e., we must have

$$\hat{J}_+ |\alpha, \beta_{max}\rangle = 0 \tag{20}$$

Similarly, there must also exist a state $|\alpha, \beta_{min}\rangle$ that cannot be lowered further by the lowering operator $\hat{J}_-$ without violating the boundedness condition of eq. (19). Therefore, we again have

$$\hat{J}_- |\alpha, \beta_{min}\rangle = 0 \tag{21}$$

Now, from eq. (20), operating on both sides with $\hat{J}_-$, we have

$$0 = \hat{J}_+ |\alpha, \beta_{max}\rangle = \hat{J}_- \hat{J}_+ |\alpha, \beta_{max}\rangle = \left(\hat{J}_1 - i\hat{J}_2\right)\left(\hat{J}_1 + i\hat{J}_2\right) |\alpha, \beta_{max}\rangle$$

$$= \left\{\hat{J}_2^1 + \hat{J}_2^2 + i\left[\hat{J}_1, \hat{J}_2\right]\right\} |\alpha, \beta_{max}\rangle = \left(\hat{\mathbf{J}}^2 - \hat{J}_3^2 - \hat{J}_3\right) |\alpha, \beta_{max}\rangle$$

$$= \left(\alpha - \beta_{max}^2 - \beta_{max}\right) |\alpha, \beta_{max}\rangle \tag{22}$$

which gives

$$\alpha = \beta_{max}(\beta_{max} + 1) \tag{23}$$

Similarly, operating on eq. (21) by $\hat{J}_+$ we have

$$0 = \hat{J}_- |\alpha, \beta_{min}\rangle = \hat{J}_+ \hat{J}_- |\alpha, \beta_{min}\rangle = \left(\hat{J}_1 + i\hat{J}_2\right)\left(\hat{J}_1 - i\hat{J}_2\right) |\alpha, \beta_{min}\rangle$$

$$= \left\{\hat{J}_1^2 + \hat{J}_2^2 - i\left[\hat{J}_1, i\hat{J}_2\right]\right\} |\alpha, \beta_{min}\rangle = \left(\hat{\mathbf{J}}^2 - \hat{J}_3^2 + \hat{J}_3\right) |\alpha, \beta_{min}\rangle$$

$$= \left(\alpha - \beta_{min}^2 + \beta_{min}\right) |\alpha, \beta_{min}\rangle \tag{24}$$

which gives

$$\alpha = \beta_{min}(\beta_{min} - 1) \tag{25}$$

Equations (23) and (25) imply

$$\beta_{min} = -\beta_{max} \tag{26}$$

Let us assume that in moving from β_{max} to β_{min}, the lowering operator $\hat{J}_-$ was applied k times on the eigenstate $|\alpha, \beta_{max}\rangle$. Then, since each operation on $|\alpha, \beta_{max}\rangle$ by $\hat{J}_-$ reduces the value of eigenvalue of $\hat{J}_3$, i.e., β by one unit in every step, we must have

$$\beta_{max} - \beta_{min} = 2\beta_{max} = k \text{ or } \beta_{max} = \frac{1}{2}k,\ k = 0, 1, 2, \ldots \tag{27}$$

i.e., k can assume any integral value. Therefore,

$$\alpha = \beta_{max}(\beta_{max} + 1) = \frac{k}{2}\left(\frac{k}{2} + 1\right),\ k = 0, 1, 2, 3, \ldots \tag{28}$$

Having obtained the eigenvalues of the operators $\hat{\mathbf{J}}^2$ and $\hat{J}_3$, we now proceed to determine their simultaneous eigenfunctions. Taking the adjoint of eq. (17), we obtain

$$\langle \alpha, \beta| \hat{J}_\mp = C_\pm^* (\alpha, \beta) \langle \alpha, \beta \pm 1| \tag{29}$$

Taking the inner product of the respective sides of eqs. (17) and (29), we obtain

$$\langle \alpha, \beta | \hat{J}_\mp \hat{J}_\pm | \alpha, \beta\rangle = C_\pm^* (\alpha, \beta)\, C_\pm (\alpha, \beta) \langle \alpha, \beta \pm 1 | \alpha, \beta \pm 1\rangle = |C_\pm (\alpha, \beta)|^2$$

or $$\langle \alpha, \beta | \hat{\mathbf{J}}^2 - \hat{J}_3^2 \pm \hat{J}_3 | \alpha, \beta\rangle = \langle \alpha, \beta | (\alpha - \beta^2 \mp \beta | \alpha, \beta)\rangle$$

$$= \alpha - \beta^2 \mp \beta = |C_{\pm}(\alpha, \beta)|^2 \tag{30}$$

Using eqs. (27) and (28), we can write eq. (30) as

$$C_{\pm}(\alpha, \beta) = \left[j(j+1) - m^2 \mp m\right]^{\frac{1}{2}} = \left[(j \mp m)(j \pm m + 1)\right]^{\frac{1}{2}} \tag{31}$$

where we have substituted $\alpha = j(j+1)$, $\beta = m, j\ \frac{k}{2} = 0, \frac{1}{2}, 1, \frac{3}{2}, \ldots$ and $m = j, j-1, j-2, \ldots, -j$.

Finally, therefore

$$\hat{J}_{\pm}\,|j, m\rangle = C_{\pm}(j, m)\,|j, m \pm 1\rangle = \left[(j \mp m)(j \pm m + 1)\right]^{\frac{1}{2}}\,|j, m \pm 1\rangle \tag{32}$$

where we have now identified the eigenfunctions by the respective numbers j, m that are related to the eigenvalues α, β by $\alpha = j(j+1)$, $\beta = m, j\ \frac{k}{2} = 0, \frac{1}{2}, 1, \frac{3}{2}, \ldots$

We can now directly write the matrix elements of $\hat{J}_1$, $\hat{J}_2$ as

$$\begin{aligned}\langle j'm'|\hat{J}_1|jm\rangle &= \frac{1}{2}\,\langle j'm'|\hat{J}_+ + \hat{J}_-|jm\rangle \\ &= \frac{1}{2}\left[C_-^*(j', m')\langle j', m'-1|jm\rangle + C_-(j, m)\langle j', m'|j, m-1\rangle\right] \\ &= \frac{1}{2}\left\{\delta_{jj'}\,\delta_{m', m+1}\left[(j-m)(j+m+1)\right]^{\frac{1}{2}}\right. \\ &\quad \left. + \delta_{jj'}\,\delta_{m', m-1}\left[(j+m)(j-m+1)\right]^{\frac{1}{2}}\right\}\end{aligned} \tag{33}$$

and similarly

$$\begin{aligned}\langle j'm'|\hat{J}_2|jm\rangle &= \frac{1}{2i}\,\langle j'm'|\hat{J}_+ - \hat{J}_-|jm\rangle \\ &= \frac{1}{2i}\left[C_-^*(j', m')\langle j', m'-1|jm\rangle - C_-(j, m)\langle j', m'|j, m-1\rangle\right] \\ &= \frac{1}{2i}\left\{\delta_{jj'}\,\delta_{m', m+1}\left[(j-m)(j+m+1)\right]^{\frac{1}{2}}\right. \\ &\quad \left. + \delta_{jj'}\,\delta_{m', m-1}\left[(j+m)(j-m+1)\right]^{\frac{1}{2}}\right\}\end{aligned} \tag{34}$$

Of course

$$\langle j'm'|\hat{J}_3|jm\rangle = m\,\langle j'\,m'|\,jm\rangle = m\,\delta_{jj'}\,\delta_{mm'} \tag{35}$$

$$\langle j'm'|\hat{\mathbf{J}}^2|jm\rangle = j\,(j+1)\,\langle j'm'|\,jm\rangle = j\,(j+1)\,\delta_{jj'}\,\delta_{mm'} \tag{36}$$

11.3 EIGENVALUE PROBLEM OF $\hat{\mathbf{L}}^2$, $\hat{\mathbf{L}}_3$

We now study the eigenvalue problem of the orbital angular momentum operator. As has been explained earlier, the orbital angular momentum does not commute with the Dirac Hamiltonian, e.g., eq. (5) and hence, we cannot have a simultaneous set of eigenfunctions of $\hat{H}$ and $\hat{\mathbf{L}}$. However, $\hat{\mathbf{L}}^2$, $\hat{L}_3$ do commute with each other and in the eigenvalue problem, we shall obtain the simultaneous eigenfunctions of these two operators. For this purpose, we introduce the operators

$$\hat{a} = \frac{1}{\sqrt{2}}\left(\hat{p}_1 - ix^1\right),\ \hat{b} = \frac{1}{\sqrt{2}}\left(\hat{p}_2 - ix^2\right) \text{ and}$$

$$\hat{c} = \hat{b} + i\hat{a} = \frac{1}{\sqrt{2}}\left(\hat{p}_2 - ix^2\right) + \frac{i}{\sqrt{2}}\left(\hat{p}_1 - ix^1\right)$$

$$= \frac{i}{\sqrt{2}}\left[\left(\hat{p}_1 - i\hat{p}_2\right) - i\left(x^1 - ix^2\right)\right] \tag{37}$$

We then have the following commutation relations for these operators

$$\left[\hat{a},\hat{a}^\dagger\right] = \left[\frac{1}{\sqrt{2}}\left(\hat{p}_1 - ix^1\right), \frac{1}{\sqrt{2}}\left(\hat{p}_1 + ix^1\right)\right]$$

$$= \frac{1}{2}\left\{\left[\hat{p}_1, \hat{p}_1\right] + i\left[\hat{p}_1, x^1\right] - i\left\{x^1, \hat{p}_1\right\} + \left[x^1, x^1\right]\right\}$$

$$= i\left[\hat{p}_1, x^1\right] = \left[\partial_1, x^1\right] = \partial_1\left(x^1.\right) - x^1\,\partial_1$$

$$= x^1\partial_1 + 1 - x^1\partial_1 = 1 \tag{38}$$

Similarly,

$$\left[\hat{b},\hat{b}^\dagger\right] = 1 \tag{39}$$

$$\left[\hat{b},\hat{a}\right] = \left[\frac{1}{\sqrt{2}}\left(\hat{p}_2 - ix^2\right), \frac{1}{\sqrt{2}}\left(\hat{p}_1 - ix^1\right)\right]$$

$$= \frac{1}{2}\left\{\left[\hat{p}_2, \hat{p}_1\right] - i\left[\hat{p}_2, x^1\right] - i\left\{x^2, \hat{p}_1\right\} + \left[x^2, x^1\right]\right\} = 0 \tag{40}$$

$$\left[\hat{b}, \hat{a}^{\dagger}\right] = \left[\hat{b}^{\dagger}, \hat{a}\right] = \left[\hat{b}^{\dagger}, \hat{a}^{\dagger}\right] = 0 \tag{41}$$

$$\left[\hat{c}, \hat{c}^{\dagger}\right] = \left[\hat{b}_{+} i\hat{a}, b^{\dagger} - i\hat{a}^{\dagger}\right]$$

$$= \left\{\left[\hat{b}, \hat{b}^{\dagger}\right] + i\left[\hat{a}, \hat{b}^{\dagger}\right] - i\left[\hat{b}, \hat{a}^{\dagger}\right] + \left[\hat{a}, \hat{a}^{\dagger}\right]\right\} = 2 \tag{42}$$

$$\left[\hat{c}, \hat{a}\right] = \left[\hat{b} + i\hat{a}, \hat{a}\right] = \left\{\left[\hat{b}, \hat{a}\right] + i\left[\hat{a}, \hat{a}\right]\right\} = 0 \tag{43}$$

$$\left[\hat{c}, \hat{a}^{\dagger}\right] = \left[\hat{b} + i\hat{a}, \hat{a}^{\dagger}\right] = \left\{\left[\hat{b}, \hat{a}^{\dagger}\right] + i\left[\hat{a}, \hat{a}^{\dagger}\right]\right\} = i \tag{44}$$

$$\left[\hat{c}, \hat{b}\right] = \left[\hat{b} + i\hat{a}, \hat{b}\right] = \left\{\left[\hat{b}, \hat{b}\right] + i\left[\hat{a}, \hat{b}\right]\right\} = 0 \tag{45}$$

$$\left[\hat{c}, \hat{b}^{\dagger}\right] = \left[\hat{b} + i\hat{a}^{\dagger}, \hat{b}^{\dagger}\right] = \left\{\left[\hat{b}, \hat{b}^{\dagger}\right] + i\left[\hat{a}, \hat{b}^{\dagger}\right]\right\} = 1 \tag{46}$$

$$\left[\hat{c}^{\dagger}, \hat{a}\right] = \left[\hat{b}^{\dagger} - i\hat{a}^{\dagger}, \hat{a}\right] = \left\{\left[\hat{b}^{\dagger}, \hat{a}\right] - i\left[\hat{a}^{\dagger}, \hat{a}\right]\right\} = i \tag{47}$$

$$\left[\hat{c}^{\dagger}, \hat{a}^{\dagger}\right] = \left[\hat{b}^{\dagger} - i\hat{a}^{\dagger}, \hat{a}^{\dagger}\right] = \left\{\left[\hat{b}^{\dagger}, \hat{a}^{\dagger}\right] - i\left[\hat{a}^{\dagger}, \hat{a}^{\dagger}\right]\right\} = 0 \tag{48}$$

$$\left[\hat{c}^{\dagger}, \hat{b}\right] = \left[\hat{b}^{\dagger} - i\hat{a}^{\dagger}, \hat{b}\right] = \left\{\left[\hat{b}^{\dagger}, \hat{b}\right] - i\left[\hat{a}^{\dagger}, \hat{b}\right]\right\} = -1 \tag{49}$$

$$\left[\hat{c}^{\dagger}, \hat{b}^{\dagger}\right] = \left[\hat{b}^{\dagger} - i\hat{a}^{\dagger}, \hat{b}^{\dagger}\right] = \left\{\left[\hat{b}^{\dagger}, \hat{b}^{\dagger}\right] - i\left[\hat{a}^{\dagger}, \hat{b}^{\dagger}\right]\right\} = 0 \tag{50}$$

$$\left[\hat{a}^{\dagger}\hat{a}, \hat{b}^{\dagger}\hat{b}\right] = \hat{a}^{\dagger}\left[\hat{a}, \hat{b}^{\dagger}\right]\hat{b} + \left[\hat{a}^{\dagger}, \hat{b}^{\dagger}\right]\hat{a}\hat{b} + \hat{b}^{\dagger}\hat{a}^{\dagger}\left[\hat{a}, \hat{b}\right] + \hat{b}^{\dagger}\left[\hat{a}^{\dagger}, \hat{b}\right]\hat{a} = 0 \tag{51}$$

$$\left[\hat{a}^{\dagger}\hat{a}, \hat{c}^{\dagger}\hat{c}\right] = \hat{a}^{\dagger}\left[\hat{a}, \hat{c}^{\dagger}\right]\hat{c} + \left[\hat{a}^{\dagger}, \hat{c}^{\dagger}\right]\hat{a}\hat{c}$$

$$+ \hat{c}^{\dagger}\hat{a}^{\dagger}\left[\hat{a}, \hat{c}\right] + \hat{c}^{\dagger}\left[\hat{a}^{\dagger}\hat{c}\right]\hat{a} = -i\hat{a}^{\dagger}\hat{c} - i\hat{c}^{\dagger}\hat{a} \tag{52}$$

$$\left[\hat{b}^{\dagger}\hat{b}, \hat{c}^{\dagger}\hat{c}\right] = \hat{b}^{\dagger}\left[\hat{b}, \hat{c}^{\dagger}\right]\hat{c} + \left[\hat{b}^{\dagger}, \hat{c}^{\dagger}\right]\hat{b}\hat{c}$$

$$+ \hat{c}^{\dagger}\hat{b}^{\dagger}\left[\hat{b}, \hat{c}\right] + \hat{c}^{\dagger}\left[\hat{b}^{\dagger}, \hat{c}\right]\hat{b} = \hat{b}^{\dagger}\hat{c} - \hat{c}^{\dagger}\hat{b} \tag{53}$$

$$\left[\hat{a}^{\dagger}\hat{a} + \hat{b}^{\dagger}\hat{b}, \hat{c}^{\dagger}\hat{c}\right] = \left[\hat{a}^{\dagger}\hat{a}, \hat{c}^{\dagger}\hat{c}\right] + \left[\hat{b}^{\dagger}\hat{b}, \hat{c}^{\dagger}\hat{c}\right] = -i\hat{a}^{\dagger}\hat{c} - i\hat{c}^{\dagger}\hat{a} + \hat{b}^{\dagger}\hat{c} - \hat{c}^{\dagger}\hat{b}$$

$$= \left(\hat{b}^{\dagger} - i\hat{a}^{\dagger}\right)\hat{c} - \hat{c}^{\dagger}\left(\hat{b} + i\hat{a}\right) = \hat{c}^{\dagger}\hat{c} - \hat{c}^{\dagger}\hat{c} = 0 \tag{54}$$

$$\hat{c}^{\dagger}\hat{c} = \left\{-\frac{i}{\sqrt{2}}\left[\left(\hat{p}_1 + i\hat{p}_2\right) + i\left(x^1 + ix^2\right)\right]\right\}\left\{\frac{i}{\sqrt{2}}\left[\left(\hat{p}_1 - i\hat{p}_2\right) - i\left(x^1 - ix^2\right)\right]\right\}$$

$$= \frac{1}{2}\left\{\left(p_1^2 + p_2^2\right) - i\left(\hat{p}_1 + i\hat{p}_2\right)\left(x^1 - ix^2\right)\right)$$

$$+ i\left(x^1 - ix^2\right)(\hat{p}_1 + i\hat{p}_2) + \left[\left(x^1\right)^2 + \left(x^2\right)^2\right]\Big\}$$

$$= \frac{1}{2}\Big\{\left[p_1^2 + \left(x^1\right)^2\right] + \left[p_2^2 + \left(x^2\right)^2\right] + i\left(x^1\hat{p}_1 - \hat{p}_1 x^1\right)$$

$$+ i\left(x^2\hat{p}_2 - \hat{p}_2 x^2\right) + 2\left(x^1\hat{p}_2 - `x^2\hat{p}_1\right)\Big\}$$

$$= \frac{1}{2}\Big\{\left[p_1^2 + \left(x^1\right)^2 + 1\right] + \left[p_2^2 + \left(x^2\right)^2 + 1\right] + 2\left(x^1\hat{p}_2 - `x^2\hat{p}_1\right)\Big\}$$

$$= \hat{a}^\dagger a + \hat{b}^\dagger\hat{b} + \hat{L}_3 \tag{55}$$

so that

$$\hat{L}_3 = \hat{c}^\dagger\hat{c} - \hat{a}^\dagger a - \hat{b}^\dagger\hat{b} \tag{56}$$

Now, because $\left[\hat{a}^\dagger\hat{a} + \hat{b}^\dagger\hat{b}, \hat{c}^\dagger\hat{c}\right] = 0$, i.e., the operator $\hat{c}^\dagger\hat{c}$ commutes with the operator $\hat{a}^\dagger\hat{a} + \hat{b}^\dagger\hat{b}$, it follows that there exist simultaneous eigenfunctions of the two operators and the eigenvalues of the sum (difference) of these two operators with respect to the simultaneous eigenfunctions would be the sum (difference) of the eigenvalues. Hence, the eigenvalues of $\hat{L}_3 = \hat{c}^\dagger\hat{c} - \left(\hat{a}^\dagger a + \hat{b}^\dagger\hat{b}\right)$ will be the difference of the eigenvalues of $\hat{c}^\dagger\hat{c}$ and $\hat{a}^\dagger\hat{a} + \hat{b}^\dagger\hat{b}$.

It is easily seen that if an operator satisfies the commutation relation $\left[\hat{O}, \hat{O}^\dagger\right] = \lambda$, then the eigenvalues of the operator $\hat{O}^\dagger, \hat{O}$ are given by $k\lambda$, $k = 0, 1, 2, 3, \ldots$ for, if $|k\rangle$ is an eigenvector of $\hat{O}^\dagger, \hat{O}$ corresponding to the eigenvalue $k\lambda$, i.e., $\hat{O}^\dagger\,\hat{O}|k\rangle = k\lambda\,|k\rangle$, then we have $k\lambda\hat{O}\,|k\rangle = \hat{O}k\lambda\,|k\rangle$ $= \hat{O}\left(\hat{O}^\dagger\hat{O}|k\rangle\right) = \left(\hat{O}\hat{O}^\dagger\right)\,\hat{O}|k\rangle = \left(\hat{O}^\dagger\hat{O} + \lambda I\right)\,\hat{O}|k\rangle$ so that

$$\left(\hat{O}^\dagger\hat{O}\right)\,\hat{O}|k\rangle = (k-1)\,\lambda\hat{O}|k\rangle \tag{57}$$

showing that the state $\hat{O}|k\rangle$ is also an eigenstate of $\left(\hat{O}^\dagger\hat{O}\right)$ with eigenvalue $(k-1)\lambda$.

Similarly,

$$k\lambda\hat{O}^\dagger|k\rangle = \hat{O}^\dagger k\lambda|k\rangle = \hat{O}^\dagger\,(\hat{O}^\dagger\hat{O}|k\rangle) = \hat{O}^\dagger\,(\hat{O}^\dagger\hat{O})|k\rangle$$
$$= \hat{O}^\dagger\,(\hat{O}\hat{O}^\dagger - \lambda I)|k\rangle = \hat{O}^\dagger\hat{O}\,(\hat{O}^\dagger|k\rangle) - \lambda I\hat{O}^\dagger|k\rangle$$

so that $$\hat{O}^\dagger\hat{O}\,(\hat{O}^\dagger|k\rangle) = (k+1)\,\lambda\,(\hat{O}^\dagger|k\rangle)$$

showing that the state $\hat{O}^\dagger|k\rangle$ is also an eigenstate of $\hat{O}^\dagger\hat{O}$ with eigenvalue $(k+1)\lambda$.

It follows from the above that

$$\hat{O}|k\rangle = C_{-}(k)\,|k-1\rangle \tag{58}$$

and

$$\hat{O}^{\dagger}|k\rangle = C_{+}(k)\,|k+1\rangle \tag{59}$$

Taking adjoint of eq. (58), we have

$$\langle k|\hat{O}^{\dagger} = C_{-}^{*}(k)\,|k+1\rangle \tag{60}$$

The inner product of the respective sides of eqs. (58) and (60) gives

$$\langle k|\hat{O}^{\dagger}\hat{O}\,|k\rangle = k\lambda = |C_{-}(k)|^{2} \tag{61}$$

and similarly

$$\langle k|\hat{O}\hat{O}^{\dagger}\,|k\rangle = \langle k|\hat{O}^{\dagger}\hat{O} + \lambda|k\rangle = (k+1)\lambda = |C_{+}(k)|^{2} \tag{62}$$

Equation (61) mandates that $k\lambda \geq 0$, i.e. all eigenvalues of the operator $\hat{O}^{\dagger}$, $\hat{O}$ must be non-negative. It follows, therefore, that there must exist some minimum eigenvalue, say k_{min} and the corresponding eigenvector $|k_{\text{min}}\rangle$ such that $\hat{O}|k_{\text{min}}\rangle = 0$ which is labelled as the ground state. Since every operation by $\hat{O}^{\dagger}$ raises the eigenvalue by λ, we also conclude that the set of eigenvalues of $\hat{O}^{\dagger}\hat{O}$ is given by $k\lambda$, $k = 0, 1, 2, 3, \ldots$.

We now establish the eigenvalues of $\hat{L}_3$ as follows:

(a) From eq. (56) the eigenvalues of $\hat{L}_3 = \hat{c}^{\dagger}\hat{c} - \left(\hat{a}^{\dagger}a + \hat{b}^{\dagger}\hat{b}\right)$ will be the difference of the eigenvalues of $\hat{c}^{\dagger}\hat{c}$ and $\hat{a}^{\dagger}\hat{a} + \hat{b}^{\dagger}\hat{b}$.

(b) Since $\left[\hat{c}, \hat{c}^{\dagger}\right] = 2$, the eigenvalues of $\hat{c}^{\dagger}\hat{c}$ will be given by $2k$, $k = 0, 1, 2, 3, \ldots$.

(c) Since $\left[\hat{a}, \hat{a}^{\dagger}\right] = \left[\hat{b}, \hat{b}^{\dagger}\right] = 1$, the eigenvalues of $\hat{a}^{\dagger}\, a$ will be given by l, $l = 0, 1, 2, 3, \ldots$ and those of $\hat{b}^{\dagger}, \hat{b}$ by m, $m = 0, 1, 2, 3, \ldots$.

(d) Therefore, the eigenvalues of $\hat{L}_3 = \hat{c}^{\dagger}\hat{c} - \left(\hat{a}^{\dagger}a + \hat{b}^{\dagger}\hat{b}\right)$ will be given by $2k - l - m$, $k, l, m = 0, 1, 2, 3, \ldots$ which can take only integral values $n = 2k - l - m = \ldots -3, -2, -1, 0, 1, 2, 3, \ldots$.

11.4 EIGENVALUES OF $\hat{S}_k$

The eigenvalue problem of the spin operator $\hat{S}_k$ can be resolved in several ways, e.g.,

(a) The spin angular momentum is defined as the vectorial difference of the total angular momentum and the orbital angular momentum and hence can be written as $\hat{S} = \hat{\mathbf{J}} - \hat{\mathbf{L}}$.

Now, since $\left[\hat{J}_k, \hat{L}_k\right] = \left[\hat{L}_k + \hat{S}_k, \hat{L}_k\right] = 0$, it follows that the eigenvalues of $\hat{S}_k = \hat{\mathrm{J}}_k - \hat{L}_k$ would be the difference of the eigenvalues of $\hat{J}_k, \hat{L}_k$.

(b) We define the spin operator explicitly as $\hat{S} = \frac{1}{2}\boldsymbol{\Sigma} = \frac{1}{2}\begin{pmatrix} \sigma & \mathbf{0} \\ \mathbf{0} & \sigma \end{pmatrix}$ or, in terms of components

$$\hat{S}_k = \frac{1}{2}\boldsymbol{\Sigma}_k = \frac{1}{2}\begin{pmatrix} \sigma_k & \mathbf{0} \\ \mathbf{0} & \sigma_k \end{pmatrix}$$

which gives

$$\hat{S}_k^2 = \frac{1}{4}\boldsymbol{\Sigma}_k^2 = \frac{1}{4}\begin{pmatrix} \sigma_k^2 & \mathbf{0} \\ \mathbf{0} & \sigma_k^2 \end{pmatrix} = \frac{1}{4}\begin{pmatrix} \mathbf{I} & \mathbf{0} \\ \mathbf{0} & \mathbf{I} \end{pmatrix}$$

with eigenvalue $\frac{1}{4}$. Consequently, $\hat{S}_k$ possesses eigenvalues $\pm\frac{1}{2}$.

11.5 THE SPIN ORIENTATION OPERATOR

In non-relativistic quantum mechanics, the spin orientation (i.e., whether the spin vector is parallel or anti-parallel to the total angular momentum vector) is ascertained in terms of the eigenvalues of the operator.

$$\hat{\sigma}.\hat{\mathbf{J}} = \hat{\sigma}\left(\hat{\mathbf{L}} + \frac{1}{2}\hat{\sigma}\right) = \hat{\sigma}\hat{\mathbf{L}} + \frac{1}{2}\hat{\sigma}^2 = \hat{\sigma}\hat{\mathbf{L}} + \frac{3}{2}$$

$$= \left(\hat{\mathbf{L}} + \frac{1}{2}\hat{\sigma}\right)^2 - \hat{\mathbf{L}}^2 + \frac{3}{4} = \hat{\mathbf{J}}^2 - \hat{\mathbf{L}}^2 + \frac{3}{4}$$

The obvious relativistic generalization of this operator is $\boldsymbol{\Sigma}.\hat{\mathbf{J}}$. However, for reasons that would become apparent in the sequel, we prefer to work with the operator $\boldsymbol{\beta}\boldsymbol{\Sigma}.\hat{\mathbf{J}}$. Let us explore the non-relativistic limit of this operator. We have

$$(\boldsymbol{\beta}_{nr})_{ab}^{+} = \langle \phi_{nr}^{a} |\boldsymbol{\beta}| \phi_{nr}^{b} \rangle = (\phi_{nr}^{a*}\ \mathbf{0})\begin{pmatrix} \mathbf{1} & \mathbf{0} \\ \mathbf{0} & \mathbf{-1} \end{pmatrix}\begin{pmatrix} \phi_{nr}^{b} \\ \mathbf{0} \end{pmatrix}$$

$$= \phi_{nr}^{a*}\phi_{nr}^{b} = \langle \phi_{nr}^{a} | \phi_{nr}^{b} \rangle = \delta_{ab} \tag{63}$$

$$(\boldsymbol{\beta}_{nr})_{ab}^{-} = \langle \chi_{nr}^{a} |\boldsymbol{\beta}| \chi_{nr}^{b} \rangle = (\mathbf{0}\ \ \chi_{nr}^{a})\begin{pmatrix} \mathbf{1} & \mathbf{0} \\ \mathbf{0} & \mathbf{-1} \end{pmatrix}\begin{pmatrix} \mathbf{0} \\ \boldsymbol{\chi}_{\mathbf{nr}}^{\mathbf{b}} \end{pmatrix}$$

$$= -\chi_{nr}^{a*}\chi_{nr}^{b} = -\langle \chi_{nr}^{a} | \chi_{nr}^{b} \rangle = -\delta_{ab} \tag{64}$$

$$(\Sigma_{nr})_{ab}^{+} = \langle \phi_{nr}^{a} |\Sigma_{nr}| \phi_{nr}^{b} \rangle = (\phi_{nr}^{a*}\ \mathbf{0})\begin{pmatrix} \sigma & \mathbf{0} \\ \mathbf{0} & \sigma \end{pmatrix}\begin{pmatrix} \phi_{nr}^{b} \\ \mathbf{0} \end{pmatrix}$$

$$= \phi_{nr}^{a*}\,\boldsymbol{\sigma}\,\phi_{nr}^{b} = \left\langle \phi_{nr}^{a}\,|\sigma|\,\phi_{nr}^{b}\right\rangle = \delta_{ab}\boldsymbol{\sigma} \tag{65}$$

$$(\Sigma_{nr})_{ab}^{-} = \left\langle \chi_{nr}^{a}\,|\Sigma_{nr}|\,\chi_{nr}^{b}\right\rangle = \begin{pmatrix}\mathbf{0} & \chi_{nr}^{a}\end{pmatrix}\begin{pmatrix}\sigma & \mathbf{0}\\ \mathbf{0} & \sigma\end{pmatrix}\begin{pmatrix}0\\ \chi_{nr}^{0}\end{pmatrix}$$

$$= \chi_{nr}^{a*}\,\boldsymbol{\sigma}\,\chi_{nr}^{b} = \left\langle \chi_{nr}^{a}\,|\sigma|\,\chi_{nr}^{b}\right\rangle = \delta_{ab}\boldsymbol{\sigma} \tag{66}$$

Furthermore, the orbital angular momentum operator $\hat{\mathbf{L}} = \mathbf{r}\times\hat{\mathbf{p}}$ remains unchanged in the relativistic framework. To summarize, we have for positive energies $\beta_{nr} \to \mathbf{I}$, $\boldsymbol{\Sigma}_{nr} \to \boldsymbol{\sigma}$, $\hat{\mathbf{L}}_{nr} \to \hat{\mathbf{L}}$, so that

$$\left(\beta\,\Sigma.\hat{\mathbf{J}}\right)_{nr} = \left[\beta\,\Sigma\left(\hat{\mathbf{L}}+\hat{\mathbf{S}}\right)\right]_{nr} = \left[\beta\,\Sigma\left(\hat{\mathbf{L}}+\frac{1}{2}\Sigma\right)\right]_{nr} \to \boldsymbol{\sigma}.\left(\hat{\mathbf{L}}+\frac{1}{2}\sigma\right) = \boldsymbol{\sigma}.\hat{\mathbf{J}}$$

i.e., the operator $(\boldsymbol{\beta\Sigma}.\hat{\mathbf{J}})$ has the desired non-relativistic limit.

Now,

$$[\hat{H},\ \boldsymbol{\beta\Sigma}.\hat{\mathbf{J}}] = \boldsymbol{\beta}[\hat{H},\ \boldsymbol{\Sigma}.\hat{\mathbf{J}}] + [\hat{H},\ \boldsymbol{\beta}]\ \boldsymbol{\Sigma}.\hat{\mathbf{J}}] \tag{67}$$

$$[\hat{H},\ \boldsymbol{\Sigma}\hat{\mathbf{J}}] = [\boldsymbol{\alpha}\hat{\mathbf{p}} + \boldsymbol{\beta}m + V(\mathbf{r}),\ \boldsymbol{\Sigma}\hat{\mathbf{J}}] = [\boldsymbol{\alpha}\,\hat{\mathbf{p}},\ \boldsymbol{\Sigma}\hat{\mathbf{J}}]$$

$$= \begin{pmatrix}\mathbf{0} & \sigma.\hat{\mathbf{p}}\\ \sigma.\hat{\mathbf{p}} & \mathbf{0}\end{pmatrix}\begin{pmatrix}\sigma.\hat{\mathbf{J}} & \mathbf{0}\\ \mathbf{0} & \sigma.\hat{\mathbf{J}}\end{pmatrix} - \begin{pmatrix}\sigma.\hat{\mathbf{J}} & \mathbf{0}\\ \mathbf{0} & \sigma.\hat{\mathbf{J}}\end{pmatrix}\begin{pmatrix}\mathbf{0} & \sigma.\hat{\mathbf{p}}\\ \sigma.\hat{\mathbf{p}} & \mathbf{0}\end{pmatrix}$$

$$= \begin{pmatrix}\mathbf{0} & \sigma\hat{\mathbf{p}}.\sigma\hat{\mathbf{J}} - \sigma\hat{\mathbf{J}}.\sigma\hat{\mathbf{p}}\\ \sigma\hat{\mathbf{p}}.\sigma\hat{\mathbf{J}} - \sigma\hat{\mathbf{J}}.\sigma\hat{\mathbf{p}} & \mathbf{0}\end{pmatrix}$$

$$= \begin{pmatrix}\mathbf{0} & \hat{\mathbf{p}}.\hat{\mathbf{J}} + i\sigma\left(\hat{\mathbf{p}}\times\hat{\mathbf{J}}\right) - \hat{\mathbf{J}}.\hat{\mathbf{p}} - i\sigma\left(\hat{\mathbf{J}}\times\hat{\mathbf{p}}\right)\\ \hat{\mathbf{p}}.\hat{\mathbf{J}} + i\sigma\left(\hat{\mathbf{p}}\times\hat{\mathbf{J}}\right) - \hat{\mathbf{J}}.\hat{\mathbf{p}} - i\sigma\left(\hat{\mathbf{J}}\times\hat{\mathbf{p}}\right) & \mathbf{0}\end{pmatrix}$$

$$= \begin{pmatrix}\mathbf{0} & 2i\sigma\left(\hat{\mathbf{p}}\times\hat{\mathbf{J}}\right)\\ 2i\sigma\left(\hat{\mathbf{p}}\times\hat{\mathbf{J}}\right) & \mathbf{0}\end{pmatrix}$$

$$= 2i\begin{pmatrix}\mathbf{0} & \sigma\\ \sigma & \mathbf{0}\end{pmatrix}(\hat{\mathbf{p}}\times\hat{\mathbf{J}}) = 2i\boldsymbol{\alpha}\,(\hat{\mathbf{p}}\times\hat{\mathbf{J}}) \tag{68}$$

$$[\hat{H},\ \boldsymbol{\beta}] = [\boldsymbol{\alpha}\hat{\mathbf{p}} + \boldsymbol{\beta}m + V(\mathbf{r}),\ \boldsymbol{\beta}] = [\boldsymbol{\alpha}\hat{\mathbf{p}},\boldsymbol{\beta}]$$

$$= \alpha\hat{\mathbf{p}}\beta - \beta\alpha\hat{\mathbf{p}} = -2\beta\alpha\hat{\mathbf{p}} \tag{69}$$

$$\alpha\hat{\mathbf{p}}.\Sigma\hat{\mathbf{J}} = \begin{pmatrix} 0 & \sigma.\hat{\mathbf{p}} \\ \sigma.\hat{\mathbf{p}} & 0 \end{pmatrix}\begin{pmatrix} \sigma.\hat{\mathbf{J}} & 0 \\ 0 & \sigma.\hat{\mathbf{J}} \end{pmatrix} = \begin{pmatrix} 0 & \sigma\hat{\mathbf{p}}.\sigma\hat{\mathbf{J}} \\ \sigma\hat{\mathbf{p}}.\sigma\hat{\mathbf{J}} & 0 \end{pmatrix}$$

$$= \begin{pmatrix} 0 & \hat{\mathbf{p}}.\hat{\mathbf{J}} + i\sigma\left(\hat{\mathbf{p}}\times\hat{\mathbf{J}}\right) \\ \hat{\mathbf{p}}.\hat{\mathbf{J}} + i\sigma\left(\hat{\mathbf{p}}\times\hat{\mathbf{J}}\right) & 0 \end{pmatrix}$$

$$= \begin{pmatrix} \mathbf{0} & \mathbf{I} \\ \mathbf{I} & \mathbf{0} \end{pmatrix}\hat{\mathbf{p}}.\,\hat{\mathbf{J}} + i\begin{pmatrix} \mathbf{0} & \sigma \\ \sigma & \mathbf{0} \end{pmatrix}(\hat{\mathbf{p}}\times\hat{\mathbf{J}})$$

$$= -\gamma^5\,\hat{\mathbf{p}}.\,\hat{\mathbf{J}} + i\alpha\,(\hat{\mathbf{p}}\times\hat{\mathbf{J}}) \tag{70}$$

From eqs. (67)-(70), we have

$$[\hat{H}, \beta\Sigma.\hat{\mathbf{J}}] = \beta\,[\hat{H}, \Sigma.\hat{\mathbf{J}}] + [\hat{H}, \beta\,]\,\Sigma.\hat{\mathbf{J}} = 2i\beta\alpha(\hat{\mathbf{p}}\times\hat{\mathbf{J}}) - 2\beta\alpha\hat{\mathbf{p}}\Sigma.\hat{\mathbf{J}}$$

$$= 2i\beta\alpha(\hat{\mathbf{p}}\times\hat{\mathbf{J}}) - 2\beta\,[-\gamma^5\hat{\mathbf{p}}.\,\hat{\mathbf{J}} + i\alpha\,(\hat{\mathbf{p}}\times\hat{\mathbf{J}})]$$

$$= 2\beta\gamma^5\hat{\mathbf{p}}.\hat{\mathbf{J}} = 2\beta\gamma^5\hat{\mathbf{p}}.\,(\hat{\mathbf{L}} + \hat{S}) = 2\beta\gamma^5\hat{\mathbf{p}}.\hat{S}$$

$$= \beta\gamma^5\hat{\mathbf{p}}.\Sigma = \beta\hat{\mathbf{p}}.\gamma^5\Sigma = -\beta\hat{\mathbf{p}}\alpha = -\beta\alpha\hat{\mathbf{p}}\,\frac{1}{2}[\hat{H}, \beta\,] \tag{71}$$

so that

$$\left[\hat{H}, \beta\Sigma.\hat{\mathbf{J}} - \frac{1}{2}\beta\right] = 0 \tag{72}$$

Thus, the operator $\hat{K} = \beta\Sigma.\,\hat{\mathbf{J}} - \frac{1}{2}\,\beta$ commutes with the Hamiltonian of the system.

Furthermore,

$$[\hat{K},\,\hat{\mathbf{J}}] = \left[\beta\Sigma.\hat{\mathbf{J}} - \frac{1}{2}\beta, \hat{\mathbf{J}}\right] = [\beta\Sigma.\,\hat{\mathbf{J}},\,\hat{\mathbf{J}}] - \frac{1}{2}\,[\beta,\,\hat{\mathbf{J}}] = [\beta\Sigma.\,\hat{\mathbf{J}},\,\hat{\mathbf{J}}]$$

$$[\beta\Sigma.\,\hat{\mathbf{J}},\,\hat{\mathbf{J}}] = \beta[\Sigma.\hat{\mathbf{J}},\,\hat{\mathbf{J}}] + [\beta,\,\hat{\mathbf{J}}]\,\Sigma.\hat{\mathbf{J}} = \beta[\Sigma.\hat{\mathbf{J}},\,\hat{\mathbf{J}}] = \beta[\Sigma,\,\hat{\mathbf{J}}]\,\hat{\mathbf{J}} = 0 \tag{73}$$

Therefore, the operator $\hat{K}$ also commutes with the total angular momentum operator $\hat{\mathbf{J}}$. From its definition $\hat{K} = \beta\Sigma.\,\hat{\mathbf{J}} - \frac{1}{2}\,\beta$,we infer that the operator $\hat{K}$ is a measure of the component of spin along the direction of the total angular momentum operator. Hence, we have four mutually commuting operators, viz. $\hat{H}$, $\hat{\mathbf{J}}^2$, $\hat{J}_3$, $\hat{K}$ with respective eigenvalues identified by the indices n_r, j, m_j, κ. These indices shall also label the simultaneous eigenstates.

We can also write the operator $\hat{K}$ as

$$\hat{K} = \boldsymbol{\beta\Sigma}.\,\hat{\mathbf{J}} - \frac{1}{2}\,\boldsymbol{\beta} = \boldsymbol{\beta\Sigma}.\left(\hat{\mathbf{L}} + \frac{1}{2}\Sigma\right) - \frac{1}{2}\,\boldsymbol{\beta}$$

$$= \boldsymbol{\beta}(\boldsymbol{\Sigma}.\,\hat{\mathbf{L}} + \mathbf{I}) = \begin{pmatrix} \boldsymbol{\sigma}.\hat{\mathbf{L}} + \mathbf{I} & 0 \\ 0 & -\boldsymbol{\sigma}.\hat{\mathbf{L}} - \mathbf{I} \end{pmatrix} \tag{74}$$

We now compute the operator $\hat{K}^2$. We have

$$\hat{K}^2 = [\boldsymbol{\beta}(\boldsymbol{\Sigma}.\hat{\mathbf{L}} + \mathbf{I})]^2 = (\boldsymbol{\Sigma}.\hat{\mathbf{L}})^2 + \mathbf{I} + 2\boldsymbol{\Sigma}.\hat{\mathbf{L}} \tag{75}$$

since $[\boldsymbol{\Sigma}, \hat{\mathbf{L}}] = [\boldsymbol{\beta}, \boldsymbol{\Sigma}.\,\hat{\mathbf{L}}] = 0$ and $\boldsymbol{\beta}^2 = \mathbf{I}$. Also

$$(\boldsymbol{\Sigma}.\hat{\mathbf{L}})^2 = \begin{pmatrix} \boldsymbol{\sigma}.\hat{\mathbf{L}} & \mathbf{0} \\ \mathbf{0} & \boldsymbol{\sigma}.\hat{\mathbf{L}} \end{pmatrix} = \begin{pmatrix} \sigma\hat{\mathbf{L}}.\sigma\hat{\mathbf{L}} & \mathbf{0} \\ \mathbf{0} & \sigma\hat{\mathbf{L}}.\sigma\hat{\mathbf{L}} \end{pmatrix}$$

$$= \begin{pmatrix} \hat{\mathbf{L}}^2 + i\sigma(\hat{\mathbf{L}} \times \hat{\mathbf{L}}) & \mathbf{0} \\ \mathbf{0} & \hat{\mathbf{L}}^2 + i\sigma(\hat{\mathbf{L}} \times \hat{\mathbf{L}}) \end{pmatrix}$$

$$= \hat{\mathbf{L}}^2 + i\boldsymbol{\Sigma}\,(\hat{\mathbf{L}} \times \hat{\mathbf{L}}) \tag{76}$$

$$(\hat{\mathbf{L}} \times \hat{\mathbf{L}}) = \mathbf{e}_1(\hat{L}_2\hat{L}_3 - \hat{L}_3\hat{L}_2) + \mathbf{e}_2(\hat{L}_3\hat{L}_1 - \hat{L}_1\hat{L}_3) + \mathbf{e}_3(\hat{L}_1\hat{L}_2 - \hat{L}_2\hat{L}_1)$$

$$= \mathbf{e}_1[\hat{L}_2, \hat{L}_3] + \mathbf{e}_2[\hat{L}_3, \hat{L}_1] + \mathbf{e}_3[\hat{L}_3, \hat{L}_1]$$

$$= i\mathbf{e}_1\hat{L}_1 + i\mathbf{e}_2\hat{L}_2 + i\mathbf{e}_3\hat{L}_3 + = i\hat{\mathbf{L}} \tag{77}$$

From eqs. (75)-(77), we get

$$\hat{K}^2 = \hat{\mathbf{L}}^2 + \mathbf{I} + \boldsymbol{\Sigma}.\hat{\mathbf{L}} \tag{78}$$

Similarly,

$$\hat{\mathbf{J}}^2 = \left(\hat{\mathbf{L}} + \frac{1}{2}\Sigma\right)^2 = \hat{\mathbf{L}}^2 + \frac{3}{4}\,\mathbf{I} + \hat{\mathbf{L}}.\boldsymbol{\Sigma}$$

$$= \hat{\mathbf{L}}^2 + \frac{3}{4}\,\mathbf{I} + \boldsymbol{\Sigma}.\hat{\mathbf{L}} = \hat{K}^2 - \frac{1}{4}\,\mathbf{I} \tag{79}$$

If we assume $-\kappa$ as the eigenvalue of $\hat{K}$ and $j(j+1)$ as that of the squared total angular momentum operator, then, on operating by these operators on a simultaneous eigenfunction, we get, in terms of eigenvalues, eq. (79) as

$$\kappa = \left[j(j+1) + \frac{1}{4}\right]^{\frac{1}{2}} = \pm\left(j + \frac{1}{2}\right) \tag{80}$$

For $\kappa > 0$, the spin is assumed to be antiparallel to the total angular momentum and vice versa. Writing the Dirac wave function, that is, the simultaneous eigenfunction of the above operators in terms of two component spinors $\phi = (\phi_1\ \phi_2)^T$, $\chi = (\chi_1\ \chi_2)^T$ as $\psi = (\phi\ \chi)^T$, we obtain the eigenvalue equation for $\hat{K}$

$$\hat{K}\psi = -\kappa\psi = \begin{pmatrix} \boldsymbol{\sigma}.\hat{\mathbf{L}} + \mathbf{1} & \mathbf{0} \\ \mathbf{0} & -\boldsymbol{\sigma}.\hat{\mathbf{L}} - \mathbf{1} \end{pmatrix} \begin{pmatrix} \phi \\ \chi \end{pmatrix} = \mp \left(j + \frac{1}{2}\right) \begin{pmatrix} \phi \\ \chi \end{pmatrix} \tag{81}$$

whence

$$(\boldsymbol{\sigma}.\hat{\mathbf{L}}) \begin{pmatrix} \phi \\ \chi \end{pmatrix} = \mp \begin{pmatrix} \left(j + \frac{1}{2} \pm 1\right)\phi \\ -\left(j + \frac{1}{2} \mp 1\right)\chi \end{pmatrix} = \begin{pmatrix} \mp\left(j + \frac{1}{2} \pm 1\right)\phi \\ \pm\left(j + \frac{1}{2} \mp 1\right)\chi \end{pmatrix} \tag{82}$$

and

$$(\boldsymbol{\Sigma}.\hat{\mathbf{L}}) \begin{pmatrix} \phi \\ \chi \end{pmatrix} = \begin{pmatrix} \boldsymbol{\sigma}\hat{\mathbf{L}} & 0 \\ 0 & \boldsymbol{\sigma}\hat{\mathbf{L}} \end{pmatrix} \begin{pmatrix} \phi \\ \chi \end{pmatrix} = \begin{pmatrix} \boldsymbol{\sigma}\hat{\mathbf{L}}\phi \\ \boldsymbol{\sigma}\hat{\mathbf{L}}\chi \end{pmatrix}$$

$$= \boldsymbol{\sigma}\hat{\mathbf{L}} \begin{pmatrix} \phi \\ \chi \end{pmatrix} = \begin{pmatrix} \mp\left(j + \frac{1}{2} \pm 1\right)\phi \\ \pm\left(j + \frac{1}{2} \mp 1\right)\chi \end{pmatrix} \tag{83}$$

Thus, although ψ is not an eigenfunction of $\boldsymbol{\Sigma}.\hat{\mathbf{L}}$, ϕ, χ both are eigenfunctions of $\boldsymbol{\sigma}.\hat{\mathbf{L}}$ with different eigenvalues as shown by eq. (82). Since ψ is assumed as a simultaneous eigenfunction of $\hat{H}$, $\hat{\mathbf{J}}^2$, $\hat{J}_3$, $\hat{K}$ we also have

$$\hat{\mathbf{J}}^2 \psi = j(j+1)\psi \tag{84}$$

and

$$\hat{\mathbf{J}}_3 \psi = m_j \psi \tag{85}$$

From eq. (79), we have

$$\hat{\mathbf{J}}^2 \begin{pmatrix} \phi \\ \chi \end{pmatrix} = \hat{\mathbf{J}}^2\psi = \left(\hat{\mathbf{L}}^2 + \frac{3}{4}\mathbf{I} + \Sigma.\hat{\mathbf{L}}\right)\psi$$

$$= \hat{\mathbf{L}}^2 \begin{pmatrix} \phi \\ \chi \end{pmatrix} + \frac{3}{4} \begin{pmatrix} \phi \\ \chi \end{pmatrix} + \begin{pmatrix} \boldsymbol{\sigma}\hat{\mathbf{L}} & \mathbf{0} \\ \mathbf{0} & \boldsymbol{\sigma}\hat{\mathbf{L}} \end{pmatrix} \begin{pmatrix} \phi \\ \chi \end{pmatrix} \tag{86}$$

or

$$\hat{\mathbf{L}}^2\begin{pmatrix}\phi\\ \chi\end{pmatrix} = \hat{\mathbf{J}}^2\begin{pmatrix}\phi\\ \chi\end{pmatrix} - \frac{3}{4}\begin{pmatrix}\phi\\ \chi\end{pmatrix} - \begin{pmatrix}\sigma\hat{\mathbf{L}}\phi\\ \sigma\hat{\mathbf{L}}\chi\end{pmatrix}$$

$$= \begin{pmatrix} j(j+1)\phi - \frac{3}{4}\phi \pm \left(j+\frac{1}{2}\pm 1\right)\phi \\ j(j+1)\chi - \frac{3}{4}\chi \mp \left(j+\frac{1}{2}\mp 1\right)\chi \end{pmatrix} \tag{87}$$

Writing the eigenvalues of $\hat{\mathbf{L}}^2$ as $l(l+1)$, we have

$$l(l+1) = \begin{pmatrix} j^2 + j(1\pm 1) - \frac{3}{4} \pm \frac{1}{2} + 1 \\ j^2 + j(1\mp 1) - \frac{3}{4} \mp \frac{1}{2} + 1 \end{pmatrix} = \begin{pmatrix} \left(j\pm\frac{1}{2}\right)\left(j\pm\frac{1}{2}+1\right) \\ \left(j\mp\frac{1}{2}\right)\left(j\mp\frac{1}{2}+1\right) \end{pmatrix} \tag{88}$$

From the above, we conclude that

(a) Both, the positive energy 2-component spinor ϕ and the negative energy 2-component spinor χ are eigenstates of $\hat{\mathbf{L}}^2$ but with different eigenvalues.

(b) The eigenvalues of $\hat{\mathbf{L}}^2$ are given by $l_\pm(l_\pm + 1)$ for the positive energy 2-component spinor and $l_\mp(l_\mp + 1)$ for the negative energy 2-component spinor where $l_\pm = j \pm \frac{1}{2}$, $l_\mp = j \mp \frac{1}{2}$.

11.6 EXPLICIT EXPRESSION FOR THE EIGENSTATES OF $\hat{\mathbf{L}}^2$, $\hat{L}_3$

In order to obtain explicit expressions for the eigenfunctions of $\hat{\mathbf{L}}^2$, $\hat{L}_3$, we express the eigenvalue problem in spherical polar coordinates in view of its underlying spherical symmetry. The transformation from Cartesian coordinates to polar coordinates is achieved by the following:

$$x^1 = r\sin\theta\cos\phi \tag{89}$$

$$x^2 = r\sin\theta\sin\phi \tag{90}$$

$$x^3 = r\cos\theta \tag{91}$$

$$ds^2 = dr^2 + r^2 d\theta^2 + r^2\sin^2\theta\, d\phi^2 \tag{92}$$

$$r = \sqrt{\left(x^1\right)^2 + \left(x^2\right)^2 + \left(x^3\right)^2},\quad \theta = \tan^{-1}\frac{\sqrt{\left(x^1\right)^2 + \left(x^2\right)^2}}{x^3},\quad \phi = \tan^{-1}\frac{x^2}{x^1} \tag{93}$$

$$\frac{\partial r}{\partial x^1} = \sin\theta\cos\phi,\quad \frac{\partial r}{\partial x^2} = \sin\theta\sin\phi,\quad \frac{\partial r}{\partial x^3} = \cos\theta \tag{94}$$

$$\frac{\partial\theta}{\partial x^1} = \frac{\cos\theta\cos\phi}{r}, \frac{\partial\theta}{\partial x^2} = \frac{\cos\theta\sin\phi}{r}, \frac{\partial\theta}{\partial x^3} = -\frac{\sin\theta}{r} \tag{95}$$

$$\frac{\partial\phi}{\partial x^1} = -\frac{\sin\phi}{r\sin\theta}, \frac{\partial\phi}{\partial x^2} = \frac{\cos\phi}{r\sin\theta}, \frac{\partial\phi}{\partial x^3} = 0 \tag{96}$$

$$\frac{\partial}{\partial x^1} \equiv \sin\theta\cos\phi\frac{\partial}{\partial r} + \frac{\cos\theta\cos\phi}{r}\frac{\partial}{\partial\theta} - \frac{\sin\phi}{r\sin\theta}\frac{\partial}{\partial\phi} \tag{97}$$

$$\frac{\partial}{\partial x^2} \equiv \sin\theta\sin\phi\frac{\partial}{\partial r} + \frac{\cos\theta\sin\phi}{r}\frac{\partial}{\partial\theta} + \frac{\cos\phi}{r\sin\theta}\frac{\partial}{\partial\phi} \tag{98}$$

$$\frac{\partial}{\partial x^3} = \cos\theta\frac{\partial}{\partial r} - \frac{\sin\theta}{r}\frac{\partial}{\partial\theta} \tag{99}$$

The gradient operator in polar coordinates is given by

$$\nabla = \hat{\mathbf{r}}\frac{\partial}{\partial r} + \hat{\theta}\frac{1}{r}\frac{\partial}{\partial\theta} + \hat{\varphi}\frac{1}{r\sin\theta}\frac{\partial}{\partial\phi} \tag{100}$$

The unit vector $\hat{\mathbf{r}} = \frac{\mathbf{r}}{r}$ is directed radially outwards, $\hat{\theta} = \frac{\theta}{\theta}$ is directed along a line of longitude away from the north pole, in the direction of increasing θ and $\hat{\varphi} = \frac{\varphi}{\phi}$ points along a line of latitude in an anticlockwise direction as seen looking on the north pole, i.e., in the direction of increasing ϕ. The three unit vectors $\hat{\mathbf{r}} = \frac{\mathbf{r}}{r}, \hat{\theta}\ \frac{\theta}{\theta}$ and $\hat{\varphi} = \frac{\varphi}{\phi}$ form a local orthonormal basis with $\hat{\mathbf{r}} \times \hat{\theta} = \hat{\varphi}$ and $\hat{\mathbf{r}} \times \hat{\varphi} = -\hat{\theta}$. Therefore,

$$r \times \nabla = \hat{\varphi}\frac{\partial}{\partial\theta} - \hat{\theta}\frac{1}{\sin\theta}\frac{\partial}{\partial\phi} \tag{101}$$

In Cartesian coordinates, we can write $\hat{\varphi} = (-\sin\phi, \cos\phi, 0)$. This is established as follows:

Consider a point P with Cartesian coordinates (x^1, x^2, x^3) and corresponding polar coordinates (r, θ, ϕ), where θ is the angle between OP and the upward pointing x^3 axis measured towards the direction of increasing θ and ϕ is the angle that the intercept of the plane containing x^3 axis and the line segment OP and the x^1-x^2 plane makes with the x^1 axis, i.e., the angle that the projection of the line segment OP on the x^1-x^2 plane, i.e., OQ makes with the x^1 axis. The direction of the unit vector $\hat{\varphi}$ is perpendicular to OQ and its magnitude is, obviously, unity. Since, in Cartesian coordinates, a unit vector along OQ would be represented by $\hat{\mathbf{x}}^1\cos\phi + \hat{\mathbf{x}}^2\sin\phi$, the unit vector $\hat{\varphi}$ being perpendicular to OQ will have the representation $\hat{\mathbf{x}}^1\sin\phi + \hat{\mathbf{x}}^2\cos\phi$.

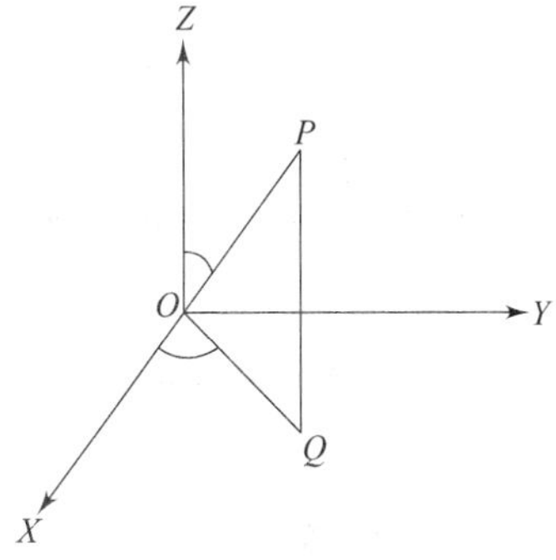

Similarly, $\hat{\theta} = (\cos\theta\cos\phi, \cos\theta\sin\phi, -\sin\theta)$ for, assuming that $\hat{\theta} = t\hat{\mathbf{x}}^1 + u\hat{\mathbf{x}}^2 + v\hat{\mathbf{x}}^3$, we have the following set of equations, viz., $t^2 + u^2 + v^2 = 1$ since $\hat{\theta}$ is a unit vector, $t\sin\theta\cos\phi + u\sin\theta\sin\phi + v\cos\theta = 0$ because $\hat{\theta}$ is perpendicular to OP and $-t\sin\phi + u\cos\phi = 0$ because $\hat{\theta}$ is also perpendicular to $\hat{\varphi} = -\hat{\mathbf{x}}^1\sin\phi + \hat{\mathbf{x}}^2\cos\phi$. Solving these sets of equations, we get $\hat{\theta} = (\cos\theta\cos\phi\ \cos\theta\sin\phi, -\sin\theta)$.

Hence,

$$\hat{\mathbf{L}} = \mathbf{r}\times\hat{\mathbf{p}} = -i\mathbf{r}\times\nabla = -i\left(\hat{\varphi}\frac{\partial}{\partial\theta} - \hat{\theta}\frac{1}{\sin\theta}\frac{\partial}{\partial\phi}\right) \tag{102}$$

$$\hat{\mathbf{L}}^2 = -\left(\hat{\varphi}\frac{\partial}{\partial\theta} - \hat{\theta}\frac{1}{\sin\theta}\frac{\partial}{\partial\phi}\right)\cdot\left(\hat{\varphi}\frac{\partial}{\partial\theta} - \hat{\theta}\frac{1}{\sin\theta}\frac{\partial}{\partial\phi}\right)$$

$$\hat{\mathbf{L}}^2 = -\left\{\left[\hat{\varphi}\frac{\partial\hat{\varphi}}{\partial\theta}\frac{\partial}{\partial\theta} + \hat{\varphi}.\hat{\varphi}\frac{\partial^2}{\partial\theta^2} - \hat{\varphi}\frac{\partial\hat{\theta}}{\partial\theta}\cdot\frac{1}{\sin\theta}\frac{\partial}{\partial\phi} - \hat{\varphi}\cdot\hat{\theta}\frac{\partial}{\partial\theta}\left(\frac{1}{\sin\theta}\frac{\partial}{\partial\phi}\right)\right]\right.$$

$$-\left[\hat{\theta}\frac{1}{\sin\theta}\frac{\partial\hat{\varphi}}{\partial\phi}\frac{\partial}{\partial\theta} + \hat{\theta}\cdot\hat{\varphi}\frac{1}{\sin\theta}\frac{\partial^2}{\partial\phi\partial\theta}\right.$$

$$\left.\left. - \hat{\theta}\frac{1}{\sin\theta}\frac{\partial\hat{\theta}}{\partial\phi}\cdot\frac{1}{\sin\theta}\frac{\partial}{\partial\phi} - \hat{\theta}\cdot\hat{\theta}\frac{1}{\sin\theta}\frac{\partial}{\partial\phi}\left(\frac{1}{\sin\theta}\frac{\partial}{\partial\phi}\right)\right]\right\}$$

$$= -\left(\frac{\partial^2}{\partial\theta^2} + \cot\theta\frac{\partial}{\partial\theta} + \frac{1}{\sin^2\theta}\frac{\partial^2}{\partial\phi^2}\right) = -\left(\frac{1}{\sin\theta}\frac{\partial}{\partial\theta}\sin\theta\frac{\partial}{\partial\theta} + \frac{1}{\sin^2\theta}\frac{\partial^2}{\partial\phi^2}\right) \tag{103}$$

where we have used the following:

(i) $\hat{\theta}.\hat{\theta} = \hat{\varphi}.\hat{\varphi} = 1.$

(ii) $\hat{\varphi}.\hat{\theta} = \hat{\theta}.\hat{\varphi} = 0.$

(iii) $\dfrac{\partial\hat{\varphi}}{\partial\theta} = 0$ since $\hat{\varphi} = (-\sin\phi, \cos\phi, 0)$ is independent of θ.

(iv) $\dfrac{\partial\hat{\theta}}{\partial\theta} = (-\sin\theta\cos\phi, -\sin\theta\sin\phi, -\cos\theta) = -\hat{\mathbf{r}}$ so that it is directed inwards along $\hat{\mathbf{r}}$, i.e., in the direction of $-\hat{\mathbf{r}}$ and, therefore, $\dfrac{\partial\hat{\theta}}{\partial\theta}$ will not contribute to the angular momentum for $\hat{\varphi}.\dfrac{\partial\hat{\theta}}{\partial\theta} = (-\sin\phi, \cos\phi, 0)\ (-\sin\theta\cos\phi, -\sin\theta\sin\phi, -\cos\theta) = 0.$

(v) $\dfrac{\partial \hat{\varphi}}{\partial \phi}$ = (–cos φ, sin φ, 0) is directed along and opposite in direction to *OQ*, i.e., the projection of $\hat{\mathbf{r}}$ on the $x^1 - x^2$ plane, for $\hat{\varphi} \cdot \dfrac{\partial \hat{\varphi}}{\partial \phi}$ = (– sin φ, cos φ, 0) (– cos φ, – sin φ, 0) = 0 so that $\dfrac{\partial \hat{\varphi}}{\partial \phi}$ is perpendicular to $\hat{\varphi}$ and is directed towards the origin in the x^1-x^2 plane. The component of $\dfrac{\partial \hat{\varphi}}{\partial \phi}$ along $\hat{\mathbf{r}}$ direction is – sin θ and that perpendicular to $\hat{\mathbf{r}}$, i.e., along the direction of $\hat{\theta}$ is – cos θ for

$$\hat{\theta} \cdot \frac{\partial \hat{\varphi}}{\partial \phi} = (\cos\theta, \cos\phi, \cos\theta\sin\phi, -\sin\theta)\,(-\cos\phi, -\sin\phi, 0) = -\cos\theta$$

(vi) Further

$$\frac{\partial \hat{\theta}}{\partial \phi} = -(\cos\theta\sin\phi, \cos\theta\cos\phi, 0) = \cos\theta\,(-\sin\phi, \cos\phi, 0) = \cos\theta\hat{\varphi}$$

Hence, $\quad \hat{\theta} \cdot \dfrac{\partial \hat{\theta}}{\partial \phi} = \cos\theta\hat{\theta} \cdot \boldsymbol{\varphi} = 0$

Now

$$\hat{L}_3 = (\mathbf{r} \times \hat{\mathbf{p}})_3 = x^1 \hat{p}_2 - x^2 \hat{p}_1 = -i\,(x^1 \partial_2 - x^2 \partial_1)$$

Using eqs. (89)-(99), we have

$$\begin{aligned}\hat{L}_3 &= -i\,(x^1 \partial_2 - x^2 \partial_1)\\ &= -i\left[r\sin\theta\cos\phi\left(\sin\theta\sin\phi\frac{\partial}{\partial r} + \frac{\cos\theta\sin\phi}{r}\frac{\partial}{\partial\theta} + \frac{\cos\phi}{r\sin\theta}\frac{\partial}{\partial\phi}\right)\right.\\ &\quad \left. - r\sin\theta\sin\phi\left(\sin\theta\cos\phi\frac{\partial}{\partial r} + \frac{\cos\theta\cos\phi}{r}\frac{\partial}{\partial\theta} - \frac{\sin\phi}{r\sin\theta}\frac{\partial}{\partial\phi}.\right)\right]\\ &= i\frac{\partial}{\partial\phi}\end{aligned} \tag{104}$$

Proceeding similarly, we have

$$\begin{aligned}\hat{L}_1 &= -i(x^2\partial_3 - x^3\partial_2) = -i\left[r\sin\theta\sin\phi\left(\cos\theta\frac{\partial}{\partial r} - \frac{\sin\theta}{r}\frac{\partial}{\partial\theta}\right)\right.\\ &\quad \left. - r\cos\theta\left(\sin\theta\sin\phi\frac{\partial}{\partial r} + \frac{\cos\theta\sin\phi}{r}\frac{\partial}{\partial\theta} + \frac{\cos\phi}{r\sin\theta}\frac{\partial}{\partial\phi}\right)\right]\end{aligned}$$

$$= i\left(\sin\phi\,\frac{\partial}{\partial\theta} + \cot\theta\cos\phi\,\frac{\partial}{\partial\phi}\right) \tag{105}$$

$$\hat{L}_2 = -i\,(x^3\partial_1 - x^1\partial_3)$$

$$= -i\left[r\cos\theta\left(\sin\theta\cos\phi\,\frac{\partial}{\partial r} + \frac{\cos\theta\cos\phi}{r}\,\frac{\partial}{\partial\theta} - \frac{\sin\phi}{r\sin\theta}\,\frac{\partial}{\partial\phi}\right)\right.$$

$$\left. - r\sin\theta\cos\phi\left(\cos\theta\,\frac{\partial}{\partial r} - \frac{\sin\theta}{r}\,\frac{\partial}{\partial\theta}\right)\right]$$

$$= -i\left(\cos\phi\,\frac{\partial}{\partial\theta} - \cot\theta\sin\phi\,\frac{\partial}{\partial\phi}\right) \tag{106}$$

The ladder operators are, therefore

$$\hat{L}_\pm = \hat{L}_1 \pm i\hat{L}_2 = i\left(\sin\phi\,\frac{\partial}{\partial\theta} + \cot\theta\cos\phi\,\frac{\partial}{\partial\phi}\right) \pm \left(\cos\phi\,\frac{\partial}{\partial\theta} - \cot\theta\sin\phi\,\frac{\partial}{\partial\phi}\right)$$

$$= \pm\,(\cos\phi \pm i\sin\phi)\left(\frac{\partial}{\partial\theta} \pm i\cot\theta\,\frac{\partial}{\partial\phi}\right) = \pm e^{\pm i\phi}\left(\frac{\partial}{\partial\theta} \pm i\cot\theta\,\frac{\partial}{\partial\phi}\right) \tag{107}$$

Now, if $\Phi_n(\phi)$ is an eigenket of $\hat{L}_3$ with eigenvalue n, then

$$\hat{L}_3\,\Phi_n(\phi) = -i\,\frac{\partial\Phi_n(\phi)}{\partial\phi}\quad n\Phi_n(\phi) \tag{108}$$

which has the solution

$$\Phi_n(\phi) = \frac{1}{\sqrt{2\pi}}\,e^{in\phi} \tag{109}$$

with $n = \ldots -3, -2, -1, 0, 1, 2, 3, \ldots$ being an integer.

Now, the operators $\hat{L}_3$, $\hat{\mathbf{L}}^2$ commute with each other, as can be explicitly seen as follows:

$$\hat{\mathbf{L}}^2\hat{L}_3 = -\left(\frac{\partial^2}{\partial\theta^2} + \cot\theta\,\frac{\partial}{\partial\theta} + \frac{1}{\sin^2\theta}\,\frac{\partial^2}{\partial\phi^2}\right)\left(-i\,\frac{\partial}{\partial\phi}\right)$$

$$= i\left(\frac{\partial^3}{\partial\theta^2\,\partial\phi} + \cot\theta\,\frac{\partial^2}{\partial\theta\partial\phi} + \frac{1}{\sin^2\theta}\,\frac{\partial^3}{\partial\phi^3}\right)$$

$$= \left(-i\,\frac{\partial}{\partial\phi}\right)\left[-\left(\frac{\partial^2}{\partial\theta^2} + \cot\theta\,\frac{\partial}{\partial\theta} + \frac{1}{\sin^2\theta}\,\frac{\partial^2}{\partial\phi^2}\right)\right] = \hat{L}_3,\ \hat{\mathbf{L}}^2 \tag{110}$$

Therefore, $\hat{L}_3$, $\hat{\mathbf{L}}^2$ have a common set of eigenfunctions. The eigenfunctions of $\hat{\mathbf{L}}^2$ must also have the same ϕ dependence and will be of the form $\Theta_l^n(\theta)\,\Phi_n(\phi)$. Substituting this solution in eq. (103) and assuming that the eigenvalues corresponding to the eigenfunctions $\Theta_l^n(\theta)\,\Phi_n(\phi)$ of $\hat{\mathbf{L}}^2$ are $l\,(l+1)$, we have

$$\hat{\mathbf{L}}^2\,\Theta_l^n(\theta)\,\Phi_n(\phi) = l\,(l+1)\,\Theta_l^n(\theta)\,\Phi_n(\phi)$$

$$= -\left(\frac{1}{\sin\theta}\frac{\partial}{\partial\theta}\sin\theta\frac{\partial}{\partial\theta} + \frac{1}{\sin^2\theta}\frac{\partial^2}{\partial\phi^2}\right)\Theta_l^n(\theta)\,\Phi_n(\phi)$$

$$= -\left[\frac{1}{\sin\theta}\frac{\partial}{\partial\theta}\left(\sin\theta\frac{\partial\Theta_l^n(\theta)}{\partial\theta}\right)\cdot\Phi_n(\phi) + \frac{1}{\sin^2\theta}\Theta_1^n(\theta)\frac{\partial^2\Phi_n(\phi)}{\partial\phi^2}\right]$$

$$= -\left[\frac{1}{\sin\theta}\frac{\partial}{\partial\theta}\left(\sin\theta\frac{\partial\Theta_1^n(\theta)}{\partial\theta}\right) - \frac{n^2}{\sin^2\theta}\Theta_l^n(\theta)\right]\Phi_n(\phi) \tag{111}$$

since $\Phi_n(\phi) = \dfrac{1}{\sqrt{2\pi}}\,e^{in\phi}$. Therefore, $\Theta_l^n(\theta)$ is a (normalized) solution of the equation.

$$\frac{1}{\sin\theta}\frac{\partial}{\partial\theta}\left(\sin\theta\frac{\partial\Theta_l^n(\theta)}{\partial\theta}\right) - \frac{n^2}{\sin^2\theta}\Theta_l^n(\theta) + l\,(l+1)\,\Theta_l^n(\theta) = 0 \tag{112}$$

We first compute the solution of the above equation for the case $n = l$ which is the highest state on the l ladder and from thereon we compute the lower states by progressively applying the "lowering" operator. If we identify the simultaneous eigenfunctions of $\hat{\mathbf{L}}^2$, $\hat{L}_3$ in terms of l, n as $|l, n\rangle$, then, for the state $n = l$, being the highest state in the l ladder, we have $\hat{L}_+|l, l\rangle = 0$ which on substituting from eq. (107) gives

$$e^{i\phi}\left(\frac{\partial}{\partial\theta} + i\cot\theta\frac{\partial}{\partial\phi}\right)|l, l\rangle = e^{i\phi}\left(\frac{\partial}{\partial\theta} + i\cot\theta\frac{\partial}{\partial\phi}\right)\Theta_l^l(\theta)\,\Phi_l(\phi)$$

$$= \frac{1}{\sqrt{2\pi}}\,e^{i\phi}\left(\frac{\partial}{\partial\theta} + i\cot\theta\frac{\partial}{\partial\phi}\right)\Theta_l^l(\theta)\,e^{il\phi}$$

$$= \frac{1}{\sqrt{2\pi}}\,e^{i\phi}\left(\frac{\partial}{\partial\theta} - l\cot\theta\right)\Theta_l^l(\theta)\,e^{il\phi} = 0 \tag{113}$$

which has the solution

$$\Theta_l^l(\theta) = N\,(\sin\theta)^l \tag{114}$$

where N is a normalization constant to be determined from the orthogonality condition

$$\int_{\theta=0}^{\pi}\int_{\phi=0}^{2\pi} Y_{l'}^{m'*}(\theta,\phi)\, Y_l^m(\theta,\phi)\sin\theta\, d\theta\, d\phi = \int Y_{l'}^{m'*}(\theta,\phi)\, Y_l^m(\theta,\phi)\, d\Omega = \delta_{l'\,l}\,\delta_{m'\,m} \tag{115}$$

where $Y_l^l(\theta,\phi) = \Theta_l^l(\theta)\,\Phi_l(\phi)$.

Since $\Phi_l(\phi) = \frac{1}{\sqrt{2\pi}} e^{il\phi}$ is already normalized, we need to normalize $\Theta_l^l(\theta) = N(\sin\theta)^l$ for which, we have, on substituting in eq. (115) from eq. (114).

$$|N|^2 \int_0^{\pi} (\sin\theta)^{2l+1}\, d\theta = 1 \tag{116}$$

The integral is standard and can be easily done to yield

$$\frac{|N|^2\, 2^{2l+1}\,(l!)^2}{(2l+1)!} = 1 \tag{117}$$

Thus, after normalization, we have

$$Y_l^l(\theta,\phi) = (-1)^l \left(\frac{(2l+1)!}{4\pi}\right)^{1/2} \frac{1}{2^l\, l!}(\sin\theta)^l\, e^{il\phi} = c_l\,(\sin\theta)^l\, e^{il\phi} \tag{118}$$

where

$$c_l = (-1)^l \left(\frac{(2l+1)!}{4\pi}\right)^{1/2} \frac{1}{2^l\, l!} \tag{119}$$

The phase factor $(-1)^l$ has been included in the normalization of the spherical harmonic $Y_l^l(\theta,\phi) = \Theta_l^l(\theta)\,\Phi_l(\phi)$ in line with the Condon Shortley convention. This facilitates the simplification of quantum mechanical operations, e.g., the application of raising and lowering operators.

Having obtained the normalized eigenfunction $|l, l\rangle$ corresponding to the highest state eigenvalue $n = l$, we can obtain eigenstates corresponding to any eigenvalue n of $\hat{L}_3$, viz., $|l, n\rangle$ by applying $\hat{L}_-$ recursively on $|l, l\rangle$. We have, in analogy with eq. (32).

$$\hat{L}_\pm |l, n\rangle = \left[(l \mp n)(l \pm n + 1)\right]^{1/2} |l, n \pm 1\rangle \tag{120}$$

so that

$$\hat{L}_- |l, l\rangle = (2l)^{\frac{1}{2}} |l, l-1\rangle \tag{121}$$

which gives

$$|l, l-1\rangle = Y_l^{l-1}(\theta,\phi) = \frac{1}{\sqrt{2l}}\, \hat{L}_- |l, l\rangle$$

$$= \frac{(-1)}{\sqrt{2l}} e^{-i\phi} \left(\frac{\partial}{\partial\theta} - i\cot\theta \frac{\partial}{\partial\phi} \right) Y_l^l (\theta, \phi) \tag{122}$$

or

$$Y_l^{l-1} (\theta, \phi) = \frac{-1}{\sqrt{2l}} c_l e^{-i\phi} \left(\frac{\partial}{\partial\theta} - i\cot\theta \frac{\partial}{\partial\phi} \right) \sin^l \theta . e^{il\phi}$$

$$= - c_l e^{i(l-1)\phi} \sqrt{2l} \sin^{l-1} \theta \cos \theta \tag{123}$$

To simplify the recursion process, we make use of the following identity:

$$\left(\sin^{1-m} \theta \frac{d}{d(\cos\theta)} \sin^m \theta \right) f(\theta) = - \sin^{1-m} \theta \left[m \sin^{m-1} \theta . \cos\theta . \frac{1}{\sin\theta} f(\theta) \right.$$

$$\left. + \sin^m \theta \left(\frac{d}{d\theta} f(\theta) \right) . \frac{1}{\sin\theta} \right]$$

$$= - m \cot \theta f(\theta) - \frac{d}{d\theta} f(\theta) \tag{124}$$

so that

$$L_- e^{im\phi} f(\theta) = - e^{-i\phi} \left(\frac{\partial}{\partial\theta} - i\cot\theta \frac{\partial}{\partial\phi} \right) e^{im\phi} f(\theta)$$

$$= e^{i(m-1)\phi} \left(-\frac{\partial}{\partial\theta} - m\cot\theta \right) f(\theta)$$

$$= \left(\sin^{1-m} \theta \frac{d}{d(\cos\theta)} \sin^m \theta \right) e^{i(m-1)\phi} f(\theta) \tag{125}$$

Hence, we have

$$Y_l^{l-1} (\theta, \phi) = \frac{1}{\sqrt{2l}} L_- Y_l^l (\theta, \phi) = \frac{c_1}{\sqrt{2l}} L_- e^{il\phi} \sin^l \theta$$

$$= \frac{c_1}{\sqrt{2l}} e^{i(l-1)\phi} \left(\sin^{1-l} \theta \frac{d}{d(\cos\theta)} \sin^l \theta \right) \sin^l (\theta) \tag{126}$$

and

$$Y_l^{l-2} (\theta, \phi) = \frac{1}{\sqrt{[2(2l-1)]}} L_- Y_l^{l-1} (\theta, \phi)$$

$$= \frac{c_l}{\sqrt{2l[2(2l-1)]}} (L_-)^2 e^{il\phi} \sin^l \theta$$

$$= \frac{c_l}{\sqrt{2l[2(2l-1)]}} L_- e^{i(l-1)\phi} \left(\sin^{1-l} \theta \frac{d}{d(\cos\theta)} \sin^l \theta \right) \sin^l (\theta)$$

$$= \frac{c_l}{\sqrt{2l[2(2l-1)]}} e^{i(l-2)\phi} \left(\sin^{2-l} \theta \frac{d^2}{d(\cos\theta)^2} \sin^l \theta \right) \sin^l (\theta)$$

$$= c_l \sqrt{\frac{[l+(l-2)]!}{(2l)![l-(l-2)]!}} e^{i(l-2)\phi} \left(\sin^{2-l} \theta \frac{d^2}{d(\cos\theta)^2} \sin^l \theta \right) \sin^l (\theta) \qquad (127)$$

and so on. To get the value of $Y_l^m(\theta, \phi)$ we need to apply the "lowering operator" recursively $(l-m)$ times to get

$$Y_l^m(\theta, \phi) = c_l \sqrt{\frac{(l+m)!}{(2l)!(l-m)!}} e^{im\phi} \sin^{-m} \theta \frac{d^{l-m}}{d(\cos\theta)^{l-m}} \sin^{2l} \theta \qquad (128)$$

We can equally well compute the above eigenfunctions by starting from the lowest eigenstates, i.e., for $n = -l$ by solving the equation $\hat{L}_- |l, -l\rangle = 0$ to obtain

$$Y_l^{-l}(\theta, \phi) = (-1)^l c_l e^{-il\phi} \sin^l \theta \qquad (129)$$

and then applying the "raising operator" recursively. We get, in this case, the equivalent expression.

$$Y_l^m(\theta, \phi) = (-1)^m c_l \sqrt{\frac{(l-m)!}{(2l)!(l+m)!}} e^{-im\phi} \sin^m \theta \frac{d^{l+m}}{d(\cos\theta)^{l+m}} \sin^{2l} \theta \qquad (130)$$

11.7 EXPLICIT EXPRESSION FOR THE EIGENSTATES OF THE SPIN OPERATOR

Our next step is to obtain the eigenfunctions of the spin operator. Since the spin angular momentum operator satisfies the same commutation relations as the "orbital angular momentum", we can, as in the case of eigenfunctions of the orbital angular momentum, identify the spin eigenfunctions in terms of the labels of eigenvalues of the spin angular momentum operators $\hat{S}^2$, $\hat{S}_3$ say s, m_s and write the eigenstates as $|s, m_s\rangle$. We work out the eigenstates for the specific case of $s = \frac{1}{2}$, i.e., spin one-half particles, for which

$$m_s = \pm \frac{1}{2}, \text{ i.e., } \hat{S}_3 \left| \frac{1}{2}, \pm \frac{1}{2} \right\rangle = \pm \frac{1}{2} \left| \frac{1}{2}, \pm \frac{1}{2} \right\rangle \qquad (131)$$

Normalization then requires

$$\left\langle \frac{1}{2}, \frac{1}{2} \middle| \frac{1}{2}, \frac{1}{2} \right\rangle = \left\langle \frac{1}{2}, -\frac{1}{2} \middle| \frac{1}{2}, -\frac{1}{2} \right\rangle = 1 \tag{132}$$

Further, since the eigenstates $\left| \frac{1}{2}, \frac{1}{2} \right\rangle$ and $\left| \frac{1}{2}, -\frac{1}{2} \right\rangle$ correspond to different eigenvalues, they must be orthogonal, i.e.

$$\left\langle \frac{1}{2}, \frac{1}{2} \middle| \frac{1}{2}, -\frac{1}{2} \right\rangle = \left\langle \frac{1}{2}, -\frac{1}{2} \middle| \frac{1}{2}, \frac{1}{2} \right\rangle = 0 \tag{133}$$

Let us define the spin "ladder operators"

$$\hat{S}_{\pm} = \hat{S}_1 \pm i\hat{S}_2 \tag{134}$$

Then, as in the case of "orbital angular momentum', we can write

$$\hat{S}_{+} \left| \frac{1}{2}, -\frac{1}{2} \right\rangle = C_{+} \left| \frac{1}{2}, \frac{1}{2} \right\rangle$$

or

$$\left\langle \frac{1}{2}, -\frac{1}{2} \right| \hat{S}_{+}^{\dagger} = \left\langle \frac{1}{2}, -\frac{1}{2} \right| \hat{S}_{-} = C_{+}^{*} \left\langle \frac{1}{2}, \frac{1}{2} \right| \tag{135}$$

and

$$\hat{S}_{-} \left| \frac{1}{2}, \frac{1}{2} \right\rangle = C_{-} \left| \frac{1}{2}, -\frac{1}{2} \right\rangle$$

or

$$\left\langle \frac{1}{2}, \frac{1}{2} \right| \hat{S}_{-}^{\dagger} = \left\langle \frac{1}{2}, \frac{1}{2} \right| \hat{S}_{+} = C_{-}^{*} \left\langle \frac{1}{2}, -\frac{1}{2} \right| \tag{136}$$

In view of the normalization condition, we obtain

$$\left\langle \frac{1}{2}, -\frac{1}{2} \right| \hat{S}_{+}^{\dagger} \hat{S}_{+} \left| \frac{1}{2}, -\frac{1}{2} \right\rangle = |C_{+}|^2 \left\langle \frac{1}{2}, \frac{1}{2} \middle| \frac{1}{2}, \frac{1}{2} \right\rangle = |C_{+}|^2 \tag{137}$$

or

$$\left\langle \frac{1}{2}, -\frac{1}{2} \right| (\hat{S}^2 - \hat{S}_3^2 - \hat{S}_3) \left| \frac{1}{2}, -\frac{1}{2} \right\rangle = [s(s+1) - m_s^2 - m_s] \left\langle \frac{1}{2}, -\frac{1}{2} \middle| \frac{1}{2}, -\frac{1}{2} \right\rangle$$

$$= \left[\frac{1}{2}\left(\frac{1}{2}+1\right) - \left(-\frac{1}{2}\right)^2 - \left(-\frac{1}{2}\right) \right] = 1 = |C_{+}|^2 \tag{138}$$

whence $C_+ = \pm 1$. Similarly, $C_- = \pm 1$. Choosing $C_+ = +1$, we get $\hat{S}_+\left|\frac{1}{2},-\frac{1}{2}\right\rangle = \left|\frac{1}{2},\frac{1}{2}\right\rangle$.

Also, since $\left|\frac{1}{2},\frac{1}{2}\right\rangle$ is the highest eigenstate of $\hat{S}_3$, we have $\hat{S}_+\left|\frac{1}{2},\frac{1}{2}\right\rangle = 0$.

Similarly, $\hat{S}_-\left|\frac{1}{2},\frac{1}{2}\right\rangle = \left|\frac{1}{2},-\frac{1}{2}\right\rangle$. Also, since $\left|\frac{1}{2},-\frac{1}{2}\right\rangle$ is the lowest eigenstate of $\hat{S}_3$, we have $\hat{S}_-\left|\frac{1}{2},-\frac{1}{2}\right\rangle = 0$.

Therefore,

$$\hat{S}_1\left|\frac{1}{2},-\frac{1}{2}\right\rangle = \frac{1}{2}(\hat{S}_+ + \hat{S}_-)\left|\frac{1}{2},-\frac{1}{2}\right\rangle = \frac{1}{2}\left|\frac{1}{2},\frac{1}{2}\right\rangle$$

$$\hat{S}_1\left|\frac{1}{2},\frac{1}{2}\right\rangle = \frac{1}{2}(\hat{S}_+ + \hat{S}_-)\left|\frac{1}{2},\frac{1}{2}\right\rangle = \frac{1}{2}\left|\frac{1}{2},-\frac{1}{2}\right\rangle$$

$$\hat{S}_2\left|\frac{1}{2},-\frac{1}{2}\right\rangle = \frac{1}{2i}(\hat{S}_+ - \hat{S}_-)\left|\frac{1}{2},-\frac{1}{2}\right\rangle = \frac{1}{2i}\left|\frac{1}{2},\frac{1}{2}\right\rangle$$

$$\hat{S}_2\left|\frac{1}{2},\frac{1}{2}\right\rangle = \frac{1}{2i}(\hat{S}_+ - \hat{S}_-)\left|\frac{1}{2},\frac{1}{2}\right\rangle = -\frac{1}{2i}\left|\frac{1}{2},-\frac{1}{2}\right\rangle$$

All these expressions lead to the following representation of the three spin operators:

$$\begin{pmatrix} <|S_1|> & \left|\frac{1}{2},\frac{1}{2}\right\rangle & \left|\frac{1}{2},-\frac{1}{2}\right\rangle \\ \left|\frac{1}{2},\frac{1}{2}\right\rangle & 0 & \frac{1}{2} \\ \left|\frac{1}{2},-\frac{1}{2}\right\rangle & \frac{1}{2} & 0 \end{pmatrix}, \begin{pmatrix} <|S_2|> & \left|\frac{1}{2},\frac{1}{2}\right\rangle & \left|\frac{1}{2},-\frac{1}{2}\right\rangle \\ \left|\frac{1}{2},\frac{1}{2}\right\rangle & 0 & -\frac{i}{2} \\ \left|\frac{1}{2},-\frac{1}{2}\right\rangle & \frac{i}{2} & 0 \end{pmatrix}$$

$$\begin{pmatrix} <|S_3|> & \left|\frac{1}{2},\frac{1}{2}\right\rangle & \left|\frac{1}{2},-\frac{1}{2}\right\rangle \\ \left|\frac{1}{2},\frac{1}{2}\right\rangle & \frac{1}{2} & 0 \\ \left|\frac{1}{2},-\frac{1}{2}\right\rangle & 0 & -\frac{1}{2} \end{pmatrix} \tag{139}$$

or, in terms of the Pauli matrices

$$\hat{S}_i = \frac{1}{2}\boldsymbol{\sigma}_i \tag{140}$$

where

$$\boldsymbol{\sigma}_1 = \begin{pmatrix} 0 & 1 \\ 1 & 0 \end{pmatrix}, \boldsymbol{\sigma}_2 = \begin{pmatrix} 0 & -i \\ i & 0 \end{pmatrix}, \boldsymbol{\sigma}_3 = \begin{pmatrix} 1 & 0 \\ 0 & -1 \end{pmatrix} \tag{141}$$

In order to obtain the eigenfunctions of $\hat{S}_3$, we consider the following eigenvalue problem:

$$\hat{S}_3 \begin{pmatrix} u_\pm \\ v_\pm \end{pmatrix} = \frac{1}{2}\begin{pmatrix} 1 & 0 \\ 0 & -1 \end{pmatrix}\begin{pmatrix} u_\pm \\ v_\pm \end{pmatrix} = \frac{1}{2}\begin{pmatrix} u_\pm \\ -v_\pm \end{pmatrix} = \pm\frac{1}{2}\begin{pmatrix} u_\pm \\ v_\pm \end{pmatrix} \tag{142}$$

which have the solutions $(1\ 0)^T$ and $(0\ 1)^T$, respectively. Thus, the eigenfunctions satisfy $\hat{S}_3 \chi_\pm = \pm \frac{1}{2}\chi_\pm$ with $\chi_+ = \begin{pmatrix} 1 \\ 0 \end{pmatrix}$ and $\chi_- = \begin{pmatrix} 0 \\ 1 \end{pmatrix}$.

These are simultaneous eigenfunctions of $\hat{S}^2$ as is easily seen, e.g.,

$$= \hat{S}^2 \begin{pmatrix} 1 \\ 0 \end{pmatrix} = (\hat{S}_1^2 + \hat{S}_2^2 + \hat{S}_3^2)\begin{pmatrix} 1 \\ 0 \end{pmatrix} = \frac{1}{4}(\boldsymbol{\sigma}_1^2 + \boldsymbol{\sigma}_2^2 + \boldsymbol{\sigma}_3^2)\begin{pmatrix} 1 \\ 0 \end{pmatrix}$$

$$= \frac{3}{4}\begin{pmatrix} 1 \\ 0 \end{pmatrix} = \frac{1}{2}\left(\frac{1}{2}+1\right)\begin{pmatrix} 1 \\ 0 \end{pmatrix} \tag{143}$$

The two spinors $\chi_+ = \begin{pmatrix} 1 \\ 0 \end{pmatrix}$ and $\chi_- = \begin{pmatrix} 0 \\ 1 \end{pmatrix}$ constitute a complete orthonormal set and hence, a basis for the 2-dimensional space in which these spinors act. We can, therefore, expand any spinor in this complete set of eigenspinors as

$$\boldsymbol{\alpha} = \alpha_+ \chi_+ + \alpha_- \chi_- = \begin{pmatrix} \alpha_+ \\ \alpha_- \end{pmatrix} = \alpha_+ \begin{pmatrix} 1 \\ 0 \end{pmatrix} + \alpha_- \begin{pmatrix} 0 \\ 1 \end{pmatrix} \tag{144}$$

with the normalization

$$1 = \langle \boldsymbol{\alpha} | \boldsymbol{\alpha} \rangle = (\alpha_+^* \ \alpha_-^*)\begin{pmatrix} \alpha_+ \\ \alpha_- \end{pmatrix} = |\alpha_+|^2 + |\alpha_-|^2 \tag{145}$$

$|\alpha_+|^2$ and $|\alpha_-|^2$ are, respectively, the probabilities of the spin operator $\hat{S}_3$ taking the values $+\frac{1}{2}$ and $-\frac{1}{2}$, respectively, when the system is in the state represented by the state $\boldsymbol{\alpha}$.

Given an arbitrary spinor $\boldsymbol{\alpha} = \alpha_+ \chi_+ + \alpha_- \chi_-$ we can correlate the coefficients $\{\alpha_+, \alpha_-\}$ to the spin axis direction. For the purpose, let us assume that the spin is pointing up along the direction of the unit vector $\mathbf{n} = (n_1, n_2, n_3) = (\sin\theta\cos\phi.\sin\theta\sin\phi, \cos\theta)$. Hence, the spinor $\boldsymbol{\alpha} = \alpha_+ \chi_+ + \alpha_- \chi_-$ is an eigenstate of the operator $\mathbf{n}.\boldsymbol{\sigma} = \begin{pmatrix} n_3 & n_1 - in_2 \\ n_1 + in_2 & -n_3 \end{pmatrix}$. Thus,

$$\begin{pmatrix} n_3 & n_1 - in_2 \\ n_1 + in_2 & -n_3 \end{pmatrix}\begin{pmatrix} \alpha_+ \\ \alpha_- \end{pmatrix} = \begin{pmatrix} n_3\,\alpha_+ + (n_1 - in_2)\,\alpha_- \\ (n_1 + in_2)\,\alpha_+ - n_3\,\alpha_1 \end{pmatrix} = \begin{pmatrix} \alpha_+ \\ \alpha_- \end{pmatrix} \tag{146}$$

which gives

$$\frac{\alpha_+}{\alpha_-} = \frac{n_1 - in_2}{1 - n_3} = \frac{\sin\theta\cos\phi - i\sin\theta\sin\phi}{1-\cos\theta} = e^{-i\phi}\cot\frac{\theta}{2} \tag{147}$$

in polar coordinates ($r = 1, \theta, \phi$), so that the spin orientation of any spinor is determined solely by the ratio $\frac{\alpha_+}{\alpha_-}$. However, since every complex number can be represented as $e^{-i\phi}\cot\frac{\theta}{2}$, it follows that, given any spinor, we can find a direction along which the spin points with probability one.

Similarly, given a particular direction represented by the unit vector $\mathbf{n} = (n_1, n_2, n_3) = (\sin\theta\cos\phi.\sin\theta\sin\phi, \cos\theta)$, we can compute the eigenstates with spin pointing up/down along the direction of this unit vector. These will be the eigenstates of the operator

$$\hat{S}.\mathbf{n} = \frac{1}{2}\,\boldsymbol{\sigma}.\mathbf{n} = \frac{1}{2}\begin{pmatrix} n_3 & n_1 - in_2 \\ n_1 + in_2 & -n_3 \end{pmatrix}$$

$$= \frac{1}{2}\begin{pmatrix} \cos\theta & e^{-i\phi}\sin\theta \\ e^{i\phi}\sin\theta & -\cos\theta \end{pmatrix}$$

with eigenvalues $\pm\frac{1}{2}$ so that

$$\hat{S}.\mathbf{n}\begin{pmatrix} u_\pm \\ v_\pm \end{pmatrix} = \frac{1}{2}\,\boldsymbol{\sigma}.\mathbf{n}\begin{pmatrix} u_\pm \\ v_\pm \end{pmatrix} = \frac{1}{2}\begin{pmatrix} \cos\theta & e^{-i\phi}\sin\theta \\ e^{i\phi}\sin\theta & -\cos\theta \end{pmatrix}\begin{pmatrix} u_\pm \\ v_\pm \end{pmatrix}$$

$$= \pm\frac{1}{2}\begin{pmatrix} u_\pm \\ v_\pm \end{pmatrix} \tag{148}$$

whence

$$\begin{pmatrix} (\cos\theta \mp 1)u_\pm + e^{-i\phi}\sin\theta v_\pm \\ e^{i\phi}\sin\theta u_\pm - (\cos\theta \pm 1)v_\pm \end{pmatrix} = \begin{pmatrix} -\sin\frac{\theta}{2}\left(\cos\frac{\theta}{2}\right)u_\pm + e^{-i\phi}\cos\frac{\theta}{2}\left(\sin\frac{\theta}{2}\right)v_\pm \\ e^{i\phi}\sin\frac{\theta}{2}\left(\cos\frac{\theta}{2}\right)u_\pm - \cos\frac{\theta}{2}\left(-\sin\frac{\theta}{2}\right)v_\pm \end{pmatrix} = 0$$

or
$$\begin{pmatrix} u_+ \\ v_+ \end{pmatrix} = \begin{pmatrix} e^{-i\frac{\phi}{2}} \cos\frac{\theta}{2} \\ e^{i\frac{\phi}{2}} \sin\frac{\theta}{2} \end{pmatrix}, \begin{pmatrix} u_- \\ v_- \end{pmatrix} = \begin{pmatrix} -e^{-i\frac{\phi}{2}} \sin\frac{\theta}{2} \\ e^{i\frac{\phi}{2}} \cos\frac{\theta}{2} \end{pmatrix} \tag{149}$$

in polar coordinates.

In Cartesian coordinates, we have

$$\hat{S}.\mathbf{n} \begin{pmatrix} u_\pm \\ v_\pm \end{pmatrix} = \frac{1}{2}\,\boldsymbol{\sigma}.\mathbf{n} \begin{pmatrix} u_\pm \\ v_\pm \end{pmatrix} = \frac{1}{2} \begin{pmatrix} n_3 & n_1 - in_2 \\ n_1 + in_2 & -n_3 \end{pmatrix} \begin{pmatrix} u_\pm \\ v_\pm \end{pmatrix}$$
$$= \pm\frac{1}{2} \begin{pmatrix} u_\pm \\ v_\pm \end{pmatrix}$$

whence

$$\begin{pmatrix} (n_3 \mp 1)\, u_\pm + (n_1 - in_2)\, v_\pm \\ (n_1 + in_2)\, u_\pm - (n_3 \pm 1) v_\pm \end{pmatrix} = 0 \text{ or } \begin{pmatrix} u_+ \\ v_+ \end{pmatrix} = \begin{pmatrix} 1 + n_3 \\ n_1 + in_2 \end{pmatrix}, \begin{pmatrix} u_- \\ v_- \end{pmatrix} = \begin{pmatrix} -n_1 + in_2 \\ 1 + n_3 \end{pmatrix} \tag{150}$$

11.8 EXPLICIT EXPRESSION FOR THE EIGENSTATES OF THE TOTAL ANGULAR MOMENTUM OPERATOR

The total angular momentum operator is the vectorial sum of the orbital angular momentum $\hat{\mathbf{L}}$ and the spin angular momentum $\hat{S}$, i.e., $\hat{\mathbf{J}} = \hat{\mathbf{L}} + \hat{S}$. Hence, our problem is reduced to finding the simultaneous eigenfunctions of these two operators which would also be the eigenfunctions of $\hat{\mathbf{J}}$ since $[\hat{\mathbf{J}}, \hat{\mathbf{L}}] = [\hat{\mathbf{L}} + \hat{S}, \hat{\mathbf{L}}] = [\hat{\mathbf{L}}, \hat{\mathbf{L}}] + [\hat{S}, \hat{\mathbf{L}}] = 0$ and similarly $[\hat{\mathbf{J}}, \hat{S}] = 0$. We shall identify the eigenstates of the orbital angular momentum as $|l, n\rangle$, where l relates to the eigenvalue $l\,(l + 1)$ of $\hat{\mathbf{L}}_2$, and n is the eigenvalue of $\hat{L}_3$, i.e., $\hat{\mathbf{L}}^2\,|l, n\rangle = l\,(l + 1)\,|l, n\rangle$ and $\hat{L}_3|l, n\rangle = n\,|l, n\rangle$. Since we are dealing here with the specific case of a spin $\frac{1}{2}$ particle, i.e., $s = \frac{1}{2}$ for which m_s can take only two values $\pm\frac{1}{2}$, the eigenstates of the spin angular momentum $|s, m_s\rangle$ can be conveniently identified as

$$\left|\pm\frac{1}{2}\right\rangle, \text{ i.e., } \hat{S}^2\left|\pm\frac{1}{2}\right\rangle = \frac{1}{2}\left(\frac{1}{2}+1\right)\left|\pm\frac{1}{2}\right\rangle = \frac{3}{4}\left|\pm\frac{1}{2}\right\rangle \text{ and } \hat{S}_3\left|\pm\frac{1}{2}\right\rangle = \pm\frac{1}{2}\left|\pm\frac{1}{2}\right\rangle$$

The common eigenstates of the operators $(\hat{\mathbf{L}}^2,\ \hat{L}_3,\ \hat{S}^2,\ \hat{S}_3)$ are denoted by $\left|l, n, \pm\frac{1}{2}\right\rangle$.

These are the products of $|l, n\rangle$ and $\left|\pm\frac{1}{2}\right\rangle$. The number of these eigenstates is $2(2l + 1)$ since n can take all integral values from $-l$ and $+l$ including 0, i.e., a total of $2l+1$ values for $m_s = +\frac{1}{2}$ and similarly another $2l+1$ values for $m_s = -\frac{1}{2}$. They transform according to the representation $D^{(l)} \otimes D^{(1/2)}$ of $SU(2)$. We can, however, reduce this representation to a sum of irreducible representations $D^{(j)}$s, where each of these $D^{(j)}$s occurs once for $\left| l - \frac{1}{2} \right| \le j \le l + \frac{1}{2}$. Each of these representations $D^{(j)}$s has $(2j+1)$ states. These are linear combinations of $\left| l, n, \pm\frac{1}{2} \right\rangle$.

Let us further identify the eigenstates of $\hat{\mathbf{J}}^2$ and $\hat{J}_3$ in terms of the numbers j, m_j such that $\hat{\mathbf{J}}^2 |j, m_j\rangle = j(j+1)|j, m_j\rangle$ and $\hat{J}_3 |j, m_j\rangle = m_j |j, m_j\rangle$.
The above situation can be explicitly summarized in the following set of equations:

$$\hat{\mathbf{L}}^2 \left| l, n, \pm\frac{1}{2} \right\rangle = l(l+1) \left| l, n, \pm\frac{1}{2} \right\rangle$$

$$\hat{L}_3 \left| l, n, \pm\frac{1}{2} \right\rangle = n \left| l, n, \pm\frac{1}{2} \right\rangle$$

$$\hat{S}^2 \left| l, n, \pm\frac{1}{2} \right\rangle = \frac{3}{4} \left| l, n, \pm\frac{1}{2} \right\rangle$$

$$\hat{S}_3 \left| l, n, \pm\frac{1}{2} \right\rangle = \pm\frac{1}{2} \left| l, n, \pm\frac{1}{2} \right\rangle$$

$$\hat{\mathbf{J}}^2 |j, m_j\rangle = j(j+1)|j, m_j\rangle$$

$$\hat{J}_3 |j, m_j\rangle = m_j |j, m_j\rangle$$

Since the operator $\hat{\mathbf{J}}^2$, $\hat{J}_3$ commute with both $\hat{\mathbf{L}}^2$ and $\hat{S}^2$, we may choose the linearly independent mutually commuting operators to be $(\hat{\mathbf{L}}^2, \hat{S}^2, \hat{\mathbf{J}}^2, \hat{J}_3)$ instead of $(\hat{\mathbf{L}}^2, \hat{L}_3, \hat{S}^2, \hat{S}_3)$ and identify $|j, m_j\rangle$ as the common eigenstates of the set of operators $(\hat{\mathbf{L}}^2, \hat{S}^2, \hat{\mathbf{J}}^2, \hat{J}_3)$ so that

$$\hat{\mathbf{J}}^2 |j, m_j\rangle = j(j+1)|j, m_j\rangle$$

$$\hat{J}_3 |j, m_j\rangle = m_j |j, m_j\rangle$$

$$\hat{\mathbf{L}}^2 |j, m_j\rangle = l(l+1)|j, m_j\rangle$$

$$\hat{S}^2 |j, m_j\rangle = \frac{3}{4} |j, m_j\rangle$$

The above two sets of equations show that the two sets of states $\left\{\left| l, n, \pm\frac{1}{2}\right\rangle\right\}$ and $\{|j, m_j\rangle\}$ constitute two different bases in the same Hilbert space. Our problem of addition of the orbital angular momentum and the spin angular momentum is, thus, equivalent to expressing the states $\{|j, m_j\rangle\}$ as linear combinations of the elements of the set of states $\left\{\left| l, n, \pm\frac{1}{2}\right\rangle\right\}$.

We express these common eigenstates $|j, m_j\rangle$ as linear combinations of the states $\left| l, n, \pm\frac{1}{2}\right\rangle$ as

$$|j, m_j\rangle = \sum_{n=-l}^{l}\left[\left|l, n, \frac{1}{2}\right\rangle C\left(l, n, \frac{1}{2}; j, m_j\right) + \left|l, n, -\frac{1}{2}\right\rangle C\left(l, n, -\frac{1}{2}; j, m_j\right)\right] \tag{151}$$

We can determine the Clebsch Gordon coefficients $C\left(l, n \pm \frac{1}{2}; j, m_j\right)$ by taking the scalar product of $|j, m_j\rangle$ with any given state, say, $\left| l, n', \pm\frac{1}{2}\right\rangle$. Because of the orthonormality of these eigenstates

$$\left\langle l, n', \frac{1}{2} \middle| j, m_j\right\rangle = C\left(l, n, \frac{1}{2}; j, m_j\right)\delta_{nn'}$$

$$\left\langle l, n', -\frac{1}{2} \middle| j, m_j\right\rangle = C\left(l, n, -\frac{1}{2}; j, m_j\right)\delta_{nn'} \tag{152}$$

Substituting from eq. (152) in (151), we obtain

$$|j, m_j\rangle = \sum_{n=-l}^{l}\left[\left| l, n, \frac{1}{2}\right\rangle\left\langle l, n, \frac{1}{2} \middle| j, m_j\right\rangle + \left| l, n, -\frac{1}{2}\right\rangle\left\langle l, n, -\frac{1}{2} \middle| j, m_j\right\rangle\right] \tag{153}$$

Operating on $|j, m_j\rangle$ by $\hat{J}_3 = \hat{\mathbf{L}}_3 + \hat{S}_3$, we obtain

$$\hat{J}_3 |j, m_j\rangle = m_j |j, m_j\rangle = m_j \sum_{n=-l}^{l}\left[\left| l, n, \frac{1}{2}\right\rangle\left\langle l, n, \frac{1}{2} \middle| j, m_j\right\rangle\right.$$

$$\left. + \left|l, n, -\frac{1}{2}\right\rangle\left\langle l, n, -\frac{1}{2}\middle| j, m_j\right\rangle\right]$$

$$= (\hat{\mathbf{L}}_3 + \hat{S}_3)\sum_{n=-l}^{l}\left[\left| l, n, \frac{1}{2}\right\rangle\left\langle l, n, \frac{1}{2} \middle| j, m_j\right\rangle + \left|l, n, -\frac{1}{2}\right\rangle\left\langle l, n, -\frac{1}{2}\middle| j, m_j\right\rangle\right]$$

$$= \sum_{n=-l}^{l}\left[\left(n + \frac{1}{2}\right)\left| l, n, \frac{1}{2}\right\rangle\left\langle l, n, \frac{1}{2} \middle| j, m_j\right\rangle + \left(n - \frac{1}{2}\right)\left| l, n, -\frac{1}{2}\right\rangle\left\langle l, n, -\frac{1}{2} \middle| j, m_j\right\rangle\right] \tag{154}$$

whence

$$\sum_{n=-l}^{l}\left[\left(m_j - n - \frac{1}{2}\right)\left| l, n, \frac{1}{2}\right\rangle\left\langle l, n, \frac{1}{2}\right| j, m_j\right\rangle$$

$$+\left(m_j - n + \frac{1}{2}\right)\left| l, n, -\frac{1}{2}\right\rangle\left\langle l, n, -\frac{1}{2}\right| j, m_j\right\rangle\Bigg] = 0 \qquad (155)$$

The linear independence of these states mandates

$$\left(m_j - n - \frac{1}{2}\right)\left\langle l, n, \frac{1}{2}\right| j, m_j\right\rangle = 0$$

and
$$\left(m_j - n + \frac{1}{2}\right)\left\langle l, n, -\frac{1}{2}\right| j, m_j\right\rangle = 0 \qquad (156)$$

whence

$$\left\langle l, n, \frac{1}{2}\right| j, m_j\right\rangle = 0 \text{ unless}$$

$$\left(m_j - n - \frac{1}{2}\right) = 0 \qquad (157)$$

and

$$\left\langle l, n, -\frac{1}{2}\right| j, m_j\right\rangle = 0 \text{ unless}$$

$$\left(m_j - n + \frac{1}{2}\right) = 0 \qquad (158)$$

Thus, in order that the Clebsch Gordon coefficients do not vanish, we must have $m_j = n \pm \frac{1}{2}$, i.e., the third component of the total angular momentum must be the sum of the third component of the orbital angular momentum and the third component of the spin. To consolidate, the various conditions for the non-vanishing of the Clebsch Gordon coefficients are

$$\left|l - \frac{1}{2}\right| \leq j \leq l + \frac{1}{2} \qquad (159)$$

$$m_j = n \pm \frac{1}{2} \qquad (160)$$

$$|m_j| \leq j \qquad (161)$$

The first of these conditions is obtained from

$$\sum_{m_s=-\frac{1}{2}}^{\frac{1}{2}} \sum_{n=-l}^{l} (n+m_s) = \sum_{n=-l}^{l}\left(n-\frac{1}{2}\right) + \sum_{n=-l}^{l}\left(n+\frac{1}{2}\right)$$

$$= \sum_{m_j=-l-\frac{1}{2}}^{l-\frac{1}{2}} m_j + \sum_{m_j=-l+\frac{1}{2}}^{l+\frac{1}{2}} m_j$$

$$= \sum_{m_j=-l-\frac{1}{2}}^{l+\frac{1}{2}} m_j - \sum_{m_j=l+\frac{1}{2}}^{l+\frac{1}{2}} m_j + \sum_{m_j=-l+\frac{1}{2}}^{l-\frac{1}{2}} m_j + \sum_{m_j=l+\frac{1}{2}}^{l+\frac{1}{2}} m_j$$

$$= \sum_{m_j=-l-\frac{1}{2}}^{l+\frac{1}{2}} m_j + \sum_{m_j=-l+\frac{1}{2}}^{l-\frac{1}{2}} m_j = \sum_{j=\left|l-\frac{1}{2}\right|}^{l+\frac{1}{2}} \sum_{m_j=-j}^{j} m_j$$

Obviously, $(m_j)_{\max} = l + \frac{1}{2} = j_{\max}$. Hence, if $j = m_j = l + \frac{1}{2} = (m_j)_{\max} = j_{\max}$ then the Clebsch Gordon coefficient corresponding to this state is $\left\langle l, n, \pm\frac{1}{2} \middle| l+\frac{1}{2}, l+\frac{1}{2}\right\rangle = \delta_{ln}\, \delta_{\pm\frac{1}{2},\frac{1}{2}}$ which is the only non-zero Clebsch Gordon coefficient corresponding to this state. It will be the one for which $n = l$ and $m_s = \frac{1}{2}$. For this state

$$\left|l+\frac{1}{2}, l+\frac{1}{2}\right\rangle = \left|l, l, \frac{1}{2}\right\rangle \left\langle l, l, \frac{1}{2} \middle| l+\frac{1}{2}, l+\frac{1}{2}\right\rangle \tag{162}$$

If we note that both the states $\left|l+\frac{1}{2}, l+\frac{1}{2}\right\rangle$ and $\left|l, l, \frac{1}{2}\right\rangle$ are normalized, then the Clebsch Gordon coefficient corresponding to this state $\left\langle l, l, \frac{1}{2} \middle| l+\frac{1}{2}, l+\frac{1}{2}\right\rangle$ is a complex number of unit magnitude conventionally taken as unity.

The matrix of Clebsch Gordon coefficients is, essentially, the transformation matrix from one basis to another in the $2(2l+1)$ dimensional Hilbert space. The matrix will be unitary provided the set of states $\left\{\left|l, n, \pm\frac{1}{2}\right\rangle\right\}$ and $\left\{\left|j, m_j\right\rangle\right\}$ are orthonormal. In the sequel, we assume that the matrix is unitary and that all the elements thereof are real so that

$$\left\langle l, n, \pm\frac{1}{2}\middle| j, m_j\right\rangle^* = \left\langle l, n, \pm\frac{1}{2}\middle| j, m_j\right\rangle \tag{163}$$

We also have the following orthogonality relations that enable us to compute the values of these Clebsch Gordon coefficients

$$\sum_{j=\left|l-\frac{1}{2}\right|}^{l+\frac{1}{2}}\ \sum_{m_j=-j}^{j}\left\langle l, n, \pm\frac{1}{2}\middle| j, m_j\right\rangle\left\langle l, n', \pm\frac{1}{2}\middle| j, m_j\right\rangle = \delta_{nn'} \tag{164}$$

$$\sum_{j=\left|l-\frac{1}{2}\right|}^{l+\frac{1}{2}}\ \sum_{m_j=-j}^{j}\left\langle l, n, \pm\frac{1}{2}\middle| j, m_j\right\rangle\left\langle l, n, \mp\frac{1}{2}\middle| j, m_j\right\rangle = 0 \tag{165}$$

$$\sum_{n=-l}^{l}\left(\left\langle l, n, \frac{1}{2}\middle| j, m_j\right\rangle\left\langle l, n, \frac{1}{2}\middle| j', m'_j\right\rangle\right.$$

$$\left.+\left\langle l, n, -\frac{1}{2}\middle| j, m_j\right\rangle\left\langle l, n, -\frac{1}{2}\middle| j', m'_j\right\rangle\right) = \delta_{m_j\, m'_j}\,\delta_{jj'} \tag{166}$$

We have already established that m_j assumes its maximum value of $l+\frac{1}{2}$ when $j = m_j = l+\frac{1}{2}$ $= (m_j)_{\max} = j_{\max}$ and, in this case, there is only one non-zero Clebsch Gordon coefficient corresponding to this state which is taken as unity, i.e., $\left\langle l, n, \frac{1}{2}\middle| l+\frac{1}{2}, l+\frac{1}{2}\right\rangle = \delta_{ln}$.

The next largest value of m_j is $l-\frac{1}{2}$. This will occur in two situations, viz. when $j = l+\frac{1}{2}$ (because m_j can take values $j, j-1, j-2,\ldots 0, -1, -2,\ldots -j$) and also when $j = \left(l+\frac{1}{2}\right) - 1 = l-\frac{1}{2}$. Furthermore, since $m_j = n\pm\frac{1}{2}$ we must also have either $n = l$ and the spin state $-\frac{1}{2}$ or $n = l-1$ and the spin state $+\frac{1}{2}$. Thus, both the final states $|j, m_j\rangle \equiv \left|l+\frac{1}{2}, l-\frac{1}{2}\right\rangle$ and $|j, m_j\rangle \equiv \left|l-\frac{1}{2}, l-\frac{1}{2}\right\rangle$ are possible and will be linear combinations of only two initial states, viz. $\left|l, l, -\frac{1}{2}\right\rangle$ and $\left|l, l-1, \frac{1}{2}\right\rangle$.

Hence, we can write

$$\left|l+\frac{1}{2}, l-\frac{1}{2}\right\rangle = \left|l, l, -\frac{1}{2}\right\rangle \left\langle l, l, -\frac{1}{2}\middle| l+\frac{1}{2}, l-\frac{1}{2}\right\rangle + \left|l, l-1, \frac{1}{2}\right\rangle \left\langle l, l-1, \frac{1}{2}\middle| l+\frac{1}{2}, l-\frac{1}{2}\right\rangle \tag{167}$$

and

$$\left|l-\frac{1}{2}, l-\frac{1}{2}\right\rangle = \left|l, l, -\frac{1}{2}\right\rangle \left\langle l, l, -\frac{1}{2}\middle| l-\frac{1}{2}, l-\frac{1}{2}\right\rangle + \left|l, l-1, \frac{1}{2}\right\rangle \left\langle l, l-1, \frac{1}{2}\middle| l-\frac{1}{2}, l-\frac{1}{2}\right\rangle \tag{168}$$

Thus, in the second and third rows and columns of the C-G coefficients matrix, there will be only two nonvanishing entries.

The next largest value of m_j is $l-\frac{3}{2}$. This will occur in situations when:

(a) $j = l+\frac{1}{2}$ (because m_j can take values $j, j-1, j-2, \ldots 0, -1, -2, \ldots -j$) and also when

(b) $j = \left(l+\frac{1}{2}\right) - 1 = l-\frac{1}{2}$.

The state $j = \left(l+\frac{1}{2}\right) - 2 = l-\frac{3}{2}$ cannot occur because it violates condition (159), i.e., $\left|l-\frac{1}{2}\right| \leq j \leq l+\frac{1}{2}$. Furthermore, since $m_j = n \pm \frac{1}{2}$ we must also have either $n = l-1$ and the spin state $-\frac{1}{2}$ or $n = l-2$ and the spin state $+\frac{1}{2}$. Thus, the final states $|j, m_j\rangle \equiv \left|l+\frac{1}{2}, l-\frac{3}{2}\right\rangle$ and $|j, m_j\rangle \equiv \left|l-\frac{1}{2}, l-\frac{3}{2}\right\rangle$ can be linear combinations of only two initial states, viz. $\left|l, l-1, -\frac{1}{2}\right\rangle$ and $\left|l, l-2, \frac{1}{2}\right\rangle$.

Let us now apply the "raising" operator $\hat{J}_+ = \hat{\mathbf{L}}_+ + \hat{S}_+$ to both sides of eq. (153). We obtain

$$\hat{J}_+ |j, m_j\rangle = \left[j(j+1) - m_j (m_j+1)\right]^{\frac{1}{2}} |j, m_j + 1\rangle$$

$$= (\hat{\mathbf{L}}_+ + \hat{S}_+) \sum_{n=-l}^{l} \left[\left| l, n, \frac{1}{2} \right\rangle \left\langle l, n, \frac{1}{2} \middle| j, m_j \right\rangle \right.$$

$$\left. + \left| l, n, -\frac{1}{2} \right\rangle \left\langle l, n, -\frac{1}{2} \middle| j, m_j \right\rangle \right]$$

$$= \sum_{n=-l}^{l} \left\{ [l(l+1) - n(n+1)]^{\frac{1}{2}} \left| l, n+1, \frac{1}{2} \right\rangle \left\langle l, n, \frac{1}{2} \middle| j, m_j \right\rangle \right.$$

$$\left. + \left[[l(l+1) - n(n+1)]^{\frac{1}{2}} \left| l, n+1, -\frac{1}{2} \right\rangle + \left| l, n, \frac{1}{2} \right\rangle \right] \left\langle l, n, -\frac{1}{2} \middle| j, m_j \right\rangle \right\} \tag{169}$$

Substituting in the left-hand side of eq. (169) for $|j, m_j + 1\rangle$ from eq. (153), we get

$$|j, m_j + 1\rangle = \sum_{n=-l}^{l} \left[\left| l, n, \frac{1}{2} \right\rangle \left\langle l, n, \frac{1}{2} \middle| j, m_j + 1 \right\rangle + \left| l, n, -\frac{1}{2} \right\rangle \left\langle l, n, -\frac{1}{2} \middle| j, m_j + 1 \right\rangle \right] \tag{170}$$

In the first and second terms of the right-hand side of eq. (169), we change the index of summation from $n \to n + 1 \equiv n$ to get

$$\sum_{n=-l+1}^{l+1} \left\{ [l(l+1) - (n-1)\,n]^{\frac{1}{2}} \left| l, n, \frac{1}{2} \right\rangle \left\langle l, n, -1, \frac{1}{2} \middle| j, m_j \right\rangle \right.$$

$$\left. + [l(l+1) - (n-1)\,n]^{\frac{1}{2}} \left| l, n, -\frac{1}{2} \right\rangle \left\langle l, n-1, -\frac{1}{2} \middle| j, m_j \right\rangle \right\}$$

$$= \sum_{n=-l}^{l} \left\{ [l(l+1) - (n-1)\,n]^{\frac{1}{2}} \left| l, n, \frac{1}{2} \right\rangle \left\langle l, n, -1, \frac{1}{2} \middle| j, m_j \right\rangle \right.$$

$$\left. + [l(l+1) - (n-1)\,n]^{\frac{1}{2}} \left| l, n, -\frac{1}{2} \right\rangle \left\langle l, n-1, -\frac{1}{2} \middle| j, m_j \right\rangle \right\} \tag{171}$$

since the terms for $n = -l$, $n = l + 1$ do not make any contribution. From eqs. (169), (170) and (171), we have

$$\left[j(j+1) - m_j(m_j+1)\right]^{\frac{1}{2}} \sum_{n=-l}^{l} \left[\left| l, n, \frac{1}{2} \right\rangle \left\langle l, n, \frac{1}{2} \middle| j, m_j + 1 \right\rangle + \left| l, n, -\frac{1}{2} \right\rangle \left\langle l, n, -\frac{1}{2} \middle| j, m_j + 1 \right\rangle \right]$$

$$= \sum_{n=-l}^{l} \left\{ [l(l+1) - (n-1)\,n]^{\frac{1}{2}} \left| l, n, \frac{1}{2} \right\rangle \left\langle l, n, -1, \frac{1}{2} \right| j, m_j \right\rangle$$

$$+ [l(l+1) - (n-1)\,n]^{\frac{1}{2}} \left| l, n, -\frac{1}{2} \right\rangle \left\langle l, n-1, -\frac{1}{2} \right| j, m_j \right\rangle$$

$$+ \left| l, n, \frac{1}{2} \right\rangle \left\langle l, n, -\frac{1}{2} \right| j, m_j \right\rangle \Bigg\} \tag{172}$$

Since the states $\left| l, n, \pm\frac{1}{2} \right\rangle$ are mutually orthogonal, equating their respective coefficients on both sides of eq. (172), we obtain

$$[j(j+1) - m_j(m_j+1)]^{\frac{1}{2}} \left\langle l, n, \frac{1}{2} \right| j, m_j + 1 \right\rangle$$

$$= [l(l+1) - (n-1)n]^{\frac{1}{2}} \left\langle l, n, -1, \frac{1}{2} \right| j, m_j \right\rangle + \left\langle l, n, -\frac{1}{2} \right| j, m_j \right\rangle \tag{173}$$

and

$$[j(j+1) - m_j(m_j+1)]^{\frac{1}{2}} \left\langle l, n, -\frac{1}{2} \right| j, m_j + 1 \right\rangle$$

$$= [l(l+1) - (n-1)n]^{\frac{1}{2}} \left\langle l, n-1, -\frac{1}{2} \right| j, m_j \right\rangle \tag{174}$$

Applying the "lowering" operator $\hat{J}_- = \hat{L}_- + \hat{S}_-$ to both sides of eq. (153) and proceeding as in the case of the "raising operator", we obtain

$$\hat{J}_-|j, m_j\rangle = [j(j+1) - m_j(m_j-1)]^{\frac{1}{2}} |j, m_j - 1\rangle$$

$$= (\hat{L}_- + \hat{S}_-) \sum_{n=-1}^{l} \left[\left| l, n, \frac{1}{2} \right\rangle \left\langle l, n, \frac{1}{2} \right| j, m_j \right\rangle + \left| l, n, -\frac{1}{2} \right\rangle \left\langle l, n, -\frac{1}{2} \right| j, m_j \right\rangle \Bigg]$$

$$= \sum_{n=-l}^{l} \left\{ [l(l+1) - n(n-1)]^{\frac{1}{2}} \left| l, n-1, \frac{1}{2} \right\rangle \left\langle l, n, \frac{1}{2} \right| j, m_j \right\rangle$$

$$+ \left| l, n, -\frac{1}{2} \right\rangle \left\langle l, n, \frac{1}{2} \right| j, m_j \right\rangle + [l(l+1) - n(n-1)^{\frac{1}{2}}$$

$$\left|l, n-1, -\frac{1}{2}\right\rangle \left\langle l, n, -\frac{1}{2}\right|\left. j, m_j \right\rangle \Bigg\} \tag{175}$$

Substituting in the left-hand side of eq. (175) for $|j, m_j - 1\rangle$ from eq. (153), we get

$$|j, m_j - 1\rangle = \sum_{n=-l}^{l} \left[\left|l, n, \frac{1}{2}\right\rangle \left\langle l, n, \frac{1}{2}\right| \left. j, m_j - 1 \right\rangle + \left|l, n, -\frac{1}{2}\right\rangle \left\langle l, n, -\frac{1}{2}\right| \left. j, m_j - 1 \right\rangle \right] \tag{176}$$

In the first and third terms of the right-hand side of eq. (175), we change the index of summation from $n \to n - 1 \equiv n$ and write these terms as

$$\sum_{n=-l-1}^{l-1} \left\{ [l(l+1) - n(n+1)]^{\frac{1}{2}} \left|l, n, \frac{1}{2}\right\rangle \left\langle l, n+1, \frac{1}{2}\right| \left. j, m_j \right\rangle \right.$$

$$\left. + [l(l+1) - n(n+1)]^{1/2} \left|l, n, -\frac{1}{2}\right\rangle \left\langle l, n+1, -\frac{1}{2}\right| \left. j, m_j \right\rangle \right\}$$

$$= \sum_{n=-l}^{l} \left\{ [l(l+1) - n(n+1)]^{\frac{1}{2}} \left|l, n, \frac{1}{2}\right\rangle \left\langle l, n+1, \frac{1}{2}\right| \left. j, m_j \right\rangle \right.$$

$$\left. + [l(l+1) - n(n+1)]^{\frac{1}{2}} \left|l, n, -\frac{1}{2}\right\rangle \left\langle l, n, +1, -\frac{1}{2}\right| \left. j, m_j \right\rangle \right\} \tag{177}$$

From eqs. (175), (176) and (177), we have

$$[j(j+1) - m_j(m_j - 1)]^{\frac{1}{2}}$$

$$\sum_{n=-l}^{l} \left[\left|l, n, \frac{1}{2}\right\rangle \left\langle l, n, \frac{1}{2}\right| \left. j, m_j - 1 \right\rangle + \left|l, n, -\frac{1}{2}\right\rangle \left\langle l, n, -\frac{1}{2}\right| \left. j, m_j - 1 \right\rangle \right]$$

$$= \sum_{n=-l}^{l} \left\{ [l(l+1) - n(n+1)]^{\frac{1}{2}} \left|l, n, \frac{1}{2}\right\rangle \left\langle l, n+1, \frac{1}{2}\right| \left. j, m_j \right\rangle + [l(l+1) - n(n+1)]^{\frac{1}{2}} \right.$$

$$\left. \left|l, n, -\frac{1}{2}\right\rangle \left\langle l, n+1, -\frac{1}{2}\right| \left. j, m_j \right\rangle + \left|l, n, -\frac{1}{2}\right\rangle \left\langle l, n, \frac{1}{2}\right| \left. j, m_j \right\rangle \right\} \tag{178}$$

which give on equating the respective coefficients

$$[j(j+1) - m_j(m_j - 1)]^{1/2} \left\langle l, n, \frac{1}{2}\right| \left. j, m_j - 1 \right\rangle$$

$$= [l(l+1) - n(n+1)]^{1/2} \left\langle l, n+1, \frac{1}{2}\right| \left. j, m_j \right\rangle \tag{179}$$

and

$$[j(j+1) - m_j(m_j-1)]^{1/2}\left\langle l, n, -\frac{1}{2}\right| j, m_j - 1\Big\rangle$$

$$= [\,l(\,l+1) - n(n+1)]^{1/2}\left\langle l, n, +1, -\frac{1}{2}\right| j, m_j\Big\rangle + \left\langle l, n, \frac{1}{2}\right| j, m_j\Big\rangle \qquad (180)$$

Using the recursion relations (173), (174), (179) and (180) and the fact that $\left\langle l, n, \frac{1}{2}\right| l+\frac{1}{2}, l+\frac{1}{2}\Big\rangle = \delta_{ln}$, we can compute all the Clebsch Gordon coefficients. For example, we compute the Clebsch Gordon coefficients for $m_j = l - \frac{1}{2}$. We substitute $n = l$, $m_j = j = l + \frac{1}{2}$ in eq. (180) to get

$$\left[\left(l+\frac{1}{2}\right)\left(l+\frac{3}{2}\right) - \left(l+\frac{1}{2}\right)\left(l-\frac{1}{2}\right)\right]^{\frac{1}{2}}$$

$$\left\langle l, l, -\frac{1}{2}\right| l+\frac{1}{2}, l-\frac{1}{2}\Big\rangle = \left\langle l, l, \frac{1}{2}\right| l+\frac{1}{2}, l+\frac{1}{2}\Big\rangle = 1 \qquad (181)$$

whence

$$\left\langle l, l, -\frac{1}{2}\right| l+\frac{1}{2}, l-\frac{1}{2}\Big\rangle = \left[2\left(l+\frac{1}{2}\right)\right]^{-\frac{1}{2}} \qquad (182)$$

Similarly, we substitute $n = l - 1$, $m_j = j = l + \frac{1}{2}$ in eq. (179) to get

$$\left[\left(l+\frac{1}{2}\right)\left(l+\frac{3}{2}\right) - \left(l+\frac{1}{2}\right)\left(l-\frac{1}{2}\right)\right]^{\frac{1}{2}}\left\langle l, l-1, \frac{1}{2}\right| l+\frac{1}{2}, l-\frac{1}{2}\Big\rangle$$

$$= [l(l+1) - (l-1)l]^{1/2}\left\langle l, l, \frac{1}{2}\right| l+\frac{1}{2}, l+\frac{1}{2}\Big\rangle$$

$$= [l(l+1) - (l-1)l]^{1/2} = (2l)^{\frac{1}{2}} \qquad (183)$$

so that

$$\left\langle l, l-1, \frac{1}{2}\right| l+\frac{1}{2}, l-\frac{1}{2}\Big\rangle = (l)^{\frac{1}{2}}\left(l+\frac{1}{2}\right)^{-\frac{1}{2}} \qquad (184)$$

Hence, we have obtained two Clebsch Gordon coefficients for $m_j = l - \frac{1}{2}$ corresponding to the state $j = l + \frac{1}{2}$. There will be two other C-G coefficients for $m_j = l - \frac{1}{2}$ corresponding to $j = l - \frac{l}{2}$. These can be obtained from the orthogonality relations (164)-(166). We obtain

$$\left\langle l, l, -\frac{1}{2}\middle| l + \frac{1}{2}, l - \frac{1}{2}\right\rangle^2 + \left\langle l, l, -\frac{1}{2}\middle| l - \frac{1}{2}, l - \frac{1}{2}\right\rangle^2 = 1 \tag{185}$$

which together with eq. (182) gives

$$\left\langle l, l, -\frac{1}{2}\middle| l - \frac{1}{2}, l - \frac{1}{2}\right\rangle = \pm\sqrt{\frac{l}{l + \frac{1}{2}}} \tag{186}$$

Similarly,

$$\left\langle l, l-1, \frac{1}{2}\middle| l + \frac{1}{2}, l - \frac{1}{2}\right\rangle^2 + \left\langle l, l-1, \frac{1}{2}\middle| l - \frac{1}{2}, l - \frac{1}{2}\right\rangle^2 = 1 \tag{187}$$

which together with eq. (184) gives

$$\left\langle l, l-1, \frac{1}{2}\middle| l - \frac{1}{2}, l - \frac{1}{2}\right\rangle = \pm\sqrt{\frac{1}{2\left(l + \frac{1}{2}\right)}} \tag{188}$$

Furthermore, we must also have

$$\left\langle l, l, -\frac{1}{2}\middle| l + \frac{1}{2}, l - \frac{1}{2}\right\rangle \left\langle l, l, -\frac{1}{2}\middle| l - \frac{1}{2}, l - \frac{1}{2}\right\rangle$$

$$+\left\langle l, l-1, \frac{1}{2}\middle| l + \frac{1}{2}, l - \frac{1}{2}\right\rangle \left\langle l, l-1, \frac{1}{2}\middle| l - \frac{1}{2}, l - \frac{1}{2}\right\rangle = 0 \tag{189}$$

whence

$$\frac{\left\langle l, l, -\frac{1}{2}\middle| l - \frac{1}{2}, l - \frac{1}{2}\right\rangle}{\left\langle l, l-1, \frac{1}{2}\middle| l - \frac{1}{2}, l - \frac{1}{2}\right\rangle} = -\frac{\left\langle l, l-1, \frac{1}{2}\middle| l + \frac{1}{2}, l - \frac{1}{2}\right\rangle}{\left\langle l, l, -\frac{1}{2}\middle| l + \frac{1}{2}, l - \frac{1}{2}\right\rangle} = -(2l)^{1/2} \tag{190}$$

Therefore, the coefficients $\left\langle l, l, -\frac{1}{2} \middle| l-\frac{1}{2}, l-\frac{1}{2} \right\rangle$ and $\left\langle l, l-1, \frac{1}{2} \middle| l-\frac{1}{2}, l-\frac{1}{2} \right\rangle$ have opposite signs. Conventionally, we take $\left\langle l, l, -\frac{1}{2} \middle| l-\frac{1}{2}, l-\frac{1}{2} \right\rangle$ as positive and $\left\langle l, l-1, \frac{1}{2} \middle| l-\frac{1}{2}, l-\frac{1}{2} \right\rangle$ as negative. The C-G coefficients for $m_j = l - \frac{1}{2}$ are tabulated below.

$m_j \rightarrow$	$l-\frac{1}{2}$	$l-\frac{1}{2}$
$j\rightarrow$ $n\downarrow$	$l+\frac{1}{2}$	$l-\frac{1}{2}$
l	$\left[2\left(l+\frac{1}{2}\right)\right]^{-\frac{1}{2}}$	$\sqrt{\frac{l}{l+\frac{1}{2}}}$
$l-1$	$(l)^{1/2}\left(l+\frac{1}{2}\right)^{-\frac{1}{2}}$	$\sqrt{\frac{1}{2\left(l+\frac{1}{2}\right)}}$

Similarly, for $m_j = l - \frac{3}{2}$ we will have C-G coefficients only for $j = l-\frac{1}{2}, l+\frac{1}{2}$ because $j = l - \frac{3}{2}$ violates condition (153). The corresponding values of n can be $n = l - 2$ in the spin state $+\frac{1}{2}$, and $n = l - 1$ in the spin state $-\frac{1}{2}$. However, the state $n = l$ will not occur because there is no spin state with $m_s = -\frac{3}{2}$. Hence, in this case we will again have a 2×2 matrix consisting of four C-G coefficients. The coefficients can be determined as in the above case. In fact, for every $m_j = l - \left(r+\frac{1}{2}\right)$, $r = 0, 1, 2, 3, \ldots$ we will have C-G coefficients for $j = l-\frac{1}{2}, l+\frac{1}{2}$. There will be only two corresponding values of n for each final state $|j, m_j\rangle$, viz. $n = l - r - 1$ in the spin state $+\frac{1}{2}$, $n = l - r$ in the spin state $-\frac{1}{2}$. In other words, each final state can be expressed as a linear combination of two initial states.

Let us now substitute $j = l + \frac{1}{2}$, $n = m - \frac{1}{2}$ $m_j = m + 1$ in eq. (179). We obtain

$$\left[\left(l+\frac{1}{2}\right)\left(l+\frac{3}{2}\right) - m(m+1)\right]^{\frac{1}{2}} \left\langle l, m-\frac{1}{2}, \frac{1}{2} \middle| l+\frac{1}{2}, m \right\rangle$$

$$= \left[l(l+1) - \left(m-\frac{1}{2}\right)\left(m+\frac{1}{2}\right)\right]^{\frac{1}{2}} \left\langle l, m+\frac{1}{2}, \frac{1}{2} \middle| l+\frac{1}{2}, m+1 \right\rangle$$

or

$$\left(l+m+\frac{3}{2}\right)^{\frac{1}{2}} \left\langle l, m-\frac{1}{2}, \frac{1}{2} \middle| l+\frac{1}{2}, m \right\rangle$$

$$= \left(l+m+\frac{1}{2}\right)^{\frac{1}{2}} \left\langle l, m+\frac{1}{2}, \frac{1}{2} \middle| l+\frac{1}{2}, m+1 \right\rangle \tag{191}$$

Similarly, on substituting $j = l + \frac{1}{2}$, $n = m + \frac{1}{2}$, $m_j = m - 1$ in eq. (174), we get

$$\left[\left(l+\frac{1}{2}\right)\left(l+\frac{3}{2}\right) - m(m-1)\right]^{\frac{1}{2}} \left\langle l, m+\frac{1}{2}, -\frac{1}{2} \middle| l+\frac{1}{2}, m \right\rangle$$

$$= \left[l(l+1) - \left(m-\frac{1}{2}\right)\left(m+\frac{1}{2}\right)\right]^{\frac{1}{2}} \left\langle l, m-\frac{1}{2}, -\frac{1}{2} \middle| l+\frac{1}{2}, m-1 \right\rangle$$

or

$$\left(l-m+\frac{3}{2}\right)^{\frac{1}{2}} \left\langle l, m+\frac{1}{2}, -\frac{1}{2} \middle| l+\frac{1}{2}, m \right\rangle$$

$$= \left(l-m+\frac{1}{2}\right)^{\frac{1}{2}} \left\langle l, m-\frac{1}{2}, -\frac{1}{2} \middle| l+\frac{1}{2}, m-1 \right\rangle \tag{192}$$

Iterating eq. (191), we get

$$\left\langle l, m-\frac{1}{2}, \frac{1}{2} \middle| l+\frac{1}{2}, m \right\rangle = \sqrt{\frac{l+m+\frac{1}{2}}{2l+1}} \left\langle l, l, \frac{1}{2} \middle| l+\frac{1}{2}, l+\frac{1}{2} \right\rangle = \sqrt{\frac{l+m+\frac{1}{2}}{2l+1}} \tag{193}$$

since $\left\langle l, l, \frac{1}{2} \middle| l+\frac{1}{2}, l+\frac{1}{2} \right\rangle = 1$.

We also have

$$\left\langle l, m+\frac{1}{2}, -\frac{1}{2} \middle| l+\frac{1}{2}, m \right\rangle = \sqrt{\frac{l-m+\frac{1}{2}}{2l+1}} \tag{194}$$

$$\left\langle l, m-\frac{1}{2}, \frac{1}{2} \middle| l-\frac{1}{2}, m \right\rangle = -\sqrt{\frac{l-m+\frac{1}{2}}{2l+1}} \tag{195}$$

$$\left\langle l, m+\frac{1}{2}, -\frac{1}{2} \middle| l-\frac{1}{2}, m \right\rangle = \sqrt{\frac{l+m+\frac{1}{2}}{2l+1}} \tag{196}$$

Hence, in the coordinate representation for the eigenstates of the orbital angular momentum $\hat{\mathbf{L}}$ and in the basis of the eigenstates of $\hat{S}_3$ for the representation of the spin operator, the eigenfunctions of the total angular momentum operator may be written (ignoring the radial part) as

$$\psi_{l+\frac{1}{2},m} = \sqrt{\frac{l+m+\frac{1}{2}}{2l+1}}\, Y_l^{m-1/2}\chi_+ + \sqrt{\frac{l-m+\frac{1}{2}}{2l+1}}\, Y_l^{m+1/2}\chi_- \tag{197}$$

and

$$\psi_{l-\frac{1}{2},m} = -\sqrt{\frac{l-m+\frac{1}{2}}{2l+1}}\, Y_l^{m-1/2}\chi_+ + \sqrt{\frac{l+m+\frac{1}{2}}{2l+1}}\, Y_l^{m+1/2}\chi_- \tag{198}$$

where $Y_l^{m-1/2}$, $Y_l^{m+1/2}$ are eigenfunctions of the orbital angular momentum and can be obtained from eq. (128) or (130) and $\chi_+ = \begin{pmatrix} 1 \\ 0 \end{pmatrix}, \chi_- = \begin{pmatrix} 0 \\ 1 \end{pmatrix}$ are the eigenstates of the spin operator in the basis of the eigenstates of $\hat{S}_3$. We now explicitly show that the functions given by eqs. (197) and (198) actually constitute eigenfunctions of $\hat{\mathbf{J}}^2$ and $\hat{J}_3$. We have

$$\hat{\mathbf{J}}^2 = \hat{\mathbf{L}}^2 + \hat{S}^2 + 2\hat{\mathbf{L}}.\hat{S} = \hat{\mathbf{L}}^2 + \hat{S}^2 + 2\hat{L}_3\hat{S}_3 + \hat{L}_+\hat{S}_- + \hat{L}_-\hat{S}_+ \tag{199}$$

Also

$$\hat{\mathbf{L}}^2\psi_{l+\frac{1}{2},m} = \hat{\mathbf{L}}^2\left(aY_l^{m-1/2}\chi_+ + bY_l^{m+1/2}\chi_-\right)$$

$$= l(l+1)\left(aY_l^{m-1/2}\chi_+ + bY_l^{m+1/2}\chi_-\right) \tag{200}$$

since
$$\mathbf{L}^2 Y_l^{m\pm 1/2} = l(l+1) Y_l^{m\pm 1/2},$$

$$\hat{S}^2 \Psi_{l+\frac{1}{2},m} = \hat{S}^2\left(aY_l^{m-1/2}\chi_+ + bY_l^{m+1/2}\chi_-\right)$$

$$= \left(aY_l^{m-1/2}\hat{S}^2\chi_+ + bY_l^{m+1/2}\hat{S}^2\chi_-\right)$$

$$= aY_l^{m-1/2}\left[\frac{1}{2}\left(\frac{1}{2}+1\right)\right]\chi_+ + bY_l^{m+1/2}\left[\frac{1}{2}\left(\frac{1}{2}+1\right)\right]\chi_-$$

$$= \frac{3}{4}aY_l^{m-1/2}\chi_+ + \frac{3}{4}bY_l^{m+1/2}\chi_- \tag{201}$$

Further

$$2\hat{L}_3\hat{S}_3\psi_{l+\frac{1}{2},m} = 2\hat{L}_3\hat{S}_3\left(aY_l^{m-1/2}\chi_+ + bY_l^{m+1/2}\chi_-\right)$$

$$= 2\hat{L}_3\left(aY_l^{m-1/2}\hat{S}_3\chi_+ + bY_l^{m+1/2}\hat{S}_3\chi_-\right)$$

$$= 2\hat{L}_3\left[aY_l^{m-1/2}\left(\frac{1}{2}\chi_+\right) + bY_l^{m+1/2}\left(-\frac{1}{2}\chi_-\right)\right]$$

$$= \hat{L}_3\left(aY_l^{m-1/2}\chi_+ - bY_l^{m+1/2}\chi_-\right)$$

$$= \left[a\left(\hat{L}_3Y_l^{m-1/2}\right)\chi_+ - b\left(\hat{L}_3Y_l^{m+1/2}\right)\chi_-\right]$$

$$= \left[a\left(m-\frac{1}{2}\right)Y_l^{m-1/2}\chi_+ - b\left(m+\frac{1}{2}\right)Y_l^{m+1/2}\chi_-\right] \tag{202}$$

$$\hat{L}_+\hat{S}_-\psi_{l+\frac{1}{2},m} = \hat{L}_+\hat{S}_-\left(aY_l^{m-1/2}\chi_+ + bY_l^{m+1/2}\chi_-\right)$$

$$= \hat{L}_+\left(aY_l^{m-1/2}\hat{S}_-\chi_+ + bY_l^{m+1/2}\hat{S}_-\chi_-\right)$$

$$= \hat{L}_+\left(aY_l^{m-1/2}\chi_-\right) = a\left(\hat{L}_+Y_l^{m-1/2}\right)\chi_-$$

$$= a\sqrt{\left(l+m+\frac{1}{2}\right)\left(l-m+\frac{1}{2}\right)}\;Y_l^{m+1/2}\chi_- \tag{203}$$

$$\hat{L}_-\hat{S}_+\psi_{l+\frac{1}{2},m} = \hat{L}_-\hat{S}_+\left(aY_l^{m-1/2}\chi_+ + bY_l^{m+1/2}\chi_-\right)$$

$$= \hat{L}_-\left(aY_l^{m-1/2}\hat{S}_+\chi_+ + bY_l^{m+1/2}\hat{S}_+\chi_-\right)$$

$$= \hat{L}_-\left(bY_l^{m+1/2}\chi_+\right) = b\left(\hat{L}_-Y_l^{m+1/2}\right)\chi_+$$

$$= b\sqrt{\left(l+m+\frac{1}{2}\right)\left(l-m+\frac{1}{2}\right)}\,Y_l^{m-1/2}\chi_+ \tag{204}$$

since $\qquad \hat{L}_+ Y_l^m = \sqrt{(l+m+1)(l-m)}\,Y_l^{m+1},$

$$\hat{L}_- Y_l^m = \sqrt{(l-m+1)(l+m)}\,Y_l^{m-1}$$

Adding eqs. (200)-(204), we get

$$\hat{\mathbf{J}}^2\,\psi_{l+\frac{1}{2},m} = \left\{a\left[l(l+1)+\frac{3}{4}+\left(m-\frac{1}{2}\right)\right]+b\sqrt{\left(l+m+\frac{1}{2}\right)\left(l-m+\frac{1}{2}\right)}\right\}Y_l^{m-1/2}\chi_+$$

$$+\left\{a\sqrt{\left(l+m+\frac{1}{2}\right)\left(l-m+\frac{1}{2}\right)}+b\left[l(l+1)+\frac{3}{4}-\left(m+\frac{1}{2}\right)\right]\right\}Y_l^{m+1/2}\chi_-$$

$$=\left\{a\left[\left(l+\frac{1}{2}\right)^2+m\right]+b\sqrt{\left(l+m+\frac{1}{2}\right)\left(l-m+\frac{1}{2}\right)}\right\}Y_l^{m-1/2}\chi_+$$

$$+\left\{a\sqrt{\left(l+m+\frac{1}{2}\right)\left(l-m+\frac{1}{2}\right)}+b\left[\left(l+\frac{1}{2}\right)^2-m\right]\right\}Y_l^{m+1/2}\chi_- \tag{205}$$

Substituting $a=\sqrt{\dfrac{l+m+\frac{1}{2}}{2l+1}}$, $b=\sqrt{\dfrac{l-m+\frac{1}{2}}{2l+1}}$ in eq. (205), we get

$$\hat{\mathbf{J}}^2\psi_{l+\frac{1}{2},m} = \left\{\sqrt{\frac{l+m+\frac{1}{2}}{2l+1}}\left[\left(l+\frac{1}{2}\right)^2+m\right]+\sqrt{\frac{l-m+\frac{1}{2}}{2l+1}}\sqrt{\left(l+m+\frac{1}{2}\right)\left(l-m+\frac{1}{2}\right)}\right\}Y_l^{m-1/2}\chi_+$$

$$+\left\{\sqrt{\frac{l+m+\frac{1}{2}}{2l+1}}\sqrt{\left(l+m+\frac{1}{2}\right)\left(l-m+\frac{1}{2}\right)}\right.$$

$$+\sqrt{\frac{l-m+\frac{1}{2}}{2l+1}}\left[\left(l+\frac{1}{2}\right)^2-m\right]\Bigg\}Y_l^{m+1/2}\chi_-$$

$$=\sqrt{\frac{l+m+\frac{1}{2}}{2l+1}}\left(l+\frac{1}{2}\right)\left(l+\frac{3}{2}\right)Y_l^{m-1/2}\chi_+ +\sqrt{\frac{l-m+\frac{1}{2}}{2l+1}}\left(l+\frac{1}{2}\right)\left(l+\frac{3}{2}\right)Y_l^{m+1/2}\chi_-$$

$$=\left(l+\frac{1}{2}\right)\left(l+\frac{3}{2}\right)\left(\sqrt{\frac{l+m+\frac{1}{2}}{2l+1}}Y_l^{m-1/2}\chi_+ +\sqrt{\frac{l-m+\frac{1}{2}}{2l+1}}Y_l^{m+1/2}\chi_-\right)$$

$$=\left(l+\frac{1}{2}\right)\left(l+\frac{3}{2}\right)\psi_{l+\frac{1}{2},m}=j(j+1)\psi_{l+\frac{1}{2},m} \tag{206}$$

Similarly, substituting $a=-\sqrt{\frac{l-m+\frac{1}{2}}{2l+1}}$, $b=\sqrt{\frac{l+m+\frac{1}{2}}{2l+1}}$ in eq. (205), we get

$$\hat{\mathbf{J}}^2\psi_{l-\frac{1}{2},m}=\left\{-\sqrt{\frac{l-m+\frac{1}{2}}{2l+1}}\left[\left(l+\frac{1}{2}\right)^2+m\right]+\sqrt{\frac{l+m+\frac{1}{2}}{2l+1}}\sqrt{\left(l+m+\frac{1}{2}\right)\left(l-m+\frac{1}{2}\right)}\right\}Y_l^{m-1/2}\chi_+$$

$$+\left\{-\sqrt{\frac{l-m+\frac{1}{2}}{2l+1}}\sqrt{\left(l+m+\frac{1}{2}\right)\left(l-m+\frac{1}{2}\right)}\right.$$

$$\left.+\sqrt{\frac{l+m+\frac{1}{2}}{2l+1}}\left[\left(l+\frac{1}{2}\right)^2-m\right]\right\}Y_l^{m+1/2}\chi_-$$

$$=-\sqrt{\frac{l-m+\frac{1}{2}}{2l+1}}\left(l-\frac{1}{2}\right)\left(l+\frac{1}{2}\right)Y_l^{m-1/2}\chi_+$$

$$+\sqrt{\frac{l+m+\frac{1}{2}}{2l+1}}\left(l-\frac{1}{2}\right)\left(l+\frac{1}{2}\right)Y_l^{m+1/2}\chi_-$$

$$=\left(l-\frac{1}{2}\right)\left(l+\frac{1}{2}\right)\left(-\sqrt{\frac{l-m+\frac{1}{2}}{2l+1}}Y_l^{m-1/2}\chi_+ + \sqrt{\frac{l+m+\frac{1}{2}}{2l+1}}Y_l^{m+1/2}\chi_-\right)$$

$$=\left(l-\frac{1}{2}\right)\left(l+\frac{1}{2}\right)\psi_{l-\frac{1}{2},m}=j(j+1)\psi_{l-\frac{1}{2},m} \qquad (207)$$

These eigenfunctions of $\hat{\mathbf{J}}^2$ are also eigenfunctions of $\hat{\mathbf{L}}.\hat{S}=\frac{1}{2}\left(\hat{\mathbf{J}}^2-\hat{\mathbf{L}}^2-\hat{S}^2\right)$ for

$$\hat{\mathrm{L}}.\hat{S}\psi_{l+\frac{1}{2},m}=\frac{1}{2}\left(\hat{\mathrm{J}}^2-\hat{\mathrm{L}}^2-\hat{S}^2\right)\psi_{l+\frac{1}{2},m}$$

$$=\frac{1}{2}\left[j(j+1)-l(l+1)-\frac{3}{4}\right]\psi_{l+\frac{1}{2},m}$$

$$=\frac{1}{2}\left[\left(l+\frac{1}{2}\right)\left(l+\frac{3}{2}\right)-(l+1)-\frac{3}{4}\right]\psi_{l+\frac{1}{2},m}=\frac{1}{2}l\psi_{l+\frac{1}{2},m} \qquad (208)$$

and

$$\hat{\mathbf{L}}.\hat{S}\psi_{l-\frac{1}{2},m}=\frac{1}{2}\left(\hat{\mathbf{J}}^2-\hat{\mathbf{L}}^2-\hat{S}^2\right)\psi_{l-\frac{1}{2},m}$$

$$=\frac{1}{2}\left[j(j+1)-l(l+1)-\frac{3}{4}\right]\psi_{l-\frac{1}{2},m}$$

$$\frac{1}{2}\left[\left(l-\frac{1}{2}\right)\left(l+\frac{1}{2}\right)-l(l+1)-\frac{3}{4}\right]\psi_{l-\frac{1}{2},m}=-\frac{1}{2}(l+1)\psi_{l-\frac{1}{2},m} \qquad (209)$$

11.9 SOLUTION OF THE DIRAC EQUATION

The Dirac equation for the hydrogen atom can be written as

$$\gamma^{\mu}[i\partial_{\mu}-eA_{\mu}(x)]\psi(x)=m\psi(x) \qquad (210)$$

For the given problem of the hydrogen atom, we have $\mathbf{A} = 0$, $A_0 = V(\mathbf{r}) = V(r) = -\dfrac{e}{r}$ so that eq. (210) becomes $[i\gamma^\mu \partial_\mu - \gamma^0 eV(r)]\,\psi(x) = m\psi(x)$ or

$$\left[\begin{pmatrix} I & 0 \\ 0 & -I \end{pmatrix}(i\partial_0 - eV(r)) + \sum_{k=1}^{3}\begin{pmatrix} \mathbf{0} & \sigma k \\ -\sigma k & \mathbf{0} \end{pmatrix}(i\partial_k) - m\begin{pmatrix} I & 0 \\ 0 & I \end{pmatrix}\right]\begin{pmatrix} \phi \\ \chi \end{pmatrix} = 0 \tag{211}$$

or

$$\left[\begin{pmatrix} i\partial_0 - eV(r) - m & 0 \\ 0 & -(i\partial_0 - eV(r) + m) \end{pmatrix}\right]\begin{pmatrix} \phi \\ \chi \end{pmatrix} = -\sum_{k=1}^{3}\begin{pmatrix} 0 & \sigma_k \hat{p}_k \\ -\sigma_k \hat{p}_k & 0 \end{pmatrix}\begin{pmatrix} \phi \\ \chi \end{pmatrix}$$

or

$$-\sum_{k=1}^{3}\begin{pmatrix} 0 & \sigma_k p_k \\ -\sigma_k p_k & 0 \end{pmatrix}\begin{pmatrix} \phi \\ \chi \end{pmatrix} = \left[\begin{pmatrix} (E - eV(r) - m) & 0 \\ 0 & -(E - eV(r) + m) \end{pmatrix}\right]\begin{pmatrix} \phi \\ \chi \end{pmatrix}$$

or

$$\boldsymbol{\sigma}.\mathbf{p}\begin{pmatrix} \chi \\ \phi \end{pmatrix} = \begin{pmatrix} (E - eV(r) - m)\phi \\ (E - eV(r) + m)\chi \end{pmatrix} \tag{212}$$

We now obtain the expression $\boldsymbol{\sigma}.\mathbf{p}$ in terms of the angular and radial components. We have

$$(\boldsymbol{\sigma}.\mathbf{r}).(\boldsymbol{\sigma}.\mathbf{r}) = \left(\sum_{i=1}^{3}\sigma_i r_i\right).\left(\sum_{j=1}^{3}\sigma_j r_j\right) = \sum_{i,j=1}^{3}\sigma_i\sigma_j r_i r_j$$

$$= \frac{1}{2}\sum_{i,j=1}^{3}(\boldsymbol{\sigma}_i\boldsymbol{\sigma}_j + \boldsymbol{\sigma}_j\boldsymbol{\sigma}_i)\, r_i r_j$$

$$= \sum_{i,j=1}^{3}\delta_{ij}\, r_i r_j = \sum_{i=1}^{3} r_i^2 = \mathbf{r}^2 = r^2 \tag{213}$$

$$(\boldsymbol{\sigma}.\mathbf{r}).(\boldsymbol{\sigma}.\mathbf{p}) = \left(\sum_{i=1}^{3}\sigma_i r_i\right).\left(\sum_{j=1}^{3}\sigma_j p_j\right) = \sum_{i,j=1}^{3}\sigma_i\sigma_j r_i p_j$$

$$= \sum_{i,j}^{3}(\delta_{ij} + i\varepsilon_{i,j,k\in\{1,2,3\}\cap\{i,j\}}\, \boldsymbol{\sigma}_{k\in\{1,2,3\}\cap\{i,j\}}) r_i p_j$$

$$= \sum_{i=1}^{3} r_i p_i + i\sum_{i,j}^{3}\varepsilon_{i,j,(k\in\{1,2,3\}\cap\{i,j\})}\, \boldsymbol{\sigma}_{k\in\{1,2,3\}\cap\{i,j\}} r_i p_j$$

$$= \mathbf{r.p} + i\boldsymbol{\sigma}(\mathbf{r}\times\mathbf{p}) = \mathbf{r.p} + i\boldsymbol{\sigma}.\mathbf{L} \tag{214}$$

From eqs. (213) and (214), we have

$$\boldsymbol{\sigma}.\mathbf{p} = \frac{1}{r^2}\ (\boldsymbol{\sigma}.\mathbf{r}).(\boldsymbol{\sigma}.\mathbf{r})\ (\boldsymbol{\sigma}.\mathbf{p})$$

$$= \frac{1}{r^2}\ (\boldsymbol{\sigma}.\mathbf{r})\ [\mathbf{r}.\mathbf{p} + i\boldsymbol{\sigma}\ (\mathbf{r} \times \mathbf{p})]$$

$$= \frac{1}{r^2}\ (\boldsymbol{\sigma}.\mathbf{r})\ (\mathbf{r}.\mathbf{p} + i\boldsymbol{\sigma}\mathbf{L})$$

or

$$\boldsymbol{\sigma}.\hat{\mathbf{p}} = \frac{1}{r^2}\ (\boldsymbol{\sigma}.\mathbf{r}).(\boldsymbol{\sigma}.\mathbf{r})(\boldsymbol{\sigma}.\hat{\mathbf{p}}) = \frac{1}{r^2}\ (\boldsymbol{\sigma}.\mathbf{r}) \left[-ir\frac{\partial}{\partial r} + i\boldsymbol{\sigma}\hat{\mathbf{L}}\right] \quad (215)$$

We also have

$$\left[\mathbf{r}\frac{\partial}{\partial r}, \frac{(\boldsymbol{\sigma}.\mathbf{r})}{\mathbf{r}}\right]\psi = \mathbf{r}\ \frac{\partial}{\partial r}\left[\frac{(\boldsymbol{\sigma}.\mathbf{r})}{\mathbf{r}}\psi\right] - \frac{(\boldsymbol{\sigma}.\mathbf{r})}{\mathbf{r}}\mathbf{r}\frac{\partial}{\partial r}\psi$$

$$= \mathbf{r}\ \frac{\partial}{\partial r}\left[\frac{(\boldsymbol{\sigma}.\mathbf{r})}{\mathbf{r}}\right].\psi + \mathbf{r}\ \frac{(\boldsymbol{\sigma}.\mathbf{r})}{\mathbf{r}}\ \frac{\partial}{\partial r}\psi - \frac{(\boldsymbol{\sigma}.\mathbf{r})}{\mathbf{r}}\ \mathbf{r}\ \frac{\partial}{\partial r}\psi$$

$$= \mathbf{r}\ \frac{\partial}{\partial r}\left[\frac{(\boldsymbol{\sigma}.\mathbf{r})}{\mathbf{r}}\right].\psi = \mathbf{r}\ \frac{\mathbf{r}\boldsymbol{\sigma} - \boldsymbol{\sigma}\mathbf{r}}{\mathbf{r}^2}\ \psi = 0 \quad (216)$$

$$[\mathbf{r}\hat{\mathbf{p}},\ \boldsymbol{\sigma}\hat{\mathbf{L}}]\psi = [\mathbf{r}\hat{\mathbf{p}}.\boldsymbol{\sigma}\ (\mathbf{r} \times \hat{\mathbf{p}}) - \boldsymbol{\sigma}\ (\mathbf{r} \times \hat{\mathbf{p}}).\mathbf{r}\hat{\mathbf{p}}]\psi$$

$$= [\boldsymbol{\sigma}\mathbf{r}\hat{\mathbf{p}}.\ (\mathbf{r} \times \hat{\mathbf{p}}) - \boldsymbol{\sigma}\ (\mathbf{r} \times \hat{\mathbf{p}}).\mathbf{r}\hat{\mathbf{p}}]\psi = 0 \quad (217)$$

Equations (216) and (217) show that $\frac{(\boldsymbol{\sigma}.\mathbf{r})}{\mathbf{r}}$ and $i\boldsymbol{\sigma}\hat{\mathbf{L}}$ both commute with $-\ i\mathbf{r}\frac{\partial}{\partial r}$. Hence, they do not act on the radial part of the state but act only on the angular momentum part.

We observe from eqs. (83), (84), (86) and (87) that ϕ, χ are both eigenfunctions of $\hat{\mathbf{J}}^2, \hat{J}_3, \hat{\mathbf{L}}^2$ with respective eigenvalues $j(j + 1)$, m_j and $l_\pm(l_\pm + 1)$ for ϕ and $j(j + 1)$, m_j and $l_\mp(l_\mp + 1)$ for χ, e.g.,

$$\hat{\mathbf{J}}^2\begin{pmatrix}\phi\\ \chi\end{pmatrix} = \hat{\mathbf{J}}^2\psi = j(j + 1)\psi = j(j + 1)\begin{pmatrix}\phi\\ \chi\end{pmatrix} = \begin{pmatrix}j(j+1)\phi\\ j(j+1)\chi\end{pmatrix} \quad (218)$$

$$\hat{J}_3\begin{pmatrix}\phi\\ \chi\end{pmatrix} = \hat{J}_3\psi = m_j\psi = m_j\begin{pmatrix}\phi\\ \chi\end{pmatrix} = \begin{pmatrix}m_j\phi\\ m_j\chi\end{pmatrix} \quad (219)$$

$$\hat{\mathbf{L}}^2\begin{pmatrix}\phi\\ \chi\end{pmatrix} = \begin{pmatrix} j(j+1)\phi - \frac{3}{4}\phi \pm \left(j + \frac{1}{2} \pm 1\right)\phi \\ j(j+1)\chi - \frac{3}{4}\chi \mp \left(j + \frac{1}{2} \mp 1\right)\chi \end{pmatrix}$$

$$\text{(since } \hat{\mathbf{L}}^2 = \hat{\mathbf{J}}^2 - \frac{3}{4} - \hat{\sigma}\hat{\mathbf{L}}, (\sigma.\hat{\mathbf{L}})\begin{pmatrix}\phi\\ \chi\end{pmatrix} = \begin{pmatrix}\mp\left(j+\frac{1}{2}\pm 1\right)\phi\\ \pm\left(j+\frac{1}{2}\mp 1\right)\chi\end{pmatrix} \text{ (eq. (82))}$$

$$= \begin{pmatrix}\left(j\pm\frac{1}{2}\right)\left(j\pm\frac{1}{2}+1\right)\phi\\ \left(j\mp\frac{1}{2}\right)\left(j\mp\frac{1}{2}+1\right)\chi\end{pmatrix} = \begin{pmatrix}l_\pm(l_\pm+1)\phi\\ l_\mp(l_\mp+1)\chi\end{pmatrix} \tag{220}$$

We can, therefore, conveniently label simultaneous eigenstates of these operators in terms of the above quantum numbers, viz. j, m_j, l and write them as $\left|j, m_j, l_k\right\rangle$, where $l_k = l_\pm = l_\phi$ for ϕ and $l_k = l_\mp = l_\chi$ for χ. We can also construct the following table of eigenvalues of the simultaneous eigenfunctions:

Operator → Eigenstate ↓	$\hat{\mathbf{J}}^2$	$\hat{J}_3$	$\hat{K}$	$\hat{\mathbf{L}}^2$	$\sigma.\hat{\mathbf{L}}$
$\psi = (\phi, \chi)^T$	$j(j+1)$	m_j	$\mp\left(j+\frac{1}{2}\right)$	Not an eigenstate	Not an eigenstate
ϕ	$j(j+1)$	m_j	$\mp\left(j+\frac{1}{2}\right)$	$l_\phi(l_\phi+1)$, $l_\phi = l_\pm = j \pm \frac{1}{2}$	$\mp\left(j+\frac{1}{2}\pm 1\right)$ $= \mp\left(l_\pm + \frac{1}{2} \pm \frac{1}{2}\right)$
χ	$j(j+1)$	m_j	$\mp\left(j+\frac{1}{2}\right)$	$l_\chi(L_\chi+1)$ $l_\chi = l_\mp = j \mp \frac{1}{2}$	$\pm\left(j+\frac{1}{2}\mp 1\right)$ $= \pm\left(l_\pm + \frac{1}{2} \mp \frac{1}{2}\right)$
Identification	j	m_j	j	$l = l_\phi, l_\chi$	l.

In conformity with eqs. (197) and (198), we can write

$$\left|j, m_j, l_k\right\rangle = \psi_{l_k,\, m_j} = a_k Y_{l_k}^{m_j-1/2}\chi_+ + b_k Y_{l_k}^{m_j+1/2}\chi_- \tag{221}$$

where $\qquad k = \phi, \chi.$

Now, from eqs. (212) and (215), we get

$$\frac{1}{r^2}(\boldsymbol{\sigma}.\mathbf{r})\left[-i r\frac{\partial}{\partial r}+i\boldsymbol{\sigma}\hat{\mathbf{L}}\right]\begin{pmatrix}\chi\\ \phi\end{pmatrix}=\begin{pmatrix}(E-eV(r)-m)\phi\\ (E-eV(r)+m)\chi\end{pmatrix} \tag{222}$$

For a fixed value of the eigenvalue κ of the operator $\hat{K}$, the orbital parities of the two-component states ϕ, χ must be opposite as shown below. The impact of the parity operator on a wave function is given by

$$\begin{aligned}\psi'(\mathbf{x}', x'^0) &= \hat{P}\psi(\mathbf{x}, x^0) = e^{i\varphi}\gamma^0\psi(-\mathbf{x}, x^0)\\ &= \psi'(-\mathbf{x}, x^0) = \hat{P}\psi(-\mathbf{x}', x'^0)\end{aligned} \tag{223}$$

where $\hat{P}\equiv e^{i\varphi}\gamma^0\hat{P}_0$ is the usual representation of the parity operator, $\hat{P}_0$ changes $\mathbf{x}\to-\mathbf{x}$ in the argument of the wave function. Since the wave function $\psi\equiv(\phi\ \chi)^T$ must have good parity, i.e. it must be an eigenfunction of the parity operator, we have

$$\psi'(\mathbf{x}', x^0) = \hat{P}\psi(\mathbf{x}, x^0) = \lambda\psi(\mathbf{x}, x^0) \tag{224}$$

Now

$$\begin{aligned}\hat{P}\psi(\mathbf{x}, x^0) &= e^{i\varphi}\begin{pmatrix}\mathbf{I} & 0\\ 0 & -\mathbf{I}\end{pmatrix}\hat{P}_0\begin{pmatrix}\phi(\mathbf{x},x^0)\\ \chi(\mathbf{x},x^0)\end{pmatrix} = e^{i\varphi}\begin{pmatrix}\hat{P}_0\phi(\mathbf{x},x^0)\\ -\hat{P}_0\chi(\mathbf{x},x^0)\end{pmatrix}\\ &= e^{i\varphi}\begin{pmatrix}\phi(-\mathbf{x},x^0)\\ -\chi(-\mathbf{x},x^0)\end{pmatrix} = e^{i\varphi}\begin{pmatrix}\phi(\mathbf{x}',x'^0)\\ -\chi(\mathbf{x}',x'^0)\end{pmatrix}\\ &= \lambda\psi(\mathbf{x}, x^0) = \lambda\begin{pmatrix}\phi(\mathbf{x},x^0)\\ \chi(\mathbf{x},x^0)\end{pmatrix}\end{aligned} \tag{225}$$

where $|\lambda| = 1$. Equation (225) shows that the parity of the two spinor ϕ must be opposite to that of the two spinor χ.

This can also be seen from eq. (212) as follows. From eq. (212), we have $\boldsymbol{\sigma}.\hat{\mathbf{p}}\chi = (E - eV(r) - m)\phi$, $\boldsymbol{\sigma}.\hat{\mathbf{p}}\phi = (E - eV(r) + m)\chi$. Since, the operator $\boldsymbol{\sigma}\,\hat{\mathbf{p}}$ causes a reversal of parity ($\hat{P}\hat{\mathbf{p}}\hat{P}^{-1} = -\hat{\mathbf{p}}$ as $\hat{P}\hat{\mathbf{x}}\hat{P}^{-1} = -\hat{\mathbf{x}}$ so that $\frac{\partial}{\partial x'^i}=\frac{\partial x^i}{\partial x'^i}\frac{\partial}{\partial x^i}=-\frac{\partial}{\partial x^i}$), the above equations imply that parity of ϕ must be opposite to that of χ. From eq. (225), we also get $\phi(-\mathbf{x}, x^0) = \pm\,\phi(\mathbf{x}, x^0)$ and $\chi(-\mathbf{x}, x^0) = \mp\,\chi(\mathbf{x}, x^0)$. Equation (212) together with eq. (215) lead to

$$\frac{1}{r^2}(\boldsymbol{\sigma}.\mathbf{r})\left[-i r\frac{\partial}{\partial r}+i\boldsymbol{\sigma}\hat{\mathbf{L}}\right]\begin{pmatrix}\chi\\ \phi\end{pmatrix}=\begin{pmatrix}(E-eV(r)-m)\phi\\ (E-eV(r)+m)\chi\end{pmatrix} \tag{226}$$

whose solutions, in terms of the eigenstates $\psi_{l+\frac{1}{2},m}$, $\psi_{l-\frac{1}{2},m}$ given by eqs. (197) and (198) are

$$\psi_1 = \begin{pmatrix} \phi_1 \\ \chi_1 \end{pmatrix} = \begin{pmatrix} F(\mathbf{r})\psi_{l+1/2,m} \\ if(\mathbf{r})\psi_{l-1/2,m} \end{pmatrix}$$

$$\psi_2 = \begin{pmatrix} \phi_2 \\ \chi_2 \end{pmatrix} = \begin{pmatrix} G(\mathbf{r})\psi_{l-1/2,m} \\ ig(\mathbf{r})\psi_{l+1/2,m} \end{pmatrix} \tag{227}$$

The radial components of the solution $F(\mathbf{r}), f(\mathbf{r}), G(\mathbf{r}), g(\mathbf{r})$ are functions of κ. The factor i has been introduced to ensure that the radial solutions are real for bound states.

Now, from eq. (74), we have

$$\hat{K}\psi = -\kappa\psi = \begin{pmatrix} \boldsymbol{\sigma}.\hat{\mathbf{L}} + \mathbf{I} & 0 \\ 0 & -\boldsymbol{\sigma}.\hat{\mathbf{L}} - \mathbf{I} \end{pmatrix}\begin{pmatrix} \phi \\ \chi \end{pmatrix}$$

$$= \begin{pmatrix} (\boldsymbol{\sigma}.\hat{\mathbf{L}} + \mathbf{I})\phi \\ -(\boldsymbol{\sigma}.\hat{\mathbf{L}} + \mathbf{I})\chi \end{pmatrix} \tag{228}$$

whence $$\boldsymbol{\sigma}.\hat{\mathbf{L}}\begin{pmatrix} \phi \\ \chi \end{pmatrix} = \begin{pmatrix} -(\kappa+1)\phi \\ (\kappa-1)\chi \end{pmatrix}.$$

Also, from eq. (82),

$$(\boldsymbol{\sigma}.\hat{\mathbf{L}})\begin{pmatrix} \phi \\ \chi \end{pmatrix} = \begin{pmatrix} \mp\left(j+\frac{1}{2}\pm 1\right)\phi \\ \pm\left(j+\frac{1}{2}\mp 1\right)\chi \end{pmatrix} \quad \text{and } l_\pm = j \pm \frac{1}{2},\ l_\mp = j \mp \frac{1}{2}$$

and from eqs. (208) and (209)

$$\hat{\mathbf{L}}.\hat{\mathbf{S}}\psi_{l+\frac{1}{2},m} = \frac{1}{2}\hat{\mathbf{L}}.\boldsymbol{\sigma}\psi_{l+\frac{1}{2},m} = \frac{1}{2}l\psi_{l+\frac{1}{2},m}$$

and $$\hat{\mathbf{L}}.\hat{\mathbf{S}}\psi_{l-\frac{1}{2},m} = \hat{\mathbf{L}}.\boldsymbol{\sigma}\psi_{l-\frac{1}{2},m} = -\frac{1}{2}(l+1)\psi_{l-\frac{1}{2},m}$$

so that we have the following combinations:

(a) for positive κ, we have $l = j + \frac{1}{2}$ and for negative κ, $l = j - \frac{1}{2}$ for ϕ.

(b) for positive κ, we have $l = j - \frac{1}{2}$ and for negative κ, $l = j + \frac{1}{2}$ for χ.

Furthermore, $\frac{1}{r}(\boldsymbol{\sigma}.\mathbf{r})$ changes parity of its operand since $\hat{P}\mathbf{r}\hat{P}^{-1} = -\mathbf{r}$, i.e., $\frac{1}{r}(\boldsymbol{\sigma}.\mathbf{r})$ is a pseudoscalar operator. The parity of the state is given by $(-1)^l$ so that $\frac{1}{r}(\boldsymbol{\sigma}.\mathbf{r})$ changes the l of a state. Hence,

$$\frac{1}{r}(\boldsymbol{\sigma}.\mathbf{r})\psi_{l+\frac{1}{2},m} = C\psi_{l-\frac{1}{2},m} \tag{229}$$

and, since
$$\left[\frac{1}{r}(\boldsymbol{\sigma}.\mathbf{r})\right]^2 = \frac{1}{r^2}(\boldsymbol{\sigma}.\mathbf{r}).(\boldsymbol{\sigma}.\mathbf{r}) = \frac{1}{r^2}(\mathbf{r}.\mathbf{r} + i\boldsymbol{\sigma}\mathbf{r}\times\mathbf{r}) = \hat{I},$$

$$\left[\frac{1}{r}(\boldsymbol{\sigma}.\mathbf{r})\right]^2 \psi_{l+\frac{1}{2},m} = \psi_{l+\frac{1}{2},m}$$

$$= C\left[\frac{1}{r}(\boldsymbol{\sigma}.\mathbf{r})\right]\psi_{l-\frac{1}{2},m} = C^2\,\psi_{l+\frac{1}{2},m} \tag{230}$$

whence $C = e^{i\delta}$. Conventionally, we take $C = -1$ giving

$$\frac{1}{r}(\boldsymbol{\sigma}.\mathbf{r})\,\psi_{l+\frac{1}{2},m} = -\psi_{l-\frac{1}{2},m} \tag{231}$$

Substituting the solution $\psi_1 = \begin{pmatrix}\phi_1\\ \chi_1\end{pmatrix} = \begin{pmatrix}F(\mathbf{r})\psi_{l+1/2,m}\\ if(\mathbf{r})\psi_{l-1/2,m}\end{pmatrix}$ in the radial eq. (226), we get

$$\frac{1}{r^2}(\boldsymbol{\sigma}.\mathbf{r})\left[-ir\frac{\partial}{\partial r} + i\boldsymbol{\sigma}\hat{\mathbf{L}}\right]\begin{pmatrix}if(\mathbf{r})\psi_{l-1/2,m}\\ F(\mathbf{r})\psi_{l+1/2,m}\end{pmatrix}$$

$$= \begin{pmatrix}(E-eV(r)-m)F(\mathbf{r})\psi_{l+1/2,m}\\ i(E-eV(r)+m)f(\mathbf{r})\psi_{l-1/2,m}\end{pmatrix} \tag{232}$$

which, using eq. (228), gives

$$\frac{1}{r^2}(\boldsymbol{\sigma}.\mathbf{r})\begin{bmatrix}\left(\mathbf{r}\frac{\partial}{\partial r} - (\kappa-1)\right)f(\mathbf{r})\psi_{l-1/2,m}\\ \left(-i\mathbf{r}\frac{\partial}{\partial r} - i(\kappa+1)\right)F(\mathbf{r})\psi_{l+1/2,m}\end{bmatrix}$$

$$= \begin{pmatrix}(E-eV(r)-m)F(\mathbf{r})\psi_{l+1/2,m}\\ i(E-eV(r)+m)f(\mathbf{r})\psi_{l-1/2,m}\end{pmatrix} \tag{233}$$

Using eq. (231), this becomes

$$\frac{1}{r}\begin{bmatrix}\left(-\mathbf{r}\frac{\partial}{\partial r}+(\kappa-1)\right)f(\mathbf{r})\psi_{l+1/2,m}\\ \left(i\mathbf{r}\frac{\partial}{\partial r}+i(\kappa+1)\right)F(\mathbf{r})\psi_{l-1/2,m}\end{bmatrix}=\begin{pmatrix}(E-eV(r)-m)F(\mathbf{r})\psi_{l+1/2,m}\\ i(E-eV(r)+m)f(\mathbf{r})\psi_{l-1/2,m}\end{pmatrix} \text{ or}$$

$$\frac{1}{r}\begin{bmatrix}\left(-\mathbf{r}\frac{\partial}{\partial r}+(\kappa-1)\right)f(\mathbf{r})\\ \left(i\mathbf{r}\frac{\partial}{\partial r}+i(\kappa+1)\right)F(\mathbf{r})\end{bmatrix}=\begin{pmatrix}(E-eV(\mathbf{r})-m)F(\mathbf{r})\\ i(E-eV(\mathbf{r})+m)f(\mathbf{r})\end{pmatrix} \text{ or}$$

$$\begin{bmatrix}\left(-\frac{\partial}{\partial r}+\frac{(\kappa-1)}{r}\right)f(\mathbf{r})\\ \left(\frac{\partial}{\partial r}+\frac{(\kappa+1)}{r}\right)F(\mathbf{r})\end{bmatrix}=\begin{pmatrix}(E-eV(r)-m)F(\mathbf{r})\\ (E-eV(r)+m)f(\mathbf{r})\end{pmatrix} \tag{234}$$

To further simplify this pair of coupled differential equations, we write

$$P=rf,\ Q=rF,\ eV=-\frac{Z\alpha}{r},$$

$$k_1=m+E,\ k_2=m-E, \tag{235}$$

whence, we get

$$\begin{bmatrix}\frac{\partial P}{\partial r}-\kappa\frac{P}{r}\\ \frac{\partial Q}{\partial r}+\kappa\frac{Q}{r}\end{bmatrix}=\begin{pmatrix}\left(k_2-\frac{Z\alpha}{r}\right)Q\\ \left(k_1+\frac{Z\alpha}{r}\right)P\end{pmatrix} \tag{236}$$

We first examine the solutions of eq. (236) near the origin, i.e. for $r\sim 0$. In this situation, we can ignore k_1, k_2 as compared to $Z\alpha$ so that the above equation becomes

$$\begin{bmatrix}\frac{\partial P}{\partial r}-\kappa\frac{P}{r}+\frac{Z\alpha}{r}Q\\ \frac{\partial Q}{\partial r}+\kappa\frac{Q}{r}-\frac{Z\alpha}{r}P\end{bmatrix}=0 \tag{237}$$

We look at a power series solution for P and Q. Since the first term of the series would dominate in the region of $r\sim 0$, we postulate the ansatz $P(r)=\sum_{m=0}^{\infty}ar^m$ and $Q(r)=\sum_{m=0}^{\infty}br^m$ as possible solutions of eq. (237). Substituting these solutions in eq. (237), we obtain

$$\begin{bmatrix} \sum_{m=0}^{\infty}(am - a\kappa + Z\alpha b)r^{m-1} \\ \sum_{m=0}^{\infty}(bm + b\kappa - Z\alpha a)r^{m-1} \end{bmatrix} = 0 \tag{238}$$

Equating powers of r on both sides, we get

$$\begin{pmatrix} (m-\kappa) & Z\alpha \\ -Z\alpha & (m+\kappa) \end{pmatrix}\begin{pmatrix} a \\ b \end{pmatrix} = 0 \tag{239}$$

For consistency, the determinant of $\begin{pmatrix} (m-\kappa) & Z\alpha \\ -Z\alpha & (m+\kappa) \end{pmatrix}$ must vanish which gives

$$\begin{aligned} m &= \pm\sqrt{\kappa^2 - Z^2\alpha^2} \\ &= \pm\sqrt{\left(j+\frac{1}{2}\right)^2 - Z^2\alpha^2} \end{aligned} \tag{240}$$

(a) The positive sign for m presents no problems since the solution is regular at the origin and the state remains normalizable.

(b) For the negative sign outside the radical, i.e., for $m = -|m|$ we have $P^2(r)+Q^2(r) \sim r^{-2|m|} = \frac{1}{r^{2|m|}}$ for $r \sim 0$ which would result in a divergent integral for the norm if $|m| \leq \frac{1}{2}$. Furthermore, if the charge Z is large enough, then the divergence is weak and the normalization integral $\int \psi^\dagger \psi d^3x = \int (P^2 + Q^2)\, dr$ becomes finite. For instance, for $\kappa^2 = 1$ and $(Z\alpha)^2 > \frac{3}{4}$, regular solutions for negative m do seem possible because $P^2 + Q^2 \sim r^{-2\sqrt{1-(Z\alpha)^2}}$ is still integrable for $r \to 0$ provided $\sqrt{1-(Z\alpha)^2} < \frac{1}{2}$ or equivalently $(Z\alpha)^2 > \frac{3}{4}$. But such solutions need to be discarded on physical considerations as discussed in the sequel. On the other hand, if the nuclear charge is small, then the divergence is too strong resulting in a non-normalizable solutions.

(c) $P^2(r) + Q^2(r) \sim r^{-2|m|}$ may still be integrable for $r \to 0$ if $2|m| = 2\sqrt{\kappa^2 - Z^2\alpha^2} < 1$. For the case of $\kappa^2 = 1$, this implies $(Z\alpha)^2 > \frac{3}{4}$. These solutions do not result in acceptable behaviour of physically meaningful quantities, e.g., the expectation value of the Coulomb energy is given by $\int \psi^\dagger$

$= \left(-\frac{Ze^2}{r}\right)\psi d^3x = \int (P^2 + Q^2)\left(-\frac{Ze^2}{r}\right) dr$. (The volume element in the integration does not contain factor r^2 because we have already substituted $P = rf$, $Q = rF$). In the limit $r \to 0$, this integrand will behave like $r^{-2|m|-1}$ and will give a finite value if $-2|m| - 1 > -1$ which gives $|m| < 0$ and hence, is not possible. Thus, we must have $m \neq -|m|$ or $m = |m|$ or $m > 0$, whence $m = +\sqrt{\kappa^2 - Z^2\alpha^2} = +\sqrt{\left(j+\frac{1}{2}\right)^2 - Z^2\alpha^2}$.

(d) From the expression $m = \sqrt{\left(j+\frac{1}{2}\right)^2 - Z^2\alpha^2}$ we find that for $j + \frac{1}{2} = 1$, real values of m can exist only for $Z \leq \frac{1}{\alpha}$. For larger values of Z, m takes imaginary values and, therefore, the states become non-normalizable.

(e) For $Z\alpha > |k|$, the real part of the wave functions show an oscillatory behavior approximating $\text{Re}(P, Q) \sim \cos(|m| \ln r)$ for small values of r.

(f) Non-acceptability of the values of $m < 0$ can also be justified on physical considerations on the following grounds:

Consider an atomic nucleus as a hollow sphere of radius R charge $(Z\alpha)^2 > \frac{3}{4}$ and $\kappa = -1$. Let us solve for the wave function separately for the inside and outside of the spherical shell.

(*i*) In the outside region of the shell, the Dirac equation will take the form of eq. (236). The wave functions corresponding to a given energy will comprise terms behaving like $r^{|m|}$ and also $r^{-|m|}$ with the term $r^{-|m|}$ dominating in the region close to the origin, i.e., $r \sim 0$. Further, for $r \to 0$, $k_1, k_2 \to 0$ so that the series solutions leading to eq. (238) hold and we have, for $\kappa = -1$ and $m = -|m|$,

$$\left.\frac{a}{b}\right|_{surface(r \to R)} = -\frac{1+|m|}{Z\alpha} = -\frac{1+\sqrt{1-Z^2\alpha^2}}{Z\alpha}.$$

(*ii*) In the region inside the shell, we have a potential of constant magnitude $\frac{Z\alpha}{R}$ so that the Dirac equation for this region for $\kappa = -1$ will become

$$\begin{bmatrix} \frac{\partial P}{\partial r} + \frac{P}{r} \\ \frac{\partial Q}{\partial r} - \frac{Q}{r} \end{bmatrix} = \begin{pmatrix} \left(m - E - \frac{Z\alpha}{R}\right)Q \\ \left(m + E + \frac{Z\alpha}{R}\right)P \end{pmatrix} \text{ or } \begin{bmatrix} \frac{\partial P}{\partial r} + \frac{P}{r} + \left(E - m + \frac{Z\alpha}{R}\right)Q \\ \frac{\partial Q}{\partial r} - \frac{Q}{r} - \left(E + m + \frac{Z\alpha}{R}\right)P \end{bmatrix} = 0$$

which is the free particle Dirac equation for a particle with energy $\varepsilon = E + \frac{Z\alpha}{R}$ and has the solution $P(r)|_{\text{inside}} = k \cos(kr + \delta) - \frac{1}{2} \sin(kr + \delta)$ and $Q(r)|_{\text{inside}} = (\varepsilon + m) \sin(kr + \delta)$ with $k^2 = \varepsilon^2 - m^2$. Thus, in the case of a constant potential, we have only one normalizable solution, which occurs for $\delta = 0$. In this case, we will have $\left.\frac{P(r)}{Q(r)}\right|_{\text{surface}} = \frac{1}{(\varepsilon + m)}\left[k\cot(kR) - \frac{1}{R}\right]$. Now we shrink the shell to a point mass for which we take the limit $R \to 0$. In this case m, $E \ll \frac{Z\alpha}{R}$ so that $k \simeq \varepsilon \simeq \frac{Z\alpha}{R}$ and we have $\left.\frac{P(r)}{Q(r)}\right|_{\text{surface}} \to -\frac{1}{Z\alpha} + \cot(Z\alpha)$.

(*iii*) Therefore, continuity at the surface of the shell mandates that $\left.\frac{a}{b}\right|_{\text{surface}(r \to R)} = \left.\frac{P(r)}{Q(r)}\right|_{\text{surface}}$ or $\frac{1 + \sqrt{1 - Z^2\alpha^2}}{Z\alpha} = \frac{1}{Z\alpha} - \cot(Z\alpha)$. Now, for $\alpha < 1$, the left-hand side of this equation is larger than $\frac{1}{Z\alpha}$ while the right-hand side is smaller than $\frac{1}{Z\alpha}$. Hence, the above equation has no solution for Z if $\alpha < 1$. Therefore, we find that there is no solution in the limit that: (a) $R \to 0$, i.e., when the spherical shell containing the nuclear charge approaches a point charge and (b) when $r^{-|m|}$ dominates the behaviour of the wave function, i.e., when $r \sim 0$. Only in the special case where the component $r^{-|m|}$ vanishes can we get a solution and it is this case that corresponds to the observed energy eigenvalues.

In order to solve eq. (236), we make the substitutions, viz. $\rho = 2\lambda r$ and $\lambda = (m^2 - E^2)^{\frac{1}{2}}$. These give $\frac{\partial \rho}{\partial r} = 2\lambda$ so that $\frac{\partial}{\partial r} = 2\lambda \frac{\partial}{\partial \rho}$ and eq. (236) reduces to

$$\begin{pmatrix} \frac{\partial P(\rho)}{\partial \rho} - \kappa \frac{P(\rho)}{\rho} \\ \frac{\partial Q(\rho)}{\partial \rho} + \kappa \frac{Q(\rho)}{\rho} \end{pmatrix} = \begin{pmatrix} \left(\frac{1}{2\lambda} k_2 - \frac{Z\alpha}{\rho}\right) Q(\rho) \\ \left(\frac{1}{2\lambda} k_1 + \frac{Z\alpha}{\rho}\right) P(\rho) \end{pmatrix} \tag{241}$$

In the asymptotic region of $\rho \to \infty$, we can neglect the terms proportional to ρ^{-1} and we get, in this region

$$\begin{pmatrix} \dfrac{\partial P(\rho)}{\partial \rho} \\ \dfrac{\partial Q(\rho)}{\partial \rho} \end{pmatrix} = \begin{pmatrix} \dfrac{1}{2\lambda} k_2 Q(\rho) \\ \dfrac{1}{2\lambda} k_1 P(\rho) \end{pmatrix} \tag{242}$$

or equivalently

$$\frac{\partial^2 Q(\rho)}{\partial \rho^2} = \frac{1}{4\lambda^2} k_1 k_2 Q(\rho) = \frac{1}{4} Q(\rho) \tag{243}$$

since $k_1 k_2 = (m + E)(m - E) = m^2 - E^2 = \lambda^2$. Equation (243) has the solution $Q(\rho) = \exp\left(\pm\frac{\rho}{2}\right)$. Now the solution $Q(\rho) = \exp\left(-\frac{\rho}{2}\right)$ is clearly normalizable as $\rho \to \infty$ and hence, is physically acceptable whereas $Q(\rho) = \exp\left(+\frac{\rho}{2}\right)$ diverges as $\rho \to \infty$. It is, therefore, not normalizable and hence, physically meaningless. It must, thus, be discarded. Hence, as a general solution for $Q(\rho)$, we consider the ansatz

$$Q(\rho) = \sqrt{k_1} e^{-\rho/2} (\phi_1(\rho) + \phi_2(\rho)) \tag{244}$$

Similarly, we propose the ansatz

$$P(\rho) = \sqrt{k_2} e^{-\rho/2} (\phi_1(\rho) - \phi_2(\rho)) \tag{245}$$

Inserting from eqs. (244) and (245) in eq. (241), we get

$$k_1^{1/2} e^{-\rho/2} \left(-\frac{1}{2} + \frac{d}{d\rho} \right) (\phi_1 + \phi_2) = -\frac{\kappa}{\rho} k_1^{1/2} e^{-\rho/2} (\phi_1 - \phi_2)$$
$$+ \left(\frac{k_1}{2\lambda} + \frac{Z\alpha}{\rho} \right) k_2^{1/2} e^{-\rho/2} (\phi_1 - \phi_2) \tag{246}$$

and

$$k_2^{1/2} e^{-\rho/2} \left(-\frac{1}{2} + \frac{d}{d\rho} \right) (\phi_1 - \phi_2) = \frac{\kappa}{\rho} k_2^{1/2} e^{-\rho/2} (\phi_1 - \phi_2)$$
$$+ \left(\frac{k_2}{2\lambda} - \frac{Z\alpha}{\rho} \right) k_1^{1/2} e^{-\rho/2} (\phi_1 + \phi_2) \tag{247}$$

or equivalently

$$\left(-\frac{1}{2}+\frac{d}{d\rho}\right)(\phi_1+\phi_2) = -\frac{\kappa}{\rho}(\phi_1+\phi_2) + \left(\frac{k_1}{2\lambda}+\frac{Z\alpha}{\rho}\right)\left(\frac{k_2}{k_1}\right)^{1/2}(\phi_1-\phi_2) \tag{248}$$

and

$$\left(-\frac{1}{2}+\frac{d}{d\rho}\right)(\phi_1-\phi_2) = \frac{\kappa}{\rho}(\phi_1-\phi_2) + \left(\frac{\kappa_2}{2\lambda}-\frac{Z\alpha}{\rho}\right)\left(\frac{k_1}{k_2}\right)^{1/2}(\phi_1+\phi_2) \tag{249}$$

Now

$$\left(\frac{k_1}{k_2}\right)^{1/2} = \left(\frac{m+E}{m-E}\right)^{1/2} = \frac{m+E}{\lambda} = \frac{k_1}{\lambda}$$

and

$$\left(\frac{k_2}{k_1}\right)^{1/2} = \frac{m-E}{\lambda} = \frac{k_2}{\lambda} \tag{250}$$

Therefore,

$$\left(-\frac{1}{2}+\frac{d}{d\rho}\right)(\phi_1+\phi_2) = -\frac{\kappa}{\rho}(\phi_1+\phi_2) + \left(\frac{k_1}{2\lambda}+\frac{Z\alpha}{\rho}\right)\frac{k_2}{\lambda}(\phi_1-\phi_2) \tag{251}$$

and

$$\left(-\frac{1}{2}+\frac{d}{d\rho}\right)(\phi_1-\phi_2) = \frac{\kappa}{\rho}(\phi_1-\phi_2) + \left(\frac{k_2}{2\lambda}-\frac{Z\alpha}{\rho}\right)\frac{k_1}{\lambda}(\phi_1+\phi_2) \tag{252}$$

Adding and subtracting eqs. (251) and (252), we get

$$2\left(-\frac{1}{2}+\frac{d}{d\rho}\right)\phi_1 = -2\frac{\kappa}{\rho}\phi_2 + \frac{k_1k_2}{\lambda^2}\phi_1 + \frac{Z\alpha}{\rho}$$

$$\left(\frac{k_2}{\lambda}-\frac{k_1}{\lambda}\right)\phi_1 - \frac{Z\alpha}{\rho}\left(\frac{k_2}{\lambda}+\frac{k_1}{\lambda}\right)\phi_2$$

or

$$\frac{d\phi_1}{d\rho} = \left(1-\frac{Z\alpha E}{\rho\lambda}\right)\phi_1 - \left(\frac{\kappa}{\rho}+\frac{Z\alpha m}{\rho\lambda}\right)\phi_2 \tag{253}$$

and

$$2\left(-\frac{1}{2}+\frac{d}{d\rho}\right)\phi_2 = -2\frac{k}{\rho}\phi_1 - \phi_2 + \frac{Z\alpha}{\rho}\left(\frac{k_1}{\lambda}+\frac{k_2}{\lambda}\right)\phi_1 + \frac{Z\alpha}{\rho}\left(\frac{k_1}{\lambda}-\frac{k_2}{\lambda}\right)\phi_2$$

or

$$\frac{d\phi_2}{d\rho} = \left(-\frac{\kappa}{\rho} + \frac{Z\alpha m}{\rho\lambda}\right)\phi_1 + \frac{Z\alpha E}{\rho\lambda}\,\phi_2 \tag{254}$$

To solve eqs. (253) and (254), we assume a power series solution of the form:

$$\phi_1 = \rho^s \sum_{r=0}^{\infty} \alpha_r\,\rho^r \text{ and } \phi_2 = \rho^s \sum_{r=0}^{\infty} \beta_r\,\rho^r \tag{255}$$

Substituting these expressions in eqs. (253) and (254), we get

$$\sum (r+s)\alpha_r \rho^{r+s-1} = \sum \alpha_r \rho^{r+s} - \frac{Z\alpha E}{\lambda} \times$$

$$\sum \alpha_r \rho^{r+s-1} - \left(\kappa + \frac{Z\alpha m}{\lambda}\right) \sum \beta_r \rho^{r+s-1} \tag{256}$$

and

$$\sum \beta_r (r+s)\rho^{r+s-1} = \left(-\kappa + \frac{Z\alpha m}{\lambda}\right)\sum \alpha_r \rho^{r+s-1} + \frac{Z\alpha E}{\lambda}\sum \beta_r \rho^{r+s-1} \tag{257}$$

On comparing coefficients of similar powers of ρ on both sides of the above equations, we get

$$\alpha_r (r+s) = \alpha_{r-1} - \frac{Z\alpha E}{\lambda}\alpha_r - \left(\kappa + \frac{Z\alpha m}{\lambda}\right)\beta_r \tag{258}$$

and

$$\beta_r (r+s) = \left(-\kappa + \frac{Z\alpha m}{\lambda}\right)\alpha_r + \frac{Z\alpha E}{\lambda}\beta_r \tag{259}$$

From eq. (259), we get

$$\frac{\beta_r}{\alpha_r} = \frac{-\kappa + Z\alpha m\lambda^{-1}}{r+s-Z\alpha E\lambda^{-1}} = \frac{\kappa - Z\alpha m\lambda^{-1}}{n'-r} \tag{260}$$

with $n' = Z\alpha E\lambda^{-1} - s$. We also have

$$\frac{\beta_0}{\alpha_0} = \frac{-\kappa + Z\alpha m\lambda^{-1}}{s - Z\alpha E\lambda^{-1}} = \frac{\kappa - (n'+s)mE^{-1}}{n'} \tag{261}$$

From eqs. (258) and (260), we get

$$\alpha_r\left[r+s+\frac{Z\alpha E}{\lambda} + \left(\kappa + \frac{Z\alpha m}{\lambda}\right)\left(\frac{\kappa - Z\alpha m\lambda^{-1}}{n'-r}\right)\right] = \alpha_{r-1} \tag{262}$$

or

$$\alpha_r\left[\left(r+s+\frac{Z\alpha E}{\lambda}\right)+(n'-r)+\kappa^2-\frac{Z^2\alpha^2m^2}{\lambda^2}\right]=\alpha_{r-1}(n'-r) \tag{263}$$

With $n' = Z\alpha E\lambda^{-1} - s$ and $s^2 = \kappa^2 - (Z\alpha)^2$, $\lambda^2 = m^2 - E^2$, eq. (263) reduces to

$$\alpha_r\left[\left(r+s+\frac{Z\alpha E}{\lambda}\right)+\left(\frac{Z\alpha E}{\lambda}-r-s\right)+\kappa^2-\frac{Z^2\alpha^2m^2}{\lambda^2}\right]$$

$$=\alpha_r\left[\left(\frac{Z\alpha E}{\lambda}\right)^2-(r+s)^2+\kappa^2-\frac{Z^2\alpha^2m^2}{\lambda^2}\right]$$

$$=\alpha_r\left[-r^2-2rs+(Z\alpha)^2+\frac{Z^2\alpha^2}{\lambda^2}(E^2-m^2)\right]$$

$$=\alpha_r\,[-r^2-2rs]=\alpha_{r-1}(n'-r) \tag{264}$$

whence

$$\alpha_r=-\frac{(n'-r)}{r(2s+r)}\,\alpha_{r-1}=\frac{(-1)^r(n'-1)\ldots(n'-r)}{r!(2s+1)\ldots(2s+r)}\,\alpha_0$$

$$=\frac{(1-n')(2-n')\ldots(r-n')}{r!(2s+1)\ldots(2s+r)}\,\alpha_0 \tag{265}$$

From eqs. (260), (261) and (265), we have

$$\beta_r=(-1)^r\,\frac{n'(n'-1)\ldots(n'-r+1)}{r!(2s+1)\ldots(2s+r)}\,\beta_0 \tag{266}$$

If we denote the confluent hypergeometric function $1+\dfrac{a}{c}\dfrac{x}{1!}+\dfrac{a(a+1)}{c(c+1)}\dfrac{x^2}{2!}+\ldots$ by $F(a, c; x)$, then

$$\phi_1=\alpha_0\rho^sF(1-n', 2s+1; \rho) \tag{267}$$

and

$$\phi_2=\beta_0\rho^sF(-n', 2s+1;\rho)=\left(\frac{\kappa-Z\alpha m\lambda^{-1}}{n'}\right)\alpha_0\rho^sF(-n', 2s+1; \rho) \tag{268}$$

Now, in order that our solution of the Dirac equation remains normalizable, the power series solution for ϕ_1 and ϕ_2 must terminate. Therefore, the hypergeometric functions must have a simple polynomial structure, i.e., n' must be a non-negative integer.

Further, from the expressions $n' = Z\alpha E\lambda^{-1} - s$ and $\lambda = (m^2 - E^2)^{\frac{1}{2}}$, we have

$$Z\alpha E(m^2 - E^2)^{-\frac{1}{2}} = n' + s = n - j - \frac{1}{2} + s \tag{269}$$

where we have defined $n = n' + |\kappa| = n' + j + \dfrac{1}{2}$ as the principal quantum number. Hence, in terms of the principal quantum number n, we have

$$\left[(Z\alpha)^2 + \left(n - j - \frac{1}{2} + s\right)^2\right]E^2 = m^2\left(n - j - \frac{1}{2} + s\right)^2 \tag{270}$$

or

$$E = \pm\frac{m\left(n - j - \frac{1}{2} + s\right)}{\left[(Z\alpha)^2 + \left(n - j - \frac{1}{2} + s\right)^2\right]^{\frac{1}{2}}} \tag{271}$$

Now $\qquad s^2 = \kappa^2 - (Z\alpha)^2 = \left(j + \dfrac{1}{2}\right)^2 - (Z\alpha)^2$ so that

$$E = \pm\frac{m}{\left\{1 + \dfrac{(Z\alpha)^2}{\left[n - j - \frac{1}{2} + \left[\left(j + \frac{1}{2}\right)^2 - (Z\alpha)^2\right]^{\frac{1}{2}}\right]^2}\right\}^{\frac{1}{2}}} \tag{272}$$

where $n = 1, 2, 3, \ldots$ and $|\kappa| = j + \dfrac{1}{2}$. It may be noted that in eq. (272), the negative sign is not acceptable because negative energies do not meet the requirement of eq. (269) for $n = 1, 2, 3, \ldots$, i.e., n is a positive integer and, therefore, the right hand-side of eq. (269) must be positive. Equation (272) represents the Sommerfeld's fine structure formula for the energy eigenvalues of a hydrogen type atom.

To obtain the complete radial wave function, we normalize the functions using $1 = \int \psi^\dagger \psi d^3x = \int_0^\infty (f^2 + F^2) r^2\, dr$, to get

$$\left.\begin{matrix} F(r) \\ f(r) \end{matrix}\right\} = \frac{\pm(2\lambda)^{3/2}}{\Gamma(2s+1)} \sqrt{\frac{(m \pm E)\Gamma(2s+n'+1)}{4m\frac{(n'+s)m}{E}\left(\frac{(n+s)m}{E} - \kappa\right)n'!}} (2\lambda r)^{s-1} \mathrm{e}^{-\lambda r}$$

$$\left\{\left(\frac{(n'+s)m}{E} - \kappa\right) \times F(-n', 2s+1; 2\lambda r) \mp n' F(1-n', 2s+1; 2\lambda r)\right\} \tag{273}$$

CHAPTER 12

The Hydrogen Atom in the Momentum Representation

12.1 CONJUGATE MOMENTA IN GENERALIZED COORDINATES

Before we discuss the transformation of the Dirac equation to the momentum representation, it is instructive to examine the issue of coordinates and conjugate momenta in generalized coordinates. In Cartesian coordinates the conjugate momentum corresponding to the coordinate operators $\hat{x}^i$, $i = 1, 2, 3$ take a particularly simple form as $\hat{p}_i$. However, in case of spherical polar and other systems of coordinates, the situation is far more involved.

Let us consider an arbitrary set of generalized coordinates $x \equiv x^i$, $i = 1, 2, 3$, describing a point in 3-space and let $p \equiv p_i$, $i = 1, 2, 3$, represent the corresponding conjugate momenta. Obviously, in quantum mechanics, all these quantities need to be Hermitian operators. In the Schrödinger representation, that we adopt here, these operators refer to a fixed instant in time and the temporal development of the system is captured by the time dependent state vector $|\rangle_{x0}$ in an appropriate Hilbert space. For the moment, we assume that each of the generalized coordinate operators possesses a continuum of eigenvalues from $-\infty$ to $+\infty$. As is conventional, we represent the squared infinitesimal distance element between two points as the quadratic form $ds^2 = g_{ij}\, dx^i\, dx^j$, where g_{ij} are the elements constituting the three-dimensional metric tensor. The determinant $g = \det\{g_{ij}\}$, $i, j = 1, 2, 3$, is abbreviated as g and must necessarily be non-singular.

Since the coordinate operators $\hat{x}^i$, $i = 1, 2, 3$, satisfy $[\hat{x}^i, \hat{x}^j] = 0$, $i, j = 1, 2, 3$ the eigenvectors of these operators can be adopted as a set of basis vectors and we can expand the state vector $|\rangle_{x0}$ in terms of the eigenvectors $|x'\rangle$ of the $\hat{x} \equiv \hat{x}^i$, $i = 1, 2, 3$, as

$$|\rangle_{x0} = \int |x'\rangle \langle x'|\rangle_{x0}\, d\omega' \tag{1}$$

where $d\omega'$ is the infinitesimal volume element

$$d\omega' = g^{1/2}(x')\, dx' \tag{2}$$

The integration needs to be carried over the entire range of eigenvalues of the coordinate operator. $\langle x'|\,\rangle_x{}^0 = \psi(x)$ is the wave function of the particle.

The normalization condition for the eigenvectors $|x'\rangle$ takes the form $\langle x'|x''\rangle \doteq \delta(x', x'')$, where $\delta(x', x'') = 0$ for $x' \neq x''$ and $\int f(x')\delta(x', x'')\, d\omega' = f(x'')$ for reasonably well-behaved functions. In terms of the Dirac delta function, we have

$$\delta(x', x'') = g^{-1/2}(x')\,\delta(x' - x'') = g^{-1/2}(x'')\,\delta(x' - x'') \tag{3}$$

which follows from the invariant volume element being $d\omega' = g^{1/2}\, dx'$ so that

$$f(x'') = \int f(x')\delta(x', x'')d\omega' = \int f(x')\delta(x', x'')g^{1/2}\, dx'$$

$$= \int f(x')\,\delta(x' - x'')\, dx'$$

whence $\delta(x', x'') = g^{-1/2}\,\delta(x' - x'')$.

Using the properties $x\delta'(x) = -\,\delta(x)$ and $x\delta(x) = 0$ of the Dirac delta function, we obtain

$$(x'^i - x''^i)\frac{\partial}{\partial x'^j}\delta(x', x'') = (x'^i - x''^i)\frac{\partial}{\partial x'^j}[g^{-1/2}\,\delta(x' - x'')$$

$$= (x'^i - x''^i)\left[\frac{\partial g^{-1/2}}{\partial x'^j}\delta(x' - x'') + g^{-1/2}\frac{\partial\delta(x' - x'')}{\partial x'^j}\right]$$

$$= (x'^i - x''^i)\, g^{-\frac{1}{2}}\frac{\partial\delta(x' - x'')}{\partial x'^j}$$

$$= -\,\delta_j^i\, g^{-\frac{1}{2}}\,\delta(x' - x'') = -\delta_j^i\,\delta(x', x'') \tag{4}$$

and also

$$\frac{\partial}{\partial x'^i}\delta(x', x'') = \frac{\partial}{\partial x'^i}[g^{-1/2}\,\delta(x' - x'')]$$

$$= -\frac{1}{2}g^{-3/2}\frac{\partial g}{\partial x'^i}\delta(x' - x'') + g^{-1/2}\frac{\partial\delta(x' - x'')}{\partial x'^i}$$

$$= -\frac{1}{2}g^{-1}\frac{\partial g}{\partial x'^i}\delta(x', x'') - g^{-1/2}\frac{\partial\delta(x' - x'')}{\partial x''^i}$$

$$= -\frac{1}{2}\frac{\partial(\ln g)}{\partial x'^i}\delta(x', x'') - g^{-1/2}\frac{\partial\delta(x' - x'')}{\partial x''^i}$$

$$= -\frac{1}{2}\frac{\partial(\ln g)}{\partial x'^i}\,\delta(x', x'') - \frac{\partial[g^{-1/2}\,\delta(x'-x'')]}{\partial x''^i}$$

$$= -\frac{1}{2}\frac{\partial(\ln g)}{\partial x'^i}\,\delta(x', x'') - \frac{\partial\delta(x', x'')}{\partial x''^i} \tag{5}$$

The matrix elements of the coordinate operators in the coordinate representation are given by

$$\langle x'|\hat{x}^i|x''\rangle = x'^i\,\delta(x', x'') = x''^i\,\delta(x', x'') \tag{6}$$

Since the conjugate momenta need to satisfy $[\hat{x}^i,\ \hat{p}_j] = i\delta^i_j$ and $[\hat{p}_i,\ \hat{p}_j] = 0$, we also have

$$i\delta^i_j\,\delta(x', x'') = \langle x'|i\delta^i_j|x''\rangle = \langle x'|[\hat{x}^i,\ \hat{p}_j]|x''\rangle$$

$$= \int d\omega'''\,\{\langle x'|\hat{x}^i|x'''\rangle\,\langle x'''|\,\hat{p}_j|x''\rangle - \langle x'|\,\hat{p}_j|x'''\rangle\,\langle x'''|\hat{x}^i|x''\rangle\}$$

$$= \int d\omega'''\,\{x'^i\,\delta(x', x''')\,\langle x'''|\,\hat{p}_j|x''\rangle - x'''^i\,\delta(x''', x'')\,\langle x'|\,\hat{p}_j|x'''\rangle\}$$

$$= (x'^i - x''^i)\,\langle x'|\,\hat{p}_j|x''\rangle \tag{7}$$

Using the identity (4), eq. (7) reduces to

$$i\delta^i_j\,\delta(x', x'') = -\,i(x'^i - x''^i)\,\frac{\partial}{\partial x'^j}\,\delta(x', x'') = (x'^i - x''^i)\,\langle x'|\,\hat{p}_j|x''\rangle \tag{8}$$

which has the general solution

$$\langle x'|\,\hat{p}_j|x''\rangle = -\,i\,\frac{\partial}{\partial x'^j}\,\delta(x', x'') + F_j\,\delta(x', x'') \tag{9}$$

where F_j is a function of the coordinates. Obviously, F_j satisfies $(x'^i - x''^i)\,F_j\,\delta(x', x'') = 0$ which follows from the property $x\delta(x) = 0$ of the Dirac delta function. Further, F_j may be evaluated at either x' or x''. Since $[\hat{p}_i,\ \hat{p}_j] = 0$, we must also have

$$0 = \int\{\langle x'|\,\hat{p}_i|x'''\rangle\,\langle x'''|\,\hat{p}_j|x''\rangle - \langle x'|\,\hat{p}_j|x'''\rangle\,\langle x'''|\,\hat{p}_i|x''\rangle\}d\omega''' \tag{10}$$

From eqs. (5) and (9) and the definition of invariant volume $d\omega' = g^{1/2}\,dx'$, substituting in eq. (10), we get

$$\int\left\{\left[-i\frac{\partial}{\partial x'^i}\delta(x', x''') + F_i\,(x')\delta(x', x''')\right]\right.$$

$$\left.\left[-i\frac{\partial}{\partial x'''^j}\delta(x''', x'') + F_j\,(x'')\delta(x''', x'')\right]\right\}$$

$$-\left\{\left[-i\frac{\partial}{\partial x'^j}\delta(x', x''') + F_j\,(x')\delta(x', x''')\right]\right.$$

$$\left[-i\frac{\partial}{\partial x'''^i}\delta(x''', x'') + F_i(x'')\delta(x''', x'')\right]\Bigg\} d\omega''' = 0 \tag{11}$$

Now

$$\frac{\partial}{\partial x'^i}\,\delta(x', x'') = \frac{\partial}{\partial x'^i}[g^{-1/2}\,\delta(x' - x'')]$$

$$= -\frac{1}{2}\,g^{-3/2}\,\frac{\partial g}{\partial x'^i}\,\delta(x' - x'') + g^{-1/2}\,\frac{\partial\delta(x' - x'')}{\partial x'^i}$$

$$= -\frac{1}{2}\,g^{-1/2}\,\frac{\partial(\ln g)}{\partial x'^i}\,\delta(x' - x'') + g^{-1/2}\,\frac{\partial\delta(x' - x'')}{\partial x'^i}$$

$$= g^{-1/2}\left[\frac{\partial\delta(x' - x'')}{\partial x'^i} - \frac{1}{2}\frac{\partial(\ln g)}{\partial x'^i}\,\delta(x' - x'')\right] \tag{12}$$

Using eq. (12), eq. (11) becomes

$$\int\left\{\begin{matrix}\left[-ig^{-1/2}\left[\frac{\partial\delta(x' - x''')}{\partial x'^i} - \frac{1}{2}\frac{\partial(\ln g)}{\partial x'^i}\delta(x' - x''')\right] + g^{-1/2}F_i(x')\delta(x' - x''')\right] \\ \left[-ig^{-1/2}\left[\frac{\partial\delta(x''' - x'')}{\partial x'''^j} - \frac{1}{2}\frac{\partial(\ln g)}{\partial x'''^j}\delta(x''' - x'')\right] + g^{-1/2}F_j(x'')\delta(x''' - x'')\right] - \\ \left[-ig^{-1/2}\left[\frac{\partial\delta(x' - x''')}{\partial x'^j} - \frac{1}{2}\frac{\partial(\ln g)}{\partial x'^j}\delta(x' - x''')\right] + g^{-1/2}F_j(x')\delta(x' - x''')\right] \\ \left[-ig^{-1/2}\left[\frac{\partial\delta(x''' - x'')}{\partial x'''^i} - \frac{1}{2}\frac{\partial(\ln g)}{\partial x'''^i}\delta(x''' - x'')\right] + g^{-1/2}F_i(x'')\delta(x''' - x'')\right]\end{matrix}\right\}$$

$$g^{1/2}\,dx''' = 0 \tag{13}$$

Obviously,

$$\left[\frac{\partial}{\partial x'^i}\delta(x' - x'''), \frac{\partial}{\partial x'^j}\delta(x''' - x'')\right] = 0 \tag{14}$$

$$\int[F_i(x')\,\delta(x' - x''')\,F_j(x'')\,\delta(x''' - x'') - F_j(x')\,\delta(x' - x''')\,F_i(x'')\,\delta(x''' - x'')]\,dx'''$$

$$= F_i(x')\,F_j(x') - F_j(x')\,F_i(x') = 0 \tag{15}$$

Also

$$\int\left[-i\frac{\partial}{\partial x'^i}\delta(x' - x''')\,F_j(x'')\delta(x''' - x'') + iF_j(x')\delta(x' - x''')\frac{\partial}{\partial x'''^i}\delta(x''' - x'')\right]dx'''$$

$$= -i\left[\frac{\partial}{\partial x'^i}\delta(x'-x'')F_j(x') - F_j(x')\frac{\partial}{\partial x'^i}\delta(x'-x'')\right]$$

$$= -i\left\{\frac{\partial}{\partial x'^i}[\delta(x'-x'')F_j(x')] - F_j(x')\frac{\partial}{\partial x'^i}\delta(x'-x'')\right\}$$

$$= -i\,\frac{\partial F_j(x')}{\partial x'^i}.\delta(x'-x'') \tag{16}$$

Similarly,

$$\int dx''' \begin{Bmatrix} [F_i(x')\delta(x'-x''')]\left[-i\dfrac{\partial}{\partial x'''^j}\delta(x'''-\delta'')\right] \\ -\left[-i\dfrac{\partial}{\partial x'^j}\delta(x'-\delta'')\right][F_i(x'')\,\delta(x'''-x'')] \end{Bmatrix}$$

$$= i\delta(x'-x'')\,\frac{\partial}{\partial x'^j}\,F_i(x') \tag{17}$$

The terms involving $\ln g$ yield exactly similar expressions and eq. (13) yields

$$ig^{-1/2}\,\delta(x'-x'')\left\{\frac{\partial}{\partial x'^j}\left[F_i(x')+\frac{i}{2}\frac{\partial(\ln g)}{\partial x'^i}\right]-\frac{\partial}{\partial x'^i}\left[F_j(x')+\frac{i}{2}\frac{\partial(\ln g)}{\partial x'^j}\right]\right\}$$

$$= i\delta(x',x'')\left\{\frac{\partial}{\partial x'^j}\left[F_i(x')+\frac{i}{2}\frac{\partial(\ln g)}{\partial x'^i}\right]-\frac{\partial}{\partial x'^i}\left[F_j(x')+\frac{1}{2}\frac{\partial(\ln g)}{\partial x'^j}\right]\right\} = 0 \tag{18}$$

Equation (18) mandates that $F_i(x') + \dfrac{i}{2}\dfrac{\partial(\ln g)}{\partial x'^i}$ must be the derivative of some function, say $F(x')$, that is independent of the indices i, j, i.e.,

$$F_i(x') + \frac{i}{2}\,\frac{\partial(\ln g)}{\partial x'^i} = \frac{\partial}{\partial x'^i}\,F(x') \tag{19}$$

so that $\langle x'|\hat{p}_j|x''\rangle = -i\dfrac{\partial}{\partial x'^j}\,\delta(x',x'') + F_j\,\delta(x',x'')$ gives

$$\langle x'|\hat{p}_j|x''\rangle = -i\frac{\partial}{\partial x'^i}\delta(x',x'') + \left[\frac{\partial}{\partial x'^i}F(x') - \frac{i}{2}\frac{\partial(\ln g)}{\partial x'^i}\right]\delta(x',x'')$$

$$= -i\frac{\partial}{\partial x'^i}\delta(x',x'') + \frac{\partial}{\partial x'^i}\,G(x')\,\delta(x',x'') \tag{20}$$

where $G(x') = F(x') - i \ln g^{1/2}$. Since $\hat{p}_i$ are Hermitian operators, we must also have $\langle x'|\hat{p}_i|x''\rangle = \langle x''|\hat{p}_i|x'\rangle^*$ which gives, on using eq. (5) and taking conjugate of eq. (20), $G^*(x') = G(x') + i \ln g^{1/2}$ for

$$\langle x''|\hat{p}_i|x'\rangle = -i\frac{\partial}{\partial x''^i}\delta(x'', x') + \left[\frac{\partial}{\partial x''^i}F(x'') - \frac{i}{2}\frac{\partial(\ln g)}{\partial x''^i}\right]\delta(x'', x') \tag{21}$$

which, using eq. (5), gives

$$\langle x''|\hat{p}_i|x'\rangle = i\frac{\partial\delta(x'', x')}{\partial x'^i} + \frac{i}{2}\frac{\partial(\ln g)}{\partial x''^i}\delta(x'', x')$$

$$+\left[\frac{\partial}{\partial x''^i}F(x'') - \frac{i}{2}\frac{\partial(\ln g)}{\partial x''^i}\right]\delta(x'', x')$$

$$= i\frac{\partial\delta(x'', x')}{\partial x'^i} + \frac{\partial}{\partial x''^i}F(x'')\,\delta(x'', x') \tag{22}$$

so that

$$\langle x''|\hat{p}_i|x'\rangle^* = -i\frac{\partial\delta(x'', x')}{\partial x'^i} + \frac{\partial}{\partial x''^i}F^*(x'')\,\delta(x'', x')$$

whence $\langle x'|\hat{p}_i|x''\rangle = \langle x''|\hat{p}_i|x'\rangle^*$ implies

$$\frac{\partial}{\partial x''^i}F^*(x'')\,\delta(x'', x') = \left[\frac{\partial}{\partial x'^i}F(x') - \frac{i}{2}\frac{\partial(\ln g)}{\partial x'^i}\right]\delta(x', x'') \tag{23}$$

or $F^*(x') = F(x') - i \ln g^{1/2}$ whence $G^*(x') = F^*(x') + i \ln g^{1/2} = F(x')$.

$= G(x') + i \ln g^{1/2}$

Now, $F^*(x') = F(x') - i \ln g^{1/2}$ implies

$$\text{Im } F(x') = \frac{i}{2}\ln g^{1/2} \text{ and so } G(x') = R(x') - \frac{i}{2}\ln g^{1/2}$$

where $R(x')$ is the real part of $F(x')$ and is a real function of the coordinates.

We can transform our basis vectors $|x'\rangle$ to a new set of eigenvectors $|x'\rangle' = \exp[-iR(x')]|x'\rangle$ as the basis. On this transformation the real function can be eliminated from consideration and we get $G(x') = -\frac{i}{2}\ln g^{1/2}$ whence the matrix elements of $\langle x'|\hat{p}_i|x''\rangle$ can be written with reference to the new basis as

$$\langle x'|\hat{p}_i|x''\rangle' = -i\frac{\partial}{\partial x'^i}\delta(x', x'') - \frac{i}{4}\frac{\partial(\ln g)}{\partial x'^i}\delta(x', x'') \tag{24}$$

In the coordinate space representation, we have $\langle x'|\hat{x}^i|\rangle_{x^0} = x^i\,\psi(x)$ and $\langle x'|\hat{p}_i|\rangle_{x^0} = p_i\psi(x)$ so that the coordinate and conjugate momentum operators have the following representation in coordinate space

$$\hat{x}^i = x'^i \text{ and } \hat{p}_i = -i\frac{\partial}{\partial x'^i} - \frac{i}{4}\frac{\partial[\ln g(x')]}{\partial x'^i}$$

$$= -i\left\{\frac{\partial}{\partial x'^i} + \frac{1}{2g^{1/2}}\frac{\partial[g^{1/2}(x')]}{\partial x'^i}\right\} \tag{25}$$

It is important, at this stage, to make certain non-trivial observations about the analysis underlying eqs. (24) and (25) as follows:

(a) The coordinate operators $\hat{x}^i$ and the conjugate momenta all have been assumed to be defined on the domain from $-\infty$ to $+\infty$. This assumption is mandatory in order that the conjugate operators of eqs. (24) and (25) be self-adjoint as

$$\int dx^\alpha\, dx^\beta\, dx^\gamma\, \psi^*\left(-i\frac{\partial\phi}{\partial x^\gamma}\right)g^{1/2} = \int dx^\alpha\, dx^\beta \int_a^b dx^\gamma\, \psi^*\left(-i\frac{\partial\phi}{\partial x^\gamma}\right)g^{1/2}$$

$$= \int dx^\alpha\, dx^\beta\, \{-i\psi^*\,\phi g^{1/2}|_a^b$$

$$-\int_a^b dx^\gamma\left[-i\left(\frac{\partial}{\partial x^\gamma} + \frac{1}{g^{1/2}}\frac{\partial g^{1/2}}{\partial x^\gamma}\right)\psi\right]^*\phi g^{1/2} \tag{26}$$

where we have performed an integration by parts using the identity

$$\int_a^b dx\,\psi^*\left(-i\frac{\partial\phi}{\partial x^\gamma}\right) = -i\psi^*\,\phi|_a^b - \int_a^b dx\left(-i\frac{\partial\psi}{\partial x^\gamma}\right)^*\phi \tag{27}$$

with $(\psi g^{1/2})^* = (g^{1/2})^*\,\psi^* = g^{1/2}\,\psi^* = \psi^*\,g^{1/2}$ as the first function.

If we assume the boundary conditions to be such that the integrated term vanishes, then it follows from eq. (26) that

$$\left(-i\frac{\partial}{\partial x^\gamma}\right)^\dagger = -i\left(\frac{\partial}{\partial x^\gamma} + \frac{1}{g^{1/2}}\frac{\partial g^{1/2}}{\partial x^\gamma}\right) \tag{28}$$

which gives

$$\left(-i\frac{\partial}{\partial x^\gamma}\right)^\dagger + \frac{i}{2g^{1/2}}\frac{\partial g^{1/2}}{\partial x^\gamma} = \left[-i\left(\frac{\partial}{\partial x^\gamma} + \frac{1}{2g^{1/2}}\frac{\partial g^{1/2}}{\partial x^\gamma}\right)\right]^\dagger$$

$$= -i\left(\frac{\partial}{\partial x^\gamma} + \frac{1}{2g^{1/2}}\frac{\partial g^{1/2}}{\partial x^\gamma}\right) \tag{29}$$

showing that $\hat{p}_i = -i\left\{\frac{\partial}{\partial x'^i} + \frac{1}{2g^{1/2}}\frac{\partial[g^{1/2}(x')]}{\partial x'^i}\right\}$ is self-adjoint. It must, however, be noted that for the integrated term to vanish in eq. (26), the function $\phi(x)$ must satisfy $\text{Lim}_{x\to-\infty}\,\phi(x) = \text{Lim}_{x\to\infty}\,\phi(x) = 0$.

Another point of importance is that the differential operator $\frac{\partial}{\partial x}$ cannot be defined in the whole of $L_2(a, b)$ because not all functions in this space are differentiable. Besides, there may also be functions which, although themselves being differentiable, their derivatives are not always in $L_2(a, b)$. Therefore, we need the domain to include continuous functions that are, at least, piecewise differentiable, i.e., the set of absolutely continuous functions whose derivatives are in $L_2(a, b)$. These functions are dense in $L_2(a, b)$. In the case of such functions, we can integrate $\int \psi\phi'\,dx$ by parts. We denote this domain by $AC(a, b)$. If we confine our domain to $\phi(x) \in AC(-\infty, \infty)$ then we have, on physical grounds, $\text{Lim}_{x\to-\infty}\,\phi(x) = \text{Lim}_{x\to\infty}\,\phi(x) = 0$ so that the integrated term of eq. (26) vanishes.

(b) According to the Hellinger-Toeplitz theorem, an everywhere dense operator $\hat{A}$ that satisfies $\langle \hat{A}\psi|\phi\rangle = \langle\psi|\hat{A}\phi\rangle$ must necessarily be a bounded operator. However, most of the operators encountered in quantum mechanics like position, momentum, energy operators are unbounded. These operators, which we want to be self-adjoint in order that they may have real eigenvalues, being unbounded, will not be self-adjoint as stated in the Hellinger Toeplitz theorem. They can only be considered self-adjoint if defined on a dense linear subspace of the Hilbert space. It, therefore, becomes necessary to define the domain of such operators and carefully examine their realm of validity before adopting them to represent meaningful physical observables.

(c) We have already considered the case when the domain of the conjugate momentum operator extends over the whole axis $(-\infty, \infty)$. We now examine the situation where the domain of $\hat{T}_i = i\,\frac{\partial}{\partial x^i}$ is restricted to the half axis, i.e., the domain consists of absolutely continuous functions on the half axis whose derivative lies in $L_2(0, \infty)$ so that Dom $(\hat{T}_i) = AC(0, \infty)$. In order that the integrated term in eq. (26) vanishes, we need to impose the additional constraint that $\phi \in AC(0, \infty)$ be such that $\phi(0) = 0$, the other boundary condition being automatically satisfied since $\text{Lim}_{x\to\infty}\,\phi(x) = 0$. The operator $\hat{T}_i$ in this case turns out to be symmetric, i.e., Dom $(\hat{T}_i) \subset$ Dom$(\hat{T}_i^\dagger)$ because, as will be shown in the sequel, while $\hat{T}_i^\dagger$ is defined $\forall\ \phi \in AC(0, \infty)$, Dom $(\hat{T}_i)$ is restricted to $\{\phi \in AC(0, \infty), \phi(0) = 0\}$. Furthermore, $\hat{T}_i$ in this case, is not self-adjoint since Dom $(\hat{T}_i) \neq$ Dom $(\hat{T}_i^\dagger)$.

(d) We now consider the case when Dom $(\hat{T}) = \{\phi \in AC(0, 1)\}$ where $\hat{T} = i\frac{\partial}{\partial x}$. For simplicity of exposition of this rather technical analysis, we restrict ourselves to the one-dimensional case, although extension to the three dimensions is straightforward.

In this case, we must additionally mandate $\phi(0) = \phi(1) = 0$ for the integrated term of eq. (26) to vanish, so that Dom $(\hat{T}) = \{\phi \in AC(0,1), \phi(0) = \phi(1) = 0\}$. Here, again $\hat{T}$ is symmetric, i.e., Dom $(\hat{T}) \subset$ Dom $(\hat{T}^{\dagger})$ but it does have self-adjoint extensions. Let us try to find $\hat{T}^{\dagger}$.

(i) For the purpose, let us assume $j(x)$ be any positive, infinitely differentiable function with support in $(-1, 1)$ so that $\int_0^1 j(x)\,dx = 1$. Let $\varphi(x) \in$ Dom $(\hat{T})$. Further define $j_\varepsilon(x) = \varepsilon^{-1} j\left(\frac{x}{\varepsilon}\right)$. Then setting $\varphi_\varepsilon(x) = J_\varepsilon\varphi = \int_0^1 j_\varepsilon(x - t)\,\varphi(t)\,dt$, we have

$$|\varphi_\varepsilon(x) - \varphi(x)| \leq \int j\varphi_\varepsilon(x - t)|\varphi(t) - \varphi(x)|dt$$

$$\leq (\text{Sup}_{(t:|x-t| \leq \varepsilon)}|\varphi(t) - \varphi(x)|) \int_{(0,1)} j_\varepsilon(x - t)dt = \sup_{(t:|x-t| \leq \varepsilon)}|\varphi(t) - \varphi(x)|$$

implying that $\varphi_\varepsilon \to \varphi$ if φ is continuous. In other words, $J_\varepsilon\,\varphi \to \varphi$ in $L^2(0,1)$ if φ is continuous (which is proved below).

We also have, $i\frac{d}{dx}\varphi_\varepsilon(x) = \int_0^1 i\frac{d}{dx} j_\varepsilon(x - t)\,\varphi(t)dt$

$$= -\int_0^1 i\left[\frac{d}{dt} j_\varepsilon(x - t)\right]\varphi(t)\,dt = \int_0^1 j_\varepsilon(x - t)\,i\frac{d}{dt}\varphi(t)\,dt$$

$$\xrightarrow{L^2(\mathbb{R})} i\frac{d}{dx}\varphi(x).$$ Thus, $\varphi_\varepsilon \in$ Dom $(\hat{T})$ for each $\varepsilon > 0$.

(ii) Secondly, each j_ε is a bounded operator of norm not greater than the one shown as follows:

Let $\psi \in L^2(0, 1)$, then

$$|\langle j_\varepsilon\,\varphi|\psi\rangle \leq \iint j_\varepsilon(x - t)|\varphi(t)||\psi(x)|dxdt$$

$$= \iint j_\varepsilon(y)|\varphi(t)||\psi(y + t)|dydt \leq \|\varphi\|\,\|\psi\| \int j_\varepsilon(y)dy = \|\varphi\|\,\|\psi\|$$

thereby giving the desired result.

(iii) We now prove that φ is continuous, i.e., a uniform limit of continuous functions is continuous so that $j_\varepsilon\varphi \xrightarrow{L^2(\mathbb{R})} \varphi$ for all $\varphi \in L^2(0,1)$. This can be done by an $\varepsilon/3$ argument as follows.

We fix $x \in (0,1)$ and $\varepsilon > 0$. We need to find $\delta > 0$ such that $|x - y| < \delta$ implies $|\varphi(x) - \varphi(y)| < \varepsilon$. We select $n \in \infty$ so that $||\varphi_n - \varphi|| < \varepsilon/3$. Now, since φ_n is continuous, we can select our $\delta > 0$ so that $|x - y| < \delta$ implies $|\varphi_n(x) < \varphi_n(y)| - \varepsilon/3$. Then $|x - y| < \delta$ implies

$$|\varphi(x) - \varphi(y)| \leq |\varphi(x) - \varphi_n(x)| + |\varphi_n(x) - \varphi_n(y)| + |\varphi_n(y) - \varphi(y)|$$

$$< \frac{1}{3}\varepsilon + \frac{1}{3}\varepsilon + \frac{1}{3}\varepsilon = \varepsilon$$

showing that φ is continuous.

(iv) Now fix $0 < \alpha < \beta < 1$ and define $f_\varepsilon^{\alpha,\beta}(x) = j_\varepsilon(x - \beta) - j_\varepsilon(x - \alpha)$ and $g_\varepsilon^{\alpha,\beta}(x) = \int_0^x f_\varepsilon^{\alpha,\beta}(t)dt$. Let $\psi \in \text{Dom}(\hat{T}^\dagger)$. If we take a sufficiently small ε, we can make $g_\varepsilon^{\alpha,\beta} \in \text{Dom}(\hat{T})$. Therefore, $\langle\psi|\hat{T}g_\varepsilon^{\alpha,\beta}\rangle = \langle\hat{T}^\dagger\psi|g_\varepsilon^{\alpha,\beta}\rangle$. Now, as $\varepsilon \to 0$, $-g_\varepsilon^{\alpha,\beta}$ converges to the characteristic function of (α, β) in $L^2(0,1)$. This gives $\langle\hat{T}^\dagger\psi|g_\varepsilon^{\alpha,\beta}\rangle \to -\int_\alpha^\beta (\hat{T}^\dagger\psi)(x)dx$.

(v) Now $\langle\psi|\hat{T}g_\varepsilon^{\alpha,\beta}\rangle = \left\langle i\frac{\partial}{\partial x}\int_0^x [j_\varepsilon(t-\beta) - j_\varepsilon(t-\alpha)]dt|\psi\right\rangle \to -i(\psi(\beta) - (\psi(\alpha))$ in mean square as $\varepsilon \to 0$ because $j_\varepsilon\varphi \xrightarrow{L^2(\mathbb{R})} \varphi$ for all $\varphi \in L^2(0,1)$. Therefore, $i(\psi(\beta)) - \psi(x)) = \int_\alpha^\beta (\hat{T}^\dagger\psi)(x)dx$, i.e., ψ is absolutely continuous because $(\hat{T}^\dagger\psi)(x) \in L^2(0,1)$ so that its integral is absolutely continuous and, further, $i\frac{d}{dx}\psi(x) = (\hat{T}^\dagger\psi)(x)$ by the "Fundamental Theorem of Calculus".

Hence, we have shown that $\psi \in AC(0,1)$ and $T^\dagger\psi = i\dfrac{d\psi(x)}{dx}$.

(vi) Conversely, for $\psi \in AC(0,1)$, we have

$$\int_0^1 \phi^*\hat{T}^\dagger\psi(x)\,dx = \phi^*\int\hat{T}^\dagger\psi(x)\,dx\Big|_0^1 - \int_0^1 (\hat{T}^\dagger\phi)^*\psi(x)\,dx$$

$$= \phi^*\int\hat{T}^\dagger\psi(x)\,dx\Big|_0^1 - \left\{\begin{array}{l}\left[\int(\hat{T}^\dagger\phi)^*dx\right]\psi(x)\Big|_0^1 \\ -\int\left[\int(\hat{T}^\dagger\phi)^*dx\right]\hat{T}\psi(x)dx\end{array}\right\}$$

$$= \phi^*(x)\psi(x)|_0^1 - \phi^*(x)\,\psi(x)|_0^1$$
$$+ \int \varphi^* \hat{T}^\dagger \psi(x)\,dx = \int \varphi^*\,\hat{T}^\dagger \psi(x)\,dx$$

so that any $\psi \in AC\,(0,1)$ is in the domain of $\hat{T}^\dagger$ and $\hat{T}^\dagger\psi = i\dfrac{d\psi(x)}{dx}$. Therefore, $\hat{T}^\dagger = i\dfrac{d}{dx}$ on Dom $(\hat{T}^\dagger) = AC(0,1)$.

(vii) Hence, while Dom $(\hat{T}) = \{\phi \in AC(0,\,1),\ \phi(0) = \phi(1) = 0\}$, Dom $(\hat{T}^\dagger) = AC(0,\,1)$ so that Dom $(\hat{T}) \subset$ Dom $(\hat{T}^\dagger)$ implying that Dom $\hat{T}$ is symmetric but not self-adjoint.

(viii) We now examine whether $\hat{T}$ has any self-adjoint extensions. Let us assume that $\hat{S}$ is a symmetric extension of $\hat{T}$. Since $\hat{S}$ is a symmetric extension of $\hat{T}$, we have Dom $(\hat{T}) \subset$ Dom $(\hat{S})$. This implies Dom $(\hat{S}^\dagger) \subset$ Dom $(\hat{T}^\dagger)$. Since Dom $(\hat{T}^\dagger) = AC(0,\,1)$, we know that functions in Dom $(\hat{S}^\dagger)$ are absolutely continuous and $\hat{S}^\dagger\,\varphi = i\dfrac{d\varphi}{dx}$. Hence, if $\varphi \in$ Dom $(\hat{S})$ and $\psi \in$ Dom $(\hat{S}^\dagger)$, integration by parts shows that $\langle\psi|\hat{S}\varphi\rangle - \langle\hat{S}^\dagger\,\psi|\varphi\rangle = \varphi^*(1)\,\psi(1) - \varphi^*(0)\,\psi(0)$ which must vanish for self-adjointness.

Now, if $\hat{S} = \hat{T}$, then $\hat{T}$ is not self-adjoint because the boundary conditions on the functions in Dom $\hat{T}$, viz. $\varphi(0) = \varphi(1) = 0$ are so strong that no boundary conditions on the functions in Dom $(\hat{T}^\dagger)$ are necessary in order that $\langle\psi|\hat{S}\varphi\rangle - \langle\hat{S}^\dagger\,\psi|\varphi\rangle$ vanishes. Therefore, Dom $(\hat{T}) \subset$ Dom $(\hat{T}^\dagger)$ and so $\hat{S} = \hat{T}$ is not self-adjoint.

What we need to do here is to extend the set of functions in Dom $(\hat{S}) \supset$ Dom $(\hat{T})$ by allowing more general boundary conditions (than those in Dom $(\hat{T})$) which are such that the equality $\langle\psi|\hat{S}\varphi\rangle - \langle\hat{S}^\dagger\,\psi|\varphi\rangle = 0$ requires the same boundary conditions on the functions in Dom $(\hat{S}^\dagger)$ as in Dom $(\hat{S})$, i.e., we want to achieve Dom $(\hat{S})$ = Dom $(\hat{S}^\dagger)$.

For this purpose, let us now assume that $\hat{S}$ is a self-adjoint extension of $\hat{T}$. Let $\varphi \in$ Dom $(\hat{S})$/ Dom $(\hat{T})$. It is instructive to write

$$\langle\psi|\hat{S}\varphi\rangle - \langle\hat{S}^\dagger\,\psi|\varphi\rangle = \varphi^*(1)\psi(1) - \varphi^*(0)\psi(0) = C$$
$$= \varphi^*(0)\psi(0) - \varphi^*(0)\psi(0) = \alpha\bar{\alpha}\varphi^*(0)\psi(0) - \varphi^*(0)\psi(0)$$

with $|\alpha| = 1$. Now, since $\varphi \in$ Dom $(\hat{S})$/Dom $(\hat{T})$, $\varphi \notin$ Dom $(\hat{T})$ and therefore $\varphi(0) \neq 0$ and we can find an α with $|\alpha| = 1$ such that $\varphi(1) - \alpha\varphi(0) = 0$. Now, if ψ is any other function in Dom $(\hat{S})$ then, because $\hat{S}$ is self-adjoint we need that $\langle\psi|\hat{S}\varphi\rangle - \langle\hat{S}^\dagger\,\psi|\varphi\rangle = 0$, which is possible if we additionally require $\psi(1) = \alpha\psi(0)$ with the same α.

In view of the above, if we write $\hat{T}_\alpha = i\dfrac{d}{dx}$ on Dom $(\hat{T}_\alpha) = \{\varphi{:}\varphi \in AC(0,1),\ \varphi(1) = \alpha\varphi(0)\}$ then Dom $(\hat{S}) \subset$ Dom $(\hat{T}_\alpha)$.

Thus, we have the following situation Dom ($\hat{S}$) $\subset$ Dom ($\hat{T}_\alpha$) so that Dom ($\hat{S}^\dagger$) $\supset$ Dom ($\hat{T}_\alpha^\dagger$), Dom $\hat{T}_\alpha \subset$ Dom ($\hat{T}_\alpha^\dagger$) because $\hat{T}_\alpha$ is symmetric and Dom ($\hat{S}$) = Dom ($\hat{S}^\dagger$) because $\hat{S}$ is desired to be self-adjoint so that Dom ($\hat{S}^\dagger$) $\supset$ Dom ($\hat{T}_\alpha^\dagger$) $\supset$ Dom ($\hat{T}_\alpha$) $\supset$ Dom ($\hat{S}$) = Dom ($\hat{S}^\dagger$) yielding Dom ($\hat{T}_\alpha$) = Dom ($\hat{S}$) so that $\hat{S} = \hat{T}_\alpha$ for some α.

(ix) To examine for what values of α, T_α are self-adjoint, we choose $\varphi \in$ Dom ($\hat{T}_\alpha$) and $\psi \in$ Dom ($T_\alpha^\dagger$). Then, we must have

$$\bar{\alpha}\varphi^*(0)\,\psi(1) - \varphi^*(0)\,\psi(0) = 0 = \bar{\alpha}\alpha\varphi^*(0)\,\psi(0) - \varphi^*(0)\,\psi(0)$$

if $\bar{\alpha}\alpha = 1$ so that $\psi(1) = \alpha\psi(0)$. Thus, $\psi \in$ Dom ($\hat{T}_\alpha$) and therefore Dom ($\hat{T}_\alpha^\dagger$) = Dom ($\hat{T}_\alpha$), whence, $\hat{T}_\alpha$ is self-adjoint for each α. Thus, the set of self-adjoint extensions of $\hat{T}$consists of the collection of operators $\{\hat{T}_\alpha: \alpha \in \mathbf{C}, |\alpha| = 1\}$.

12.2 CONJUGATE MOMENTUM OPERATORS IN SPHERICAL POLAR COORDINATES

We now elaborate on the consequences of the above analysis of conjugate momentum operators in relation to spherical polar coordinates.

The spherical polar coordinates are, classically, curvilinear coordinates in Euclidean space, because after a coordinate transformation the metric tensor g_{ij} can take the values δ_{ij}. The transformation equations between Cartesian coordinates to spherical polar coordinates are, respectively, $x^1 = r\sin\theta\cos\phi$, $x^2 = r\sin\theta\sin\phi$ and $x^3 = r\cos\theta$. The metric tensor in spherical polar coordinates is $g_{ij} = \begin{pmatrix} 1 & 0 & 0 \\ 0 & r^2 & 0 \\ 0 & 0 & r^2\sin^2\theta \end{pmatrix}$ since, the distance between two points (r, θ, ϕ) and $(r + dr, \theta + d\theta, \phi + d\phi)$ is $ds^2 = dr^2 + r^2 d\theta^2 + r^2\sin^2\theta\, d\phi^2$. Comparison with $ds^2 = g_{ij}\,dx^i\,dx^j$ yields the above expression for the metric tensor. The determinant of the metric tensor is $g = \det(g_{ij}) = r^4\sin^2\theta$.

We consider in more detail the change between Cartesian and spherical polar coordinates. Let us represent the point transformation between the two sets of coordinates as (we shall not consider here explicit time dependence) $x^k = x^k(q)$. In order that there exists an equivalence of the two sets x and q, it is necessary that each one of them describes the system completely. Therefore, the transformation $x^k = x^k(q)$ must, of necessity, be invertible with the inverse $q^k = q^k(x)$. This requirement of invertibility mandates that: (i) the transformation functions are continuous together with their partial derivatives and (ii) the Jacobian of the transformation is nonzero. Hence, we require $J^2 = g = r^4\sin^2\theta \neq 0$ for the compliance of which we have to assume that $r > 0$, $0 < \theta < \pi$ and $0 \le \phi < 2\pi$.

Additionally, we cannot determine $\phi = \tan^{-1}\left(\frac{x^2}{x^1}\right)$ when $x^1 \to 0$ and $x^2 \to 0$ together, corresponding to the origin. Also, we need to accept that $\theta = 0, \pi$ include the x^3 axis. There are, therefore, several impediments to interpreting $x^k = x^k(q)$ as a "point" transformation. Hence, it does not lead to a complete description of the system and so is not a true representation, because it is not completely invertible.

The fallout of the above is that the conjugate momenta in spherical polar coordinates obtained by applying eq. (25), viz.

$$\hat{p}_r = -i\left[\frac{\partial}{\partial r} + \frac{1}{2r^2 \sin\theta}\frac{\partial(r^2 \sin\theta)}{\partial r}\right] = -i\left[\frac{\partial}{\partial r} + \frac{1}{r}\right] \tag{30}$$

$$\hat{p}_\theta = -i\left[\frac{\partial}{\partial \theta} + \frac{1}{2r^2 \sin\theta}\frac{\partial(r^2 \sin\theta)}{\partial \theta}\right] = -i\left[\frac{\partial}{\partial \theta} + \frac{1}{2}\cot\theta\right] \tag{31}$$

$$\hat{p}_\phi = -i\left[\frac{\partial}{\partial \phi} + \frac{1}{2r^2 \sin\theta}\frac{\partial(r^2 \sin\theta)}{\partial \phi}\right] = -i\frac{\partial}{\partial \phi} \tag{32}$$

do not satisfy the requirement of being self-adjoint except for $\hat{p}_\phi = -i\frac{\partial}{\partial \phi}$ with Dom($\hat{p}_\phi$) $= AC$ ($\phi = 0,\ 2\pi$).

The above expressions for the conjugate momenta in spherical polar coordinates can also be obtained directly. For this, we consider a dynamical system with canonical coordinates (r, θ, ϕ) and the corresponding conjugate angular momenta operators $(\hat{p}_r, \hat{p}_\theta, \hat{p}_\phi)$. We shall illustrate the procedure with reference to the radial component of the angular momentum operator as follows:

(a) For the radial component we first require that the radial operator $\hat{r}$ satisfies the commutator $[\hat{r}, \hat{p}_r] = i$ and $[\hat{\theta}, \hat{p}_r] = 0$, $[\hat{\phi}, \hat{p}_r] = 0$ with its conjugate momentum operator $\hat{p}_r$. The most general form of $\hat{p}_r$ that can satisfy these commutators can be written as $\hat{p}_r = -i\left[\frac{\partial}{\partial r} + F(r, \theta, \phi)\right]$, where F is an arbitrary real function of the canonical coordinates that can be decomposed as $F(r, \theta, \phi) = f(r) + g(\theta) + h(\phi)$.

(b) Secondly, we require that $\hat{p}_r$ be Hermitian, i.e., for arbitrary complex functions ψ, ξ belonging to our Hilbert space of square integrable functions, we must have $\langle\psi|\hat{p}_r\xi\rangle - \langle\hat{p}_r\psi|\xi\rangle = 0$. The first requirement for this, as has been shown earlier (by performing an integration by parts), is that the surface term should vanish, i.e., the functions ψ, ξ should be such that their derivatives should vanish at the limits of integration.

Defining the scalar product in the usual way $\langle\psi|\xi\rangle = \int d\omega\psi^\dagger\xi$, where $d\omega = r^2\sin\theta\, dr\, d\theta\, d\phi$ is the infinitesimal volume element, we get, from the Hermiticity condition

$$-i\int d\omega\left(\psi^\dagger\frac{\partial\xi}{\partial r} + \psi^\dagger F(r,\theta,\phi)\xi + \frac{\partial\psi^\dagger}{\partial r}\xi + \psi^\dagger F(r,\theta,\phi)\xi\right) = 0 \tag{33}$$

which gives

$$\int\sin\theta\, d\theta\, d\phi\left[\int_{-\infty}^{\infty}\begin{pmatrix}\psi^\dagger\dfrac{\partial\xi}{\partial r} + \psi^\dagger F(r,\theta,\phi)\xi \\ +\dfrac{\partial\psi^\dagger}{\partial q_i}\xi + \psi^\dagger F(r,\theta,\phi)\xi\end{pmatrix} r^2\, dr\right] = 0 \tag{34}$$

or

$$\int_{-\infty}^{\infty}\left(\psi^\dagger\frac{\partial\xi}{\partial r} + \frac{\partial\psi^\dagger}{\partial r}\xi + 2\psi^\dagger F(r,\theta,\phi)\xi\right) r^2\, dr$$

$$= \int_{-\infty}^{\infty}\left[\frac{\partial(\psi^\dagger\xi)}{\partial r} + 2F(r,\theta,\phi)\psi^\dagger\xi\right] r^2\, dr = 0 \tag{35}$$

which can be written as $\int_{-\infty}^{\infty}\left[\dfrac{\partial(r^2\psi^\dagger\xi)}{\partial r}\right]dr = (r^2\psi^\dagger\xi)|_{-\infty}^{\infty}$ with the identification $f(r) = \dfrac{1}{r}$, $g(\theta) = 0$ and $h(\phi) = 0$. Hence, the Hermitcity condition for $\hat{p}_r$ reduces to $(r^2\psi^\dagger\xi)|_{-\infty}^{\infty} = 0$ which implies that ψ, ξ remain finite at $r = 0$ and go exponentially to zero when $r \to \infty$.

(c) Furthermore, with the above identification of $f(r)$, $g(\theta)$ and $h(\phi)$, we have $\hat{p}_r = -i\left(\dfrac{\partial}{\partial r} + \dfrac{1}{r}\right)$.

(d) Proceeding on exactly similar lines, we have $\hat{p}_\theta = -i\left(\dfrac{\partial}{\partial\theta} + \dfrac{1}{2}\cot\theta\right)$ because

$$\int_0^\pi\left[\frac{\partial(\psi^\dagger\xi)}{\partial\theta} + 2F(r,\theta,\phi)\psi^\dagger\xi\right]\sin\theta d\theta$$

$$= \int_0^\pi\left[\frac{\partial(\sin\theta\,\psi^\dagger\xi)}{\partial\theta} - \cos\theta\psi^\dagger\xi + 2F(r,\theta,\phi)\sin\theta\psi^\dagger\xi\right]d\theta = 0$$

giving the identification $F(r, \theta, \phi) = \dfrac{1}{2}\cot\theta$ so that $\hat{p}_\theta = -i\left[\dfrac{\partial}{\partial\theta} + \dfrac{1}{2}\cot\theta\right]$.

(e) Similarly, $\int_0^{2\pi}\left[\frac{\partial(\psi^\dagger\xi)}{\partial\phi}+2F(r,\theta,\phi)\psi^\dagger\xi\right]d\phi = 0$ whence $F(r, \theta, \phi) = 0$ and $\hat{p}_\phi = -i\frac{\partial}{\partial\phi}$.

(f) The requirement of Hermiticity of $\hat{p}_\theta$ and $\hat{p}_\phi$ require $(\sin\theta\psi^\dagger\xi)|_0^\pi$, i.e., ψ, ξ remain finite at $\theta = 0, \pi$ for Hermiticity of $\hat{p}_\theta$, and $(\psi^\dagger\xi)|_0^{2\pi} = 0$ for the Hermiticity of $\hat{p}_\phi$, i.e., ψ, ξ satisfy the periodicity requirement.

12.3 THE ANGLE VARIABLES

Actually, for any allowed r and θ, there need to be periodic boundary conditions associated with ϕ as can be shown by expanding an arbitrary wave function ψ in a Fourier series as $\psi = \frac{1}{\sqrt{2\pi}}\sum_{-\infty}^{\infty} C_n e^{in\phi}$. The matrix representation of $[\hat{p}_\phi, \phi]$ takes the form $([\hat{p}_\phi, \phi])_{mn} = -i[\delta_{mn} - (-1)^{m-n}]$ so that we recover the standard commutator on the subset of $L_2(-\pi, \pi)$ consisting of those functions for which $\sum(-1)^n C_n = 0$, i.e., those which vanish at $+\pi$ and $-\pi$.

In fact, $\hat{p}_\phi = -i\frac{\partial}{\partial\phi}$ is self-adjoint with spectrum consisting of the points $m \in \mathbb{Z}$, if and only if, the domain is taken to be the space of absolutely continuous functions on $(-\pi, \pi)$, i.e., $\psi \in AC(-\pi, \pi)$ whose derivatives are in $L_2(-\pi, \pi)$ and also satisfy $\psi(\pi) = \psi(-\pi)$. Therefore, if ψ is in this domain, $\phi\psi$ is also in it, if and only if, $\psi(\pi) = 0$.

We conclude that $\hat{p}_\phi = -i\frac{\partial}{\partial\phi}$ is Hermitian only in $\{\psi\}_P$, i.e., the space of periodic functions with period 2π. This is because, as shown earlier, with ψ being periodic, we have, explicitly

$$\langle\hat{p}_\phi\psi_1|\psi_2\rangle = \int_{-\pi}^{\pi}\left(-i\frac{\partial\psi_1}{\partial\phi}\right)^*\psi_2\,d\tau = -i\psi_1^*\psi_2|_{-\pi}^{\pi} - \int_{-\pi}^{\pi}\psi_1^*\left(-i\frac{\partial\psi_2}{\partial\phi}\right)^* d\tau$$

which will be equal to $\langle\psi_1|\hat{p}_\phi\psi_2\rangle$ only on the vanishing of the surface term $-i\psi_1^*\psi_2|_{-\pi}^{\pi}$ for which the periodicity of ψ with period 2π is mandated.

But as per the standard definition $\phi = \tan^{-1}\frac{x^2}{x^1}$ and so ϕ itself is not periodic and can take values from $-\infty$ to $+\infty$. Therefore, if we adopt this definition for ϕ, then $\hat{p}_\phi$ will not be Hermitian and therefore, the uncertainty relation $\Delta\hat{p}_\phi\,\Delta\hat{\phi} \geq \frac{1}{2}$ will not follow in this case from the commutator $[\hat{p}_\phi, \phi] = -i$.

There is another intriguing observation regarding the Hermiticity of $\hat{p}_\phi$. On taking the matrix elements of the commutator relation $[\phi, \hat{p}_\phi] = i$, we obtain $\langle lm'|[\phi, \hat{p}_\phi]|lm\rangle = i\langle lm'|lm\rangle$ or $(m - m')$

$\langle lm'|\phi|lm\rangle = i\delta_{mm'}$ which leads to the absurdity 0 = 1 for $m' = m$. The flaw, obviously, lies in supposing that, in $\langle lm'|\hat{p}_\phi\ \phi|lm\rangle$, $\hat{p}_\phi$ can operate to the left as well, on the assumption that it is Hermitian.

The genesis of the above discussion is that the non-periodicity of $\phi|lm\rangle$, or alternately that of ϕ not being Hermitian with respect to $\hat{p}_\phi|lm\rangle$, needs to be accounted for in defining the conjugate momentum $\hat{p}_\phi$ in spherical polar coordinates.

In this context, it is qualitatively well accepted that fluctuations in ϕ, greater than 2π have little physical meaning. We can easily choose wave functions for which $\Delta\hat{p}_\phi \le \frac{1}{4}$ so that $\Delta\phi$ is greater than 2π. In fact, we can even choose $\Delta\hat{p}_\phi = 0$ by selecting eigenfunctions of $\hat{p}_\phi$, the spherical harmonics $Y_{lm}(\theta, \phi)$, denoted by $|lm\rangle$ as the wave functions. The uncertainty relation is then clearly violated.

One way out of this quagmire is to take as angle observable in quantum mechanics multiplication, not by ϕ, the usual angle of trigonometry which runs from $-\infty$ to ∞ but instead by

$$Y(\phi) \equiv \phi - 2\pi \sum_{n=0}^{\infty} H[\phi - (2n+1)\pi] + 2\pi \sum_{n=0}^{\infty} H[-\varphi - (2n+1)\pi]$$

where $H(x)$ is the unit step function. Thus, $Y(\phi)$ is taken as ϕ (mod 2π) in the range $(-\pi, \pi)$ which is also the range of integration in the scalar product and $\hat{p}_\phi = -i\frac{\partial}{\partial\phi}$ remains Hermitian with regard to the domain consisting of $Y(\phi)$ or any function of it since they are periodic (and are, therefore, in the domain $\hat{p}_\phi$ of as a self-adjoint operator). The commutator, in this case, becomes $[Y(\phi), \hat{p}_\phi]$ $= i\left\{1 - 2\pi\sum_{-\infty}^{\infty}\delta[\phi - (2n+1)\pi]\right\}$.

However, one of the delta functions in the right-hand side of this equation would contribute in the physical interval of 2π and, therefore, the uncertainty relation $[Y(\phi), \hat{p}_\phi] = i$ will not hold.

Actually, the difference between the classical and quantum variables ϕ and $Y(\phi)$ arises from the fact that a classical trajectory enables and implies continuous observation. One can measure the angle that is turned through by the radius vector of a classical particle as it moves from one position to another — this angle might be many multiples of 2π. In quantum mechanics such a measurement is impossible and meaningless.

Another approach that appears in the literature to obviate the problem of non-periodicity of the angle variable is to use continuous periodic variables to locate the azimuthal position, e.g., to adopt sin ϕ and cos ϕ instead of ϕ. We can, then construct any other physical (periodic) function using Fourier series. We then have the commutation relations $[\sin\phi, \hat{p}_\phi] = i\cos\phi$ and $[\cos\phi, \hat{p}_\phi] = -i\sin\phi$. The uncertainty relations become

$$(\Delta\hat{p}_\phi)^2 (\Delta\sin\phi)^2 \ge \frac{1}{4}\langle\cos\phi\rangle^2 \text{ and } (\Delta\hat{p}_\phi)^2 (\Delta\cos\phi)^2 \ge \frac{1}{4}\langle\sin\phi\rangle^2$$

or $\dfrac{(\Delta\hat{p}_\phi)^2 (\Delta\sin\phi)^2}{\langle\cos\phi\rangle^2} \ge \dfrac{1}{4}$ and $\dfrac{(\Delta\hat{p}_\phi)^2 (\Delta\cos\phi)^2}{\langle\sin\phi\rangle^2} \ge \dfrac{1}{4}$

since $\hat{p}_\phi$ is now Hermitian.

We can also write the above pair of uncertainty relations as one relation symmetrical in sin ϕ and cos ϕ as

$$U \equiv \frac{(\Delta\hat{p}_\phi)^2\left[(\Delta\cos\phi)^2 + (\Delta\sin\phi)^2\right]}{\langle\sin\phi\rangle^2 + \langle\cos\phi\rangle^2} \ge \frac{1}{4}$$

When the distribution of ϕ in the wave function is sufficiently localized, we recover the canonical commutation relations. We construct the probability distribution $P(\phi)$ by integrating out all variables except ϕ and assume that $P(\phi)$ is peaked at ϕ_0 with an essentially symmetrical distribution so that by expanding cos ϕ and sin ϕ up to second order in $\phi - \phi_0$, we obtain

$$\cos\phi = \cos\phi_0\left[1 - \frac{1}{2}(\delta\phi)^2\right] - \sin\phi_0\,\delta\phi$$

and
$$\sin\phi = \sin\phi_0\left[1 - \frac{1}{2}(\delta\phi)^2\right] + \cos\phi_0\,\delta\phi$$

where $\delta\phi = \phi - \phi_0$. On taking expectations,

$$\langle\cos\phi\rangle = \cos\phi_0\left[1 - \frac{1}{2}\langle(\delta\phi)^2\rangle\right],$$

$$\langle\sin\phi\rangle = \sin\phi_0\left[1 - \frac{1}{2}\langle(\delta\phi)^2\rangle\right],$$

and
$$\langle\cos^2\phi\rangle = \cos^2\phi_0\,[1 - \langle(\delta\phi)^2\rangle] + \sin^2\phi_0\,\langle(\delta\phi)^2\rangle,$$
$$\langle\sin^2\phi\rangle = \sin^2\phi_0\,[1 - \langle(\delta\phi)^2\rangle] + \cos^2\phi_0\,\langle(\delta\phi)^2\rangle,$$

whence $\langle\Delta\cos\phi\rangle^2 \equiv \langle\cos^2\phi\rangle - \langle\cos\phi\rangle^2 = \sin^2\phi_0\,\langle(\delta\phi)^2\rangle$ and $\langle\Delta\sin\phi\rangle^2 \equiv \langle\sin^2\phi\rangle - \langle\sin\phi\rangle^2 = \cos^2\phi_0$ $\langle(\delta\phi)^2\rangle$ upto second order.

Since $\langle\cos\phi\rangle^2 \approx \cos^2\phi_0$ to lowest order, we further have $\langle\Delta\cos\phi\rangle^2 \equiv \langle\sin\phi\rangle^2\,\langle(\delta\phi)^2\rangle$ and similarly $\langle\Delta\sin\phi\rangle^2 \equiv \langle\cos\phi\rangle^2\,\langle(\delta\phi)^2\rangle$

so that

$$\langle\Delta\hat{p}_\phi\rangle^2\,\langle\Delta\sin\phi\rangle^2 \ge \frac{1}{4}\langle\cos\phi\rangle^2 \Rightarrow (\Delta\hat{p}_\phi)^2\,\langle\cos\phi\rangle^2\,\langle(\delta\phi)^2\rangle$$

$$\ge \frac{1}{4}\langle\cos\phi\rangle^2 \Rightarrow (\Delta\hat{p}_\phi)^2\,\langle(\delta\phi)^2\rangle \ge \frac{1}{4}$$

which are the canonical commutation relations with the identification $\langle(\delta\phi)^2\rangle = (\Delta\phi)^2$.

Thus, to use an angle variable ϕ to locate a geometrical point in the $x^1 - x^2$ plane leads to problems with the Hermiticity of the conjugate angular momentum operator $\hat{p}_\phi$ if the canonical commutation relations are to be maintained. A possible way is to use a pair of continuous periodic variables cos ϕ and sin ϕ or a combination thereof. Then $\hat{p}_\phi$ is Hermitian, and we also recover the canonical commutation relations to lowest order.

We examine a third approach towards resolving this intriguing problem. Let a Hilbert space H be defined to consist of square integrable functions on the segment $-\pi \leq \phi \leq \pi$ with the usual scalar product given by $\langle u|v\rangle = \int_{-\pi}^{\pi} u^*(\phi)\, v(\phi)\, d\phi$. The domain of the operator $\hat{p}_\phi \equiv -i\frac{\partial}{\partial\phi}$ consists of all the differentiable functions $v(\phi)$.

Since, we have $\langle u|\,\hat{p}_\phi\, v\rangle - \langle\,\hat{p}_\phi^\dagger u|v\rangle = -i[u^*(\pi)\, v(\pi) - u^*(-\pi)\, v(-\pi)]$, it follows that the operator $\hat{p}_\phi \equiv -i\frac{\partial}{\partial\phi}$ is not a self-adjoint operator. Since $v(\pi)$ and $v(-\pi)$ are arbitrary the above expression will vanish, if and only if, $u^*(\pi) = u^*(-\pi) = 0$ implying that the domain of Dom $(\hat{p}_\phi^\dagger) \subset$ Dom $(\hat{p}_\phi)$, in this case. This may be contrasted with the case discussed earlier wherein we had Dom $(\hat{p}_\phi) \subset$ Dom $(\hat{p}_\phi^\dagger)$. The difference is due to the fact that in the present case, the boundary restrictions $u^*(\pi) = u^*(-\pi) = 0$ are imposed on the adjoint operator whereas in the earlier case, restrictions were imposed on the primary operator i.e., we had $v(\pi) = v(-\pi) = 0$.

Let us now consider the domain $u(\phi) \in \mathrm{H}_\alpha \Leftrightarrow u(\pi) = e^{2\pi i\alpha}\, u(-\pi)$ with $0 \leq \alpha \leq 1$ and define the family of operators $\hat{p}_\phi^\alpha \equiv -i\frac{\partial}{\partial\phi}$ on this domain. They are self-adjoint. In fact, for each $0 \leq \alpha \leq 1$, they constitute self-adjoint extensions of $\hat{p}_\phi$.

But the problem does not end here and while we have found self-adjoint extensions of $\hat{p}_\phi$, the angle operator $\hat{\phi}$ corresponding to azimuthal angle in spherical polar coordinates needs some discussion. The angle operator, obviously satisfies $\hat{\phi}u(\phi) = \phi u(\phi)$. The problem is that the domain of $\hat{\phi}$ is not H_α. This is because the spectrum of $\hat{\phi}$ must be restricted to a finite interval, e.g., $-\pi \leq \phi \leq \pi$. Hence, we cannot find a common domain for $\hat{p}_\phi$ and $\hat{\phi}$.

A possible way to resolve this problem, that has also found mention in the literature, is to define a series of differentiable functions $f_\varepsilon(\phi) \in \mathrm{H}_{\alpha=0}$ such that the function $f_\varepsilon(\phi)$ takes the value $f_\varepsilon(\phi) = \phi$ for $-\pi \leq \phi < \pi - \varepsilon$ and $f_\varepsilon(\phi)$ goes smoothly to the value $f_\varepsilon(-\pi)$ for $\phi \geq \pi - \varepsilon$ in the limit $\phi \to \pi$. This implies that $f_\varepsilon(-\pi) = f_\varepsilon(\pi)$ with $\mathrm{Lim}_{\varepsilon\to 0} f_\varepsilon(\phi) \to \phi$. With this trick the family of operators $\hat{\phi}_\varepsilon \equiv f_\varepsilon(\hat{\phi})$ all belong to H_α and they also constitute good approximations of the angle operator $\hat{\phi}$ in the limit of small ε.

Because the canonical commutation relation $[\hat{p}_\phi^\alpha, \hat{\phi}_\varepsilon] = -if'_\varepsilon(\hat{\phi}_\varepsilon)$ does not exist at $\varepsilon = 0$ due to the discontinuity of the function $f_{\varepsilon=0}(\phi)$ at the point $\phi = \pi$, we write the canonical commutation relation in the form $[\hat{p}^{\alpha}_{\phi}, \hat{\phi}_{\varepsilon=0}] = -i[1 - 2\pi\delta(\hat{\phi} - \pi)]$, where $\hat{\phi}_{\varepsilon=0} \neq \hat{\phi}$ only at the point $\phi = \pi$.

Let an eigenstate of $\hat{p}_{\phi}$ corresponding to the eigenvalue m be represented by $|m\rangle$. Then, taking the matrix elements of both sides of $[\hat{p}^{\alpha}_{\phi}, \hat{\phi}_{\varepsilon=0}] = -i[1 - 2\pi\delta(\hat{\phi} - \pi)]$, we get $(m - m') \langle m|\hat{\phi}_{\varepsilon=0}|m'\rangle = - i\{\delta_{m,\, m'} - \exp[i(m' - m)\pi]\}$ both sides of which identically vanish for $m = m'$.

By adopting this approach, we are able to ensure that although the angle operator ϕ does not belong to H_{α}, any periodic function of it is a self-adjoint operator in H_{α}.

Another radical approach questions the very need of requiring quantum mechanical obervables to be represented by "Hermitian" operators. Attempts have been made to develop a self-consistent theory of quantum mechanics based on "non-Hermitian" operators. In order to appreciate the nuances of this approach, we first need to understand the rationale behind the restriction of representing observables only by Hermitian operators in traditional quantum mechanics. For the sake of completeness, we briefly trace the logic behind this mandate.

In quantum mechanics, it is traditional to represent phase space variables like position, momentum, etc. and other physical properties of systems by operators. The result of a particular measurement of a physical property is then construed as the value of the eigenvalue of the relevant operator in the prescribed state of the system. Since such results of measurements must necessarily take real values, it is natural to demand that the eigenvalue spectrum of such operators be "real" in order that they may constitute "well meaning" representations of physical quantities. Hermitian operators, thus, were an automatic choice to represent physical properties in quantum mechanics since they have real eigenvalues.

Such Hermitian operators possess another invaluable attribute – their eigenvectors constitute a complete orthogonal set so that an operator can be most conveniently represented in the basis constituted by its eigenvectors, e.g., $\hat{X} = \sum_k x_k|x_k\rangle\langle x_k|$ or, in the continuous case $\hat{X} = \int_{-\infty}^{\infty} dx x\,|x\rangle\langle x|$.

It, therefore, follows that any operator and consequently, the physical quantity that it represents can be constructed and/or identified through a knowledge of its set of eigenvectors with the corresponding eigenvalues, i.e., its spectrum provided, that the above series converges in the desired domain of existence of the operator.

There may be cases where the series is non-convergent over the entire infinite dimensional Hilbert space of the system but convergence may be achieved by suitably defining the domain of the operator to a desired truncation of the said Hilbert space compatible with the physical quantity proposed to be represented.

If the above recipe does not work and convergence is not achieved, we can still obtain knowledge about of the eigenvalue spectrum by defining, for example, a bounded function f of the operator $\hat{X}$, say, $f(\hat{X})$. We choose this function as being such as to permit an expansion of the corresponding operator $\hat{f}$ in the basis of the eigenvectors of $\hat{X}$ to obtain $\hat{f} = \sum_k f(x_k)|x_k\rangle\langle x_k|$ so that a knowledge of the eigenvalues of $\hat{f}$ together with the eigenvectors would enable us to ascertain the eigenvalues of $\hat{X}$. Furthermore, if $f(\hat{X})$ is a bounded function so that $\hat{f}$ is a bounded operator a spectral decomposition of $\hat{f}$ is then, guaranteed.

The lack of Hermiticity of an operator would have two immediate consequences that need to be addressed as a prerequisite to any consistent quantum theory, viz.

(a) that the eigenvalues of the operator can no longer be mandated to be real.

(b) the eigenvectors would not constitute an orthogonal set and hence cannot be adopted as a basis for the spectral resolution of the operator (unless the operator is, otherwise, unitary).

We first address (b). If the eigenvectors do not constitute an orthogonal set, the set or eigenvectors would not constitute a complete basis and we may or may not end up with a spectral resolution of the identity with the set of eigenvectors. To explore the situation, let us assume that a set of state vectors exists that permits a resolution of the identity as $I = \int dx|x\rangle\langle x|$. Another fallout of (b) is that there may or may not exist well-defined operators representing the decomposition $\hat{f} = \int dx f(x)|x\rangle\langle x|$ in terms of the eigenvectors $|x\rangle$. In the event that $\hat{f}$ does permit the decomposition, we have, as pointed by Levy-Leblond, for any normalized state $|\psi\rangle$,

$$\langle\psi|\hat{f}|\psi\rangle = \langle\psi|\int dx f(x)|x\rangle\langle x||\psi\rangle = \int dx f(x)|\langle x|\psi\rangle|^2$$

$$1 = \langle\psi|\psi\rangle = \langle\psi|\int dx|x\rangle\langle x||\psi\rangle = \int dx|\langle x|\psi\rangle|^2$$

and

$$|\psi\rangle = \boldsymbol{I}|\psi\rangle = \int dx|x\rangle\langle x|\psi\rangle$$

so that the fundamental axioms of quantum mechanics are still met.

However, we would need to generalize some of the conventional definitions of expectation values and the spreads about expectation values when dealing with non-Hermitian operators. For instance, Levy-Leblond defines the spectral extension of a non-Hermitian operator in a normalized state $|\psi\rangle$ by

$$(\Delta\hat{F})^2_\psi = \|(\hat{F} - \langle\hat{F}\rangle I)|\psi\rangle\|^2 = [(\hat{F} - \langle\hat{F}\rangle I)|\psi\rangle]^\dagger(\hat{F} - \langle\hat{F}\rangle I)|\psi\rangle$$

$$= \langle\psi|(\hat{F} - \langle\hat{F}\rangle I)^\dagger(\hat{F} - \langle\hat{F}\rangle I)|\psi\rangle = \psi\langle\hat{F}^\dagger\hat{F}|\psi\rangle - \langle\hat{F}\rangle^2$$

$$= \langle\psi|\hat{F}^\dagger\hat{F}|\psi\rangle - |\langle\psi|\hat{F}|\psi\rangle|^2$$

This may be contrasted with the conventional definition (for Hermitian operators) $(\Delta\hat{F})^2_\psi = \langle\psi|\hat{F}^2|\psi\rangle - |\langle\psi|\hat{F}|\psi\rangle|^2$ the difference between the two definitions being the consequence of the fact that the eigenvectors of $\hat{F}$ may not be orthogonal if $\hat{F}$ is non-Hermitian and, therefore, $|\hat{F}|^2 \neq \hat{F}^\dagger\hat{F}$. We obviously recover the latter definition from the former for Hermitian $\hat{F}$ with the identification $\hat{F} = \hat{F}^\dagger$. For a unitary operator $\hat{F}^\dagger\hat{F} = I$, the above definition reduces to $(\Delta\hat{F})^2 = 1 - |\langle\hat{F}\rangle|^2$. Application of Schwarz's inequality ensures that $\langle\psi|\hat{F}^\dagger\hat{F}|\psi\rangle| \geq \langle\psi|\hat{F}|\psi\rangle|^2$ so that $(\Delta\hat{F})^2_\psi \geq 0$ with the equality holding, if and only if, $|\psi\rangle$ is an eigenstate of $\hat{F}$ for, in such a case if f is the corresponding eigenvalue, we have $\langle\psi|\hat{F}^\dagger\hat{F}|\psi\rangle = f^*f = |f|^2 = |\langle\psi|\hat{F}|\psi\rangle|^2$.

The issue of the eigenvalues of non-Hermitian operators not being real and thereby measurements of physical quantities represented by such operators not yielding real values can also be addressed by the above prescription, i.e., by defining a suitable function of the relevant operator mapping the eigenvalues (including imaginary ones) to the measured ones.

The implications of representing physical quantities by non-Hermitian operators are intriguing, particularly insofar as measurements of physical quantities of quantum systems are concerned. The cardinal point is that the eigenvectors of Hermitian operators are orthogonal. Now, the result of measurement of a physical property (represented by an operator) of a quantum system is the eigenvalue of that operator in the given eigenstate. If the property is represented by a Hermitian operator, then, because the eigenvectors of such an operator are orthogonal, i.e., $\langle x_i|x_j\rangle = \delta_{ij}$, the system cannot be construed to be in any other eigenstate as the probability thereof is $|\langle x_i|x_j\rangle|^2 = 0$ for $i \neq j$. Hence, any successive measurements will yield the same value provided the system remains in the same state. However, if the physical property is represented by non-Hermitian operators, then the corresponding eigenvectors are not orthogonal so that $|\langle x_i|x_j\rangle|^2 \neq 0$ for $i \neq j$. Thus, there exists a finite nonzero probability that successive measurements of the same property of the system in the same state will not yield the same result. This philosophy seems to be finding endorsement with the development of quantum field theories on non-commutative manifolds.

In view of the aforesaid observations on the angle variables and the conjugate momenta, we define the unitary operator $v = e^{i\phi}$ as a proxy for the azimuthal angle ϕ and define the conjugate momentum as $\hat{p}_\phi = v\frac{\partial}{\partial v} = e^{i\phi}\frac{\partial \phi}{\partial v}.\frac{\partial}{\partial \phi} = -i\frac{\partial}{\partial \phi}$. The commutation relation v and $\hat{p}_\phi$ is $[\hat{p}_\phi, v] = v\frac{\partial}{\partial v}(v.) - v.v\frac{\partial}{\partial v} = v$. Now $\hat{p}_\phi$ is Hermitian over the whole range of v. We use the Levy-Leblond definition of a spectral extension, viz. $(\Delta\hat{F})^2 = \langle\alpha|\hat{F}^\dagger\hat{F}|\alpha\rangle - |\langle\alpha|\hat{F}|\alpha\rangle|^2$ which, in the case of $\hat{F}$ being a unitary operator becomes $(\Delta\hat{F})^2 = 1 - |\langle\hat{F}\rangle|^2$ which gives for $\hat{F} = v = e^{i\phi}$, $(\Delta v)^2 = 1 - |\langle \exp i\phi\rangle|^2 = 1 - \langle\cos\phi\rangle^2 - \langle\sin\phi\rangle^2$. Now, as has been proved in the foregoing, $[\hat{p}_\phi, \cos\phi] = i\sin\phi$ and $[\hat{p}_\phi, \sin\phi] = -i\cos\phi$ and the corresponding uncertainty relations become $(\Delta\hat{p}_\phi)^2\,(\Delta\cos\phi)^2 \geq \frac{1}{4}\langle\sin\phi\rangle^2$ and $(\Delta\hat{p}_\phi)^2\,(\Delta\sin\phi)^2 \geq \frac{1}{4}\langle\cos\phi\rangle^2$, whence we get

$$(\Delta\hat{p}_\phi)^2\,(\Delta v)^2 = (\Delta\hat{p}_\phi)^2\,[(\Delta\cos\varphi)^2 + (\Delta\sin\varphi)^2] \geq \frac{1}{4}\,(\langle\cos\phi\rangle^2 + \langle\sin\phi\rangle^2)$$

$$= \frac{1}{4}|\langle e^{i\phi}\rangle|^2 = \frac{1}{4}\,[1 - (\Delta v)^2] \tag{36}$$

It is pertinent to note here that the above statement of the uncertainty principle is intuitively more appealing insofar as, given a wave function that is an eigenstate of the angular momentum operator, the conventional uncertainty relation, $(\Delta\hat{p}_\phi)^2\,(\Delta\phi)^2 \geq \frac{1}{4}$, would correspond to an infinite spread in the angle variable (since $\Delta \hat{p}_\phi = 0$ if the wave function is an eigenstate of $\hat{p}_\phi$) conveying no physical meaning. However, in the Levy-Leblond framework the maximal spread of the angle variable is $\Delta v = 1$.

It is instructive to examine here the implications of the discrete spectrum of the angular momentum operator. For the purpose, we introduce the unitary operator $\hat{U}_m = e^{im\phi}$ with $\hat{p}_\phi = -i\dfrac{\partial}{\partial\phi}$.

We then have $[\hat{p}_\phi, \hat{U}_m] = m\hat{U}_m$ and the corresponding uncertainty relation in the Levy-Leblond framework becomes $(\Delta \hat{p}_\phi)^2 (\Delta\hat{U}_m)^2 \geq \dfrac{m^2}{4}[1-(\Delta\hat{U}_m)^2]$. Now we have two situations, viz.

(a) ϕ is a continuous parameter (i.e., the spectrum of $\hat{p}_\phi = -i\dfrac{\partial}{\partial\phi}$ is continuous) so that the set of operators $\hat{U}_m = e^{im\phi}$ constitute a one parameter group of unitary operators generated by the Hermitian generator ϕ. Expanding $\hat{U}_m$ as a power series to second order, viz. $\hat{U}_m = e^{im\phi}I + im\phi - \dfrac{1}{2}m^2\phi^2 + O(m^3)$, we obtain

$$(\Delta\hat{U}_m)^2 = 1 - |\langle\hat{U}_m\rangle|^2 = 1 - \left|\left\langle I + im\phi - \frac{1}{2}m^2\phi^2 + O(m^3)\right\rangle\right|^2$$

$$= m^2(\langle\phi^2\rangle - \langle\phi\rangle^2) + O(m^4) = m^2(\Delta\phi)^2 + O(m^4)$$

considering terms to lowest order. Furthermore, from this equality, we also get $\Delta\phi = \text{Lim}_{m\to 0}\dfrac{1}{m}(\Delta\hat{U}_m)$ whence

$$(\Delta\hat{p}_\phi)^2(\Delta\phi)^2 = \text{Lim}_{m\to 0}(\Delta\hat{p}_\phi)^2\left(\frac{1}{m}\Delta\hat{U}_m\right)^2 \geq \text{Lim}_{m\to 0}\frac{1}{4}[1-(\Delta\hat{U}_m)^2] \simeq \frac{1}{4}$$

where we have used

$$\text{Lim}_{m\to 0}(\Delta\hat{U}_m)^2 = \text{Lim}_{m\to 0}[1 - |\langle\hat{U}_m\rangle|^2] = \text{Lim}_{m\to 0}[1 - \langle e^{im\phi}\rangle^2] = 0$$

and we recover the standard Heisenberg inequality.

(b) In the event that m takes only integral values corresponding to the discrete spectrum of the angular momentum operator, the operators $\hat{U}_m$ constitute a discrete group. Furthermore, since, in this case, we cannot invoke the limit taking process in m (because the group is not continuous), there does not exist any Hermitian generator of the group. In this case, we get a set of Heisenberg type inequalities corresponding to each permissible value of m and they would need to be simultaneously considered in order to obtain a full characterization of the angular spectral extension. In this case, we cannot recover the standard Heisenberg inequality as a limiting case since no limit process is allowed.

12.4 THE RADIAL VARIABLE

We now examine the issue of Hermiticity of the conjugate radial momentum in the light of the observations made above with regard to the self-adjointness of unbounded operators. We consider a Hilbert space H_r consisting of square integrable functions $u(r)$ with $0 \le r \le \infty$ vanishing at infinity, i.e., $u(r = \infty) = 0$ and define the inner product by $\langle u|v\rangle = \int_0^\infty u^*(r)\, v(r) dr$. We then have, on integration by parts

$$\langle u|\, \hat{p}_r\, v\rangle = \left\langle u\left| -i\frac{\partial v}{\partial r}\right.\right\rangle = \int_0^\infty u^*(r)\left(-i\frac{\partial v(r)}{\partial r}\right) dr = iu^*(0)\, v(0) - \int_0^\infty \left(-i\frac{\partial u(r)}{\partial r}\right)^\dagger v(r) dr$$

$$= iu^*(0)\, v(0) + \langle p_r^\dagger u|v\rangle$$

The radial momentum $\hat{p}_r = -i\dfrac{\partial}{\partial r}$ is, therefore, not a self-adjoint operator. Further, since $\langle u|\, \hat{p}_r\, v\rangle - \langle p_r^\dagger\, u|v\rangle = iu^*(0)\, v(0)$ and because $v(0)$ is arbitrary, in order that $\hat{p}_r^\dagger = \hat{p}_r$, the domain of $\hat{p}_r^\dagger$ must be restricted by the second boundary condition $u(0) = 0$. Further, if $u(r)$ belongs to the domain of $\hat{p}_r^\dagger$, then $\hat{p}_r^\dagger = \hat{p}_r$ also mandates that $0 = \langle \hat{p}_r\, u|\, \hat{p}_r\, v\rangle - \langle \hat{p}_r\, u|\, \hat{p}_r\, v\rangle = \langle \hat{p}_r\, u|\, \hat{p}_r\, v\rangle - \langle \hat{p}_r^\dagger\, \hat{p}_r\, u|v\rangle = i\,\hat{p}_r\, u^*(0)\, v(0)$ which implies that $\hat{p}_r u(0) = -iu'(0) = 0$ and by induction $u^{(n)}(0) = 0$. In other words, a self-adjoint conjugate radial momentum operator can be defined only on a subspace of the given space consisting of all differentiable functions $u(r) \in H$ such that $u^{(n)}(0) = 0\ \forall n \in \mathbb{N}$. The truncation of the Hilbert space through this boundary condition renders it of very limited utility for physical applications since most wave functions do not belong to it.

Besides this, objections as to $\hat{p}_r = -i\left(\dfrac{\partial}{\partial r} + \dfrac{1}{r}\right)$ being an "observable" are also levelled on the ground that there exist no solution to the differential equation $\hat{p}_r\, \psi(r) = -i\left(\dfrac{\partial}{\partial r} + \dfrac{1}{r}\right)\psi(r) = \lambda_r \psi(r)$ that satisfy the boundary conditions stipulated above (which can be conveniently merged as a single boundary condition $\mathrm{Lim}_{r \to 0}\, r\psi(r) = 0$).

An approach that has been advocated towards enlarging the domain of the radial momentum operator is to define two subspaces of the Hilbert space H_r according as $u(r) \in H_r^+ \Leftrightarrow u^{(2m+1)}(0) = 0\ \forall m = 0, 1, 2, \ldots$ and $u(r) \in \mathrm{H}_r^- \Leftrightarrow u^{(2m)}(0) = 0\ \forall m = 0, 1, 2, \ldots$. Thus, H_r^+ includes all square integrable functions whose odd derivatives vanish at the origin and H_r^- includes all square integrable functions whose even derivatives vanish at the origin, i.e. the former includes all even functions and the latter, all odd functions. However, the two subspaces are not disjoint, e.g., functions whose all order derivatives vanish at the origin would appear in both subspaces. Then, for all real and even functions of $\hat{p}_r$, we can find two self-adjoint extensions with domains H_r^+ and H_r^-. For example, let $\hat{p}_r^2 = -\dfrac{\partial^2}{\partial r^2}$ then $\langle u|\,\hat{p}_r^2 v\rangle - \langle(\hat{p}_r^2)^\dagger u|v\rangle = u^*(0)\left.\dfrac{\partial v}{\partial r}\right|_{r=0} - \left.\dfrac{\partial u^*}{\partial r}\right|_{r=0} v(0)$. Thus, if $\hat{p}_r^2$ is to be a self-adjoint operator, we must require that $u^*(0) = v(0) = 0$ or $\left.\dfrac{\partial v}{\partial r}\right|_{r=0} = \left.\dfrac{\partial u^*}{\partial r}\right|_{r=0} = 0$. But if the domain of $\hat{p}_r^2 = -\dfrac{\partial^2}{\partial r^2}$ and its adjoint is so restricted, then, by iteration, as shown above $-\dfrac{\partial^2 u(r)}{\partial r^2} = -u''(r)$ must also satisfy the same boundary conditions, i.e., $u''(0) = 0$ or $u'''(0) = 0$. Thus, $\hat{p}_r^2$ becomes a self-adjoint operator if its domain is either H_r^+ or H_r^-. Furthermore, any real function of $\hat{p}_r^2$ also has the same self-adjoint extension and any odd function of $\hat{p}_r$ does not have a self-adjoint extension.

We can expand a real and even function $u(r) \in H_r^+$ and an odd function $v(r) \in H_r^-$ in a Fourier series as $u(r) = \int_{-\infty}^{\infty} d\chi U_\chi \cos(\chi r)$ and $v(r) = \int_{-\infty}^{\infty} d\chi V_\chi \sin(\chi r)$, where U_χ and V_χ are the Fourier components. The inner product is, as usual, given by $\int u^*(r)\, v(r) dr$. We can span the subspace H_r^+ by the basis vectors $e_\chi^+(r) = \cos(\chi r)$ that get normalized as $\int_0^{\infty} d\chi\, e_\chi^+(r) e_\chi^+(r') = \delta(r - r')$ and also $\int_{-\infty}^{\infty} dr e_\chi^+(r) e_{\chi'}^+(r) = \delta(\chi - \chi')$. An eigenstate of any real and even function of $\hat{p}_r$ in this subspace can be represented as $|\chi\rangle = \int_{-\infty}^{\infty} dr \cos(r\chi)|r\rangle$. Similarly, the space H_r^- is spanned by the basis $e_\chi^+(r) = \sin(\chi r)$ and we can write the eigenstate of the absolute value of $\hat{p}_r$ in this as $|\chi\rangle = \int_{-\infty}^{\infty} dr \sin(r\chi)|r\rangle$.

12.5 THE CLASSICAL-QUANTUM CORRESPONDENCE FOR SPHERICAL COORDINATES

In Cartesian coordinates, a transition can be achieved from a classical function of the phase space variables $F(x_i, p_{xi})$ to its quantum analogue directly by making the substitutions $\hat{p}_{x_i} = -i\dfrac{\partial}{\partial x_i}$ to obtain the operator valued function $\hat{F}(\hat{x}_i, \hat{p}_{x_i})$. However, the same correspondence cannot be applied in the case of spherical polar coordinates because of several issues that have been elaborated in the preceding content, particularly the absence of Hermiticity of the corresponding conjugate momenta.

Towards obtaining the correspondence between classical phase space variables and their quantum counterparts, we write $\hat{p}_r$, $\hat{p}_\theta$, $\hat{p}_\phi$ as

$$\hat{p}_r = -i\left(\frac{\partial}{\partial r} + \frac{1}{r}\right) = \hat{p}_r^0 - i\frac{1}{r},\ \hat{p}_\theta = -i\left(\frac{\partial}{\partial \theta} + \frac{1}{2}\cot\theta\right)$$

$$= \hat{p}_\theta^0 - i\frac{1}{2}\cot\theta,\ \hat{p}_\phi = -i\frac{\partial}{\partial \phi} = \hat{p}_\phi^0$$

The fact that the conjugate momenta are linear combinations of the corresponding canonical coordinates and the derivative with reference to the said coordinate suggests that the correspondence rule may be written as

$$\hat{F}(\hat{q}_i, \hat{p}_i) \equiv \hat{F}(\hat{r}, \hat{\theta}, \hat{\phi}, \hat{p}_r, \hat{p}_\theta, \hat{p}_\phi)$$

$$= \hat{G}(\hat{r}, \hat{\theta}, \hat{\phi}, \hat{p}_r, \hat{p}_\theta, \hat{p}_\phi) + Q(r, \theta, \phi) \tag{37}$$

Expanding the operator valued quantum function $\hat{G}(\hat{r}, \hat{\theta}, \hat{\phi}, \hat{p}_r, \hat{p}_\theta, \hat{p}_\phi)$ as a Taylor series about $(\hat{p}_r^0, \hat{p}_\theta^0, \hat{p}_\phi^0)$, we obtain

$$\begin{aligned}
\hat{G}(\hat{r}, \hat{\theta}, \hat{\phi}, \hat{p}_r, \hat{p}_\theta, \hat{p}_\phi) &= \hat{G}(\hat{r}, \hat{\theta}, \hat{\phi}, \hat{p}_r^0, \hat{p}_\theta^0, \hat{p}_\phi^0) \\
&+ \left[(\hat{p}_r - \hat{p}_r^0)\frac{\partial \hat{G}}{\partial \hat{p}_r}\bigg|_{\hat{p}_r = \hat{p}_r^0} + (\hat{p}_\theta - \hat{p}_\theta^0)\frac{\partial \hat{G}}{\partial \hat{p}_\theta}\bigg|_{\hat{p}_\theta = \hat{p}_\theta^0}\right. \\
&\left. + (\hat{p}_\phi - \hat{p}_\phi^0)\frac{\partial \hat{G}}{\partial \hat{p}_\phi}\bigg|_{\hat{p}_\phi = \hat{p}_\phi^0}\right] \\
&+ \frac{1}{2!}\left[(\hat{p}_r - \hat{p}_r^0)^2\frac{\partial^2 \hat{G}}{\partial \hat{p}_r^2}\bigg|_{\hat{p}_r = \hat{p}_r^0} + (\hat{p}_\theta - \hat{p}_\theta^0)^2\frac{\partial^2 \hat{G}}{\partial \hat{p}_\theta^2}\bigg|_{\hat{p}_\theta = \hat{p}_\theta^0}\right. \\
&\left. + (\hat{p}_\phi - \hat{p}_\phi^0)^2\frac{\partial^2 \hat{G}}{\partial \hat{p}_\phi^2}\bigg|_{\hat{p}_\phi = \hat{p}_\phi^0}\right] + \ldots \\
&= \hat{G}(\hat{r}, \hat{\theta}, \hat{\phi}, \hat{p}_r^0, \hat{p}_\theta^0, \hat{p}_\phi^0) - i\left(\frac{1}{r}\frac{\partial \hat{G}}{\partial \hat{p}_r}\bigg|_{\hat{p}_r = \hat{p}_r^0} + \frac{\cot\theta}{2}\frac{\partial \hat{G}}{\partial \hat{p}_\theta}\bigg|_{\hat{p}_\theta = \hat{p}_\theta^0}\right) \\
&- \frac{1}{2!}\left[\frac{1}{r^2}\frac{\partial^2 \hat{G}}{\partial \hat{p}_r^2}\bigg|_{\hat{p}_r = \hat{p}_r^0} + \left(\frac{\cot\theta}{2}\right)^2\frac{\partial^2 \hat{G}}{\partial \hat{p}_\theta^2}\bigg|_{\hat{p}_\theta = \hat{p}_\theta^0}\right] + \ldots
\end{aligned} \tag{38}$$

Since the operator $\hat{G}$ and its derivatives may include the conjugate momentum operators resulting in operator products that may or may not be Hermitian, we can ensure their Hermiticity by replacing $\hat{A}\,\hat{B} = \dfrac{\hat{A}\hat{B} + \hat{B}\hat{A}}{2}$. The above expression then becomes

$$\hat{G}(\hat{r},\,\hat{\theta},\,\hat{\phi},\,\hat{p}_r,\,\hat{p}_\theta,\,\hat{p}_\phi) = \hat{G}(\hat{r},\,\hat{\theta},\,\hat{\phi},\,\hat{p}_r^0,\,\hat{p}_\theta^0,\,\hat{p}_\phi^0)$$

$$-\frac{i}{2}\left[\left[\left(\frac{1}{r}\frac{\partial \hat{G}}{\partial \hat{p}_r}\bigg|_{\hat{p}_r = \hat{p}_r^0} + \frac{\partial \hat{G}}{\partial \hat{p}_r}\right]_{\hat{p}_r = \hat{p}_r^0}\frac{1}{r}\right)\right.$$

$$\left.+\left(\frac{\cot\theta}{2}\frac{\partial \hat{G}}{\partial \hat{p}_\theta}\bigg|_{\hat{p}_\theta = \hat{p}_\theta^0} + \frac{\partial \hat{G}}{\partial \hat{p}_\theta}\bigg|_{\hat{p}_\theta = \hat{p}_\theta^0}\frac{\cot\theta}{2}\right)\right]$$

$$-\frac{1}{2.2!}\left\{\left(\frac{1}{r^2}\frac{\partial^2 \hat{G}}{\partial \hat{p}_r^2}\bigg|_{\hat{p}_r\,\hat{p}_r^0} + \frac{\partial^2 \hat{G}}{\partial \hat{p}_r^2}\bigg|_{\hat{p}_r = \hat{p}_r^0}\frac{1}{r^2}\right)\right.$$

$$\left.+\left[\left(\frac{\cot\theta}{2}\right)^2\frac{\partial^2 \hat{G}}{\partial \hat{p}_\theta^2}\bigg|_{\hat{p}_\theta = \hat{p}_\theta^0} + \frac{\partial^2 \hat{G}}{\partial \hat{p}_\theta^2}\bigg|_{\hat{p}_\theta = \hat{p}_\theta^0}\left(\frac{\cot\theta}{2}\right)^2\right]\right\} + \ldots \tag{39}$$

Furthermore, we know that if $\hat{A}$ is a differential operator then $(\hat{A}\,\hat{B} + \hat{B}\,\hat{A})\psi = (\hat{A}\,\hat{B}.\psi + \hat{B}\,\hat{A}\psi) + \hat{B}\,\hat{A}\psi = \hat{A}\,\hat{B}.\psi + 2\hat{B}\,\hat{A}\psi$ for any arbitrary function ψ. Hence, we may write the above equation as

$$\hat{G}(\hat{r},\,\hat{\theta},\,\hat{\phi},\,\hat{p}_r,\,\hat{p}_\theta,\,\hat{p}_\phi) = \hat{G}(\hat{r},\,\hat{\theta},\,\hat{\phi},\,\hat{p}_r^0,\,\hat{p}_\theta^0,\,\hat{p}_\phi^0)$$

$$-\frac{i}{2}\left[\left(\frac{2}{r}\frac{\partial \hat{G}}{\partial \hat{p}_r}\bigg|_{\hat{p}_r = \hat{p}_r^0}\;\frac{\partial \hat{G}}{\partial \hat{p}_r}\bigg|_{\hat{p}_r = \hat{p}_r^0}\frac{1}{r}\right) + \left(\cot\theta\frac{\partial \hat{G}}{\partial \hat{p}_\theta}\bigg|_{\hat{p}_\theta = \hat{p}_\theta^0} + \frac{\partial \hat{G}}{\partial \hat{p}_\theta}\bigg|_{\hat{p}_\theta = \hat{p}_\theta^0}\frac{\cot\theta}{2}\right)\right]$$

$$-\frac{1}{2.2!}\left\{\left(\frac{2}{r^2}\frac{\partial^2 \hat{G}}{\partial \hat{p}_r^2}\bigg|_{\hat{p}_r = \hat{p}_r^0} + \frac{\partial^2 \hat{G}}{\partial \hat{p}_r^2}\bigg|_{\hat{p}_r = \hat{p}_r^0}\frac{1}{r^2}\right) + \left[\begin{array}{l} 2\left(\dfrac{\cot\theta}{2}\right)^2\dfrac{\partial^2 \hat{G}}{\partial \hat{p}_\theta^2}\bigg|_{\hat{p}_\theta = \hat{p}_\theta^0} \\ + \dfrac{\partial^2 \hat{G}}{\partial \hat{p}_\theta^2}\bigg|_{\hat{p}_\theta = \hat{p}_\theta^0}\left(\dfrac{\cot\theta}{2}\right)^2 \end{array}\right]\right\} + \ldots \tag{40}$$

Now, substituting

$$\frac{\partial \hat{G}}{\partial \hat{p}_r} = \frac{\partial \hat{G}}{\partial \hat{p}_r^0}\cdot\frac{\partial \hat{p}_r^0}{\partial \hat{p}_r} = \frac{\partial \hat{G}}{\partial \hat{p}_r^0} \text{ and } \frac{\partial \hat{G}}{\partial \hat{p}_\theta} = \frac{\partial \hat{G}}{\partial \hat{p}_\theta^0}\cdot\frac{\partial \hat{p}_\theta^0}{\partial \hat{p}_\theta} = \frac{\partial \hat{G}}{\partial \hat{p}_\theta^0}$$

$$\hat{G}(\hat{r}, \hat{\theta}, \hat{\phi}, \hat{p}_r, \hat{p}_\theta, \hat{p}_\phi) = \hat{G}(\hat{r}, \hat{\theta}, \hat{\phi}, \hat{p}_r^0, \hat{p}_\theta^0, \hat{p}_\phi^0)$$

$$-\frac{i}{2}\left[\left(\frac{2}{r}\frac{\partial \hat{G}}{\partial \hat{p}_r^0} + \frac{\partial \hat{G}}{\partial \hat{p}_r^0}\frac{1}{r}\right) + \left(\cot\theta\frac{\partial \hat{G}}{\partial \hat{p}_\theta^0} + \frac{\partial \hat{G}}{\partial \hat{p}_\theta^0}\frac{\cot\theta}{2}\right)\right]$$

$$-\frac{1}{2.2!}\left[\left(\frac{2}{r^2}\frac{\partial^2 \hat{G}}{\partial (\hat{p}_r^0)^2} + \frac{\partial^2 \hat{G}}{\partial (\hat{p}_r^0)^2}\frac{1}{r^2}\right) + \left[\begin{matrix} 2\left(\frac{\cot\theta}{2}\right)^2 \frac{\partial^2 \hat{G}}{\partial (\hat{p}_\theta^0)^2} \\ + \frac{\partial^2 \hat{G}}{\partial (\hat{p}_\theta^0)^2}\left(\frac{\cot\theta}{2}\right)^2 \end{matrix}\right]\right] + \ldots. \tag{41}$$

Collecting terms that contain the conjugate momenta and those that do not, we get

$$\hat{G}(\hat{r}, \hat{\theta}, \hat{\phi}, \hat{p}_r, \hat{p}_\theta, \hat{p}_\phi) = \hat{G}(\hat{r}, \hat{\theta}, \hat{\phi}, \hat{p}_r^0, \hat{p}_\theta^0, \hat{p}_\phi^0)$$

$$-\left[\frac{i}{2}\left(\frac{2}{r}\frac{\partial \hat{G}}{\partial \hat{p}_r^0} + \cot\theta\frac{\partial \hat{G}}{\partial \hat{p}_\theta^0}\right) + \frac{1}{2.2!}\left(\frac{2}{r^2}\frac{\partial^2 \hat{G}}{\partial (\hat{p}_r^0)^2} + 2\left(\frac{\cot\theta}{2}\right)^2\frac{\partial^2 \hat{G}}{\partial (\hat{p}_\theta^0)^2}\right) + \ldots\right]$$

$$-\left[\frac{i}{2}\left(\frac{\partial \hat{G}}{\partial \hat{p}_r^0}\frac{1}{r} + \frac{\partial \hat{G}}{\partial \hat{p}_\theta^0}\frac{\cot\theta}{2}\right) + \frac{1}{2.2!}\left(\frac{\partial^2 \hat{G}}{\partial (\hat{p}_r^0)^2}\frac{1}{r^2} + \frac{\partial^2 \hat{G}}{\partial (\hat{p}_\theta^0)^2}\left(\frac{\cot\theta}{2}\right)^2\right) + \ldots\right] \tag{42}$$

where the last square bracket contains terms that are non-differential operators and its preceding square brackets consists of differential operator containing terms. Furthermore, if the Hamiltonian contains $\hat{p}_i$ upto degree n, then the coefficient of $\frac{\partial^n \hat{G}}{\partial (\hat{p}_r^0)^n}\cdot\frac{\partial^n \hat{G}}{\partial (\hat{p}_\theta^0)^n}$ will vanish in the first square brackets. In the second square brackets, the normal coefficient will be multiplied by a factor of 2.

Since the momentum operators must take the form of differential operators, the non-differential operators in $F(\hat{r}, \hat{\theta}, \hat{\phi}, \hat{p}_r, \hat{p}_\theta, \hat{p}_\phi) = \hat{G}(\hat{r}, \hat{\theta}, \hat{\phi}, \hat{p}_r, \hat{p}_\theta, \hat{p}_\phi) + Q(r, \theta, \phi)$ must aggregate to zero so that

$$Q(r, \theta, \phi) - \frac{i}{2}\left(\frac{\partial \hat{G}}{\partial \hat{p}_r^0}\frac{1}{r} + \frac{\partial \hat{G}}{\partial \hat{p}_\theta^0}\frac{\cot\theta}{2}\right) - \frac{1}{2.2!}\left[\begin{matrix} \frac{\partial^2 \hat{G}}{\partial (\hat{p}_r^0)^2}\frac{1}{r^2} \\ + \frac{\partial^2 \hat{G}}{\partial (\hat{p}_\theta^0)^2}\left(\frac{\cot\theta}{2}\right)^2 \end{matrix}\right] + \ldots. = 0 \tag{43}$$

or

$$Q(r, \theta, \phi) = \frac{i}{2}\left(\frac{\partial \hat{G}}{\partial \hat{p}_r^0}\frac{1}{r} + \frac{\partial \hat{G}}{\partial \hat{p}_\theta^0}\frac{\cot\theta}{2}\right) + \frac{1}{2.2!}\begin{bmatrix} \dfrac{\partial^2 \hat{G}}{\partial(\hat{p}_r^0)^2}\dfrac{1}{r^2} \\ + \dfrac{\partial^2 \hat{G}}{\partial(\hat{p}_\theta^0)^2}\left(\dfrac{\cot\theta}{2}\right)^2 \end{bmatrix} - \ldots \tag{44}$$

whence

$$F(\hat{r}, \hat{\theta}, \hat{\phi}, \hat{p}_r, \hat{p}_\theta, \hat{p}_\phi) = \hat{G}(\hat{r}, \hat{\theta}, \hat{\phi}, \hat{p}_r^0, \hat{p}_\theta^0, \hat{p}_\phi^0)$$

$$-\frac{i}{2}\left[\left(\frac{2}{r}\frac{\partial \hat{G}}{\partial \hat{p}_r^0} + \frac{\partial \hat{G}}{\partial \hat{p}_r^0}\frac{1}{r}\right) + \left(\cot\theta\frac{\partial \hat{G}}{\partial \hat{p}_\theta^0} + \frac{\partial \hat{G}}{\partial \hat{p}_\theta^0}\frac{\cot\theta}{2}\right)\right]$$

$$-\frac{1}{2.2!}\left\{\left(\frac{2}{r^2}\frac{\partial^2 \hat{G}}{\partial(\hat{p}_r^0)^2} + \frac{\partial^2 \hat{G}}{\partial(\hat{p}_r^0)^2}\frac{1}{r^2}\right) + \begin{bmatrix} 2\left(\dfrac{\cot\theta}{2}\right)^2 \dfrac{\partial^2 \hat{G}}{\partial(\hat{p}_\theta^0)^2} \\ + \dfrac{\partial^2 \hat{G}}{\partial(\hat{p}_\theta^0)^2}\left(\dfrac{\cot\theta}{2}\right)^2 \end{bmatrix}\right\} + \ldots$$

$$+\frac{i}{2}\left(\frac{\partial \hat{G}}{\partial \hat{p}_r^0}\frac{1}{r} + \frac{\partial \hat{G}}{\partial \hat{p}_\theta^0}\frac{\cot\theta}{2}\right) + \frac{1}{2.2!}\begin{bmatrix} \dfrac{\partial^2 \hat{G}}{\partial(\hat{p}_r^0)^2}\dfrac{1}{r^2} \\ + \dfrac{\partial^2 \hat{G}}{\partial(\hat{p}_\theta^0)^2}\left(\dfrac{\cot\theta}{2}\right)^2 \end{bmatrix} - \ldots$$

$$= \hat{G}(\hat{r}, \hat{\theta}, \hat{\phi}, \hat{p}_r^0, \hat{p}_\theta^0, \hat{p}_\phi^0)$$

$$-\left[\frac{i}{2}\left(\frac{2}{r}\frac{\partial \hat{G}}{\partial \hat{p}_r^0} + \cot\theta\frac{\partial \hat{G}}{\partial \hat{p}_\theta^0}\right) + \frac{1}{2.2!}\left(\frac{2}{r^2}\frac{\partial^2 \hat{G}}{\partial(\hat{p}_r^0)^2} + 2\left(\frac{\cot\theta}{2}\right)^2\frac{\partial^2 \hat{G}}{\partial(\hat{p}_\theta^0)^2}\right) + \ldots\right] \tag{45}$$

which represents the quantum-classical correspondence rule in spherical polar coordinates.

12.6 THE HAMILTONIAN IN SPHERICAL POLAR COORDINATES

We know the Hamiltonian of a classical particle in a force free field is given by

$$H = \frac{1}{2m}\left(p_r^2 + \frac{1}{r^2}p_\theta^2 + \frac{1}{r^2\sin^2\theta}p_\phi^2\right) \tag{46}$$

The operator valued quantum mechanical Hamiltonian of a free particle corresponding to this classical Hamiltonian can be directly obtained by using the above correspondence by making the following substitutions:

$$\hat{G}(\hat{r}, \hat{\theta}, \hat{\phi}, \hat{p}_r, \hat{p}_\theta, \hat{p}_\phi) = \frac{1}{2m}\left[(\hat{p}_r)^2 + \frac{1}{r^2}(\hat{p}_\theta)^2 + \frac{1}{r^2 \sin^2\theta}(\hat{p}_\phi)^2\right] \tag{47}$$

$$\hat{G}(\hat{r}, \hat{\theta}, \hat{\phi}, \hat{p}_r^0, \hat{p}_\theta^0, \hat{p}_\phi^0) = \frac{1}{2m}\left[(\hat{p}_r^0)^2 + \frac{1}{r^2}(\hat{p}_\theta^0)^2 + \frac{1}{r^2 \sin^2\theta}(\hat{p}_\phi^0)^2\right] \tag{48}$$

$$\frac{\partial \hat{G}}{\partial \hat{p}_r^0} = \left.\frac{\partial \hat{G}}{\partial \hat{p}_r}\frac{\partial \hat{p}_r}{\partial \hat{p}_r^0}\right|_{\hat{p}_r = \hat{p}_r^0} = \left.\frac{\partial \hat{G}}{\partial \hat{p}_r}\right|_{\hat{p}_r = \hat{p}_r^0} = \frac{1}{m}\hat{p}_r^0 \tag{49}$$

$$\frac{\partial \hat{G}}{\partial \hat{p}_\theta^0} = \left.\frac{\partial \hat{G}}{\partial \hat{p}_\theta}\frac{\partial \hat{p}_\theta}{\partial \hat{p}_\theta^0}\right|_{\hat{p}_\theta = \hat{p}_\theta^0} = \left.\frac{\partial \hat{G}}{\partial \hat{p}_\theta}\right|_{\hat{p}_\theta = \hat{p}_\theta^0} = \frac{1}{mr^2}\hat{p}_\theta^0 \tag{50}$$

$$\frac{\partial \hat{G}}{\partial \hat{p}_\phi^0} = \left.\frac{\partial \hat{G}}{\partial \hat{p}_\phi}\frac{\partial \hat{p}_\phi}{\partial \hat{p}_\phi^0}\right|_{\hat{p}_\phi = \hat{p}_\phi^0} = \left.\frac{\partial \hat{G}}{\partial \hat{p}_\phi}\right|_{\hat{p}_\phi = \hat{p}_\phi^0} = \frac{1}{mr^2\sin^2\theta}\hat{p}_\phi^0 \tag{51}$$

$$\frac{\partial^2 \hat{G}}{\partial(\hat{p}_r^0)^2} = \frac{1}{m}, \quad \frac{\partial^2 \hat{G}}{\partial(\hat{p}_\theta^0)^2} = \frac{1}{mr^2} \text{ and } \frac{\partial^2 \hat{G}}{\partial(\hat{p}_\phi^0)^2} = \frac{1}{mr^2\sin^2\theta} \tag{52}$$

All the higher order derivatives of $\hat{G}(\hat{r}, \hat{\theta}, \hat{\phi}, \hat{p}_r, \hat{p}_\theta, \hat{p}_\phi)$ with respect to the conjugate momenta vanish.

Expanding $\hat{G}(\hat{r}, \hat{\theta}, \hat{\phi}, \hat{p}_r, \hat{p}_\theta, \hat{p}_\phi)$ as a Taylor's series about $(\hat{p}_r^0, \hat{p}_\theta^0, \hat{p}_\phi^0)$ using symmetrized product of operators, we obtain

$$\begin{aligned}\hat{G}(\hat{r}, \hat{\theta}, \hat{\phi}, \hat{p}_r, \hat{p}_\theta, \hat{p}_\phi) &= \frac{1}{2m}\left[(\hat{p}_r^0)^2 + \frac{1}{r^2}(\hat{p}_\theta^0)^2 + \frac{1}{r^2\sin^2\theta}(\hat{p}_\phi^0)^2\right] \\ &- \frac{i}{2m}\left[\left(\hat{p}_r^0\frac{1}{r} + \frac{1}{r}\hat{p}_r^0\right) + \frac{1}{r^2}\left(\hat{p}_\theta^0\frac{\cot\theta}{2} + \frac{\cot\theta}{2}\hat{p}_\theta^0\right)\right] \\ &- \frac{1}{2.2!}\left\{\left(\frac{1}{m}\frac{1}{r^2} + \frac{1}{r^2}\frac{1}{m}\right) + \frac{1}{m}\left[\left(\frac{\cot\theta}{2}\right)^2\frac{1}{r^2} + \frac{1}{r^2}\left(\frac{\cot\theta}{2}\right)^2\right]\right\}\end{aligned} \tag{53}$$

wherein we have used

$$\hat{p}_r = -i\left(\frac{\partial}{\partial r} + \frac{1}{r}\right) = \hat{p}_r^0 - i\frac{1}{r},\ \hat{p}_\theta = -i\left(\frac{\partial}{\partial \theta} + \frac{1}{2}\cot\theta\right)$$

$$= \hat{p}_\theta^0 - i\frac{1}{2}\cot\theta, \hat{p}_\phi = -i\frac{\partial}{\partial\phi} = \hat{p}_\phi^0$$

Noting that $\hat{p}_r^0$, $\hat{p}_\theta^0$, $\hat{p}_\phi^0$ are all differential operators and so $(\hat{A}\hat{B} + \hat{B}\hat{A})\psi = (\hat{A}\hat{B}.\psi + \hat{B}\hat{A}\psi) + \hat{B}\hat{A}\psi = \hat{A}\hat{B}.\psi + 2\hat{B}\hat{A}\psi$, we obtain

$$\hat{G}(\hat{r}, \hat{\theta}, \hat{\phi}, \hat{p}_r, \hat{p}_\theta, \hat{p}_\phi) = \frac{1}{2m}\left[(\hat{p}_r^0)^2 + \frac{1}{r^2}(\hat{p}_\theta^0)^2 + \frac{1}{r^2\sin^2\theta}(\hat{p}_\phi^0)^2\right]$$

$$-\frac{i}{2m}\left[\left(\frac{2}{r}\hat{p}_r^0 - \frac{1}{r^2}\right) + \left(\cot\theta\frac{1}{r^2}\hat{p}_\theta^0 - \frac{1}{2r^2}\operatorname{cosec}^2\theta\right)\right]$$

$$-\frac{1}{2.2!}\left(\frac{2}{mr^2} + \frac{1}{2mr^2}\cot^2\theta\right) \tag{54}$$

Equating the sum of all non-differential operators to zero, we get

$$-\frac{i}{2m}\left[\left(-\frac{1}{r^2}\right) + \left(-\frac{1}{2r^2}\operatorname{cosec}^2\theta\right)\right]$$

$$-\frac{1}{2.2!}\left(\frac{2}{mr^2} + \frac{1}{2mr^2}\cot^2\theta\right) + Q(r, \theta, \phi) = 0 \tag{55}$$

whence

$$Q(r, \theta, \phi) = \frac{i}{2m}\left[\left(-\frac{1}{r^2}\right) + \left(-\frac{1}{2r^2}\operatorname{cosec}^2\theta\right)\right] + \frac{1}{2.2!}\left(\frac{2}{mr^2} + \frac{1}{2mr^2}\cot^2\theta\right) \tag{56}$$

Hence,

$$\hat{H}(\hat{r}, \hat{\theta}, \hat{\phi}, \hat{p}_r, \hat{p}_\theta, \hat{p}_\phi) = \frac{1}{2m}\left[(\hat{p}_r^0)^2 + \frac{1}{r^2}(\hat{p}_\theta^0)^2 + \frac{1}{r^2\sin^2\theta}(\hat{p}_\phi^0)^2\right]$$

$$-\frac{i}{2m}\left(\frac{2}{r}\hat{p}_r^0 + \cot\theta\frac{1}{r^2}\hat{p}_\theta^0\right) \tag{57}$$

Now

$$\hat{p}_r^2 = (\hat{p}_r^0)^2 - \frac{2i}{r}\hat{p}_r^0,\ \hat{p}_\theta^2 = (\hat{p}_\theta^0)^2 - i\cot\theta\,\hat{p}_\theta^0 + \frac{1}{2\sin^2\theta} - \frac{\cos^2\theta}{4\sin^2\theta} \tag{58}$$

whence

$$\hat{H}(\hat{r}, \hat{\theta}, \hat{\phi}, \hat{p}_r, \hat{p}_\theta, \hat{p}_\phi) = \frac{1}{2m}\left[\hat{p}_r^2 + \frac{1}{r^2}\hat{p}_\theta^2 + \frac{1}{r^2 \sin^2\theta}\hat{p}_\phi^2\right] - \frac{1}{8mr^2\sin^2\theta}(1 + \sin^2\theta) \qquad (59)$$

In spherical polar coordinates, we also have directly from eq. (57)

$$\begin{aligned}\hat{H}(\hat{r}, \hat{\theta}, \hat{\phi}, \hat{p}_r, \hat{p}_\theta, \hat{p}_\phi) &= \frac{1}{2m}\left[(\hat{p}_r^0)^2 + \frac{1}{r^2}(\hat{p}_\theta^0)^2 + \frac{1}{r^2\sin^2\theta}(\hat{p}_\phi^0)^2\right] \\ &\quad - \frac{i}{2m}\left(\frac{2}{r}\hat{p}_r^0 + \cot\theta\frac{1}{r^2}\hat{p}_\theta^0\right) \\ &= -\frac{1}{2m}\left[\left(\frac{\partial^2}{\partial r^2} + \frac{2}{r}\frac{\partial}{\partial r}\right) + \frac{1}{r^2}\left(\frac{\partial^2}{\partial\theta^2} + \cot\theta\frac{\partial}{\partial\theta}\right)\right. \\ &\quad \left. + \frac{1}{r^2\sin^2\theta}\left(\frac{\partial^2}{\partial\phi^2}\right)\right] \\ &= -\frac{1}{2m}\left[\frac{1}{r^2}\frac{\partial}{\partial r}\left(r^2\frac{\partial}{\partial r}\right) + \frac{1}{r^2\sin\theta}\frac{\partial}{\partial\theta}\left(\sin\theta\frac{\partial}{\partial\theta}\right) + \frac{1}{r^2\sin^2\theta}\frac{\partial^2}{\partial\phi^2}\right] \\ &= -\frac{1}{2m}\nabla^2 \qquad (60)\end{aligned}$$

as is well known, thereby vindicating the procedure outlined above.

12.7 SOLUTION OF THE DIRAC EQUATION IN THE MOMENTUM REPRESENTATION

The transformation from coordinate space to momentum space takes the form of a unitary transformation operator $\hat{S}(p, q)$ under which the wave function in coordinate space $\psi(q)$ transforms as $\varphi(p) = \frac{1}{\sqrt{2\pi}}\int dq\, \hat{S}(p, q)\, \psi(q)$. The "unitary" nature of the transformation operator is mandated by the requirement that the probability amplitude $\psi^\dagger\psi$ be preserved. Hence, we must have $\int dq\, \hat{S}^\dagger(p', q)\, \hat{S}(p, q) = 2\pi\delta(p' - p)$.

In momentum space, functions of "position" coordinates are replaced by the corresponding operators $q \to i\dfrac{d}{dp}$ and $q^{-1} \to -i\int dp$ while p is retained as a variable. In Cartesian coordinate systems, this one-dimensional picture can easily be extended to three-space dimensions with $\varphi(\mathbf{p}) = \dfrac{1}{(2\pi)^{\frac{3}{2}}}\int d^3x\hat{S}(\mathbf{p},\mathbf{x})\,\psi(\mathbf{x}),\ x \to i\nabla_{\mathbf{p}}$ and $\mathbf{x}^{-1} \to -i\int d^3\,\mathbf{p}$.

12.8 CHOICE OF VARIABLES IN "MOMENTUM SPACE" FOR THE DIRAC EQUATION

Radial Variable

Despite the reservations expressed above about the suitability of $\hat{p}_r = -i\left(\dfrac{\partial}{\partial r} + \dfrac{1}{r}\right)$ as the momentum conjugate to the radial vector, we retain it as our conjugate radial momentum because functions of $\hat{p}_r$ (such as the square of the total momentum) are observables.

Angle Variables

As stated above, we define the unitary operator $v = e^{i\phi}$ as a proxy for the azimuthal angle ϕ and define the conjugate momentum as $\hat{p}_\phi = v\dfrac{\partial}{\partial v} = e^{i\phi}\dfrac{\partial\phi}{\partial v}.\dfrac{\partial}{\partial\phi} = -i\dfrac{\partial}{\partial\phi}$. The commutation relation between v and $\hat{p}_\phi$ is $[\hat{p}_\phi, v] = v\dfrac{\partial}{\partial v}(v.) - v.v\dfrac{\partial}{\partial v} = v$. Now $\hat{p}_\phi$ is Hermitian over the whole range of $v(|v| \leq 1)$.

In lieu of θ, we shall adopt $t = e^{i\pi\cos\theta} = e^{i\pi u}$ as the second angle variable, t is obviously unitary and $\hat{p}_\phi = \pi t\dfrac{\partial}{\partial t} = \pi e^{i\pi\cos\theta}\dfrac{\partial\theta}{\partial t}\dfrac{\partial}{\partial\theta} = \dfrac{i}{\sin\theta}\dfrac{\partial}{\partial\theta} = -i\dfrac{\partial}{\partial(\cos\theta)} = -i\dfrac{\partial}{\partial u}$. The commutation relation between $\hat{p}_\theta$ and t becomes $[\hat{p}_\theta, t] = \pi t$ and the corresponding uncertainty relation is $(\Delta\hat{p}_\theta)^2(\Delta t)^2 \geq \dfrac{1}{4}\pi^2[1-(\Delta t)^2]$ in the Levi-Leblond framework.

Thus, $\hat{r}, \hat{t}, \hat{v}$ constitute our "*position*" operators and $\hat{p}_r, \hat{p}_\theta, \hat{p}_\phi$ are their associated momentum operators.

The Hamiltonian for the hydrogen atom $\hat{H} = \dfrac{\hat{\mathbf{p}}^2}{2m} - \dfrac{Ze^2}{r} = -\dfrac{1}{2m}\nabla^2 - \dfrac{Ze^2}{r}$ may be written in terms of the standard spherical polar coordinates as

$$\hat{H} = -\frac{1}{2m}\left[\frac{1}{r^2}\frac{\partial}{\partial r}\left(r^2\frac{\partial}{\partial r}\right) + \frac{1}{r^2\sin\theta}\frac{\partial}{\partial\theta}\left(\sin\theta\frac{\partial}{\partial\theta}\right) + \frac{1}{r^2\sin^2\theta}\frac{\partial^2}{\partial\phi^2}\right] - \frac{Ze^2}{r}$$

$$= -\frac{1}{2m}\left[\frac{\partial^2}{\partial r^2} + \frac{2}{r}\frac{\partial}{\partial r} + \frac{1}{r^2}\frac{\partial^2}{\partial\theta^2} + \frac{\cot\theta}{r^2}\frac{\partial}{\partial\theta} + \frac{1}{r^2\sin^2\theta}\frac{\partial^2}{\partial\phi^2}\right] - \frac{Ze^2}{r} \tag{61}$$

We transform this in the coordinates r, $u = \cos\theta$, $v = e^{i\phi}$ as

$$\frac{\partial}{\partial\theta} = \frac{\partial u}{\partial\theta}\frac{\partial}{\partial u} = -\sin\theta\frac{\partial}{\partial u} = -(1-u^2)^{1/2}\frac{\partial}{\partial u} \tag{62}$$

$$\frac{\partial^2}{\partial\theta^2} = \frac{\partial u}{\partial\theta}\frac{\partial}{\partial u}\left[-(1-u^2)^{1/2}\frac{\partial}{\partial u}\right] = -(1-u^2)^{1/2}$$

$$\left[-(1-u^2)^{1/2}\frac{\partial^2}{\partial u^2} + \frac{u}{(1-u^2)^{1/2}}\frac{\partial}{\partial u}\right] = (1-u^2)\frac{\partial^2}{\partial u^2} - u\frac{\partial}{\partial u} \tag{63}$$

$$\frac{\partial}{\partial\phi} = \frac{\partial v}{\partial\phi}\frac{\partial}{\partial v} = ie^{i\phi}\frac{\partial}{\partial v} = iv\frac{\partial}{\partial v} \tag{64}$$

$$\frac{\partial^2}{\partial\phi^2} = \frac{\partial v}{\partial\phi}\frac{\partial}{\partial v}\left[iv\frac{\partial}{\partial v}\right] = iv\left[iv\frac{\partial^2}{\partial v^2} + i\frac{\partial}{\partial v}\right] = -v^2\frac{\partial^2}{\partial v^2} - v\frac{\partial}{\partial v} \tag{65}$$

Substituting these values, we get

$$\hat{H} = -\frac{1}{2m}\left(\frac{\partial^2}{\partial r^2} + \frac{2}{r}\frac{\partial}{\partial r} + \frac{\wp^2}{r^2}\right) - \frac{Ze^2}{r} \tag{66}$$

where

$$\wp^2 = (1-u^2)\frac{\partial^2}{\partial u^2} - 2u\frac{\partial}{\partial u} - \frac{1}{1-u^2}\left(v^2\frac{\partial^2}{\partial v^2} + \frac{\partial}{\partial v}\right) \tag{67}$$

is the square of the total angular momentum.

Let us assume that the solution of the eigenvalue equation

$$\hat{H}\psi(r, u, v) = E\psi(r, u, v) \tag{68}$$

can be expressed through a separation of variables as

$$\psi(r, u, v) = R(r)\,P(u),\,Z(v) \tag{69}$$

Substituting eq. (69) in (68) and using eq. (66), we get three one-dimensional equations

$$v^2\frac{\partial^2 Z(v)}{\partial v^2} + v\frac{\partial Z(v)}{\partial v} = m^2 Z(v) \tag{70}$$

$$\wp^2 P(u) = -\, l(l+1)\, P(u) \tag{71}$$

$$\left(\frac{\partial^2}{\partial r^2} + \frac{2}{r}\frac{\partial}{\partial r} - \frac{l(l+1)}{r^2} + \frac{2mZe^2}{r} + 2mE\right) R(r) = 0 \tag{72}$$

where m^2 and $l(l + 1)$ are separation constants.

We now assume that the momentum wave function $\varphi(p_r, p_\theta, p_\phi)$ corresponding to the coordinate space wave function $\psi(r, u, v)$ is also separable in momentum space, i.e.,

$$\varphi(p_r, p_\theta, p_\phi) = \alpha(p_r)\, \beta(p_\theta)\, \rho(p_\phi) \tag{73}$$

As mentioned earlier, the momentum function can be obtained from the position wave function by the unitary transformation operator $\hat{S}\,(p, \mathbf{x})$ as

$$\varphi(\mathbf{p}) = \frac{1}{(2\pi)^{\frac{3}{2}}} \int d^3x \hat{S}\,(\mathbf{p}, \mathbf{x})\, \psi(\mathbf{x})$$

which in spherical polar coordinates becomes

$$\varphi(p_r, p_\theta, p_\phi) = \frac{1}{(2\pi)^{\frac{3}{2}}} \int d\tau\, \hat{S}\,(p_r, p_\theta, p_\phi, r, \theta, \phi)\, \psi(r, \theta, \phi) \tag{74}$$

Equation (74) may be transformed to the set of variables (r, u, v) by introducing the volume element

$$d\tau = r^2 \sin\theta dr d\theta\, d\phi = -r^2(1-u^2)^{1/2}$$

$$dr(1-u^2)^{-1/2}\, du(iv)^{-1}\, dv = ir^2\, drdud(\ln v) \tag{75}$$

where $\ln v$ is chosen to be the principal value of the natural logarithm. This gives

$$\varphi(p_r, p_\theta, p_\phi) = (2\pi)^{-3/2} \int_{\substack{r=0\\ u=-1\\ v=1}}^{\substack{r=\infty\\ u=+1\\ v=e^{i2\pi}}} \hat{S}\,(p_r, p_\theta, p_\phi, r, u, v)\, \psi(r, u, v) ir^2\, drdud(\ln v) \tag{76}$$

We choose the transformation $\hat{S}\,(p_r, p_\theta, p_\phi, r, u, v)$ as

$$\hat{S}\,(p_r, p_\theta, p_\phi, r, u, v) = \sqrt{\frac{2}{\pi}} \cdot \frac{1}{r} \exp\,(irp_r + ip_\theta u + p_\phi \ln_v) \tag{77}$$

which is, obviously, unitary.

The transformation of eq. (77) may be written as a product of three separate one-dimensional transforms

$$\hat{S}\,(p_r, r) = \frac{\sqrt{2}}{r} \exp\,(irp_r) \tag{78}$$

$$\hat{S}(p_\theta, u) = \sqrt{\frac{1}{\pi}} \exp(ip_\theta u) \tag{79}$$

$$\hat{S}(p_\phi, v) = \exp(p_\phi \ln v) \tag{80}$$

Our next objective is to transform Eqs. (70), (71) and (72) to momentum space. We first consider the angular equations, viz. eqs. (70) and (71). To achieve the transformation to momentum space, we make the following transformations:

$$Z(v) \to \rho(p_\phi) \tag{81}$$

$$-iv\frac{d}{dv} \to p_\phi \tag{82}$$

On substitution in eq. (70), we get

$$p_\phi^2 \, \rho(p_\phi) = m^2 \, \rho(p_\phi) \tag{83}$$

which has the solution

$$\rho_m(p_\phi) = \delta(p_\phi + m) \tag{84}$$

where δ is the Dirac delta function.

The matrix elements of the commutator $[p_\phi, v] = v$ are given by

$$\langle m|[p_\phi, v]|m'\rangle = \langle m|v|m'\rangle \tag{85}$$

whence

$$\langle m|[p_\phi, v] - v|m'\rangle = \langle m|(p_\phi - 1)v - vp_\phi|m'\rangle = 0 \tag{86}$$

As p_ϕ is Hermitian, we get $m' = m - 1$ which shows that successive eigenvalues of p_ϕ differ by unity. Further, the minimum value of m^2 is zero so that the spectrum of p_ϕ consists of all positive and negative integers, including zero.

Equation (71), read with eq. (67) and (70), may be written as

$$\left[(1-u^2)\frac{d^2}{du^2} - 2u\frac{d}{du} - \frac{m^2}{1-u^2}\right] P(u) = -l(l+1)\, P(u) \tag{87}$$

Substituting

$$P(u) = (1 - u^2)^{-m/2}\, Q(u) \tag{88}$$

we get

$$(1-u^2)\frac{d^2 Q}{du^2} + 2(m-1)u\frac{dQ}{du} - m(m-1)Q = -\, l(l+1)Q \tag{89}$$

Transformation to the momentum representation is achieved by the substitutions

$$Q(u) \to \omega(p_\theta),\ u \to i\frac{d}{dp_\theta} \text{ and } \frac{d}{du} \to ip_\theta \tag{90}$$

Substituting these expressions in eq. (89), we get

$$\left[p_\theta^2\frac{d^2}{dp_\theta^2} + 2(m+1)\,p_\theta\frac{d}{dp_\theta} + p_\theta^2 + m\,(m+1) - l(l+1)\right] \omega(p_\theta) = 0 \tag{91}$$

Equation (91) has the solution

$$\omega_l^m (p_\theta) = N_\theta p_\theta^{-(m+1/2)} J_{l+1/2}(p_\theta) \tag{92}$$

where N_θ is a normalization constant given by

$$N_\theta^2 = \frac{2^{2m}(m!)^2(l+m)!}{(2m)!(l-m-1)!} \tag{93}$$

and $J_{l+1/2}(p_\theta)$ are the Bessel functions. In order to prevent $\omega(p_\theta)$ from diverging at $p_\theta = 0$, we need to impose a further condition $|m| \leq l$. We now obtain expression for $P(u)$ using $P(u) = (1-u^2)^{-m/2} Q(u)$. Since, $|u| = |\cos\theta| \leq 1$, we can expand $(1-u^2)^{-m/2}$ as a power series

$$(1-u^2)^{-m/2} = 1 + \frac{1}{2}mu^2 + \frac{1}{2!}.\frac{1}{2}m\left(\frac{1}{2}m+1\right)u^4 + \ldots \tag{94}$$

which, on transformation to momentum space using eq. (90), becomes

$$(1-u^2)^{-m/2} = 1 - \frac{1}{2}mD^2 + \frac{1}{2!}.\frac{1}{2}m\left(\frac{1}{2}m+1\right)D^4 + \ldots \tag{95}$$

where $D \equiv \dfrac{d}{dp_\theta}$. The equivalent of eq. (88) in momentum space may then be written as

$$\beta_l^m (p_\theta) = (1 + D^2)^{-m/2}\, \omega_l^m(p_\theta) \tag{96}$$

with $\omega_l^m(p_\theta)$ being given by eq. (92). From the properties of Bessel functions and the fact that $|m| \leq l$, we can write $\beta_l^m(p_\theta)$ as a finite polynomial in ω_j^k e.g.,

$$\beta_l^m = \sum_{j=E(m/2)}^{E(l/2)} a_j\, \omega_{j-m/2}^{m/2} \tag{97}$$

which on substituting from eq. (92), becomes

$$\beta_l^m (p_\theta) = \sum_{j=E(m/2)}^{E(l/2)} a_j\, p_\theta^{-(m+1)^2} J_{j-(m-1)/2}(p_\theta) \tag{98}$$

where the normalization constant is absorbed in the coefficients.

As shown in Chapter 11, the four components of the Dirac wave function may be written as

$$u_1 = \hat{F}(r)\left[\frac{l+m+\frac{1}{2}}{2l+1}\right]^{1/2} Y_{l,\, m-1/2(\theta,\, \phi)} \tag{99}$$

$$u_2 = -\hat{F}(r)\left[\frac{l-m+\frac{1}{2}}{2l+1}\right]^{1/2} Y_{l,\, m+1/2(\theta,\, \phi)} \tag{100}$$

$$u_3 = -i\hat{f}(r)\left[\frac{l-m+3/2}{2l+3}\right]^{1/2} Y_{l+1,\,m-1/2\,(\theta,\,\phi)} \tag{101}$$

$$u_4 = -i\hat{f}(r)\left[\frac{l+m+3/2}{2l+3}\right]^{1/2} Y_{l+1,\,m+1/2\,(\theta,\,\phi)} \tag{102}$$

We have already obtained the solutions of the angular part of the wave function in momentum space in the foregoing. The radial part of the wave function satisfies the following equations in coordinate space

$$\frac{d\hat{F}}{dr} + (1+\kappa)\frac{\hat{F}}{r} = \left[m + E + \frac{Z\alpha}{r}\right]\hat{f} \tag{103}$$

and

$$\frac{d\hat{f}}{dr} + (1-\kappa)\frac{\hat{f}}{r} = \left[m - E - \frac{Z\alpha}{r}\right]\hat{F} \tag{104}$$

where $\kappa = -(l+1)$ for $j = l + \frac{1}{2}$ and $\kappa = l$ for $j = l - \frac{1}{2}$. In order to transform eqs. (103) and (104) to momentum space, we make the transformations

$$\hat{f}(r) \to f(p_r),\ \hat{F}(r) \to F(p_r),\ i\left(\frac{d}{dr} + \frac{1}{r}\right) \to p_r,\ \frac{1}{r} \to -i\int dp_r \equiv I \tag{105}$$

to get

$$p_r F + i\kappa \underline{I} F = [i(m+E) + Z\alpha i \underline{I}] f \tag{106}$$

and

$$p_r f - i\kappa \underline{I} f = [i(m-E) - Z\alpha i \underline{I}] F \tag{107}$$

Let us take the $(\gamma - 1)^{\text{th}}$ derivative of eqs. (106) and (107) where γ is not necessarily an integer, i.e., we take the $(\gamma - 1)^{\text{th}}$ fractional derivative of eqs. (106) and (107). The product rule for fractional derivatives is

$$D^{\alpha} u(z)\, v(z) = \sum_{n=0}^{\infty} \frac{\Gamma(\alpha+1) D^{\alpha-n} u(z) D^{n} v(z)}{\Gamma(\alpha+1-n)\Gamma(n+1)} \tag{108}$$

For example, taking $u(p_r) \equiv F(p_r)$ and $v(p_r) \equiv p_r$, we get

$$D^{\gamma-1} p_r F(p_r) = D^{\gamma-1} F(p_r)\, p_r = \sum_{n=0}^{\infty} \frac{\Gamma(\gamma) D^{\gamma-1-n} F(p_r) D^{n} p_r}{\Gamma(\gamma-n)\Gamma(n+1)} \tag{109}$$

$$= p_r D^{\gamma-1} F(p_r) + (\gamma-1) D^{\gamma-2} F(p_r) \tag{110}$$

Let us define $G = F^{(\gamma-2)}$ and $g = f^{(\gamma-2)}$. Substituting these in eqs. (106) and (107), we get

$$p_r G' + (\gamma - 1 + \kappa)\, G = i(m + E)g' + Z\alpha g \tag{111}$$

and

$$p_r g' + (\gamma - 1 + \kappa)\, g = i(m - E)G' - Z\alpha G \tag{112}$$

To simplify calculations, we make further substations

$$g = \left(1 - \frac{E}{m}\right)^{1/2} (\xi_1 + \xi_2) \tag{113}$$

and

$$G = \left(1 + \frac{E}{m}\right)^{1/2} (\xi_1 - \xi_2) \tag{114}$$

Substituting these values from eqs. (113) and (114) in eqs. (111) and (112), alternately adding and subtracting the resulting equations, we get

$$[p_r - im(1 - E^2 m^{-2})^{1/2}]\, \xi_1' + \left[\gamma - 1 + \frac{Z\alpha E m^{-1}}{(1 - E^2 m^{-2})^{1/2}}\right] \xi_1$$

$$= \left[\kappa + \frac{Z\alpha}{(1 - E^2 m^{-2})^{1/2}}\right] \xi_2 \tag{115}$$

and

$$[p_r + im(1 - E^2 m^{-2})^{1/2}]\, \xi_2' + \left[\gamma - 1 - \frac{Z\alpha E m^{-1}}{(1 - E^2 m^{-2})^{1/2}}\right] \xi_2$$

$$= \left[\kappa - \frac{Z\alpha}{(1 - E^2 m^{-2})^{1/2}}\right] \xi_1 \tag{116}$$

We now make the following substitutions in eqs. (115) and (116):

$$\lambda = m\left(1 - \frac{E^2}{m^2}\right)^{1/2} \quad \text{and} \quad N = \frac{Z\alpha m}{\lambda} = Z\alpha\left(1 - \frac{E^2}{m^2}\right)^{-1/2} \tag{117}$$

to get

$$(p_r - i\lambda)\xi_1' + (\gamma - 1 + NEm^{-1})\xi_1 = (\kappa + N)\xi_2 \tag{118}$$

$$(p_r + i\lambda)\xi_2' + (\gamma - 1 - NEm^{-1})\,\xi_2 = (\kappa - N)\,\xi_1 \tag{119}$$

Equations (118) and (119) constitute a pair of coupled first order differential equations with the solution

$$\xi_1 = C\frac{(p_r + i\lambda)^{n_1}}{(p_r - i\lambda)^{n_2}} \tag{120}$$

and

$$\xi_2 = \frac{1}{C}\frac{(p_r - i\lambda)}{(p_r + i\lambda)}\xi_1 \tag{121}$$

where C is a constant. Substitution from eqs. (120) and (121) in eqs. (118) and (119), gives

$$\left(n_1 - n_2 + \gamma - 1 + NEm^{-1} - \frac{\kappa + N}{C}\right)p_r$$

$$+\left(-n_1 - n_2 + \gamma - 1 + NEm^{-1} + \frac{\kappa + N}{C}\right)i\lambda = 0 \tag{122}$$

and

$$[n_1 - n_2 + \gamma - 1 - NEm^{-1} - (\kappa - N)C]p_r$$

$$+ [-n_1 + n_2 + 2) - (\gamma - 1 - NEm^{-1}) - (\kappa - N)C]i\lambda = 0 \tag{123}$$

Solving eqs. (122) and (123), we get

$$n_1 = \frac{\kappa + N}{C} \quad \text{and} \quad N = \frac{m}{E}(n_1 + \gamma) \tag{124}$$

Substituting from eq. (124) in eqs. (118) and (119), we get

$$(p_r - i\lambda)\xi'_1 + (2\gamma - 1 + n_1)\xi_1 = (\kappa + N)\xi_2 \tag{125}$$

and

$$(p_r + i\lambda)\xi'_2 + (1 + n_1)\xi_2 = (\kappa - N)\xi_1 \tag{126}$$

Using eqs. (120) and (121) in eqs. (125) and (126) and eliminating C by using eq. (124), we get

$$[(n_1 - n_2)p_r - (n_1 + n_2)i\lambda + (2\gamma - 1 + n_1)(p + i\lambda)]\xi_1$$

$$= (\kappa + N)\,(p_r + i\lambda)\xi_2 \tag{127}$$

$$[(n_1 - n_2)p_r - (n_1 + n_2 + 2)i\lambda - (n_1 + 1)(p_r - i\lambda)]\xi_2$$

$$= (\kappa - N)(p_r - i\lambda)\xi_1 \tag{128}$$

Equations (127) and (128) have the solution

$$n_1 = n_2 - 2\gamma + 1 \tag{129}$$

and

$$N^2 - \kappa^2 = (n_2 + 1)(n_2 - 2\gamma + 1) \tag{130}$$

Writing

$$n' = n_1 = n_2 - 2\gamma + 1 \tag{131}$$

we obtain from eq. (124)

$$\frac{E}{m} = \frac{(n' + \gamma)}{N} \tag{132}$$

Furthermore, we also have from eqs. (117), (130), (131) and (132)

$$\begin{aligned} Z^2 \alpha^2 &= N^2 \left(1 - \frac{E^2}{m^2}\right) = N^2 - N^2 \frac{E^2}{m^2} \\ &= [\kappa^2 + n_1(n_2 + 1)] - (n_1 + \gamma)^2 \\ &= [\kappa^2 + n_1(n_1 + 2\gamma)] - (n_1 + \gamma)^2 = \kappa^2 - \gamma^2 \end{aligned} \tag{133}$$

The solutions of eqs. (125) and (126) can now be written as

$$\xi_1 = \frac{\kappa + N}{n'} \frac{(p + i\lambda)^{n'}}{(p - i\lambda)^{n' + 2\gamma - 1}} \tag{134}$$

with ξ_2 given by eq. (121).

Having obtained expressions for ξ_1, ξ_2, we can obtain expressions for g, G using eqs. (113) and (114) and thereafter of f, F by fractional integration $(\gamma - 2)$ times. We finally get

$$f = (1 - Em^{-1})^{1/2} \frac{(-1)^{3\gamma}}{(i\lambda)^{2\gamma+2}} \sum_{k=0}^{n'+1} \left[\frac{\kappa + N}{n'} \frac{n' - k + 1}{n' + 1} + 1\right] \frac{a_k (i\lambda)^k}{(p - i\lambda)^{\gamma + k + 1}} \tag{135}$$

and

$$F = (1 + Em^{-1})^{1/2} \frac{(-1)^{3\gamma}}{(i\lambda)^{2\gamma+2}} \sum_{k=0}^{n'+1} \left[\frac{\kappa + N}{n'} \frac{n' - k + 1}{n' + 1} - 1\right] \frac{a_k (i\lambda)^k}{(p - i\lambda)^{\gamma + k + 1}} \tag{136}$$

where

$$a_k = \frac{2^k \,\Gamma(n' + 2)\Gamma(\gamma + k + 1)}{\Gamma(k + 1)\Gamma(n' - k + 2)\Gamma(2\gamma + k + 1)} \tag{137}$$

The quantum condition n' = an integer then emerges from the requirement that $\mathrm{Lim}_{p \to \infty} f$, $F = 0$ as $p_r \to \infty$.

Bibliography

Abramowicz, M. and Stegun, I.A., 1965, Handbook of Mathematical Functions, Dover.

Aharonov, Y. and Bohm, D., 1959. *Phys. Rev.*, 115, 485.

Aharonov, Yakir, Anandan, J. and Vaidman, L., 1993, *Phys. Rev. A*, 47, 4616.

Akhiezer, A. and Berestetskii, V.B., 1963, Quantum Electrodynamics, Wiley.

Albert, David Z., 1992, Quantum Mechanics and Experience, Harvard University Press.

Albert, David Z., 1994, "Bohm's Alternative to Quantum Mechanics", *Scientific American*, May, 58-67.

Alvarez-Estrada, R.F., Fernandez, F., Sanchez-Gomez, J.L. and Vento, V., 1986, Models of Hadron Structure Based on Quantum Chromodynamics, Springer.

Anandan, J., 1993, "Protective Measurement and Quantum Reality", *Found. Phys. Lett.*, 6, 503.

Anandan, J., 1995, "Reality and Geometry of States and Observables in Quantum Theory", Proceedings of the International Conference on Non-Accelerator Particle Physics, Bangalore, Jan. 2-9, 1995, Ed. R. Cowsik, World Scientific.

Aspect, Alain, Phillipe, Grangier & Gerard Roger, 1982, "Experimental Realization of the Einstein-Podolsky-Rosen Gedanken Experiment: A New Violation of Bell's Inequalities", *Phys. Rev. Lett.*, 49, 91; "Experimental Tests of Bell's Inequalities Using Time-Varying Analyzers", 1804.

Atkatz, David, 1994, "Quantum Cosmology for Pedestrians", *Am. J. Phys.*, 62(7), 619.

Ballentine, L.E., 1970, "The Statistical Interpretation of Quantum Mechanics", *Rev. Mod. Phys.*, 42(4), pp. 358-381.

Ballentine, L.E., 1988, "What Do We Learn about Quantum Mechanics from the Theory of Measurement?", *Int. J. Theor. Phys.*, 27(2), 221-218.

Ballentine, Leslie E., Quantum Mechanics, Englewood Cliffs, Prentice Hall.

Bargmann, V. and Wigner, E.P., 1948, *Proc. Nat. Acad. Sci.* USA, 34, 211.

Bell, J.S., 1964, Physics 1, 195.

Bell, John S., 1966, *Rev. Mod. Phys.* 38, pp. 447-452.

Bell, J.S., 1971, "Introduction to the hidden-variables question", Foundations of Quantum Mechanics, Ed. B. d'Espagnet, Academic.

Bell, J.S., 1982, "On the impossible pilot wave", *Found. Phys.* 12, 989-999.

Bell, J.S., 1987, Speakable and Unspeakable in Quantum Mechanics, Cambridge University Press.

Berestetskii, V.B., Lifshitz, E.M. and Pitaevskii, L.P., 1979, Relativistic Quantum Theory, Pergamon.

Berg, R.A., 1965, J. Math. Phys., 6, 34.

Bethe, H.A. and Salpeter, E.E., 1957, Quantum Mechanics of One and Two Electron Atoms, Springer.

Bjorken, J.D. and Drell, S.D., 1964, Relativistic Quantum Mechanics, McGraw-Hill.

Bjorken, J.D. and Drell, S.D., 1965, Relativistic Quantum Fields, McGraw-Hill.

Blatt, J.M. and Weisskopf, V.F., 1952, Theoretical Nuclear Physics, Chapman,

Bleuler, K., 1950, Helv. Phys. Acta, 23, 567.

Bloch, F. and Nordsiek, A., 1937, Phys. Rev., 52, 54.

Bogoliubov, N.N. and Shirkov, D.V., 1959, Introduction to the Theory of Quantized Field, Interscience.

Bogoliubov, N.N., Logunov, A.A. and Todorov, I.T., 1975, Axiomatic Quantum Field Theory, Benjamin.

Bohm, David, 1951, Quantum Theory, Englewood Cliffs, Prentice Hall.

Bohm, David, 1952, "A Suggested Interpretation of Quantum Theory in Terms of 'Hidden Variables' I and II", Phys. Rev. 85, 166.

Bohm, D. and Hiley, B.J., 1993, The Undivided Universe: An Ontological Interpretation of Quantum Mechanics, Routledge.

Bohr, N. and Rosenfeld, L., 1950, Phys. Rev., 78, 794.

Born, Max, 1949, Natural Philosophy of Cause and Chance, Clarendon Press.

Born, M., Ed. 1971, The Born-Einstein Letters, Macmillan.

Bouchiat, M.A. and Bouchiat, C, 1974, Phys. Rev. B, 28, 111.

Brown, G.E. and Jackson, A.D., 1976, The Nucleon-Nucleon Interaction, North Holland.

Capra, Fritjof, 1975, The Tao of Physics, Shambhala.

Carter, B., 1983, Phil. Trans. Roy. Soc. Lond. A, 310, 347.

Carruthers, P. and Nieto, M.M., 1968, Rev. Mod. Phys., 40, 411.

Chakrabarti, A., 1988, Fortschritte der Physik, 36, 863.

Chambers, R.G., 1960, Phys. Rev. Lett. 5, 3.

Chiao, Raymond Y., Kwiat, Paul, G. and Steinberg, A.M., 1993, "Faster than Light?" Scientific American, August, 52.

Clauser, J.F. and Horne, M.A., 1974, Phys. Rev. D, 10, 526.

Clauser, John F. and Shimony, Abner, 1978, Rep. Prog. Phys., 41, 1881-1927.

Coleman, S., Hartle, J.B., Piran, T. and Weinberg, S., Eds., 1991, Quantum Cosmology and Baby Universes, World Scientific.

Condon, E.U. and Shortley, G.H., 1967, The Theory of Atomic Spectra, Cambridge.

Cramer, John G., 1986, "The Transactional Interpretation of Quantum Mechanics" Rev. Mod. Phys., 58, 647.

Cramer, John G., 1988, Int. J. Theo. Phys., 27, 227.

Cushing, J. and McMullin, E., Ed., 1989, Philosophical Consequences of the Quantum Theory: Reflections on Bell's Theorem, University of Notre Dame Press.

Das, T.P., 1973, Relativistic Quantum Mechanics of Electrons, Harper and Row.

Datta, A., Home, D. and Raychaudhuri, A., 1987, Phys. Lett. A, 123, 4.

Davies, P.C.W., 1974, The Physics of Time Asymmetry, University of California Press.

D'Espagnat, Bernard, 1989, Reality and the Physicist: Knowledge, Duration and the Quantum World, Cambridge University Press.

Deutsch, D., 1985, "Quantum Theory, the Church-Turing Principle and the Universal Quantum Computer", Proc. Royal Soc. Lond. A, 400, 97.

DeWitt, Bryce S. and Neill, Graham, Eds., The Many-Worlds Interpretation of Quantum Mechanics, Princeton University Press.

DeWitt, B.S., 1952, *Phys. Rev.*, 85(4), 653.

Dicke, R.H., 1961, "Dirac's Cosmology and Mach's Principle", *Nature* 192, 440.

Dirac, P.A. M., 1951, *Nature*, 168, 906.

Dobyns, York H., 1992, *J. Scientific Exploration*, 6, 23.

Dobyns, York H., 1993, *J. Scientific Exploration*, 7, 259.

Domingos, J.M. and Caldeira, M.H., 1984, *Found. Phys.*, 14(2), 147.

Dowker, Fay and Kent, Adrian, 1994, "On the Consistent Histories Approach to Quantum Mechanics", gr-qc/9412067.

Dunbar, D.N.F., Pixley, R.E., Wenzel, W.A. and Whaling W., 1953, *Phys. Rev.*, 92, 649.

Dunne, Brenda J. and Jahn, Robert J., 1992, *J. Scientific Exploration* 6, 311.

Eberhard, P., 1978, Nuovo Cimento B, 46, 392.

Eberhard, Phillippe H. and Ross, Ronald R., 1989, *Found. Phys. Lett.*, 2, 127.

Einstein, A., Podolsky, B. and Rosen, N., 1935, "Can the Quantum Mechanical Description of Physical Reality be Considered Complete?", *Phys. Rev.*, 47, 777.

Elton, L.R. B., 1959, Introductory Nuclear Theory, Pitman.

Everett III, Hugh, 1957, *Rev. Mod. Phys.*, 29, 454.

Fermi, E., 1932, *Rev. Mod. Phys.*, 4, 87.

Feynman, R.P., 1948, "Spacetime Approach to Non-relativistic Quantum Mechanics", *Rev. Mod. Phys.*, 20, 367.

Feynman, R.P., 1949, "The Theory of Positrons", *Phys. Rev.*, 76, 749.

Feynman, R.P., 1949, "Spacetime Approach to Quantum Electrodynamics", *Phys. Rev.*, 76, 769.

Feynman, R.P. and Hibbs, A.R., 1965, Quantum Mechanics and Path Integrals, McGraw-Hill.

Feynman, Richard P., 1965, The Character of Physical Law, Cox and Wyman.

Feynman, Richard P., 1965, "The Development of the Space-time View of Quantum Electrodynamics" Nobel Lectures Physics 1963-1970, Elsevier, 1992.

Feynman, Richard P., 1985, Surely You're Joking, Mr. Feynman, W.W. Norton.

Feynman, R.P., 1986, "Quantum Mechanical Computers", *Found. Phys.*, 16, 507.

Foldy, L.L., 1956, Phys. Rev., 102, 568.

Franck, Philipp, 1947, Einstein – His Life and Times, Knopf.

French, A.P. and Kennedy, P.J., Eds., 1985, Niels Bohr: a Centenary Volume, Harvard University Press.

Fushchich, W.I. & Nikitin, A.G., 1994, Symmetries of the Equations of Quantum Mechanics, Alberton Press, New York and references therein.

Galindo, A. and Pascual, P., 1978, Quantum Mechanics, Vols. I, II, Springer.

Gel'fand, I.M. & Minlos, R.A., 1963, Representations of the rotation and lorentz groups and their applications, Pergamon Press.

Gell-Mann, Murray and Hartle, James P., 1991, "Time Symmetry and Asymmetry in Quantum Mechanics and Quantum Cosmology" in the Proceedings of the 1st International A.D. Sakarov Conference on Physics, Moscow, May 27-31, 1991 and in the Proceedings of the Nato Workshop on the Physical Origin of Time Asymmetry, Mazagon, Spain, September 30-October 4, 1991, Ed. by Haliwell, J., Perez-Mercader, J. and Zurek, W., Cambridge University Press, 1992.

Gell-Mann, Murray, 1994, The Quark and the Jaguar: Adventures in the Simple and the Complex, W.H. Freeman.

Ghirardi, G.C., Rimini, A. and Weber, T., 1986, *Phys. Rev. D*, 34, 470.

Gisin, N., 1990, *Phys. Lett.* A, 143, 1.

Goldberger, M.L. and Watson, K., 1965, Collision Theory, Wiley.

Goldstein, H., 1965, Classical Mechanics, Addison-Wesley.

Gottfried, K., 1966, Quantum Mechanics, Benjamin.

Gour, G., Khanna, F.C. and Revzen, M., 2004, *Phys. Rev. A*, 69, 014101.

Greenberger, D.M., Horne, M.A., Shimony, A. and Zeilinger, A., 1990, *Am. J. Phys.*, 58, 1131.

Greiner, W. et al., Relativistic Quantum Mechanics and other Volumes in the Series, Springer.

Griffiths, R.J., 1984, *J. Stat. Phys.*, 36, 219.

Griffiths, Robert J., 1986, "Correlations in Separated Quantum Systems: A Consistent History Analysis of the EPR Problem" *Am. J. Phys.*, 55(1), 11.

Gupta, S.N., 1950, *Proc. Roy. Soc. Lond.* A, 63, 681.

Guth, A., 1981, "Inflationary Universe: A Possible Solution to the Horizon and Flatness Problems", *Phys. Rev.* D, 23, 347.

Guth, A. and Steinhardt, P., 1984, "The Inflationary Universe", *Scientific American*, 215(5), 116.

Halpern, F.R., 1968, Special Relativity & Quantum Mechanics, Englewood Cliffs.

Hartle, J.B., 1968, "Quantum Mechanics of Individual Systems", *Am. J. Phys.*, 36(8), 704.

Hawking, S.W., 1985, *Phys. Rev.* D, 32, 2989.

Hawking, Stephen W., 1988, A Brief History of Time: From the Big Bang to Black Holes. Bantam.

Hegerfeldt, G.C., 1985, *Phys. Rev. Lett.*, 54, 2395.

Heisenberg, W., 1927, *Z. Phys*, 43, 172.

Heisenberg, W., 1958, Physics and Philosophy, Harper and Row.

Heisenberg, W., 1958, Daedalus, 87, 95.

Herbert, N., 1982, *Found. Phys.*, 12, 1171.

Hiley, B.J. and David Peat, F., Eds., 1987, Quantum Implications: Essays in Honor of David Bohm, Routledge and Kegan Paul.

Home, D. and Whitaker, M.A.B., 1992, "Ensemble Interpretations of Quantum Mechanics, A Modern Perspective", *Phys. Rep.* 210, 213-317.

Itzykson, C. and Zuber, J.B., 1980, Quantum Field Theory, McGraw-Hill.

Jammer, Max, 1974, The Philosophy of Quantum Mechanics: The Interpretations of Quantum Mechanics in Historical Perspective, John Wiley.

Jauch, J.M. and Rohrlich, F., 1959, The Theory of Photon and Electrons, Addison-Wesley.

Jefferys, William H., 1992, J. *Scientific Exploration*, 6, 47.

Jordan, T.F., 1980, J. Math. Phys., 21, 2028.

Judge, D. and Lewis, J.T., 1963, *Phys. Lett.*, 5(3), 190.

Kafatos, Menas, Ed., 1989, Bell's Theorem, Quantum Theory and Conceptions of the Universe, Kluwer Academic.

Kaplan, David M. and Wu, F.Y., 1971, *Chinese J. Phys.*, 9(1), 31.

Kauffman, Stuart A., 1993, The Origins of Order: Self-Organization and Selection in Evolution, Oxford University Press.

Kauffman, Louis H. and Pierre Noyes, H., 1994, "Discrete Physics and the Derivation of Electromagnetism from the Formalism of Quantum Mechanics", SLAC-PUB-6697.

Karplus, R. and Klein A., 1952, *Phys. Rev.*, 87, 848.

Kazanas, D., 1980, *Astrophysical J.*, 241, L59.

Kiesling, C., 1988, Tests of the Standard Model of Electroweak Interactions, Springer.

Kochen, S. and Specker, E.P., 1967, *J. Math. Mech.*, 17, 59.

Kolb, Edward W. and Turner, Michael S., 1990, The Early Universe, Addison-Wesley.

Landau, L.D. and Lifshitz, E.M., 1975, Classical Mechanics, Pergamon.

Landau, L.D. and Lifshitz, E.M., 1951, The Classical Theory of Fields, Pergamon.

Landau, L.D. and Lifshitz, E.M., 1958, Quantum Mechanics, Pergamon.

Levi-Leblond, Jean-Marc, 1976, *Ann. Phys.*, 101, 319.

Linde, A.D., 1982, *Phys. Lett.* B, 108, 389.

Linde, Andre, 1987, "Particle Physics and Inflationary Cosmology", *Phys. Today*, 40, 61.

Linde, Andre, 1990, Particle Physics and Inflationary Cosmology, Academic Press.

Linde, Andre, 1994, "The Self-Reproducing Inflationary Universe", *Scientific American*, November, 48.

Lombardi, John R., 1980, *Phys. Rev.* A, 22(3), 797.

Lombardi, John R., 1983, *Phys. Rev.* A, 27(3), 1275.

Magnus, W., Oberhettinger, F. and Soni, R.P., 1966, Formulas and Theorems of the Special Functions of Mathematical Physics, Springer.

Marlow, A.R., Ed., 1978, The Mathematical Foundations of Quantum Mechanics, Academic Press.

Marshall, T.W., Santos, E. and Selleri, F., 1983, *Phys. Lett. A*, 98, 5.

Mermin, N.D., 1990, *Am. J. Phys.*, 58, 731.

Naimark, M.A., 1964, Linear Representations of the Lorentz Group, Pergamon, New York.

Newton, T.D. and Wigner, E.P., 1949, *Rev. Mod. Phys.*, 21,400.

Ohnuki, Y., Unitary Representations of the Poincare Group & Relativistic Wave Equations, 1988, World Scientific.

Omnès, R.J., 1988, *J. Stat. Phys.*, 53, 893, 933, 957.

Omnès, R.J., 1992, *Rev. Mod. Phys.*, 64, 339.

Omnès, Roland J., 1994, The Interpretation of Quantum Mechanics, Princeton University Press.

O'Raifeartaigh, L., Straumann, N. and Wipf, A., 1991, Comments Nucl. Phys. 20, 15.

Page, D., 1985, *Phys. Rev. D.*, 32, 2496.

Pagels, Heinz R., 1985, *Phys. Today*, January, 97.

Pais, A., 1979, *Rev. Mod. Phys.* 51, 863.

Park, James L., 1968, "Nature of Quantum States", *Am. J. Phys.*, 36, 21.

Penrose, Roger, 1979, in General Relativity: An Einstein Century Survey, Ed. by S. Hawking and W. Israel, Cambridge University Press.

Penrose, Roger, 1989, The Emperor's New Mind: Concerning Computers, Minds, and the Laws of Physics, Oxford University Press.

Penrose, Roger, 1994, Shadows of the Mind: A Search for the Missing Science of Consciousness, Oxford University Press.

Peres, Asher, 1995, Quantum Theory-Concepts & Methods, Springer.

Polchinski, Joseph, 1991, *Phys. Rev.*, 66, 397.

Prigogine, Ilya and Stengers, Isabella, 1984, Order out of Chaos, Bantam.

Przibam, K., Ed., 1967, Letters on Wave Mechanics, Philosophical Library.

Pryce, M.H.L., 1948, Proc. Royal Soc. London, Series A, 195, 62.

Rae, Alstair, 1986, Quantum Physics: Illusion or Reality?, Cambridge University Press.

Rao, K. N. Srinivasa, 1988, The rotation and lorentz groups and their representations for physicists, John Wiley.

Rarita, M.E., 1961, Relativistic Electron Theory, Wiley.

Redhead, Michael, 1987, Incompleteness, Nonlocality, and Realism, Clarendon Press.

Reed, M. & Simon B, Methods of Modern Mathematical Physics, Vols. I & II, Academic Press.

Ritus, V.I., 1961, Soviet Phys. - JETP, 13, 240.

Selleri, Franco, Ed., 1988, Quantum Mechanics Versus Local Realism: The Einstein-Podolsky-Rosen Paradox, Plenum Press.

Sakurai, J.J., 1967, Advanced Quantum Mechanics, Addison-Wesley.

Schiff, L.J., 1968, Quantum Mechanics, McGraw-Hill.

Schweber, S.S., 1961, Relativistic Quantum Field Theory, Row, Peterson.

Schwinger, J., 1951, *Phys. Rev.*, 82, 664.

Shirokov, J M., 1958, A group theoretical consideration of the basis of relativistic quantum mechanics, I,II,III, IV, Soviet Phys. - JETP, 6, 664, 918, 929, ibid. 7, 493.

Sommers, Paul, 1994, "The Role of the Future on Quantum Theory" gr-qc.xxx.lanl.gov/ 9404022.

Soo, Chopin & Lin, Cyrus C.Y., arXiv: quant-ph/0307107v4.

Stapp, H.P., 1971, Phys. Rev. D, 3, 1303.

Stapp, Henry Pierce, 1972, "The Copenhagen Interpretation", *Am. J. Phys.*, 40, 1098.

Stapp, Henry P., 1985, *Am. J. Phys.* 54, 306.

Stapp, Henry P., 1993, Mind, Matter, and Quantum Mechanics, Springer-Verlag.

Stapp, Henry P., 1994, *Phys. Rev.* A, 50, 18.

Stefanovich, Eugene V., 2007, Relativistic Quantum Mechanics, arXiv:physics/0504062v7

Strocchi, F., 1967, *Phys. Rev.*, 162, 1429, *Phys Rev. D*, 2, 2334 (1970).

Temple, G., 1934, An Introduction to Quantum Theory, D. Van Nostrand.

Thirring, W., 1950, *Phil. Mag.*, 41,113.

Tung, Wu-Ki, 1984, Group Theory in Physics, *World Scientific*.

Von Neumann, John, 1955, The Mathematical Foundations of Quantum Mechanics, Princeton University Press.

Weinberg, S., 1964, in Brandeis Lectures on Particles and Field Theory, Vol 2, Prentice Hall.

Weinberg, S., 1989, *Ann. Phys.* (N.Y.), 194, 336.

Weinberg, Steven, 1992, Dreams of a Final Theory: The Search for the Fundamental Laws of Nature, Pantheon Books.

Weinberg, Steven, 1995, The Quantum Theory of Fields, Cambridge University Press.

Wheeler, J.A., Wojciech, H. and Zurek, W.H., Eds., 1983, Quantum Theory and Measurement, Princeton University Press.

Wightman, A.S., 1960, in Dispersion Relations, Les Houches Lectures, Wiley.

Wigner, E.P., 1939, *Ann. Math.*, 40, No. 1.

Wigner, E.P., 1959, Group Theory, Academic Press.

Yndurain F.J., 1971, Nuovo Cimento A, 2, 104.

Zhan, Youbang, 1988, *Phys. Lett.* A, 128(9), 451.

Zurek, W.H., 1981, *Phys. Rev. D*, 24, 1516.

Zurek, W.H., 1982, *Phys. Rev. D*, 26, 1862.

Zurek, Wojciech H., 1991, "Decoherence and the Transition from Quantum to Classical", *Phys. Today* 36, 36.

Zwanziger, D., 1964, *Phys. Rev. B*, 113, 1036.

Zwanziger, D., 1964, in Lectures in Theoretical Physics, Vol. VII, University of Colorado Press.

Index